Handbook of Research on Microbial Tools for Environmental Waste Management

Vinay Mohan Pathak
Gurukul Kangri University, India

Navneet
Gurukul Kangri University, India

A volume in the Advances in Environmental Engineering and Green Technologies (AEEGT) Book Series

IGI Global
DISSEMINATOR OF KNOWLEDGE

Published in the United States of America by
IGI Global
Engineering Science Reference (an imprint of IGI Global)
701 E. Chocolate Avenue
Hershey PA, USA 17033
Tel: 717-533-8845
Fax: 717-533-8661
E-mail: cust@igi-global.com
Web site: http://www.igi-global.com

Library of Congress Cataloging-in-Publication Data

Names: Pathak, Vinay Mohan, 1990- | Navneet, 1963-
Title: Handbook of research on microbial tools for environmental waste management / Vinay Mohan Pathak and Navneet, editors.
Description: Hershey, PA : Engineering Science Reference, [2018]
Identifiers: LCCN 2017018918| ISBN 9781522535409 (hardcover) | ISBN 9781522535416 (ebook)
Subjects: LCSH: Bioremediation. | Microorganisms--Industrial applications.
Classification: LCC TD192.5 .M3257 2018 | DDC 628.5--dc23 LC record available at https://lccn.loc.gov/2017018918

This book is published in the IGI Global book series Advances in Environmental Engineering and Green Technologies (AEEGT) (ISSN: 2326-9162; eISSN: 2326-9170)

British Cataloguing in Publication Data
A Cataloguing in Publication record for this book is available from the British Library.

For electronic access to this publication, please contact: eresources@igi-global.com.

Advances in Environmental Engineering and Green Technologies (AEEGT) Book Series

Sang-Bing Tsai
University of Electronic Science and Technology of China
Zhongshan Institute, China
Ming-Lang Tseng
Lunghwa University of Science and Technology, Taiwan
Yuchi Wang
University of Electronic Science and Technology of China
Zhongshan Institute, China

ISSN:2326-9162
EISSN:2326-9170

Mission

Growing awareness and an increased focus on environmental issues such as climate change, energy use, and loss of non-renewable resources have brought about a greater need for research that provides potential solutions to these problems. Research in environmental science and engineering continues to play a vital role in uncovering new opportunities for a "green" future.

The **Advances in Environmental Engineering and Green Technologies (AEEGT)** book series is a mouthpiece for research in all aspects of environmental science, earth science, and green initiatives. This series supports the ongoing research in this field through publishing books that discuss topics within environmental engineering or that deal with the interdisciplinary field of green technologies.

Coverage

- Renewable energy
- Cleantech
- Sustainable Communities
- Green Technology
- Air Quality
- Electric Vehicles
- Radioactive Waste Treatment
- Biofilters and Biofiltration
- Pollution Management
- Contaminated Site Remediation

IGI Global is currently accepting manuscripts for publication within this series. To submit a proposal for a volume in this series, please contact our Acquisition Editors at Acquisitions@igi-global.com or visit: http://www.igi-global.com/publish/.

The Advances in Environmental Engineering and Green Technologies (AEEGT) Book Series (ISSN 2326-9162) is published by IGI Global, 701 E. Chocolate Avenue, Hershey, PA 17033-1240, USA, www.igi-global.com. This series is composed of titles available for purchase individually; each title is edited to be contextually exclusive from any other title within the series. For pricing and ordering information please visit http://www.igi-global.com/book-series/advances-environmental-engineering-green-technologies/73679. Postmaster: Send all address changes to above address.

Titles in this Series

For a list of additional titles in this series, please visit: www.igi-global.com/book-series

Food Systems Sustainability and Environmental Policies in Modern Economies
Abiodun Elijah Obayelu (Federal University of Agriculture – Abeokuta (FUNAAB), Nigeria)
Engineering Science Reference • copyright 2018 • 371pp • H/C (ISBN: 9781522536314) • US $195.00 (our price)

Microbial Biotechnology in Environmental Monitoring and Cleanup
Pankaj (G. B. Pant University of Agriculture and Technology, India) and Anita Sharma (G. B. Pant University of Agriculture and Technology, India)
Engineering Science Reference • copyright 2018 • 427pp • H/C (ISBN: 9781522531265) • US $235.00 (our price)

Handbook of Research on Renewable Energy and Electric Resources for Sustainable Rural Development
Valeriy Kharchenko (Federal Scientific Agroengineering Center VIM, Russia) and Pandian Vasant (Universiti Teknologi PETRONAS, Malaysia)
Engineering Science Reference • copyright 2018 • 672pp • H/C (ISBN: 9781522538677) • US $345.00 (our price)

Green Production Strategies for Sustainability
Sang-Binge Tsai (University of Electronic Science and Technology of China (Zhongshan Institute), China & Civil Aviation University of China, China) Bin Liu (Shanghai Maritime University, China) and Yongian Li (Nankai University, China)
Engineering Science Reference • copyright 2018 • 325pp • H/C (ISBN: 9781522535379) • US $235.00 (our price)

Innovative Strategies and Frameworks in Climate Change Adaptation Emerging Research and Opportunities
Alexander G. Flor (University of the Philippines Open University, Philippines) and Benjamina Gonzalez Flor (University of the Philippines, Philippines)
Engineering Science Reference • copyright 2018 • 165pp • H/C (ISBN: 9781522527671) • US $165.00 (our price)

Utilizing Innovative Technologies to Address the Public Health Impact of Climate Change Emerging Research and Opportunities
Debra Weiss-Randall (Florida Atlantic University, USA)
Engineering Science Reference • copyright 2018 • 295pp • H/C (ISBN: 9781522534143) • US $185.00 (our price)

Economical and Technical Considerations for Solar Tracking Methodologies and Opportunities for Energy Management
S. Soulayman (Higher Institute for Applied Sciences and Technology, Syria)
Engineering Science Reference • copyright 2018 • 647pp • H/C (ISBN: 9781522529507) • US $245.00 (our price)

IGI Global
DISSEMINATOR OF KNOWLEDGE

701 East Chocolate Avenue, Hershey, PA 17033, USA
Tel: 717-533-8845 x100 • Fax: 717-533-8661
E-Mail: cust@igi-global.com • www.igi-global.com

Editorial Advisory Board

List of Contributors

Table of Contents

Section 2
Bioremediation of Environmental Waste and Pollutants

Section 3
Removal of Toxic Product and Hazardous Waste

Detailed Table of Contents

This chapter contends that bioethanol has received the most attention over other fuels due to less emission of greenhouse gases and production from renewable sources. It is mainly produced from sugar containing feedstocks. Since feedstocks are utilized as food for humans, its consumption in bioethanol production creates a food crisis for the entire world. Bioethanol derived from agriculture waste, which is most abundant at global level, is the best option. Agriculture wastes contain lignin, cellulose and hemicelluloses which creates hindrances during conversion to ethanol. Pretreatment of agriculture wastes remove lignin, hemicelluloses and then enzymatically hydrolyzed into sugars. Both pentose and hexose sugars are fermented to bioethanol. There are still various problems for developing an economically feasible technology but a major one is the resistance to degradation of the agricultural material. Use of two or more pretreatment methods for delignification and the use of genetically modified agricultural biomass can be developed for economically feasible ethanol production.

Cellulose is the most abundant polymer in plants and the microbial conversion of cellulose is a subject of active research. Currently, cellulase is commonly used in many industrial applications, especially in animal feed, textile, waste water, brewing and wine making. A challenging strategy for the efficient utilization of this renewable resource is to use it as a base material for the production of desired metabolites. This

This chapter describes how pollutants are increasing in the environment due to the rapid industrialization all over the world. The environment has been contaminated with large number of organic and inorganic pollutants. The organic pollutants are largely anthropogenic and are introduced to the environment in many ways. Soil contamination with toxic metals, such as Cd, Pb, Cr, Zn, Ni, etc., as a result of worldwide industrialization has increased noticeably within the past few years. Bioremediation is a process for reclaiming the environment which has been polluted with the help of living forms. It is an option that offers the possibility to destroy various contaminants using natural biological activity and to degrade the environmental contaminants into less toxic forms. It is also applicable for the heavy metal hazards. It has proven to be cheap and efficient than other techniques. This chapter focuses on the possible trends in the remediation of environment pollutants with the help of plants as well as microbes.

Chapter 6

Composting an Emerging Technology for Solid Waste Management in India................................. 105

Mansi Rastogi, Maharshi Dayanand University, India

Meenakshi Nandal, Maharshi Dayanand University, India

This chapter describes how "Solid waste" (SW) is a term usually applied to an assorted collection of wastes produced in urban areas, the nature, characteristics and quantity of which varies from region to region. Waste generation is influenced not only by lifestyle of the region's inhabitants, but also by copiousness and type of the region's natural resources. Composting of municipal solid waste is, therefore, a simple and cost-effective technology for treating the organic fraction of municipal solid waste. As the composting proceeds there occurs drastic change in microbial communities, and resemblance to the initial community is lost with process. Generally, three categories of microorganisms play a vital role in composting process: bacteria, actinomycetes and fungi. It is primarily the mesophilic (> 45 °C) and specifically the thermophilic bacteria (< 60 °C), that plays a major role by rising the temperature of the piles. This chapter mainly focused on composting process for Municipal waste stabilization and role of microbes in the process.

Chapter 7

Valorization of Tannery Solid Waste Materials Using Microbial Techniques: Microbes in Tannery Solid Waste Management... 127

Thazeem Basheer, Bharathiar University, India

Mridul Umesh, Bharathiar University, India

This chapter describes how environmental pollution is the major problem associated with rapid industrialization, due to which the quantity of solid waste generated from the industries is increasing substantially. Solid wastes generated in leather industries are briefly classified into pre-tanned and post-tanned wastes. Although landfills, composting, anaerobic digestion and thermal incineration are available for disposal of these solid wastes, they do again pose severe environmental and financial burdens to the tanners. Microbe-mediated tannery solid waste management points to the recovery of value-added products from these wastes. Microbial valorization of tannery solid wastes for the production of saleable products would be a convincing, challenging and eco-friendly opening for its utilization when compared to that of chemical and thermal hydrolysis. Exploitation of traditional chemicals could be reduced and innovative products could be recovered, enabling sustainable solid waste management. This would ultimately alleviate the solid waste disposal problems and financial crisis faced by the tanners.

Nowadays municipal wastewater (MWW) treatment by phytoremediation techniques goes as an emerging technique in the USA and many European countries. Cleaning wastewater with constructed wetland (CW) is an advanced type of phytoremediation. Low concentration of hazardous metallic elements in this major wastewater caused its capability for treatment by CW. This treatment method highly depends on the presence of macrophytes, media, and operating factors which have high influences in the efficiency of this technology. This chapter will discuss on engineering controls that traditionally are available and practically could be used in the commonly CW. Recirculation, dry to wet duration, artificial aeration, absorbent application, dilution and carbon source addition through this eco-friendly decontamination method. The concentration of this manuscript will be on Cyperus alternifolius as a well-known rapid growth plant species which often known as an ornamental plant. This review on try to illustrate practical ways to enhancing efficiency of decontamination of MWW in CW

This chapter describes how exposure to pesticides has become one of the most prominent components of current agriculture. Pesticides have high impact on farmer's economy as they prevent and minimize agricultural losses due to pests (insects, fungi, nematode), improve yield and quality food in terms of superficial appeal. However, these synthetic pesticides have serious environmental concerns and create a variety of toxic effects in living beings. In this regard, the bioremediation approach can be a suitable alternative over the conventional treatment methods for the treatment of pesticides-contaminated sites. Bioremediation is the use of microorganisms or their enzymes to degrade and detoxify the pesticide from the environment. Microbial interactions greatly influence natural bio remedial processes occurring in soil. Microorganisms involved various biochemical mechanisms for the degradation of pesticides, such as oxidation and reduction reactions. This chapter provides an overview on the nature, characteristics and toxicity of pesticides in environment and role of microbes in bioremediation of pesticides.

Soil, though often overlooked, is a critical component of the earth and is essential for human existence. Soil pollution has deteriorated large areas of agricultural land around the globe. Human health is at risk due to the high concentration of pollutants found in soils and food produced in polluted lands. Soil pollution may arise from a wide range of sources that can be discrete point sources, or diffuse sources. It is usually due to domestic, municipal, industrial, mining, and agricultural wastes as well as agrochemicals such as fertilizers and pesticides. Each source has specific characteristics, briefly presented below. The most dangerous categories of soil pollutants are polycyclic aromatic hydrocarbons, polychlorinated biphenyl, pesticides, heavy metals, and radionuclides. New soil pollutants, such as veterinary medicines, have emerged with industrialization and globalization. This chapter deals with sources of soil pollution, pollutants generated from these sources and their possible adverse effects on the environment.

Type of industry and its raw material decides quantity and quality of the emerged wastes including both biodegradable as well as non-biodegradable. Among non-biodegradable wastes, copper, chromium, nickel, cadmium, etc. are widespread contaminants of soil, water, and these are most common heavy metals. Several heavy metals such as cadmium, mercury, and lead are highly poisonous and fatal to human as well as animals. Several plants as well as microbes respond to heavy metals by diverse biological processes like biosorption to their cell wall and entrapment in their capsule, oxidation and reduction, precipitation, complexation, etc. These responses may help significantly in the remediation of heavy metals from the contaminated sites.

With its ever-increasing demand, the enormous production of low density polyethylene (LDPE) is leading to its huge accumulation in the environment. LDPE remains durable and inert to natural degradation and deterioration. This chapter focuses on various microbial effects and approaches on biodegradation of LDPE. Biodegradation takes place through several different strategies, such as assimilation, biodeterioration, biofragmentation, etc. Few microorganisms have so far been isolated which can grow on LDPE. Degradation by microbial enzyme is considered to be one of the most powerful tools to study the biodegradation of LDPE. Some blends and composites of LDPE with natural polymers have been found to be biodegradable; however, their manufacturing is costly. Thorough analysis of the microbial degradation of LDPE helps us realize the overall mechanism involved. In essence, this chapter aims to objectify the in situ biodegradation of LDPE via development of microbial biofilm on the polymer surface.

Halogenated organic compounds having many beneficial applications, both in industries and agriculture sectors. Basically, the uses are as pesticides, solvents, surfactants, and plastics. However, their large, widespread uses throughout the world have resulted the negative impact on the environment. Considering their treatment process are widely accepted by using the bioreactor systems. The large variety of microorganisms present in the bioreactor and their interaction is the key to the effective treatment and removal of these compounds. Usually the microbes produce the enzymes known as dehalogenase to remove the halogen form the compounds to make it non-toxic. Many of the different steps and about the microbial groups in degradation process of halogenated compounds are well understood, but more details concerning the microbial community are yet to be discovered. This chapter describes about the different dehalogenation systems available in microbes and their ultimate application in different bioreactor systems for the degradation analysis of several harmful halogenated compounds.

Azo dyes are considered as xenobiotic compounds, which are often recalcitrant to be biodegraded. Many dyes are relevant for staining in industries, especially textile and food related. Among other environmental problems, one obvious issue is the coloring effect on effluents from industrial sites and thus the release of xenobiotics into nature. Microorganisms (algae, fungi, yeast, and bacteria) have been found to decolorize a number of azo dyes. Decolorization of azo dyes by microbial cultures is summarized and this is mostly linked to initial activation or even cleavage of the azo bond (e.g., by azoreductases). However, it does not necessarily mean that the compound is degraded to non-toxic products. Various mechanisms of microbial decolorization processes were discovered, including adsorption, enzymatic degradation, or a combination of both. Oxidases and reductases were found to be involved, which contain azoreductase, lignin peroxidase, Mn peroxidase, laccase, tyrosinase, and so on. A focus is on the azoreductases including classification, activity, and applicability.

Proper waste management in developing countries is increasingly becoming popular, especially the use of engineered sanitary landfills. In this study, the classification, design, and operation of sanitary landfills are stated and suggestions made. The landfills were classified in accordance to waste type and population size. Leachate control mechanisms were put in place in a bid to avoid surface and underground water pollution. Several liner materials such as clay and geotextile material are recommended to avoid leachate perforating into the ground. Waste management is encouraged to be done through the cell system which must be compacted and covered on a daily basis. Resource recovery of biogas for power generation is recommended to recovery value from the waste as well as to lower landfill operation costs. Post-closure monitoring of the landfill must be done and recreation centers can be developed on the landfill as a rehabilitation strategy. Land filling is essential for proper waste management and is also governed by local acts.

Bioinformatics is the new area of science to study various living organisms, plants and animals. Bioremediation is the removal of toxic pollutants from the environment using microorganisms. Many of the databases are available online and can be used for the study of the microorganism genomics, transcriptomics, proteomics. This chapter mainly focuses the uses of recent databases and tools for the exploration of microorganisms.

Petroleum and petroleum-based products are highly beneficial for our daily life system. Functioning of several industries and machineries directly depends on different forms of petroleum. There are strong possibilities of the release of such petroleum and petroleum-based compounds like aliphatic as well as aromatic hydrocarbons during their refining process, usage, storage, transport, and other activities due to the accidents, leakages, or just a little lack of awareness. Their exposure to soils or water causes serious problems for aquatic as well as human beings. The efficient removal or detoxification of such pollutants is the demand of the present time. The use of microbial sources for the bioremediation of such petroleum wastes may be promising technique because it does not require any drastic conditions for detoxification process and by-products produced by them are also harmless unlike chemical and other techniques. In this regard, this chapter discusses the enzymatic role in detoxification or bioremediation of petroleum-based hydrocarbons.

Preface

In current scenario we have reached to huge level of industrialization. These progressive development and exploration of industrialization showing major pressure on our environment by introducing hazardous pollutants. Industrial wastes include liquid, solid and gaseous waste that are produced mainly from fermentation, petroleum, chemical processing, textile, paper and pulp industries etc. Improper waste management strategy leading towards the accumulation of increased solid waste in the natural environment. The occurrence of hazardous compound in an elevated amount leading to increased land and water pollution are of great concern. Such pollutants are responsible for lethal effect (lung and heart diseases, lung cancers and chest infections caused by air pollutant) on human after prolonged exposure. Bioremediation, phytoremediation serves as most prominent tool to remediate polluted environment in an eco-friendly manner. Biological agents and their metabolic enzymes can be exploited as a potent tool for degradation of hazardous pollutants. Bacterial and fungal species are the most abundant biological agents found in nature and have distinct degradation abilities for xenobiotic degradation. These techniques serve as an economical alternative to conventional approaches for waste management. This book is a compilation of scientific data on various advanced and eco-friendly approaches for sustainable environment. Biodegradation of natural polymers and xenobiotic compounds (poly aromatics hydrocarbons, dyes etc.) have been discussed in these chapters and contributed by the leading scientists drawn from all over the world.

The first chapter of this having the information on agricultural waste management for bioethanol production. Agriculture waste serves as cheap and abundant biomass for bioethanol production. Agriculture waste can be converted into bioethanol by different pretreatments (physical, chemical, physio-chemical and biological). Biological pretreatments should be used because it is safe, environmentally friendly and less energy consuming as compared to other pretreatments. Pretreated biomass is used for enzymatic hydrolysis followed by fermentation using microbes such as bacteria or yeast. Bioethanol have received most attention over other fuels due to less emission of greenhouse gases and production from renewable sources. Fossil energy sources such as coal, petrol etc. are consumed by the society to produce fuels and chemicals. Bioethanol is mainly produced from feedstocks such as sugarcane, sugar beet, rice and corn grain, but the supply of these feedstocks is limited. This limitation will create the food crisis if these feedstocks utilized for bioethanol production.

The second chapter of this having the information on microbial cellulase: characteristics, sources and application. Cellulose serves as leading bio-residue from agricultural sector is the predominant renewable biopolymer in the world which comprises of celluloses, hemicelluloses and lignin. Microbial conversion of cellulosic biomass is a promising strategy focussing towards low-cost biomass processing. The biological degradation of cellulose has been studied for many years, and a number of cellulolytic enzymes, especially cellulases produced by fungi and bacteria. The catalytic domains of cellulases and related xylanases have been identified. According to the mechanism of cellulose degradation, cellulases are subdivided into either non-processive cellulases (endocellulases) or processive cellulases. Endocellulases can randomly cleave the cellulose at exposed positions and produce new reducing ends; however, the processive cellulases remain attached to the chain and release cellobiose units. During the degradation of cellulose, non-processive cellulases and processive cellulases have been found to work synergistically. Cellulase is an important enzymes and find applications in industries of starch processing, grain alcohol fermentation, malting and brewing, extraction of fruit and vegetable juices, pulp and paper industry, and textile industry. Therefore, it is of considerable economic interest to develop processes for effective treatment and utilization of cellulosic wastes as inexpensive carbon sources. Because of the biomass potential as a source of renewable fuels, there has been a rapid increase in the amount of research in this area and our scientific approach towards the cellulose degradation should continue to increase. Cellulolytic bacteria can be exploited for cellulase production which serves wide applications in industries, pharmaceuticals and further, use of these cellulose degrading bacteria (CDB) as bio-inoculants can be incorporated to enhance organic matter decomposition in soil to increase soil fertility and to minimize the applications to fertilizer. It also finds wide applications in reducing the environmental pollution and promotes sustainable agriculture.

Third chapter of this book explain bioprocessing requirements for bioethanol: sugarcane vs. sugarcane bagasse. Authors discuss alternative energy sources and the advantages of biofuels over fossil fuels. It outlines the main steps of bioethanol production and suggests some alternative sources as potential feedstock. The core focus of this chapter is to examine new research which considers the use of agricultural waste as a feedstock for bioethanol production rather than conventional feedstocks such as sugarcane and corn. The main advantages of the use of waste materials are overcoming the problem of greenhouse gases, the renewability of feedstock, economic benefits including more employment opportunities and to sustain global transport systems. As the world demand for energy alternatives continues to increase, lignocellulosic material/waste seems a viable possibility as a renewable feedstock for the production of biofuels.

The fourth chapter of this book discusses the microbe mediated bioconversion of fruit waste into value added products: microbes in fruit waste management. This chapter summarizes microbe mediated fermentative utilization of fruit waste, for the production of value added products like organic acid, single cell protein, bioplastics, enzymes and biogas. Increasing concern about environment pollution due to careless discharge of waste called for development of sustainable waste management strategies for production of value added products. Microbes have served and continue to serve as imperial candidates in both production and management strategies. Microbe mediated techniques has emerged as ecofriendly and sustainable alternative to their synthetic counterparts. Fruit waste with high content of carbohydrates are easily available cheap raw material for the production of a wide range of products like organic acids, single cell protein, bioplastics, enzymes and biogas. Fruit based industries produces large volumes of solid and liquid wastes contributing to increase in pollution load. Disposal of these waste not only represent loss of valuable biomass but also leads to substantial increase in Biological Oxygen Demand (BOD) and Chemical Oxygen Demand (COD). Utilization of biomass through microbial fermentation serves as sustainable alternative for chemical synthesis process for the manufacture of products of commercial interest. Proper planning, execution, financial assistance from public and private sector for both research and product development can take this initiative in to promising domains in the future emphasizing the concept of 'Wealth from Waste'.

The fifth chapter of this book on bioremediation of environmental pollutants. This Chapter will focuses on the possible trends in the remediation of environment pollutants with the help of plants as well as microbes. Pollutants are increasing in the environment due to the rapid industrialization all over the world. The environment has been contaminated with large number of organic and inorganic pollutants. The Organic pollutants are largely anthropogenic and are introduced to the environment in many ways. Soil contamination with toxic metals, such as Cd, Pb, Cr, Zn, Ni etc., as a result of worldwide industrialization has increased noticeably within the past few years. Bioremediation is a process for reclaiming the environment which has been polluted with the help of living forms. It is an option that offers the possibility to destroy various contaminants using natural biological activity and to degrade the environmental contaminants into less toxic forms. It is also applicable for the heavy metal hazards. It has proven to be cheap and efficient than other techniques. Bioremediation is a technology, utilizing the life forms to kill or expel contamination from waste. It takes the advantages of the microbes, plants, and also the natural symbiosis with mycorrhiza and root associated natural microbial flora for the degradation of the harmful contaminants. The studies described above represent the plant microbes interaction to remediate the heavy metals which would be an important tool for scientist as the pollutants are increasing in the environments day by day due to anthropogenic and other activity on earth. These systems could be an interesting tool to further improve and develop bioremediation into a widely accepted technique.

The sixth chapter of this book on composting an emerging technology for solid waste management in India. This chapter mainly focused on composting process for Municipal waste stabilization and role of microbes in the process. In developing countries like India, the feasible and productive municipal solid waste management treatment and disposal options have been constraint by finances, technology and qualified personnel. All these issues have narrowed down composting as only option that best fits within the available resources in developing countries. A major feature that renders composting as a suitable technology is its adaptability to a broad range of situations. This process has many advantages like lower equipment and operating costs, environment friendly and useful product production. As the composting proceeds there occurs drastic change in microbial communities, and resemblance to the initial community is lost with process. Generally, three categories of microorganisms play a vital role in composting process: bacteria, actinomycetes and fungi. For successful composting the major prerequisites include understanding and implementation of the basic principles to the process. This knowledge about the process enables a rational evaluation of compost process and utilization of these technologies to broader aspects to resolve the waste related problems.

Seventh chapter of this book discusses the valorization of tannery solid waste materials using microbial techniques: microbes in tannery solid waste management. Microorganisms play a vital role in tannery solid waste management, replacing traditional disposal methods and harsh chemicals which have been utilized by the tanners regularly till now to get rid of their unhealthy situation in tanneries. Proteolytic, lipolytic and probiotic microbes have a great potential in managing environmental wastes with recovery of marketable products. They act as a major tool for reducing environmental pollution caused by adverse disposal systems. Microbial mediated tannery solid waste management not only alleviates tanners' problems faced during waste disposal, but also profits the industry greatly in terms of fermentation end products that could be beneficially recycled and reused by the tanneries itself in conversion of raw hides into finished leather. Much more initiations and advancements have to be provoked among scientists and researchers to work with microorganisms in managing tannery solid waste and its threats that may reduce the burden of the tanners.

Eighth chapter on phytoremediation of nitrogen and phosphorus in municipal wastewater by *Cyperus alternifolius* planted constructed wetland. This chapter on operating elements tries to illustrate practical ways to enhancing efficiency of decontamination of municipal wastewater by planted constructed wetland in Malaysia. Municipal wastewater treatment by phytoremediation techniques goes as an emerging technique in USA and many European countries. Cleaning wastewater with constructed wetland is an advanced types of phytoremediation. Low concentration of hazardous metallic elements and wide range of numerous contaminants in this major wastewater caused its capability for treatment by constructed wetland. This treatment method highly depends to presence of macrophytes, media, and operating factors which have high influences in efficiency of this green technology. This chapter also discuss on engineering controls that traditionally are available and practically could be used in common constructed wetland. Recirculation, dry to wet duration, artificial aeration, absorbent application, dilution and carbon source addition through this eco-friendly decontamination method. Wastewater engineering is a type of managing these responses based on the environmental circumstances, pollutants concentration environmental discharge regulations and cost efficiency of treatment. Results of this chapter clearly illustrate municipal wastewater could be effectively fully treated by constructed wetland even with a limited engineering techniques.

In ninth chapter information on role of microorganisms in bioremediation of pesticides. This chapter provides an overview on the nature, characteristics and toxicity of pesticides in environment and role of microbe in bioremediation of pesticides. Microorganisms involve various biochemical mechanisms for degradation of pesticides such as oxidation and reduction reactions. Exposure to pesticides has become one of the most prominent components of current agriculture. Pesticides have high impact on farmer's economy as they prevent and minimize agricultural losses due to pests (insects, fungi, nematode), improve yield and quality food in terms of superficial appeal. Pesticides are often considered a quick, easy, and inexpensive solution for controlling weeds and insect pests in urban landscapes. However, pesticide use comes at a significant cost. Pesticides have contaminated almost every part of our environment. Pesticide residues are found in soil and air, and in surface and ground water across the countries, and urban pesticide uses contribute to the problem. Bioremediation is the use of microorganisms or their enzymes to degrade and detoxify the pesticide from the environment. Microbial interactions greatly influence natural bio remedial processes occurring in soil. Bioremediation is a characteristic procedure for the remediation of polluted site and to expel the contamination from nature environment collaborative use of biological agents (bacteria, fungi, plants). For bioremediation to be effective, the pesticides must be subject to microbial attack or microbial transformation, the metabolic product must be safe and the process must not cause adverse ecological side effects.

The tenth chapter in this book on sources of soil pollution and their impact on environment. This chapter deals with sources of soil pollution, pollutants generated from these sources and their possible adverse effects on the environment. Soil, though often overlooked, is a critical component of the earth and is essential for human existence. Soil pollution has deteriorated large areas of agricultural land around the globe. There are many anthropogenic sources, which cause important environmental risks. It may arise from a wide range of sources that can be discrete point sources, or diffuse sources. It is usually due to domestic, municipal, industrial, mining and agricultural wastes as well as agrochemicals such as fertilizers and pesticides. Each source has specific characteristics, briefly presented below. The most dangerous categories of soil pollutants are polycyclic aromatic hydrocarbons, polychlorinated biphenyl, pesticides, heavy metals and radionuclide. New soil pollutants, such as veterinary medicines, have emerged with industrialization and globalization. Appropriate measures need to be taken to control levels of organic and inorganic pollutants and their sources particularly in cultivatable soil. This will in turn result in enhancement of life quality and human health.

The eleventh chapter in this book on sources of air pollution and their impact on environment. This chapter provides an overview on air pollution sources as well as the ways in which pollutants can affect human health and the environment. Air pollution is a global concern that creates a need for international strategies and agreements-reflections of the social demands of the way we live. Numerous potential air pollutants may enter the environment from many natural and anthropogenic sources. They are known to have negative effects on pulmonary and cardiovascular functioning. Air pollutants cause serious health problems together with adverse environmental and ecological effects. The most common are carbon monoxide (the most abundant air pollutant), ammonia, NOx, SOx and VOCs. They are emitted from either stationary sources or mobile sources. It is well known that emissions from transportation sources are a major component of the total ambient air pollution concentration in large urban cities. Thus, developments of alternative fuels that can reduce emissions of air pollutants as well as the use of public transport are necessary. From a policy perspective, the close association between traffic patterns and air pollution leads to the development of transport policies and regulations with population health improvements as a primary objective. Other important sources of air pollution are power stations. Emissions from transportation and power generation sources have begun to decline in recent years owing to the introduction of efficient abatement measures. Urban inventories are important for identifying these sources and their impact on the environment. The information related to air pollution presented in this chapter is intended to increase awareness of the problem and of the need for prevention and control. More drastic solutions will be needed so that air quality can continue to improve. Economic attempts to reduce energy use and to advance and expand the use of pollutant-reducing technologies must be increased. Improved scientific understanding of the links between air pollution and health effects will help both to sustain this progress and to improve it by better targeting the most toxic emissions.

Chapter 12 discusses on anaerobic digestion enhancement with microbial electrolysis cells: is biomethane production the direction to go for commercialization. This chapter on the bioprocessing of biomethane using AD and MECs and addresses important issues, namely the lack of pilot-scale studies. Downstream processing techniques are discussed, as well as a novel suggestion of further utilising MECs in the purification process. The engineering of replacements for crude oil is a priority within industrial biotechnology. Biogas, produced by anaerobic digestion (AD) during organic waste degradation, has been used for electricity generation and heating. Microbial electrolysis cells (MECs) are an emerging technology which when combined with AD can produce higher yields of such energy while simultaneously treating waste water and sludge. MECs are bioelectrochemical systems which utilize the metabolism of microbes to oxidize organics. The majority of the research has been focused on biohydrogen production, despite associated issues, which has resulted in poor commercialization prospects. Consequently, scientists are now suggesting that methane production should be the focus of MEC technology. In addition, this is a relatively new technology and in future years it is likely MECs will be an additional tool for scientists faced with tackling the energy crisis.

Chapter 13 in this book on biosorption of heavy metals: biological approach to control the industrial waste. Large scale production of commodities for mankind by industries did huge damage to the environment. Industrial waste contains lots of toxic materials including heavy metals were drained to water bodies like river, lakes, ponds etc. These effluents drastically ruin water quality as well as the soil fertility. Type of industry and its raw material decides quantity and quality of the emerged wastes including both biodegradable as well as non-biodegradable. Among non-biodegradable wastes copper, chromium, nickel, cadmium etc. are widespread contaminants of soil, water and these are most common heavy metals. Several heavy metals such as cadmium, mercury and lead are highly poisonous and fatal to human as well as animals. Several plants as well as microbes respond to heavy metals by diverse biological processes like biosorption to their cell wall and entrapment in their capsule, oxidation and reduction, precipitation, complexation etc. These responses may help significantly in the remediation of heavy metals from the contaminated sites. Researchers are paying attention on the economy of the biosorbents, source and the type of biosorbents plays an important role in the economy of the process. Most economic biosorbents includes several fungi, algae which can be successfully grown for large scale because of their minimal requirement of nutrients.

Fourteenth chapter of this book on biodegradation of low density polyethylene films. LDPE remains durable and inert to natural degradation and deterioration. This chapter aims to objectify the *in situ* biodegradation of LDPE via development of microbial biofilm on the polymer surface. This chapter focuses on various microbial effects and approaches on biodegradation of LDPE. Biodegradation takes place through several different strategies, such as assimilation, biodeterioration, biofragmentation etc. Few microorganisms have so far been isolated which can grow on LDPE. Degradation by microbial enzyme is considered to be one of the most powerful tools to study the biodegradation of LDPE. Some blends and composites of LDPE with natural polymers have been found to be biodegradable; however, their manufacturing is costly. Thorough analysis of the microbial degradation of LDPE helps us realize the overall mechanism involved. This chapter also described the *in situ* biodegradation of LDPE, which depends on several factors like hydrophilicity or hydrophobicity, thickness of the film, types of microbes engaged for biodegradation, etc. It appeared that when hydrophilicity is more, the rate of biodegradation is also more. Again, the microbes which are oxidative produce more effective biodegradation rate. Therefore, *in situ* biodegradation process renders the film hydrophilic; thereby, forming biofilm on the surface by the microbes. This results in a faster rate of degradation.

Fifteenth chapter on microbial bioreactor systems for dehalogenation of organic pollutants. This chapter describes about the different dehalogenation systems available in microbes and their ultimate application in different bioreactor systems for the degradation analysis of several harmful halogenated compounds.

Halogenated organic compounds having many beneficial applications, both in industries and agriculture sectors. Basically the uses are as pesticides, solvents, surfactants and plastics. However, their large, widespread uses throughout the world have resulted the negative impact on the environment. Considering their treatment process the biological methods are widely accepted by using the bioreactor systems. The large variety of microorganisms present in the bioreactor and their interaction is the key to the effective treatment and removal of these halogenated compounds. Usually the microbes produce the enzymes known as dehalogenase, to remove the halogen form the compounds to make it non-toxic. The reduction of toxic halogenated compounds in the contaminated material be of economic interest. In spite of several chemical based and other methods, the biological (microorganism mediated) is preferred due to its environmental friendly nature.

In sixteenth chapter information on microbial degradation of azo dyes, the role of azoreductase to initiate degradation. Azo dyes are considered as xenobiotic compounds, which are often recalcitrant to be biodegraded. Microorganisms (algae, fungi, yeast, and bacteria) have been found to decolorize a number of azo dyes. Decolorization of azo dyes by microbial cultures is summarized and this is mostly linked to initial activation or even cleavage of the azo bond, e.g. by azoreductases. However, it does not necessarily mean that the compound is degraded to non-toxic products. Various mechanisms of microbial decolorization processes were discovered, including adsorption, enzymatic degradation or a combination of both. Oxidases and reductases were found to be involved, which contain azoreductase, lignin peroxidase, Mn peroxidase, laccase, tyrosinase etc.

The seventeenth chapter of this book on design and development of a sanitary landfill for low income countries for optimal waste management. In this study, the classification, design and operation of sanitary landfills were understated and suggestions made. Proper waste management in developing countries is increasingly become popular especially the use of engineered sanitary landfills. The landfills were classified in accordance to waste type and population size. Leachate control mechanisms were put in place in a bid to avoid surface and underground water pollution. Several liner materials such as clay and geotextile material are recommended to avoid leachate perforating into the ground. Waste management is encouraged to be done through the cell system which must be compacted and covered on a daily basis. Resource recovery of biogas for power generation is recommended to recovery value from the waste as well as to lower landfill operation costs. Post closure monitoring of the landfill must be done and recreation centers can be developed on the landfill as a rehabilitation strategy. It is essential for developing counties to adopt the engineered sanitary landfill technology. The environmental regulatory authorities must also ensure that the landfills are being constructed to specification according to the landfill design guidelines.

The eighteenth chapter of this book discusses on insilico tools to study the bioremediation in microorganisms. This chapter mainly focuses the uses of recent databases and tools for the exploration of depth of our understanding among microorganisms. Bioinformatics is the new area of science to study the various living organisms, plants and animals. Due to their large scale use for accurate results as compare to laboratory experiment. Bioremediation is the removal of toxic pollutants from the environment using microorganisms. There are many databases which can be used for the study of the microorganism genomics, transcriptomics and proteomics. On the basis of known sequences from the database we can analyze the sequences with respect to bioremediation.

The nineteenth chapter on enzymatic treatment of petroleum based hydrocarbons. This chapter deals the brief discussions on enzymatic role in detoxification or bioremediation of petroleum based hydrocarbons. Petroleum and petroleum based products are highly beneficial for our daily life system. Functioning of several industries and machineries directly depend on different forms of petroleum. There are strong possibilities of the release of such petroleum and petroleum based compounds like aliphatic as well as aromatic hydrocarbons due to the accidents, leakages or just a little lack of awareness during their refining process, usage, storage, transport and other activities. Their exposure to soils or water causes serious problems for aquatic environment as well as human beings. The efficient removal or detoxification of such pollutants is the demand of the present time. The use of microbial sources for the bioremediation of such type of petroleum wastes may be promising technique because it does not require any drastic conditions for detoxification process and by-products produced by them are also harmless unlike chemical and other techniques.

Vinay Mohan Pathak
Gurukul Kangri University, India

Navneet
Gurukul Kangri University, India

Acknowledgment

The completion of this book could not have been possible without the contribution and support we get from those around us, as always, our team and mentors. It is a pleasant aspect and expresses our gratitude for all of them.

Dr. Jitender Laura, Dr. Raj K. Salar, Dr. Neelam, Dr. Sanjay, Dr. Shiv Shankar, Dr. Reshu Chauhan, Dr. Mayank Pandey, Dr. Tarun Bisht, Dr. Pankaj Bhatt, Mr. Ajeet Singh, Mrs. Aparna, Miss. Minakshi Rajput, Mrs. Sanyogita Kumari, Mr. Akash, Mr. Neelesh Babu, and Mr. Ramveer Singh for their endless support, kind and understanding spirit during the completion of book.

We would also like to express our deep gratitude to publisher IGI global and all those people whose name could not figure out in this acknowledgement but who knowingly or unknowingly helped us.

Vinay Mohan Pathak
Gurukul Kangri University, India

Navneet
Gurukul Kangri University, India

Section 1
Biodegradation of Agricultural Waste Products

Chapter 1
Agricultural Waste Management for Bioethanol Production

Miss Priyanka
Lucknow University, India

Dileep Kumar
Lucknow University, India

Uma Shankar
Lucknow University, India

Anurag Yadav
S K Nagar Dantiwada Agricultural University, Gujarat, India

Kusum Yadav
University of Lucknow, India

ABSTRACT

This chapter contends that bioethanol has received the most attention over other fuels due to less emission of greenhouse gases and production from renewable sources. It is mainly produced from sugar containing feedstocks. Since feedstocks are utilized as food for humans, its consumption in bioethanol production creates a food crisis for the entire world. Bioethanol derived from agriculture waste, which is most abundant at global level, is the best option. Agriculture wastes contain lignin, cellulose and hemicelluloses which creates hindrances during conversion to ethanol. Pretreatment of agriculture wastes remove lignin, hemicelluloses and then enzymatically hydrolyzed into sugars. Both pentose and hexose sugars are fermented to bioethanol. There are still various problems for developing an economically feasible technology but a major one is the resistance to degradation of the agricultural material. Use of two or more pretreatment methods for delignification and the use of genetically modified agricultural biomass can be developed for economically feasible ethanol production.

DOI: 10.4018/978-1-5225-3540-9.ch001

INTRODUCTION

In the 20th century, the entire world was mostly depended on fossil energy sources, such as petroleum, coal and natural gas to produce fuels, chemicals, materials and power. Fossil fuel is mainly utilized in the transport and agricultural sectors. However, it contributes environmental pollution and global warming. In the past few decades, the harmful effects of fossil fuels on the environment such as increase in the global warming due to greenhouse gases (GHGs) emission (e.g. CO, CO_2, CH_4 and NO_2), continuous energy requirement, decrease in the energy supply sources and unavailability of stable oil market have attracted the for finding the alternative fuels (Sarkar et al., 2011). Various health diseases are also increased because of rapid increment in the GHG emissions. Transport sector contributes about 22% of worldwide GHG emissions. According to the data of The International Energy Agency (IEA) that global GHG emissions will increase by 92% between 1990 and 2020. Additionally, these GHG emissions are released in the atmosphere is expected to reach 8.6 billion metric tons from 2020 to 2035. This increment in GHG emission will increase the global temperature by 2°C which may cause the death of hundreds of millions of people of the world (Pimentel & Patzek, 2005).

Biofuels are the alternative source which can reduce the dependence on fossil energy sources. Many countries have put their target to develop biofuels as they have potential to reduce more than 80% of GHG emissions (Walker, 2011). These alternative fuels have already covered 2% of the total transport sector and the expectation is that it will be more promoted and utilized in the near future with technology and researches development. The alternative fuels have several benefits such as reduction of the environmental pollution, decreased emission of GHG, highly abundance of raw materials (Du et al., 2016).

In 1970s, after the oil crisis problems many countries have started searching for an abundant and low cost fuel. Continuous efforts towards the solution of the energy supply and the environmental impacts caused by the transportation and agricultural sector have led to the innovation of the alternative fuels through extensive research activities and search for renewable energy sources which have become a matter of widespread attention. Renewable energy sources like biofuels are the main alternatives to overcome the depletion of non-renewable energy (like petroleum, coal, or natural gas to produce fuels, chemicals, materials and power), increasing environmental pollution and increasing greenhouse gases. In biofuel includes biomethanol bioethanol, biohydrogen, biodiesel, vegetable oils, bio-oil, bio-char, biogas and bio-synthetic gas (Ayadi et al., 2016). Production of bioethanol from edible agricultural products may cause rise of cost of these crops leading to food insecurity and enhancement in the GHGs. Non-edible agricultural products must be investigated to overcome these problems.

Ethanol is considered as most potential alternative fuel due to fast depletion in the world reserves of petroleum (Prasad et al., 2007b). Ethanol is the most important alcohol that can be produced by converting the starchy material into alcohol along with the production of carbon dioxide (CO_2) (Duff & Murray, 1996). The fermentation process is an anaerobic catalyzed by enzymes produced from bacteria and fungi. In this process, yeast and heat are used to break down the starch and sugar containing materials into fermentable sugars and producing ethanol. The equation for conversion of glucose to ethanol is shown below:

$$C_6H_{12}O_6 \rightarrow 2C_2H_5OH + 2CO_2$$

During the fermentation process, the sugar is fermented by the yeast cells and converted into ethanol and carbon dioxide (Demirbas, 2005).

HISTORY OF BIOETHANOL

In ancient time, bioethanol had been prepared from sugar containing feedstocks using fermentation process. In 12-14th century, pure ethanol production started using distillation. Ethanol was mostly used for making the medical drugs and color pigments for painting in the middle ages. In the 12th century, starch containing feedstock was first used for ethanol production in Ireland. In 19th century, the high bioethanol production at industrial level was started due to the low cost of distilling process (Roehr et al., 2000).

German inventor Nicholas Otto had used ethanol as a fuel in an internal combustion engine in 1860. After many years, Henry Ford constructed his first automobile engine that was based on ethanol. In 1908, Ford used T engine equipped with carburettors that could be used ethanol, gasoline or a mixture of ethanol and gasoline (Solomon et al., 2007). First time, Ford told ethanol as "the future fuel". In 1925, he told that the future fuel can be produced from fruit, weeds, sawdust and vegetables. Fossil fuels were mostly used in automobile industry due to their lower production cost. Hydrous ethanol can be an alternative source than gasoline used in engines. On other hand, anhydrous ethanol has power to increase the octane number when it is mixed with gasoline (Turner et al., 2011). In the starting of the 20th century, ethanol used in various combustion engines, mostly for automobiles. Various ethanol production processes were developed based on sugar cane, beet molasses in 20th century (Roehr et al., 2000).

PHYSICOCHEMICAL CHARACTERISTIC OF ETHANOL

Bioethanol is a transparent, colorless liquid with pleasant odor. The taste of ethanol is varies according to its concentration; when it is diluted has sweet flavor and has a burning taste if it is concentrated. It is the second member alcoholic group which contain hydroxyl group. The melting and boiling points of ethanol are -114.1°C and 78.5°C, respectively. The density of ethanol is 0.789 g/mL at 20°C. It forms homogeneous mixture with both types of solvents i.e. polar as well as non-polar solvents. It is also used as organic solvent and utilized in perfumes, paints, lacquer and explosive industry. Ether is formed after the dehydration of ethanol. It can be further oxidized to acetaldehyde and then into acetic acid. The detailed physico-chemical properties of bioethanol are shown in the Table 1. Alcoholic solutions containing non-volatile substances are called tinctures and solution having volatile substances is called spirit (Kwon et al., 2012).

BIOETHANOL PRODUCTION POLICY OF SOME COUNTRIES

Brazil is known as the leading country in production as well as import of bioethanol in the world. In 1970s, the government of Brazil launched The National Ethanol Fuel Program at the time of oil crisis; series of measures were introduced to improve bioethanol production in order to reduce the oil import and to deal with the decrease of the national sugar price (Tan et al., 2015). Three important stages had been experienced by this program; between 1970 and 1990, the government started to encourage the developing of aqueous ethanol to handle with the situation of oil crisis. The stability in oil prices was recorded in the period from 1990 to 2000. Brazil succeeded in developing anhydrous ethanol with initial volume ratio of 20% ethanol mixed with gasoline, then 22% in 1993 and 25% in 2002. From 2000 to today, about 98% of ethanol is used in transport sector. Therefore, ethanol blended gasoline cars are

Table 1. Physicochemical characteristics of ethanol

Parameter	Characteristics
Molecular formula	C_2H_5OH
Molecular mass	46.07 g/mol
Appearance	Colorless liquid
Water solubility	Between –117°C and 78°C
Density	0.789 kg/l
Boiling temperature	78.5°C
Freezing point	–117°C
Flash point	12.8°C
Ignition temperature	lowest temperature of ignition
Explosion limits	Lower 3.5% (v/v) Upper 19%(v/v)
Vapour pressure at 38°C	50 mm Hg
Higher heating value (at 20°C)	29,800 KJ/kg
Lower heating value (at 20°C)	21,090 KJ/kg
Specific heat	Kcal/Kg 60°C
PKa	15.9
Viscosity	1.200 mPa.s (20°C)
Refractive index (*n*D)	1.36 (25°C)
Octane number	99

(Walker, 2011)

emerged in Brazilian market for transport. Bioethanol blended fuels are used in more than 80% of vehicles (Ayadi et al., 2016).

In Colombia, the government has made the biofuel policies in which it has been mentioned that it is mandatory blending a 10% bioethanol in cities whose population is above 5,00,000 (USDA, 2009). The law was applied in 2005 and this aim had become true after four years, when 75% of total consumed gasoline had 10% ethanol content. Colombian government aims to enhance the ethanol content up to 25% by 2020. Colombia offers many facilities to support and to encourage bioethanol production such as prices of sugarcane which used for production of the bioethanol are decided by the Colombian government and in addition, bioethanol is exempted from the VAT food processing with large quantities (Torres-Jimenez et al., 2011).

India is a developing country and has a considerable growth in the economy. The crisis for energy security would not end in India until alternative fuels developed from renewable sources to replace traditional fossil fuel. At present, molasses is mostly used for ethanol production in India. By 2017, Indian policy is proposed the goal, to blend 20% of bioethanol and biodiesel in gasoline. It has been already made compulsory to mix bioethanol in the gasoline which has been operative from October 2008. The Oil Marketing Companies has accepted the 5% combination of ethanol with gasoline in 20 States and four Union Territories of India. The permission has been granted by sugar industry to produce bioethanol directly from sugarcane juice to fulfil the requirement of bioethanol (Ministry of New and Renewable Energy (MNRE), 2009). Additionally, the sugar industry has also encouraged increasing bioethanol productivity to meet the requirements of bioethanol for blending in gasoline that are recommended from time to time by the Government of India. But it has been promised that this does not make any problem with the supply of sugar and requirement of alcohol for utilization in industry (Ministry of New and Renewable Energy, India 2009).

WORLDWIDE BIOETHANOL DISTRIBUTION

Brazil and the US are the two leading bioethanol producers in world. An about 62% world bioethanol produced by these countries. Sugarcane is the major crop biomass used for bioethanol production in Brazil. Corn grain is the major biomass for bioethanol production in the United States. Sugar or starch containing crops are the main source of ethanol production. Lignocellulosic biomass includes agricultural wastes (e.g. corn stover, crop straw, sugar cane bagasse, peels of banana, orange, forestry wastes and wastepaper etc.) are other potential sources for bioethanol production (Foyle et al., 2007). The use of lignocellulosic materials for bioethanol production is still under developed in some countries. The worldwide distribution of bioethanol production in different countries is shown in Table 2.

In the year 2014, the Table 2 scenario has been changed that global ethanol fuel production was increased up to 24,570 million gallons where USA (14,300) and Brazil (6,190) were the two leading bioethanol production countries.

COMPARATIVE ACCOUNT OF ETHANOL AND GASOLINE

Gasoline is a hydrocarbon used as fuel in transport sector. It is the by-product of crude oil which obtained after fractional distillation. It is hydrophobic and has a flash point of approximately -45°F, varying with octane rating. Vapour density lies between 3 and 4. It has a specific gravity of 0.72-0.76 and insoluble in water, due to these properties it floats on top of water. The gasoline internal combustion point lies from 280°C to 456°C.The boiling point of gasoline is varied from 37°C to 204°C. The high level of gasoline exposure for long time produced harmful effects to respiratory system (Ameri et al., 2008).

Bioethanol is an oxygenated fuel having 35% oxygen. Ethanol reduces particulate and nitrogen oxides (NOX) emission from combustion. Pure ethanol is polar solvent that is water soluble and has 55°F flash point. It is heavier than air vapour due to density of 1.59. The specific gravity of ethanol is 0.79, which means it is lighter than water. It is slightly soluble in water. The toxicity of bioethanol is less than gasoline and after blended with gasoline, utilized as automobile fuels. It reduces use of petroleum oil consequently it increases the oxygen in the fuel, which improve the combustion of gasoline consequently it produces less amount of greenhouse gas emission (Wang et al., 1999). The concentration of blending ethanol with gasoline is 10% bioethanol to 90% gasoline which is known as "E10" and commonly called "gasohol."

FEEDSTOCKS FOR BIOETHANOL

On the basis of feedstocks, bioethanol is classified into two types. Ethanol which is derived from feedstocks containing sugar and starch such as sugarcane (Figure 1), rice and wheat etc. is known as first generation bioethanol. These food sources are mainly utilized as food by the human population and they

Table 2. Bioethanol producing countries and amount of bioethanol produced in million gallons

Country	Ethanol Production (Million Gallons)		
	In 2004	In 2005	In 2006
Brazil	3,989	4,227	4,491
USA	3,535	4,264	4,855
China	964	1,004	1,017
India	462	449	502
France	219	240	251
Russia	198	198	171
South Africa	110	103	102
UK	106	92	74
Saudi Arabia	79	32	52
Spain	79	93	122
Thailand	74	79	93
Germany	71	114	202
Ukraine	66	65	71
Canada	61	61	153
Poland	53	58	66
Indonesia	44	45	45
Argentina	42	44	45
Italy	40	40	43
Australia	33	33	39
Rest of the World Total	545	909	477
World Total	10,770	12,150	13,489

(Balat, 2009)

are less abundant if these are used in bioethanol production it can cause a serious problem to the world population. So, there is a need to move on the alternative sources that do not interfere with the food security and produces less GHGs emission as compared to that of first generation ethanol production. If the ethanol is produced from feedstocks containing lignocelluloses such as crop residues, grasses, sawdust, wood chips and solid animal waste material, it is known as second generation bioethanol (Sun & Cheng, 2002). Advantages of second generation bioethanol include the reduction of GHGs emissions and highly abundance of waste rich in lignocelluloses. There are potentially 1×10^{10} metric ton lignocelluloses produced annually worldwide and it is the most plentiful renewable biomass. Hence, second generation bioethanol has the enormous potential to substitute fossil fuels without affecting the food-supply (Goh et al., 2010).

Figure 1. Schematic process of sugarcane to produce ethanol and sugar
(Dutton, n.d.)

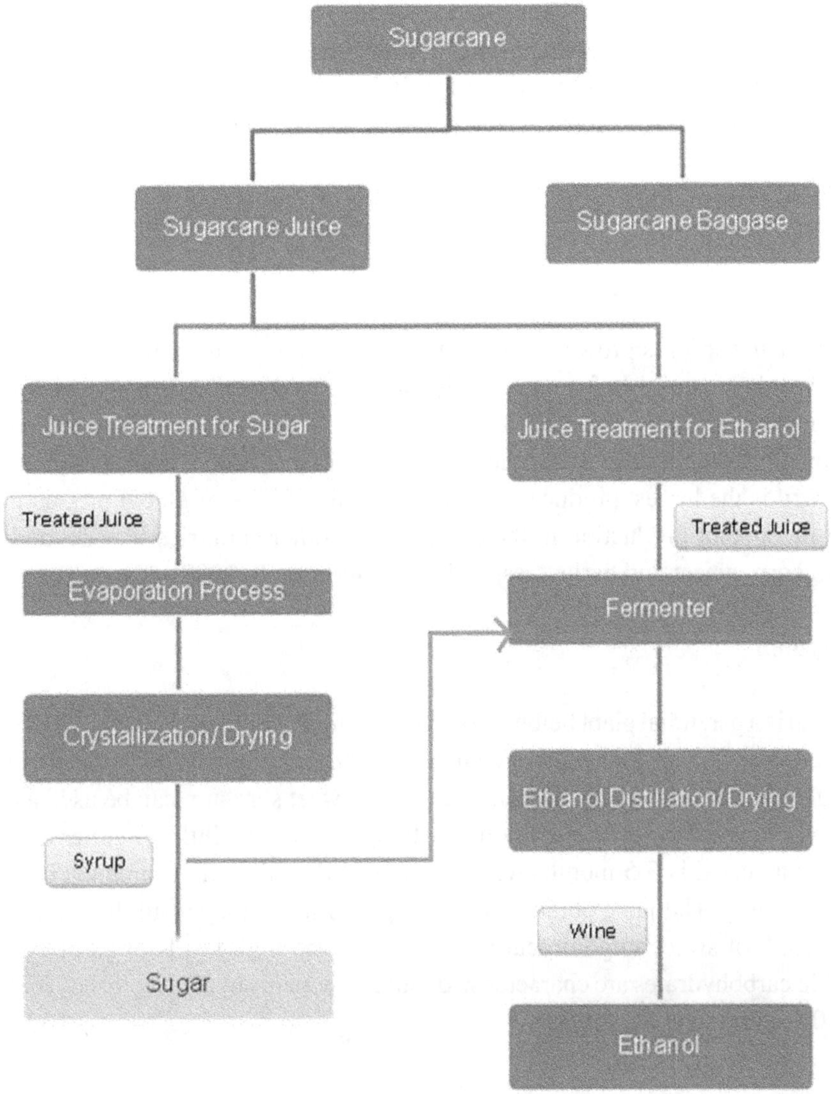

First Generation Bioethanol

Bioethanol produced from natural sources (e.g. sugarcane, sugarbeet and corn wheat etc.) is called first generation bioethanol. The first-generation bioethanol does not need any pretreatment before fermentation because their structure is simple. It is cheaper method for bioethanol production (Sun & Cheng, 2002).

Sources for First Generation Bioethanol Production

Production of biofuels requires new and cheap carbohydrate sources. Currently, biofuels made from crops, such as corn, sugar cane, wheat, and soybeans, for use as renewable energy sources. Though it may seem beneficial to use renewable plant materials for biofuel but the use of crop residues and other biomass for biofuels cause many problems, including food shortages and destruction of vital soil resources (Balat et al., 2009). Different types of biomass used in the production of biofuels are shown in Table 3.

Bioethanol feedstocks roughly divided into three major groups: (1) sucrose-containing feedstocks (e.g. sugar beet, sweet sorghum, sugar cane and fruits), (2) starch materials (e.g. wheat, rice, potatoes and corn) and (3) lignocellulosic biomass (e.g. grasses and wood etc).

Sugar Containing Plant Crops

1. Sugarcane

Sugarcane (*Saccharum* spp.) is grown in tropical and sub-tropical areas throughout the world. It is an important crop worldwide not only for sugar production, but also increasingly as a bioenergy crop due to its phenomenal dry matter production capacity. Most of worldwide bioethanol is produced from sugarcane and remaining bioethanol produced from other crops such as sugarbeet, sorghum, wheat, rice etc. (Dufey et al., 2006). Brazil is the largest producer of sugarcane (about 27% of global production) which is the main source of bioethanol production in the country. The other countries like US and Europe mainly uses starch from corn, wheat and barley, respectively (Linde et al., 2008).

2. Sweet Sorghum

Sweet sorghum is a perennial plant belongs to Poaceae family (Ratnavathi et al., 2010). Sweet sorghum reduces carbon emissions. Sorghum cultivated in temperate, subtropical and tropical climates. All components of the plant have economic value; the grain from sweet sorghum can be used as food, the leaves for forage, and the stalk (along with the grain) for fuel, the fiber (cellulose) either as mulch or animal feed. Its maturation period is 3-5 months which is shorter than that of sugarcane (10-12 months). It is also tolerance to salinity. The juice obtained from sorghum stalks may result high ethanol productivity (Figure 2). The stalk of sweet sorghum cultivars contains the high levels of carbohydrates (15-23%). Total fermentable carbohydrates are characterized into three main sugars; sucrose (70%), glucose (20%) and fructose (10%) (Prasad et al., 2007a).

Table 3. Different type of biomass used in the production of biofuels

S. No.	Biomass/ Feedstock	Biofuel	Country
1.	Sugarcane	Bioethanol	Brazil
2.	Corn	Bioethanol	Estados Unidos da America (EUA), Canada
3.	Sugar beet	Bioethanol	EUA
4.	Cassava	Biodiesel	China
5.	Wheat	Bioethanol	Canada
6.	Sweet Sorghum	Bioethanol	EUA
7.	Jatropha	Biodiesel	China
8.	Soybean oil	Biodiesel	EUA, Brazil

Figure 2. Schematic representation of ethanol production from sweet sorghum
(Almodares & Hadi, 2009)

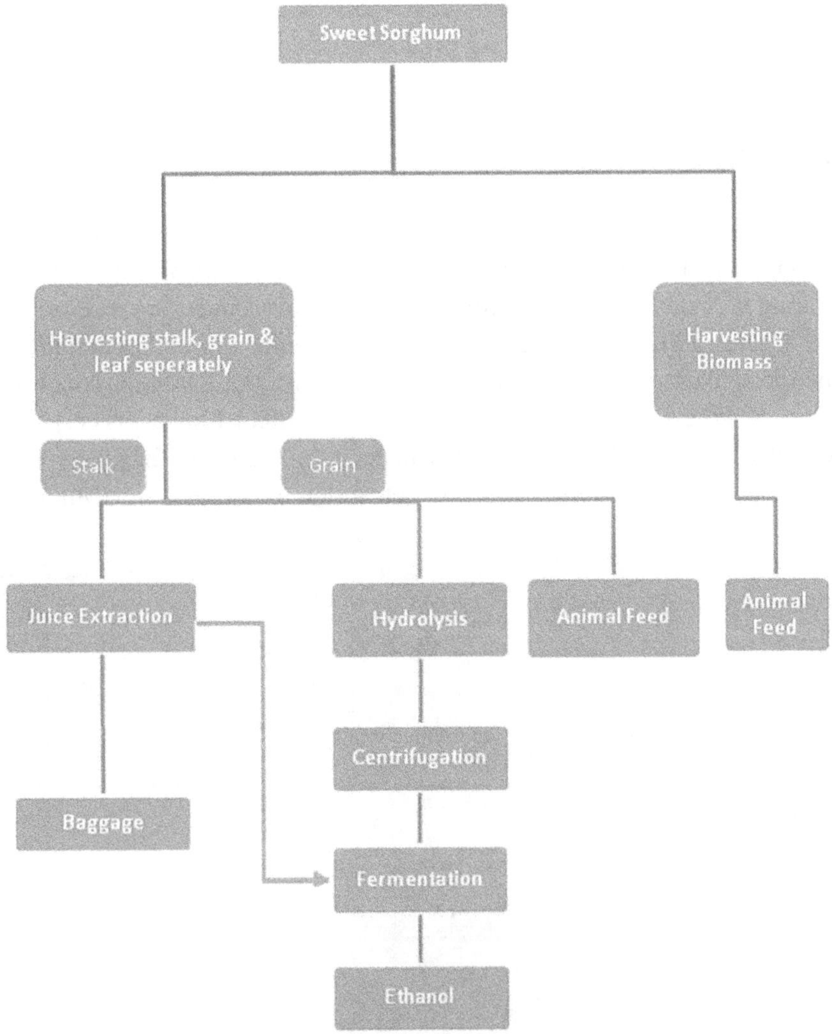

3. Sugarbeet

In European countries, beet molasses is the most utilized sucrose containing feedstock for bioethanol production. Sugar beet crop yields high amount of bioethanol than sweet sorghum and wheat. The advantage of sugar beet is shorter duration of crop production, high yield, high tolerance to broad range of climatic changes (e.g. drought and flood, etc.), low requirement of water and fertilizer (Balat et al., 2009).

Starch Containing Plant Crops

Among biofuels feedstocks, corn, wheat, and barley are examples of starch containing materials (Balat et al., 2008). Starch materials are pretreated with acid and/or enzyme to produce a high sugar concentration for biofuel production (Figure 3). It is a polymer made up of D-glucose units. Hydrolysis breaks down the starch into fermentable sugars. Bioethanol production from starch is required to hydrolyze the polysaccharide chains and produce glucose syrup which can be converted into bioethanol by the yeasts. Corn is the most common starch containing feedstock used for bioethanol production followed by wheat (Cardona et al., 2007). United States produced 14% corn-based bioethanol in comparison to other countries.

Second Generation Bioethanol

Bioethanol which is produced from agriculture waste is called second generation bioethanol. It is the best option for production of bioethanol. Conversion of these wastes into bioethanol would reduce global waste and generating a valuable fuel. Agricultural residues or byproducts are renewable and abundantly available (Sun & Cheng, 2002). Second generation biofuels emit low carbon, enhanced efficiency of energy and reduce dependency on energy (Taherzadeh & karimi, 2008). In some countries where cultivation of energy crops for bioethanol production is difficult, agriculture biomass is best option. Agriculture waste include wheat straw, barley husks, corn cobs, paper pulp, sugar cane bagasse, banana peel, orange peels and pineapple peels are examples of second generation feedstocks. Agricultural residues have a

Figure 3. Bioethanol production from starch feedstock
(Balat et al., 2009)

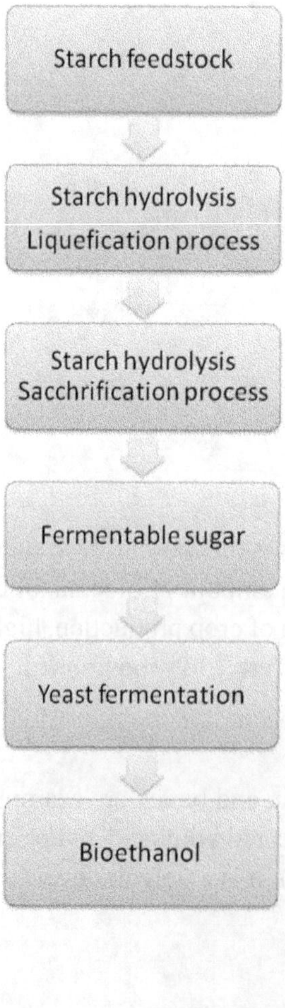

high potential because they are easily available in higher amounts at global level (Koltermann et al., 2014). Polysaccharides such as cellulose, hemicelluloses and lignin present in agricultural materials are of major interest as feedstocks for production of second generation ethanol.

Composition of Agricultural Waste

Agricultural waste is composed of three main components: cellulose (35-50%) of the total dry weight, hemicellulose (20-35%) and lignin (10-25%). Cellulose and hemicelluloses are joined to lignin through hydrogen and covalent bonds which makes it resistant to degradation (Verma et al., 2011). The composition of cellulose, hemicelluloses and lignin of common biomass is shown in Table 4.

Cellulose

Cellulose $(C_6H_{10}O_5)_n$ is a major component of agriculture biomass and found in plant cell wall. It is a linear, crystalline homopolymer polysaccharide made up of approximately 500-15,000 repeating units of D-glucose linked by β-1,4-glycosidic bonds (Figure 4). The structure of cellulose is rigid and compact so harsh treatment is required to obtain glucose (Gray et al., 2006). Cellulose chains are joined together by hydrogen and vanderwaal bonds and packed into microfibrils which make it crystalline in nature. Due to their crystalline nature cellulose becomes resistant to degradation into fermentable sugars (glucose, mannose etc.) (Singhania et al., 2010; Agbor et al., 2011).

Hemicellulose

Hemicellulose $(C_5H_8O_4)_n$ is second major component of agriculture matter. It is a highly branched, long chain heteropolymer consist of pentose sugar (xylanose and arabinose) and hexose sugar (galactose and

Table 4. Composition of different components of common biomass

Agriculture Waste Feedstock	Cellulose (%)	Hemicellulose (%)	Lignin (%)
Leaves	15-20	80-85	0
Nut shells	25-30	25-30	30-40
Softwoods stems	45-50	25-35	25-35
Grasses	25-40	35-50	10-30
Corn stover	40	29.60	23
Rice husks	36.70	20.05	21.30
Wheat straw	30	50	15
Barley husks	21.40	36.62	19.20
Rye straw	41.10	30.20	22.90
Oat straw	39.40	27.10	17.50
Rice straw	36.20	19.00	9.90
Corn stalks	35.00	16.80	7.00
Sugarcane bagasse	40.00	27.00	10.00

(Sun & Cheng, 2002)

Figure 4. Structure of cellulose

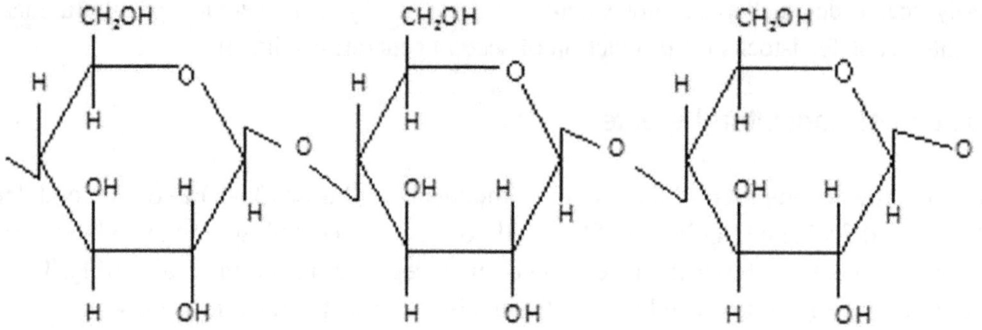

mannose) along with acetate units in its sidechains (Kumar et al., 2008) (Figure 5). It is less complex and easily converted into fermentable sugars such as pentose sugars and hexose sugars (Saha et al., 2007).

Lignin

Lignin ($C_6H_{11}O_2$) is the third major component of agriculture biomass. It is an aromatic polymer of phenyl propane units (p-coumaryl, coniferyl and sinapyl alcohol) are joined by C–O–C and C–C bonds (Figure 6) (Howard et al., 2003). It is highly resistant to degradation. It is tightly linked to cellulose and hemicelluloses in agriculture biomass. Therefore, this strong linkage affects the enzyme degradation. Lignin is an amorphous heteropolymer which makes the cell wall water-resistant and protects from microbial attack. This property of lignin creates the hindrance for the conversion agriculture waste to bioethanol. Therefore, delignification of agriculture biomass should be done for efficient conversion into ethanol (Vanholme et al., 2010).

CONVERSION OF AGRICULTURAL WASTE TO ETHANOL

Agriculture waste conversion to ethanol is typically done in four steps: (1) pretreatment of waste, (2) enzyme hydrolysis of pretreated waste to fermentable sugars, (3) fermentation of sugars to ethanol and (4) ethanol recovery (Figure 7). Pretreatment of agricultural biomass before hydrolysis can significantly

Figure 5. Structure of hemicellulose

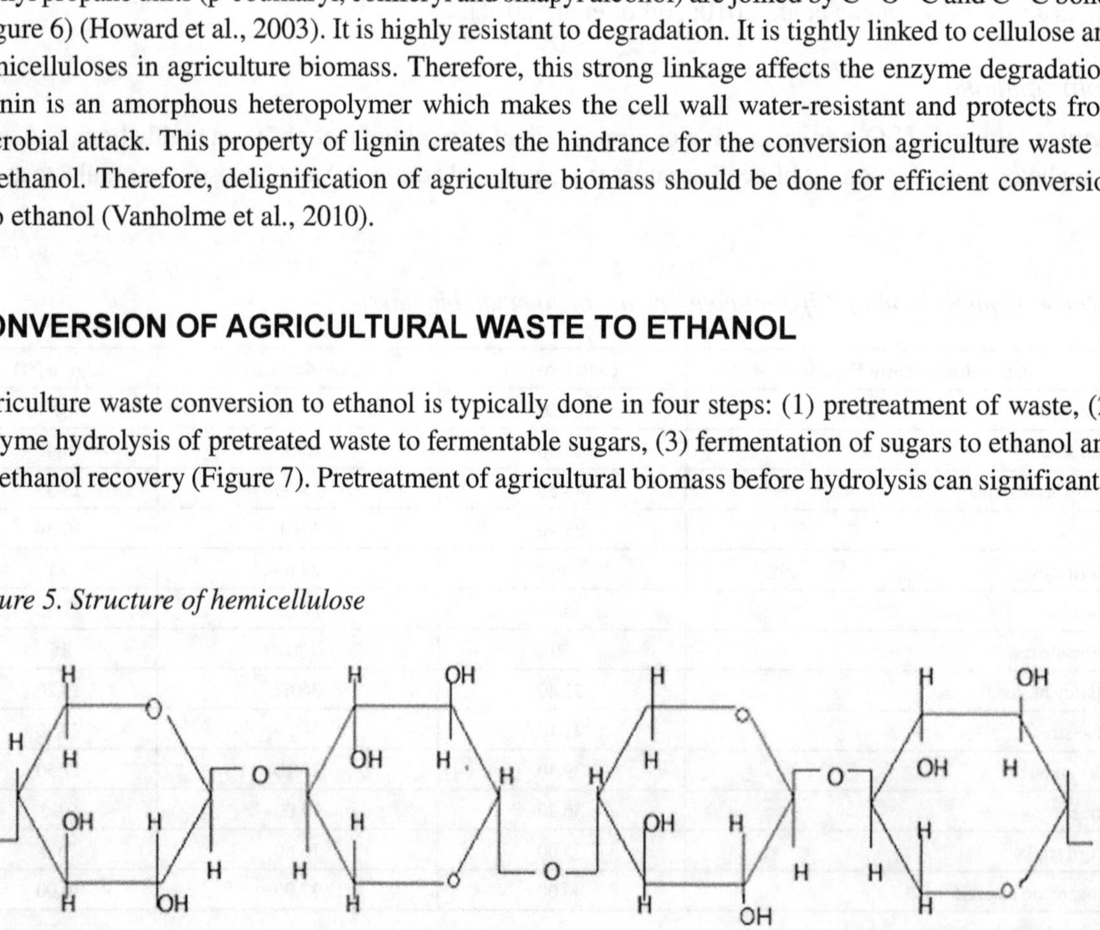

Figure 6. Structure of lignin

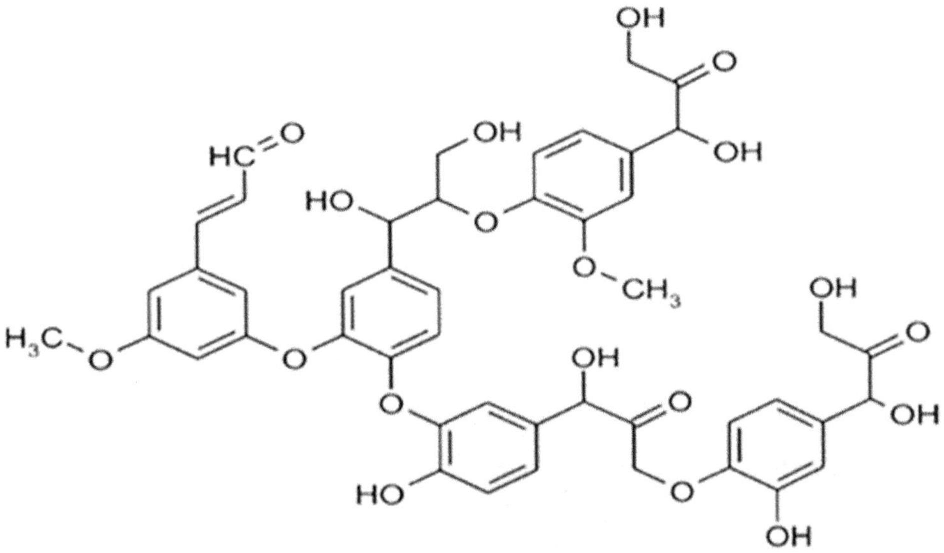

enhanced the hydrolysis efficiency by removal of lignin and hemicellulose; decrease cellulose crystallinity and increase of porosity (Sun & Cheng, 2002; Kumar et al., 2009). The enzymatic hydrolysis is performed by using cellulase enzymes followed by fermentation is carried out by microbes such as yeast and bacteria (Taherzadeh & Karimi, 2008).

Pretreatment

Pretreatment methods involve the separation and solubilization of the components of agriculture materials. Cellulose and lignin is the main components of agriculture biomass. Pretreatment is essential for the following reasons: (i) to break hemicelluloses and remove lignin barrier from agriculture materials, (ii) break the crystalline structure of cellulose, (iii) increase the available surface area and pore volume of the substrate, (iv) avoid the loss of sugars formed, (v) avoid the formation of inhibitors which inhibit the fermentation process, (vi) to minimize energy demands and (vii) to reduce costs. These changes in agriculture biomass make it easier for enzymatic hydrolysis, results in higher levels of fermentable sugars (glucose, xylose, arabinose, mannose and galactose) and will have a major impact on the overall process (Sun & Cheng, 2002; Yang & Wyman, 2008).

An efficient and economical pretreatment method should fulfill the following conditions (i) simple operation, (ii) limited use of energy, process water and chemicals, (iii) low equipment corrosion, (iv) high recovery of valuable hemicelluloses and lignin derived products, (v) decrease production of unwanted degradation products and (vi) reduced generation of wastes (Taherzadeh & Karimi, 2008).

Four different pretreatment have been used for the bioethanol production from agriculture waste: (1) physical (e.g. milling, grinding, pyrolysis and irradiation), (2) chemical (e.g. alkali, dilute, concentrated acid, oxidizing agents and organic solvents), (3) physicochemical (e.g. steam explosion pretreatment, ozonolysis, liquid hot water and wet oxidation) and (4) biological (fungi and bacteria) (Taherzadeh &

Figure 7. Flow chart of second generation ethanol production from agricultural waste
(Joshi et al., 2011)

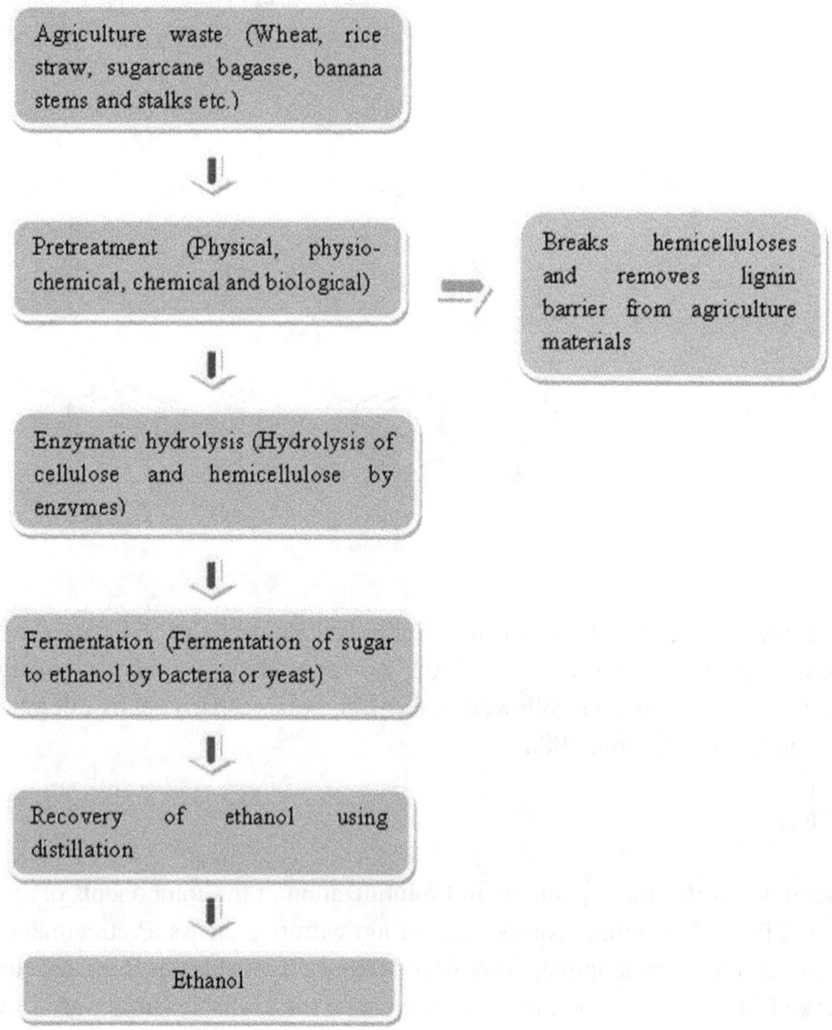

Karimi, 2008; Kumar et al., 2009; Gnanambal et al., 2012; Mohanty et al., 2016) which are summarized in Figure 8.

Physical Methods

Various types of mechanical and non-mechanical methods can be used for the physical pretreatment of agricultural waste. Mechanical methods include milling, grinding and chipping of agriculture materials. In non-mechanical methods includes Irradiation e.g., gamma rays, electron beam, microwaves pretreatment to hydrolyze the agricultural waste (Zheng et al., 2009). The main advantage of physical pretreatment is that it is chemicals free, fermentation inhibitors are not produced and is environment friendly. However, these methods are very time consuming and require good skills and mechanical power (Hendriks et al., 2009).

Figure 8. Different pretreatment methods for agricultural waste
(Kumar & Sharma, 2017)

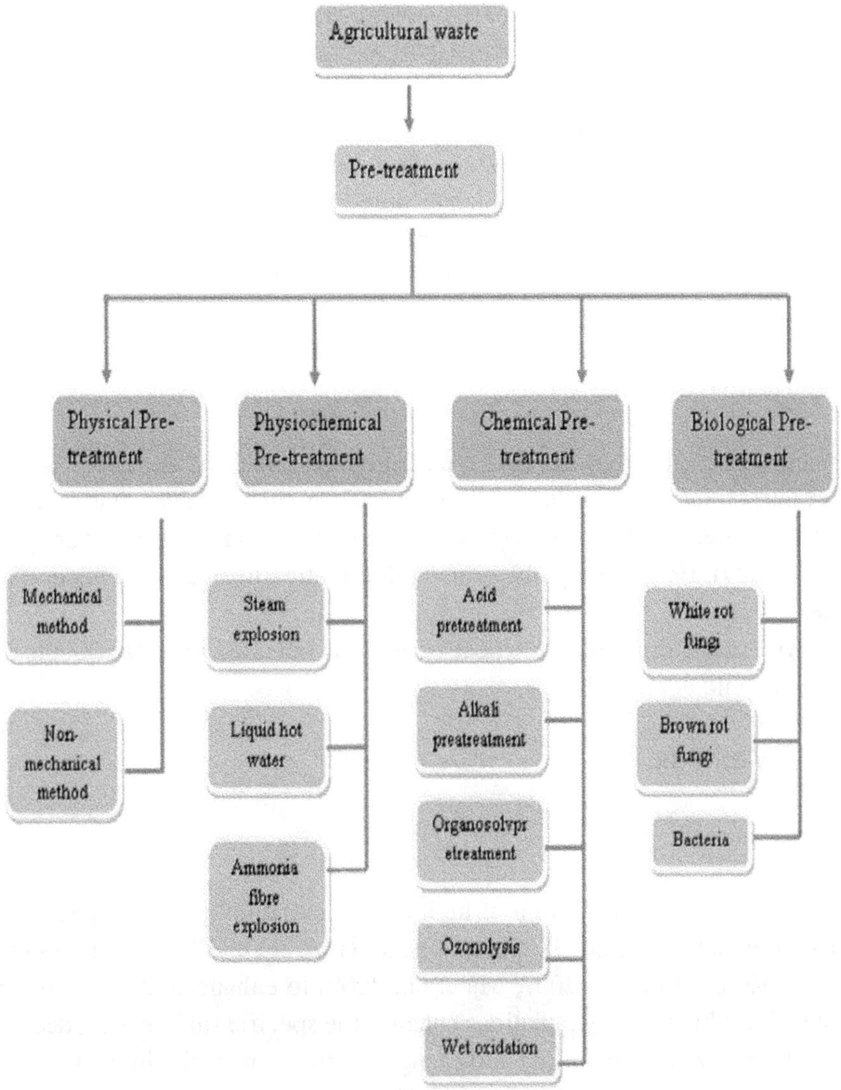

Mechanical Methods

In this method involves the combination of chipping, grinding, shearing, or milling, which decreases the particle size and increases surface area thus facilitating the cellulases to attack on biomass surface and enhanced the conversion of cellulose to ethanol.

1. Milling

Milling is a type of mechanical pretreatment of agricultural biomass. This pretreatment decreases particle size and increase surface area of agricultural material in this way cellulose becomes more accessible to enzymatic hydrolysis (Harmsen et al., 2010). Different types of milling such as ball milling, hammer milling, two-roll milling, disk milling and colloid milling are used for bioethanol production from agriculture biomass. The size of final particle obtained depends on the type of physical treatment method for e.g. size of particle obtained up to 10-30 mm after chipping and 0.2-2 mm after milling, respectively (Sun & Cheng, 2002). The main disadvantage of milling is the requirement of high energy.

Ball milling (BM) and wet disk milling (WDM) pretreatment on sugarcane bagasse and straw were compared. It has been seen that BM decreases the crystallinity resulting in enhanced enzymatic hydrolysis, while WDM remove fibres from waste which favoured enzymatic conversion (da Silva et al., 2010). Centrifugal roller mill pretreatment were given to corn straw which resulted in increase in surface area by a factor of 4.8, decrease in cellulose crystallinity and enhanced the yield of low molecular carbohydrates by a factor of 2 (Bychkov et al., 2014).

2. Pyrolysis or Thermal Cracking

In this treatment, waste is treated with high temperature of 300°C. It breaks down cellulose and produces gaseous products for e.g. hydrogen, carbon monoxide gas and residual char. At lower temperature (<300°C), the cellulose decomposition becomes slower and less volatile products are produced (Prasad et al., 2007b; Mtui, 2009). However, pyrolysis method of pretreatment is more expensive due to use of high temperature.

In a study, 80 wt% of the dry lignin of wheat straw was converted into 40-60% bio-oil and 30-40% biochar by pyrolysis (Wild et al., 2012).

Non Mechanical Methods

1. Irradiation Pretreatment

In this pretreatment involves the utilization of high energy radiation including γ-ray (Chunping et al., 2008), ultrasound (Rehman et al., 2013), electron beam (Bak et al., 2009; Henniges et al., 2013) and microwave heating (Keshwani et al., 2007; Ma et al., 2009) to enhanced the enzymatic hydrolysis of agricultural waste. These high energy radiations enhance the specific surface area, decrease the crystallinity of cellulose, hydrolyze hemicellulose and change the structure of lignin. However, these methods are expensive at large scale (Taherzadeh & Karimi, 2008).

Microwave irradiation is one of the most common pretreatment method for agriculture waste. It has several advantages includes (i) easy and simple operation, (ii) less energy requirement, (iii) generation of low level of inhibitors and (iv) degrades or disrupt the structure of cellulose (Kumar & Sharma, 2017). Pretreatment of sweet sorghum bagasse using microwave radiation was evaluated at different lime doses (0, 0.1, 0.15, and 0.2 g/g sorghum bagasse), water content of 10 or 20 ml/g sorghum bagasse, at exposure times of 2, 4, and 6 min. At optimum conditions of 0.1 g lime, 10 ml water/g sorghum bagasse in 4 min, sugar yield of 32.2g/100g sorghum bagasse was produced (Choudhary et al., 2012).

Chemical Methods

Chemical pretreatment involves the use of different chemical agents such as oxidizing agents (hydrogen peroxide and ozone), acids (H_2SO_4, HCl and organic acids etc.), alkalis (NaOH, Na_2CO_3, $Ca(OH)_2$ and

NH_3 etc.), organic solvents, SO_2, CO_2 and other chemicals which degrades the hemicellulose and removes lignin from lignocellulosic waste materials (Nwosu-Obieogu et al., 2016).

Acid Pretreatment

Acid pretreatment is a method in which agricultural waste is treated with dilute or concentrated acids (e.g. sulphuric acid, hydrochloric, peracetic acid or nitric acid). This method hydrolyzes the polysaccharides, especially hemicelluloses to monosaccharides resulting in increased accessibility of cellulose to enzymatic hydrolysis. It can be done either under low concentration of acid and high temperature or under higher concentration of acid and lower temperature (Taherzadeh & Karimi, 2008).

$$\underset{\text{Lignocelluloses}}{\left(C_6H_{10}O_5\right)} \xrightarrow[\text{Conc. } H_2SO_4]{+H_2O} \underset{\text{Reducing sugars}}{C_6H_{12}O_6}$$

Acid pretreatment have been utilized for many agriculture wastes. For instance, sorghum was pretreated with (0.5-4%) H_2SO_4. After enzymatic hydrolysis and pretreatment, 0.408g reducing sugars are produced (Akansha et al., 2014). In another study, durian seed waste was pretreated with 0.6% sulphuric acid followed by enzymatic hydrolysis resulted in 50.0944g/L of glucose content (Ghazalia et al., 2016). The disadvantage of using acid is causing corrosion to the equipment and formation of inhibitors (furans, carboxylic acids and phenolic compounds) which inhibits the fermentation process (Singh & Trivedi, 2013).

Alkaline Pretreatment

Alkaline pretreatment is a process in which agricultural waste is pretreated with bases, such as sodium, potassium, calcium and ammonium hydroxide. Alkali pretreatment results in (i) breaks down the ester bonds between lignin, hemicellulose and cellulose, (ii) disruption of the structure of lignin, (iii) decrease the crystallization of cellulose and (iv) enhanced the accessibility of enzyme to cellulose (Singh et al., 2013). Alkali method has many advantages such as requires low temperatures, pressure, removes hemicelluloses, lignin, and increases accessible surface area. The limitation of this method includes long residence times and high concentration of alkalis is required. Among NaOH and $Ca(OH)_2$, $Ca(OH)_2$ pretreatment is most favored because it is cheaper, safer and can be easily removed from the hydrolysate by reaction with CO_2 as compared to NaOH (Mosier et al., 2005b). Several studies have been done using alkaline pretreatment. Wheat straw was pretreated with 0.18g of NaOH and 0.06g of lime followed by enzymatic hydrolysis. 62.5% delignification and 93.1% conversion of cellulose to glucose after enzymatic hydrolysis were achieved. A total of $80.3 \pm 1.2\%$ yield of monosaccharide sugars (glucose plus xylose and arabinose) from cellulose and hemicellulose of wheat straw were obtained (Jaisamut et al., 2013). Pretreatment of coffee pulp was done with 4% (w/v) NaOH for 25 min. After NaOH treatment, 25.19% delignification was achieved and 38.13g/L of reducing sugars were produced (Menezes et al., 2013).

Organosolv Pretreatment

In this method, agriculture waste is pretreated with organic or aqueous solvents such as methanol, ethanol, acetone, ethyleneglycol, triethylene glycol and tetrahydrofurfuryl alcohol at temperatures of 150-200°C.This pretreatment can be performed with or without using catalysts such as oxalic, salicylic, acetylsalicylic, hydrochloric and sulphuric acid (Taherzadeh & Karimi, 2008). These solvents break the lignin and hemicellulose bonds (Joshi et al., 2011). Five solvents namely, ethanol, methanol, diethylene glycol, acetone and butanol were examined for organosolv pretreatment of wheat straw. The biorefinery process was catalyzed by sulphuric acid at 160 °C for 20 min. Out of these solvents; butanol removed maximum lignin (Sidiras et al., 2015). Pretreatment of wheat straw were successfully done by using solvents and ionic liquids, such as formic acid (formalin process), concentrated phosphoric acid (CPA), N-methylmorpholine-N-oxide (NMMO) and 1-allyl-3-methylimidazolium chloride ([AMIM]Cl). CPA post treatment showed the best efficacy. Among these solvents, CPA produced best results and obtained an ethanol concentration of 41.6g/L with 91.2% of yield (Li et al., 2017). This treatment having the advantage that organic solvents can be recovered and recycle due to easy process of distillation (Harmsen et al., 2010). The disadvantages are that formation of toxic inhibitors, corrosion occurs due to the requirement of organic acids and the process is expensive due to use of high pressure and temperature (Bensah & Mensah, 2013).

Ozonolysis Pretreatment

The process in which agricultural waste is treated with ozone is called ozonolysis. Ozone is a powerful oxidant, soluble in water and easily available. This pretreatment efficiently degrades lignin and increases the biomass digestibility. Hemicellulose is partially degraded by ozone while the cellulose remains unaffected (Travaini et al., 2016). Ozone has been used to degrade lignin in feedstocks such as bagasse (Eqra et al., 2014), maize stover (Li et al., 2015), wheat, rye straw (Garcia-Cubero et al., 2009; Aljibouri et al., 2015) and poplar sawdust (Vidal & Molinier, 1988).The advantage of ozone pretreatment includes harmful inhibitors does not produced and the reactions are carried out under room temperature and pressure (Vidal & Molinier, 1988). Ozone is required in high amount which makes this process costly. Therefore, this method is not efficient to use at industry level (Kumar et al., 2009).

Wet Oxidation Pretreatment

Agricultural waste is treated with water or air at elevated temperature (120°C for 30 min) in the presence of oxygen (Varga et al., 2003). This method is used for waste with low lignin amount, because the efficiency of delignification decreases with increasing the lignin amount. Wet oxidation pretreatment has been used in several agricultural feedstocks such as rape straw (Arvaniti et al., 2012), sugarcane bagasse, rice hulls, cassava stalks and peanut shells (Banerjee et al., 2012; Martin et al., 2007). This treatment has following advantages: (i) degrade hemicellulose and lignin under lower temperature, (ii) recovery of chemical is not required and (iii) higher enzymatic hydrolysis efficiency. However, this process is expensive due to supply of high pressure oxygen and chemicals (Biswas et al., 2015).

Physicochemical Methods

This category includes mixture of pretreatment methods of physical and chemical methods.

Steam Explosion (Autohydrolysis) Pretreatment

Steam pretreatment is one of the most commonly used methods for pretreatment of agriculture waste. This pretreatment involves the use of hot steam (180 to 240°C) under pressure (1 to 3.5 MPa) and pressure is instantly decreased which causes the explosive decomposition of agricultural biomass (Sun & Cheng, 2002). This method was earlier known as "steam explosion" because it is based on explosive action on the cellulose fibres which makes the materials accessible to hydrolysis (Mosier et al., 2005). This method has been used very efficiently to remove cellulose fibres in meadow hay (Tutt et al., 2014). The advantages of steam explosion pretreatment include the less energy requirement (Sun & Cheng, 2002; Kumar et al., 2009). Limitations of steam explosion include incomplete cleavage of the lignin-carbohydrate matrix, destroy a part of the xylan fraction, the generation of inhibitors products (e.g. furaldehyde, 5-hydroxymethyl-2-furaldehyde and phenolic compounds) that may inhibit enzymatic hydrolysis and fermentation (Garcia-Aparicio et al., 2006).

Liquid Hot Water Pretreatment (LHW)

In this method, hot liquid water having pressure above the saturation point is used to treat the agricultural waste to hydrolyze hemicelluloses. The main advantage of this process has high xylose recovery, no use of chemicals or catalyst which making this process eco-friendly in nature (Hamelinck et al., 2005). Limitation of LHW includes the production of inhibitors which inhibit the growth of microorganisms during fermentation process (Taherzadeh & Karimi, 2008). LHW has been successfully utilized for the pretreatment of rye straw, rice straw, corn stover and sugarcane bagasse for hydrolysis of hemicelluloses (Ingram et al., 2009; Imman et al., 2012; Li et al., 2014; Gurgel et al., 2014).

Ammonia Fibre Explosion (AFEX)

It is based on steam explosion process and the use of liquid ammonia at medium temperature (60-120°C) and high pressure (1.72-2.06 MPa) for several minutes (>30 min) (Kumar et al., 2009). Rapid release of ammonia gas breaks the lignin-carbohydrate complex resulting increased enzymatic digestibility of agricultural waste (Chundawat et al., 2007). AFEX has been used for pretreatment of corn stover. This treatment increased metabolic yield and ethanol production (Lau & Dale, 2009). It has been also used to pretreat sugarcane bagasse and cane leaf residues. The AFEX pretreatment increased the accessibility of hemicellulose and cellulose to enzyme hydrolysis (Krishnan et al., 2010). AFEX method has following advantages: (i) inhibitors such as furans are not produced, (ii) high depolymerisation of biomass is achieved, (iii) cheaper and less energy requirement (Chundawat et al., 2007; Sanchez, 2009).

Biological Pretreatment

Agricultural waste can be treated with either lignin-degrading microorganism, such as white, brown-rot and soft-rot fungi or enzymes (Lee, 1997). It has been found that brown-rot fungi attack on cellulose

(Ray et al., 2010). On the other hand, white-rot and soft rot fungi mainly degrade the lignin, cellulose and hemicellulose component of agricultural waste and enhancing enzymatic hydrolysis of waste. Various fungi such as *Trichoderma reesei* and *Aspergillus niger* produce several extracellular cellulolytic enzymes including endoglucanases, cellobiohydrolases (exoglucanases) and β-glucosidases which efficiently degrade cellulole residues. Basidiomycetes (e.g. *Phanerochaete chrysosporium)* degrade agriculture materials (Dashtban et al., 2009; Sanchez, 2009). Biological pretreatment using white rot fungi and *Streptomyces* have been studied in pulp for removal of lignin. In soft wood10.5% and hard wood 23.5% loss of lignin was observed (Saritha et al., 2012).

Several studies using biological treatment have been performed on agricultural waste using enzymes produced by micro-organisms. 85.6% delignification was achieved (55.2 g/L) from sawdust treated with cellulose enzyme derived from *Trichoderma/Hypocrea* (Saravana kumar & Kathiresan, 2014). The rice straw was biologically pretreated with white rot fungi, *Pleurotusostreatus* for 44 days to remove lignin. This pretreatment degraded 27.85% of lignin and 24.30% of cellulose in 24 days (Balasubramaniam & Rajarathinam, 2013).

The advantage of this pretreatment involves no use of chemicals, less energy needed, high specificity to lignin, mild operating conditions, no inhibitors generation and environment friendly (Sanchez, 2009). The main drawback is requirement of long residence time due to lower hydrolysis rate by fungi (Sun & Cheng, 2002). The limitations of the biological pretreatment make it less attractive at pilot scale (Eggeman & Elander, 2005) (Table 5).

Enzymatic Hydrolysis

Enzymatic hydrolysis is a process in which agriculture biomass hydrolytically break down into fermentable sugars by enzymes. Enzymatic hydrolysis is becoming more attractive as compared to acid hydrolysis because it avoids equipment corrosion, needed less energy and mild environment conditions, while fewer fermentation inhibitors are formed (Sanchez et al., 2004).

Three major enzymes namely, endoglucanase, exoglucanase and cellobiase are used for the hydrolysis of cellulose. The endoglucanases attack and cleaved the cellulose chains into glucose, cellobiose and cellotriose. The exoglucanases attack the non-reducing end of cellulose and form the cellobiose units. Cellobiase converts cellobiose units into D-glucose (Joshi et al., 2011). Hemicelluloses structure is more complex therefore, several enzymes are needed for their degradation. For example, for xylan hydrolysis, the action of various enzymes is required such as endo-β-1,4-xylanase, β-xylosidase, and accessory enzymes like α-L-arabinofuranosidase, α-glucuronidase, α-galactosidase, acetylxylan esterase and feruloyl esterase (Moreira & Filho, 2016).

Hydrolytic enzymes obtained from both bacterial and fungal sources are used for hydrolysis of agricultural materials. Enzymes from the bacterial sources such as *Clostridium, Cellulomonas, Bacillus, Thermomonospora, Ruminococcus, Bacteriodes, Erwinia, Acetovibrio, Microbispora* and *Streptomyces* genera have been used extensively. Fungal enzymes obtained from *P. chrysosporium, Trichoderma, Aspergillus, Schizophyllum, Humicola, Acremonium* and *Penicillium* have been studied for production of hemicellulase and cellulases. Due to the limitations of bacterial species such as anaerobic nature of some bacteria and the generally low yields obtained from bacteria which attracted the scientists to the use of fungal strains for commercial enzyme preparations (Sun & Cheng, 2002).

Table 5. Methods used for pretreatment of agricultural waste

Pretreatment Methods	Operating Conditions	Advantages	Limitations and Disadvantages	Biomass Treated
		Physical		
Milling	Room temperature Energy input < 30Kw/ton biomass	Reduces crystallinity of cellulose	High power consumption	Corn straw, Sugarcane bagasse and straw
Pyrolysis	High temperature of 300°C	Maximum conversion of cellulose to fermentable sugars	Expensive	Wheat straw
Irradiation	γ-ray, ultrasound, electron beam and microwave heating	Enhanced the specific surface area, decrease the crystallinity of cellulose, hydrolyze hemicelluloses and change the structure of lignin.	Expensive	Sorghum bagasse
		Physio-Chemical		
Steam Explosion	160-260°C (0.69- 4.83MPa) for 5-15 min, followed by decompression until atm. pressure	Low energy requirement	Incomplete cleavage of the lignin-carbohydrate matrix, destroy a part of the xylan fraction, formation of inhibitors that may inhibit hydrolysis and fermentation process	Meadow hay
Liquid Hot Water (LHW)	Hot liquid waterhaving pressure above the saturation point	High xylose recovery, no use of chemicals or catalyst	The formation of inhibitors which inhibit the growth of microorganisms during fermentation process	Rye straw, rice straw, corn stover and sugarcane bagasse
Ammonia Fiber Explosion (AFEX)	Anhydrous ammonia 60-100 °C 250-300 psi followed by decompression	removes lignin and hemicellulose, increases accessible surface area, no toxic inhibitors are generated	Not recommended for waste with high amount of lignin	Corn stover, sugarcane bagasse and cane leaf residues
Wet Oxidation	148-200°C for 30 min	Efficiently removed lignin and low inhibitors formation	Expensive due to use of oxygen and acid catalyst	Rape straw, sugarcane bagasse, rice hulls, cassava stalks and peanut shells
Ozonolysis	Room temperature	Decrease lignin content, does not produce toxic products	Costly due to requirement of large amount of ozone	Bagasse, maize stover, wheat, rye straw and poplar sawdust
		Chemical		
Acid Hydrolysis	Dilute (0.5-3%) H_2SO_4, HCl or HNO_3, at 130-200°C/3-15 atm pressure. Conc. (10-30%) H_2SO_4, 170-190°C	Hydrolyzes hemicellulose to xylose and other sugars, change lignin structure	Expensive, equipment corrosion, toxic substances are formed	Sorghum and durian seed waste
Alkaline Hydrolysis	Low temperature; long time, high concentration of the base	Lower temperatures and pressures than other pretreatment techniques, removes hemicelluloses and lignin, increases accessible surface area	Long residence times and high concentration of alkalis are required,	Wheat straw and coffee pulp
Organosolv Pretreatment	150-200°C with or without addition of catalysts (oxalic, salicylic and acetylsalicylic acid)	Hydrolyzes lignin and hemicelluloses	Need to recovered organic solvents from the reactor, expensive	Wheat straw
Biological Pretreatment	Several fungi (brown-, white- and soft-rot fungi	Inexpensive, degrades lignin and hemicelluloses, low energy requirements	Lower hydrolysis rate and time consuming	Sawdust and rice straw

(Kumar et al., 2009)

Enzymatic hydrolysis is most costly steps in production of bioethanol from agriculture waste because of requirement of high amount of enzymes and the price of cellulases is high. Thus, it becomes necessary to reduce enzyme loading. It has been observed that certain additives such as surfactants had been used and showed a positive effect on enzymatic hydrolysis of agriculture biomass. The use of additives increased hydrolytic efficiency and lower enzyme dosage is required (Qing et al., 2010; Cui et al., 2011).

Fermentation

During fermentation, both pentose and hexose sugars (glucose) are fermented to bioethanol and carbon dioxide under aerobic/anaerobic conditions:

$$\underset{\text{Glucose}}{C_6H_{12}O_6} \rightarrow \underset{\text{Ethanol}}{2C_2H_5OH} + \underset{\text{Carbon dioxide}}{2CO_2}$$

Saccharomyces cerevisiae (baker's yeast) is most commonly used for bioethanol production by fermenting the sugars (e.g. glucose, fructose, maltose and sucrose) obtained from first generation feedstocks. The main limitations of baker's yeast that it is not capable of fermenting the pentose sugars (Walker, 2010). Therefore, genetically modified *S. cerevisiae* and non Saccharomyces yeasts are used for fermenting pentose sugars (e.g. xylose, arabinose) obtained from second generation feedstocks (Matsushika et al., 2009; Oreb et al., 2012; Sinumvayo et al., 2015). Some other microorganisms such as *Pichiastipitis, Candida shehatae and Pachysolantannophilus* are also capable of fermenting both hexose and pentose sugars (Lin & Tanaka, 2006).

Thermophilicbacteria (e.g. *Thermoanaerobacterium saccharolyticum, Thermoanaerobacter ethanolicus*) are of recent interest for bioethanol production from agricultural residues. These bacteria have several advantages for ethanol production such as (i) improved solubility of substrates, (ii) enhanced mass transfer due to decreased viscosity, (iii) increased rate of diffusion, (iv) high bioconversion rates, (v) use variety of low-cost biomass feed stocks, (vi) low risk of contamination and (vii) ferment both pentose and hexose sugars found in agriculture hydrolysates. These bacteria also produce cellulase and hemicellulase enzymes which ferment biomass to ethanol without the addition of external hydrolytic enzymes results in decreased cost (Ahmad & Qazi, 2014).

Various systems are employed for processing and fermenting the agricultural waste. These include batch, fed batch, simultaneous saccharification and fermentation (SSF) (Olofsson et al., 2008), simultaneous saccharification and co-fermentation (SSCF), separate hydrolysis and fermentation (SHF), consolidated bioprocessing (CBP) (Lynd et al., 2005). The features, advantages and limitations of these systems are summarized in Table 6.

Separation and Distillation

Distillation separates ethanol from water on the basis of the differences of volatilities. First, the mixture containing lignin, enzymes, unreacted cellulose and hemicellulose, yeast and various salts in the fermentation broth can be added to a distillation (beer) column to concentrate the ethanol in the overhead product and water. Solids removed from the bottom of the distillation device. Ethanol can then pass to a rectification column for concentration of the ethanol-water mixture to the azeotrope composition of

Table 6. Various systems used for processing and fermenting the agricultural waste

Fermentation Process	Features	Advantages	Disadvantages
Batch Fermentation	The batch process is a closed system in which initially limited amount of nutrients provided and then inoculated with microorganisms for fermentation (Olsson & Han-Hagerdal, 1996)	Very simple method	Low productivity with high labour costs, low cell density
Fed Batch Fermentation	The microorganism works at low substrate concentration with an increasing ethanol concentration during the course of fermentation process.	Gives higher yield and productivities than batch cultures, maximum viable cell concentration and less time is required	Labor intensive
Continuous Fermentation	It is a system in which substrate, culture medium and nutrients is added continuously in agitated culture vessel	Gives higher productivity than batch fermentation at low dilution rates	Long growth periods increased the risk of contamination
Simultaneous Saccharification and Fermentation (SSF)	The enzymatic hydrolysis and fermentation is performed simultaneously.	Cheaper because less number of vessels needed for combined enzymatic hydrolysis and fermentation and higher overall ethanol yield, less enzyme requirement and a shorter process time	The temperature needed for enzymatic hydrolysis (45-50°C) is higher than that of fermentation (30°C) and the yeast cannot be reused in an SSF process due to the problems of recovering the yeast from the lignin after fermentation
Separate Enzymatic Hydrolysis and Fermentation (SHF)	Pretreated agricultural wastes are first enzymatically hydrolyzed to glucose followed by fermentation into ethanol.	The temperature for hydrolysis and fermentation can be optimized	Microbial contamination, hydrolysis process is long and end product inhibited when glucose and cellobiose accumulate
Simultaneous Saccharification and Cofermentation	Fermentation of both pentoses and hexoses sugars to ethanol occur in single bioreactor with single microorganism.	Cheaper, short processing time, high ethanol yield	-
Consolidated Bioprocessing (CBP) or Direct Microbial Conversion (DMC)	Production of ethanol along with all of the required enzymes in single bioreactor and single microorganism.	Cheaper	Low yield of ethanol due to formation of byproducts (acetate and lactate), less tolerance of microorganism to ethanol

about 95% by weight (Hamelinck et al., 2005). The remaining product is added to the stripping column to remove extra water from this product. The bottoms solid from the first column can be further concentrated by centrifugation. If the ethanol is used as a hydrous fuel, the azeotropic mixture would be used. If anhydrous ethanol is blended with gasoline then utilize benzene or cyclohexane in distillation column. The main disadvantage associated with distillation is that impurity with similar boiling points to ethanol also recovered and high cost of this process (Onuki et al, 2008).

ENVIRONMENT ASPECTS OF BIOETHANOL

Combustion of the bioethanol produces CO_2 gas which would be assimilated again by plants. Almost no net CO_2 is produced by using bioethanol generated from agriculture biomass (Marszalek & Kaminski,

2008). It reduces GHGs emission. The exposure of humans to bioethanol carried out mostly by inhalation of vapour and by body contact is harmless. Degradation of bioethanol in the atmosphere is also fast. Use of ethanol along with diesel decrease octane number, increase heating value, aromatics fractions and changes distillation temperatures. These properties facilitate the complete burning of ethanol and less emissions of GHGs (Chandel et al., 2007). It has also low atmospheric photochemical reactivity and reducing the formation of ozone (Wyman, 1994).

FUTURE SCOPE

Conversion of agricultural waste to bioethanol has great potential due to highly abundance and relatively low cost of agriculture materials. There are still various problems on the way of developing an economically feasible technology, due to the complex structure and resistance to degradation of the agriculture material (Sun & Cheng, 2002). Single pretreatment method is not sufficient for complete delignification of biomass. Therefore, combined pretreatment methods (e.g. physical with chemical, chemical with biological) should be used for efficient hydrolysis (Kumar & Sharma, 2017). Agriculture biomass components are resistant to degradation. Therefore, genetically modified agricultural biomass can be used for ethanol production. Genetic engineering generates the plants having new genes which provide biomass with modified composition. Genetically modified agricultural biomass can increase ethanol yield and decreasing the cost of enzyme hydrolysis (Wang & Zhu, 2010).

CONCLUSION

Fossil energy sources such as coal, petrol etc. are consumed by the society to produce fuels and chemicals. These fossil fuels emit GHGs resulting increased global warming and creates environment pollution. Therefore, it is a demand to search an ecofriendly and less costly biofuel. Bioethanol is best option as an alternative fuel. It emits less greenhouse gases and produced from renewable sources. Bioethanol is mainly produced from feedstocks such as sugarcane, sugar beet, rice and corn grain, but the supply of these feedstocks is limited. This limitation will create the food crisis if these feedstocks utilized for bioethanol production. Agriculture waste serves as cheap and abundant biomass for bioethanol production.

Agriculture waste can be converted into bioethanol by different pretreatments (physical, chemical, physio-chemical and biological). Biological pretreatments should be used because it is safe, environmentally friendly and less energy consuming as compared to other pretreatments. Pretreated biomass is used for enzymatic hydrolysis followed by fermentation using microbes such as bacteria or yeast. Bioethanol production from agriculture waste will help decrease emission of greenhouse gases, environmental pollution and serve as a sustainable solid waste management strategy.

REFERENCES

Agbor, V. B., Cicek, N., Sparling, R., Berlin, A., & Levin, D. B. (2011). Biomass pretreatment: Fundamentals toward application. *Biotechnology Advances*, *29*(6), 675–685. doi:10.1016/j.biotechadv.2011.05.005 PMID:21624451

Ahmad, Q., & Qazi, J. I. (2014). Thermophilic fermentations of lignocellulosic substrates and economics of biofuels: Prospects in Pakistan. *International Journal of Energy and Environmental Engineering*, *5*(2-3), 2–3. doi:10.100740095-014-0094-4

Akansha, K., Prasad, A., Sukumaran, R. K., Nampoothiri, K. M., & Pandey, A. (2014). Dilute acid pretreatment and enzymatic hydrolysis of sorghum biomass for sugar recovery: A statistical approach. *Indian Journal of Experimental Biology*, *52*, 1082–1089. PMID:25434103

Aljibouri, A. K., Turcotte, G., Wu, J., & Cheng, C. (2015). Ozone pretreatment of humid wheat straw for biofuel production. *Energy Science & Engineering*, *3*(6), 541–548. doi:10.1002/ese3.93

Almodares, A., & Hadi, M. R. (2009). Production of bioethanol from sweet sorghum: A review. *African Journal of Agricultural Research*, *4*(9), 772–780.

Ameri, M., Ghobadian, B., & Baratian, I. (2008). Technical comparison of a CHP using various blends of gasohol in an IC engine. *Renewable Energy*, *33*(7), 1469–1474. doi:10.1016/j.renene.2007.09.015

Arvaniti, E., Bjerre, A. B., & Schmidt, J. E. (2012). Wet oxidation pretreatment of rape straw for ethanol production. *Biomass and Bioenergy*, *39*, 94–105. doi:10.1016/j.biombioe.2011.12.040

Ayadi, M., Sarma, S. J., Pachapur, V. L., Brar, S. K., & Cheikh, R. B. (2016). History and Global Policy of Biofuels. In C.R. Soccol, S.K. Brar, C. Faulds et al. (Eds.), Green Fuels Technology. Switzerland: Springer International Publishing. doi:10.1007/978-3-319-30205-8_1

Bak, J. S., Ko, J. K., Han, Y. H., Lee, B. C., Choi, I., & Kim, K. H. (2009). Improved enzymatic hydrolysis yield of rice straw using electron beam irradiation pretreatment. *Bioresource Technology*, *100*(3), 1285–1290. doi:10.1016/j.biortech.2008.09.010 PMID:18930388

Balasubramaniam, M. K., & Rajarathinam, R. (2013). Implementation of white rot fungal pretreated rice straw for sustainable bioethanol production by *Saccharomyces cerevisiae*. *International Journal of Engineering Research & Technology*, *2*(11), 4047–4053.

Balat, M. (2009). Gasification of biomass to produce gaseous products. *Energy Sources*, *31*(6), 516–526. doi:10.1080/15567030802466847

Balat, M., Balat, H., & Cahide, O. Z. (2008). Progress in bioethanol processing. *Progress in Energy & Combustion*, *34*(5), 551–573. doi:10.1016/j.pecs.2007.11.001

Banerjee, S., Sen, R., Morone, A., Chakrabarti, T., Pandey, R. & Mudliar, S. (2012). Improved wet air oxidation pretreatment for enhanced enzymatic hydrolysis of rice husk for bioethanol production. *Dynamic biochemistry, process biotechnology and molecular biology, 6*, 43-45.

Bensah, E. C., & Mensah, M. (2013). Chemical Pretreatment methods for the production of cellulosic ethanol: Technologies and innovations. *International Journal of Chemical Engineering*, *2013*, 1–21. doi:10.1155/2013/719607

Biswas, R., Uellendahl, H., & Ahring, B. K. (2015). Wet Explosion: A universal and efficient pretreatment process for lignocellulosic biorefineries. *BioEnergy Research*, *8*(3), 1101–1116. doi:10.100712155-015-9590-5

Bychkov, A. L., Buchtoyarov, V. A., & Lomovsky, O. L. (2014). Mechanical pretreatment of corn straw in a centrifugal roller mill. *Cellulose Chemistry and Technology, 48*(5-6), 545–551.

Cardona, C. A., & Sanchez, O. J. (2007). Fuel ethanol production: Process design trends and integration opportunities. *Bioresource Technology, 98*(12), 2415–2457. doi:10.1016/j.biortech.2007.01.002 PMID:17336061

Chandel, A.K., E.S. C., Rudravaram, R., Narasu, M.L., Rao, L.V., & Ravindra, P. (2007). Economics and environmental impact of bioethanol production technologies: an appraisal. *Biotechnology & Molecular Biology Review, 2*(1), 014-032.

Choudhary, R., Umagiliyage, A. L., Liang, Y., Siddaramu, T., Haddock, J., & Markevicius, G. (2012). Microwave pretreatment for enzymatic saccharification of sweet sorghum bagasse. *Biomass and Bioenergy, 39*, 218–226. doi:10.1016/j.biombioe.2012.01.006

Chundawat, S. P., Venkatesh, B., & Dale, B. E. (2006). Effect of particle size based separation of milled corn stover on AFEX pretreatment and enzymatic digestibility. *Biotechnology and Bioengineering, 96*(2), 219–231. doi:10.1002/bit.21132 PMID:16903002

Chunping, Y., Zhiqiang, S., Guoce, Y. U., & Jianlong, W. (2008). Effect and after effect of gama radiation pretreatment on enzymatic hydrolysis of wheat straw. *Bioresource Technology, 99*(14), 6240–6245. doi:10.1016/j.biortech.2007.12.008 PMID:18248982

Cui, L., Liu, Z., Hui, L. F., & Si, L. F. (2011). Effect of cellobiase and surfactant supplementation on enzymatic hydrolysis of pretreated wheat straw. *BioResources, 6*(4), 3850–3858.

Da Silva, A. S., Inoue, H., Endo, T., Yano, S., & Bon, E. P. (2010). Milling pretreatment of sugarcane bagasse and straw for enzymatic hydrolysis and ethanol fermentation. *Bioresource Technology, 101*(19), 7402–7409. doi:10.1016/j.biortech.2010.05.008

Dashtban, M., Schraft, H., & Qin, W. (2009). Fungal bioconversion of lignocellulosic residues; opportunities & perspectives. *International Journal of Biological Sciences, 5*(6), 578–595. doi:10.7150/ijbs.5.578 PMID:19774110

Demirbas, A. (2005). Bioethanol from Cellulosic Materials: A Renewable Motor Fuel from Biomass. *Energy Sources, 27*(4), 327–337. doi:10.1080/00908310390266643

Demirbas, A. (2008). The importance of bioethanol and biodiesel form biomass. *Energy Source. Part B, 3*, 177–185.

Du, H., Kommalapati, R. R., & Huque, Z. (2016). Assessment of Bioethanol applications on transportation vehicles in houston. *JFundam Renewable Energy App, 6*, 207.

Dufey, A. (2006). *Biofuels production, trade and sustainable development: emerging issues*. London: International Institute for Environment and Development.

Duff, S. J. B., & Murray, W. D. (1996). Bioconversion of forest products industry waste cellulosics to fuel ethanol: A review. *Bioresource Technology, 55*(1). doi:10.1016/0960-8524(95)00122-0

Dutton, J. A. (n.d.). 7.2 Sugarcane Ethanol Production. Retrieved from https://www.e-education.psu.edu/egee439/node/647

Eggeman, T., & Elander, R. T. (2005). Process and economic analysis of pretreatment technologies. *Bioresource Technology*, *96*(18), 2019–2025. doi:10.1016/j.biortech.2005.01.017 PMID:16112490

Foyle, T., Jennings, L., & Mulcahy, P. (2007). Compositional analysis of lignocellulosic materials: Evaluation of methods used for sugar analysis of waste paper and straw. *Bioresource Technology*, *98*(16), 3026–3036. doi:10.1016/j.biortech.2006.10.013 PMID:17142038

García-Aparicio, M. P., Ballesteros, I., González, A., Oliva, J., Ballesteros, M., & Negro, M. (2006). Effect of inhibitors released during steam-explosion pretreatment of barley straw on enzymatic hydrolysis. *Applied Biochemistry and Biotechnology*, *129*(1-3), 278–288. doi:10.1385/ABAB:129:1:278 PMID:16915647

García-Cubero, M. T., González-Benito, G., Indacoechea, I., Coca, M., & Bolado, S. (2009). Effect of ozonolysis pretreatment on enzymatic digestibility of wheat and rye straw. *Bioresource Technology*, *100*(4), 1608–1613. doi:10.1016/j.biortech.2008.09.012 PMID:18951781

Ghazali, K. A., Salleh, S. F., Riayatsyah, T. M., Aditiya, H. B., & Mahlia, T. M. (2016). The effect of dilute acid pre-treatment process in bioethanol production from durian (*Durio zibethinus*) seeds waste. In *Proceedings of IOP Conference Series: Earth and Environmental Science*. Malaysia: IOP Publishing. 10.1088/1755-1315/32/1/012058

Goh, C. S., Tan, K. T., Lee, K. T., & Bhatia, S. (2010). Bioethanol from lignocellulose: Status, perspectives and challenges in Malaysia. *Bioresource Technology*, *101*(13), 4834–4841. doi:10.1016/j.biortech.2009.08.080 PMID:19762229

Gray, K. A., Zhao, L., & Emptage, M. (2006). Bioethanol. *Current Opinion in Chemical Biology*, *10*(2), 141–146. doi:10.1016/j.cbpa.2006.02.035 PMID:16522374

Gurgel, L. V., Pimenta, M. T., & Curvelo, A. A. (2014). Enhancing liquid hot water (LHW) pretreatment of sugarcane bagasse by high pressure carbon dioxide (HP-CO_2). *Industrial Crops and Products*, *57*, 141–149. doi:10.1016/j.indcrop.2014.03.034

Hamelinck, C. N., Hooijdonk, G. V., & Faaij, A. P. (2005). Ethanol from lignocellulosic biomass: Techno-economic performance in short-, middle- and long-term. *Biomass and Bioenergy*, *28*(4), 384–410. doi:10.1016/j.biombioe.2004.09.002

Harmsen, P., Huijgen, W., Bermudez, L., & Bakker, R. (2010). Literature review of physical and chemical pretreatment processes for lignocellulosic biomass (Tech. Rep. 1184). Biosynergy, Wageningen UR Food & Biobased Research.

Hendriks, A., & Zeeman, G. (2009). Pretreatments to enhance the digestibility of lignocellulosic biomass. *Bioresource Technology*, *100*(1), 10–18. doi:10.1016/j.biortech.2008.05.027 PMID:18599291

Henniges, U., Hasani, M., Potthast, A., Westman, G., & Rosenau, T. (2013). Electron beam irradiation of cellulosic materials-opportunities and limitations. *Materials (Basel)*, *6*(5), 1584–1598. doi:10.3390/ma6051584 PMID:28809230

Howard, R., Abotsi, E., Jansen, V. R., & Howard, S. (2003). Lignocellulose biotechnology: Issues of bioconversion and enzyme production. *African Journal of Biotechnology*, *2*(12), 602–619. doi:10.5897/AJB2003.000-1115

Imman, S., Arnthong, J., Laosiripojana, N., & Champreda, V. (2012). Liquid hot water pretreatment of rice straw for enzymatic hydrolysis. In *Proceedings of the4th International Conference on Sustainable Energy and Environment: A Paradigm Shift to Low Carbon Society, Bangkok.*

Ingram, T., Rogalinski, T., Bockemühl, V., Antranikian, G., & Brunner, G. (2009). Semi-continuous liquid hot water pretreatment of rye straw. *The Journal of Supercritical Fluids*, *48*(3), 238–246. doi:10.1016/j.supflu.2008.10.023

Jaisamut, K., Paulova, L., Patakova, P., Rychtera, M., & Melzoch, K. (2013). Optimization of alkali pretreatment of wheat straw to be used as substrate for biofuels production. *Plant, Soil and Environment*, *59*(12), 537–542. doi:10.17221/7129-PSE

Joshi, B., Bhatt, M. R., Sharma, D., Joshi, J., Malla, R., & Sreerama, L. (2011). Lignocellulosic ethanol production: Current practices and recent developments. *Biotechnology & Molecular Biology Review*, *6*(8), 172–182.

Keshwani, D. R., Cheng, J. J., Burns, J. C., Li, L., & Chiang, V. (2007). Microwave pretreatment of switchgrass to enhance enzymatic hydrolysis. *Paper presented at the 2007 ASAE Annual Meeting.*

Koltermann, A., Kraus, M., Rarbach, M., Reisinger, C., Zavrel, M., & Soltl, Y. (2014). Cellulosic ethanol from agricultural residues–an advanced biofuel and biobased chemical platform. *JSM Biotechnology Bioengineering*, *2*(1), 1024.

Krishnan, C., Sousa, L. D., Jin, M., Chang, L., Dale, B. E., & Balan, V. (2010). Alkali-based AFEX pretreatment for the conversion of sugarcane bagasse and cane leaf residues to ethanol. *Biotechnology and Bioengineering*, *107*(3), 441–450. doi:10.1002/bit.22824 PMID:20521302

Kumar, A. K., & Sharma, S. (2017). Recent updates on different methods of pretreatment of lignocellulosic feedstocks: A review. *Bioresources and Bioprocessing*, *4*(1), 7. doi:10.118640643-017-0137-9 PMID:28163994

Kumar, P. D., Barrett, D. M., Delwiche, M. J., & Stroeve, P. (2009). Methods for pretreatment of lignocellulosic biomass for efficient hydrolysis and biofuel production. *Industrial & Engineering Chemistry Research*, *48*(8), 3713–3729. doi:10.1021/ie801542g

Kumar, R., Singh, S., & Singh, O. (2008). Bioconversion of lignocellulosic biomass: Biochemical and molecular perspectives. *Journal of Industrial Microbiology & Biotechnology*, *35*(5), 377–391. doi:10.100710295-008-0327-8 PMID:18338189

Kwon, E. E., Jean, Y. J., & Yi, H. (2012). New candidate for biofuel feedstock beyond terrestrial biomass for thermo-chemical process (pyrolysis/gasification) enhanced by carbon dioxide (CO_2). *Bioresource Technology*, *123*, 673–677. doi:10.1016/j.biortech.2012.07.035 PMID:22939597

Lau, M. W., & Dale, B. E. (2009). Cellulosic ethanol production from AFEX-treated corn stover using *Saccharomyces cerevisiae* 424A(LNH-ST). *Proceedings of the National Academy of Sciences of the United States of America, 106*(5), 1368–1373. doi:10.1073/pnas.0812364106 PMID:19164763

Lee, J. (1997). Biological conversion of lignocellulosic biomass to ethanol. *Journal of Biotechnology, 56*(1), 1–24. doi:10.1016/S0168-1656(97)00073-4 PMID:9246788

Li, C., Wang, L., Chen, Z., Li, Y., Wang, R., Luo, X., & Lu, J. (2015). Ozonolysis pretreatment of maize stover: The interactive effect of sample particle size and moisture on ozonolysis process. *Bioresource Technology, 183*, 240–247. doi:10.1016/j.biortech.2015.01.042 PMID:25746300

Li, T., Fang, Q., Chen, H., Qi, F., Ou, X., Zhao, X., & Liu, D. (2017). Solvent-based delignification and decrystallization of wheat straw for efficient enzymatic hydrolysis of cellulose and ethanol production with low cellulase loadings. *RSC Advances, 7*(17), 10609–10617. doi:10.1039/C6RA28509K

Li, X., Lu, J., Zhao, J., & Qu, Y. (2014). Characteristics of corn stover pretreated with liquid hot water and fed-batch semi-simultaneous saccharification and fermentation for bioethanol production. *PLoS ONE, 9*(4).

Lin, Y., & Tanaka, S. (2005). Ethanol fermentation from biomass resources: Current state and prospects. *Applied Microbiology and Biotechnology, 69*(6), 627–642. doi:10.100700253-005-0229-x PMID:16331454

Linde, M., Galbe, M., & Zacchi, G. (2008). Bioethanol production from non-starch carbohydrate residues in process streams from a dry-mill ethanol plant. *Bioresource Technology, 99*(14), 6505–6511. doi:10.1016/j.biortech.2007.11.032 PMID:18281213

Lynd, L., Zyl, W., Mcbride, J., & Laser, M. (2005). Consolidated bioprocessing of cellulosic biomass: An update. *Current Opinion in Biotechnology, 16*(5), 577–583. doi:10.1016/j.copbio.2005.08.009 PMID:16154338

Ma, H., Liu, W., Chen, X., Wu, Y., & Yu, Z. (2009). Enhanced enzymatic saccharification of rice straw by microwave pretreatment. *Bioresource Technology, 100*(3), 1279–1284. doi:10.1016/j.biortech.2008.08.045 PMID:18930389

Marszalek, J., & Kaminski, W. (2008). Environmental impact of bioethanol production. In *Proceedings of ECOpole*. Wolczanska.

Menezes, E. G., Do Carmo, J. R., Alves, J. G., Menezes, A. G., Guimaraes, I. C., Queiroz, F., & Pimenta, C. J. (2014). Optimization of alkaline pretreatment of coffee pulp for production of bioethanol. *Biotechnology Progress, 30*(2), 451–462. doi:10.1002/btpr.1856 PMID:24376222

Ministry of New and Renewable Energy, Government of India. (2009). *Annual report.*

Mohanty, B., & Abdullahi, I. I. (2016). Bioethanol production from lignocellulosic waste-a review. *Biosciences Biotechnology Research Asia, 13*(2), 1153–1161. doi:10.13005/bbra/2146

Moiser, N. (2005b). Features of promising technologies for pretreatment of lignocellulosic biomass. *Bioresource Technology, 96*(6), 673–686. doi:10.1016/j.biortech.2004.06.025 PMID:15588770

Moreira, L. R., & Filho, E. X. (2016). Insights into the mechanism of enzymatic hydrolysis of xylan. *Applied Microbiology and Biotechnology*, *100*(12), 5205–5214. doi:10.100700253-016-7555-z PMID:27112349

Mtui, G.Y.S. (2009). Recent advances in pretreatment of lignocellulosic wastes and production of value added products. *African Journal of Biotechnology, 8*(8).

Mussatto, S. I., & Teixeira, J. A. (2010). Lignocellulose as raw material in fermentation processes. In A. Mendez-Vilas (Ed.), *Current Research, Technology and Education, Topics in Applied Microbiology and Microbial Biotechnology* (pp. 897–907). Badajoz: Formatex Research Center.

Nwosu-Obieogu, K., Chiemenem, L. I., & Adekunle, K. F. (2016). Utilization of agricultural waste for bioethanol production- a review. *International Journal of Current Research & Review*, *8*(19).

Olofsson, K., Bertilsson, M., & Liden, G. (2008). A short review on SSF – an interesting process option for ethanol production from lignocellulosic feedstocks. *Biotechnology for Biofuels*, *1*(1), 7. doi:10.1186/1754-6834-1-7 PMID:18471273

Olsson, L., & Hahn-Hagerdal, B. (1996). Fermentation of lignocellulosic hydrolysates for ethanol production. *Enzyme and Microbial Technology*, *18*(5), 312–331. doi:10.1016/0141-0229(95)00157-3

Onuki, S., Koziel, J. A., Leeuwen, V. J., Jenks, W. S., Grewell, D., & Cai, L. (2008). Ethanol production, purification, and analysis techniques: a review. In *Proceedings of the* ASABE Annual International Meeting, *Rhode Island*.

Oreb, M., Dietz, H., Farwick, A., & Boles, E. (2012). Novel strategies to improve co-fermentation of pentoses with D-glucose by recombinant yeast strains in lignocellulosic hydrolysates. *Bioengineered*, *3*(6), 347–351. doi:10.4161/bioe.21444 PMID:22892590

Pimentel, D., & Patzek, T. W. (2005). Ethanol Production Using Corn, Switch grass and wood; biodiesel production using soybean and sunflower. *Natural Resources Research*, *14*(1), 65–76. doi:10.100711053-005-4679-8

Prasad, S., Singh, A., Jain, N., & Joshi, H. C. (2007a). Ethanol production from sweet sorghum syrup for utilization as automotive fuel in India. *Energy & Fuels*, *21*(4), 2415–2420. doi:10.1021/ef060328z

Prasad, S., Singh, A., & Joshi, H. C. (2007b). Ethanol as an alternative fuel from agricultural, industrial and urban residues. *Conservation & Recycling*, *50*(1). doi:10.1016/j.resconrec.2006.05.007

Qing, Q., Yang, B., & Wyman, C. E. (2010). Impact of surfactants on pretreatment of corn stover. *Bioresource Technology*, *101*(15), 5941–5951. doi:10.1016/j.biortech.2010.03.003 PMID:20304637

Ratnavathi, C. V., Suresh, K., Kumar, B. S. V., Pallavi, M., Komala, V. V., & Seetharama, N. (2010). Study on genotypic variation for ethanol production from sweet sorghum juice. *Biomass and Bioenergy*, *34*(7), 947–952. doi:10.1016/j.biombioe.2010.02.002

Ray, M. J., Leak, D. J., Spanu, P. D., & Murphy, R. J. (2010). Brown rot fungal early stage decay mechanism as a biological pretreatment for softwood biomass in biofuel production. *Biomass and Bioenergy*, *34*(8), 1257–1262. doi:10.1016/j.biombioe.2010.03.015

Rehman, M. S. U., Kim, I., Chisti, Y., & Han, J. I. (2013). Use of ultrasound in the production of bioethanol from lignocellulosic biomass. *Energy Education Science and Technology, Part A. Energy Science and Research*, *30*(2), 1391–1410.

Roehr, M. (2000). *Biotechnology of Ethanol, Classic and future application. Publication.* Wiley. doi:10.1002/3527602348

Saha, B. C., & Cotta, M. A. (2007). Enzymatic saccharification and fermentation of alkaline peroxide pretreated rice hulls to ethanol. *Enzyme and Microbial Technology*, *41*(4), 528–532. doi:10.1016/j.enzmictec.2007.04.006

Sanchez, C. (2009). Lignocellulosic residues: Biodegradation and bioconversion by fungi. *Biotechnology Advances*, *27*(2), 185–194. doi:10.1016/j.biotechadv.2008.11.001 PMID:19100826

Sanchez, G., Pilcher, L., Roslander, C., Modig, T., Galbe, M., & Liden, G. (2004). Dilute-acid hydrolysis for fermentation of the bolivian straw material pajabrava. *Bioresource Technology*, *93*(3), 249–256. doi:10.1016/j.biortech.2003.11.003 PMID:15062819

Saravanakumar, K., & Kathiresan, K. (2013). Bioconversion of lignocellulosic waste to bioethanol by *Trichoderma* and yeast fermentation. *Biotech*, *4*(5), 493–499. PMID:28324381

Saritha, M., Arora, A., & Lata. (2011). Biological pretreatment of lignocellulosic substrates for enhanced delignification and enzymatic digestibility. *Indian Journal of Microbiology*, *52*(2), 122–130. doi:10.100712088-011-0199-x PMID:23729871

Sarkar, N., Ghosh, S. K., Bannerjee, S., & Aikat, K. (2011). Bioethanol production from agricultural wastes: An overview. *Renewable Energy*, *37*(1), 19–27. doi:10.1016/j.renene.2011.06.045

Sidiras, D. K., & Salapa, L. S. (2015). Organosolv pretreatment as a major step of lignocellulosic biomass refining. In *Proceedings of the International Engineering Conferences.*

Singh, D. P., & Trivedi, R. K. (2013). Acid and Alkaline pretreatment of lignocellulosic biomass to produce ethanol as biofuel. *International Journal of Chemtech Research*, *5*(2), 727–734.

Singhania, R. R., Sukumaran, R. K., Patel, A. K., Larroche, C., & Pandey, A. (2010). Advancement and comparative profiles in the production technologies using solid-state and submerged fermentation for microbial cellulases. *Enzyme and Microbial Technology*, *46*(7), 541–549. doi:10.1016/j.enzmictec.2010.03.010

Sinumvayo, J. P., Ishimwe, N., Komera, I., & Niyomukiza, S. (2015). Ethanol biofuel production from lignocellulosic biomass by Engineered *Saccharomyces cerevisiae. Journal of Academia & Industrial Research*, *3*(10), 463–471.

Soccol, C. R., Vandenbarghe, L. P. S., Medeiros, A. B. P., Krap, S. G., Buckeridge, M., Ramos, L. P., ... Torres, F. A. G. (2010). Bioethanol from lignocelluloses: Status and perspectives in Brazil. *Bioresource Technology*, *101*(13), 4820–4825. doi:10.1016/j.biortech.2009.11.067 PMID:20022746

Sun, Y., & Cheng, J. (2002). Hydrolysis of lignocellulosic materials for ethanol production: A review. *Bioresource Technology*, *83*(1). doi:10.1016/S0960-8524(01)00212-7 PMID:12058826

Taherzadeh, M. J., & Karimi, K. (2008). Pretreatment of lignocellulosic wastes to improve ethanol and biogas production: A review. *International Journal of Molecular Sciences*, *9*(9), 1621–1651. doi:10.3390/ijms9091621 PMID:19325822

Tan, L., Sun, Z. Y., Okamoto, S., Takaki, M., Tang, Y. Q., Morimura, S., & Kida, K. (2015). Production of ethanol from raw juice and thick juice of sugar beet by continuous ethanol fermentation with flocculating yeast strain KF-7. *Biomass and Bioenergy*, *81*, 265–272. doi:10.1016/j.biombioe.2015.07.019

Torres-Jimenez, E., Jerman, M. S., Gregorc, A., Lisec, I., Dorado, M. P., & Kegl, B. (2011). Physical and chemical properties of ethanol-diesel fuel blends. *Fuel*, *90*(2), 795–802. doi:10.1016/j.fuel.2010.09.045

Travaini, R., Martin-Juarez, J., Lorenzo-Hernando, A., & Bolado-Rodriguez, S. (2016). Ozonolysis: An advantageous pretreatment for lignocellulosic biomass revisited. *Bioresource Technology*, *199*, 2–12. doi:10.1016/j.biortech.2015.08.143 PMID:26409859

Turner, D., Xu, H., Cracknell, R. F., Natarajan, V., & Chen, X. (2011). Combustion performance of bioethanol of various blend ratios in a gasoline direct injection engine. *Fuel*, *90*(5), 1999–2006. doi:10.1016/j.fuel.2010.12.025

Tutt, M., Kikas, T., Kahr, H., Pointner, M., Kuttner, P., & Olt, J. (2014). Using steam explosion pretreatment method for bioethanol production from floodplain meadow hay. *Agronomy Research (Tartu)*, *12*(2), 417–424.

United States Department of Agriculture. (2009). Colombia: a new ethanol producer on the rise. Economic Research Service: USDA.

Vanholme, R., Demedts, B., Morreel, K., Ralph, J., & Boerjan, W. (2010). Lignin biosynthesis and structure. *Plant Physiology*, *153*(3), 895–905. doi:10.1104/pp.110.155119 PMID:20472751

Varga, E., Schmidt, A. S., Reczey, K., & Thomsen, A. B. (2003). Pretreatment of corn stover using wet oxidation to enhance enzymatic digestibility. *Applied Biochemistry and Biotechnology*, *104*(1), 37–50. doi:10.1385/ABAB:104:1:37 PMID:12495204

Verma, A., Kumar, S., & Jain, P. K. (2011). Key pretreatment technologies on cellulosic ethanol production. *Journal of Scientific Research*, *55*, 57–63.

Vidal, P. F., & Molinier, J. (1988). Ozonolysis of lignin improvement of in vitro digestibility of poplar sawdust. *Biomass*, *16*(1), 1–17. doi:10.1016/0144-4565(88)90012-1

Walker, G. M. (2011). Fuel alcohol: Current production and future challenges. *Journal of the Institute of Brewing*, *117*(1), 3–22. doi:10.1002/j.2050-0416.2011.tb00438.x

Wang, M., Saricks, C., & Santini, D. (1999). *Effect of fuel ethanol use on fuel-cycle energy and greenhouse gas emission*. Argonne, IL: Argonne National Laboratory.

Wang, Q., & Zhu, S. (2010). Genetically modified lignocellulosic biomass for improvement of ethanol production. *BioResources*, *5*(1), 3–4.

Wilda, P. J. D., Huijgena, W. J. J., & Heeres, H. J. (2012). Pyrolysis of wheat straw-derived organosolv lignin. *Journal of Analytical and Applied Pyrolysis*, *93*, 95–103. doi:10.1016/j.jaap.2011.10.002

Wyman, C. E. (1994). Ethanol from lignocellulosic biomass: Technology, economics, and opportunities. *Bioresource Technology*, *50*(1), 3–15. doi:10.1016/0960-8524(94)90214-3

Zheng, Y. I., Pan, Z., & Zhang, R. (2009). Overview of biomass pretreatment for cellulosic ethanol production. *International Journal of Agricultural and Biological Engineering*, *2*(3), 51–68.

KEY TERMS AND DEFINITIONS

Enzyme Hydrolysis: A process in which enzymes facilitate the conversion of pretreated biomass into fermentable sugars.

Feedstock: Raw material used for fuel production.

Fermentation: A biological process which converts sugars such as glucose, fructose, and sucrose into ethanol and carbon dioxide using yeast or bacteria.

First Generation Bioethanol: Bioethanol which is derived from sugar containing feedstocks.

Pretreatment: Treatment of biomass to remove lignin and hemicelluloses.

Second Generation Bioethanol: Bioethanol which is derived from agriculture waste.

Chapter 2
Cellulolytic Microorganisms: A Review

Amritha Govindrao Kulkarni
Karnatak University, India

Ankala Bassappa Vedamurthy
Karnatak University, India

ABSTRACT

Cellulose is the most abundant polymer in plants and the microbial conversion of cellulose is a subject of active research. Currently, cellulase is commonly used in many industrial applications, especially in animal feed, textile, waste water, brewing and wine making. A challenging strategy for the efficient utilization of this renewable resource is to use it as a base material for the production of desired metabolites. This chapter therefore focuses on exploring the cellulase producing bacteria and optimizing the parameters for the enzyme cellulase under varied conditions. Cellulolytic bacteria can be exploited for cellulase production which serves wide applications in industries, pharmaceuticals and further, use of these CDB as bio-inoculants can be incorporated to enhance organic matter decomposition in soil to increase soil fertility and to minimize the fertilizer application. It finds wide applications in reducing the environmental pollution and promote sustainable agriculture.

INTRODUCTION

Anselme Payen (1795-1871) coined the term cellulose and introduced it to scientific literature in 1839 after isolating a fibrous substance mostly found in wood, cotton and other plants (Payen, 1838). Cellulose, the leading bio-residue from agricultural sector is the predominant renewable biopolymer in the world which comprises of celluloses, hemicelluloses and lignin. A promising strategy for the efficient utilization of this renewable resource is to use it as a base material for the production of desired metabolites (Kulkarni and vedamurthy 2016). Cellulose is probably the most abundant Biomass/biological compound on earth specially, on terrestrial and aquatic ecosystem and is the main component of plant biomass. Cellulose is probably one of the most ubiquitous and abundant polymers on the planet, given its widespread industrial use in the present age, but also in the past for ropes, sails, paper, timber for

DOI: 10.4018/978-1-5225-3540-9.ch002

housing and many other applications. By far the most commercially exploited natural resource containing cellulose is wood (Nishida et al., 2007).

It occurs in almost pure form in cotton fiber and in combination with other materials, such as lignin and hemicelluloses, in wood, plant leave stalks, etc. It has already been used in processing of coffee, in textile industry and in laundry detergents. Cellulases cause hydrolysis of the individual cellulose fibres to break it into smaller sugars units & finally producing glucose molecules (Shaheen *et al*, 2010).

Cellulose is probably one of the most ubiquitous and abundant polymers on the planet, given its widespread industrial use in the present age, but also in the past for ropes, sails, paper, timber for housing and many other applications. By far the most commercially exploited natural resource containing cellulose is wood (Echhiron *et al*, 2009). It occurs in almost pure form in cotton fiber and in combination with other materials, such as lignin and hemicelluloses, in wood, plant leave stalks, etc. It has already been used in processing of coffee, in textile industry and in laundry detergents. Cellulases cause hydrolysis of the individual cellulose fibres to break it into smaller sugars units & finally producing glucose molecules (Muhammed, 2012).

It is the dominant waste material from agricultural industry in the form of stalks, stems and husk, there has been great interest in utilizing cellulose as an energy resource and feed (Karmakar & Ray, 2011). The cellulose is composed of D-glucose units linked together to form linear chain via ß-1, 4-glycosidic linkages (Abdelnasser & Ahmed, 2007). It is the primary product of photosynthesis in terrestrial environments and the most abundant renewal bio-resource produced in the bio-sphere. Cellulose in natural substrates is persistent in the environment and remains as an environmental pollutant (Singh, 2007). Cellulosic materials can, however, be a valuable and vast resource replenished by photosynthesis. Apart from the production of value-added products, its bioconversion offers an effective solution for the abatement of pollution due to solid-waste and their utilization, which would act at low sustainable process and products. It is well known that plants are the most common source of renewable carbon and energy on the earth. Lignocellulosic crop residues, such as cereal straw, provide the principal input of cellulose to arable soils (Lynch et al., 2008).

Structure and Classification of Cellulose

The cellulose molecule is an unbranched linear polymer of glucose. This is shown by its empirical formula ($C_6 H_{10}O_5$) n, and its structural formula shown in Figure 1.

Figure 1. Structure of Cellulose

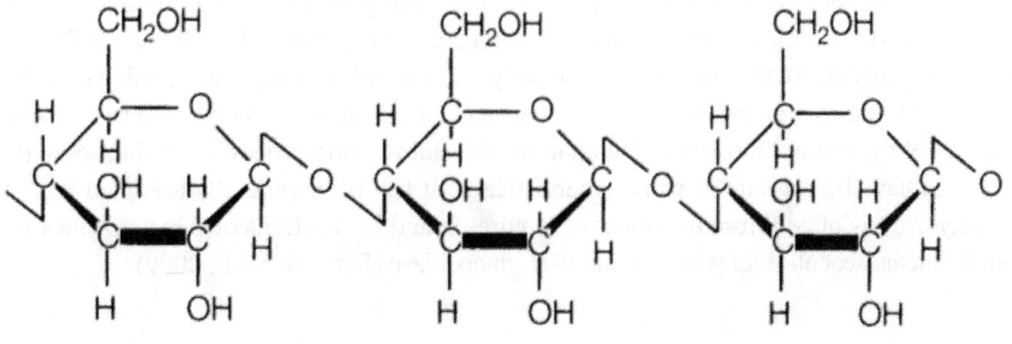

Cellulose is linear condensation polymer consisting of D anhydroglucopyranose joined together by β-1, 4-glycosidic bonds with a degree of polymerization (DP) from 100 to 20,000 (Zhang & Lynd, 2004). Coupling of adjacent cellulose chains and sheets of cellulose by hydrogen bonds and Vander Waal's forces results in a parallel alignment and a crystalline structure with straight, stable supra-molecular fibres of great tensile strength and low accessibility (Gomashe et al, 2013). The glucose units are linked at their 1- and 4-carbon positions by glycosidic bonds into molecular chains varying in length from a few glucose units to as many as 3,000 or more. From this we have the more technical name for cellulose, 1, 4- p-polyanhydroglucopyranose. The linear molecules of cellulose are bound laterally by hydrogen bonds or Van der Wall forces into linear fibrils of 50 to 100 chains. The cellulose is composed of D-glucose units linked together to form linear chain via ß-1, 4-glycosidic linkages It is the primary product of photosynthesis in terrestrial environments and the most abundant renewal bio-resource produced in the bio-sphere (Abdelnaser & Ahmed, 2007).

They annually produce about $9x10^4$ tons of cellulose which is a highly stable polymer consisting of β-1,4-linked glycosyl residues, along with other polysaccharides (Tabao et al., 2010). The potential of these biological resources are the possible substitutes for diminishing fossil energy resources and hence becoming increasingly important. Types of cellulose constitute Cotton Cellulose: Cotton is the purest form of cellulose found in nature. Since it contains only minor amounts of extraneous materials such as pectins, waxes, and others, it has been adopted as a standard for research work. Wood Cellulose: Wood cellulose is structurally identical with cotton cellulose, except that it is associated with polymers of basic sugar units other than glucose and has a lower proportion of crystalline material. However, many organisms virulently cellulolytic against cotton are unable to utilize wood for food. This is primarily due to the nature of association of the cellulose with the lignin of wood. A chemical bond between cellulose and lignin has been postulated but evidence for its existence remains inconclusive. Modified celluloses: natural celluloses, as described above, may be changed in susceptibility to enzymatic degradation by physical and chemical modification. Much of our present understanding of the mechanism of action of cellulolytic enzymes has resulted from the use of cellulosic substrates that have been modified in specific known ways. Modified cellulosic materials may be prepared from natural celluloses by five general methods (a) hydrolysis, (b) substitution, (c) regeneration, (d) mechanical degradation and (e) irradiation (Pew, 1957). The biological degradation of cellulose has been studied for many years and a number of cellulolytic enzymes, especially cellulases produced by fungi and bacteria, have been isolated and characterized. These enzymes, which cleave the β-1,4 bond of cellulose, belong to the large family of glycosyl hydrolases (Zhang & Lynd, 2004). On the basis of sequence comparisons and hydrophobic cluster analysis, the catalytic domains of all glycosyl hydrolases have been classified into 63 families of homologous folds. The catalytic domains of cellulases and related xylanases have been identified in 14 of these families. According to their mechanism of cellulose degradation, cellulases are subdivided into either non-processive cellulases (endocellulases) or processive cellulases (Lynch et al. 1981). Endocellulases can randomly cleave the cellulose at exposed positions and produce new reducing ends, however, the processive cellulases remain attached to the chain and release mainly cellobiose or cellotetraose units from one end of the chain. During the degradation of cellulose, non-processive cellulases and processive cellulases have been found to work synergistically. It has been generally accepted that effective biological hydrolysis of cellulose into glucose requires synergistic actions of three enzymes including endo-β-1,4-glucanase, cellobiohydrolase and β-*D*-glucosidase (Saratale et al., 2009).

Sources of Cellulose

Cellulose is the most abundant plant biomass on earth with an estimated annual production of 4.0×10^7 units (Bhat, 2000). This abundance confers to cellulose an immense potential as a renewable source of energy. Cellulose is also the most common organic compound on earth. It is well known that plants are the most common source of renewable carbon and energy on the earth. (Yakubu et al., 2011). Cellulose is basically the structural component of the primary cell wall of green plants, many forms of algae and the oomycetes. Cellulose is the major component of plant biomass. Plants produce 4×10^9 tons of cellulose annually. It is also considered as one of the most important sources of carbon on this planet and its annual biosynthesis by both land plants and marine occurs at a rate of 0.85×10^{11} tons per annum. Higher plant tissues such as trees, cotton, flax, cereal straw represent the main sources of cellulose i.e. it makes up 35-50% of dry plant weight (Lynd et al., 1999). Algae such as *Valoniaventricosa* and *Microdicyan* are representatives of lower plants that synthesize cellulose (Petre et al., 1999). In addition to plants, non-photosynthetic organisms such as bacteria i.e. aerobic *Acetobacter xylinum*, marine invertebrates from the ascite family i.e. tunicates, fungi, slime moulds and amoebae also produce cellulose (Lynd et al., 2002). Cellulose is used in the development of energy, and reduction in the load of waste and greenhouse gas to the environment (Chiu et al., 2008). Kitchen Waste forms a significant part of domestic waste. Food waste is an unwanted raw or cooked food discarded during or after food preparation that is no longer fit for consumption (Boonmee, 2009). Plants cellulose are the most common source of renewable carbon and energy on the earth. It has no taste, is odorless and is hydrophilic (Ghosal et al., 2011). Cellulose is derived from D-glucose units, which condense through β (1-4) glycosidic bonds (Yakubu et al., 2011). These linkages being the primary structural material of plant cell wall, is the most abundant carbohydrate in nature (Saha et al., 2006). Compost is a fertilizing mixture of partially decomposed organic matter from plant and animal origin.

Cellulose Degrading Microorganisms

Various biological studies have been carried out to identify the microbiological agents responsible for biodegradation of cellulose (Lyand et al., 2002). Microorganisms having the ability to degrade cellulosic compounds are of great importance from the municipal waste and are potent enough to produce different biological and ecological point of view. The cellulose degrading ability of fungal and bacterial species has provided a broad platform for research in determining the physico-biochemical properties of these microorganisms as well as their use in different biotechnological processes (Ariffin et al., 2006). Among bacteria, *Bacillus* species can produce numbers of extracellular polysaccharide hydrolyzing enzyme. However, carboxymethyl cellulase (endoglucanase) cannot hydrolyze crystalline cellulose. These polysaccharide hydrolyzing enzymes also include alkaline cellulase with high potential as laundry detergent additives (Petre et al., 1999). However, there are some studies on *Bacillus* and fungal endoglucanases which had shown detectable activity on microcrystalline cellulose. A cellulase producing *Bacillus* strains have isolated from fermented food and optimized the culture conditions for enzyme production. In this study, the cellulase was purified and characterized from *Bacillus subtilis*. Even though bulk quantity of cellulosic residues gets accumulated in the terrestrial ecosystem and contribute in maintaining the carbon cycle (Kadam & Wyman, 1996; Denis et al., 2000). Lignocellulose -degrading micro-organisms and their enzymes (cellulose, xylanase, peroxidase, etc.) have been well studied (Liszka et al., 2012); and several related applications have been developed for textile, food, and paper-pulp processing. However, it is hard to utilize these micro-organisms and enzymes to process natural cellulosic material without pre-treatment and/or sterilization. Cellulose and hemicelluloses each account for about 40% (w/w) of straw and they are degraded at about equal rates and much more rapidly than the lignin component (about 14%, w/w) under field conditions (Harper & Lynch, 1981; Rathnan et al., 2013). Cellulolytic fungi including, *Trichoderma* sp, *Geocladium* sp, *Chaetomium* sp and *Penicillium* sp are known to play an important role in agriculture by facilitating enhanced seed germination, rapid plant growth and flowering, improved root system and increased crop yields (Immanuel, 2007). Muhammed et al., (2012), also reported the cellulytic activity of *Pseudomonas spp, Bacillus* sp, *Bacillus polymyxa,* and *Bacillus brevis* from soil of Bhitarakanika, Odisha. A number of organisms have been isolated for cellulose degradation such as *Bacillus subtilis, Bacillus licheniformis, Bacillus cereus* and *Paenibacillus.* In 2012, Irfan et al., observed *Cellulomonas* sp. and Farzaneh et al., (2011) had found various anaerobic hydrolytic and methanogenic bacteria active in cattle dung biogas plants including *Syntrophobacter wolinii* a propionate degrading bacterium in co-culture with hydrogen utilizing methanogen *viz., Methanobacterium formicicum.* Five cellulose degrading bacteria were reported *Sphingomonas* sp., *Pseudomonas* sp.M1, *Achromobacter* sp., *Pseudomonas* sp. M2, and *Stenotrophomonas* sp., and Dubey et al., (2014) observed among the 15 isolates one belonged to *Pseudomonas spp.*, one to *Aeromonas spp.*, one to *Pasteurella* spp., two belonged to *Staphylococcus spp.* and then to *Bacillus* genus. Among them, *Bacillus spp.* has shown a remarkable ability of cellulose hydrolysis. Kaur and Arora (2012) reported the successful full-scale composting of the primary and secondary sludge bulked with bark. Comparing the product with other commercial composts and peat, Fitzpartick (1989) reported that Dwarf oleander growth was significantly better in this compost than in the peat and at least one other commercial comport blend. In recent years, more attention has been directed toward screening of novel microbial strains that have broad spectrum of enzyme activities. The high cost of cellulase production and low enzyme activities limit their Industrial use. So, efforts are to be taken to economize and increase the yield of cellulase production by media optimization, and hence isolation, characterization and media optimization for cellulase producing bacteria remain to be an important area of research. So, we focused on the isolation of novel cellulase producing bacteria with an industrial perspective (Kulkarni and vedamurthy, 2015).

Cellulose Degradation Pathway

The cellulose molecule is very stable, with half-life of 5-8 million years for β-glucosidic bond cleavage at 25°C (Yin, 2010), while the much faster enzyme-driven cellulose biodegradation process is vital to return the carbon in sediments to atmosphere (Harita et al., 2009). The biodegradation of cellulose has important consequences for the cultivation of arable soil, since under condition of restricted oxygen supply, microbial population and metabolism change significantly with the result that phytotoxic concentration of acetic acid accumulate (Lynch & Gunn, 1980). Under aerobic conditions, plant growth can be promoted by the products of cellulolysis (Lynch, 1977), but the precise nature of the population changes is poorly understood.

An understanding of the overall cellulolytic activity of pure or mixed microbial cultures depends on knowledge of the activity of the three groups of enzymes involved in cellulose degradation which act synergistically (Yin, 2010). Activity against Avicel, carboxy methyl cellulose and cellobiose (for cellubiase) provides this information. Primary hydrolysis that occurs on the surface of solid substrates releases soluble sugars with a degree of polymerization (DP) up to 6 into the liquid phase upon hydrolysis by endoglucanases is the rate-limiting step for the whole cellulose hydrolysis process. Secondary

Figure 2. Cellulose degradation by the combine action of endoglucanase, exoglucanase and glucosidase

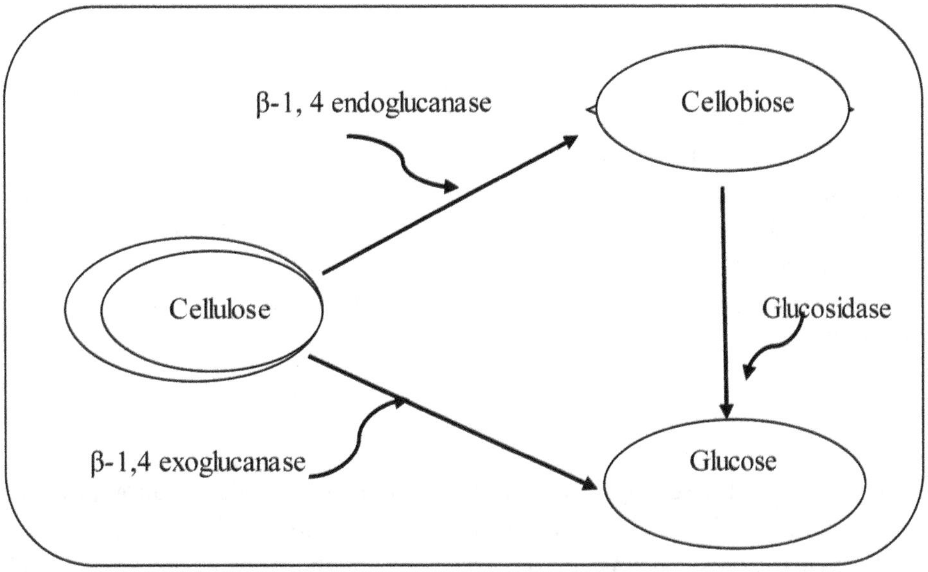

hydrolysis that occurs in the liquid phase involves primarily the hydrolysis of cellobiose to glucose by β-glucosidases, although some β-glucosidases also hydrolyse longer cellodextrin (Zhang & Lynd, 2004). During cellulose hydrolysis, the solid substrate characteristic vary, including - Changes in the cellulose chain end number resulting from generation by endoglucanase and consumption by exoglucanases (Kleman-Leyer et al., 1992, 1994 & 1996; Srisodsuk et al., 1998; Kongruang et al., 2004; Zhang & Lynd, 2005) and Changes in cellulose accessibility resulting from substrate consumption and cellulose fragmentation (Wang et al., 2003). The combined actions of endoglucanases and exoglucanases modify the cellulose surface characteristics (topography) over time, resulting in rapid changes in hydrolysis rates.

The widely accepted mechanism for enzymatic cellulose hydrolysis involves synergistic actions by endoglucanase (EC 3.2.1.4), exoglucanase or cellobiohydrolase (EC 3.2.1.91) and β-glucosidase (EC 3.2.1.21) (Lynd et al., 2002; Zhang & Lynd, 2004). Endoglucanase hydrolyze accessible intramolecular β-1, 4- glucosidic bonds of cellulose chains randomly to produce new chain ends; exoglucanases processively cleave cellulose chains at the ends to release soluble cellobiose or glucose; and β-glucosidase hydrolyze cellobiose to glucose in order to eliminate cellobiose inhibition.

The complicated interactions among endoglucanases, exoglucanases, and the changing substrate characteristics during hydrolysis have been stimulated by a new functionally mathematical model, applying a set of enzymatic parameters for endoglucanase I, cellobiohydrolases I and II to a variety of substrates with two important substrate properties: the fraction of β-glucosidic bond accessible to cellulase and the degree of polymerization (DP) (Zhang & Lynd, 2004).

Hydrolysis of Cellulose

A key step in the global carbon cycle is the hydrolysis of the cellulose in plant cell walls, which is the most abundant source of carbon on land. The half-life of cellulose at neutral pH in the absence of enzymes is estimated to be several million years so that microbial activity is responsible for most of the turnover of the carbon in cellulose although fire also plays a role. Even when ruminants, shipworms or termites utilize cellulose as an energy source, microorganisms usually are involved in its degradation. In the case of ruminants, all of their cellulose hydrolysis is carried out by anaerobic rumen bacteria, fungi, or protozoa, though for some termites both microbial enzymes and termite enzymes are utilized for cellulose hydrolysis (Lee et al., 2011). Since cellulose is mostly present in plant cell walls, which are very difficult to degrade, only a small fraction of all microorganisms that are specialized for plant cell wall degradation can hydrolyze cellulose (Dubey et al., 2014). Since bacteria and fungi are unable to engulf particles, these organisms need to secrete their cellulases and most anaerobic bacteria that produce cellulosomes attach them to their outer surface. Because of the recalcitrance of plant cell wall some cellulolytic microorganisms secrete up to 50% of their total protein during growth on biomass or cellulose (David, 2011).

Cellulase

Cellulolytic enzymes or cellulase play an important role in natural biodegradation processes in which plant lignocellulosic materials are efficiently degraded by cellulolytic fungi, bacteria, actinomycetes and protozoa. In industry, these enzymes have found novel applications in the production of fermentable sugars and ethanol, organic acids, detergents and other chemicals. Originally it was thought that only microorganisms produced cellulases, but it is now clear that some insects, molluscs, nematodes, and protozoa also produce cellulases. Cellulases provide a key opportunity for achieving tremendous benefits of biomass utilization (Wen et al., 2005).

The conversion of cellulose into glucose is now known to consist of two steps in the enzyme system. In the first step, beta-1,4glucanase breaks the glucosidic linkage to cellobiose, which is a glucose dimer with a beta-1, 4 bond as opposed to maltose, a counterpart with an alpha-1, 4 bond. Subsequently, this beta-1,4glucosidic linkage is broken by beta-glycosidase:

b-1,4 glucanase b-glucosidase

Cellulose -------------------> Cellobiose -----------------> Glucose 4

The kinetics of cellulose hydrolysis has been widely studied and Michaelis-Menten types of rate expressions with substrate or product inhibition terms have been proposed to describe the observed reaction kinetics.

Cellulose is commonly degraded by an enzyme called cellulase. Cellulases are the inducible bioactive compounds produced by microorganisms during their growth on Cellulosic matters (Lee et al., 2001). Increasing knowledge of mode of action of Cellulase; they were used in enzymatic hydrolysis of cellulosic substances (Guber et al., 1993). Cellulase enzymes, which can hydrolyze cellulose forming glucose and other commodity chemicals, can be divided into three types: endoglucanase (endo-1, 4-β-D-glucanase); cellobiohydrolase or exoglucanase (exo-1, 4-β-D-glucanase) and β-glucosidase (1,4-β-D-glucosidase) (Li et al., 2009; Malherbe et al., 2011). Endo-glucanases make nicks within the cellulose biomolecule there by exposing their reducing and non-reducing ends, cellobiohydrolases release cellobiose units—a disaccharide of two glucose molecules linked by a β-1,4 linkage—the repeating units of cellulose from the chain ends; and finally, β-glucosidases act on cellobiose to liberate glucose (Sukumaran et al., 2005). Cellulase hydrolyses the β-1, 4-glycosidic bonds in the polymer to release glucose units. Various biological studies have been carried out to identify the microbiological agents responsible for biodegradation of cellulose (Kulkarni & Vedamurthy, 2015). Native crystalline cellulose is degraded to its glucose units through the synergistic actions. Cellulose is a linear polysaccharide of glucose residues with β-1, 4-glycosidic linkages. *Insitu*, numerous parallel poly-glucose chains form insoluble complex of crystalline cellulose fibre through hydrogen bonds and Vander Waal's forces. Cellulose (exo-β-1, 4-glucanase, exocellulase) split off cellubiose, a glucose dimer from the non-reducing end of the oligosaccharide chain, while amorphous and soluble cellulose may be degraded through the chain of endocellulose alone, the degradation of crystalline cellulose requires the activity of an exocellulose (Wood and Garciacampaya) at least in the case of common extra cellular cellulases cellubiose (β-Glucosidase) cleaves cellubiose into two glucose moieties. Possibly, glucohydrolase is also involved in cellulose degradation as a component of the exoglucanase splitting off glucose instead of cellubiose from the non-reducing end of poly and oligosaccharides. The transfer of carbon and energy from detritus to animal consumers appears to occur via grazing of easily digestible and highly nutritive microbial biomass resulting from bacterial and fungal transformation of the lignocellulosic detritus. Thus most of the bacteria and fungi degrade this cellulosic material by producing enzyme called cellulase (Nishida et al., 2007). Currently, cellulase is commonly used in many industrial applications, especially in animal feed, textile, waste water, brewing and wine making. To utilize these materials and to avoid waste pollution, one of the most important approaches is to find applicable cellulase and hemicellulase to hydrolyze the lignocellulosic biomass (Farzaneh, 2011). Accordingly, a highly stable, efficient and also economic cellulase or hemicellulase for the bio-fuel industry has been attracted many scientist's concerns. Applications of this enzyme such as the production of animal feed, formulation of detergents, juice clarification, paper industry and wine production. Cellulase contribute to 8% of the worldwide industrial enzyme demands and the demand is expected to increase by 100% within 2014 (Costa et al., 2008; Saha et al., 2012). Cellulases are among the industrially important hydrolytic enzymes and are of great significance in present day biotechnology (Kulkarni & Vedamurthy 2015). Cellulases are widely used in the food, feed, textile and pulp industries. The potential of these biological resources are possible substitutes for diminishing fossil energy resources and becoming increasingly important.

Applications of Cellulase

For years, cellulases have been a target for academic and industrial research and are currently being applied in many industries (Singh et al., 2007). Research has been conducted to find out industry beneficiary strain to increase the yield of cellulase production.

Cellulase in the Textile Industry

The textile industry has been revolutionized by introduction of enzymes that are slowly replacing the conventional chemical processes, which are generally severe and lead to fiber damage (Bhat, 2000; Kuhad et al., 2011). Cellulases have the ability to modify cellulosic fiber in a controlled and desired manner thus improving the fabric quality (Mojsov, 2012). They are mostly used during wet processing to improve fabric properties. Processes that involve cellulase activity include biostoning of jeans and biopolishing cellulosic fibers with increased productivity and a safe working environment (Christian et al., 2006; Karmakar et al., 2011). Cellulase mixtures usually rich in endoglucanases are used in this process to remove the small protruding fibers from the fabric surface without using chemicals. The fabric attains a smooth and a glossy appearance, improved brightness and uniformity. Biopolishing is a key procedure in the production of high quality garments (Bhat, 2000; Kuhad et al., 2011).

Cellulase in the Wine and Brewing Industry

Enzyme technology plays a crucial role in the beer and wine industries. Wine making requires the extraction of juice from grapes and subsequent fermentation by yeast while beer brewing involves malting of barley and fermentation of the resulting wort (Bhat, 2000). Cellulase can be used for production of alcohol from cellulose biomass, because hydrolysis of cellulose produces reducing sugar as end product. Cellulase play important role in beverage industry for making fruit juice by degrading the cell wall which made up of cellulose, along with increasing the glucose concentration (Cherry & Fidantsef, 2003). In wine making, pectinases, β glucanases and hemicellulases comprise the main exogenous enzymes added. These enzymes give a better skin maceration, improved colour extraction, easy must clarification and filtration and improved wine quality and stability (Benitez et al., 1998). In order to improve the wine's aroma, β glucosidase is added to modify glycosylated precursors that are naturally present (Sun & Cheng, 2006).

Cellulase in the Detergent Industry

Recent innovations in the detergent industry have seen the incorporation of enzymes such as cellulases, proteases and lipases in detergents (Singh et al., 2007). Cellulase containing detergents are capable of degrading the cellulose micro fibrils to restore a smooth surface and original color to the garment. In addition, the degradation softens the fabric and removes dirt particles trapped in the micro fibril network (Sukumaran et al., 2005). Cellulases active under alkaline conditions increase the cleaning capacity of detergents by selective contraction fibers hence facilitating the removal of oil from inter fiber space (Karmakar & Ray, 2011).

Cellulase in Pulp and Paper Industry

Application of enzyme preparations comprising cellulases, xylanases and lignases in the pulp and paper industry has increased in the last decade (Mai et al., 2004). Bio pulping using cellulases and allied enzymes reduces the energy required to achieve the desired strength and freeness of the pulp hence it's a better alternative to mechanical pulping. Cellulases containing enzyme mixtures are also useful in the hydrolysis of fine small particles produced during refining of primary or secondary fibers. Deinking process is crucial during paper recycling. Enzymatic deinking using cellulases reduces the need for deinking chemicals and also results to little or no loss in paper strength. Enzymatic deinking is usually combined with mechanical agitation in order to improve the efficacy of the process (Aygan, Karcioglu & Arikan, 2011).

Cellulase in Agriculture

Cellulases and related enzymes from certain fungi -are capable of degrading the cell wall of plant pathogens hence controlling plant diseases. The β-1, 3- glucanase and N-acetyl- glucosaminidase from *Trichoderma harzianum* were reported to synergistically inhibit the spore germination and germ tube elongation of *B. Cinerea* (Cavaco-Paulo, 1998; Bhat, 2000). The β- 1, 3- glucanase from *Trichoderma harzianum* CECT 2413 induced morphological changes such as hyphal tip swelling, cytoplasm leakage and the formation of numerous septae and also inhibited the growth of *R. Solani* and *Fusarium sp* (Benitez et al.,1998). Cellulases are also important in soil quality improvement. In order to reduce overreliance on mineral fertilizers, farmers incorporate straw in soil. Microbial routes to hasten straw decomposition were using organisms such as *Aspergillus*, *Chaetomium* and *Trichoderma* and *actinomycetes* have shown promising results (Ortiz Escobar & Hue, 2008; Tejada, et al., 2008).

Cellulase in the Food Industry

Research on industrially suitable enzymes such as cellulases, hemicellulases and pectinases from food grade microorganisms such as *Aspergillusniger, Trichoderma sp* and increased knowledge of fruit components led to the overcoming of these challenges and led to improved methods of extraction, clarification and stabilization (Singh et al., 2007). Cellulases along with xylanases and pectinases are the macerating enzymes that serve to increase the yield and process performance without any additional cost. Macerating enzymes are usually used in two steps; after crushing, the fruit pulp is macerated to either partial or complete liquefaction. After the extraction, pectinases are then used for its clarification and this lowers viscosity of fruit juice prior to its concentration and further increases the filtration rate and the stability of the juice.

There is a growing demand for natural pigments for food colorants such as carotenoids. In their natural state, carotenoids remain bound to proteins thus preventing pigment oxidation. When solvents are used to extract carotenoids, they disrupt that association thus making the pigments insoluble in water and oxidation. This can be prevented by use of enzymatic methods. Cellulases hydrolyze cellulose in the cell walls hence the structural rigidity is interfered with exposing intracellular materials for extraction. These pigments remain bound to proteins and are more stable than those obtained through traditional methods that involve use of solvents (Bassi et al., 1993).

Cellulase in the Bio-Refinery

Bioconversion of lignocellulosic biomass to produce biofuel is the most popular area of cellulase application being investigated recently (Sukumaran et al., 2005). Potential lignocellulosic feedstocks sources include agricultural crop residues such as straw, the perennial prairie grass, municipal waste, packaging and construction debris, agricultural or forest processing by products e.g. food processing residues, pulping liquor from paper mills and forest woody biomass either logging residues from conventional harvest operations or removal of excess biomass from timberlands. Lignocellulosic biomass consists of cellulose tightly linked to lignin and hemicellulose (Kuila et al., 2011).

Cellulase as Bioenergy

Interest in bio energy has been sharply increasing in recent years due to the necessity of sustainable economies and clean environments (Sun et al., 2010). Cellulose and hemicellulose are the most abundant biomasses on earth, and therefore have the greatest potential to resolve both the energetic and environmental demands of bio energy (Kim et al., 2008). Cellulose is a linear polysaccharide which is constructed from monomer of glucose bound together with 1-4 glucosidal linkage. Cellulose, the leading bio-residue from agricultural sector is the predominant renewable biopolymer in the world which comprises of celluloses, hemicelluloses and lignin. A promising strategy for the efficient utilization of this renewable resource is to use it as a base material for the production of desired metabolites. Apart from the production of value-added products, its bioconversion offers an effective solution for the abatement of pollution due to solid-waste and their utilization, which would at low sustainable process and products well known that plants are the most common source of renewable carbon and energy on the earth (Kulkarni & Vedamurthy, 2015). Compost is a fertilizing mixture of partially decomposed organic matter from plant and animal origin. Cellulose degradation from an initial 40% (dry basis) to a final 19% was observed during composting primary sludge, bark and nitrogen fertilizers as amendment (Farzaneh et al., 2011). In other study, Sangharak et al. (2013) reported composting a blend of bleached primary sludge's, ash and paunch (cattle stomach contents as nitrogen source) from a slaughterhouse. The biological degradation of cellulose has been studied for many years, and a number of cellulolytic enzymes, especially cellulases produced by fungi and bacteria, have been isolated and characterized. These enzymes, which cleave the β-1, 4 bond of cellulose, belong to the large family of glycosyl hydrolases (Gomashe et al., 2013). The bioconversion of cellulosic materials is now a subject of intensive large-scale conversion process beneficial to mankind. Considering this huge significance of cellulase which contribute 8% of the world wide industrial enzyme demand and the demand was expected to increase by 100% within 2014 (Muhammad et al., 2012). Surprisingly very little is known about the bacterial and fungal cellulase in compost, swamps and adjacent environments. Keeping the above in vision, the present review is aimed at isolation, screening, characterization and thorough understanding of cellulolytic micro-organisms.

CONCLUSION

Despite the exciting new findings and studies regarding cellulose degrading organisms, there is still much we do not understand about the mechanism of cellulose degradation and the microbial communities that carry it out. Cellulase is an important enzymes and find applications in industries of starch processing, grain alcohol fermentation, malting and brewing, extraction of fruit and vegetable juices, pulp and paper industry, and textile industry. Therefore, it is of considerable economic interest to develop processes for effective treatment and utilization of cellulosic wastes as inexpensive carbon sources. Because of the biomass potential as a source of renewable fuels, there has been a rapid increase in the amount of research in this area and our scientific approach towards the cellulose degradation should continue to increase.

REFERENCES

Ariffin, H., Abdullah, N., Umi Kalsom, M. S., Shirai, Y., & Hassan, M. A. (2006). Production and characterisation of cellulase by bacillus pumilus. *IACSIT International Journal of Engineering and Technology, 3*(1), 47–53.

Aygan, A., Karcioglu, L., & Arikan, B. (2011). Alkaline thermostable and halophilic endoglucanase from Bacillus licheniformis C108. *African Journal of Biotechnology, 10*(5), 789–796.

Bhat, M. K. (2000). Cellulases and related enzymes in biotechnology. *Biotechnology Advances, 18*(5), 355–383. doi:10.1016/S0734-9750(00)00041-0 PMID:14538100

Wanderley, K. J., Torres, F. A., Moraes, L. M., & Ulhoa, C. J. (2004). Biochemical characterization of Î±-amylase from the yeast Cryptococcus flavus. *Microbiology Letters, 231*(2), 165–169. doi:10.1016/S0378-1097(03)00955-8

Boonmee, A. (2009). Screening of rice straw degrading microorganisms and their cellulose activities. *KKU Sci. j., 37*, 83-88.

Cavaco-Paulo, A. (1998). Mechanism of cellulose action in textile processes. *Carbohydrate Polymers, 37*(3), 273–277. doi:10.1016/S0144-8617(98)00070-8

Cherry J. R & Fidantsef A. L (2003). Directed evolution of industrial enzymes: an update. *Curr. Opin. Biotechnol., 14*, 438-443.

Dubey, S. K., Meena, R. K., Sao, S., Pate, J., Thakur, S., & Shukla, P. (2014). Isolation and Characterization of Cellulose Degrading Bacteria from Biogas Slurry and their RAPD profiling. *Current Research in Microbiology and Biotechnology.*, *2*(4), 416–442.

Farzenah, R., Joh, L. D., Kashima, H., Reddy, A. P., & VanderGheynst, J. S. (2011). Selection of condition for cellulase and xylanase extraction from switchgrass colonized by Acidothermus cellulolyticus. *Applied Biochemistry and Biotechnology, 164*(6), 793–803. doi:10.100712010-011-9174-6 PMID:21318368

Gomashe, A. V., Gulhane, P. A., & Bezalwar, P. M. (2013). Isolation and Screening of Cellulose Degrading Microbes from Nagpur Region Soil. *International Journal of Life Sciences (Kathmandu), 1*(4), 291–293.

Immanuel, G., Bhagavath, C. M., Raj, I. P., Esakkiraj, P., & Palavesam, A. (2007). Production and Partial Purification of Cellulase by Aspergillus niger and A. fumigates Fermented in Coir waste and Sawdust. *The Internet Journal of Microbiology, 3*(1). doi:10.5580/49

Irfan, M., Safdar, A., Syed, Q., & Nadeem, M. (2012). Isolation and screening of cellulolytic bacteria from soil and optimization of cellulose production and activity. *Turkish Journal of Biochemistry–Turk. Journal of Biochemistry, 37*(3), 287–293.

Karmakar, M., & Ray, R. (2011). Current trends in research and application of microbial cellulases. *Research Journal of Microbiology, 6*(1), 41–53. doi:10.3923/jm.2011.41.53

Kaur, M., & Arora, S. (2012). Isolation and Screening of Cellulose Degrading Bacteria in Kitchen Waste and Detecting Their Degrading Potential. *IOSR Journal of Mechanical and Civil Engineering, 1*(2), 33–35.

Kuhad, R. C., Gupta, R., & Singh, A. (2011). Microbial cellulases and their industrial applications. *Enzyme Research.* doi:.10.4061/2011/280696

Kuila, A., Mukhopadhyay, M., Tuli, D. K., & Banerjee, R. (2011). Production of ethanol from lignocellulosics: An enzymatic venture. *EXCLI Journal, 10*, 85–96. PMID:27857667

Kulkarni, A. G., & Vedamurthy, A. B. (2015). Isolation and Characterization of Cellulolytic Bacteria from Soil. *International Journal of Environmental Biology, 5*(3), 57–65.

Li, L. L., McCorkle, S. R., Monchy, S., Taghavi, S., & van der Lelie, D. (2009). Bioprospecting metagenomes: Glycosyl hydrolases for converting biomass. *Biotechnology for Biofuels, 128*, 231–236. PMID:19450243

Lin, L., Kan, X., Yan, H., & Wang, D. (2012). Characterization of extracellular cellulose degrading enzymes from Bacillus thuringiensis strains. *Electronic Journal of Biotechnology, 15*(3). doi:10.2225/vol15-issue3-fulltext-1

Liszka, M. J., Clark, M. E., Schneider, E., & Clark, D. S. (2012). Nature Versus Nurture: Developing Enzymes That Function Under Extreme Conditions. *Annual Review of Chemical and Biomolecular Engineering*, *3*(1), 77–102. doi:10.1146/annurev-chembioeng-061010-114239 PMID:22468597

Lu, W.-J., Wang, H.-T., Yang, S.-J., Wang, Z.-C., & Nie, Y.-F. (2005). Isolation and characterization of mesophilic cellulose degrading bacteria from flower stalks-vegetable waste co-composting system. *The Journal of General and Applied Microbiology*, *51*(6), 353–360. doi:10.2323/jgam.51.353 PMID:16474195

Lyand, L. R., Weimer, P. J., Vanzyl, W. H., & Pretorious, I. S. (2002). Microbial cellulose utilization. Fundamentals and Biotechnology. *Microbiology and Molecular Biology Reviews*, *66*(3), 506–577. doi:10.1128/MMBR.66.3.506-577.2002 PMID:12209002

Lynch, J. M., Slater, J. H., Bennett, J., & Harper, S. H. T. (1981). Cellulase activities of some aerobic microorganisms isolated from soil. *Journal of General Microbiology*, *127*, 231–236.

Malherbe, S., & Cloete, T. E. (2003). Lignocellulose biodegradation: Fundamentals and applications. *Reviews in Environmental Science and Biotechnology*, *1*(2), 105–114. doi:10.1023/A:1020858910646

Muhammad, H., Iqbal, N., Ahmed, I., Zia, M. A., & Irfan, M. (2012). Purification and characterization of the kinetic parameters of cellulase produced from wheat straw by Trichoderma viride under SSF and its detergent compatibility. *Advances in Bioscience and Biotechnology*, *2*, 149–156.

Petre, M., Zarnea, G., Adrian, P., & Gheorghiu, E. (1999). Biodegradation and bioconversion of cellulose wastes using bacterial and fungal cells immobilized in radio polymerized hydrogels. *Resources, Conservation and Recycling*, *27*(4), 309–332. doi:10.1016/S0921-3449(99)00028-2

Pew, J. C. (1957). Properties of powdered wood and isolation of lignin by cellulolytic enzymes. *TAPPI*, *40*, 553–558.

Saha, S., Roy, R., Sen, S. K., & Ray, A. K. (2006). Characterization of cellulase-producing bacteria from the digestive tract of tilapia, Oreochromis mossambica (Peters) and grass carp, Ctenopharyngodon idella (Valenciennes). *Aquaculture Research*, *37*(4), 380–388. doi:10.1111/j.1365-2109.2006.01442.x

Saha, S., Roy, R., Sen, S. K., & Ray, A. K. (2006). Characterization of cellulase-producing bacteria from the digestive tract of tilapia, Oreochromis mossambica (Peters) and grass carp, Ctenopharyngodon idella (Valenciennes). *Aquaculture Research*, *37*(4), 380–388. doi:10.1111/j.1365-2109.2006.01442.x

Sangkharak, K., Vangsirikul, P., & Janthachat, S. (2013). Isolation of novel cellulase from agricultural soil and application for ethanol production. *International Journal of Advanced Biotechnology and Research*, *2*(2), 230-239.

Singh, C. P. (1987). Preparation of high grade compost by enrichment on organic matter decomposition. *Bio. Agric. Host*, *5*, 41–49.

Sukumaran, R. K., Singhania, R. R., & Pandey, A. (2005). Microbial cellulases- Production, Applications and challenges. *Journal of Scientific and Industrial Research*, *64*, 832–844.

Sukumaran, R. K., Singhania, R. R., & Pandey, A. (2005). Microbial Cellulases Production, Applications and Challenges. *Journal of Scientific and Industrial Research*, *64*(11), 832–844.

Wilson, D. B. (2011). Microbial diversity of cellulose hydrolysis. *Current Opinion in Microbiology*, *14*(3), 1–5. doi:10.1016/j.mib.2011.04.004 PMID:21531609

Wyman, C. E. (2007). What is (and is not) vital to advancing cellulosic ethanol. *Trends in Biotechnology*, *25*(4), 153–157. doi:10.1016/j.tibtech.2007.02.009 PMID:17320227

Yin, L. J., Huang, P. S., & Lin, H. H. (2010). Isolation of Cellulase-Producing Bacteria and Characterization of the Cellulase from the Isolated Bacterium *Cellulomonas* Sp. YJ5. *Journal of Agricultural and Food Chemistry*, *58*(17), 9833–9837. doi:10.1021/jf1019104 PMID:20687562

Yin, L. J., Lin, H. H., & Xiao, Z. R. (2010). Purification and characterization of a cellulase from Bacillus subtilis YJ1. *Journal of Marine Science and Technology*, *18*(3), 466–471.

Zhang, Y. H., Himmel, M. E., & Mielenz, J. R. (2006). Outlook for cellulase improvement: Screening and selection strategies. *Biotechnology Advances*, *24*(5), 452–481. doi:10.1016/j.biotechadv.2006.03.003 PMID:16690241

Chapter 3
Bioprocessing Requirements for Bioethanol:
Sugarcane vs. Sugarcane Bagasse

Sophie Anderson
Teesside University, UK

Pattanathu K.S.M. Rahman
Teesside University, UK

ABSTRACT

This chapter discusses alternative energy sources and the advantages of biofuels over fossil fuels. It outlines the main steps of bioethanol production and suggests some alternative sources as potential feedstock. The core focus of this chapter is to examine new research which considers the use of agricultural waste as a feedstock for bioethanol production rather than conventional feedstocks such as sugarcane and corn. The advantages of sugarcane bagasse as a feedstock are discussed in detail and the bioprocessing requirements are studied in comparison to traditional methods that use sugarcane as the feedstock. The chapter concludes by briefly outlining further research that could potentially improve these processes.

INTRODUCTION

The world demand for energy and the depletion of oil reserves motivates the research for alternative energy sources. Biofuels can be defined as solid, liquid or gaseous fuels predominantly produced from biomass. There is a large amount of interest regarding biofuels due to the advantages over fossil fuels; biofuels are easily accessible from common biomass sources, they have a potential to be more environmentally friendly, there are benefits to the environment, economy and consumers in using biofuels and they are biodegradable and sustainable (Demirbas, 2008). Biofuels are claimed to have a lower 'carbon footprint' than fossil fuels and they contribute less to the greenhouse effect due to the CO_2 neutral conversion owing to the renewability of biomass. However, biofuels do not come without concerns and it is a worry that biofuel crops are replacing food crops. This has adverse implications on food security, especially in poor countries (Vassilev et al., 2015).

DOI: 10.4018/978-1-5225-3540-9.ch003

Ethanol is one of the most widely used liquid biofuels for vehicles and so is one of the most produced biofuels worldwide. The need for bioethanol is increasing due to depletion of crude oil supplies and the need for a sustainable fuel alternative (Sarkar et al., 2012). The largest producers of ethanol are the United States and Brazil, together producing 85% of the world's ethanol (Alternative Fuels Data Center, 2016). The vast majority of U.S. ethanol is produced from the starch from corn, while Brazil primarily uses sugarcane. However, these conventional crops are unable to meet the global demand of bioethanol production due to their primary value of food and feed. New research is considering the use of agricultural waste as a feedstock for bioethanol production. It is claimed that it is a cost effective, renewable and abundant alternative and it avoids the 'food vs. fuel' argument. A study by Kim and Dale (2004) estimated that 491 billion litres of bioethanol could be produced per year using agricultural waste – 16 times higher than the actual world bioethanol production at that time. Conventional raw materials such as corn or sugarcane contribute to 40-70% of production costs, so by using waste products the cost of feedstock could be reduced (Cardona et al., 2010). The following discussion will study and analyse two different methods of bioethanol production: (1) sugarcane as feedstock and (2) sugarcane bagasse (waste) as feedstock. The study will discuss the similarities and differences of production and if using waste as a feedstock is beneficial.

BIOETHANOL PRODUCTION

Sources and Applications

The main sources of biomass to produce bioethanol are conventionally sugarcane and corn feedstock, but as discussed above, these sources are barely sufficient to meet the demand (Vohra et al., 2014). Lignocellulosic biomass, also known as agrowaste, is a non-food alternative to this. The possible sources include crop residues, grasses, sawdust and woodchips. More recent research has also explored other possible feedstocks – Byadgi and Kalburgi (2016) considered the use of waste newspapers due to lower feedstock costs and to minimise the newspaper load on municipal waste. Others include bioethanol production form waste potatoes and waste money bills (Sheikh et al., 2013; Ali & Jiyad, 2015)

However, there are four major sources of agrowaste that are widely studied and are most favourable due to their abundant availability. These sources are rice straw, wheat straw, corn straw and bagasse (the dry pulpy residue left after the extraction of juice from sugar cane). As well as the production of bioethanol, a small amount of these feedstocks are also used for animal feed and bedding or fuel for heating and electricity (Sarkar et al., 2012).

Table 1. Worldwide potential bioethanol production from agricultural wastes

Agrowaste	Potential Bioethanol Production (Billion Litres Annually)
Rice straw	205
Wheat Straw	104
Corn Straw	58.6
Sugarcane bagasse	51.3

(Sarkar et al., 2012)

Sugarcane as a Feedstock

The process of bioethanol production depends on the materials used. There are commonly three major steps in its production: (1) extraction of fermentable sugars, (2) conversion of sugars into ethanol by fermentation and (3) separation and purification of ethanol usually by distillation and dehydration (Vohra et al., 2014). The steps prior to fermentation to obtain the fermentable sugars, are the main difference between conventional ethanol production from corn or sugarcane and ethanol production from agrowaste. Figure 1 outlines the main processes for the production of bioethanol from sugarcane.

Cleaning of Sugarcane, Extraction of Sugars and Juice Treatment

Sugarcane is harvested from fields. The stalks contain most of the sugars so are most desirable for industry. A dry-cleaning system is used to remove ~70% of dirt which is dragged along from harvesting. The extraction of the fermentable sugars is carried out by mills. Water is used to maximize sugar recovery (~96%) by a process called imbibition. Sugarcane juice and bagasse are obtained from the mills at this stage. In a study carried out by Dias et al. (2009) the recovery of sugars is ~96%. The bagasse

Figure 1. Process flow diagram of bioethanol product from sugarcane

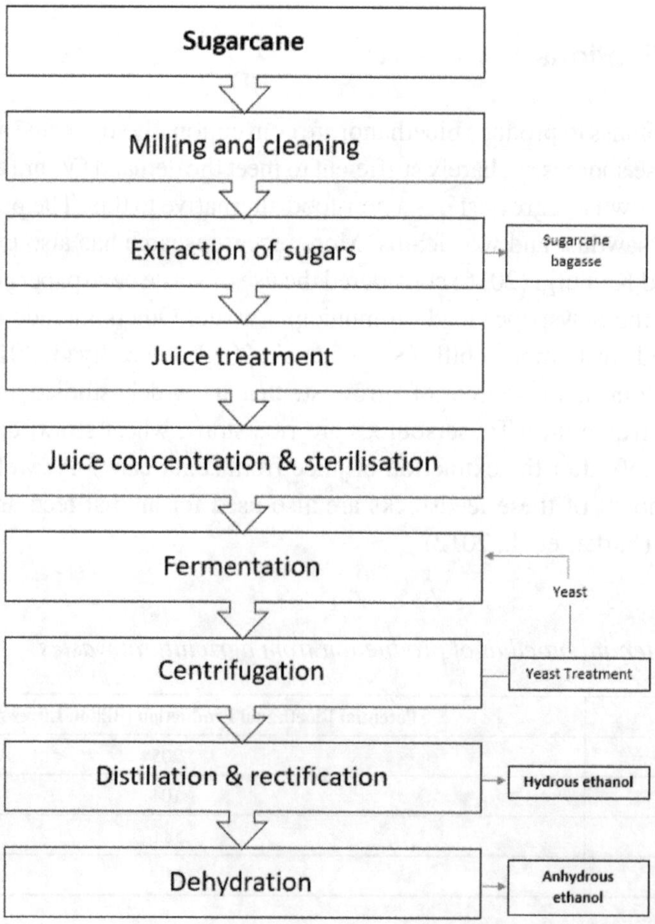

produced here can be burnt for the generation of steam and energy or it can be used as a raw material for bioethanol production.

The sugarcane juice contains impurities that need to be removed for efficient use as a raw material. Hydrocyclone removes physical dirt particles, phosphoric acid is added to enhance impurity removal and increase phosphate content. The juice is then initially heated to 30-70°C and then a secondary heated at 105°C.

Juice Concentration and Sterilisation

The juice must then be concentrated prior to fermentation and this is carried out by a five-stage multiple effector evaporator (MEE). Table 2 shows the pressure and temperature stages of juice concentration and sterilisation.

Fermentation

Yeast suspension (*Saccharomyces cerevisiae)* is added to the fermentation reactor along with the sterilized juice. Yeast is a facultative anaerobe. In an aerobic environment, it converts sugars into CO_2 and water whereas in an anaerobic environment, it converts sugars into CO_2 and ethanol. So for ethanol production anaerobic conditions are necessary. Optimal conditions are temperature of 28°C and pH 4.5 (Dias et al., 2009). Asif et al. (2015) used similar parameters i.e. temperature 30°C and pH 4.6 using *S. cerevisiae*. In contrast, Wong and Sanggari (2014) stated that the optimum temperature from their findings was 35°C but with the same pH of 4.5. Most of these studies used a batch fermentation method with a known amount of yeast suspension and juice.

During fermentation, sucrose is hydrolysed into fructose and glucose which are converted into ethanol and $CO_{2.}$ (wine and gases). Gases are collected through a column to recover evaporated ethanol and wine is centrifuged to recover and recycle yeast cells. The centrifuged wine is then fed to the distillation unit.

Distillation and Dehydration

The bioethanol from fermentation requires further separation and purification of ethanol from water (Limayem & Rick, 2012). Distillation is used to produce hydrous ethanol using a series of distillation and rectification columns operating under atmospheric pressure. It is the most recognized technique of purifying ethanol. The basic principle is that the wine from the fermenter is heated and the components

Table 2. Pressure and temperature on each stage of MEE.

Stage	Pressure (kPa)	Temperature (°C)
1	169.6	115.5
2	135.4	108.8
3	101.0	100.6
4	52.9	83.6
5	20.0	64.6

Data from Dias et al. (2009)

with a low boiling point vaporises. These vapours are then condensed to obtain more concentrated, less volatile products in the form of a liquid. So, ethanol and water are separated using these columns. To produce anhydrous ethanol, an extractive separation process is used involving two columns; extractive column and recovery column. Hydrous ethanol vapours are added to the extraction column along with a suitable solvent and anhydrous ethanol is obtained from the top and the solvent solution exits at the bottom. The recovery column is needed to produce a pure solvent that can be recycled back in to the extractive column. (Dias et al., 2011). The columns operate at high temperatures and pressures. Figure 2 shows a schematic of the distillation and dehydration columns (although the extraction and recovery columns are not shown here).

Typical Yield

Anhydrous ethanol has a purity of ~99.5% compared to hydrous ethanol which is between 93.8 – 93.5 wt.% ethanol. Both types can be used as a fuel source. However, only anhydrous ethanol is used when combined with fossil fuels. The study carried out by Dias et al. (2009) obtained a total yield of 102.5L of anhydrous ethanol per tonne of sugarcane. That included 83.7 (L/t) from sugarcane and 18.8 (L/t) from bagasse.

Sugarcane Bagasse as a Feedstock

The process of producing bioethanol from bagasse has some major differences. Figure 3 outlines the main methods involved. Lignocellulosic materials like sugarcane bagasse are mainly composed of cellulose, hemicellulose and lignin. The conversion of these materials into bioethanol or other biofuels is

Figure 2. Distillation and dehydration columns (Campo, 2012)

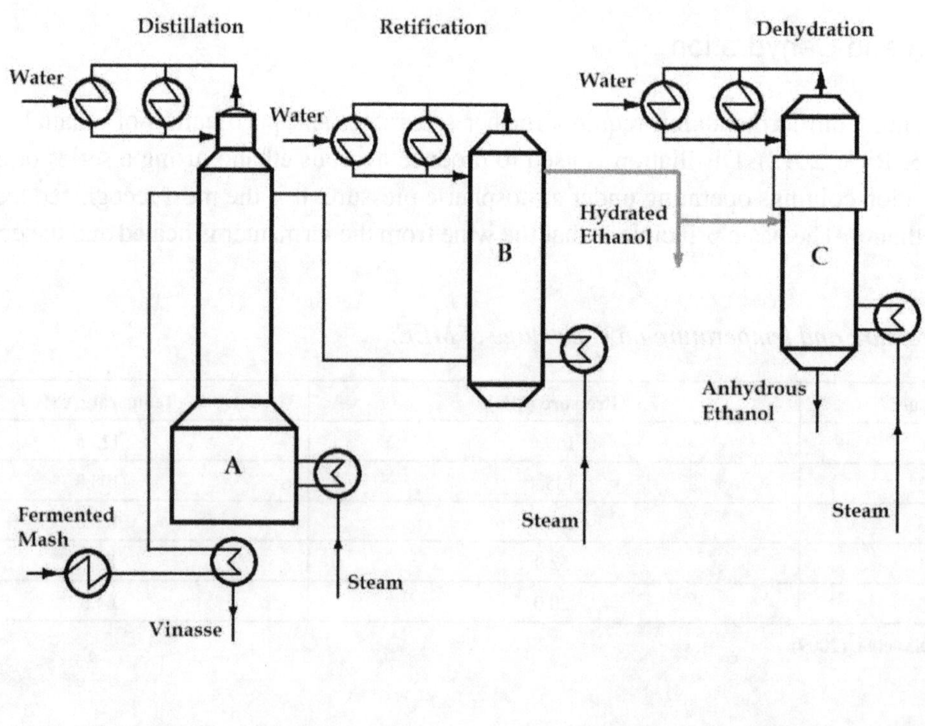

Figure 3. Process flow diagram of bioethanol production from sugarcane bagasse (Limayem & Ricke, 2012)

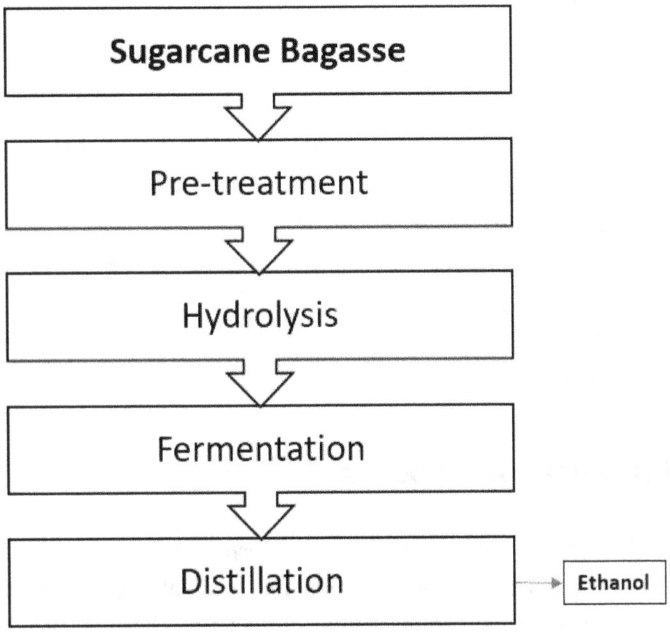

complex and needs pre-treatment and hydrolysis stages to increase the efficiency of the product (Eberhard et al., 2017).

Pre-Treatment and Hydrolysis

This process differs in that pre-treatment methods are needed to break down lignocellulosic biomass into its three main components; hemicellulose, lignin and cellulose. By breaking these down, it makes the remaining solid biomass more accessible for further treatment i.e. the next step of hydrolysis (Sarkar et al., 2012). This step is fundamental for successful hydrolysis and downstream processes. Unfortunately, these pre-treatment steps are the most expensive part of the process.

The major pre-treatment methods are classified as physical, chemical or biological. Physical methods include mechanical size reduction (such as milling), steam treatment and hydro-thermolysis. For chemical methods, acid/alkali is added to the bagasse to remove lignin, which increases the surface area and makes the material more suitable for fermentation. For biological methods, fungi can be used to enzymatically delignify the material. Table 3 outlines some advantages and disadvantages of these methods.

The next step is hydrolysis to convert the separated cellulose to sugar monomers for the fermentation to bioethanol. This process can be acidic or enzymatic. Enzymatic hydrolysis can be carried out using enzymes from certain fungi. Talebnia et al. (2010) suggested the use of *Trichoderma reesei* or *Aspergillus niger*. Acidic hydrolysis can be subdivided into dilute or concentrated hydrolysis. Concentrated hydrolysis is the preferred method as it isn't followed by high concentrations of inhibitors and produces a large yield of sugars. Unfortunately, it requires large amounts of acid and is an expensive process so it is less attractive commercially (Limayem & Ricke, 2012).

Table 3. Advantages and disadvantages of pre-treatment methods

Method	Process	Advantages	Disadvantages
Physical	Milling	• Increases accessible surface area of material • Decreases cellulose crystallinity • Highly efficient method	High energy demand
Chemical	Acid/Alkali	• Increases accessible surface area of material • Decreases cellulose crystallinity • Delignifies • Hydrolysis of hemicellulose • Most effective process • Rapid rate of treatment	Some methods can generate toxic inhibitors
Biological	Fungi	• Delignifies • Partial hydrolysis of hemicellulose • Low energy requirement	• Very low rate of treatment • Unsure on quality for industrial use

(Limayem & Ricke 2012; Eberhard et al., 2017)

Fermentation and Downstream Process

Once the sugar substrates from hydrolysis are available for fermentation, these next steps of fermentation and distillation/separation are the same as the methods discussed above for sugarcane.

The industrial potential for *S. cerevisiae* fermentation has already been proven and its capability is utilised in the industry. Despite its advantages, *S. cerevisiae* is not able to ferment sugars other than hexose. Unfortunately, lignocellulosic material such as bagasse contains mainly pentose sugars so new research is interested in developing genetically engineered microorganisms capable of fermenting hexose and pentose sugars simultaneously. The genetic improvement of the conventional *S. cerevisia* strain is gaining increasing research interest since this strain is already the most optimally adapted to bioethanol fermentation conditions (Limayem & Ricke, 2012).

FUTURE RESEARCH DIRECTIONS

Further research to improve these processes of producing bioethanol involve the genetic improvement of the conventional *S. cerevisiae* to be able to ferment both sugar types. There are already a wide range of bacteria able to ferment these sugars but they are not all capable of working under the fermentation conditions and result in low ethanol yields. If research uncovers a way to resolve these issues, the use of lignocellulosic biomass could be an attractive alternative to the use of non-renewable, polluting fossil fuels for fuel.

CONCLUSION

The main advantages of the use of waste materials are overcoming the problem of greenhouse gases, the renewability of feedstock, economic benefits including more employment opportunities and to sustain global transport systems (Eberhard et al., 2017).

As the world demand for energy alternatives continues to increase, lignocellulosic material/waste seems a viable possibility as a renewable feedstock for the production of biofuels. Despite advances in research, the process of producing this from waste is still expensive and challenging. The availability of these materials is still an issue although studies by Kim and Dale (2004) concluded that sugarcane bagasse is an attractive feedstock due to its abundant available. A significant barrier is the cost of the pre-treatment process which is estimated to account for ~33% of the total cost of production from lignocellulosic material (Talebnia et al., 2010). Until a cost-effective method is discovered, the cost of bioethanol will be high.

REFERENCES

U.S. Dept. of Energy. (2016). Alternative fuels data center: Maps and data. Retrieved 1 February 2017 from http://www.afdc.energy.gov/data/

Campo, A. B. (2012). PID Control Design, MATLAB - A Fundamental Tool for Scientific Computing and Engineering Applications (Vol. 1). InTech. Retrieved 1 February 2017 from http://www.intechopen.com/books/matlab-a-fundamental-tool-for-scientific-computing-and-engineering-applications-volume-1/pid-control-design

Asif, H. K., Ehsan, A., Kashaf, Z., Abeera, A. A., Azra, N., & Muneeb, Q. (2015). Comparative study of bioethanol production from sugarcane molasses by using *Zymomonas mobilis* and *Saccharomyces cerevisiae*. *African Journal of Biotechnology*, *14*(31), 2455–2462. doi:10.5897/AJB2015.14569

Byadgi, S. A., & Kalburgi, P. B. (2016). Production of bioethanol from waste newspaper. *Procedia Environmental Sciences*, *35*, 555–562. doi:10.1016/j.proenv.2016.07.040

Cardona, C. A., Quintero, J. A., & Paz, I. C. (2010). Production of bioethanol from sugarcane bagasse: Status and perspectives. *Bioresource Technology*, *101*(13), 4754–4766. doi:10.1016/j.biortech.2009.10.097 PMID:19945863

Demirbas, A. (2008). Biofuels sources, biofuel policy, biofuel economy and global biofuel projections. *Energy Conversion and Management*, *49*(8), 2106–2116. doi:10.1016/j.enconman.2008.02.020

Dias, M. O. S., Ensinas, A. V., Nebra, S. A., Maciel Filho, R., Rossell, C. E. V., & Maciel, M. R. W. (2009). Production of bioethanol and other bio-based materials from sugarcane bagasse: Integration to conventional bioethanol production process. *Chemical Engineering Research & Design*, *87*(9), 1206–1216. doi:10.1016/j.cherd.2009.06.020

Kandasamy, M., Hamawand, I., Bowtell, L., Seneweera, S., Chakrabarty, S., Yusaf, T., ... Eberhard, F. (2017). Investigation of ethanol production potential from lignocellulosic material without enzymatic hydrolysis using the ultrasound technique. *Energies*, *10*(1), 62. doi:10.3390/en10010062

Kim, S., & Dale, B. E. (2004). Global potential bioethanol production from wasted crops and crop residues. *Biomass and Bioenergy*, *26*(4), 361–375. doi:10.1016/j.biombioe.2003.08.002

Sheikh, M. M. I., Kim, C., Lee, J., Kim, S., Kim, G., Shim, S., & Kim, J. (2013). Production of bioethanol from waste money bills – A new cellulosic material for biofuels. *Food and Bioproducts Processing*, *91*(1), 60–65. doi:10.1016/j.fbp.2012.09.001

Limayem, A., & Ricke, S. C. (2012). Lignocellulosic biomass for bioethanol production: Current perspectives, potential issues and future prospects. *Progress in Energy and Combustion Science*, *38*(4), 449–467. doi:10.1016/j.pecs.2012.03.002

Mohammed, A. K., Ali, S. A., & Jiyad, J. H. (2015). Production of bioethanol from waste potatoes. Al-Khawarizmi Engineering Journal, 10(3), 62-67.

Sarkar, N., Ghosh, S. K., Bannerjee, S., & Aikat, K. (2012). Bioethanol production from agricultural wastes: An overview. *Renewable Energy*, *37*(1), 19–27.

Talebnia, F., Karakashev, D., & Angelidaki, I. (2010). Production of bioethanol from wheat straw: An overview on pretreatment, hydrolysis and fermentation. *Bioresource Technology*, *101*(13), 4744–4753. doi:10.1016/j.biortech.2009.11.080 PMID:20031394

Vassilev, S. V., Vassileva, C. G., & Vassilev, V. S. (2015). Advantages and disadvantages of composition and properties of biomass in comparison with coal: An overview. *Fuel*, *158*, 330–350. doi:10.1016/j.fuel.2015.05.050

Vohra, M., Manwar, J., Manmode, R., Padgilwar, S., & Patil, S. (2014). Bioethanol production: Feedstock and current technologies. *Journal of Environmental Chemical Engineering*, *2*(1), 573–584. doi:10.1016/j.jece.2013.10.013

Wong, Y. C., & Sanggari, V. (2014). Bioethanol production from sugarcane bagasse using fermentation process. *Oriental Journal of Chemistry*, *30*(2), 507–513. doi:10.13005/ojc/300214

de Albuquerque Wanderley, M. C., Martín, C., de Moraes Rocha, G. J., & Gouveia, E. R. (2013). Increase in ethanol production from sugarcane bagasse based on combined pretreatments and fed-batch enzymatic hydrolysis. *Bioresource Technology*, *128*, 448–453. doi:10.1016/j.biortech.2012.10.131

KEY TERMS AND DEFINITIONS

Agrowaste: Also known as agricultural waste, is waste produced from agricultural operations.

Bagasse: The dry pulpy residue waste from the extraction of juice from sugar cane.

Biofuels: Fuels produced directly from biomass.

Biomass: Renewable organic matter such as agricultural crops or waste, often used for fuel.

Carbon Footprint: The amount of carbon dioxide released into the atmosphere because of human activity.

Feedstock: Raw material to supply or fuel a machine or process (such as fermentation).

Fermentation: The chemical breakdown of a substance by bacteria, yeasts, or other microorganisms.

Chapter 4
Microbe Mediated Bioconversion of Fruit Waste Into Value Added Products:
Microbes in Fruit Waste Management

Mridul Umesh
Bharathiar University, India

Thazeem Basheer
Bharathiar University, India

ABSTRACT

Biosynthetic capabilities of microbes have solved several hurdles in the human welfare. Microbes have served and continue to serve as imperial candidates in both production and management strategies. Microbe mediated techniques has emerged as ecofriendly and sustainable alternative to their synthetic counterparts. Fruit based industries produces large volumes of solid and liquid wastes contributing to increase in pollution load. Disposal of these waste not only represent loss of valuable biomass but also leads to substantial increase in Biological Oxygen Demand (BOD) and Chemical Oxygen Demand (COD). However, in spite of their pollution and hazard aspects, in many cases, fruit processing wastes have a promising potential for being chief raw materials for secondary industries. This chapter summarizes microbe mediated fermentative utilization of fruit waste, for the production of value added products like organic acid, single cell protein, bioplastics, enzymes and biogas.

INTRODUCTION

The exponential growth of the human population has led to the accumulation of huge amounts of biodegradable and non-biodegradable waste materials across the world (Rivard et al., 1995). Living conditions in the biosphere are therefore changing dramatically, in such a way that the presence of waste residues is affecting the potential survival of many species. Fruit peels are a class of agro wastes that may be regarded as a non-product flow of raw materials having economic values less than the cost of collection

DOI: 10.4018/978-1-5225-3540-9.ch004

and recovery for reuse. Fruit based industries produces large volumes of wastes, both solid and liquid; these wastes pose increased disposal and pollution (High BOD or COD) problems and represents a loss of valuable biomass and nutrients. In spite of their hazardous effects contributing to environmental pollution, fruit peels have a good potential for conversion into useful products of higher value as by-product, or even as raw materials for other industries. Fruit waste contain an appreciable amount of carbohydrates which could be utilized by microorganisms producing economically important Biopolymers, Single Cell Proteins, Organic acids, Enzymes, Biogas etc. with potential application in food, fuel, agriculture, packaging, and pharmaceutical industries.

Bioplastics especially polyhydroxyalkanoates (PHA's) are exclusively synthesized as intracellular carbon and energy storage compounds by wide range of microorganisms and are reported to be completely degraded in to benign compounds both aerobically and anaerobically (Lemoigne, 1926). Their material properties closely resemble synesthetic plastics and thus are emerging to be the best alternatives to their synthetic counterparts. The ability to be synthesized from a wide variety of waste substrates along with their fully degradable nature is an added advantage to this kind of biopolymers.

SCPs are protein rich microbial biomass or total proteins extracted from microbial cell that could be used as protein supplements in food and feed (Gour et al., 2015). Their nutritious value and growth promoting essential amino acids content makes them ideal for tackling the global issues associated with Protein Calorific Malnutrition (PCM). Production of organic acids from fruits and fruit wastes dates back to several centuries. It is still an important industry that sustains the economy of many developed, developing and under developed countries. Acetic acid, butyric acid, lactic acid, citric acid and tartaric acid are the well-known examples of microbe mediated 'organic acids'. Enzyme production from fruit wastes using microbes could pave as a sustainable method for reducing cost of production. Fruit peels are used for the commercial production of enzymes like amylase, protease, and laccase. Microbes could also be exploited as tools for production of bioethanol and biogas from fruit peels. This serves as a promising alternative for increasing energy crisis all over the world.

Fruit peel wastes could be thus considered as valuable by-product if appropriate technical means are used to increase the value of the subsequent products to exceed the cost of reprocessing. This chapter highlights easily adoptable microbiological methodologies for recycling, reprocessing and eventual utilization of fruit peel wastes for the production of value added products rather than their discharge to the environment.

Background of the Study

Fruit wastes represent an important class of food processing waste that has been turned to be a menace for all the countries due to problems associated with their disposal and treatment. Majorly it consists of large fractions of solid or semi-solid waste generated during the separation of desired products from undesired ones in early stages of processing. Undesirable constituents discharged from fruit-based industries include fruit peels, seeds, extracted pulp and pits. Usages of these wastes either as feed ingredient for livestock or application on land are two conventional methods of fruit waste management. Major characteristic of fruit processing industries are wide range of waste water and organic load. Biological Oxygen Demand (BOD) and Total Suspended Solids (TSS) accounts for the chief pollutants in the fruit-based waste. Fruit based industries are conservative and reluctant to invest capital in waste management principles. The fruit processing industry produces 12 million tons of fruit annually. This in turn generates approximately 39 billion gal of waste water and 4.7 million tons of solid residuals. Fruit processing waste majorly contain mostly biodegradable organic matter in both soluble and insoluble forms (Omoregie et al., 2013). Due to the presence of high volume of suspended solids and BOD, fruit wastes can lead to serious environmental concerns when discharged in to the main streams. Microorganisms metabolize the organic matter present in the waste water thereby reducing the oxygen concentration that leading to the death of aquatic organisms. Continuous oxygen depletion triggers the anaerobic decomposition of organic matter and producing foul odor and stream discoloration.

MAJOR FOCUS OF THE CHAPTER

The thrust for development of integrated production strategies in food industries along with substantial byproduct utilization has paved way for valorization of fruit waste rather than their usage as feed or composting materials. Microbial mediated bioconversion has always been excelled as an important tool in waste management and product development (Bayer et al., 2007).

Organic Acid

Microbes have been exploited as the important biological agents for the production of organic acid from a wide variety of substrates. This trend continues even during the 21st century due to their robust nature to support fermentation and ability to utilize even the very cheap agricultural waste serves as feed stock for organic acid production. Microbial fermentation has advantages over chemical synthesis of organic acid which includes their ability to use waste as substrate for acid production, ease in recovery process, cost of production is low and problems associated with disposal of waste chemicals is nullified. These factors make them ideal and affordable even for developing and under developed countries. The major organic acids produced from fruit waste along with their applications in various sectors are illustrated in Figure 1.

Lactic Acid

Lactic acid classified as GRAS (generally regarded as safe) for use as a food additive by the US FDA (Food and Drug Administration), can be produced by either microbial fermentation or chemical synthesis. There are two optical isomers of lactic acid: L (+) -lactic acid and D (-) –lactic acid. Depending upon organism, metabolic pathways differ when glucose is the main carbon source: homofermentative bacteria yield two lactic acid molecules from a glucose molecule whereas the heterofermentative bacteria transform a glucose molecule in to lactate, ethanol and carbon dioxide. Lactic Acid Bacteria (LAB) consist of a number of bacterial genera within the phylum *Firmicutes*. The genera *Coronobacterium, Enterococcus, Lactobacillus, Lactococcus, Lactosphaera, Leuconostoc, Melissococcus, Oenococcus, Pediococcus, Streptococcus, Tetragenococcus, Vagococcus and Weissella* are recognized as the Lactic Acid Bacteria (Ercolini et al., 2001). Lactic acid production from fruit waste was reported in many literatures (Mridul & Preethi, 2014; Pushparani et al., 2012) Production of lactic acid from fruit waste through microbial fermentation can be depicted through the flow diagram:

Fruit Waste → Acid hydrolysis/ Enzymatic treatment →Fermentable Carbohydrates → Microbial Fermentation → Recovery and Purification

Figure 1. Important organic acid and their uses

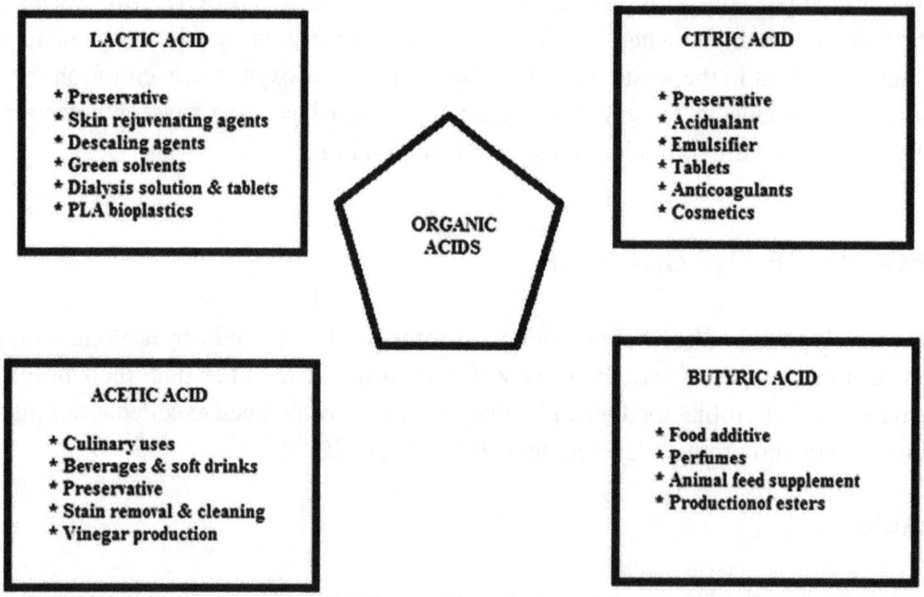

Lactic acid is also used in pharmaceutical industry as an electrolyte in many parenteral and intra venal solutions that are intended to replenish the body fluids or electrolytes. Example includes Lactated Ringer's or Hartmann's solutions, CAPD (continuous ambulatory peritoneal dialysis) solution, and dialysis solution for conventional artificial kidney machines. Moreover, lactic acid is used in preparation of tablets, prostheses, surgical sutures and controlled drug delivery systems (Pushparani et al., 2012).

Lactic acid offers natural ingredients for cosmetic applications. Although primarily used as moisturizers and pH regulators, they possess multiple other properties such as antimicrobial activity, skin lightening and skin hydration. The moisturizing effect is related directly to lactate's water retaining capacity and the skin lightening action of lactic acid is produced by the suppression of the formation of tyrosinase. Since they are naturally ingredients of the human body, lactic acid and its salt fit perfectly in to the modern trend towards natural and safer formulations, and they produce such effect as skin lightening and rejuvenation which makes them very useful as active ingredients in cosmetics.

Lactic acid and its salts are used increasingly in various types of chemical products and processes. In this category of application, lactic acid function as a descaling agent, pH regulator, neutralizer, chiral intermediate, solvent, cleaning agent, metal complexing agent, antimicrobial agent, and humectants. Natural lactic acid serves as an excellent and clean solvent as compared to their synthetic counterparts for cleansing application. Due to the high solvency power and solubility of lactic acid, it is an excellent remover of polymer and resins. It is available with an isomeric purity of 98%, and is suitable as a starting material in the production of herbicides or pharmaceuticals. Since lactic acid offers better descaling properties than conventional organic descalers do, it is often used in many decalcification products, such as bathroom cleaners, coffee machines, and toilets. Ethyl lactate is used in many anti-acne preparations, because it combines excellent solvency power against oils and polymeric stains, with no environmental impact and toxicological effects.

Lactic acid is considered the most potential monomer for chemical conversion, because it contains two reactive functional groups, a carboxylic group and a hydroxyl group. Lactic acid can undergo a variety of chemical conversions into potentially useful chemicals, such as propylene oxide (via hydrogenation), acetaldehyde (via decarboxylation), acrylic acid (via dehydration), propanoic acid (via reduction), and dilactide (via self–esterification). Lactic acid has recently received a great deal of attention as a feedstock monomer for the production of Poly Lactic Acid (PLA), which serves as a biodegradable commodity plastic. The optically pure lactic acid can be polymerized in to a high molecular mass PLA through serial reactions of polydecondensation, depolymerization, and ring-opening polymerization. The resultant polymer, PLA, has numerous uses in a wide range of application, such as protective clothing, food packaging, mulch film, trash bags, rigid containers, shrink wrap and short life trays. The huge growth of the PLA market will stimulate future demands on lactic acid considerably.

Success of fermentative production of lactic acid relies on the availability of cheap raw materials, because polymer producers and other industrial users usually require large quantities of lactic acid at a relatively low cost. Raw materials for lactic acid production should have the following characteristics: cheap, low levels of contaminating microbes, rapid production rate, high yield, little or no by-product formation, ability to be fermented with little or no pretreatment and year-round availability. When refined materials are used for production the costs for production and the costs for product purification should be significantly reduced. However still it is not economically favorable as refined carbohydrates are so expensive that they eventually result in higher production costs. Therefore, there have been many attempts to screen for cheap raw material for economical production of lactic acid (John et al., 2007).

Cheap raw materials, such as starch and cellulosic materials, whey, and molasses have been used for lactic acid production. Among these, starch and cellulosic materials are currently receiving a great deal of attention, because they are cheap and abundant renewable source. The starch material used for lactic acid production includes sweet sorghum, wheat, corn, cassava, potato, rice, rye and barley. These materials have to be hydrolyzed in to fermentable sugars before fermentation, because they consist of mainly α (1, 4) and α (1, 6) linkage between glucose molecules. This hydrolysis can be simultaneously with fermentation (Ligouri et al., 2013).

Cellulosic materials have been used for lactic acid production in similar ways as starch materials. These materials consist mainly of (1, 4) – glucan and often contain xylan, arabinan, galactan and lignin. The utilization of corncob, waste paper and wood has been reported as well. Sreenath et al., (2001) investigated the production of lactic acid from agricultural residues such as alfalfa fiber, wheat bran, corn stover and wheat straw. They suggested that, during SSF of alfalfa fiber, lactic acid production was enhanced by adding pectinase and cellulase together. Garde et al., (2002), used hemicellulose hydrolyzate from wheat straw for lactic acid production by co-culture of *L. brevis* and *L. pentosus*. Fermentation of lignocellulosic hydrolyzate is inhibited usually by inhibitory compounds, such as furfural, 5-hydroxymethyl furfural and acetic acid, which are generated during pre-treatment of lignocellulose. Most studies on methods to decrease this inhibition have been focused on the chemical and physical detoxification of the hydrolyzate. Wee et al., (2005) however, reported that the inhibition of fermentation caused by wood hydrolyzate was reduced to a slight degree by direct adaptation of LAB to the wood hydrolyzate-based medium.

Some industrial waste products, such as whey and molasses are of interest for common substrates for lactic acid production. Whey is a major by-product of the dairy industry and it contains lactose, protein, fat and mineral salts. For complete utilization of whey lactose, it is necessary to supplement whey with an additional nitrogen source. Amrane and Prigent (1998), Kulozik and Wilde (1999), and Schepers et al., (2002) supplemented whey with yeast extract for rapid production of lactic acid with *L. helveticus*. According to Fitzpatrick and O'Keeffe (2001) the addition of whey protein hydrolyzate to whey medium would make the fermentation more economically viable and would also reduce the amount of unused nutrients left during fermentation. There were several attempts to produce lactic acid from whey by batch culture of *L. casei*. Molasses is a waste product from the sugar manufacturing process and it usually contains a large amount of sucrose. *L. delbrueckii* and *E. faecalis* have recently been used for lactic acid production from molasses. Shukla et al., (2004) also reported D (–)-lactic acid production from molasses with recombinant strains of *E. coli*.

It is necessary to supplement the fermentation media with sufficient nutrients for rapid lactic acid production. The most common nutrient for lactic acid production is yeast extract, but this may contribute significantly to an increase in production costs. As an alternative to yeast extract, corn steep liquor, a by-product from the corn steeping process, has been used successfully for lactic acid production. Since it is derived from corn, 85% of its total nitrogen content is composed of proteins, peptides and amino acids. Yun et al., (2004) suggested that rice bran and wheat bran play important roles as effective nutrients for lactic acid production, because they usually contain several nutritional factors as well as fermentable carbohydrates. Kurbanoglu (2003) demonstrated that ram horn waste was an effective supplement for lactic acid production. Similarly, Bustos et al., (2004) proposed that vinification lees could be used for the formulation of low-cost media for lactic acid production. According to Wee et al., (2004) wastewater from electrodialyzed fermentation broth still contained some nutrients that could be available to LAB. Their result indicated that, if small amounts of other nutrients were supplemented to electro dialysis wastewater, then the efficiencies of fermentation would be improved significantly (Yun et al., 2004).

Agro wastes may be regarded non product flows of raw materials whose economic values are less than the cost of collection and recovery for reuse; and are therefore discarded.These wastes could be considered valuable by-product if appropriate technical means are used to increase the value of the subsequent products to exceed the cost of reprocessing. Recycling, reprocessing and eventual utilization of food processing residues offer potential of returning these by-product to beneficial uses rather than their discharge to the environment which might cause detrimental environmental effects. Food industries produces large volumes of wastes both solids and liquids; these wastes pose increasing disposal and pollution (High BOD or COD) problem and represent a loss of valuable biomass and nutrients. However, in spite of their pollution and hazard aspects, in many cases, food processing wastes have a good potential for conversion into useful products of higher value as by-product, or even as raw materials for other industries. Organic acids are example of such valuable by-product of the fermentation of high carbohydrate containing industrial substrates. Fruit peels contain an appreciable amount of carbohydrates which could be utilized by microorganisms producing intermediate volume of high value organic acid like lactic acid.

Lactic acid, an intermediate – volume specialty chemical is under increasing demand in food, pharmaceutical and chemical industries and for the production of Poly lactic acid polymers, which possess excellent biomedical applications. The global production of this organic acid is estimated to be 100 million pounds/year and is expected to grow by 8.6% annually. In India, the annual production capacity of Lactic acid is 6000t and estimated gaps of 2300t in supply by the year 2015 have been predicted, if the present level of production is not increased. Starch rich waste substrates could be used for meeting this growing demand for lactic acid, if appropriate biotechnological interventions are used and specific sectors amongst the Indian food processing industry are targeted.

The biotechnological processes for the production of lactic acid from cheap raw materials should be improved further to make them competitive with the chemically derived one.

Acetic Acid

Acetic acid is formed through multistep process involving conversion of starch to sugar by amylases, anaerobic conversion of sugars to ethanol by yeast fermentation, conversion of ethanol to hydrated acetaldehyde and dehydrogenation to acetic acid by aldehyde dehydrogenase. The last two steps are performed aerobically with the aid of acetic acid forming bacteria. Vinegar bacteria also called acetic acid bacteria, are members of the genus *Acetobacter* characterized by their ability to convert ethyl alcohol into acetic acid, by oxidation as shown below:

Fruit waste → Acetic acid bacteria/ Yeast → Anaerobic fermentation

Ethanol → Acetic acid bacteria → Oxidation → Acetic acid

Earlier processes used for making acetic acid were the Orleans process (which is also known as the slow process), the quick process (which is also called the generator process) and the submerged culture process. The quick process and submerged culture process were developed and are used for commercial acetic production today. Acetic acid can be produced from fruit peel waste containing appreciable amount of carbohydrates through fermentation process (Vikas & Mridul, 2014).

- **Citric Acid:** Citric acid is an intermediate in Tri Carboxylic Acid cycle and is a commercially valuable product with wide spread applications. It is produced through submerged fermentation in commercial scale from sucrose rich media. In small scale citric acid can be produced through solid state fermentation. In recent years, citric acid is produced from fruit wastes like grape pomace, kiwi fruit peel, carob pod, pineapple peels, orange peels and mosami peels. The microbial strains used in commercial production of citric acid from carbohydrate rich waste include *Aspergillus niger, Candida sp.* and *Bacillus licheniformis* (Kumar et al., 2003; Imandi et al., 2008).

Butyric Acid

Butyric acid belongs to the class of short chain fatty acids widely used in the manufacture of textile fibers, heat and temperature resistant films. Salts of butyric acid are used as flavoring agents in food. It is used extensively in the production of drugs used for treatment of gastrointestinal disorders and cancer (Bouallagui et al., 2004). Butyric acid is commercially produced through chemical synthesis using crude oil or through microbial fermentation approach. Despite of its high cost of production, microbial fermentation of butyric acid is unavoidable for some specific application. Butyric acid is produced by anaerobic microorganism belonging to the genera – *Clostridium, Fusobacterium* and *Sarcina.* Microbial production of butyric acid from fruit waste could be out lined in Figure 2.

Single Cell Proteins

Single cell proteins are dried microbial biomass or total proteins extracted from microbes that could be used as protein supplement in both food and feed. It is also called as bioprotein or microbial protein. Apart from high protein content (about 60-82% of dry cell weight) SCP also contains fats, carbohydrates, nucleic acids, vitamins and minerals. Another advantage with SCP is that it is rich in certain essential

Figure 2. Butyric acid production pathway

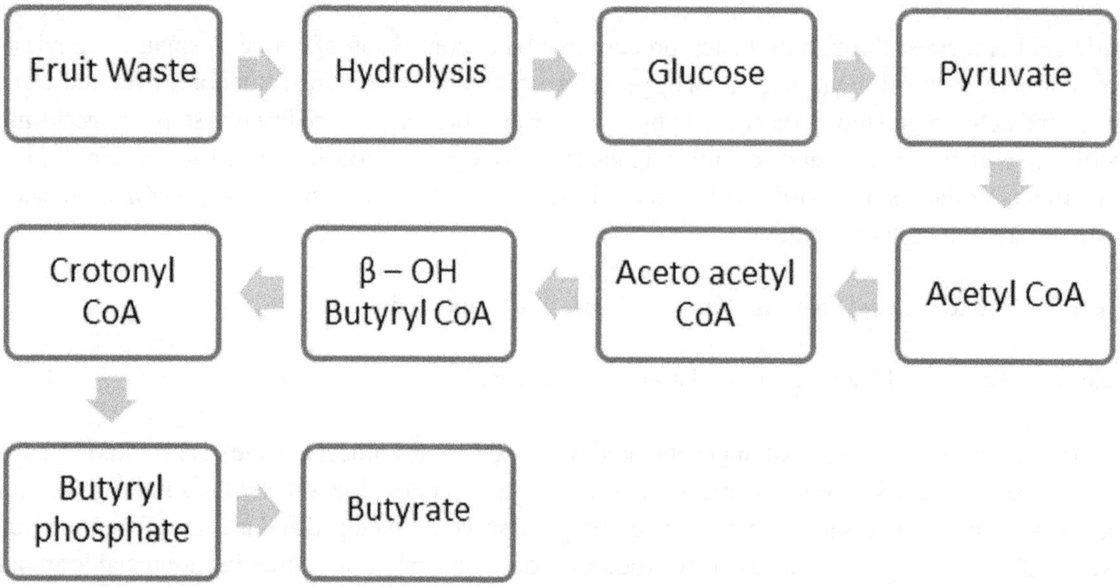

amino acids like lysine and methionine which are limiting in most plant and animal foods (Gour et al., 2015). Microbes like yeast, fungi, bacteria and algae have been used for SCP production. Yeast is another source of Single Cell Protein, and has been produced since a long time ago. Algae had drawn the attention of scientist to bridge the protein gap. Some of the algae like *Chloralla, Soenedesmus, Coelastrum* and *Spirulina* have been found to suitable for mass cultivation and utilization. The advantages in using algae include high protein and nutrient content, simple cultivation, effective utilization of solar energy and faster growth.

Many species of bacteria have been investigated for use in single cell protein production. *Methylophilus methylotrophus* with a generation time of about 2 hours and is usually and mainly used in animal feed as bacteria, in general produce a more favorable protein composition than yeast or fungi. Therefore, the large quantities of SCP animal feed can be produced using bacteria. The major factors that make bacteria ideal candidate for SCP production: their rapid growth and short generation times most can double their cell mass in 20 minutes to 2 hours. Capable of growing on a variety of raw materials, ranging from carbohydrates such as starch and sugar, hydrocarbons as methane, ethane and petrochemicals (Solomons et al., 1985; Nigam et al.,2000).

Importance of Microbial Protein

Microbial protein or SCP has various benefits over animal and plant proteins in that its requirement for growth are neither seasonal or climate dependent; it can be produced all around the year. It does not require a large expanse of land and it has high protein content with wide amino acid spectrum, low fat content and higher protein - carbohydrate ratio than forages. It can be grown on waste and it is environmental friendly as it helps in recycling waste (Gour et al., 2004). Beside nutritional value a protein should have desirable functional properties for its incorporation as food. Microbial protein has found to meet all the requirements for its inclusion as diet supplement for both human and livestock especially in the developing countries of Africa (Adedayo et al., 2011). Genetic manipulation of microbes may improve chemical composition and nutritional value and also production of microbial protein requires less labour than does agriculture production.

Benefits of Production of SCP From Microorganisms

1. Microorganisms have a high rate of multiplication and hence rapid succession of generations (algae: 2–6 hours, yeast: 1–3 hours, bacteria: 0.5–2 hours)
2. They can be easily genetically modified for varying the amino acid composition.
3. Protein content accounts for 43–85% in the dry mass.
4. They can utilize a broad spectrum of raw materials as carbon sources, which include waste products. Thus, they help in the removal of pollutants also.
5. Strains with high yield and good composition can be selected or produce relatively easily.
6. SCP production is independent of seasonal variation and usually consistent.
7. Land requirement is low and is ecologically beneficial.
8. A high solar-energy-conversion efficiency per unit area.
9. Solar energy conversion efficiency can be maximized and yield can be enhanced by easy regulation of physical and nutritional factors.
10. Algal culture can be done in space that is normally unused and so there is no need to compete for land (Rashad et al., 1990; Mateles et al., 1968; Miller et al., 1976; Parajo et al., 1995).

Limitations of SCP

Limitations of microbial proteins are enlisted below:

1. Production of non-targeted toxic end product can occur that may seriously affect health of the livestock when used as feed or may be toxic to human health.
2. Problem with digestion may occur sometimes.
3. Occurrence of allergic response in humans upon consumption of SCP may occur rarely.
4. Presence of high nucleic acid content may make SCP undesirable for human consumption and may often lead to the formation of gout and kidney stones.
5. The high nucleic acid content of many types of microbial biomass may lead to poor digestibility, gastrointestinal problem and also some skin reactions in humans.
6. Single cell protein production is a very expensive procedure as it needs high level of sterility control in the production unit or in the laboratory.

Fruit Waste as Substrate for SCP Production

Fruit waste containing an appreciable amount of complex carbohydrates need to be subjected to chemical or physical pretreatments to release the monomeric sugars that are easily fermentable (Figure 3). Apart from the carbon source, SCP production requires nitrogen and mineral supplementation depending on the targeted microorganism. Aseptic conditions need to be maintained throughout the process to prevent contamination. The starter culture is inoculated in its pure state and fermentation is carried

Figure 3. SCP production from fruit waste

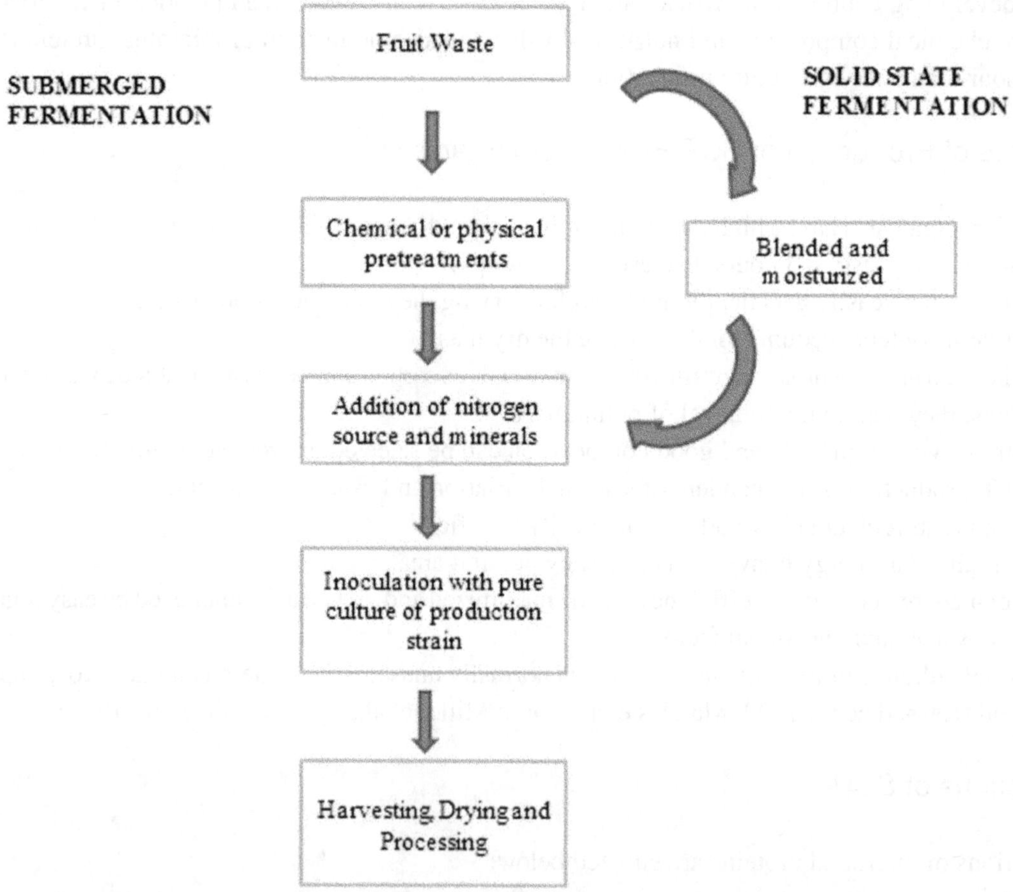

out in aerobic conditions. Provisions for ensuring minimizing the heat generation during SCP production should be monitored precisely for maximizing the yield. The microbial biomass is harvested from the medium, dried and processed. Commercial production of SCP basically occurs through submerged fermentation using *Saccharomyces cerevisiae, Aspergillus oryzae, Rhizopus oligosporus* and *Candida utilis* (Nasseri et al., 2011).

Enzymes

Enzymes are important class of bioactive compounds that has a great industrial significance in day to day life. Using of synthetic media for enzyme production results in the overall increase in the cost of production of enzymes. Global demand for increased production of enzymes paved way for the utilization of cheap agro waste substrates like fruit waste as commercially valuable substrates for enzyme production. Fruit wastes are primarily composed of complex polysaccharide that enhances the microbial growth for the production of industrial enzymes. The important enzymes produced from fruit waste along with their potential application are enlisted below:

Pectinases

Pectinases are pectin degrading enzyme produced majorly from *Aspergillus niger, Penicillium sp., Rhizobium sp.* especially using citrus peel waste. Pectinases are broadly divided in to two:

1. Depolymerizing enzyme – breaks α 1, 4 linkage between pectin.
2. Demethoxylating enzyme – converts pectin to pectic acid.
   Production of pectinase from fruit waste is outlined in Figure 4

Amylase

It is a class of starch degrading enzyme basically produced from fungi, plants and animals. Bacterial amylases are commercially produced using *Bacillus sp.* and are highly compatible due to thermal stability. Important steps in amylase production is outlined in Figure 5.

Biogas Production From Fruit Waste

The technique of using organic materials for biogas production dates back to several decades. Methane accounts for the major portion of the biogas produced by anaerobic digestion of organic matter. Biogas is basically a flammable gas produced when organic materials are fermented under anaerobic condition. It contains methane and carbon dioxide along with traces of hydrogen sulfide and water vapour. It usually burns with a pale blue flame and has a calorific value between 25.9 – 30 J/m^3 (Weiland, 2010).

The formation of biogas from fruit waste basically occurs through three major steps:

Figure 4. Pectinase production from fruit waste

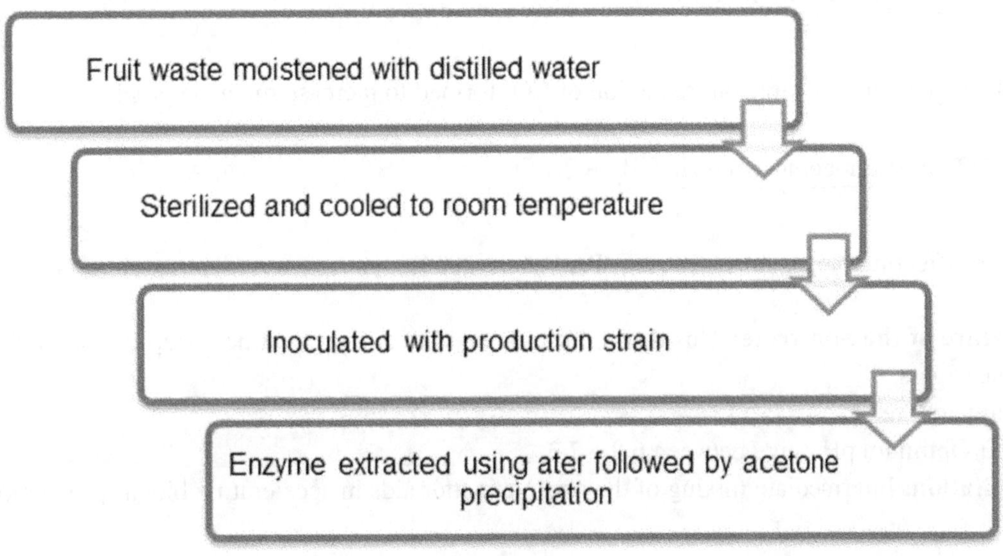

Figure 5. Amylase production from fruit waste

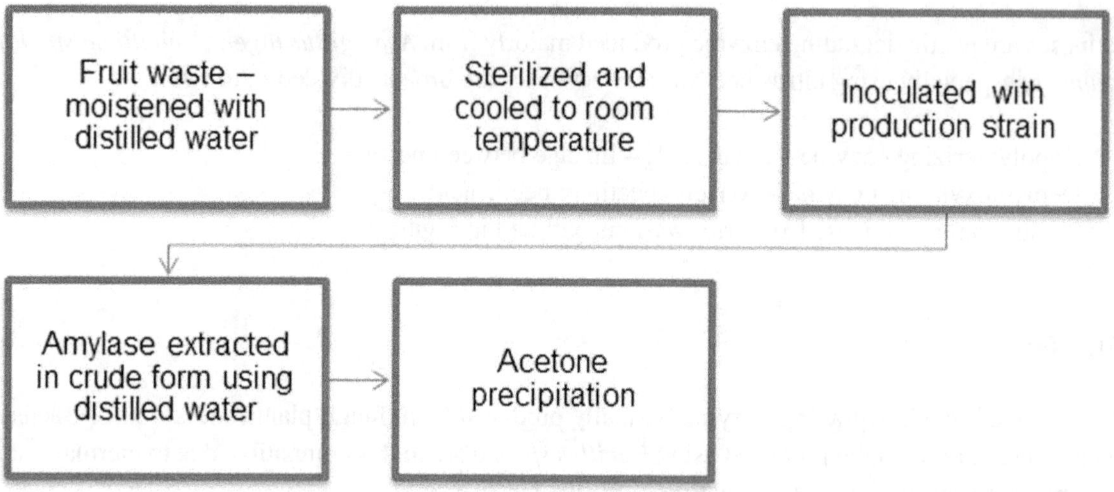

1. Complex organic molecules present in the fruit waste are decomposed by microorganisms in to simple molecules.

$n\,(C_6H_{10}O_5) + nH_2O\ n(C_6H_{12}O_6)$

2. These simple organic molecules are then converted in to organic acids by acid forming bacteria.

$n(C_6H_{12}O_6)\ \xrightarrow{\text{Acid forming bacteria}}\ CH_3COOH\ \xrightarrow{\text{Amylolytic bacteria}}$

3. These organic acids are then converted in to methane by methanogenic bacteria.

$CH_3COOH\ \xrightarrow{\text{Methanogenic bacteria}}\ CH_4 + CO_2$

Methanogens further carry out reduction of CO_2 formed to increase methane yield.

$CO_2 + 4H2\ \xrightarrow{\text{Methanogenic bacteria}}\ CH_4 + 2H_2O$

Factors affecting biogas formation are (Paul et al., 2014):

- **Nature of the Substrate:** This determines the need of any pretreatment step to release fermentable sugars.
- Temperature
- **pH:** Optimum pH range between 6.8 – 7.3
- **Agitation:** Intermediate mixing of the organic matter aids in accelerating biogas production.

- **Nutrient Supply:** Addition of nutrient sources to accelerate microbial growth can inturn increase biogas production.
- Slurry Concentration
- Digester Configuration
- **C/N Ratio:** Optimum concentration between 15-25
- **Retention Time:** At least 10 days
- Intitial feeding
- Total volatile acids
- COD
- Total Solid

The basic methodology involved in production of biogas from fruit waste is outlined below. Preparation of biogas from fruit waste preliminary involves drying of the waste followed by preparation of slurry for inoculation in to the digester of varying concentration. It is followed by inoculating the slurry with inoculum from previous batch or using cow dung containing substantial amount of methanogens.

BIOPLASTIC PRODUCTION

Plastics material is one of the most popular materials and indispensable in the present world. The word plastic comes from the Greek word "plastikos", which means 'able to be molded into different shapes' (Joel, 1995). Plastics are manmade long chain polymeric molecules (Scott, 1999). Plastics have become technologically significant since the 1940s and since then they have come to replace glass, wood, masonry and other constructional materials, and even metals in many industrial, domestic, commercial and environmental applications (Cain, 1992). Synthetic plastics are made from inorganic and organic raw materials, such as carbon, silicon, hydrogen, nitrogen, oxygen and chloride. The basic materials used for making plastics are extracted from oil, coal and natural gas (Seymour, 1989). They have versatile qualities of strength, lightness, durability and resistance to degradation. They have become an important commodity to enhance the comfort and quality of life. Accumulation of recalcitrant plastics in the environment has become a world-wide problem. During the combustion of plastic waste, hydrogen cyanide can be formed from acrylonitrile-based plastics and may cause potential health hazards are also causing other serious environmental problems due to their non-biodegradability. The sorting of wide variety of discarded plastic material is also a very time-consuming process. In such a scenario, biodegradable plastics offer the best solution to the environmental hazard posed by conventional plastics.

Synthetic Polymers and Their Impact on the Environment

The properties of plastics such as low cost, durability, lightweight, ease in processing and high resistance to chemical and biological degradation have made them an integral part of everyday life. Their chemical structure can be easily manipulated so that they can be molded into almost any desired shape and therefore are used in the manufacture of many durable, disposal goods and as packaging materials (Rivard et al., 1995; Yu, 2007). The four major commodity thermoplastic resins are, polyethylene, polypropylene, polystyrene and polyvinyl chloride. The major features of plastics that have contributed immensely towards their popularity are causing grave environmental and societal concerns. Plastics being xenobiotic in nature and are recalcitrant to microbial degradation (Flechter, 1993). Their large molecular size is responsible for the resistance of plastics to biodegradation (Reddy et al., 2003). Since their presence in nature is increased enormously during the recent years, new enzyme structures capable of degrading synthetic polymers are yet to be evolved and nature's in-built mechanisms are unable to degrade these novel unfamiliar pollutants (Reddy et al., 2003; Mueller, 2006). Approximately, 140 million tonnes per year of synthetic petroleum-based plastics are produced and several thousand tonnes of it is discarded into the environment as industrial waste products (Shimao, 2001). The accumulation of these discarded plastics in the environment is reducing the aesthetic qualities of cities, forests, water bodies as well as endangering the flora and fauna especially in the marine ecosystem (Moore, 2008; Ojumu et al., 2004). This dramatic increase in production and recalcitrant property of the synthetic polymers is imposing a strain on nature due to their persistence in the environment for centuries (Albertsson et al., 1987). Disposal of these xenobiotics is currently achieved through landfilling, incineration or recycling. Unfortunately, only a small fraction of the discarded plastic wastes reach the disposal sites while the rest litters the landscape or is blown off into the sea posing severe threat to the marine life (Moore, 2008). Incineration of plastic waste results in the generation of large amounts of carbonaceous material and undesirable pollutants such as carbon dioxide, carbon monoxide, furans, dioxins, hydrogen chloride, hydrogen cyanide, nitrogen oxides and benzopyrene which are highly corrosive and disease inducing (Jayasekara et al., 2005; Johnston, 1995; Reddy *et al.,* 2003). Recycling of synthetic plastics is disadvantageous due to alteration in the material properties, limiting its further application range and thereby inevitably increasing the cost of the recycled plastics as compared to the original form of plastics. The persistence of the discarded plastics is a global threat due to their potential adverse impacts on the environment (Mohanty et al., 2002).

Bioplastics

A fully biodegradable polymer is defined as a polymer that is completely converted by living organisms (usually microorganisms) to carbon dioxide, water and humic material. They can be completely degraded in landfills, composters or sewage treatment plants by action of naturally occurring microorganisms. Fully degradable plastics leave no toxic, visible or distinguishable residues following degradation. Bioplastics are a special type of biomaterial. They are polyesters produced by a range of microbes cultured under different nutrient and environmental conditions. These polymers are storage granules rich in lipids, are accumulated as storage materials (in the form of mobile, amorphous, liquid granules) allowing microbial survival under stress conditions. The rate of bioplastic formation, material and physicochemical properties vary depending on the production strain. They can be observed intracellularly as light-refracting granules or as electronlucent bodies that, in overproducing mutants, cause a striking alteration of the bacterial shape.

Types of Bioplastics

Biodegradable plastics can be divided into three categories:

1. **Chemically Synthesised Polymers:** Polyglycollic acid, polylactic acid, poly (ε-caprolactone), polyvinyl alcohol, poly (ethylene oxide) fall into this category. These are susceptible to enzymic or microbial attack.

2. **Starch-Based Biodegradable Plastics:** In this type, starch is added as filler and cross-linking agent to produce a blend of starch and plastic (for example, starch– polyethylene). Soil micro-organisms degrade the starch easily, thus breaking down the polymer matrix. This results in significant reduction of degradation time. But such plastics are only partially degradable. The fragments left after starch removal are recalcitrant and remain in the environment for a long time.
3. **Polyhydroxyalkanoates (PHAs):** These are the only 100% biodegradable polymers. They are polyesters of various hydroxyalkanoic acids which are synthesised by numerous micro-organisms as energy reserve materials when an essential nutrient such as nitrogen or phosphorus is available only in limiting concentrations in the presence of excess carbon source. Polyhydroxyalkanoates (PHAs) are a family of biopolyesters synthesized and accumulated by a wide range of microorganisms as reserve food material. These are the only plastics produced exclusively by microorganisms and hence are completely degraded to benign compounds (Anderson and Dawes, 1990). They are non-polluting as they do not need catalysts or additives to promote their degradation. Plastics produced from PHAs have been reported to be truly biodegradable in both aerobic and anaerobic environments, unlike many of the "so-called" biodegradable plastics made synthetically. PHAs are composed mainly of poly-betahydroxybutyric acid (PHB) and poly-beta hydroxyvaleric acid (PHV). More than 80 different forms of PHAs have been detected in bacteria (Lee, 1996). These copolymers have better film forming and mechanical properties quite similar to low-density polyethylene. PHB was first discovered in bacteria (Lemoigne, 1926). It is a unique intracellular polymer accumulated under stress conditions but with excess carbon source.
    a. **Occurrence and Biosynthesis of PHA:** There is widespread occurrence of PHA-producing organisms in the environment. The main candidates for commercial-scale PHA production are bacteria and transgenic plants. Bacterial cells are capable of accumulating PHA to levels as high as 90% (w/w) of the dry cell weight (DCW) (Steinbuchel & Lutke-Eversloh, 2003) thus emphasizing its importance as a potential candidate for PHA production. A wide variety of bacteria are able to synthesize PHA. Unbalanced nutrient supply triggers the bacteria to store the excess carbon and energy in the form of PHA. The PHA polymer is deposited in the cell cytoplasm as discrete insoluble granules which vary in number depending on the bacterial strain (Anderson & Dawes, 1990). Polyhydroxybutyrate (PHB), the first PHA to be discovered, is produced via the classical PHB biosynthetic pathway in three steps catalyzed by the PHA biosynthetic enzymes. These are polyesters containing monomers of medium chain length (mcl PHAs, C5–C14) or long-chain length (lcl PHAs, >C14). Although PHAs are structurally related to PHBs (short-chain length, scl PHAs), the microbes that synthesise PHBs usually fail to make PHAs. However, recombinant organisms containing mixed catabolic pathways are able to synthesise either polymers (or co-polymers) containing scl, mcl monomers, or both. PHA production starts in response to stress imposed on cells, usually by nitrogen or phosphorus limitation, although in the presence of abundant carbon source. Under these conditions (PHA accumulation phase), the cells do not grow or divide but instead divert their metabolites towards the biosynthesis of hydoxyalkyl-CoA (HA-coA). HA-CoA is polymerized by the enzymatic action of PHA synthase to form PHA polymer (Figure 6). Being insoluble in water, PHA begins to form amorphous and nearly spherical granules that gradually fill the cells and force them to expand (Lee et al., 1999).
    b. **Chemical Structure of Polyhydroxyalkanoates:** Polyhydroxyalkanotes are hydroxyalkanoic acids linked to each other by an ester linkage. Polyhydroxybutyrate (PHB) is the most com-

monly occurring and hence most studied PHA. The composition of the synthesized polymer is influenced by the bacterial strain as well as type and relative quantity and quality of carbon sources supplied to the growth medium (Steinbuchel et al., 1993).

c. **General Structure of PHA:** The polyhydroxyalkanoates can be divided into three groups depending on the carbon chain length of the monomeric units (Figure 7):
   i. Short-Chain-Length (scl-PHA), which consist of 3 to 5 carbon atoms
   ii. Medium-Chain-Length (mcl-PHA), which consist of 6 to 14 carbon atoms
   iii. Long-Chain-Length (1c1-PHA), which consist of 17 to 18 carbon atoms (Volova, 2004). Biosynthetic pathway for the synthesis of PHA is illustrated in Figure 5

Figure 6. PHA production from fruit waste

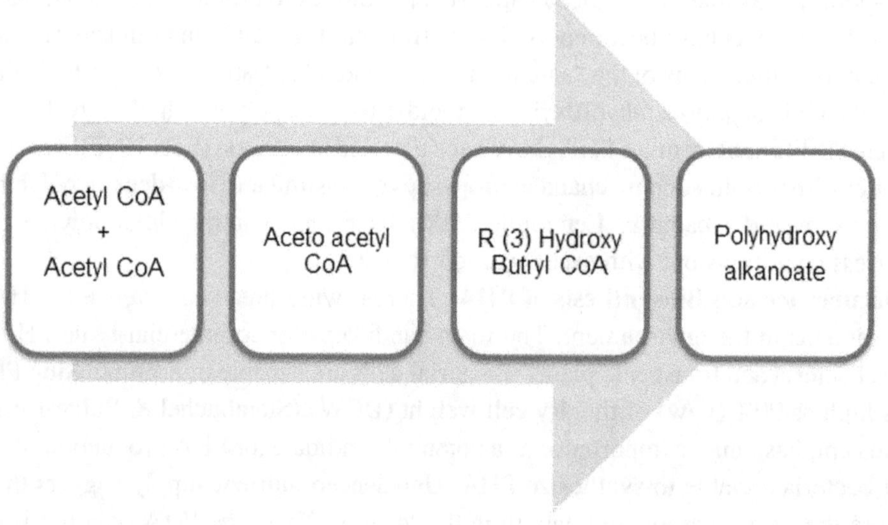

Figure 7. General structure of PHA

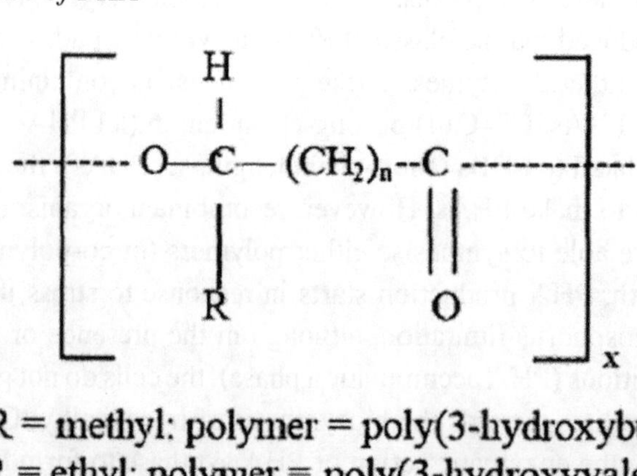

n = 1 R = methyl: polymer = poly(3-hydroxybutyrate)
R = ethyl: polymer = poly(3-hydroxyvalerate)
n = 2 R = hydrogen: polymer = poly(4-hydroxybutyrate)
n = 3 R = hydrogen: polymer = poly(5-hydroxybutyrate)
x = 100-30000

Bioplastic Production From Waste Substrate

The use of plastics has grown rapidly over the past few decades. An interesting alternative to decrease the environmental impacts of plastics is to replace conventional petroleum-derived polymers with biodegradable ones. The most important factors affecting the overall economics of PHA production are PHA productivity, PHA content and PHA yield which depend mainly upon the carbon source used, the raw material cost and the recovery methods. In the manufacturing process the raw material cost, especially the carbon source greatly influences the overall cost of the final product. Synthesis of PHA using expensive carbon sources such as glucose renders the process economically non-viable. Alternatively, inexpensive and easily available raw materials such as wastes generated during agricultural and industrial processes are now being considered as potential carbon substrates (Castilho et al., 2009). These wastes contain large amounts of organic matter which can be utilized as carbon feedstock for PHA production. This will not only ensure the reduction in the raw material cost but also simultaneously save energy and decrease the costs associated with its disposal. The synthesis of value-added products such as PHA from waste carbon sources coupled with their biodegradability makes them increasingly attractive in the pursuit of sustainable development. The biological polyesters known as polyhydroxyalkanoates (PHAs) are mainly produced by microbial fermentation processes, and there is a major challenge to reduce their production costs (Nath et al., 2008). Since the traditional approach for PHA production has been based on submerged fermentation (SMF) processes, most studies reported in the literature are limited either to liquid wastes or to liquid by-products. Wastes generated in the agricultural sector are being reviewed as potential feedstock for polymer production. In a wheat-based biorefinery strategy for PHB production, accumulation of PHB occurred when wheat hydrolysate and fungal extract were provided as carbon and nitrogen sources (Koutinas et al., 2013). A large number of industrial processes generate organic wastes such as molasses and corn steep liquor which are rich in carbon. These wastes are being used for the synthesis of value-added products namely; lactic acid, ethanol, enzymes and antibiotics. Wastes generated by the fast food manufacturing units can also serve as ideal feedstock due to their organic content.

FUTURE RESEARCH DIRECTIONS

The immense fermentative potential of fruit waste along with its global availability throughout the year makes it a versatile for important fermentation process involving microbes. Microbe mediated bioconversion continues to be the most effective tool for waste management due to the inherent ability of microbial enzymes to break the most complex organic molecules. Development of integrated process for simultaneous production of more than one microbial metabolite utilizing fruit waste will be the major focus for research in the upcoming years. Research aimed at utilization of carbohydrate rich waste like fruit peels can spread light on the concept of 'Reduce, Reuse and Recycle'. Development in genetic engineering and microbial fermentation process in the upcoming years could act as promising milestones in the development of novel compounds of commercial interest from cheap raw material like fruit wastes.

CONCLUSION

Increasing concern about environment pollution due to careless discharge of waste called for development of sustainable waste management strategies for production of value added products. Fruit waste with high content of carbohydrates are easily available cheap raw material for the production of a wide range of products like organic acids, single cell protein, bioplastics, enzymes and biogas. Utilization of biomass through microbial fermentation serves as sustainable alternative for chemical synthesis process for the manufacture of products of commercial interest. Proper planning, execution, financial assistance from public and private sector for both research and product development can take this initiative in to promising domains in the future emphasizing the concept of 'Wealth from Waste'.

REFERENCES

Adedayo, M. R., Ajiboye, E. A., Akintunde, J. K., & Odaibo, A. (2011). Single cell proteins: As nutritional enhancer. *Advances in Applied Science Research*, *2*(5), 396–409.

Albertsson, A. C., Andersson, S. O., & Karlsson, S. (1987). The mechanism of biodegradation of polyethylene. *Polymer Degradation & Stability*, *18*(1), 73–87. doi:10.1016/0141-3910(87)90084-X

Amrane, A., & Prigent, Y. (1998). Lactic acid production rates during the different growth phases of *Lactobacillus helveticus* cultivated on whey supplemented with yeast extract. *Biotechnology Letters*, *20*(4), 379–383. doi:10.1023/A:1005331430943

Anderson, A. J., & Dawes, E. A. (1990). Occurrence, metabolism, metabolic role, and industrial uses of bacterial polyhydroxyalkanoates. *Microbiological Reviews*, *54*(4), 450–472. PMID:2087222

Bayer, E. A., Lamed, R., & Himmel, M. E. (2007). The potential of cellulases and cellulosomes for cellulosic waste management. *Current Opinion in Biotechnology*, *18*(3), 237–245. doi:10.1016/j.copbio.2007.04.004 PMID:17462879

Berezinaa, N., Nysb, J., & Yadaa, B. (2013). Method for the Analysis of Grafted Cellulosic Materials. *Chemical Engineering (Albany, N.Y.)*, 32.

Bouallagui, H., Torrijos, M., Godon, J. J., Moletta, R., Cheikh, R. B., Touhami, Y., & Hamdi, M. (2004). Two-phases anaerobic digestion of fruit and vegetable wastes: Bioreactors performance. *Biochemical Engineering Journal*, *21*(2), 193–197. doi:10.1016/j.bej.2004.05.001

Buss, S. E., Butler, A. P., Sollars, C. J., Perry, R., & Johnston, P. M. (1995). Mechanisms of leakage through synthetic landfill liner materials. *Water and Environment Journal: the Journal / the Chartered Institution of Water and Environmental Management*, *9*(4), 353–359. doi:10.1111/j.1747-6593.1995.tb00952.x

Bustos, G., Moldes, A. B., Alonso, J. L., & Vázquez, M. (2004). Optimization of D-lactic acid production by *Lactobacillus coryniformis* using response surface methodology. *Food Microbiology*, *21*(2), 143–148. doi:10.1016/S0740-0020(03)00061-3

Cain, R. B. (1992). Microbial degradation of synthetic polymers. *Journal of Microbial control of pollution.*

Carmo, I. M. T. D. D. (2013). *Food waste valorization through the production of polyhydroxyalkanoates by mixed microbial cultures* [Doctoral dissertation]. Faculdade de Ciências e Tecnologia.

Castilho, L. R., Mitchell, D. A., & Freire, D. M. (2009). Production of polyhydroxyalkanoates (PHAs) from waste materials and by-products by submerged and solid-state fermentation. *Bioresource Technology, 100*(23), 5996–6009. doi:10.1016/j.biortech.2009.03.088 PMID:19581084

Chirila, T. V., Constable, I. J., Crawford, G. J., Vijayasekaran, S., Thompson, D. E., Chen, Y. C., & Griffin, B. J. (1993). Poly (2-hydroxyethyl methacrylate) sponges as implant materials: In vivo and in vitro evaluation of cellular invasion. *Biomaterials, 14*(1), 26–38. doi:10.1016/0142-9612(93)90072-A PMID:7678755

Daramwal, N., & Gaur, R. (2004). Single cell protein. *Environmental microbiology and biotechnology, 218.*

Fitzpatrick, J. J., & O'keeffe, U. (2001). Influence of whey protein hydrolysate addition to whey permeate batch fermentations for producing lactic acid. *Process Biochemistry, 37*(2), 183–186. doi:10.1016/S0032-9592(01)00203-5

García, I. L., Lopez, J. A., Dorado, M. P., Kopsahelis, N., Alexandri, M., Papanikolaou, S., & Koutinas, A. A. (2013). Evaluation of by-products from the biodiesel industry as fermentation feedstock for poly (3-hydroxybutyrate-co-3-hydroxyvalerate) production by *Cupriavidus necator. Bioresource Technology, 130*, 16–22. doi:10.1016/j.biortech.2012.11.088 PMID:23280181

Garde, A., Jonsson, G., Schmidt, A. S., & Ahring, B. K. (2002). Lactic acid production from wheat straw hemicellulose hydrolysate by *Lactobacillus pentosus* and *Lactobacillus brevis. Bioresource Technology, 81*(3), 217–223. doi:10.1016/S0960-8524(01)00135-3 PMID:11800488

Goldberg, S. Z., Eisenberg, R., Miller, J. S., & Epstein, A. J. (1976). Tetrakis (methyl isocyanide) palladium (II) tetrakis (7, 7, 8, 8-tetracyano-p-quinodimethane), [Pd (CNMe) 4](TCNQ) 4.2 MeCN: Synthesis, structure, and physical properties. *Journal of the American Chemical Society, 98*(17), 5173–5182. doi:10.1021/ja00433a020

Imandi, S. B., Bandaru, V. V. R., Somalanka, S. R., Bandaru, S. R., & Garapati, H. R. (2008). Application of statistical experimental designs for the optimization of medium constituents for the production of citric acid from pineapple waste. *Bioresource Technology, 99*(10), 4445–4450. doi:10.1016/j.biortech.2007.08.071 PMID:17936623

Jayasekara, R., Harding, I., Bowater, I., & Lonergan, G. (2005). Biodegradability of a selected range of polymers and polymer blends and standard methods for assessment of biodegradation. *Journal of Polymers and the Environment, 13*(3), 231–251. doi:10.100710924-005-4758-2

Jendrossek, D., Knoke, I., Habibian, R. B., Steinbüchel, A., & Schlegel, H. G. (1993). Degradation of poly (3-hydroxybutyrate), PHB, by bacteria and purification of a novel PHB depolymerase from *Comamonas sp. Journal of Polymers and the Environment, 1*(1), 53–63.

John, R. P., Sukumaran, R. K., Nampoothiri, K. M., & Pandey, A. (2007). Statistical optimization of simultaneous saccharification and L (+)-lactic acid fermentation from cassava bagasse using mixed culture of lactobacilli by response surface methodology. *Biochemical Engineering Journal, 36*(3), 262–267. doi:10.1016/j.bej.2007.02.028

Ko, P. C., & Yu, Y. (1968). Production of SCP from hydrocarbons: Taiwan. *Single Cell Protein*, 225.

Kulozik, U., & Wilde, J. (1999). Rapid lactic acid production at high cell concentrations in whey ultrafiltrate by *Lactobacillus helveticus. Enzyme and Microbial Technology*, *24*(5), 297–302. doi:10.1016/S0141-0229(98)00122-7

Kumar, D., Jain, V. K., Shanker, G., & Srivastava, A. (2003). Utilisation of fruits waste for citric acid production by solid state fermentation. *Process Biochemistry*, *38*(12), 1725–1729. doi:10.1016/S0032-9592(02)00253-4

Kurbanoglu, E. B., & Kurbanoglu, N. I. (2003). Utilization for lactic acid production with a new acid hydrolysis of ram horn waste. *FEMS Microbiology Letters*, *225*(1), 29–34. doi:10.1016/S0378-1097(03)00472-5 PMID:12900017

Lee, S. Y. (1996). Bacterial polyhydroxyalkanoates. *Biotechnology and Bioengineering*, *49*(1), 1–14. doi:10.1002/(SICI)1097-0290(19960105)49:1<1::AID-BIT13.0.CO;2-P PMID:18623547

Lee, S. Y., Choi, J. I., & Wong, H. H. (1999). Recent advances in polyhydroxyalkanoate production by bacterial fermentation: Mini-review. *International Journal of Biological Macromolecules*, *25*(1), 31–36. doi:10.1016/S0141-8130(99)00012-4 PMID:10416647

Lemoigne, M. (1926). Produits de deshydration et de polymerisation de l'acide boxybutyrique. *Biological science, 8*, 770-782.

Liguori, R., Amore, A., & Faraco, V. (2013). Waste valorization by biotechnological conversion into added value products. *Applied Microbiology and Biotechnology*, *97*(14), 6129–6147. doi:10.100700253-013-5014-7 PMID:23749120

Mohanty, A. K., Misra, M., & Drzal, L. T. (2002). Sustainable bio-composites from renewable resources: Opportunities and challenges in the green materials world. *Journal of Polymers and the Environment*, *10*(1), 19–26. doi:10.1023/A:1021013921916

Moore, C. J. (2008). Synthetic polymers in the marine environment: A rapidly increasing, long-term threat. *Environmental Research*, *108*(2), 131–139. doi:10.1016/j.envres.2008.07.025 PMID:18949831

Umesh, M., & Preethi, K. (2014). Fermentative Utilization of fruit peel waste for Lactic Acid production by *Lactobacillus plantarum. Indian Journal of Applied Research*, *4*(9), 449–451.

Mudaliyar, P., Sharma, L., & Kulkarni, C. (2012). Food waste management-lactic acid production by *Lactobacillus species. International Journal of Advanced Biological Research*, *2*(1), 34–38.

Mueller, R. J. (2006). Biological degradation of synthetic polyesters—enzymes as potential catalysts for polyester recycling. *Process Biochemistry*, *41*(10), 2124–2128. doi:10.1016/j.procbio.2006.05.018

Nasseri, A. T., Rasoul-Amini, S., Morowvat, M. H., & Ghasemi, Y. (2011). Single cell protein: Production and process. *American. Journal of Food Technology*, *6*(2), 103–116. doi:10.3923/ajft.2011.103.116

Nath, A., Dixit, M., Bandiya, A., Chavda, S., & Desai, A. J. (2008). Enhanced PHB production and scale up studies using cheese whey in fed batch culture of Methylobacterium sp. ZP24. *Bioresource Technology*, *99*(13), 5749–5755. doi:10.1016/j.biortech.2007.10.017 PMID:18032031

Nigam, J. N. (2000). Cultivation of Candida langeronii in sugar cane bagasse hemicellulosic hydrolyzate for the production of single cell protein. *World Journal of Microbiology & Biotechnology*, *16*(4), 367–372. doi:10.1023/A:1008922806215

Oh, H., Wee, Y. J., Yun, J. S., Han, S. H., Jung, S., & Ryu, H. W. (2005). Lactic acid production from agricultural resources as cheap raw materials. *Bioresource Technology*, *96*(13), 1492–1498. doi:10.1016/j.biortech.2004.11.020 PMID:15939277

Oikeh, E. I., Oriakhi, K., & Omoregie, E. S. (2013). Proximate analysis and phytochemical screening of *Citrus sinensis* fruit wastes. *The Bioscientist*, *1*(2), 164–170.

Ojumu, T. V., Yu, J., & Solomon, B. O. (2004). Production of polyhydroxyalkanoates, a bacterial biodegradable polymers. *African Journal of Biotechnology*, *3*(1), 18–24. doi:10.5897/AJB2004.000-2004

Pandey, A., Soccol, C. R., Nigam, P., & Soccol, V. T. (2000). Biotechnological potential of agro-industrial residues. I: Sugarcane bagasse. *Bioresource Technology*, *74*(1), 69–80. doi:10.1016/S0960-8524(99)00142-X

Parajó, J. C., Santos, V., Domínguez, H., & Vázquez, M. (1995). NH 4 OH-Based pretreatment for improving the nutritional quality of single-cell protein (SCP). *Applied Biochemistry and Biotechnology*, *55*(2), 133–149. doi:10.1007/BF02783554

Preethi, K., & Maha Lakshmi, G., Mridul Umesh, Priyanka, K., & Thazeem, B. (2017). Fruit peels: A potential substrate for acetic acid production using *Acetobacter aceti. International Journal of Applied Research*, *2*(4), 286–291.

Rashad, M. M., Moharib, S. A., & Jwanny, E. W. (1990). Yeast conversion of mango waste or methanol to single cell protein and other metabolites. *Biological Wastes*, *32*(4), 277–284. doi:10.1016/0269-7483(90)90059-2

Reddy, C. S. K., Ghai, R., & Kalia, V. (2003). Polyhydroxyalkanoates: An overview. *Bioresource Technology*, *87*(2), 137–146. doi:10.1016/S0960-8524(02)00212-2 PMID:12765352

Rivard, C., Moens, L., Brigham, K. R., & Kelley, S. (1995). Starch esters as biodegradable plastics: Effects of ester group chain length and degree of substitution on anaerobic biodegradation. *Enzyme and Microbial Technology*, *17*(9), 848–852. doi:10.1016/0141-0229(94)00120-G

Schepers, A. W., Thibault, J., & Lacroix, C. (2002). *Lactobacillus helveticus* growth and lactic acid production during pH-controlled batch cultures in whey permeate/yeast extract medium. Part I. multiple factor kinetic analysis. *Enzyme and Microbial Technology*, *30*(2), 176–186. doi:10.1016/S0141-0229(01)00465-3

Scott, D. S., & Legge, R. L. (1999). Bioplastics. CIGR handbook of agricultural engineering, 5, 305-310.

Seymour, R. B. (1989). Polymer science before & after 1899: Notable developments during the lifetime of Maurtis Dekker. *Journal of Macromolecular Science. Chemistry*, *26*(8), 1023–1032. doi:10.1080/00222338908052032

Shimao, M. (2001). Biodegradation of plastics. *Current Opinion in Biotechnology*, *12*(3), 242–247. doi:10.1016/S0958-1669(00)00206-8 PMID:11404101

Shukla, V. B., Zhou, S., Yomano, L. P., Shanmugam, K. T., Preston, J. F., & Ingram, L. O. (2004). Production of d (−)-lactate from sucrose and molasses. *Biotechnology Letters, 26*(9), 689–693. doi:10.1023/B:BILE.0000024088.36803.4e PMID:15195965

Solomons, G. L. (1985). Production of biomass by filamentous fungi. *Comprehensive biotechnology: the principles, applications, and regulations of biotechnology in industry, agriculture, and medicine.*

Sreenath, H. K., Moldes, A. B., Koegel, R. G., & Straub, R. J. (2001). Lactic acid production from agriculture residues. *Biotechnology Letters, 23*(3), 179–184. doi:10.1023/A:1005651117831

Steinbüchel, A., & Lütke-Eversloh, T. (2003). Metabolic engineering and pathway construction for biotechnological production of relevant polyhydroxyalkanoates in microorganisms. *Biochemical Engineering Journal, 16*(2), 81–96. doi:10.1016/S1369-703X(03)00036-6

Vikas, O.V., & Mridul, U. (2014). Bioconversion of papaya peel waste in to vinegar by *Acetobacter aceti. International Journal of Scientific Research, 3*(11), 409–411.

Volova, T. G. (2004). *Polyhydroxyalkanoates--plastic materials of the 21st century: production, properties, applications.* Nova publishers.

Wee, Y. J., Kim, J. N., Yun, J. S., & Ryu, H. W. (2004). Utilization of sugar molasses for economical L (+)-lactic acid production by batch fermentation of Enterococcus faecalis. *Enzyme and Microbial Technology, 35*(6), 568–573. doi:10.1016/j.enzmictec.2004.08.008

Weiland, P. (2010). Biogas production: Current state and perspectives. *Applied Microbiology and Biotechnology, 85*(4), 849–860. doi:10.100700253-009-2246-7 PMID:19777226

Wong, R. C., & Solomon, A. R. (1985). Acquired dermal smooth-muscle hamartoma. *Cutis, 35*(4), 369–370. PMID:3996041

Yu, W. W., Chang, E., Falkner, J. C., Zhang, J., Al-Somali, A. M., Sayes, C. M., & Colvin, V. L. (2007). Forming biocompatible and nonaggregated nanocrystals in water using amphiphilic polymers. *Journal of the American Chemical Society, 129*(10), 2871–2879. doi:10.1021/ja067184n PMID:17309256

KEY TERMS AND DEFINITIONS

Amylase: Starch degrading enzyme produced by wide variety of microbes.

Biogas: Biogas is basically a flammable gas produced when organic materials are fermented under anaerobic condition.

Bioplastics: A fully biodegradable polymer is defined as a polymer that is completely converted by living organisms, usually microorganisms, to carbon dioxide, water and humic material.

Fermentation: Metabolic process mediated by microbes that involves conversion of organic matter in to value added products.

Organic Acid: Acids derived from organic source preferably plant or animals.

Polyhydroxyalkanoates: Polyhydroxyalkanotes are hydroxyalkanoic acids linked to each other by an ester linkage

Single Cell Protein: Live or dried protein supplement produced from microbes.

Section 2
Bioremediation of Environmental Waste and Pollutants

Chapter 5
Bioremediation of Environmental Pollutants

Akash
Gurukul Kangri University, India

Vinay Mohan Pathak
Gurukul Kangri University, India

Neelesh Babu
Gurukul Kangri University, India

Navneet
Gurukul Kangri University, India

ABSTRACT

This chapter describes how pollutants are increasing in the environment due to the rapid industrialization all over the world. The environment has been contaminated with large number of organic and inorganic pollutants. The organic pollutants are largely anthropogenic and are introduced to the environment in many ways. Soil contamination with toxic metals, such as Cd, Pb, Cr, Zn, Ni, etc., as a result of world-wide industrialization has increased noticeably within the past few years. Bioremediation is a process for reclaiming the environment which has been polluted with the help of living forms. It is an option that offers the possibility to destroy various contaminants using natural biological activity and to degrade the environmental contaminants into less toxic forms. It is also applicable for the heavy metal hazards. It has proven to be cheap and efficient than other techniques. This chapter focuses on the possible trends in the remediation of environment pollutants with the help of plants as well as microbes.

INTRODUCTION

Bioremediation is a method which utilizing the life forms to kill or expel contamination from waste. It is vital to comprehend that this type of waste remediation utilizes no poisonous chemicals, in spite of the fact that it might utilize a living organism that can be detrimental in specific situations (Mathiyazhagan et al., 2011; Obreque et al., 2015; Akpomie et al., 2016; Satyapal et al., 2016, Ramakrishnan, 2016; Xe-

DOI: 10.4018/978-1-5225-3540-9.ch005

nia et al., 2016; Kiraye et al., 2016). In other words, we can say that bioremediation is the use of living microorganisms to degrade environmental pollutants or to prevent environment from pollution. It is an important technology for removing pollutants from the environment and thus restoring the original natural surroundings so that the environment can be prevent from pollution. It can be divided into three phases or levels. Firstly, by natural attenuation in which the contaminants are reduced by native microorganisms without any human augmentation, secondly the biostimulation is employed where nutrients and oxygen are applied in the systems to improve their effectiveness and to accelerate the process of biodegradation. Finally, during bioaugmentation microorganisms are added in the systems. The techniques of bioremediation are now widely being applied to remove the pollutants or contaminants from the environment.

Types of Bioremediation

There are much more than nine types of bioremediation, yet the accompanying are the most well-known ways in which it is utilized (Wilson et al., 1994, Griffiths, 2003, Lai et al., 1995, Edens,1984,). The list of some of the bioremediation techniques are as follows:

- **Phytoremediation:** Phytoremediation is a process of utilization of plants to expel contaminants. The plants can draw the contaminants from the soil and water into their structures and clutch them, adequately expelling them from soil or water.
- **Bioventing:** Bioventing is an *in situ* remediation technique that includes blowing of air through soil to expand oxygen rates in the waste. It works well with light contaminants as which can evaporate easily. It is a very efficient approach to neutralize certain oxygen sensitive metals or chemicals with the help of microorganisms.
- **Bioleaching:** This technique includes expelling metals from soil with the help of living life forms especially bacteria. Certain sorts of life forms attract to heavy metals and other type of contaminants and ingest them. This technique has widely being used to remove the metals like lead, zinc, cobalt, lead, gold etc.
- **Landfarming:** This technique involves placing the contaminants soil in a biocell which is consisting of a linear surrounded by a berm. After that the soil placed on the linear and turned periodically to help the microorganism to breakdown the pollutants.
- **Bioreactor:** The utilization of uniquely outlined containers to hold the waste while bioremediation happens.
- **Composting:** Containing waste so a characteristic decay and remediation process happens.
- **Bioaugmentation:** Adding microorganisms and living organisms to fortify the same in waste to permit them to assume control and purify the region.
- **Rhizofiltration:** The utilization of plants to expel metals in water.
- **Biostimulation:** The utilization of organisms intended to expel contamination connected in a medium to the waste.

Classes of Bioremediation

There are two classes of bioremediation which are as follows:

- ***In-situ:*** The term *'In situ'* means when contaminated waste material is treated at the site of its origin. For instance, there might be soil that is contaminated. Rather than to expel the dirt from its point of root, it is treated right where it is. The main advantage to *in situ* treatment is that it prevents the spread of contaminated waste material during its transportation and displacement. (Bitter et al., 1984; Eden et al., 1984; Sambrook et al., 1989; Mulbry et al., 1989; WHO, 1990; Looman et al., 1993; Wilson et al., 1994; Parekh et al., 1996; lai et al., 1999; Di Sioudi et al., 1999; Griffiths et al., 2003; Geitz et al., 2007; Schofield et al., 2010;). It is very superior method for removing the contaminants/pollutants from the environment as it is very cheaper and uses harmless microorganism to degrade the chemicals compounds.
- ***Ex-situ:*** The term *'Ex situ'* involves the contaminants of soil and water once has been excavated out of the location or area at which it was found. *Ex situ* bioremediation is of two types depending on the state of contaminants which are solid phase and slurry phase. The soil might be expelled and transported to the zone where the process of bioremediation might be applied. The main advantage of this technique is to contain and control the bioremediation products. This technique can be faster, easier and has usually more able to degrade contaminants.

BIOREMEDIATION OF SOIL AND WATER POLLUTANTS

There are so many methods of the bioremediation of the contaminants but the specific methods are Composting, Intrinsic bioremediation, land farming and slurry bioreactor. These all method are *ex situ* biological processes and shows advantages and disadvantages at the time of their using. Soil scrubbing is also an *ex situ* chemical/physical remediation process known which uses mechanical energy to separate the pollutants from the soil and in each batch the quantity of the soil treated so that the pollutants can be removed. After that he soil is crushed and then separated via sieving which ensures that the soil sample is homogeneous or not. The soil is then dispersed into the water. Sometime additives are also added along with water to dissolve the pollutants. Generally, the additives are used to overcome the bonding forces between the pollutants or contaminants and the soil particles. After all these steps the soil is separated on the basis of their density which is now low density and high-density solids. The highly contaminated particles are separated and removed out and dewatered and then particles rinsed with uncontaminated

Table 1. Some of the important factors required for bioremediation

Factors	Condition Required
Environmental Factors	Temperature, pH, Oxygen content and Electron acceptor/donor
Type of soil	Low clay or silt content
Natural Biological processes of microorganism	Catabolism and Anabolism
Microorganisms	Aerobic or Anaerobic
Nutrients	Carbon, Nitrogen and Oxygen etc
Soil Moisture	25-28% of water holding capacity

(Khan, 2014)

Table 2. Some of the important environmental conditions required for bioremediation

Environmental Factor	Optimum Conditions	Condition for Microbial Activity
Available soil moisture	25-85% water holding capacity	25-28% of water holding capacity
Nutrients	C:N:P= 120: 10:1 mol. Ratio	N, P for microbial growth
Redox potential	Eh > 50 mill volts	
pH	6.5 to 8.0	5.5 - 8.5
Heavy metals	700ppm	Total content 2000ppm
Temperature	20-30 °C	15-45°C
Oxygen	>0.2 mg/L DO, >10% air-filled pore space for aerobic degradation	Aerobic, minimum air-filled pore space of 10%
Contaminants	Hydrocarbon app. 5-10% of dry weight of soil	Not much toxic

(Vidali, 2001)

water. The exhaust air and the wastewater are then purified. It is the one of the most effective method for the removal of BTEX, PAH, PCB, TPH and heavy metals and dioxins (Koning et al., 2000).

Dibble and Bertha (1979) introduces land farming into scientific literature by an article which described that disposal by biodegradation of oily sludge in soil. It is a technique, based on the principles applied in the agriculture which aims at controlling and bicycling of natural compounds. This method is inexpensive, has simple procedure and self-heating system which are the important features than other methods. It can be optimized by the dilution of contaminated soil with clean soil and tilling of the contaminated soil to reduce the initial toxicity

Generally, Composting is process in which the solid waste is converted into humus like material. This process is driven by aerobic microbes. In this technique degradation of organic material and condition are control to help thermophillic microorganism transform organic material into soil like product. It can be used to stabilize and decrease industrial wastes, yard wastes, sewage sludge and municipal wastes (Boopathy, 2000). Sometime the rate of the decomposition is very slow in composting but overall it is very good, inexpensive and self-heating technique. For industrial purposes, optimization of microbial growth is done to increases the rate. This means optimizing moisture content, carbon to nitrogen ratio and particle size, oxygen concentration, ph. After the degradation of organic matter, the temperature increases to 50-60 °C. So, at this temperature the thermophiliic bacteria grow easily (Thassitou et al., 2001).

The technique of slurry bioreactor has fast degradation rate, good control over parameters and its incubation periods are approximately one week. In this method, contaminate soil is excavated and mixed with the liquid (water) to form slurry which is mechanically aerated in a reactor vessel then the reactor contents are agitated to facilitate the breakdown of soil aggregated, enhances desorption of contaminate from soil, increases contact with the wastes and microorganism and enhances oxygenation of the slurry (Baker, 1994). Sometime substances like dispersants, surfactants and microorganism growth area added to slurry so that the biodegradation capability can be increased (Thassitou et al., 2017). On the other hand, intrinsic bioremediation manages the innate capability of naturally occurring microorganism to degrade environmental contaminants or pollutants. The effectiveness of intrinsic bioremediation must be proven with a site-monitoring regime that routinely analyzes contaminants concentration. It requires

Table 3. List of Some common atmospheric pollutants and their source

Category	Source of Pollutants	Emitted Pollutants
Agriculture	Open burning	CO, SPM and VOC
Mining and quarrying	Coal mining	SO_2, SPM and NOx,
	Petroleum and natural gas production	So_2
	Non-ferrous ore mining	Pb and SPM
	Stone quarrying	SPM
	Beverages and tobacco	VOC, CO and H_2S
	Textiles and leather industries	VOC and SPM,
	Wood products	VOC, SPM,
	Paper products, printing	SPM, VOC, CO, H_2S and SO_2
Chemicals manufacturing	Phallic anhydride	CO_2, SPM and SO_2
	Chlor-alkali	Cl_2
	Hydrochloric acid	HCl
	Hydrofluoric acid	SiF_4 and HF
	Sulphuric acid	SO_3 and SO_2
	Nitric acid	NOx
	Phosphoric acid	F_2 and SPM
	Lead oxide and pigments	Pb and SPM
	Ammonia	NOx, SPM, SO_2, VOC and NH_3
	Sodium carbonate	NH_3 and SPM
	Calcium carbide	SPm
	Adipic acid	SPM, VOC, NOx and CO
	Alkyl lead	Pb
	terephtalic acid, Maleic anhydride	VOC and CO
	Pesticide production, Fertilizers	NH_3 and SPM
	Ammonium nitrate	SPM, NH_3 and HNo_3,
	Ammonium sulphate	VOC
	plastic, Syntheticresins, material, fibres	SPM, CS_2, VOC and H_2S,
	Varnishes, Paints, lacquers	VOC and SPM
	Soap	SPM
	Carbon black and printing ink	SO_2, VOC, NOx, CO, H_2s and SPM,
	Trinitrotoluene	NOx, SO_2, SPM, SO_3 and HNO_3
Petroleum refineries	Products of petroleum and coal	VOC, F2, SPM, SO_2 and CO_2
	clay products	
	Cement lime and plaster	SO_2, SPM and NOx,
Music metal industries	Iron and steel	NOx, SO_2, CO, VOC and Pb
	Non -ferrous industries	SO_2, F, Pb and SPM,
Wholesale and retail trade	Gas, Steam Electricity	NOx, CO, SO_3, Pb and SO_2
Transport	filling operation	VOC
Community services	Municipal incinerators	Pb, SPM, SO_2, NOx, CO and VOC,

SPM: Suspended particulate matter; VOC: Volatile organic compounds (Economopoulos, 1993)

documentation of the role of native microorganisms in the eliminating contaminants with the help of test performed at the field sites or on site derived samples of soil, sediment or water.

The *in situ* remediation includes various techniques such as bioventing, bioslurping, biosparging and phytoremediation along with physical, chemical, and other thermal processes. One of the most important advantages of *In situ* remediation is that it is less costly due to the lack of excavation and transportation costs but sometime these remediation techniques are less controllable and less effective. (Koning et al., 2000). So many works is going on *In situ* thermal processes. It involves injecting a steam-air mixture at 60-100 °C into the soil. Sometime the steam air must stay in temperature range to avoid the transport of pollutants to the groundwater. The volatile and semi-volatile compounds transport from soil to the gas phase after injection, using the soil vapour the gases are removed from surface then treated. The process of *In situ* technique is used for only soil with low organic content and homogeneous soils with high permeability. This process can be stripped in the lower temperature (e.g. BTEX) (Koning et al., 2000). These processes are also known as pump and treat processes because these pumps water into the subsurface to draw out the pollutants. To increase the solubility of the pollutants Surfactants are added and the water is treated by standard wastewater treatment techniques. Sometime chemical oxidation is done to destroy pollutants like trichloroethylene (TCE) and PAHs (Koning et al., 2000). Some other chemicals like peroxide, permanganate, ozone injected into the soil to promote the toxic organic compounds. (Van Deuren et al., 2002). In soil vapour extraction other *in situ* chemical/physical process are also being used. The vacuum blowers are used to extract pollutants which are volatile through perforated pipes for soil. After that the pollutants are treated at the site using compost filters or compost filters. Some of the soil characters like permeability, moisture content, temperature on which the effectiveness of soil is dependent, play important role in this process. According to Van Deuren et al., (2002) high percentage of fine soil can also hinder the effectiveness of soil vapour extraction. In *in situ* process, bioventing is an effective technique which allows treatment of unsaturated soil. This technique is not effective when water table is within several feet of the surface. (Van Deuren et al., 2002). When vacuum is used, the atmospheric oxygen flows into the subsurface due to pressure gradient then aerobic contaminant decomposition process starts. Nitrogen salt is an additive by sprinkling a nutrient solution on top of the soil or by injecting into the contaminated zone (Held and Dörr, 2000). In this system sufficient air is required, if the concentration of pollutants is high then clogging of the soil pores maybe occur, in that situation soil vapour extraction is needed. Bioventing is effective for the removal of non-volatile hydraulic oils, hydrocarbons, aromatic hydrocarbons (Held and Dörr, 2000), but it is effective in area having high temperature (Van Deuren et al., 2002).

PLANTS AS PHYTOREMEDIATORS

The term 'Phytoremediation' refers to the use of plants along with the associated microorganisms to partially or completely remediate selected contaminants/pollutants from soil, sediments, wastewater and ground water. It can also be used for removal of radionuclides, heavy metals and organic pollutants (Ali, 2013). This technique utilizes a variety of plant processes, physical characteristics of plants which help in the remediation of contaminated zone.

'Phytoremediation' (phyto means plant, and remediation means correct evil) is a term which was coined in 1991.This process utilizes the metabolism of plants for the removal of the contaminants. Today conventionally techniques are extremely expansive, so phytoremediation is a best technique to solve many issues. The applications of this process can be classified on the basis of contaminant fate: degradation, containment, extraction and combination of these. The mechanism includes extraction of contaminants or pollutants from soil or groundwater, concentration of contaminants in plant tissue, degradation of contaminants/pollutants by various processes, transpiration of volatile contaminants from plants to the air, volatilization or immobilization of contaminants or pollutants in the root zone of plants, hydraulic control of contaminated groundwater, and control of runoff, erosion.

In other words, it consists of a set of innovative technologies for cleaning of environments that takes advantage of the metabolic capabilities of plants to remove the contaminants of the surrounding area or zone. Phytoremediation presents clear benefits over traditional methods due to its applicability, ecological value and cost-effectiveness etc. The organic compounds and other contaminants can be easily degraded into less toxic forms by plants or even mineralized and reduces the volume of the polluted material necessary for disposal.

Past data from two decades has strengthened that plants may be an interesting alternative for the remediation of environmental pollutants (Pilon-Smits, 2005). Plant-based processes use solar energy which makes these techniques easy, cheap and most important is that it is eco-friendly. Traditional methods that are employed for cleaning the contaminants like heavy metal of the soil disrupt the soil structure and also reduce the soil productivity whereas this technique cleans up the soil without causing harm to the soil and do not reduce the soil quality.

The set of the technologies remove or degrade the environmental pollutants/contaminants from landfills, groundwater, soil, wastewater, etc., with the help of wild, genetically modified plants are referred to as phytoremediation in scientific literature since 1990 (Salt et al., 1998; McCutcheon & Schnoor, 2003). Phytoremediation includes phytoextraction, phytofiltration, phytostabilization, phytovolatilization and phytodegradation (Alkorta, 2004). These different techniques can lead to contaminants degradation, removal by accumulation or dissipation or immobilization. The chief application of phytoremediation is for lightly contaminated soils and waters where the material to be treated is at a shallow or medium depth and the area to be treated is large so that this technique would be economic. This technique can be applied *in situ* for the remediation of the shallow soil and ground water and can be used in surface water bodies. Plants have the ability to remove the contamination of the soil by up taking into their roots, binding of these contaminants into their roots and transfer these into the shoot so that it inhibits the contaminants leaching out of the soil. The plants also grow well in these conditions but if the plant die, the contaminants will return to the soil. There are so many plant species used in the phytoremediation which are alfalfa, poplar trees, forage kochia, Kentucky bluegrass, water hyacinth, sugarcane, rice, wheat, maize etc. For complete removal of contaminants from the soil, the plant should be cut and disposed elsewhere in a better way. At present many industries, companies are funding to the scientist to test the different plant to observe their activity against different types of environmental pollutants, heavy metals, organic pollutants etc. Scientists also favour *Brassica juncea* and *B. olearacea*, the species of the mustard family, for phytoremediation because these plants have the ability to remove large quantities of Ni, Cu, Pb, Cr from the soil. The different plant species interact with the pollutants in the environment in different way like indigo bush (accumulate lead), water velvet (absorb metal), water hyssop (accumulate metal), hydrilla (hyper accumulate metal) and so on. It is now full known that plants are capable of removing the organic and inorganic compounds from soil by their roots, stomata, shoot etc.

Methods of Phytoremediation

Traditional methods of remediating lead contaminated sites include a variety of physical, thermal and chemical treatments (Berti & Cunningham, 2000; Chen et al., 2000). Recently, also phytoremediation technologies have been recognized as quite efficient and cost-effective method for remediating sites contaminated with toxic metals (Salt et al., 1995; Raskin et al., 1997; Meagher, 2000).

Phytoextraction (Phytoaccumulation)

Phytoextraction is a process in which the contaminants uptake by the roots of plants with the subsequent accumulation of that contaminants in the plant tissue. This process is also known as phytoaccumulation, phytoabsorbtion and phytosequestration. It is a promising cleanup technique which includes the use of plants to remove the variety of metal containing soil by accumulation of these in their tissue, (Kumar, 1995; Cunningham, 1997). This technique helps in saving the remediation cost by accumulating low level of the contaminants in a big area. The important feature of the advantage of this technique is, it is eco-friendly and compared to other traditional techniques it does not affect the soil quality as well as the nature of the soil. Some plants showing the phytoextraction are *Rumex crispus, Sedum alfredii, Viola baoshanensis* (Macek, 2000; Zhuang, 2007). Metal such as Mn, Ni, Cu, Zn, Cu, are essential micronutrient. The plants which are hyper accumulator do not accumulate these micronutrients but absorb significant amount of non-essential metals like Cd (Alloway, 1996). The absorption of heavy metal is governed by the various soil characteristics such as organic matter content, pH so that the amount of heavy metal in the soil do not always show similar concentration in plants. The plants which grow in the soil with heavy metal concentration have increased heavy metal ion content (Alloway, 1996). Some cultivated plant takes up the metal as mobile ion or through root, foliar absorption from the soil.

Rhizofiltration

It is also known as Phytofiltration. The concept of rhizofiltration is similar to phytoextraction but the main difference is, in rhizofiltration, accumulation can occur in roots or in the portion of the above plant water but in phytoextraction above ground not in the root So, in rhizofiltration the contaminants are either adsorbed onto the root surface of plants or are absorbed by the roots. This technique is helpful in treating large volume of water with low contaminants/pollutants concentration. In rhizofiltration the plants firstly are acclimated to the pollutants for *in situ* than they are grown hydroponically in clean water until the root has developed. After developing the roots of the plants, the water supply is substituted for a polluted water supply to acclimatise the plant. When the plants become acclimatised they are planted in polluted area where roots uptake the contaminated water. They are harvested when roots become saturated. This technique is effective when it is repeated several times

Phytostabilisation

Phytostabilisation is a technique in which contaminants are treated by the modification of the chemicals, biological and physical condition the soil. It is also known as in-place inactivation or phtyoimmobilization. It includes the use of certain plants to immobilize soil and water contaminants. The contaminant or pollutants are absorbed and accumulated by roots, and precipitated in the rhizosphere. This technique

Table 4. Some of the plant species involved in the heavy metal remediation

Species	Metal	References
Pteris vittata	Zn, As Cu and Ni	Ma et al. (2001)
Thlaspi,Eichhornia, Hydrocotyle spp.	Cd, Zn, Ni, Pb and Cu	Prasad and Freitas (2003)
Brassica juncea	Cd and Se	Banuelos et al. (2005)
Helianthus annuus	Cd	Mani et al. (2007)
Populus Sp.	Hg	Lyyra et al. (2007)
Husk of Oryza sativa and Zea mays	Pb	Abdel-Ghani et al. (2007)
Brassica juncea and Alyssum sp.	Ni and Pb	Chaney et al., (2007)
Brassica napus	Cd	Selvam and Wong (2008)
Typha latifolia	Pb	Tiwari et al. (2008)
Nelumbo nucifera	Cu, Ni, Zn, Co, Cd and Pb	Kumar et al. (2008a)
Amaranthus viridis	Cr	Liu et al. (2008)
Helianthus annuus	Zn, Pb, As, Cd, Ni, Cu,Hg and	Jadia and Fulekar (2008)
Trifolium pretense	Cs	Wu and Tang (2009)
Spinacia oleracea	Pb and Zn	Nouri et al., (2009)
Vetiveria zizanioides	Pb and Cd	Danh et al., (2009)
Nicotiana tabacum	Cd	Wojas et al., (2009)
Alyssum lesbiacum	Ni	Kramer (2010)
Lupinus luteus	Pb, Cu and Cd	Dary et al., (2010)
Paspalum,Brassica, Ammania, Vetiveria	As, Pb, Cu, Zn and Cd	Mukhopadhyay and Maiti (2010)
Populus tremula	Zn, Cd and Cu	Ruiz et al., (2011)
Beta maritime	Pb	De la Fuente et al. (2011)
Sedum plumbizincicola	Zn, Cu and Cd	Wu et al., (2012)
Pistia stratiotes	Pb, Cd and Zn	Vesely et al., (2012)
Spinacea oleracia	Cd and Pb	Mani et al., (2012a, c)
Helianthus annuus	Zn and Cr	Mani et al., (2012b)
Gmelina arborea	Al	Dudhane et al., (2012)
Arabidopsis thaliana	As and Cd	Guo et al., (2012b)

(Mani et al., 2014)

reduces the mobility of the contaminants preventing migration into the groundwater or air, and also reduces the bioavailability of the contaminants so that preventing the spread through the food chain.

Phytoremediation Strategies

- **Phytodegradation (Phytotransformation)**

Phytodegradation is an important degradation process which include breakdown of organic contaminants by metabolic processes driven by the plant. These metabolic processes of plants hydrolyse organic compounds into smaller substances that can be absorbed by the plants and some of the contaminants can be absorbed by the plant and then broken down by plant enzymes. These contaminants are used by plants as metabolites, thus incorporating into the tissue of the plants.

- **Rhizodegradation**

It is also called phytostimulation, and is a plant assisted bioremediation which include the breakdown of organic contaminants in the soil by soil dwelling microbes which is enhanced by the rhizosphere's presence. Microorganism such as bacteria, yeast and fungi use the contaminants of pollutants as their source of energy in the rhizosphere. Many hazards chemicals or pollutants are converted into harmless substances. Plant root exudates some products like alcohols, organic acids, sugars, which used by the soil microflora to enhances microbial growth and activity. The roots of the plants also loosen the soil and transport water to the rhizosphere thus enhancing microbial activity.

- **Phytovolatilization**

Phytovolatilization is a natural process in which plants uptake contaminants or pollutants and transpire them or release them into the atmosphere. Sometime the pollutants become modified when the water travel along with the vascular system of the roots to the leaves. Now it well known to all that *Poplar, Brassica* and many plant species had shown phytovolatilization.

Phytoremediation of Soil and Water Pollutants

The process of phytoremediation depends on the absorption, translocation and metabolism of organic pollutants in plants. So, it is the process of using green plants to reduce the contaminants of the environmental problem without the need to excavate the pollutants. There are many organic pollutants which enter the plant cell by penetrating the cell membrane. Plants like *Medicago sativa, Tagetes patula* show activity against soil pollutants like PAEs and PAHs (Hemdi et al., 2012; Sun et al., 2011; Fud, 2012). Study also indicates that phytoremediation of *M. sativa* exhibits the largest bio- concentration factors (BCFs) of PAEs (Ma, 2012). Plants use their own metabolism to remove, stabilize and transfer the contaminants of soil and water. Simple diffusion in an important process which driven the uptake of organic contaminants from soil by plant roots on the basis of their chemical properties and bioavailability (Nwoko, 2010). To assist the process of phytoremediation of organic pollutants, many amendments like methylated-β-cyclodextrins, citric and oxalic acids, Tween 80, biochar have been applied (Bessely et al., 2010, Mitton et al., 2012; Shen, 2009). At the time of the entrance of the pollutants further translocated and metabolized in plants.

The diffusion of organic contaminants from root symplast into xylem apoplast is necessary for their translocation from roots to shoots and can be calculated by transpiration stream concentration factor (TSCF) (Nowoko., 2010; Verkleji et al., 2009). Several detoxification enzymes like cytochrome P450 enzymes play important role in the oxidative process for emulsifying highly hydrophobic pollutants (Page, 2009). Other enzymes like glutathione-S-transferases (GSTs) catalyze the conjugation between toxic organic contaminants and glutathione (GSH). GST-pollutants conjugates then transported and sequestrated from cytosol to vacuoles in plant cells (Cummins et al., 2011). As it is well known to us that plant require macro (N, P, K, Ca, Mg, S) as well as micronutrients (B, Cl, Cu, Fe, Mn, Mo, Zn,

Co, Ni, Se, Si, V). Sometime the nutrient uptake pathway in plants can take up contaminants which are similar to the nutrients in terms of chemical form. Nutrients like calcium and zinc which are similar to cadmium so plant uptake them from roots but poplar leaves did not accumulate cadmium (Pierzynski et al., 1994). So it is not necessary that all plant species accumulate the same contaminants as others. Plant nutrient like arsenic taken up by plants because of its similarity to the arsenic. Chemicals like arsenic by *Helianthus annus* (Marchiol et al., 2007), cadmium by *Salix viminalis* (Greger et al., 1999), caesium-137 and strontium-90 by *Helianthus annus* (Adler et al., 1996), the harmful contaminants are removed the phytoremediation process. The chemical must be in a solution either in ground water or in the soil for plant uptake. The roots tissue of plants absorbs water from soil solution then the water moves through epidermis to the casparian strip and after that to the endodermis where it can be absorbed, bound or metabolized. The metabolite or chemicals through the endodermis reaches to the xylem and then transported in the transpiration stream or sap. These chemicals might react with plant tissue, or may be released to the atmosphere through stomatal pores (Paterson et al., 1990; Shimp et al., 1993). There are many factors on which the uptake of pollutant by plant is depending which are physical and chemical characteristics of the soil, type of plant, age of the contaminant. Paterson et al., (1990) identified that about 70 organic pollutants were taken up by 88 species of plants and trees. The exact conclusion cannot be made for a particular chemicals or contaminants, for example PCP a chemical; when spike into soil was shown that 21% was found in roots and 15% in shoots after 155 days in the presence of grass (Qiu et al., 1994). So, phytoremediation is totally environmental friendly as it is govern by plant own metabolism and it does not causes harm to the other surrounding area or zone.

Plant Microbes Interaction for the Remediation of Pollutants

The interaction between plant and microbes in the agricultural, natural ecosystem has great importance for growth, productivity and removal of toxic substances in the environment. The interaction has undoubtedly major effects on the components of ecosystem and this has so many advantages for the living community. The two types of interaction i.e. positive and negative not occur only between microorganisms but also occur between plant and microorganism. The mutualistic and communal interaction of plant and microorganism is rhizosphere. The rhizospheric microorganism degrade so many pollutants of their surrounding as they use the food source namely, the root exudates in soil. The degradation of the pollutants by microorganism in rhizosphere zone has positive effects on the plants because due to the degradation of the pollutants near the roots of plants by microbes provide better growing environment. The first study was carried out on the degradation of herbicide and pesticide in the rhizosphere pesticides (Hoagland et al., 1994; Jacobsen, 1997; Zablotowicz et al., 1994) and it suggested that plants are protected against these contaminants by degrading bacteria. Trial and error research has made attention on fast growing and well-developed root system in *Brassica* sp has great capability to remediate the chemicals. It has been also noticed that *Populus deltoids, Morus alba* remove chlorophenols and chlorinated solvent trichloroethylene (TCE) (Stomp et al., 1993). Leguminous plants also have shown rhizoremediation (Kuiper et al., 2001, 2004). There are so many environmental factors which affect the plant microbe interaction, like nutrients, particles size, sunlight, unavailability of the sunlight and so on. Symbiosis is also a plant miocrobes association between leguminous plant and rhizobia. A study revealed that *Astragalus bisulcatus, A. racemosus, A. praelongus* are accumulator of selenium and this is not ony possible but also common for heavy metal remediation (Lin et al., 2011; Alford et al., 2012). Another study showed that the tolerance of bacteria and fungi inhabiting the root of Zn-hyper accumulating *Thlaspi caerulescens* to heavy metals while ecotype microflora did not accumulate Zn (Dechamps et al., 2005). Plant growth promoting rhizobacteria are the resistant rhizospheric microbes which are capable to stimulate plat growth in direct way (Zhuang et al., 2007). The process involves auxins release, N_2 fixation etc. Study also revealed that the growth rate and accumulation of nickel ions by *Alyssum serpylifolium* and *Thlaspi goesingense* can be stimulated by auxins production, IAA production (Alford et al., 2010; Wenzel et al., 2009; Arshad et al., 2007). Plant produce hormone like ethylene, stress hormone against abiotic stresses like heavy. Plant growth contributes gain in biomass thus remove the metal from soil in well manner (Li et al., 2007). The enzyme ACC deaminases transform ethylene precursor, aminocyclopropane-1-carboxylic acid, into alpha-ketobutyric acid and ammonia (Glick et al., 2007). PGPR also make phosphorus available to the plant by solubilisation process. The metal uptake is not only facilitated by bacteria but also by some symbiotic fungi which enhances phytoextraction. The mcorrhizal association gain organic compounds by the process of photosynthesis by plant and supply phodpjphorus, and metal ion like cooper and zinc to plant (Bonfante et al.,2009). This results in the increase of metal uptake by plants with the help of these mycorrhizal fungi like, *Berkheya coddi* mycorrhized with *Rhizophagus intraradices* can accumulate up to 20 times more nickel in comparison with non-mycorrhized plants (Orłowska et al., 2011). Root exudates also attract plant pathogen and helps i the remediation of heavy metals. The rate of the root exudates changes with the age of plant (Chaudhry et al., 2005). The quantity, timing and nature of exudates are important for rhizoremediation. Plants change the composition of root exudates in response to the chemical stress according to the metabolic activities of rhizosphere microorganisms (Chaudhry et al., 2005). Some organic carbon which is present in root exudates serves N_2 and carbon for the growth and survival of microorganism that are capable of degrading organic pollutants (Pilon-Smits, 2005; Salt et al., 1998; Anderson et al., 1993). The biological, chemical, and physical properties of the root- associated soil are responsible for variation in the microorganism diversity and metabolic activities of microorganisms in the rhizosphere this is known as rhizosphere effect (Barea et al., 2005; Kuiper et al., 2004; Pilon-Smits, 2005; Salt et al., 1998). The effects of temperature, pH, and organic matter content, soil moisture, aeration, on the biodegradation of pesticides have been investigated in many studies (Bending et al., 2006; Charnay et al., 2005; Rasmussen et al., 2005).

So, phytoremediation along with the microbes has showed great opportunities which remediate the environmental pollutants along with heavy metals without causing any harm to the surrounding environment. Rhizoremediation takes the advantages of the mycorrhiza, natural symbiosis and with the association of the microorganism helps in the degradation of the contaminants/pollutants in the rhizosphere. Now it is well known that a single species of plant along with the associated microbes can degrade so many chemicals compounds in their surrounding environment. It has been perceived to be more eco-friendly and low tech alternative remedial method which has no destructive impact on the soil as well as on the surrounding environment.

Advantages of Bioremediation

- Microbes have ability to degrade the contaminant/ pollutants from the soil and water. The residue of the treatment is generally harmless which contains Co_2, water and cell biomass etc.
- The technique of bioremediation is helpful for the complete degradation of the variety of the pollutants/contaminants. Today, any compounds which are considered to be hazard can be treated and transformed into harmless products.

Table 5. Plant species promote microbial degradation

Plant Species	Pollutants	References
Kolchia	Atrazine,metolachlor and trifluralin	Anderson et al.,1994
Rye Grass	Chlorpyrifos	Korade and Fulekar., 2010
Zinnia anguistifolia	mefenoxam	Pai et al., 2001
Penninsetum pedicellatum	Chlorpyrifos and Cypermethrie, Fenvalerate	Dubey and Fulekar., 2011 a
Alfalfa and alpine blue grass	Hexadecane and PAHs	Nichols et al.,1997
Sugar beet (cv.Rex)	Polychlorinated biphenyls(PCBS)	Brazil et al., 1995
Hordeum vulgare	2,4-D	Jacobsen, 1997
Triticum aestivum (Wheat)	2,4-D	Kingley et al., 1994
Populus deltoids nigra	1,4-dioxane	Schnoor et al., 1998
Poplar root extract	1,4-dioxane	Kelley *et al.,* 2001
Zea mays (mays)	3-methylbenzoate	Ronchel et al., 2001
Astragalus sinicus	Cd	Sriprang et al.,2002
Azolla pinnata (Fern)	Diesel fuel	Cohen et al.,2002
Sugarcane	2,4-D	Sandmann and Loose,1984
Rice	Benthiocarb	Sato, 1989
Corn	Atrazine	Seibert et al.,1981

(Modified from Kala, 2014)

- The advantage of bioremediation is that it is carried out on site without causing too much disturbance in the surrounding and it also eliminates the need of the transport quality of waste off site and the various problems arises at the time of transportation.
- This technique is less expensive than other techniques which are used for cleanup of hazardous waste.
- Maximum contaminants are destroyed and not transferred to other environments.

In this technique pollutants are converted into innocuous product.

Disadvantages of Bioremediation

- At present day the technique of bioremediation considered to a boon for the environment because of the pollutants and contamination are increasing day by day but it is also causing some problems. The additives which we used to enhance the activity of fungi, bacterium and other microorganisms *in situ*, may cause disturbances in the surrounding environment (Vidali, 2001).
- This process is limited for those pollutants/ compounds that are biodegradable. All compounds are not susceptible to complete degradation.
- Sometimes it is very difficult to control and also take too much time than other treatment option like removal of soil or incineration and excavation.
- Its future effectiveness is difficult to predict because of it is a dynamic process.
- Sometimes it may require more extensive monitoring.

- Sometime when the genetically modified organism is released into the environment it is difficult to remove them from that particular environment.

FUTURE RESEARCH DIRECTIONS

There are so many works which is still going on the bioremediation, phytoremediation and plant microbes interaction to remediate the environment pollutants and others harmful contaminants. These techniques are easy, low cost, efficient that is why, now widely applied by the research and some modification on these techniques are still going on so that all the weakness of these techniques can be overcome. The technique of phytoremediation will most likely focus on improving the speed and efficiency of the plant species, now currently being employed to remediate contaminated sites, either through genetic engineering or traditional methods. Some of the targets for enhancement may be characteristics such as root spread, accumulation, tolerance, and rate of transpiration. Increasingly more companies are investing in bioremediation and uses of microbes and plant so that the research would be environmentally friendly approach to the remediation of polluted soils.

CONCLUSION

Bioremediation is a technology, utilizing the life forms to kill or expel contamination from waste. It takes the advantages of the microbes, plants, and also the natural symbiosis with mycorrhiza and root associated natural microbial flora for the degradation of the harmful contaminants. The studies described above represent the plant microbes' interaction to remediate the heavy metals which would be an important tool for scientist as the pollutants are increasing in the environments day by day due to anthropogenic and other activity on earth. These systems could be an interesting tool to further improve and develop bioremediation into a widely accepted technique.

REFERENCES

Abd El Haleem, D., Von Wintzingerode, F., Moter, A., Moawad, H., & Göbel, U. B. (2000). Phylogenetic analysis of rhizosphere-associated β-subclass proteobacterial ammonia oxidizers in a municipal wastewater treatment plant based on rhizoremediation technology. *Letters in Applied Microbiology*, *31*(1), 34–38. doi:10.1046/j.1472-765x.2000.00760.x PMID:10886611

Abdel-Ghani, N. T., Hefny, M., & El-Chaghaby, G. A. (2007). Removal of lead from aqueous solution using low cost abundantly available adsorbents. *International Journal of Environmental Science and Technology*, *4*(1), 67–73. doi:10.1007/BF03325963

Adler, T. (1996). Botanical cleanup crews. *Science News*, *150*(3), 42–43. doi:10.2307/3980349

Akpomie, O. O., & Ejechi, B. O. (2016). Bioremediation of Soil Contaminated with Tannery Effluent by Combined Treatment with Cow Dung and Microorganisms Isolated from Tannery Effluent. *Journal of Bioremediation & Biodegradation*, *7*, 354.

Ali, H., Khan, E., & Sajad, M. A. (2013). Phytoremediation of heavy metals—concepts and applications. *Chemosphere, 91*(7), 869–881. doi:10.1016/j.chemosphere.2013.01.075 PMID:23466085

Alford, É. R., Pilon-Smits, E. A., & Paschke, M. W. (2010). Metallophytes—a view from the rhizosphere. *Plant and Soil, 337*(1-2), 33–50. doi:10.100711104-010-0482-3

Alford, É. R., Pilon-Smits, E. A., Fakra, S. C., & Paschke, M. W. (2012). Selenium hyperaccumulation by *Astragalus* (Fabaceae) does not inhibit root nodule symbiosis. *American Journal of Botany, 99*(12), 1930–1941. doi:10.3732/ajb.1200124 PMID:23204487

Alkorta, I., Hernández-Allica, J., Becerril, J. M., Amezaga, I., Albizu, I., & Garbisu, C. (2004). Recent findings on the phytoremediation of soils contaminated with environmentally toxic heavy metals and metalloids such as zinc, cadmium, lead, and arsenic. *Reviews in Environmental Science and Biotechnology, 3*(1), 71–90. doi:10.1023/B:RESB.0000040059.70899.3d

Alloway, B. J. (1996). *Heavy Metal in Soils*. London: Halsted Press, John Wiley & Sons Inc.

Anderson, T. A., Kruger, E. L., & Coats, J. R. (1994). Enhanced degradation of a mixture of three herbicides in the rhizosphere of a herbicide-tolerant plant. *Chemosphere, 28*(8), 1551–1557. doi:10.1016/0045-6535(94)90248-8

Anderson, T. A., Guthrie, E. A., & Walton, B. T. (1993). Bioremediation in the rhizosphere. *Environmental Science & Technology, 27*(13), 2630–2636. doi:10.1021/es00049a001

Arshad, M., Saleem, M., & Hussain, S. (2007). Perspectives of bacterial ACC deaminase in phytoremediation. *Trends in Biotechnology, 25*(8), 356–362. doi:10.1016/j.tibtech.2007.05.005 PMID:17573137

Baker, K. H. (1994). Bioremediation of surface and surfaces soils. In K.H. Baker & D.S. Herson (Eds.), Bioremediation (pp. 9-60). New York, USA: McGraw-Hill.

Bañuelos, G., Terry, N., LeDuc, D. L., Pilon-Smits, E. A., & Mackey, B. (2005). Field trial of transgenic Indian mustard plants shows enhanced phytoremediation of selenium-contaminated sediment. *Environmental Science & Technology, 39*(6), 1771–1777. doi:10.1021/es049035f PMID:15819237

Barea, J. M., Pozo, M. J., Azcon, R., & Azcon-Aguilar, C. (2005). Microbial co-operation in the rhizosphere. *Journal of Experimental Botany, 56*(417), 1761–1778. doi:10.1093/jxb/eri197 PMID:15911555

Barfield, B. J., Blevins, R. L., Flofle, A. W., Madison, C. E., Inamder, S., Carey, D. I., & Evangelou, V. P. (1992). Water quality impacts of natural riparian grasses: Empirical studies. American Society of Agricultural Engineers, St. Joseph, MI.

Beesley, L., Moreno-Jiménez, E., & Gomez-Eyles, J. L. (2010). Effects of biochar and green waste compost amendments on mobility, bioavailability and toxicity of inorganic and organic contaminants in a multi-element polluted soil. *Environmental Pollution, 158*(6), 2282–2287. doi:10.1016/j.envpol.2010.02.003 PMID:20219274

Bending, G. D., Lincoln, S. D., & Edmondson, R. N. (2006). Spatial variation in the degradation rate of the pesticides isoproturon, azoxystrobin and diflufenican bioremediation and Phytoremediation of Pesticides in soil and its relationship with chemical and microbial properties. *Environmental Pollution, 139*(2), 279–287. doi:10.1016/j.envpol.2005.05.011 PMID:16043274

Berti, W. R., & Cunningham, S. D. (2000). *Phytostabilizat ion of metals* (pp. 71–88). New York: Wiley.

Bellin, C. A., & O'Connor, G. A. (1990). Plant uptake of pentachlorophenol from sludgeamended soils. *Journal of Environment Qualality*, *19*(3), 598–602. doi:10.2134/jeq1990.00472425001900030040x

Boopathy, R.(2000). Factors limiting bioremediation technologies. *Bioresource technology.74.63*

Bonfante, P., & Anca, I. A. (2009). Plants, mycorrhizal fungi, and bacteria: A network of interactions. *Annual Review of Microbiology*, *63*(1), 363–383. doi:10.1146/annurev.micro.091208.073504 PMID:19514845

Brazil, G. M., Kenefick, L., Callanan, M., Haro, A., De Lorenzo, V., Dowling, D. N., & O'gara, F. (1995). Construction of a rhizosphere pseudomonad with potential to degrade polychlorinated biphenyls and detection of bph gene expression in the rhizosphere. *Applied and Environmental Microbiology*, *61*(5), 1946–1952. PMID:7646029

Bitter, G. A., & Egan, K. M. (1984). Expression of heterologous genes in *Saccharomyces cerevisiae* from vectors utilizing the glyceraldehyde-3-phosphate dehydrogenase gene promoter. *Gene*, *32*(3), 263–274. doi:10.1016/0378-1119(84)90002-7 PMID:6099318

Economopoulos, A. P. (1993). *Assessment of Sources of Air Water and Land Pollution. A Guide to Rapid Source Inventory Techniques and Their Use in Formulating Environmental Control Strategies. Part One: Rapid Inventory Techniques in Environmental Pollution. Part Two: Approaches for Consideration in Formulating Environmental Control Strategies.* WHO.

Chaney, R. L., Angle, J. S., Broadhurst, C. L., Peters, C. A., Tappero, R. V., & Sparks, D. L. (2007). Improved understanding of hyperaccumulation yields commercial phytoextraction and phytomining technologies. *Journal of Environmental Quality*, *36*(5), 1429–1443. doi:10.2134/jeq2006.0514 PMID:17766822

Charnay, M. P., Tuis, S., Coquet, Y., & Barriuso, E. (2005). Spatial variability in 14C-herbicide degradation in surface and subsurface soils. *Pest Management Science*, *61*(9), 845–855. doi:10.1002/ps.1092 PMID:16003827

Chaudhry, Q., Blom-Zandstra, M., Gupta, S. K., & Joner, E. (2005). Utilising the synergy between plants and rhizosphere microorganisms to enhance breakdown of organic pollutants in the environment. *Environmental Science and Pollution Research International*, *12*(1), 34–48. doi:10.1065/espr2004.08.213 PMID:15768739

Chen, H., & Cutright, T. (2001). EDTA and HEDTA effects on Cd, Cr, and Ni uptake by *Helianthus annuus*. *Chemosphere*, *45*(1), 21–28. doi:10.1016/S0045-6535(01)00031-5 PMID:11572587

Cohen, M. F., Williams, J., & Yamasaki, H. (2002). Biodegradation of diesel fuel by an *Azolla*-derived bacterial consortium. *Journal of Environmental Science and Health. Part A*, *37*(9), 1593–1606.

Comans, R. N., Haller, M., & De Preter, P. (1991). Sorption of cesium on illite: Non-equilibrium behaviour and reversibility. *Geochimica et Cosmochimica Acta*, *55*(2), 433–440. doi:10.1016/0016-7037(91)90002-M

Cunningham, S. D., Shann, J. R., Crowley, D. E., & Anderson, T. A. (1997). Phytoremediation of contaminated water and soil. In: E. Kruger et al. (Ed.), Phytoremediation of soil and water allocainants, (pp 2–17). ACS Symp. Ser. 664. Am. Chemical Societ, Washington, DC. doi:10.1021/bk-1997-0664.ch001

Dary, M., Chamber-Pérez, M. A., Palomares, A. J., & Pajuelo, E. (2010). "In situ" phytostabilisation of heavy metal polluted soils using Lupinus luteus inoculated with metal resistant plant-growth promoting rhizobacteria. *Journal of Hazardous Materials*, *177*(1), 323–330. doi:10.1016/j.jhazmat.2009.12.035 PMID:20056325

Dary, M., Chamber-Pérez, M. A., Palomares, A. J., & Pajuelo, E. (2010). "In situ" phytostabilisation of heavy metal polluted soils using Lupinus luteus inoculated with metal resistant plant-growth promoting rhizobacteria. *Journal of Hazardous Materials*, *177*(1), 323–330. doi:10.1016/j.jhazmat.2009.12.035 PMID:20056325

Davies, B. E. (1983). A graphical estimation of the normal lead content of some British soils. *Geoderma*, *29*(1), 67–75. doi:10.1016/0016-7061(83)90031-9

Dechamps, C., Roosens, N. H., Hotte, C., & Meerts, P. (2005). Growth and mineral element composition in two ecotypes of *Thlaspi caerulescens* on Cd contaminated soil. *Plant and Soil*, *273*(1), 327–335. doi:10.100711104-005-0099-0

Danh, L. T., Truong, P., Mammucari, R., Tran, T., & Foster, N. (2009). Vetiver grass, Vetiveria zizanioides: A choice plant for phytoremediation of heavy metals and organic wastes. *International Journal of Phytoremediation*, *11*(8), 664–691. doi:10.1080/15226510902787302 PMID:19810597

De la Fuente, C., Clemente, R., Martínez-Alcalá, I., Tortosa, G., & Bernal, M. P. (2011). Impact of fresh and composted solid olive husk and their water-soluble fractions on soil heavy metal fractionation; microbial biomass and plant uptake. *Journal of Hazardous Materials*, *186*(2), 1283–1289. doi:10.1016/j.jhazmat.2010.12.004 PMID:21216095

Dibble, J. T., & Bartha, R. (1979). *Effects of environmental parameters on the biodegradation of oil sludges*. Applied Environmental Biology.

Di Sioudi, B. D., Miller, C. E., Lai, K., Grimsley, J. K., & Wild, J. R. (1999). Rational design of organophosphorus hydrolase for altered substrate specificities. *Chemico-Biological Interactions*, *119*, 211–223. doi:10.1016/S0009-2797(99)00030-7 PMID:10421455

Dubey, K. K., & Fulekar, M. H. (2011). Effect of pesticides on the seed germination of *Cenchrus setigerus* and *Pennisetum pedicellatum* as monocropping and co-cropping system: Implications for rhizospheric bioremediation. *Romanian Biotechnological Letters*, *16*(1), 5908–5918.

Dudhane, M., Borde, M., & Jite, P. K. (2012). Effect of aluminium toxicity on growth responses and antioxidant activities in Gmelina arborea Roxb. inoculated with AM fungi. *International Journal of Phytoremediation*, *14*(7), 643–655. doi:10.1080/15226514.2011.619230 PMID:22908633

Edens, L., Bom, I., Ledeboer, A. M., Maat, J., Toonen, M. Y., Visser, C., & Verrips, C. T. (1984). Synthesis and processing of the plant protein thaumatin in yeast. *Cell*, *37*(2), 629–633. doi:10.1016/0092-8674(84)90394-5 PMID:6327079

Fu, D., Teng, Y., Luo, Y., Tu, C., Li, S., Li, Z., & Christie, P. (2012). Effects of alfalfa and organic fertilizer on benzo [a] pyrene dissipation in an aged contaminated soil. *Environmental Science and Pollution Research International*, *19*(5), 1605–1611. doi:10.100711356-011-0672-4 PMID:22134861

Scientific Research. (2000). Phytoremediation: An Environmentally Sound Technology for Pollution Prevention, Control and Remediation: An Introductory Guide to Decision-Makers. United Nation Environment Program.

Gietz, R. D., & Schiest, R. H. (2007). High-efficiency yeast transformation using the LiAc/SS carrier DNA/PEG method. *Nature Protocols*, *2*(1), 31–34. doi:10.1038/nprot.2007.13 PMID:17401334

Glick, B. R., Cheng, Z., Czarny, J., & Duan, J. (2007). Promotion of plant growth by ACC deaminase-producing soil bacteria. *European Journal of Plant Pathology*, *119*(3), 329–339. doi:10.100710658-007-9162-4

Greger, M., & Landberg, T. (1999). Use of willow in phytoextraction. *International Journal of Phytoremediation*, *1*(2), 115–123. doi:10.1080/15226519908500010

Griffiths, A. D., & Tawfik, D. S. (2003). Directed evolution of an extremely fast phosphotriesterase by in vitro compartmentalization. *The EMBO Journal*, *22*(1), 24–35. doi:10.1093/emboj/cdg014 PMID:12505981

Guo, J., Xu, W., & Ma, M. (2012 a). The assembly of metals chelation by thiols and vacuolar compartmentalization conferred increased tolerance to and accumulation of cadmium and arsenic in transgenic Arabidopsis thaliana. *Journal of Hazardous Materials*, *199*, 309–313. doi:10.1016/j.jhazmat.2011.11.008 PMID:22119299

Guo, J., Xu, W., & Ma, M. (2012b). The assembly of metals chelation by thiols and vacuolar compartmentalization conferred increased tolerance to and accumulation of cadmium and arsenic in transgenic *Arabidopsis thaliana. Journal of Hazardous Materials*, *199*, 309–313.

Hamdi, H., Benzarti, S., Aoyama, I., & Jedidi, N. (2012). Rehabilitation of degraded soils containing aged PAHs based on phytoremediation with alfalfa (Medicago sativa L.). *International Biodeterioration & Biodegradation*, *67*, 40–47.

Held, T., & Dörr, H. (2000). *In Situ* Remediation. *Biotechnology*, *11b*, 350–370.

Hoagland, R. E., Zablotowicz, R. M., & Locke, M. A. (1993). Propanil metabolism by rhizosphere microflora. In J. A. Anderson & J. R. Coats (Eds.), *Bioremediation through Rhizosphere Technology* (pp. 160–183). Washington, D.C.: American Chemical Society.

Jacobsen, C. S. (1997). Plant protection and rhizosphere colonization of barley by seed inoculated herbicide degrading Burkholderia (*Pseudomonas*) cepacia DBO1 (pRO101) in 2, 4-D contaminated soil. *Plant and Soil*, *189*(1), 139–144. doi:10.1023/A:1004296615446

Jadia, C. D., & Fulekar, M. H. (2008). Vermicomposting of vegetable waste: A biophysicochemical process based on hydro-operating bioreactor. *African Journal of Biotechnology*, *7*(20), 3723–3730.

Jesus, H. E., Peixoto, R. S., & Rosado, A. S. (2015). Bioremediation in Antarctic Soils. *Journal of Petroleum & Environmental Biotechnology*, *6*(06), 248. doi:10.4172/2157-7463.1000248

Jeyaratnam, J. (1990). Acute pesticide poisoning: A major global health problem. *World Health Statistics Quarterly*, *43*(3), 139–144. PMID:2238694

Kala. (2014). Rhizoremediation: A Promising Rhizosphere Technology. *IOSR-JESTFT, 8*(8), 23-27.

Kumar, P. B. A. N., Dushenkov, V., Motto, H., & Raskin, I. (1995). Phytoextraction: The use of plants to remove heavy metals from soils. *Environmental Science & Technology*, *29*(5), 1232–1238. doi:10.1021/es00005a014 PMID:22192016

Kelley, S. L., Aitchison, E. W., Deshpande, M., Schnoor, J. L., & Alvarez, P. J. (2001). Biodegradation of 1, 4-dioxane in planted and unplanted soil: Effect of bioaugmentation with *Amycolata* sp. CB1190. *Water Research*, *35*(16), 3791–3800. doi:10.1016/S0043-1354(01)00129-4 PMID:12230161

Kingsley, M. T., Fredrickson, J. K., Metting, F. B., & Seidler, R. J. (1994). Environmental restoration using plant-microbe bioaugmentation. In Bioremediation of chlorinated and polyaromatic hydrocarbon compounds (pp. 287-292). Boca Raton, FL: Lewis Publishers.

Kiraye, M., John, W., & Gabriel, K. (2016). Bioremediation Rate of Total Petroleum Hydrocarbons from Contaminated Water by *Pseudomonas aeruginosa* Case Study: Lake Albert, Uganda. *Journal of Bioremediation & Biodegradation*, *7*, 335.

Khan, M.A. (2014). Microbiological Solution to Environmental Problems- A Review on Bioremediation. *International Journal of Pure and Applied Bioscience, 2*(6), 295-303.

Koning, M., Hupe, K., & Stegmann, R. (2008). *Thermal processes, scrubbing/extraction, bioremediation and disposal. In Biotechnology Set* (2nd ed., pp. 304–317).

Korade, D. L., & Fulekar, M. H. (2009). Rhizosphere remediation of chlorpyrifos in mycorrhizospheric soil using ryegrass. *Journal of Hazardous Materials*, *172*(2), 1344–1350. doi:10.1016/j.jhazmat.2009.08.002 PMID:19720454

Kramer, U. (2010). Metal hyperaccumulation in plants. *Annual Review of Plant Biology*, *61*(1), 517–534. doi:10.1146/annurev-arplant-042809-112156 PMID:20192749

Kuiper, I., Bloemberg, G. V., & Lugtenberg, B. J. (2001). Selection of a plant-bacterium pair as a novel tool for rhizostimulation of polycyclic aromatic hydrocarbon-degrading bacteria. *Molecular Plant-Microbe Interactions*, *14*(10), 1197–1205. doi:10.1094/MPMI.2001.14.10.1197 PMID:11605959

Kuiper, I., Lagendijk, E. L., Bloemberg, G. V., & Lugtenberg, B. J. (2004). Rhizoremediation: A beneficial plant-microbe interaction. *Molecular Plant-Microbe Interactions*, *17*(1), 6–15. doi:10.1094/MPMI.2004.17.1.6 PMID:14714863

Kumar, R., Joshi, S. R., & Acharya, C. (2008b). Metal tolerant Bacillus and Pseudomonas from uranium rich soils of Meghalaya. In *Proceedings of ISBT 2008* (pp. 345–350).

Lai, K., Stolowich, N. J., & Wild, J. R. (1995). Characterization of PS bond hydrolysis in organophosphorothioate pesticides by organophosphorus hydrolase. *Archives of Biochemistry and Biophysics, 318*(1), 59–64. doi:10.1006/abbi.1995.1204 PMID:7726573

Lyyra, S., Meagher, R. B., Kim, T., Heaton, A., Montello, P., Balish, R. S., & Merkle, S. A. (2007). Coupling two mercury resistance genes in Eastern cottonwood enhances the processing of organomercury. *Plant Biotechnology Journal, 5*(2), 254–262. doi:10.1111/j.1467-7652.2006.00236.x PMID:17309680

Li, W. C., Ye, Z. H., & Wong, M. H. (2007). Effects of bacteria on enhanced metal uptake of the Cd/Zn-hyperaccumulating plant, *Sedum alfredii. Journal of Experimental Botany, 58*(15-16), 4173–4182. doi:10.1093/jxb/erm274 PMID:18039737

Lin, H. (2011). Three principles of soil change and pedogenesis in time and space. *Soil Science Society of America Journal, 75*(6), 2049–2070. doi:10.2136ssaj2011.0130

Liu, D., Zou, J., Wang, M., & Jiang, W. (2008). Hexavalent chromium uptake and its effects on mineral uptake, antioxidant defence system and photosynthesis in Amaranthus viridis L. *Bioresource Technology, 99*(7), 2628–2636. doi:10.1016/j.biortech.2007.04.045 PMID:17570658

Looman, A. C., & Kuivenhoven, J. A. (1993). Influence of the three nucleotides upstream of the initiation codon on expression of the *Escherichia coli* lacZ gene in *Saccharomyces cerevisiae. Nucleic Acids Research, 21*(18), 4268–4271. doi:10.1093/nar/21.18.4268 PMID:8414981

Macek, T., Mackova, M., & Káš, J. (2000). Exploitation of plants for the removal of organics in environmental remediation. *Biotechnology Advances, 18*(1), 23–34. doi:10.1016/S0734-9750(99)00034-8 PMID:14538117

Marchiol, L., Fellet, G., Perosa, D., & Zerbi, G. (2007). Removal of trace metals by *Sorghum bicolor and Helianthus annuus* in a site polluted by industrial wastes: A field experience. *Plant Physiology and Biochemistry, 45*(5), 379–387. doi:10.1016/j.plaphy.2007.03.018 PMID:17507235

Ma, L. Q., Komar, K. M., Zhang, W., Cai, Y., & Kennelley, E. D. (2001). A fern that hyperaccumulates arsenic. *Nature, 409*(6820), 579. doi:10.1038/35054664 PMID:11214308

Ma, T., Luo, Y., Christie, P., Teng, Y., & Liu, W. (2012). Removal of phthalic esters from contaminated soil using different cropping systems: A field study. *European Journal of Soil Biology, 50*, 76–82. doi:10.1016/j.ejsobi.2011.12.001

Mani, D., Sharma, B., & Kumar, C. (2007). Phytoaccumulation, interaction, toxicity and remediation of cadmium from *Helianthus annuus* L. (sunflower). *Bulletin of Environmental Contamination and Toxicology, 79*(1), 71–79. doi:10.100700128-007-9153-3 PMID:17549427

Mani, D., Sharma, B., Kumar, C., Pathak, N., & Balak, S. (2012b). Phytoremediation potential of Helianthus annuus L. in sewageirrigated Indo-Gangetic alluvial soils. *International Journal of Phytoremediation, 14*(3), 235–246. doi:10.1080/15226514.2010.498844 PMID:22567708

Mani, D., Sharma, B., Kumar, C., & Balak, S. (2012c). Cadmium and lead bioaccumulation during growth stages alters sugar and vitamin C content in dietary vegetables. *Proceedings of the National Academy of Sciences. India. Section B, Biological Sciences*, *82*(4), 477–488. doi:10.100740011-012-0057-6

Mathiyazhagan, N., & Natarajan, D. (2011). Bioremediation on effluents from Magnesite and Bauxite mines using *Thiobacillus* Spp and *Pseudomonas* Spp. *Journal of Bioremediation & Biodegradation*, *2*(01), 115. doi:10.4172/2155-6199.1000115

McCutcheon, S. C., & Schnoor, J. L. (2003). Overview of phytotransformation and control of wastes. *Phytoremediation: Transformation and control of contaminants, 358.*

Meagher, R. (2000). Phytoremediation of toxic elements and organic pollutants. *Current Opinion in Plant Biology*, *3*(2), 153–162. doi:10.1016/S1369-5266(99)00054-0 PMID:10712958

Menzel, R. G. (1965). Soil-plant relationships of radioactive elements. *Health Physics*, *11*(12), 1325–1332. doi:10.1097/00004032-196512000-00009 PMID:5884542

Minton, N. A., & Murray, V. S. (1988). A review of organophosphate poisoning. *Medical Toxicologyand Adverse Drug Experiment*, *3*, 350–375. PMID:3057326

Mitton, F. M., Gonzalez, M., Peña, A., & Miglioranza, K. S. (2012). Effects of amendments on soil availability and phytoremediation potential of aged p, p′-DDT, p, p′-DDE and p, p′-DDD residues by willow plants (*Salix* sp.). *Journal of Hazardous Materials*, *203*, 62–68. doi:10.1016/j.jhazmat.2011.11.080 PMID:22188788

Mukhopadhyay, S., & Maiti, S. K. (2010). Phytoremediation of metal mine waste. *Applied Ecology and Environmental Research*, *8*(3), 207–222.

Mulbry, W. W., & Karns, J. S. (1989). Parathion hydrolase specified by the Flavobacterium opd gene: Relationship between the gene and protein. *Journal of Bacteriology*, *171*(12), 6740–6746. doi:10.1128/jb.171.12.6740-6746.1989 PMID:2556372

Nichols, T. D., Wolf, D. C., Rogers, H. B., Beyrouty, C. A., & Reynolds, C. M. (1997). Rhizosphere microbial populations in contaminated soils. *Water, Air, and Soil Pollution*, *95*(1), 165–178. doi:10.1007/BF02406163

Nouri, J., Khorasani, N., Lorestani, B., Karami, M., Hassani, A. H., & Yousefi, N. (2009). Accumulation of heavy metals in soil and uptake by plant species with phytoremediation potential. *Environmental Earth Sciences*, *59*(2), 315–323. doi:10.100712665-009-0028-2

Nwoko, C. O. (2010). Trends in phytoremediation of toxic elemental and organic pollutants. *African Journal of Biotechnology*, *9*(37), 6010–6016.

Obreque-Contreras, J., Pérez-Flores, D., Gutiérrez, P., & Chávez-Crooker, P. (2015). Acid mine drainage in Chile: An opportunity to apply bioremediation technology. *Hydrology: Current Research*, *6*(3), 1. doi:10.4172/2157-7587.1000215

Orłowska, E., Przybyłowicz, W., Orlowski, D., Turnau, K., & Mesjasz-Przybyłowicz, J. (2011). The effect of mycorrhiza on the growth and elemental composition of Ni-hyperaccumulating plant, *Berkheya coddii Roessler. Environmental Pollution, 159*(12), 3730–3738. doi:10.1016/j.envpol.2011.07.008 PMID:21835516

Page, V., & Schwitzguébel, J. P. (2009). The role of cytochromes P450 and peroxidases in the detoxification of sulphonated anthraquinones by rhubarb and common sorrel plants cultivated under hydroponic conditions. *Environmental Science and Pollution Research International, 16*(7), 805–816. doi:10.100711356-009-0197-2 PMID:19529968

Pai, S. G., Riley, M. B., & Camper, N. D. (2001). Microbial degradation of mefenoxam in rhizosphere of *Zinnia angustifolia. Chemosphere, 44*(4), 577–582. doi:10.1016/S0045-6535(00)00368-4 PMID:11482644

Paterson, S., Mackay, D., Tam, D., & Shiu, W. Y. (1990). Uptake of organic chemicals by plants: A review of processes, correlations and models. *Chemosphere, 21*(3), 297–331. doi:10.1016/0045-6535(90)90002-B

Pierzynski, G. M., Schnoor, J. L., Banks, M. K., Tracy, J. C., Licht, L. A., & Erickson, L. E. (1994). Vegetative remediation at Superfund Sites. Mining and It's Environ. Impact. *Environmental Science & Technology, 1*, 49–69.

Pilon-Smits, E. (2005). Phytoremediation. *Annual Review of Plant Biology, 56*(1), 15–39. doi:10.1146/annurev.arplant.56.032604.144214 PMID:15862088

Prasad, M. N. V. (2007). Phytoremediation in India. *Phytoremediation: Methods and Reviews*, 435-454.

Qiu, X., Shah, S. I., Kendall, E. W., Sorensen, D. L., Sims, R. C., & Engelke, M. C. (1994). Grass-enhanced bioremediation for clay soils contaminated with polynuclear aromatic hydrocarbons. In T.A. Anderson & J.R. Coats (Eds.), Bioremediation Through RhizosphereTechnology (pp. 142-157). American Chemical Society, Washington, DC. 10.1021/bk-1994-0563.ch013

Ramakrishnan, B. (2016). *Bioremediation with Simultaneous Recovery and Reuse of Resources. Journal of Bioremediation and Biodegradation, 7*, 172.

Raskin, I., Smith, R. D., & Salt, D. E. (1997). Phytoremediation of metals: Using plants to remove pollutants from the environment. *Current Opinion in Biotechnology, 8*(2), 221–226. doi:10.1016/S0958-1669(97)80106-1 PMID:9079727

Ronchel, M. C., & Ramos, J. L. (2001). Dual system to reinforce biological containment of recombinant bacteria designed for rhizoremediation. *Applied and Environmental Microbiology, 67*(6), 2649–2656. doi:10.1128/AEM.67.6.2649-2656.2001 PMID:11375176

Ruiz, O. N., Alvarez, D., Gonzalez-Ruiz, G., & Torres, C. (2011). Characterization of mercury bioremediation by transgenic bacteria expressing metallothionein and polyphosphate kinase. *BMC Biotechnology, 11*(1), 82. doi:10.1186/1472-6750-11-82 PMID:21838857

Salt, D. E., Smith, R. D., & Raskin, I. (1998). Phytoremediation. *Annual Review of Plant Biology, 49*(1), 643–668. doi:10.1146/annurev.arplant.49.1.643 PMID:15012249

Sambrook, J., Fritsch, E. F., & Maniatis, T. (1989). *Molecular cloning: a laboratory manual.* New York: Cold Spring Harbor Press.

Sandmann, E. R. I. C., & Loos, M. A. (1984). Enumeration of 2, 4-D-degrading microorganisms in soils and crop plant rhizospheres using indicator media; high populations associated with sugarcane (*Saccharum officinarum*). *Chemosphere, 13*(9), 1073–1084. doi:10.1016/0045-6535(84)90066-3

Sato, K. (1989). In F. Kunc (Ed.), *Interrelationships between Microorganisms and Plants in Soil. Vancura, V* (pp. 335–342). New York: Elsevier. doi:10.1016/S0166-2481(08)70234-4

Satyapal, G. K., Rani, S., & Kumar, M. (2016). Potential Role of Arsenic Resistant Bacteria in Bioremediation: Current Status and Future Prospects. *Journal of Microbial & Biochemical Technology, 8*(3), 256–258. doi:10.4172/1948-5948.1000294

Schnoor, J. L., Aitchison, E. W., Kelley, S. L., & Alvarez, P. J. J. (1998). Phytoremediation and bioaugmentation of 1, 4-dioxane. In *Abstr. Adv. Biol. Syst. Rem. Conf.* (pp. 87-91).

Schofield, D. A., & DiNovo, A. A. (2010). Generation of a mutagenized organophosphorus hydrolase for the biodegradation of the organophosphate pesticides malathion and demeton-S. *Journal of Applied Microbiology, 109*(2), 548–557. PMID:20132373

Seibert, K., Fuehr, F., & Cheng, H. H. (1981). In *Theory and Practical Use of Soil-Applied Herbicides Symposium (*pp. 137-146). European Weed Resource Society, Paris.

Selvam, A., & Wong, J. W. C. (2008). Phytochelatin systhesis and cadmium uptake of *Brassica napus. Environmental Technology, 29*(7), 765–773. doi:10.1080/09593330801987079 PMID:18697518

Shen, C., Tang, X., Cheema, S. A., Zhang, C., Khan, M. I., Liang, F., ... Chen, Y. (2009). Enhanced phytoremediation potential of polychlorinated biphenyl contaminated soil from e-waste recycling area in the presence of randomly methylated-β-cyclodextrins. *Journal of Hazardous Materials, 172*(2), 1671–1676. doi:10.1016/j.jhazmat.2009.08.064 PMID:19748180

Shimp, J. F., Tracy, J. C., Davis, L. C., Lee, E., Huang, W., Erickson, L. E., & Schnoor, J. L. (1993). Beneficial effects of plants in the remediation of soil and groundwater contaminated with organic materials. *Critical Reviews in Environmental Science and Technology, 23*(1), 41–77. doi:10.1080/10643389309388441

Sriprang, R., Hayashi, M., Yamashita, M., Ono, H., Saeki, K., & Murooka, Y. (2002). A novel bioremediation system for heavy metals using the symbiosis between leguminous plant and genetically engineered rhizobia. *Journal of Biotechnology, 99*(3), 279–293. doi:10.1016/S0168-1656(02)00219-5 PMID:12385715

Stomp, A., Han, K.-H., Wilbert, S., Gordon, M. P., & Cunningham, S. D. (1994). Genetic strategies for enhancing phytoremediation. *Annals of the New York Academy of Sciences, 721*(1), 481–491. doi:10.1111/j.1749-6632.1994.tb47418.x PMID:8010696

Sun, Y., Zhou, Q., Xu, Y., Wang, L., & Liang, X. (2011). Phytoremediation for co-contaminated soils of benzo [a] pyrene (B [a] P) and heavy metals using ornamental plant *Tagetes patula*. *Journal of Hazardous Materials, 186*(2), 2075–2082. doi:10.1016/j.jhazmat.2010.12.116 PMID:21269763

Tamura, T., & Jacobs, D. G. (1960). Structural implications in cesium sorption. *Health Physics, 2*(4), 391–398. doi:10.1097/00004032-195910000-00009 PMID:13836791

Thassitou, P.K., & Arvanitonyannis, I S. (2001). Bioremediation: A novel approach to food waste management. *Trends in Food Science & Technology, 12*(185).

Tiwari, S., Kumari, B., & Singh, S. N. (2008). Evaluation of metal mobility/immobility in fly ash induced by bacterial strains isolated from the rhizospheric zone of Typha latifolia growing on fly ash dumps. *Bioresource Technology, 99*(5), 1305–1310. doi:10.1016/j.biortech.2007.02.010 PMID:17382536

Uqab, B., Mudasir, S., & Qayoom, A. (2016). Bioremediation: A Management Tool. *Journal of Bioremediation & Biodegradation, 7*, 331.

Uqab, B., Mudasir, S., & Nazir, R. (2016). Review on Bioremediation of Pesticides. *Journal of Bioremediation & Biodegradation, 7*, 343.

Van Deuren, J., Lloyd, T., Chhetry, S., Raycharn, L., & Peck, J. (2002). Remediation Technologies Screening Matrix and Reference Guide. In *Federal Remediation Technologies Roundtable.*

Verkleij, J. A., Golan-Goldhirsh, A., Antosiewisz, D. M., Schwitzguébel, J. P., & Schröder, P. (2009). Dualities in plant tolerance to pollutants and their uptake and translocation to the upper plant parts. *Environmental and Experimental Botany, 67*(1), 10–22. doi:10.1016/j.envexpbot.2009.05.009

Veselý, T., Tlustoš, P., & Száková, J. (2012). Organic acid enhanced soil risk element (Cd, Pb and Zn) leaching and secondary bioconcentration in water lettuce (Pistia stratiotes L.) in the rhizofiltration process. *International Journal of Phytoremediation, 14*(4), 335–349. doi:10.1080/15226514.2011.620650 PMID:22567715

Vidali, M. (2001). Bioremediation. an overview. *Pure and Applied Chemistry, 73*(7), 1163–1172. doi:10.1351/pac200173071163

Wenzel, W. W. (2009). Rhizosphere processes and management in plant-assisted bioremediation (phytoremediation) of soils. *Plant and Soil, 321*(1-2), 385–408. doi:10.100711104-008-9686-1

Wilson, J. H., Leung, W. Y., Bosco, G., Dieu, D., & Haber, J. E. (1994). The frequency of gene targeting in yeast depends on the number of target copies. *Proceedings of the National Academy of Sciences of the United States of America, 91*(1), 177–181. doi:10.1073/pnas.91.1.177 PMID:8278360

Wojas, S., Hennig, J., Plaza, S., Geisler, M., Siemianowski, O., Skłodowska, A., & Antosiewicz, D. M. (2009). Ectopic expression of Arabidopsis ABC transporter MRP7 modifies cadmium root-to-shoot transport and accumulation. *Environmental Pollution, 157*(10), 2781–2789. doi:10.1016/j.envpol.2009.04.024 PMID:19467746

World Health Organization. (1990). *Public health impact of pesticides used in agriculture.*

Wu, H., Tang, S., Zhang, X., Guo, J., Song, Z., Tian, S., & Smith, D. L. (2009). Using elevated CO 2 to increase the biomass of a Sorghum vulgare× Sorghum vulgare var. sudanense hybrid and Trifolium pratense L. and to trigger hyperaccumulation of cesium. *Journal of Hazardous Materials*, *170*(2), 861–870. doi:10.1016/j.jhazmat.2009.05.069 PMID:19523760

Wu, L., Li, Z., Akahane, I., Liu, L., Han, C., Makino, T., ... Christie, P. (2012). Effects of organic amendments on Cd, Zn and Cu bioavailability in soil with repeated phytoremediation by Sedum plumbizincicola. *International Journal of Phytoremediation*, *14*(10), 1024–1038. doi:10.1080/15226514.2011.649436 PMID:22908662

Xenia, M. E., & Refugio, R. V. (2016). Microorganisms Metabolism during Bioremediation of Oil Contaminated Soils. *Journal of Bioremediation & Biodegradation*, *7*(340), 2.

Zablotowicz, R. M., Hoagland, R. E., & Locke, M. A. (1994). Glutathione S-transferase activity in rhizosphere bacteria and the potential for herbicide detoxification. In T. A. Anderson & J. R. Coats (Eds.), *Bioremediation Through Rhizosphere Technology* (pp. 184–198). Washington, DC: American Chemical Society. doi:10.1021/bk-1994-0563.ch015

Zhuang, P., Yang, Q. W., Wang, H. B., & Shu, W. S. (2007). Phytoextraction of heavy metals by eight plant species in the field. *Water, Air, and Soil Pollution*, *184*(1-4), 235–242. doi:10.100711270-007-9412-2

Zhuang, X., Chen, J., Shim, H., & Bai, Z. (2007). New advances in plant growth-promoting rhizobacteria for bioremediation. *Environment International*, *33*(3), 406–413. doi:10.1016/j.envint.2006.12.005 PMID:17275086

Chapter 6
Composting an Emerging Technology for Solid Waste Management in India

Mansi Rastogi
Maharshi Dayanand University, India

Meenakshi Nandal
Maharshi Dayanand University, India

ABSTRACT

This chapter describes how "Solid waste" (SW) is a term usually applied to an assorted collection of wastes produced in urban areas, the nature, characteristics and quantity of which varies from region to region. Waste generation is influenced not only by lifestyle of the region's inhabitants, but also by copiousness and type of the region's natural resources. Composting of municipal solid waste is, therefore, a simple and cost-effective technology for treating the organic fraction of municipal solid waste. As the composting proceeds there occurs drastic change in microbial communities, and resemblance to the initial community is lost with process. Generally, three categories of microorganisms play a vital role in composting process: bacteria, actinomycetes and fungi. It is primarily the mesophilic (> 45 °C) and specifically the thermophilic bacteria (< 60 °C), that plays a major role by rising the temperature of the piles. This chapter mainly focused on composting process for Municipal waste stabilization and role of microbes in the process.

INTRODUCTION

"Solid waste" (SW) is a term usually applied to an assorted collection of wastes produced in urban areas, the nature, characteristics and quantity of which varies from region to region. Waste generation is influenced not only by lifestyle of the region's inhabitants, but also by copiousness and type of the region's natural resources. Considering population India is the second largest nation in the world with a population of 1.21 billion, accounting for nearly 18% of world's human population with inadequate resources and adequate systematic methodology to treat its solid wastes. Thus, enormous generation of

DOI: 10.4018/978-1-5225-3540-9.ch006

solid waste is being caused by such a high population growth rate posing a serious threat to environmental quality and human health (Balasubramanian & Birundha, 2012). The urban population in India grew at a rate of 31.8% during the last decade to 377 million, which is greater than the entire population of US, the third largest country in the world according to population (Census of India, 2011). The improper disposal of this waste results in contamination of environment and spread of diseases. Management of these waste related problems has become a major problem and the impacts of these wastes on the economy cannot be ignored.

Solid waste management has evolved greatly since its early days and it is now considered an inter-related series of options that aim at waste reduction at source, recycling methods, treatment and final disposal methods. It is one of the most challenging issues in India than elsewhere at the global level and gigantic steps have to be taken to fill these gaps. Hence, approach for a system analysis has become necessary considering many options available and a system model that is desirable because of the interactions between many factors within a waste management system. As per MSW is concerned, it includes street sweeping, sanitation residues, household garbage and rubbish, trade and non-hazardous industrial refuse, construction and demolition debris. MSW management is of global concern for both the developed and under-developing countries. There is lack of proper municipal solid waste (MSW) disposal systems to tackle the ever-expanding quantity of wastes. The current SWM services are inept to sustain the heavy expenditure and could cause a potential threat to the environmental quality and public health. In India about 80-85% fraction of the solid waste is characterized to be organic. In India waste contains a large part of organic matter, composting looks as an attractive and feasible option for handling the degradable part of MSW.

Waste Generation in India

There has been tremendous increase in the waste generation with improved life style and social status of the populations in urban centers. Each Generation of municipal solid waste (MSW) in India increased, between 1991 and from 23.86 million tons/year to more than 39 million tons/year (Figure 1). In Per capita terms, the growth of MSW in Indian cities has been estimated to be 1-1.33% and within next two decades, the annual waste generation is debris, sanitation residues, and trade and non-hazardous estimated to increase by more than five times than that of the present level (Singhal, 2000).

Due to the inadequate transport capacity and less workforce, collection efficiency in the Indian cities has been in the range of 70-73% (Gupta, Prasad, & Gupta, 1998). Thus, only one-third of the total waste remains uncollected even though the municipal bodies allocate 85-90% of their total budget for collection and transportation activities. On comparison of the per capita waste generation, being 0.44 kg/day in 2001 to 0.5 kg/day in 2011, stimulated by changing and improved urban lifestyle. There is 50% increase in the waste generation due to increased population growth in Indian cities. About 68.8 million tons per year (TPY) or 188,500 tons per day (TPD) of MSW is generated by urban India, including 53 cities with a million plus population together generating 86,000 TPD (31.5 million tons per year) at a 500 grams/day per capita waste generation rate (Annepu, 2012). Presently, the urban population in India generates 1.15 lakh tones of MSW per day that amounts to 42 million tons of garbage per annum (Asnani, 2006). Such an abrupt increase in waste generation has resulted in acute stress on all available natural, budgetary and infrastructural resources. The predominant mode of disposal has been open dumping as 94% and only 5% of the total waste is taken for composting (Nunan, 2000). The average waste collection in the Indian cities is 70% (Sharholy, Ahmad, Vaishya, & Gupta, 2007). About 70 - 90% of MSW generated is

Figure 1. Municipal solid waste generation in Indian cities

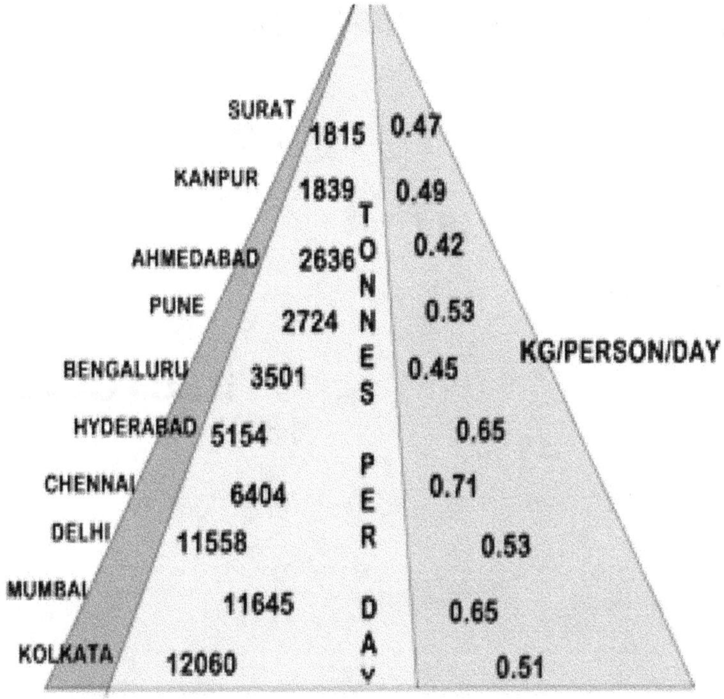

collected in big cities, whereas collection in smaller cities and towns is less than 50%. The 85-91% of the collected MSW is conventionally disposed of by land filling on open lands and dumps with about 10% being burnt on open areas or caught in landfill fires (Kumar, 2010; Annepu, 2012). Such open burning of MSW and landfill fires results as increase in environmental pollution and the pollutants released includes carbon monoxide (CO), nitrogen oxides (NOx), carcinogenic hydro carbons (HC) (includes dioxins and furans), sulfur dioxide (SO_2) and particulate matter (PM) as shown in Figure 2 (Annepu, 2012).

In India all technological solutions that were attempted to tackle this enormous ever-expanding waste disposal issue encountered initial failures. In 1975-1976 around ten MBT (mechanical biological treatment) composting facilities were built in 2003 two RDF (refuse derive fuel) plants built near Hyderabad and Vijayawada and in 1985 the WTE (waste to energy) facility built in Delhi. With all of these plants being non-functional and the ten MBT and the WTE plant completely closed. The Major reasons that could be figured out for their failures were allocation of funds for plant maintenance was ignored, the plants were designed for handling more waste than could be acquired and local conditions were not considered while importing the technology. The present favorable outcomes of MBT in India could be a result of the lessons learned from such failures (Annepu, 2012).

Solid Waste Management (SWM) In India

Solid waste can be defined as non-liquid material that no longer has any value to the person who is responsible for it. The words rubbish, garbage, trash or refuse are often used as synonyms when talking

Figure 2. Solid waste composition of urban India

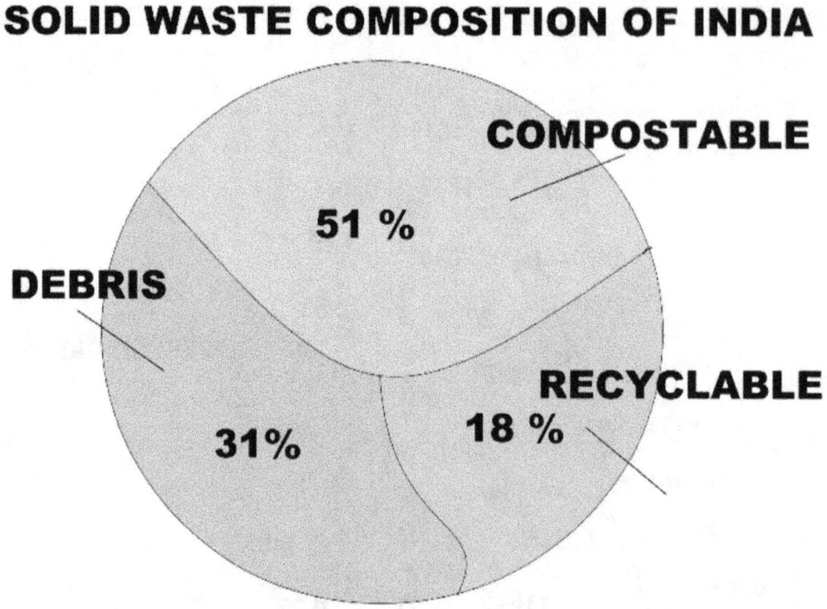

about solid waste. A typical waste management system in a low- or middle-income country like India includes the following elements:

- Waste generation and storage
- Segregation, reuse and recycling at the household level
- Primary waste collection and transport to a transfer station or community bin
- Street sweeping and cleansing of public places
- Management of the process
- Secondary collection and transport to the waste disposal site
- Waste disposal in landfills

As we can see that disposal is the last functional element in the solid waste management system and the ultimate fate of all the waste that are of no further value. Implementation of waste disposal facilities in the country is found to be far from satisfaction. Most of cities/towns are facing problem in identification of landfill sites for construction of sanitary landfills. This is due to public resistance, rapid growth of urban areas, escalating land prices and not having proper master plan.

Technologies for Solid Waste Management

The main technological options available for processing and disposal of MSW are composting, vermicomposting, anaerobic digestion, incineration, gasification and pyrolysis, plasma pyrolysis. It also includes production of Refuse Derived Fuel (RDF) known as pelletization and sanitary land filling/ landfill gas recovery. Over 50% of an average developing country city's municipal solid waste stream could be readily composted. Based on a committee report the municipal solid waste management and handling rules (MSWMH) drafted in 1999 and effective from 2000 made composting mandatory for all the municipal divisions. This chapter focus on composting process for municipal waste stabilization.

Composting is a technology known in India since times immemorial. This process is suitable for organic biodegradable fraction of MSW, yard (or garden) waste/waste containing high proportion of lignocelluloses materials, which do not readily degrade under anaerobic conditions. Composting of MSW is, therefore, the simplest and cost-effective technology for treating the organic fraction of MSW. The compost is thereafter used for improvement in soil texture and augmenting of micronutrient deficiencies. It also increases moisture-holding capacity of the soil and helps in maintaining soil health. Although there are certain issues with composting process like land required for open compost plants is relatively large. Also issue with methane emission, odour and pathogens from badly managed open compost plants remain. At the operational level, waste segregation at source if is not properly carried out there is possibility of toxic material entering the stream of municipal solid waste. It is essential that compost produced to be safe for application but still it is an old age effective practice for waste disposal.

OVERVIEW OF COMPOSTING PROCESS

Composting is a cornerstone of sustainable development, yet it is often neglected within integrated municipal solid waste management programs. Composting refers to the controlled decomposition of organic (or carbon- containing) matter by using microorganisms (mainly bacteria and fungi) into a stable humus material that is dark brown or black in color and has an earthy smell. The process is controlled with the aim of accelerating decomposition, optimizing efficiency and minimizing any potential environmental or nuisance problems that could develop. During the composting process, feedstock is placed in a pile or windrow (an elongated pile) where decomposition takes place. The rate of decomposition depends on the level of technology used as well as on such physical, chemical and biological factors as microorganisms, oxygen levels, moisture content and temperature (Figure 3). Composting works best when these factors are carefully monitored and controlled. The end products of a well-run composting process are a humus-like material, heat, water and carbon dioxide. The end product compost is organic material that can be used as a soil amendment or as a medium to grow plants. Mature compost contains a stable carbon-rich material called humus that is dark brown or black in color with a soil-like earthy smell. In India aerobic composting is the most widely employed SWM technology. It is estimated that up to 6% of MSW collected is composted in various MBT (mechanical biological treatment) facilities. There are more than 80 MBT plants in India treating mixed MSW, most of them located in the states of Maharashtra (CityMayors, 2007), Himachal Pradesh (UNEP, 2004), Chhattisgarh (MoUD, 2000) and Orissa (CityMayors.com, 2007). More than 26 new MBT plants are proposed in different cities and towns across India. Even though composting of mixed wastes is a better solution as compared to land filling or openly burning of waste material (Kant, 2011). Compost coming from MBT facilities was found to be of low quality and to contain toxic heavy metals which could enter human food chain by being used in agriculture practices. Considering MSWM (municipal solid waste management), composting is extensively used technique in India, with above 70 composting plants in more than 43 cities (Sinha, 1997; DOE, 2009; Annepu, 2012). Most of the composting facilities are managing around 100 – 1000 TPD of

MSW. Composting is a low-cost procedure and involves less infrastructural framework with production of marketable by product, compost. The capital investment for building of waste processing composting plant is about INR 200,000) the compost being sold at INR 2,000 – 2,200 (Annepu, 2012). Additional government aid and increasing entrepreneurial interest resulted in inflation in the number of composting facilities nationwide. MSW Composting can be done by two methods, a) Windrow composting, b) Vermicomposting (UNEP, 2004).

Types of Composting

Composting may be divided into two categories by the nature of the decomposition process.

Aerobic Composting

Aerobic composting takes place in the presence of ample oxygen. In this process, organic matter is decomposed by aerobic microorganisms with release of stable end products, carbon dioxide (CO_2), water, ammonia, humus and heat. Some intermediate compounds such as organic acids are also produced which are further decomposed by aerobic micro-organisms. This process has less phytotoxicity risk. The heat released fastens the breakdown of proteins, fats and complex carbohydrates such as cellulose and hemicellulose. This leads to shortening of the processing time. Moreover, if the process gets sufficiently high temperature it causes destruction of many weeds, seeds and human or plant pathogenic micro-organisms. Hence, in comparison to anaerobic composting the aerobic composting process is considered more efficient and reliable for agricultural manure production.

Figure 3. Components of composting process

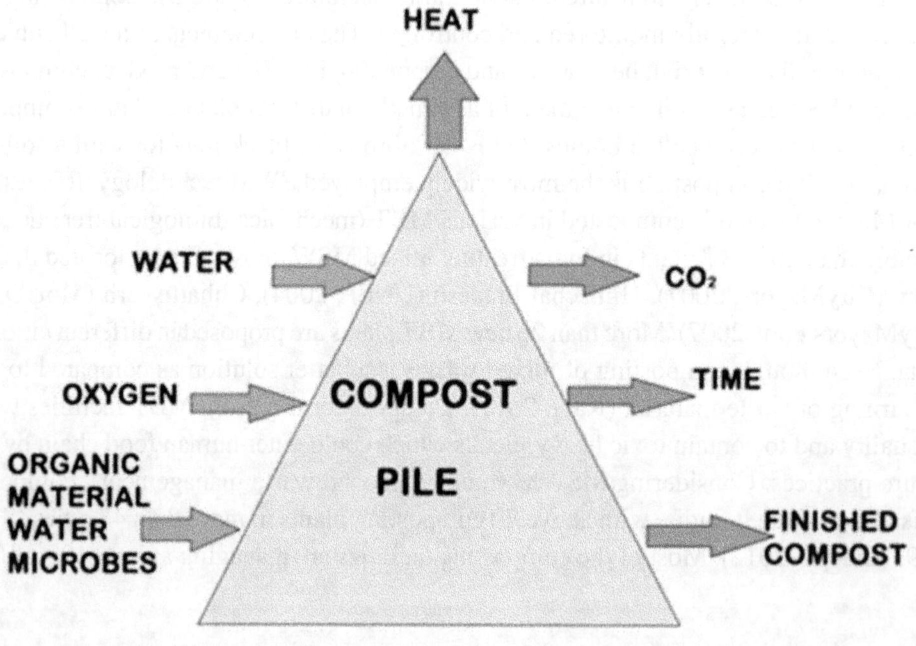

Anaerobic Composting

In this process, organic compounds are broken down with aid of living anaerobic organisms in absence of oxygen. The anaerobic composting process is practiced in large composting systems that consist of well packed stacks. The moisture content ranges between 40%-75% with little or no oxygen penetration, while in liquid suspension system 80% to 99% moisture is maintained for the feedstock material. Similar to the aerobic process here also the microorganisms use nitrogen, phosphorus and other nutrients for cell protoplasm formation. However, unlike aerobic decomposition, organic nitrogen is reduced to ammonia and organic acids and carbon is released in form of methane gas (CH_4) but a small portion of carbon may be respired as carbon dioxide. Being a reduction process the final product humus and also undergoes some aerobic oxidation. The temperature for the putrefying material raises high due to liberation of heat energy but it is not enough to destroy the pathogens, this is a drawback of this process. Another problem may occur due to the odour nuisance. However, if the volume of water in the feed material is maintained accurately, gases gradually dissolve in the water and released slowly into the atmosphere.

Factors Influencing the Composting Process

Aeration/Oxygen

A large amount of Oxygen is required in the aerobic composting process at earlier stages. This requirement of oxygen is indispensable for aerobic composting and is fulfilled by the provision of aeration. The oxygen concentration within the pile is to be maintained between 5-15% (ambient air- 21%). By mixing, turning or by using forced aeration systems in the pile appropriate oxygen concentration can be maintained. With insufficient supply of O_2 there is limited growth of aerobic micro-organisms that results in decomposition at slow rate. Aeration removes water vapor, gases and excessive heat generated in the pile. Proper bulking agents could be added which would create void spaces which needs to be filled with air to support aerobic microbial activity.

Moisture

It is an essential parameter which supports the metabolic activity of the microorganisms. A 40-65% of moisture content works best with the composting process. If the pile is too dry due to lack of moisture, composting proceeds slowly while if it exceeds 65% then anaerobic conditions prevail. In practice composting a moisture content of 50-60% is optimum, while the mature compost ends with 30% moisture content.

Nutrient Level

The microorganisms that basically play an important role in the composting process require specific nutrients in an adequate concentration, available form and proper ratio as an energy source of carbon (C), nitrogen (N), phosphorus (P) and potassium (K). These are the essential macronutrients and are required by microorganisms in relatively large amounts. The ideal C: N ratio of feedstock is between 25:1 and 30:1. A higher ratio limits the growth of microorganisms resulting in a longer composting time. While a lesser one leads to underutilization of N along with odour problems and the excess may be lost to the atmosphere as nitrous oxide or ammonia. This may also be toxic to microorganisms and delay the composting process. The C: N ratio of the mature product should range in 10:1 to 15:1.

Particle Size

Particle size of the feedstock is an important parameter that affects the composting process. A smaller amount feedstock material has greater surface area/ volume ratio and so, there is more of the particle surface available as homogeneous mixture for direct microbial action with better decomposition and improve insulation in the initial stages of composting. This also helps to maintain the optimum temperatures in the composting pile. If the particle size is too small or the surface is too compact this would result in loss of oxygen from void spaces.

Temperature

It is a vital factor for regulating the rate at which the feedstock would be decomposed in a composting pile. There are two temperature ranges mesophilic and thermophilic. At initial composting stage the ideal temperature of 20-45°C, while the process is preceded by thermophilic organisms with an ideal temperature range between 50-70°C. Microbe's works best at high temperature ranges. The critical point for elimination of pathogens is above 55 °C, while the weed seeds is 62°C. So, the maintenance of temperatures at the highest level possibly will result in most effective compost managing without affecting the rate of microbial decomposition (FAO, 2003).

Lignin Content

Lignin is one of the main elements of plant cell walls being highly resistant to microbial degradation due to the complex chemical. It serves as a porosity enhancer and helps in improving the aeration but it also reduces the actual C: N ratio lower than what is normally required as it blocks the bioavailability of the other cell-wall constituents.

PH/Alkalinity Value

Composting process has a unique buffering capacity but the optimum pH for composting process is 7-8. Efficacy of the composting process is increased as the pH proceeds towards neutrality (pH 7). The pH ranging between 5.5 and 9 is best for decomposition (Gray, Sherman, & Biddlestone, 1971; Rynk, 1992;). As the composting process starts there is formation of organic acids and acidic conditions develop with a pH of about 5. At this point the acid-tolerating fungi play a significant role in decomposition. As the breakdown procedure further proceeded by microorganisms the acids are broken down. A gradual approach of pH towards neutrality or as high as 8.5 indicates good compost maturity with generation of ammonia gas that is eventually lost to atmosphere.

Size of Compost System

An optimum compost size would prevent rapid dissipation of heat and moisture and would allow good air circulation. Generally, the piles should be at least 1 m^3 in size to ensure sufficient heat and moisture retention. A small heap would lead to less aeration and a very high temperature which would hinder the process (Robert, 2002). The optimal conditions for composting have been illustrated in Figure 4.

Process of Composting

Process of composting involves three stages: 1) mesophilic phase, 2) thermophilic phase, 3) the cooling phase or curing phase as shown in Figure 5.

1. **Mesophilic Stage:** Being the initial stage of the composting process it is an essential step. In this phase, the microbes present in the compost produce carbon dioxide and energy using carbon and oxygen. Microorganisms use a part for growth and reproduction giving off rest as heat. As the mesophilic stage escalates, there is a gradual rise temperature up to 44 °C (111 °F) in the composting mass but with the mesophilic bacterial population starts declining after a point with this rise in temperature as they are taken over by thermophilic bacteria which proliferate in temperature range of 44 °C- 52 °C (111 °F- 125.6 °F)
2. **Active or Thermophilic Phase:** It is the most active stage and major part of disintegration and simplification of materials occurs in this stage of composting. At such a high temperature (above 45°C) the aerobic, thermophilic organisms e.g. Actinomycetes, fungi and protozoa thrive and decompose the waste material. Oxygen is consumed as energy source for breakdown of feedstock materials and carbon dioxide is respired. Unsuitable temperature range (40-60 °C) and moisture content 50-60% will reduce the biological activity due to reduction in oxygen or a very high rise in temperature. If the temperature rises and the heap gets too hot turning could be considered to cool it down. Mixing up or turning of the contents is more important in the first 14 days. For evaluating the time to be taken by composting method the feedstock material and ambient air temperatures are deciding factors.
3. **Maturing or Curing Phase:** As the temperature lowers down in this stage there is a shift in the microbial populations to those that prefer lower temperatures. Compost should remain aerobic though turning is not needed. In the curing phase, compost is not yet mature. It is a time dependent process and it will take the same time by the compost to reach maturity. It indicates the intensity of humification or the transformation of organic matter into humic substances. Generally, it takes less than 3 months for curing to occur. Various lab tests or germination tests could confirm the

Figure 4. Optimal conditions for composting process

Parameter	Active composting	Curing	Product storage
Oxygen concentration	13 to 18%		
FAS	40 to 60%		
Particle size	A mixture of particles between 3 and 50 mm		
C:N ratio	25:1 to 30:1	18:1 to 23:1	15:1 to 20:1
Moisture content	55 to 65%	45 to 55%	40 to 45%
Temperature	55 to 60°C	Less than 50°C	Ambient
pH	6.5 to 8		

Figure 5. Stages in composting process

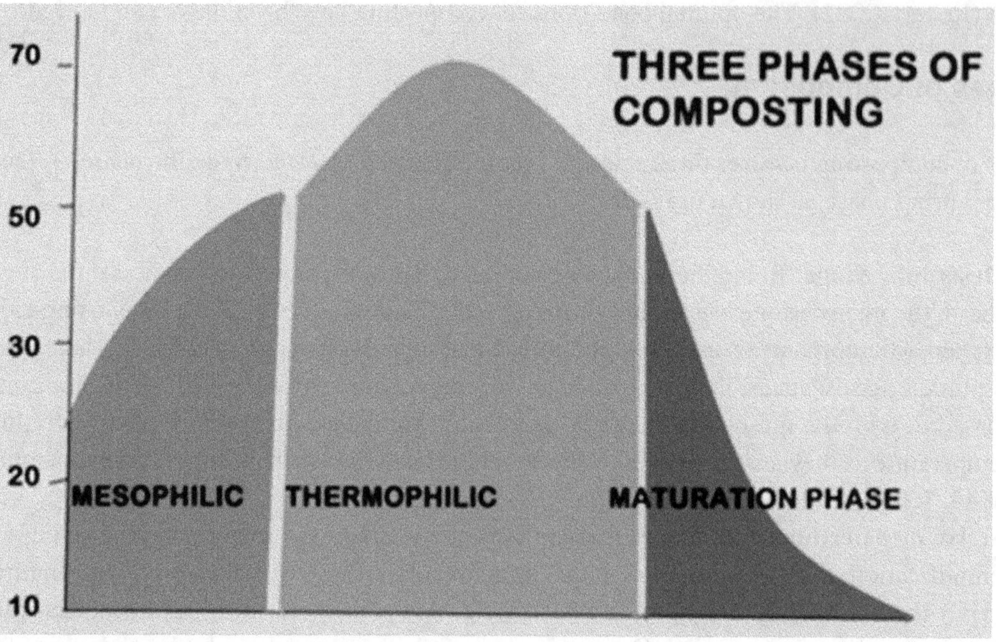

compost maturity e.g., the "squeeze-test" if the feedstock material is low in organic matter (e.g. 30% OM), it is sufficiently wet to about 30 to 40% moisture content. A high organic content, typical of a fresh compost mix should be ideally moistened to 65%. Water holding capacity decreases as the biodegradation proceeds due to loss of organic content and so on comparing the moisture content one can determine if the compost has matured or not. Injury or plant death could be caused by immature compost as higher levels of organic acids, high salt contents, extreme pH values and high C: N ratios could be found in it that could lead to plant death if applied as a fertilizer. So, maturity of compost plays a vital role and the compost should be utilized as soon as possible without longer storage period.

Microbial Diversity in Composting Process

A number of microorganisms have been found to be involved in the process of composting, but the role they play precisely is seldom known. The compost microbes are affected by the substrate composition and temperature of the compost pile. Also, diverse microorganisms have an additive influence on each other. In the initial stage of composting, there is a drastic increase in microbial activity leading to a temperature rise. As composting proceeds fungus replaces the bacterial dominance during compost maturation. The factors like nutrient, oxygen and water or moisture affecting the microbial processes may vary according to input substrate, the compost piles size, turning frequency, particle size, aeration and moisture content. Also selected microbial strain inoculation is done in waste to produce composts. The beneficial microbes are competent to harmful ones during the composting process with a positive effect on crops when applied on field. The microbial introduction in the beginning of the composting process depends on the original matrix of organic fraction (Fuchs, 2010).

As the composting proceeds there occurs drastic change in microbial communities and resemblance to the initial community is lost with process. Generally, three categories of microorganisms play a vital role in composting process: bacteria, actinomycetes and fungi. It is primarily the mesophilic (> 45 °C) and specifically the thermophilic bacteria (< 60 °C), that plays a major role by rising the temperature of the piles. Actinomycetes are considered intermediary between bacteria and fungi because they fall in category of bacterial classification and morphologically look similar to fungi. They proliferate in the last phase of composting and are generally thought to replace the thermophilic bacteria in succession. During the later stage, fungus prevails in the composting process. There are at least 100,000 known species of fungi, the over- whelming numbers of them being microscopic. Only thermophilic fungi being heat tolerant can at or above 50 °C. These microbes are temperature dependent as above 60 °C fungi tend to be absent in compost and above 70 °C actinomycetes do not survive. Finally, as the temperature rises above 82 °C biological activity ceases. To summarize, the microbial population's existing during different phases of compost are defined by dynamic complex interactions among the microorganisms and environment. Also, high temperature depending on the availability of substrate influences the microbial community. Composting process starts with a drastic increase in microbial biomass (Hellmann, Zelles, Palojarvi & Bai, 1997; Narihiro, Takebayashi, & Hiraishi, 2004). Klamer and Baath (1998) studied on composting of shredded straw of *Miscanthus* amended with pig slurry depicted a six-fold increase on the first day. They concluded that there are significant changes in microbial community and that the microbial multiplication is not equally fast and are a complex procedure among individual species. There is a change in chemical and physical properties of the substrate as the initially active microorganisms produce metabolites after degrading the original substrate creating an advanced physico-chemical environment to be used by next set of microorganisms (Ryckeboer et al., 2003a). Also, different enzymes are secreted by different microorganisms to degrade different substrates (Hu & Van Bruggen, 1997; Tuomela, Vikman, Hatakka, & Itavaara, 2000; Ryckeboer et al., 2003b).

PROCESSING TECHNIQUES

There are three subsequent stages of MSW composting; preprocessing, processing and post processing.

Preprocessing

Before composting, feedstock is preprocessed (Figure 6). The quality of the final compost product highly depends on preprocessing procedure and the processing speed. Generally, three procedures are typically performed during preprocessing have been listed below (US EPA, 1994):

1. Sorting/segregation of feedstock material
2. Reduction in size of feedstock particle
3. Treatment of feedstock

Figure 6. Typical MSW composting operation

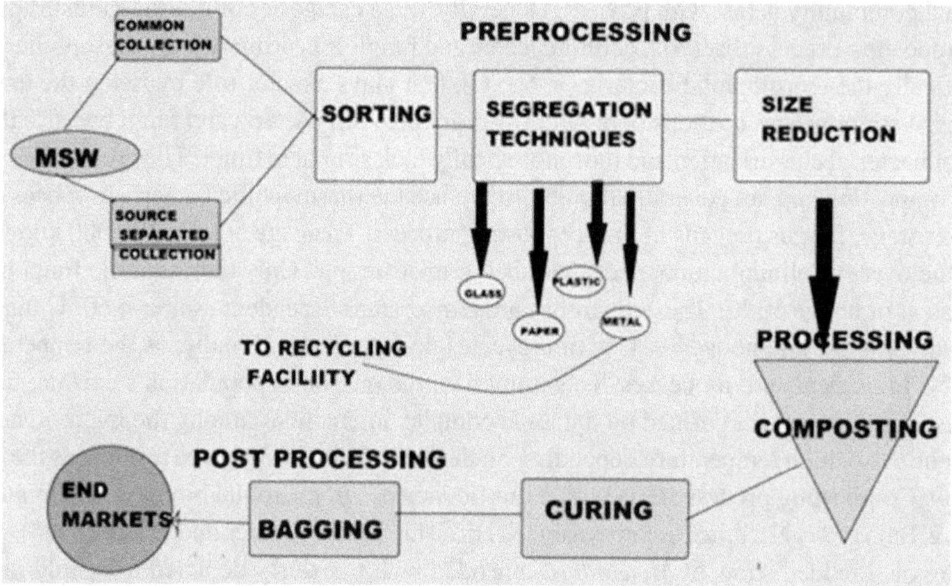

Sorting

The effort required for sorting to segregate and separate undesirable materials from the feedstock depends on diversification of feedstock material. MSW (which comprises heterogeneous materials) can require copious sorting and segregation. If the end use requires higher quality compost then a high level of sorting effort will be required. For example, compost used on food crops will be more polished and have lower levels of unwanted materials and then compost to be used as landfill cover.

Sorting Techniques

MSW is extremely diverse in size, nutrient and moisture contents with organic fractions possessing contrasting non-compostable and possibly hazardous material. The chemical and physical contents found in the feedstock affect the compost and can negatively impact the marketability of the finished product along with the heavy expenses reoccurring in their removal. MSW can be sorted both manually and by mechanical techniques to remove unwanted items. The removal of recyclables, such as plastic bottles, aluminum cans, ferrous materials etc. will prevent their pulverization and mixing into the feed material, which further could prevent various problems.

The manual separation of the waste can be done by three different ways:

- Directly at the source,
- During the collection
- At the disposal site

Mechanical methods are listed below:

- **Screens:** As size of the particle plays an important role in composting process, screening is done to separate materials in different categories based on size. MSW composting facilities to govern the maximum size of feedstock that the facility can control. The type of screen used depends on the moisture content, density, particle shape, viscosity, heterogeneity and of the segregated feedstock (Figure 7).
- **Magnetic-Based Separators:** In these separators magnetic fields is generated along conveyers that leads to attraction mechanism and remove ferrous metals from the feedstock stream. The speed and the quantity of materials being processed through the conveyer define the efficacy of magnetic separators.
- **Eddy-Current Machines:** Eddy-current machines generate electromagnetic field emitting high-energy that produces an electrical charge and separate the aluminum and nonferrous metals from MSW (Figure 8).
- **Air Classifiers:** In air classifiers difference in weight of the feedstock is the main factor. They work well with the feedstock that has been reduced in size.
- **Weight Separation Technology:** In this technology material separation depends upon density and in this process air is replaced by water as the floating medium.
- **Ballistic or Inertial Separation:** The elasticity and density differences in this technology separate the inert and organic constituents. The feedstock falls upon a rotating drum or whirling cone and due to density and elasticity of stones, glass, metal and other material throws the materials at different lengths in trajectories away from the compost feedstock.

Particle Size of the Feedstock

After separation of non-compostable part from the feedstock reduction in size of particles is performed. To increase the surface area to volume ratio mainly size reduction is conducted. This is done to enhance the decomposition rate to increase the surface area on which microbes can act to compost the materials. It also maximizes the efficiency of composting process and creates a balance between the reduced particle size and aerobic conditions. The hammermill, rotating drums and shear shredders are few of the size-reduction devices that are available:

- **Hammermills:** In hammermill, the sets of swinging hammers counter rotate and convert the feedstock into smaller sized particles (Figure 9).
- **Rotating Drums:** In Rotating drums gravity and cascading effect is used to trip down materials into a rotating cylinder (Figure 10).
- **Shear Shredders:** In Shear shredders a pair of oppositely placed knives or hooks rotate at a slow speed with high torque.

Processing

Processing of compost feedstock involves diverse methods for decomposition of the feedstock materials and their transformation to a finished and ethical product. These methods should maximize the pace of the process and minimize the negative effects like odour release and leachate runoff.

Figure 7. Trommel

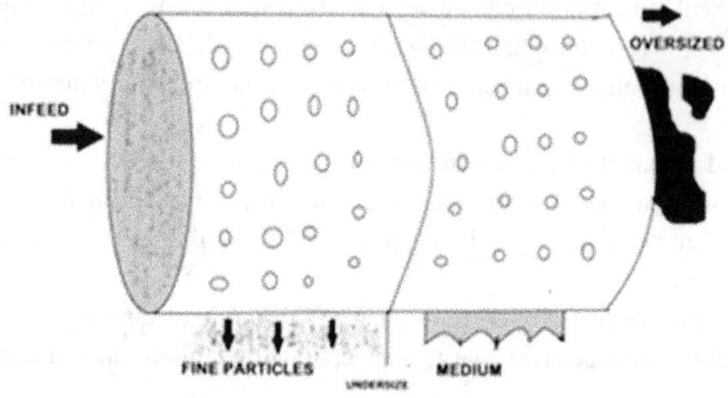

Figure 8. Eddy current separator

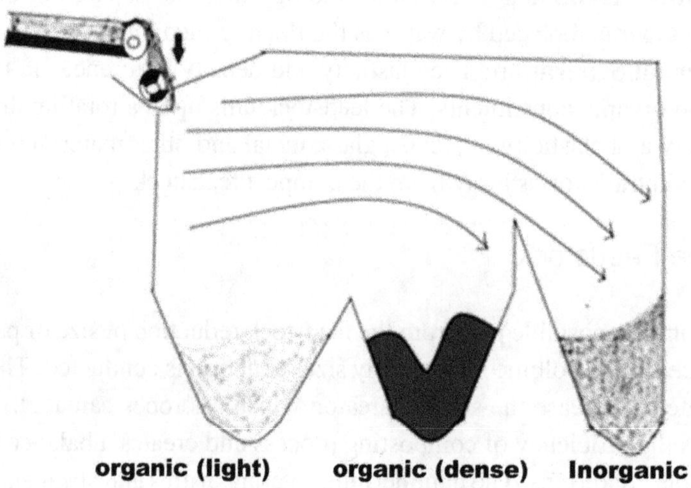

Methods Employed in MSW Composting Process

1. Turned windrows

This method epitomizes as a less technological and medium labor-intensive approach producing uniform compost. Waste is placed in windrows provided with mechanical aeration. Aeration is provided in the compost mix by a windrow turner which could be self-powered/propelled or by a farm tractor. As a general rule, the height of the windrows should be less than 2.5 m (8 ft) or width less than 3.6 m (12 ft) otherwise could lead to reduction in air flow in the windrow. Typically, for dense materials the height of the windrows should be 90 cm high to 360 cm high for light, voluminous materials such as leaves. The width may vary from 300 to 600 cm. Thorough turning ensures uniform mixing of the feedstock materials and overheating or accumulation of high moisture materials which could slow the waste degradation process. Generally, as the composting proceeds due to turning the moisture content decreases which could be prevented by provision of adequate space between windrows or water can be added. A

Figure 9. A hammer mill

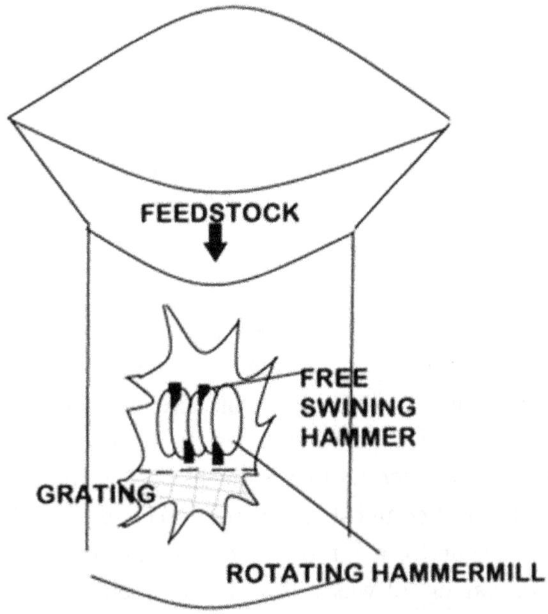

Figure 10. A rotating drum

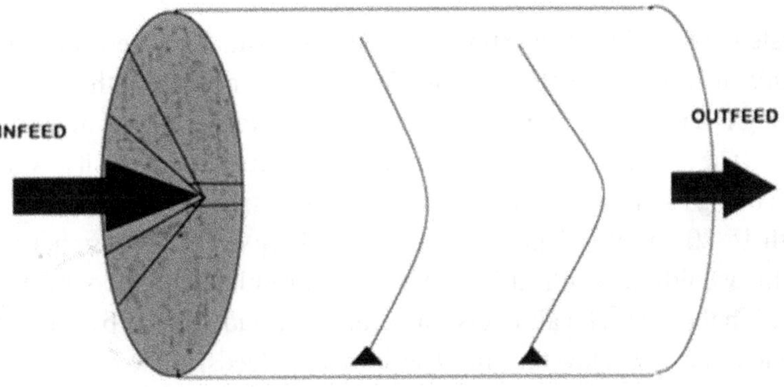

covering semi-permeable layer could lower down evaporation rate during less rain in a windrow. To shed the rainwater windrow should be peaked up in a triangular pattern. Turning of waste is an important component and when done at regular intervals in the initial 2-3 weeks of composting helps to keep the temperatures above 55°C for 10-15 days (Makan, Assobhei, & Mountadar, 2013). This kills most pathogens and weed seeds along with increase in oxygen level which would stabilize after 1-3 hrs. This can be done by one of following equipment systems:

1. tractor-loader
2. tractor + windrow turner
3. tractor + spreader +tractor-loader

2. Enclosed Aerated Windrow

An enclosed aerated windrow is basically a blend of windrow and an aerated static pile. In this the composting process accelerates and materials are physically broken down by forced air circulation with controlled oxygen and temperatures levels along with pile agitation. A building comprising of a hall floor with aerated trenches layered by grates with biofilter installed to transfer air out of the building is used as a facility for this purpose. Generally, initial processing in rotary drums is followed by aerated windrow (with a retention time of three-day in the drums) or by in-vessel containers (FAO, 2003).

3. Passive Windrow

This composting process is a very low-cost, less labor and capital requiring technique with large land requirement. Typically, feedstock material is collected and quickly piled into windrows consisting of more or less triangular elongated pile of material. It can take six months to two years for this process to complete. As illustrated in Figure 11, typically in a windrow length could vary according to quantity of feedstock width about 3 meters (10 feet) and 1.5 meters (5 feet) height. As the pile heats up fresh air enters the pile. A finished layer of compost can be used to cover the pile to reduce odour problems, prevent moisture loss and produce uniform compost. The dimensions for large passive windrows could be of any length depending on amount of waste as wide as 7 meters (24 feet) and as high as 4 meters (12 feet). In such a large heap turning is necessary to supply oxygen to the centre part otherwise it will become anaerobic and will exude an unpleasant odour (FAO, 2003).

4. Aerated Static Piles

In early 1970s aerated static pile composting method was primarily used for composting municipal sewage sludge. In this method composting is done in piles or windrows with mechanical aeration. This method of composting is a low labor approach with an average technology which sometimes results in a non-uniform product. The piling of waste is done according to material porosity, weather conditions and the equipment reach being used for building the pile with an initial height of about 150-250 cm. The pile may be layered with 15-20 cm of finished compost or bulking agent as extra height helps in retaining heat (Figure 12). This would prevent heat loss, drying and insulate the pile with check on the possible odours produced within the pile. The air ducts for aeration provide the air by blowing or drawing air in windrow or pile is and is located below the pile. The amount of feedstock to be composted is an important decisive factor for optimum sizes of pile along with the air flow efficiency and the handling equipment. The aeration pipe runs along the length below the ridge of the pile. It consists of comparatively simple aeration systems with fans, electrical motors, ducting with sensors and alarms. To improve pile porosity, control temperatures and enhance the airflow wood chips can be used as bulking agent. The constant movement of air, timing and duration are important. Aeration could also be mechanical provided near the end of the active compost period.

5. In-Vessel Systems

In-vessel compost systems involve mixing of compost mechanically. These are high rate, environmentally controlled aeration systems. In comparison to windrow method this is a low labor approach producing a uniform product and processing large quantities of waste in less space. It is designed in a way to maintain optimal conditions for waste to be composted in silos, channels or drums with properly regulated aeration system. The aerating machines built into concrete floors provide air flow from ducts by steady agitation with use of concrete bays and/or fans. In addition, it can accommodate virtually any type of organic waste with production of very little odour and minimal leachate (FAO, 2003). Three main categories of in- vessel system are as follows:

- **Rotating Drum:** The popularity of rotary drums is mainly because they are multipurpose when used. They could be used to blend, screen and reduce size without shredding. In a rotating drum mounted on large bearings having a daily capacity of about 60 tones being about 40 m in length and 4.5 m in diameter, aeration is provided by feeding air into the drum through a bull gear aeration. During exit the material is biofiltered and is passed through a screen to remove the contaminants. When considering high oxygen demanding materials this is a speedy and efficient waste degradation process. Sometimes the waste decomposition process is followed by a second phase of composting generally wind-rows or aerated static piles and the rotating drum merely acts as a mixing device.
- **Horizontal (Rectangular/Cylindrical) or Vertical Silos:** In vertical silos passive aeration composts the municipal organics without any forced aeration. To enable airflow through the material it is placed in vertical, wire-mesh "cages". The cages are few feet in width but can be long and tall (e.g. 3.8 m to 4.4 m high). Everyday an auger is used to remove the composted residue from the bottom of the silo and a mixture of fresh raw materials is added to top. Air is blown from the lower part of the silo throughout the composting materials. Exhaust air can be collected for odour treatment at the top of the silo. Typically, in 14 days this process is completed and so daily removal and replacement of one-fourteenth of the total silo volume has to be done.
- **Channels:** These comprise of a series of channels provided with a turner clambered on wheels. At the top of the channel tracks are placed on the turners which move and aids as a mixing device. For regulating this forced aeration being continuous or intermittent a perforated bottom is present in channel.

EM: Inoculation Based Composting Process

EM stands for Effective Microorganisms not being specific type of microorganisms. It consists of mixed liquid culture solution of aerobic and anaerobic micro-organisms being common and food-grade like photosynthetic bacteria, actinomycetes, lactobacillus, yeast and streptomyces etc. The strains of the micro-organisms make about 5 families, 10 genera and 80 different species and are easily available in the microbe banks or could be isolated from the environment (Higa, 1991). Being mutually compatible with one another they can coexist in liquid culture for extended periods. Increase in microbial diversity of soil could improve soil quality, increase growth and enhance the yield with quality of crops (Higa & Parr, 1994; Kishore, 2000). As investigated in various studies, the beneficial aspects of integrated recycling of urban organic waste with EM technology inoculation causes an increase in the polycondensed structures and the presence of more stable organic matter in the mature compost, which may be good indicators of humification processes of MSW composting with inoculation (Wididana & Higa, 1999). Inoculation may enhance the process of composting with increased degree of humification especially in the first thermophilic stage.

Figure 11. Passive windrow method

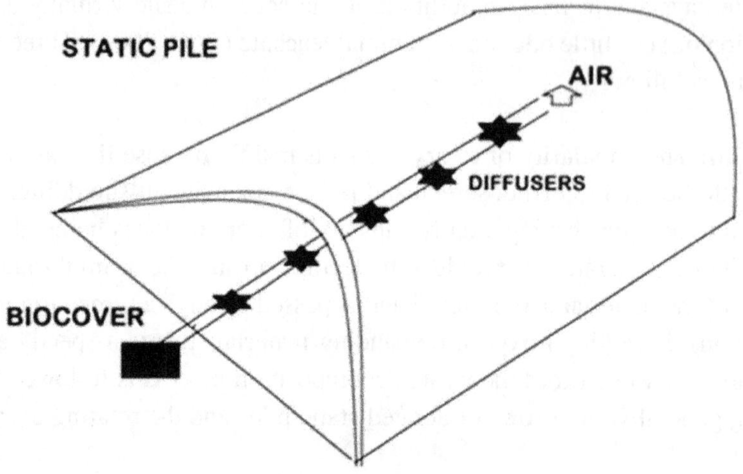

Figure 12. Aerated static pile layout

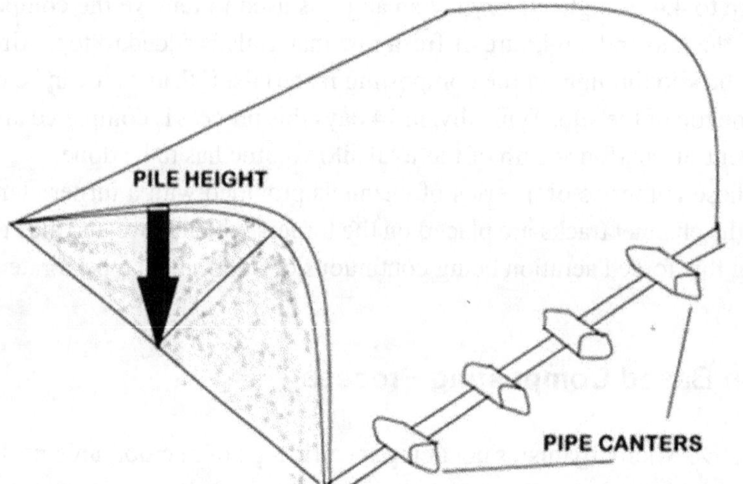

Post Processing

Post processing is generally done to polish the compost product though being optional is important for good rating for end-use specifications or market requirements. To ensure complete stabilization waste is examined in post processing. Further testing for chemical or pathogenic impurities and to ascertain nutrient levels, removal of undesired material, size sorting, screened, size reduction, blending with other materials, stored and bagged is done. For ascertaining high quality compost without any unwanted large particles and aesthetically pleasing; sorting, screening and removal operations can be done. A suitable storage ensures the quakiness of the compost product. General storage problems include saturation of compost due to defective drainage controls. Excessive wet compost becomes difficult to handle, smelly and bulky. The storage area should be designed to hold about 25% of total volume of product.

Also post processing ensures that compost maturity or curing by proper testing. This could be done by assessing parameters that indicate the compost maturity level or seed germination tests. Laboratory or experimental analyses of important compost parameters such as total nitrogen, oxygen consumption and C: N ratio could be done. After determining the nutrient levels and contaminants results can be unified into compost labels. This will ensure the acceptable range of composts nutrient for the receiving end according to their specific needs. Finally bagging of compost can be before its market distribution. Bagging should be proceeded only if is secured by buying party and the cost of bagging can be adjusted with expected revenues. So, bagging facilitates like transporting, labelling and marketing of compost turn out to be labour intensive as well as costly.

Benefits of Process of Composting

- It increases overall waste diversion from final disposal, especially since as much as 80% of the waste stream in low- and middle-income countries is compostable and enhances recycling and incineration operations by removing organic matter from the waste stream.
- It produces a valuable soil amendment as it contains a full spectrum of essential plant nutrients, helps to bind clusters of soil particles called aggregates, which provide good soil structure and enhances the effectiveness of fertilizer application along with water holding capacity.
- Integral to sustainable agriculture and promotes environment friendly practices such as waste stabilization, solving waste disposal problem and the reduction in methane gas generation at sanitary landfills.
- It is easier for execution at different levels, from small, medium to large-scale centralized facilities with integration of present informal sectors involved in waste management i.e. collection, separation and recycling of wastes.
- It includes involvement of less capital and operating costs with provision of an excellent opportunity to improve overall waste management plan.

CONCLUSION

In developing countries like India, the feasible and productive municipal solid waste management treatment and disposal options have been constraint by finances, technology and qualified personnel. All these issues have narrowed down composting as only option that best fits within the available resources in developing countries. A major feature that renders composting as a suitable technology is its adaptability to a broad range of situations. This process has many advantages like lower equipment and operating costs, environment friendly and useful product production. For successful composting the major prerequisites include understanding and implementation of the basic principles to the process. This knowledge about the process enables a rational evaluation of compost process and utilization of these technologies to broader aspects to resolve the waste related problems.

REFERENCES

Annepu, R. K. (2012, January 10). Report on Sustainable Solid Waste Management in India- Open Dumping of Solid Waste in India. Retrieved from http://www.seas.columbia.edu/earth/wtert/sofos/Sustainable%20 Solid%20Waste%20 Management %20in%20India_Final.pdf

Asnani, P. U. (2006). *Solid waste Management: India Infrastructure Report, 3iNetwork*. New Delhi: Oxford University Press.

Balasubramanian, M., & Dhulasi Birundha, V. (2012). An Economic Analysis of Solid Waste Management in Madurai District, Tamil Nadu. *Applied Journal of Hygiene, 1*(1).

Census of India. (2011). [Data file] Retrieved from www.censusindia.gov.in/2011census/PCA/PCA_Highlights/pca_highlights_file/India/Chapter-1.pdf

City Mayors Statistics. (2007). The Largest Cities in the World by Land Area, Population and Density. [Data file]. Retrieved from www.citymayors.com/statistics/largest-cities-density-125.html

Department of Economic Affairs, Ministry of Finance, Government of India. (2009, November). Position Paper on the Solid Waste Management Sector in India. *Public Private Parternships in India*. Retrieved from http://www.seas.columbia.edu/earth/wtert/sofos/Sustainable%20Solid%20Waste%20Management%20in%20India_Final.pdf

FAO. (2003). Report on State of the World's Forests, Rome. Retrieved from http://www.fao.org/docrep/005/ Y7581E/y7581e04.htm

Fuchs, J. G. (2010). Interactions Between Beneficial and Harmful. Microorganisms: From the Composting Process to Compost Application. In Insam et al... (Eds.), *Microbes at Work. From waste to resources* (pp. 213–230). Heidelberg: Springer-Verlag. doi:10.1007/978-3-642-04043-6_11

Gray, K. R., Sherman, K., & Biddlestone, A. J. (1971). A review of composting Part 2- The practical process. *Process Biochemistry*, *6*(10), 22–28.

Gupta, S., Mohan, K., Prasad, R., Gupta, S., & Kansal, A. (1998). Solid Waste Management in India, Options and Opportunity. *Resources, Conservation and Recycling*, *24*(2), 115–137. doi:10.1016/S0921-3449(98)00033-0

Hellmann, B., Zelles, L., Palojarvi, A., & Bai, Q. (1997). Emission of climate-relevant trace gases and succession of microbial communities during open-window composting. *Applied and Environmental Microbiology*, *63*, 1011–1018. PMID:16535535

Higa, T. (1991). Effective Microorganisms: A biotechnology for mankind. In J.F. Parr, S.B. Hornick, & C.E. Whitman (Eds.), *Proceedings of the First International Conference on Kyusei Nature Farming* (pp.8-14). U.S. Department of Agriculture, Washington, DC.

Higa, T., & Parr, J. F. (1994). *Beneficial and Effective Microorganisms for a Sustainable Agriculture and Environment* (p. 16). Atami, Japan: International Nature Farming Research Center.

Hu, S., & Van Bruggen, A. H. C. (1997). Microbial dynamics associated with multiphasic decomposition of 14C-labeled cellulose in soil. *Microbial Ecology*, *33*(2), 134–143. doi:10.1007002489900015 PMID:9052647

Kant, R. (2011). *Report on Sustainable Solid Waste management in India*. Ramky Enviro Engineers Ltd. Retrieved from ramkyenviroengineers.com/.../enviro/enviro%20engeneering%20brochure_latest.pdf

Kishore, S. (2000). Response of Effective Micro-organisms (EM) and other Nutrients to Rice and Wheat under Field conditions at Khumaltar, Nepal. *World Journal of Emergency Medicine*, *1*(1), 40–44.

Klamer, M., & Baath, E. (1998). Microbial community dynamics during composting of straw material studied using phospholipid fatty acid analysis. *FEMS Microbiology Ecology*, *27*, 9–20. doi:10.1111/j.1574-6941.1998.tb00521.x

Kumar, S., & Chakrabarti, T. (2010). Effective Municipal Solid Waste Management in India, Waste Management. InTech. Retrieved from http://www.intechopen.com/books/waste-management/effective-municipal-solid-waste-management-in-india

Makan, A., Assobhei, O., & Mountadar, M. (2013). Effect of initial moisture content on the in-vessel composting under air pressure of organic fraction of municipal solid waste in Morocco. *Iranian Journal of Environmental Health Sciences & Engineering*, *10*(1), 3. doi:10.1186/1735-2746-10-3 PMID:23369502

MoUD. (2000). Annual Report on Municipal Solid Waste Management on a Regional Basis, Government of India [Data file]. Retrieved from www.asean.org/uploads/archive/files/UNEP.pdf

Narihiro, T., Takebayashi, S., & Hiraishi, A. (2004). Activity and phylogenetic composition of proteolytic bacteria in mesophilic fedbatch garbage composters. *Microbes and Environments*, *19*(4), 292–300. doi:10.1264/jsme2.19.292

Nunan, F. S. L. (2000). Urban Organic Waste Markets: Responding to Change in Hubli-Dharwad, India. *Habitat International*, *24*(3), 347–360. doi:10.1016/S0197-3975(00)00002-3

Ryckeboer, J., Mergaert, J., Coosemans, J., Deprins, K., & Swings, J. (2003a). Microbiological aspects of bio waste during composting in a monitored compost bin. *Journal of Applied Microbiology*, *94*(1), 127–137. doi:10.1046/j.1365-2672.2003.01800.x PMID:12492933

Ryckeboer, J., Mergaert, J., Vaes, K., Klammer, S., De Clerq, D., Coosemans, J., ... Swings, J. (2003b). Survey of bacteria and fungi occurring during composting and self-heating processes. *Annals of Microbiology*, *53*(4), 349–410.

Rynk, R. (Ed.). (1992). *On-Farm Composting Handbook*. Ithaca, NY: Northeast Regional Agricultural Engineering Service Cooperative Extension.

Sharholy, M., Ahmad, K., Vaishya, R. C., & Gupta, R. D. (2007). Municipal Solid Waste Characteristics and Management in Allahabad, India. *Journal of Waste Management*, *27*(4), 490–496. doi:10.1016/j.wasman.2006.03.001 PMID:16766176

Singhal, S. (2000). Solid Waste Management in India, Status and Future direction, TERI. *Information Monitor on Environ. Sci*, *6*(1), 1–4.

Sinha, C. (1997). Open burning of municipal solid waste: State level Analysis. *TERI Monitoring on Environmental Science*, *2*(2), 71–75.

Tuomela, M., Vikman, M., Hatakka, A., & Itavaara, M. (2000). Biodegradation of lignin in a compost environment: A review. *Bioresource Technology*, *72*(2), 169–183. doi:10.1016/S0960-8524(99)00104-2

UNEP. (2004). Report on State of Waste Management in South East Asia, Types of Wastes - Sources and Composition [Data file]. Retrieved from www.asean.org/uploads/archive/files/UNEP.pdf

U.S. Environment Protection Agency. (1994). *Composting Yard Trimmings and Municipal Solid Waste*. Solid Waste and Emergency Response (Report No. EPA530-R- 94-003).

Ward, M. M. (2002). Composting: A beginner's guide. The Robert A. Macoskey Center for Sustainable Systems Education and Research, Slippery Rock University, PA.

Wididana, G. N., & Higa, T. (1999). Integrated Recycle System of Organic Urban Waste with EM-Technology. *Presented in 6th International Conference on Kyusei Nature Farming*, South Africa.

Chapter 7
Valorization of Tannery Solid Waste Materials Using Microbial Techniques:
Microbes in Tannery Solid Waste Management

Thazeem Basheer
Bharathiar University, India

Mridul Umesh
Bharathiar University, India

ABSTRACT

This chapter describes how environmental pollution is the major problem associated with rapid industrialization, due to which the quantity of solid waste generated from the industries is increasing substantially. Solid wastes generated in leather industries are briefly classified into pre-tanned and post-tanned wastes. Although landfills, composting, anaerobic digestion and thermal incineration are available for disposal of these solid wastes, they do again pose severe environmental and financial burdens to the tanners. Microbe-mediated tannery solid waste management points to the recovery of value-added products from these wastes. Microbial valorization of tannery solid wastes for the production of saleable products would be a convincing, challenging and eco-friendly opening for its utilization when compared to that of chemical and thermal hydrolysis. Exploitation of traditional chemicals could be reduced and innovative products could be recovered, enabling sustainable solid waste management. This would ultimately alleviate the solid waste disposal problems and financial crisis faced by the tanners.

DOI: 10.4018/978-1-5225-3540-9.ch007

INTRODUCTION

Globally, leather is one of the most widely traded products. Leather products are the greatest export earners. The drastic transformation of leather industries over years from a mere supplier of raw and tanned leather into a leading exporter of finished leather goods is due to the demand driven by several other industries for fashion and commercial purposes. Leather industries mark a prominent place in India's economy. Tanning process is highly complicated and could be briefly classified into beam house and tanyard operations that involve various steps such as curing, soaking, dehairing, fleshing, deliming, pickling, tanning, wringing, splitting, shaving, re-tanning, fat-liquoring, buffing and finishing. Raw hides undergo these step-wise processes to transform into final leather product (Kanagaraj et al., 2006).

Leather processing industries are also well known for their disagreeable odor emanating from the effluents and solid wastes, causing severe environmental pollution. The drastic growth of tanneries is coupled with issues in discharging and dumping of tannery effluents and solid wastes respectively, enriched with organic, inorganic and chemical load. Protein and fat are the major components of these wastes. Due to the lack of proper waste management facilities and inferior infrastructure technologies, tanneries aggravate pollution problems day by day (Ahamed & Kashif, 2014).

Common effluent treatment plants and effluent treatment plants regularly monitor and help in safe discharge of treated tannery effluents into the disposal sites. Major problem of the industry that dominates wastewater/effluent treatment is the unhealthy disposal of solid wastes emanating from the tanning processes. These wastes remain unutilized or underutilized. They are found stagnated at a particular area within the industry, undergoing decomposition with a familiar putrefying smell of sulfide and ammonia gases. Available methods to dispose these wastes create secondary environmental pollution in terms of soil, water and air contamination. Eco-friendly strategies to minimize the tannery solid wastes and maximize their recycling are scarce. With this perspective, the objective of this chapter lies at featuring the role of microbes and related techniques in tannery solid waste management. Products acquired from microbial waste management and their possible utilization as a feed for several other industries has also being discussed (Thazeem et al., 2017b).

BACKGROUND

Leather industries aim at the transformation of putrescible (liable to decompose) animal raw hides and skins into a stable non-putrescible material (leather), ultimately resulting in the generation of solid as well as liquid wastes. An average of 250 kg non-tanned waste and 50,000 kg waste water is produced during one metric ton of raw hide processing (Sundar et al., 2011). Tanneries in India alone produce 150,000 tonnes of unutilized solid wastes out of the whole tanning process, which result in severe environmental threat (Rai et al., 2010). Many biological methods have been implemented to reduce environmental pollution emerging through tannery effluents (Bhaskar et al., 2007). Although numerous cleaner technologies and effluent treatment plants support in the alleviation of liquid waste generation, there remains inadequacy in the healthy disposal of solid waste among the tanners. One of the major issues faced by the leather industry lies at the improper disposal of solid waste generated during leather processing, with maximum waste emerging at the pre-tanning operations. From 1000 kg of raw hide, nearly 850 kg of solid wastes are generated which infers that only 150 kg of raw material is converted into finished leather. Tannery solid wastes consist of: fleshing waste (56-60%); skin trimmings (5-7%); chrome shaving, splits and buffing dust (35-40%) and hair (2-5%) that are generated predominantly during beam house operations (80%) followed by tanning (19%) and finishing (1%) processes (Kanagaraj et al., 2006). Briefly, tannery solid waste could be classified into tanned and untanned collagenous wastes (Figure 1).

Untanned wastes consist of raw hide trimmings, green fleshings, limed fleshings, limed trimmings and hair waste.

- **Trimmings:** Collagen rich pieces of neck, tail, head and belly regions of hides that are cut off during the earlier stages of tanning to provide a perfect configuration to the final leather (6% wet weight of raw hides). If not removed, it may harm the efficacy of the tanning operations.
- **Green Fleshings:** Fleshing operation of raw hides before they undergo washing and soaking result in the yield of green fleshings (10% wet weight of raw hides), that are rendered with other non-edible offal (meat and bone) as they do not undergo any chemical processing.
- **Limed Fleshings:** Fleshing operation of limed hides either manually or through a fleshing machine results in the yield of proteinaceous limed fleshings that is made up of cutaneous and subcutaneous muscle tissues (35% wet weight of raw hides).
- **Hair Waste:** During liming of raw hides, calcium hydroxide and sodium sulfide are employed along with large mechanical rotators to remove hair and wool from hides. This traditional process could be done manually via painting the lime solution over the flesh side of skins or through dipping the hides into the lime solution, enabling generation of hair and wool waste.
- **Limed Trimmings:** Are ragged pieces removed from limed hides after fleshing process in order to make them into uniform layers without any torn surfaces (2-7% wet weight of raw hides).

Chrome trimmings, chrome splits, chrome shavings, crust leather trimmings, buffing dust and finished leather trimmings are categorized under tanned wastes.

- **Chrome Trimmings:** Pleats developed at the edges of hide during chrome tanning are trimmed out during removal of excess water, before splitting process and shaving. If not trimmed, they may undesirably interfere in the further processes (5.5-6% wet weight of hides).

Figure 1. Solid waste generation from leather industries

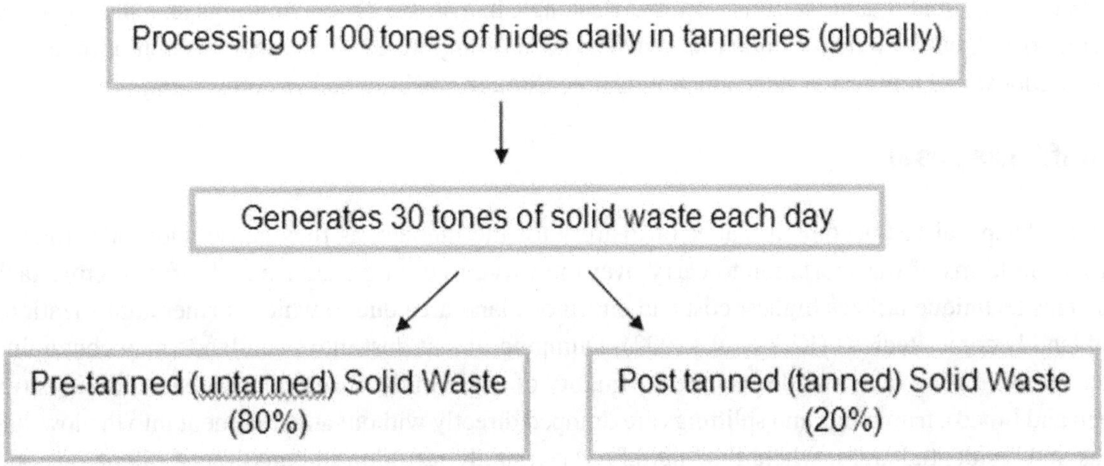

- **Chrome Splits:** 4-6% of the total tannery solid waste generated during post tanning operations consist of splits that are very thin substances obtained during chrome splitting process.
- **Chrome Shavings:** Fibrous waste obtained during shaving process of tanned leather constitutes chrome shavings which are around 1-2% of raw hide weight. They are usually found in shredded nature.
- Crust leather trimmings, buffing dust and finished leather trimmings are the lightweight unenviable leather pieces obtained during the final tanning operations to contribute aesthetic value to the finally finished leather (Fathima et al., 2011).

All the wastes generated from the whole tanning process are collagenous in nature with high protein, moisture and lipid content. Disposal of these wastes in an unhealthy manner indirectly correlates to the point that bulk amount of nutritious matter is disposed off without its value-recognition and proper utilization.

POSSIBLE TANNERY SOLID WASTE DISPOSAL METHODS

Composting

Composting is considered as an aerobic, cost-effective biotechnological method for solid waste management, through which microbes convert waste material into a more stabilized form. Carbonaceous and nitrogenous solid waste compounds are subsequently converted into a less complex organic matter through successive activities of different microbial populations during composting, allowing wastes to be disposed of after reduction in their volume and mass. In case of tannery solid waste, only negligible proportions of the nitrogen and carbon were exploited by microbes during bacterial composting, leaving the complex molecules unutilized. It took around 70-90 days for composting of tannery waste into manure, with high emission of greenhouse gases. Vermin composting of tannery solid waste along with cow dung and leaf litters have already been performed using *Eisenia foetida* (Sekaran et al., 2007). Composting of tannery solid waste with petrochemical sludge was done using *Eudrilus euginea* (Banu et al., 2005). Biotransformation of fleshings into organic compost using *Selenomonas ruminantium* HM000123, followed by mixing with leaf litter and cow dung was accomplished by Ravindran and Sekaran (2010) to generate composted organic fertilizer. Limed fleshings, hair waste, limed splittings and shavings could be composted, but with major limitations such as need of high residence time, emission of toxic gases from compost and removal of huge amount of lime through washing before composting.

Landfill Disposal

Landfill Disposal has become a matter of distress for the tanners, as they cause additional financial burdens in terms of transportation to carry over the waste to a long distance and infrastructure facilities. This technique utilizes highest cost and enormous land area due to which further authorization of new land becomes tedious (Kirk et al., 2002). Dumping of salt dust into open lands may contaminate groundwater during rains and deteriorate the quality of water in the surrounding areas. Fleshings (raw, green and limed), trimmings and splittings are dumped directly without any treatment into the low-lying areas and residential areas, where they putrefy very rapidly resulting in emission of obnoxious odors (landfill gases), ultimately resulting in air pollution and greenhouse gases emission. This may also result in epidemics, fires and bird menace around the landfill areas. Thus, landfill disposal method is discouraged by environmental scientists.

Anaerobic Digestion (Bio-Methanization)

Anaerobic digestion of tannery solid waste is a bio-chemical conversion process that consists of breakdown of biodegradable organic material by microorganisms under anaerobic conditions. This involves anaerobic, facultative, acetogenic and methanogenic bacteria. Methane rich biogas and sludge (effluent) are the products of anaerobic digestion. The methane can be used to produce heat and electricity. The residues could be used as organic manure after stabilization. Careful monitoring and need for technical expertise are the negative aspects of anaerobic digesters.

Thermal Incineration

Incineration is a thermo-chemical conversion method that involves burning of solid wastes at temperatures around 800°C and above, with the presence of oxygen, resulting in heat liberation and ash formation. Attempts were made for the first time in a laboratory scale at the Institute of Inorganic Chemistry and Technology of the Cracow University of Technology in the 1990s. Major advantage with incinerating the solid waste is that the total volume of the combustibles could be drastically reduced (80-95%).

Gasification and Pyrolysis techniques are the same except that the former needs negligible amount of oxygen whereas the latter allows none. Both are costlier methods and the drawback here is the issues such as the release of halogenated organic compounds, poly aromatic hydrocarbons etc. into the environment (European IPPC report, 2001).

Leather industries are confronted with two better options, either to minimize the quantity of solid waste generation or to maximize the quantity of value added products out of the waste (Alexandar et al., 1992). As numerous legal arrangements all over the world are enforcing the tanneries to reuse the solid wastes generated during leather processing into useful byproducts, there remains a great demand for techniques that could be implemented for the recovery of reusable byproducts instead of the above-mentioned disposal methods.

ROLE OF MICROBES IN TANNERY SOLID WASTE MANAGEMENT

Bioremediation is one of the recent technologies utilized for recycling industrial wastes. The role of microorganisms in recycling and recovery of value added products from wastes is gaining significance due to their well-known versatile nature. Microbial fermentation processes are familiar from ancient times and it has been a prominent practice of using fermented products and food. Biotechnological and microbiological techniques are gaining importance and popularity nowadays due to their feasibility and eco-friendly nature. Fermentation process could adequately replace thermal and chemical treatments, thus reducing the need of high capital and energy sources. Fermentation end products are potentially usable and consumable. Microorganisms may undergo various biosynthetic pathways and metabolisms depending upon the substrate. Enormous microbes have been employed by the researchers to manage tannery solid waste and the pollution caused by them in and around the tanneries (Figure 2). Microorgan-

Figure 2. Role of different microorganisms in tannery solid waste management

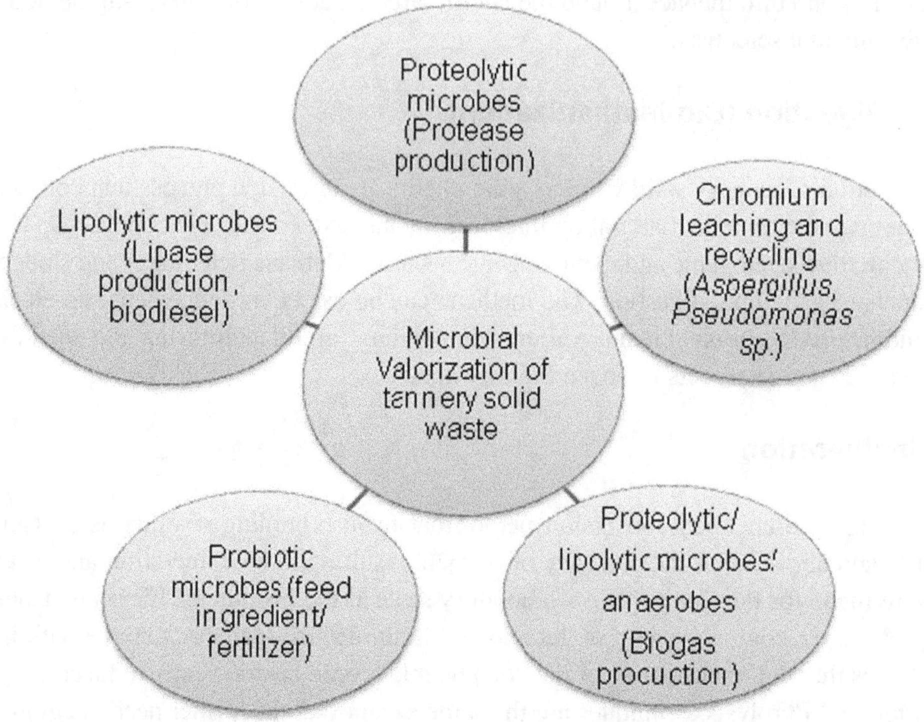

isms, their sources, fermentation products and application of the fermented residue in different valuable streams is as follows:

Proteolytic Microbes in Tannery Solid Waste Management

Microbes are extraordinary source of proteolytic enzymes. Among microbial proteases, bacterial proteases are relatively used in a wider scale than fungal and animal proteases (Kumar et al., 2008b). Animal proteases are less preferred due to their heterogeneous nature and need for purification procedures. About 50% of industrial enzymes constitute of proteases that act as industrial process aids worldwide. Proteases may be intracellular or extracellular products. Mostly, extracellular proteases are of commercial importance which is employed in assisting protein degradation. Among proteases, acid and alkaline proteases stand out for their exceptional usage in harsh industrial environments and bioremediation purposes. They also find application mainly at laundry, textile, food and pharmaceutical industries. The quantity and activity of the proteases depend upon the fermentation substrates and is directly proportional to the amount of protein available for hydrolysis. As limed animal fleshings are predominant among tannery solid waste and highly proteinaceous in nature, several works on protease production utilizing fleshings as a substrate has been carried out my many researchers. A fungi named *Paecilomuces lilacinus* was able to produce alkaline protease (heat stable) enzyme under solid state fermentation, solubilizing fleshings which was assessed by Kjeldahl analysis (Chakraborty & Sarkar 1998).

A study involving strains of *Bacillus* sp. isolated from Morocco compost soil was performed to identify collagenase producers that could completely digest tannery solid waste (Zerdani et al., 2004). Among the eight strains isolated, *Bacillus subtilis* and *Bacillus licheniformis* worked more potently in digesting protein of tannery solid waste and their growth were exceptional in the substrate when compared to reference strain *Bacillus subtilis* ATCC 6633. This beneficiary attempt of isolation of collagenase producers and hydrolysis of tannery solid waste into protein hydrolysates paved way for their subsequent usage as bio-fertilizer or nutrient source in medium for microbial growth. Research on screening for an efficient microorganism from a consortium for digestion of tannery fleshings as their biological treatment was carried out successfully by Ghosh et al., (2004).

Enzymatic treatment of shavings, trimmings and splittings of leather processing was performed using *Bacillus subtilis* isolated from tannery waste (Aftab et al., 2006). Under submerged fermentation, the organism produced extracellular alkaline protease enzyme, resulting in subsequent hydrolysis of the leather waste. Degradation was measured by release of amino acid hydroxyproline. The process was quite economical as it does not required energy. Leather waste that consisted of 80% of protein was hydrolyzed, for their consequent use as poultry feed ingredient/feed additive or as a fertilizer after proper purification.

Kumar et al., (2008b) were successful in hydrolyzing animal fleshings microbiologically using acid protease produced by *Synergistes* sp isolated from an anaerobic ecosystem used for the treatment of tannery solid waste. This method was a promising alternative to usual chemical and thermal hydrolysis of animal fleshings. Minced limed fleshings were delimed with 25% v/v ammonia solution for 3-4 hours to remove the adhered calcium salts. The microbe isolated from anaerobic digester was capable hydrolyzing gelatin and was identified via 16SrDNA sequencing to be *Synergistes jonesii* (85.6% similarity). Maximum enzyme production of 350 - 420 U/ml at pH-6.0 was obtained along with hydrolysis of animal fleshings substrate which was confirmed by scanning electron microscopy, gas chromatography-mass spectrometry and nuclear magnetic resonance techniques. Enzyme purified with 80% ammonium sulphate had a molecular weight of 60kDa. Particle size of the fleshings greatly influenced the enzyme production. Smaller the particle size, larger the enzyme production. It was a remarkable attempt for the implementation of biotechnological methods in tannery solid waste management, yielding mesophilic acid protease at stationary growth phase of the microbe.

Alkaline proteases, classified under serine proteases were primarily used as detergent additives. They also play a vital role in tanneries, food and feed industries, chemical, silver and medical industries. Utilization of animal fleshings for production of industrially important alkaline protease enzyme using *Pseudomonas aeruginosa* proved the tremendous potential of fleshings for industrial level production of the enzyme (Kumar et al., 2008c). *Pseudomonas aeruginosa,* isolated from tannery waste water was capable of secreting extracellular alkaline protease enzyme with maximum activity of 1160-1175 U/ml at 20 hours of incubation during stationary growth phase. Animal fleshings medium acted as a sole carbon and nitrogen source for the organism's biomass production and enzyme secretion.

Kumar et al., (2008d) studied the effect of nonionic surfactant Tween-80 on the enhancement of anaerobic fermentation and tannery fleshings hydrolysis to obtain maximum activity of protease and deaminase enzymes. Under anaerobic fermentation, proteins present in animal fleshings were converted into smaller peptides and amino acids, followed by the formation of volatile fatty acids, ammonia and carbon dioxide (McInerney 1988). 0.15% of Tween 80 served as an emulsifier to facilitate enhanced production of fermentative enzymes with maximum enzyme activities of 338-360 U/ml after 72 hours (protease), 23-31 U/m at 24 hours (lipase) and 187-206 U/ml after 72 hours (deaminase).

Acidogenic fermentation of animal fleshings was performed by inoculating anaerobic sludge in to a reactor consisting of smaller pieces of fleshings (Kumar et al., 2008a). Anaerobic sludge primarily constitutes of acidogenic bacteria *Clostridium* sp. Aliphatic amino acids were released initially which evidenced degradation of globular proteins, followed by release of aromatic amino acids.

Albeit various substrates such as mustered oil cake, rice bran, wheat bran, green gram husk, *Imperata cylindrical* grass, potato peels, used tea leaves, banana leaves and lentil husk have been exploited for alkaline protease production using *Bacillus* sp, alkaline protease production through solid state fermentation from animal fleshing as a substrate, using *Bacillus cereus* 1173900 isolated from tannery wastewater added note to the already explored phenomena that Bacillus genus are great producers of different proteases (Ravindran et al., 2011). Crude extract of the fermentation exhibited increased activity of 12,310 U/g at 60 hours resulting in the yield of extracellular alkaline metalloprotease enzyme (molecular weight-66 kDa) when animal fleshings delimed with 1.25% w/v of ammonium sulphate served as solid substrate.

Sumathi et al., (2012b) successfully reported the production of different proteases by microorganisms isolated from the fish *Labeo rohita* fed tannery fleshings feed for 15 days. The fish were fed with raw fleshings incorporated diet and sacrificed to isolate microbes from gut. Three bacterial strains that exhibited vibrant zone of hydrolysis of skim milk agar were further sequenced and identified to be *Bacillus megaterium* (ANFLR1), *Serratia marcescens* (NPLR1) and novel *Pontibacter* sp. (PROLR15) with 99%, 99% and 92% similarity respectively. Maximum protease activities under optimized conditions were 420 U/ml, 550 U/ml and 530 U/ml for enzyme produced by ANFLR1, NPLR1 and PROLR15 respectively. Alkaline metalloprotease production by novel *Pontibacter* sp. and aspartate protease production by *Serratia marcescens* exploiting tannery fleshings substrate remained unexplored in the study. Among the three, *Pontibacter* sp. PROLR15, a Gram-negative coccobacillus established its potential to degrade the fleshings efficiently, enabling its industrial scale use.

As proteinaceous tannery solid waste – animal fleshings are hydrolyzed through thermal and chemical processes that consume more energy and cost, an attempt for anaerobic digestion with acid cysteine protease production using *Selenomonas ruminantium* HM000123 was accomplished by Ravindran et al., (2012). Monomeric enzyme with molecular weight of around 53 kDa was obtained from the fleshings incorporated minimal medium optimized through response surface methodology that could act as a biocatalyst for various industrial processes.

Chrome shaving wastes from tanneries were hydrolyzed enzymatically using proteolytic fungus – *Aspergillus tamarii* under solid state fermentation conditions, producing alkaline protease enzyme. 6 units of the enzyme were required to completely hydrolyze 1 gram of chrome shaving waste (Kumar et al., 2012). After hydrolyzing, proteins were separated through filtration that could be used in animal feed and pharmaceutical industries. The settled chromium was air-dried to obtain a chromium cake which could be recycled and reused for tanning processes. Thus, the biological method of chromium recovery from waste is highly adaptable than that of chemical and thermal methods, with subsequent yield of protein hydrolysates.

Untanned solid waste – animal fleshings was explored as a substrate for alkaline protease production by various *Bacillus* sp. such as *Bacillus licheniformis, Bacillus coagulans, Bacillus polymxa* and *Bacillus amyloliquefaciens* isolated from soil samples (Ahmad & Ansari, 2013). Among the four, *Bacillus licheniformis* was the most efficient strain for fleshings hydrolysis and alkaline protease production, which could meet the needs of industry with elaboration of the study.

Probiotic Microbes in Tannery Solid Waste Management

Fermentation with probiotic microorganisms has grown adversely as an effective alternative to conventional acid ensilaging using strong acids. Organic acids obtained as the fermentation end products play a vital role in all emerging fields of research and industries. Among organic acids, lactic acid has a long history and it plays an exemplary role in food-based as well as non-food-based industries. It is primarily used as a preservative that could ensilage meat and meat products. Lactic acid fermentation favors valorization of renewable substrates with less cost and energy. It is one among environmentally safe and viable microbial techniques.

Bhaskar et al., (2007) worked on preparation of protein hydrolysates through bio-stabilization using lactic acid fermentation. Wet fleshings from goat and sheep skins were procured and delimed with 0.1% H_2O_2 (1:10 v/v), followed by 0.2 N HCl (1:10 w/v) washes. *Lactobacillus plantarum* B4496, *Lactobacillus lactis* B634, *Lactobacillus acidophilus* B4495 and *Pediococcus acidolactici* CFR2181 were assessed for their fermentation capacities. Fermentation mixture consisting of 19.5% v/w inoculum, 20% w/w sugar and 2% w/w common salt was held at $30 \pm 2°$ C for 60 days. Degree of hydrolysis was measured under optimized conditions followed by enzymatic hydrolysis using Protex-6L. Among the four probiotic microorganisms, *Pediococcus acidolactici* exhibited lowest pH (5.26 ± 0.20) with maximum conversion of sugars into lactic acid. Protein hydrolysates obtained after fermentation had a higher protein content of 12.4% when compared to that of unfermented fleshings (6.8%). Quality of protein is determined on the basis of essential amino acids profile. As the protein hydrolysates of tannery fleshings were rich in tyrosine, arginine and phenylalanine, they had a higher potential to be utilized as a livestock/aquaculture feed ingredient. Bio-conversion of animal fleshings into hydrolysates through lactic acid fermentation and protease digestion could alleviate solid waste disposal problems faced by the tanners.

A native lactic acid bacterium was isolated from tannery fleshings to check for its fermentation efficiency by Rai et al., (2009). 50 discrete colonies were isolated from wet limed fleshings, delimed fleshings and fleshings fermented under natural conditions through plating on Deman, Rogosa, and Sharpe (MRS) medium. Isolate CTF2 was characterized and identified to be *Enterococcus faecium* HAB01 through schemes of Bergey's manual of Systematic Bacteriology and 16S rRNA gene sequencing respectively. *Enterococcus faecium* HAB01 exhibited better antibacterial activity (strong inhibition against *Listeria monocytogenes* Scott A). The bacteriocin responsible for antibacterial activity was thermostable with molecular weight of ~5 kDa. The fermented liquor exhibited good antioxidant properties which may be due to the presence of protein peptides as the end products of fermentation. Fermented liquor could be subsequently dried and used in aquaculture and animal feed industries, cutting off feed production rates in terms of commercial antibiotics and antioxidants addition. Antibacterial and antioxidative property of the fermented fleshings hydrolysates obtained using *Enterococcus faecium* HAB01 paved way for the strong recycling of tannery solid waste using microbiological method. Rai et al., (2010) further optimized the fermentation conditions for better degree of protein hydrolysis of delimed fleshings using *Enterococcus faecium* HAB01. Minced delimed fleshings mixed with distilled water in the ratio of 1:1 was subjected to fermentation along with 17.5% w/w glucose, 12.5% v/w inoculum and 2% w/w sodium chloride for 96 hours under shaking conditions. Total antioxidant activity and DPPH radical scavenging activity of the fermented powder (spray dried) was proportional to the degree of protein hydrolysis as they all are inter-connected. Under optimized conditions, fermented fleshings powder exhibited relatively higher antioxidative property that may be contributed by the peptides and amino acids of the fleshings. Peptides or amino acids in the form of antioxidants are of great demand now in food industries due to their safety and efficacy (Hattori et al., 1998). Fermented protein powders with such antioxidant activities can be supplemented in animal feeds, as hydrolyzed proteins are well ingested and digested in animal systems than the intact ones and they could relieve or prevent oxidative stress in animals (Ziegler et al., 1998).

Sumathi and Sekaran, (2010) were successful in formulating diets for major Indian carp, *Labeo rohita* using raw, chemical treated and *Bacillus megaterium* LR1 fermented proteinaceous tannery fleshings as a sole protein source. Microbial digestion of solid waste is considered as an adaptable biotechnological method to breakdown complex proteins into simpler forms. *Bacillus megaterium* LR1 isolated from the gut of *Labeo rohita,* exhibited proteolytic activity due to which feed formulated with the fermented fleshings had higher protein content (36%) than the control feed (35%). Fish groups fed with fermented fleshings had greater growth performance and protein utilization values than that of raw, hydrogen peroxide treated and control feeds. This study proved the significance of microbial fermentation in terms of easier uptake of nutrients by fish due to the availability of protein hydrolysates and therefore proper assimilation and utilization of nutrients by the fish. This study also showed that the proteinaceous pre-tanned tannery fleshings could sufficiently replace fish meal in aqua feeds. *Bacillus megaterium* ANFLR1 isolated from tannery fleshings fed *Labeo rohita* gut was further inoculated for 48 hours in tannery fleshings broth, which was centrifuged and filtered to get a clear supernatant (Sumathi et al., 2012a). It was used to assess antimicrobial activity against Gram positive, Gram negative bacterial pathogens and fungal pathogenic strains. Maximum zone of inhibition was observed against *Salmonella typhii, Microsporum canis* and *Aspergillus fumigatus*. Researchers were excited to discover its antibacterial and antifungal property and further studies on the active metabolite will pave way for its use as an antibiotic as well as a probiotic.

A comparative study on *in vitro* antioxidant activity and antibacterial activity of hydrolyzed tannery fleshings proteins through fermentation and acid hydrolysis was performed by Balakrishnan et al., (2011). Acid hydrolysis consisted of mixing fleshings with 20% acid mixture of propionic acid: formic acid (1:1) and incubation for 9 days followed by spray drying. A native isolate of tannery fleshings named *Enterococcus faecium* NCIM5335 was used for fermentation which resulted in the yield of fermented hydrolysate powder. Fermented powder had higher antioxidant property than acid hydrolysed powder. It exhibited maximum antibacterial activity against *Salmonella typhii* FB231. This eminent antioxidant and antibacterial nature of fermented fleshings product was due to the bacteriocin produced by the lactic acid bacterium and the peptides formed during fermentation, enabling their usage in feed industries as relatively low-cost ingredients.

Acute and sub-acute toxicological studies were performed by Jini et al., (2016) to evaluate the safety of incorporating tannery fleshings products in feed industries. Acid hydrolyzed and fermented (using *Enterococcus faecium* MTCC 5691) delimed tannery fleshings were used separately for different feed formulations and diets were fed to weanling male wistar rats. No mortality and abnormal behaviors were observed and the animals were healthy. Acute toxicity study for 15 days and sub-acute study for 30 days showed that both acid-hydrolyzed and fermented tannery fleshings powder could be supplemented up to 15% in animal diets. Feed intake, organ weights, biochemical and histological parameters of rats fed test diets was comparable to the control groups and this work emphasized a novel form of protein based feed ingredient from tannery solid waste material that could be used in poultry feeds.

Thazeem et al., (2017b) worked on lactic acid fermentation of delimed bovine tannery fleshings using homo-fermentative *Lactobacillus plantarum* (NCDC685). Wet limed fleshings were delimed with 1.5% formic acid solution [1:10 (w/v)] containing 0.1% v/v hydrogen peroxide. Delimed fleshings were steam cooked and lactic acid fermentation was carried out using 5% lactose (w/w) and 10% tapioca flour (w/w) as carbon sources. Lactic acid production increased with increase in incubation time up to 96 hours of fermentation under shaking conditions. Maximum yield of 10.8 g/L of lactic acid was obtained with subsequent hydrolysis of the fleshings medium, which upon air-drying resulted in the formation of fermented tannery fleshings flour. Five iso-nitrogenous diets for *Labeo rohita* were prepared, where fermented tannery fleshings flour replaced fish meal by 25, 50, 75 and 100%. Superior growth parameters, food indices and carcass composition ($P < 0.05$) were observed in experimental fish group fed with feed formulated by replacing fish meal with 75% fermented tannery fleshings flour than other groups and control diets. Fish were healthy and feed-related mortality was not observed during the entire feeding period of 60 days. Lactic acid production for the first time valorizing proteinaceous untanned tannery solid waste material-bovine fleshings as a relatively low-cost substrate remained unique in this study. Such a novel probiotic fermented offal product could act as a successful aqua feed ingredient and its incorporation in feeds could curtail feed production costs as fish meal is expensive and by doing so, antimicrobial drug resistance among fish pathogens could be prevented as probiotic microorganisms are well documented for their beneficial metabolites such as organic acids, hydrogen peroxide and bacteriocins production in the fermentation medium as end products.

Lipase Producers in Tannery Solid Waste Management

Lipase enzymes are popularly known for their biocatalyst role in hydrolyzing fats and oils. Both solid state as well as submerged fermentation is suitable for production of lipases by bacteria, fungi and yeast cells. In a study conducted by Dayanandan et al., (2013), lipases were produced by solid state fermentation using wheat bran, gingili oil cake and gingili oil as substrates by *Aspergillus tamarii* MTCC 5152. Crude extract of the enzyme was concentrated by ultra-filtration. Concentrated lipase enzyme was checked on tannery fleshings for its fat solubility ability. Maximum fat solubility of around 92% was observed when 3.0% lipase enzyme (750 U/mL) was used at a reaction time of 6 hours. This indirectly showed that 92% of the fleshings fat could be dissolved by lipase of *Aspergillus tamarii* MTCC 5152. The hydrolyzed fat could be further used for biodiesel production. The residual protein fraction could be used in animal feed formulations.

Biodiesel, commonly considered as green fuel stands out for its biodegradable, non-toxic and renewable nature, enabling successful replacement of fossil fuels. Biodiesel is produced by trans-esterification of oils and fats. Tannery solid waste rich in animal fats could be utilized for biodiesel production as a sustainable way of recycling. Animal fleshings were utilized for solid state fermentation, through which the intact fats were fermented into fatty acids esters (long chain and short chain fatty acids) by the microorganism *Microbacterium* sps (Krishnan et al., 2016). The organism was involved in this pre-treatment method, which was followed by trans-esterification reactions. *Microbacterium* sps was isolated from tannery soak effluent and immobilized on acid treated silica. Animal fleshings were inoculated with *Microbacterium* immobilized silica sand for 3-4 days. Lipase produced by the organism helped in the hydrolysis of fleshings fats into smaller fatty acids esters, which was then followed by methanol treatment. These fatty acid methyl esters obtained from tannery fleshings could be used as a feed stock for biodiesel production. *Microbacterium* sps from soak liquor was able to produce maximum lipase enzyme after 36 hours of fermentation and this was evident through gas chromatography mass spectrometry analysis of the fermented substrate. After biodiesel production, the residue could be further fermented to get bio ethanol. The fermented broth may then be utilized for bio methane or bio hydrogen production. Thus, tannery solid waste materials could be converted into an eco-friendly and sustainable fuel substance with the interventions of microbes.

ROLE OF MICROBES IN BIOGAS PRODUCTION USING TANNERY SOLID WASTE

Vasudevan and Ravindran (2007, 2009) isolated bacterial strains with lipolytic and proteolytic activity from different locations such as tannery polluted soil, agricultural soil, compost, beam house and sludge stored soil. Tannery fleshings was completely digested (liquefied) by *Bacillus cereus* and *Pseudomonas aeruginosa* (lipolytic bacterial strains) and *Bacillus subtilis* and *Pseudomonas fluorescence* (proteolytic bacterial strains). 5% of the proteolytic and lipolytic strains were used as inoculum at log phase of their growth. Results showed that biologically liquefied tannery fleshings with proteolytic and lipolytic bacteria were well suited for bio-methanization process. Liquefied tannery fleshings along with sludge and cow dung were checked for their biogas producing efficiency for 30-35 days in a batch reactor (digester). This pre-treatment method of biological digestion of organic solid waste before introducing into methanogenic reactor was highly appreciable. Three-fold increase in the gas production was observed with liquefied tannery fleshings than the control. Higher gas production with drastic COD reduction and greater methane yield was the ultimate result. After this process, the residual slurry could be utilized as organic manure. Thus, microbial liquefaction of tannery solid waste is safe and viable recycling technique for its dual end products such as biogas and nutrient rich slurry that could be used for agriculture or horticulture purposes. Not only tannery fleshings, but also other proteinaceous tannery solid waste such as trimmings, shavings, splittings and sludges were also capable of undergoing bio-methanization and generation of residual slurry for other commercial purposes.

Limed tannery fleshings were co-digested with municipal solids' biodegradable fractions to enhance biogas production and reduce ammonia liberation under anaerobic conditions. This process of co-digestion helped in maximizing the recycling of tannery solid waste and municipal waste together into a renewable energy, with negligible ammonia liberation (Shanmugam & Horan, 2009). Here the role of methanogenic bacteria stands out for their consortial work, yielding enhanced amount of bio methane than separate anaerobic digestion systems.

Selenomonas Ruminantium HM000123 in Tannery Solid Waste Management

As non-tanned proteinaceous animal fleshings generated during initial stages of leather industry are rich sources of carbon, nitrogen and as they are not likely to be polluted by carcinogens and heavy metals, proteolytic bacterium *Selenomonas ruminantium* HM000123 was chosen for hydrolysis of fleshings through anaerobic fermentation to convert it into an organic fertilizer. Generally, nitrogen rich solid wastes could act as a great source of organic fertilizers (Cordovil et al., 2006). Hydrolyzed animal fleshings were tested for their efficiency to promote growth of tomato seedlings. Fleshings were completely hydrolyzed during the termination of fermentation period (5 days). End products of fermentation such as amino acids and fatty acids acted as growth promoters for the germination of tomato seedlings. Neutral pH (6.6) of the fermented fleshings was apt for tomato plants that favored its growth and seed germination. Incorporation of hydrolysed fleshings significantly influenced the height, weight, biochemical parameters such as chlorophyll, carotenoid, protein content and fruit yield of the plant in an increasing trend. Nutritional quality of the fruit in terms of minerals, vitamins and protein were exceptional among treatment plants supplemented with hydrolyzed fleshings. Thus, underutilized tannery solid waste material could act as a potent plant growth promoting material, performing equal to that of the existing fertilizers. *Selenomonas ruminantium* HM000123 fermented tannery fleshings served as a nutrient-rich fertilizer for tomato (*Lycopersicon esculentum* L.) plants (Ravindran et al., 2013). The mode of action of these fertilizers from fleshings is very similar to that of organic manure in slow release of nitrogen into the soil. They could also be added to the soil directly before planting.

Aspergillus Carbonarius UACC 92 in Tannery Solid Waste Management

Recycling of chromium through solid state fermentation using *Aspergillus carbonarius* UACC 92 was performed by Katsifas et al., (2003). Chromed tannery solid wastes are of major concern as they are highly toxic and unsafe for disposal. Need for chromium recycling from these wastes and their re-utilization in tanning process is the need of the hour. Microbial fermentation and therefore hydrolysis of chrome shaving waste is gaining popularity as the use of harsh acid and alkaline conditions result in severe environmental threat again. The ability of microorganisms to grow at high chromium enriched sites and its recovery points to the utilization of those microbes in tannery solid waste management, especially in the treatment of post-tanned solid waste. With this aim in mind, a study was designed in Greece for chromium recovery and reuse by microbiological means. *Aspergillus carbonarius* was able to degrade polyphenolic substances and was able to produce collagenase enzyme that could break the collagen cross-linking fibers. *Aspergillus carbonarius* UACC 92 degraded autoclaved chromium shaving within 12 days at 30°C to the maximum extent. From long term fermentation experiments, it was observed that out of each 500 ml of fermented liquid, 68 g of chromium precipitate sludge and 400 ml of proteinaceous supernatant was the ultimate yield. This proteinaceous supernatant may be valorized into a feed ingredient or used as fertilizer or in silage production. Chromium sulfate solution could be reused for tanning hides or could be sold for other domestic purposes. This microbe mediated minimization of hazardous tannery solid waste material paves way to a relatively low-cost technique for chrome shavings utilization. *Aspergillus fumigates* and few *Pseudomonas* sp. are also well known to leach out chrome from wastes and effluents. Mechanism indulged by the microbes here is biosorption. It is a process of interaction (non-specific binding) of the chromium metals with polysaccharides of the microorganisms via enzymatic mode, resulting in adsorption and accumulation of the metal. *Bacillus* sp. and *Rhodococcus* sp. are widely known for their biosorption property.

MICROBIAL ENZYMES REPLACING TRADITIONAL CHEMICALS OF TANNERIES

Microbial and animal enzymes have been widely used in various processes of the leather industry, as conventional chemicals create an unhealthy platform for the workers in tannery. For certain economic reasons, microbial enzymes are preferred than animal enzymes. Moreover, animal enzymes consist of mixture of one or more enzymes (for example-trypsin, chymotrypsin and amylase as secondary enzymes). Mostly, proteases, lipases and amylases are of commercial interest in leather industries. The following table shows various enzymes produced by microorganisms that are being used in the leather industry during pre-tanning operations.

Table 1. Microbial enzymes used in pre-tanning operations at tanneries (Kamini et al., 1999)

Process	Enzyme	Microorganism
Soaking	Protease	*Aspergillus parasiticus, A. flavus, A. oryzae, Bacillus subtilis* and *Rhizopus rhizopodiformis.*
	Carbohydrases	Aspergillus awamori
Dehairing	Protease	*Aspergillus flavus, Aspergillus sp. Bacillus subtilis, Lactobacillus sp. Conidiobolus sp. B. amyloliquefaciens Streptomyces griseus, S. fradiae, S. moderatus*
Bating	Protease	*A. parasiticus, S.rimosus and B. licheniformis, B. subtilis, Penicillium janthinellium*
Degreasing	Lipase	*Rhizopus nodosus, A. oryzae and A. flavus*

Proteases and carbohydrases are used in soaking process of beam house operation in tannery, which favors loosening and opening of fiber structures, resulting in the reduction of wrinkles in the leather. It softens the hides making them amenable for further processing.

Dehairing, one of the major processes in beam house operation is traditionally done using the application of lime and sodium sulphide onto the hides with mechanical agitation favoring removal of hair from the hides. This process is the main contributor of COD and alkalinity in tannery effluents with emission of noxious odors due to hydrogen sulphide and ammonia emission. Keratinases are of potent interest here as they enable uniform hair removal with good quality and strength that could be reused. Enzymatic dehairing is comparatively fast and eco-friendly method than chemical dehairing, leading to conductive ambience for the tanners to work further. Preethi et al., (2015) were successful in isolating keratinase producing *Bacillus* sp. from tannery aeration tank effluent and identified their ability to dehair hides. Isolates were identified as *Bacillus cereus, Bacillus vallismortis* and *Bacillus mojavensis* through biochemical and MALDI-TOF analysis. All the three isolates had good dehairing activity and feather degrading ability too. Microbial proteases have been well developed for this process and three ways to apply were designed as dip, paint and spray method. Out of the three, dip method proved to be an eminent method for its feasibility and efficacy. An alkaline serine protease from *Aspergillus flavus* named as Clarizyme has been developed by Central Leather Research Institute, Adyar, Chennai, India for dehairing of hides and skins (Kamini et al., 1999). Dehairing is followed by bating that also indulges protease enzymes to soften the hides further. Thazeem et al., (2016) assessed the dehairing ability of proteolytic *Bacillus pumilus* isolated from tannery lime effluent through dip method. Thazeem et al., (2017a) performed a comparative study on the dehairing activity of three different alkaline protease producing *Bacillus* sp. through dip, paint and spray method. Dehairing activity of *Bacillus licheniformis* was exceptional via dip method, followed by *Bacillus cereus* M-1 and *Bacillus subtilis*. Moderate results were observed when hides were sprayed and painted with enzymes, enabling dip method to be the best out of three.

Microbial enzymes find application in another tanning process named degreasing, where removal of excess fats from oilier skins are performed using acid or alkaline lipases that could replace solvents used in this step. Lipases from *Rhizopus nodosus* and *Aspergillus niger* have been reported to be utilized in the degreasing process that could solubilize fats present in the interfibrillar layers of hides to give a complete finishing to the final product.

FUTURE RESEARCH DIRECTIONS

Albeit various roles of different microorganisms in tannery solid waste management has been discussed and detailed, only few researches have been commercialized and implemented by the leather industries due to technical and economic reasons. More studies on the stability of microbial enzymes for industrial scale application are necessary as they do not perform similarly in the laboratory and scale up processes. Improvisation of crude enzyme activities and its industrial utilization has to be encouraged, as purification step needs more cost, time and labor. Pathogenecity of the microbes has to be focused thoroughly, as the residual part of hydrolysis/fermentation has fertilizer and feed applications. As probiotic microorganisms are used for biostabilization and fermentation of tannery fleshings, incorporation of such probiotic fermented products into animal feeds may influence the immunity of the experimental animal. Further studies on the effect of such products on the immunity and oxidative stress tolerance of the animal system have to be emphasized. Many works on bioconversion of tannery solid waste materials into bio fuel and bio gas has to be encouraged and funded as there is a great need of such a sustainable energy in the present world. Residual nutrients from tannery fleshings have to be applied for commercial crops on fields and checked for the growth and productivity. As chromium recovery and recycling is a matter of distress in tanneries, upcoming researchers could concentrate on microbes that have the ability to leach out chromium from solid wastes and effluents for its successful recycling in tanning process.

CONCLUSION

The present chapter concludes with the point that microorganisms play a vital role in tannery solid waste management, replacing traditional disposal methods and harsh chemicals which have been utilized by the tanners regularly till now to get rid of their unhealthy situation in tanneries. Proteolytic, lipolytic and probiotic microbes have a great potential in managing environmental wastes with recovery of marketable products. They act as a major tool for reducing environmental pollution caused by adverse disposal systems. Microbial mediated tannery solid waste management not only alleviates tanners' problems faced during waste disposal, but also profits the industry greatly in terms of fermentation end products that could be beneficially recycled and reused by the tanneries itself in conversion of raw hides into finished leather. Much more initiations and advancements have to be provoked among scientists and researchers to work with microorganisms in managing tannery solid waste and its threats that may reduce the burden of the tanners.

REFERENCES

Aftab, M. N., & Hameed, A., & Ikram-ul-Haq, C. H. E. N. (2006). Biodegradation of leather waste by enzymatic treatment. *The Chinese Journal of Process Engineering*, *6*(3), 462–465.

Ahamed, M. I. N., & Kashif, P. M. (2014). Safety disposal of tannery effluent sludge: Challenges to researchers—a review. *International Journal of Pharma Sciences and Research*, *5*, 733–736.

Ahmad, J., & Ansari, T. A. (2013). Alkaline protease production using proteinaceous tannery solid waste. *Journal of Petroleum & Environmental Biotechnology*, *4*(1). doi:10.4172/2157-7463.1000136

Alexander, K. T. W., Donohue, V. J., Cory, N. J., & Sykes, R. L. (1992). Environmental and safety issues-clean technology and environmental auditing. *Journal of the Society of Leather Technologists and Chemists*, *76*, 17–23.

Balakrishnan, B., Prasad, B., Rai, A. K., Velappan, S. P., Subbanna, M. N., & Narayan, B. (2011). In vitro antioxidant and antibacterial properties of hydrolysed proteins of delimed tannery fleshings: Comparison of acid hydrolysis and fermentation methods. *Biodegradation*, *22*(2), 287–295. doi:10.100710532-010-9398-0 PMID:20680665

Banu, J. R., Esakkiraj, S., Nagendran, R., & Logakanthi, S. (2005). Biomanagement of petrochemical sludge using an exotic earthworm Eudrilus eugineae. *Journal of Environmental Biology*, *26*(1), 43–47. PMID:16114460

Bhaskar, N., Sakhare, P. Z., Suresh, P. V., Gowda, L. R., & Mahendrakar, N. S. (2007). Biostabilization and preparation of protein hydrolysates from delimed leather fleshings. *Journal of Scientific and Industrial Research*, *66*, 1054–1063.

Chakraborty, R., & Sarkar, S. K. (1998). Enzyme hydrolysis of solid tannery wastes: Solid state enzyme production. *Journal of the Society of Leather Technologists and Chemists*, *82*(2), 56–58.

Cordovil, C. M., Cabral, F., & Coutinho, J. (2007). Potential mineralization of nitrogen from organic wastes to ryegrass and wheat crops. *Bioresource Technology*, *98*(17), 3265–3268. doi:10.1016/j.biortech.2006.07.014 PMID:16905311

Dayanandan, A., Rani, S., Shanmugavel, M., Gnanamani, A., & Rajakumar, G. S. (2013). Enhanced production of Aspergillus tamarii lipase for recovery of fat from tannery fleshings. *Brazilian Journal of Microbiology*, *44*(4), 1089–1095. doi:10.1590/S1517-83822013000400010 PMID:24688497

European Commission. Directorate General JRC, Joint Research Centre, Institute for prospective technological Studies, Technologies for sustainable development, European IPPC Bureau, Integrated Pollution Prevention and Control. (2001). Reference document on best available techniques for the tanning of hides and skins (Final draft). Retrieved from http://eippcb.jrc.es.2001

Fathima, N., Rao, R., & Nair, B. U. (2012). Tannery solid waste to treat toxic liquid wastes: A new holistic paradigm. *Environmental Engineering Science*, *29*(6), 363–372. doi:10.1089/ees.2010.0445

Ghosh, A., Chakroborti, A., Das, F., Chattopadhyay, D., & Chakrabarti, K. (2004). Use of bacteria for utilization of solid waste from poultry and leather industry. *Journal of Indian Leather Technologists'. Association.*, *55*, 619–623.

Hattori, M., Yamaji-Tsukamoto, K. A., Kumagai, H., Feng, Y., & Takahashi, K. (1998). Antioxidative activity of soluble elastin peptides. *Journal of Agricultural and Food Chemistry*, *46*(6), 2167–2170. doi:10.1021/jf970484r

Jini, R., Bijinu, B., Baskaran, V., & Bhaskar, N. (2016). Utilization of Solid Wastes from Tanneries as Possible Protein Source for Feed Applications: Acute and Sub-acute Toxicological Studies to Assess Safety of Products Prepared from Delimed Tannery Fleshings. *Waste and Biomass Valorization*, *7*(3), 439–446. doi:10.100712649-015-9463-3

Kamini, N. R., Hemachander, C., Mala, J. G. S., & Puvanakrishnan, R. (1999). Microbial enzyme technology as an alternative to conventional chemicals in leather industry. *Current Science*, *77*(1), 80–86.

Kanagaraj, J., Velappan, K. C., Babu, N. K., & Sadulla, S. (2006). Solid wastes generation in the leather industry and its utilization for cleaner environment-A review. *Journal of Scientific and Industrial Research*, *65*, 541–548.

Katsifas, E. A., Giannoutsou, E., Lambraki, M., Barla, M., & Karagouni, A. D. (2004). Chromium recycling of tannery waste through microbial fermentation. *Journal of Industrial Microbiology & Biotechnology*, *31*(2), 57–62. doi:10.100710295-004-0115-z PMID:14767674

Kirk, D. W., Chan, C. C., & Marsh, H. (2002). Chromium behavior during thermal treatment of MSW fly ash. *Journal of Hazardous Materials*, *90*(1), 39–49. doi:10.1016/S0304-3894(01)00328-4 PMID:11777591

Krishnan, C. S., Sakinah, A. M. M., Singh, L., & Wahid, Z. A. (2016). Production of biodiesel using tannery fleshing as a feedstock via solid state fermentation. *International Journal of Chemical, Molecular, Nuclear. Materials and Metallurgical Engineering*, *10*(5), 591–596.

Kumar, A. G., Kamatchi, P., Umashankari, J., Vidhya, S., Murthy, P. S., & Sekaran, G. (2008a). Acidogenic fermentation of proteinaceous solid waste and characterization of different bioconversion stages and extracellular products. *Biodegradation*, *19*(4), 535–543. doi:10.100710532-007-9159-x PMID:17952609

Kumar, A. G., Nagesh, N., Prabhakar, T. G., & Sekaran, G. (2008b). Purification of extracellular acid protease and analysis of fermentation metabolites by Synergistes sp. utilizing proteinaceous solid waste from tanneries. *Bioresource Technology*, *99*(7), 2364–2372. doi:10.1016/j.biortech.2007.05.001 PMID:17601727

Kumar, A. G., Swarnalatha, S., Sairam, B., & Sekaran, G. (2008c). Production of alkaline protease by Pseudomonas aeruginosa using proteinaceous solid waste generated from leather manufacturing industries. *Bioresource Technology*, *99*(6), 1939–1944. doi:10.1016/j.biortech.2007.03.025 PMID:17481889

Kumar, A. G., Venkatesan, R., Kirubagaran, R., Prabhakar, T. G., & Sekaran, G. (2008d). Effects of nonionic surfactant on hydrolysis and fermentation of protein rich tannery solid waste. *Biodegradation*, *19*(5), 739–748. doi:10.100710532-008-9178-2 PMID:18288576

Kumar, P. S., Sounderraj, L., Sivakumar, K., & Rajakumar, S. (2012). Enzymatic hydrolyses of chrome shaving waste from tannery through Aspergillus tamerii. *Journal of Biological and Environmental Sciences*, *6*(18), 285–292.

Preethi, K., Anand, M., & Thazeem, B. (2015). Isolation and Identification of keratinolytic bacteria from tannery effluent: A Study on their biodegradative and dehairing activity. *International Journal of Multidisciplinary Research and Development*, *2*(10), 227–234.

Rai, A. K., Bhaskar, N., Halami, P. M., Indirani, K., Suresh, P. V., & Mahendrakar, N. S. (2009). Characterization and application of a native lactic acid bacterium isolated from tannery fleshings for fermentative bioconversion of tannery fleshings. *Applied Microbiology and Biotechnology*, *83*(4), 757–766. doi:10.100700253-009-1970-3 PMID:19333593

Rai, A. K., General, T., Bhaskar, N., Suresh, P. V., Sakhare, P. Z., Halami, P. M., ... Mahendrakar, N. S. (2010). Utilization of tannery fleshings: Optimization of conditions for fermenting delimed tannery fleshings using Enterococcus faecium HAB01 by response surface methodology. *Bioresource Technology*, *101*(6), 1885–1891. doi:10.1016/j.biortech.2009.10.015 PMID:19892548

Ravindran, B., Gayathri, V., Priya, D. M., Kanimozhi, S., & Sekaran, G. (2012). Hydrolysis of proteinaceous tannery solid waste for the production of extracellular acidic protease by Selenomonas ruminantium. *African Journal of Biotechnology*, *11*(56), 11978–11990. doi:10.5897/AJB11.2484

Ravindran, B., Kumar, A. G., Bhavani, P. A., & Sekaran, G. (2011). Solid-state fermentation for the production of alkaline protease by Bacillus cereus 1173900 using proteinaceous tannery solid waste. *Current Science (Bangalore)*, *100*(5), 726–730.

Ravindran, B., Nagalakshmi, C., Devi, S. G., Subramanian, K. S., Mandal, A. B., & Sekaran, G. (2013). Efficiency of hydrolyzed tannery solid waste for the growth and yield potentials on tomato. *Journal of Plant Nutrition*, *36*(5), 795–810. doi:10.1080/01904167.2012.756893

Ravindran, B., & Sekaran, G. (2010). Bacterial composting of animal fleshing generated from tannery industries. *Waste Management (New York, N.Y.)*, *30*(12), 2622–2630. doi:10.1016/j.wasman.2010.07.013 PMID:20727727

Sekaran, G., Swarnalatha, S., & Srinivasulu, T. (2007). Solid waste management in leather sector. *Journal on Design and Manufacturing Technologies*, *1*(1), 47–52. doi:10.18000/ijodam.70008

Shanmugam, P., & Horan, N. J. (2009). Optimising the biogas production from leather fleshing waste by co-digestion with MSW. *Bioresource Technology*, *100*(18), 4117–4120. doi:10.1016/j.biortech.2009.03.052 PMID:19395254

Sumathi, C., Jayashree, S., & Sekaran, G. (2012a). Screening and evaluation of antimicrobial activity of tannery fleshing utilizing fish gut bacteria. *International Journal of Pharmacology*, *8*(5), 373–381. doi:10.3923/ijp.2012.373.381

Sumathi, C., Mohanapriya, D., Mandal, A. B., & Sekaran, G. (2012b). Production of different proteases from fish gut microflora utilizing tannery fleshing. *Engineering in Life Sciences*, *12*(2), 223–237. doi:10.1002/elsc.201100097

Sumathi, S., & Sekaran, G. (2010). Nutritional Evaluation of Animal Fleshing as a Fish Meal Replacer in Labeo rohita. *Journal of Aquaculture Feed Science and Nutrition*, *2*(2-4), 6–10.

Sundar, V. J., Gnanamani, A., Muralidharan, C., Chandrababu, N. K., & Mandal, A. B. (2011). Recovery and utilization of proteinous wastes of leather making: A review. *Reviews in Environmental Science and Biotechnology*, *10*(2), 151–163. doi:10.100711157-010-9223-6

Thazeem, B., Beryl, G. P., & Umesh, M. (2017a). A comparative study on alkaline protease production from *Bacillus* spp. and their biodegradative, dehairing and destaining activity. *International Journal of Academic Research and Development*, *2*(2), 74–79.

Thazeem, B., Preethi, K., Umesh, M., & Radhakrishnan, S. (2017b). Nutritive Characterization of Delimed Bovine Tannery Fleshings for Their Possible Use as a Proteinaceous Aqua Feed Ingredient. *Waste and Biomass Valorization*.

Thazeem, B., Umesh, M., & Vikas, O. V. (2016). Isolation and identification of proteolytic *Bacillus pumilus* from tannery lime effluent and its dehairing activity. *International Journal of Academic Research and Development, 1*(2), 5-7.

Vasudevan, N., & Ravindran, A. D. (2007). Biotechnological process for the treatment of fleshing from tannery industries for methane generation. *Current Science*, *93*(11), 1492–1494.

Vasudevan, N., & Ravindran, A. D. (2009). Bioconversion of tannery solid waste into energy. *International Journal of Integrated Energy Systems*, *1*(2), 97–101.

Zerdansi, I., Faid, M., & Malki, A. (2004). Digestion of solid tannery wastes by strains of Bacillus species isolated from compost in Morocco. *International Journal of Agriculture and Biology*, *6*(5), 758–761.

Ziegler, F., Nitenberg, G., Coudray-Lucas, C., Lasser, P., Giboudeau, J., & Cynober, L. (1998). Pharmacokinetic assessment of an oligopeptide-based enteral formula in abdominal surgery patients. *The American Journal of Clinical Nutrition*, *67*(1), 124–128. doi:10.1093/ajcn/67.1.124 PMID:9440386

KEY TERMS AND DEFINITIONS

Dehairing: Process of removing hairs from raw hides and skins by lime treatment and mechanical agitation followed by gentle scrapping manually.

Deliming: Process of removing calcium hydroxide and sodium sulfide from hides with the help of commercial ammonium salts, weak acid or alkali solutions.

Fleshings: Undesirable cutaneous and sub-cutaneous layers of muscles scrapped out from the animal hides during beam house operation using fleshing machine.

Lactic Acid Fermentation: Process of conversion of available substrate sugars into lactic acid by lactic acid producing bacteria through oxidation.

Probiotics: Microorganisms generally considered as safe that exert beneficial effects in the host system when consumed either in viable or attenuated form.

Proteases: A group of enzymes that degrade (hydrolyze) complex proteins into simpler peptide units.

Proteinaceous: A substance which is highly enriched with single or a group of proteins.

Putrefaction: Stage of decomposition of any organic material due to the action of microorganisms with emission of notorious smell.

Tanning: Process of conversion of animal raw hides into wet-blue uniform leather pieces using vegetable oil or chromium.

Valorization: Transformation of any unutilized substance into a value added saleable substance.

Chapter 8
Phytoremediation of Nitrogen and Phosphorus in Municipal Wastewater by Cyperus alternifolius Planted Constructed Wetland

Hossein Farraji
University Saiens Malaysia (USM), Malaysia

ABSTRACT

Nowadays municipal wastewater (MWW) treatment by phytoremediation techniques goes as an emerging technique in the USA and many European countries. Cleaning wastewater with constructed wetland (CW) is an advanced type of phytoremediation. Low concentration of hazardous metallic elements in this major wastewater caused its capability for treatment by CW. This treatment method highly depends on the presence of macrophytes, media, and operating factors which have high influences in the efficiency of this technology. This chapter will discuss on engineering controls that traditionally are available and practically could be used in the commonly CW. Recirculation, dry to wet duration, artificial aeration, absorbent application, dilution and carbon source addition through this eco-friendly decontamination method. The concentration of this manuscript will be on Cyperus alternifolius as a well-known rapid growth plant species which often known as an ornamental plant. This review on try to illustrate practical ways to enhancing efficiency of decontamination of MWW in CW

INTRODUCTION

Municipal wastewater fundamentally divided into two types black water and gray water. Gray water is from kitchen wastewater, sink, washing machine and bathroom. This wastewater can reach up to 75% of the total wastewater generated in the household (Hernández et al., 2011). Black water comes from toilets with high concentrations of nitrogen and phosphorus (Zeeman et al., 2008). This kind of domestic

DOI: 10.4018/978-1-5225-3540-9.ch008

wastewater should be considered as one of critical environmental problem since through the quality and quantity, it can contaminate the environment, interfere human health and other creatures. An increasing number of domestic wastewater causes negative impact on environment quality since it generally discharge directly to natural water like as river and lakes so it can cause increasing pollutant load on the water bodies. Constructed wetlands provide a greater level of pollutant control to discharge effluent to surface waters (Kadlec & Wallace 2008). The sewage discharge from the domestic sewage treatment plant is typically subjected to the local, state and federal regulations and standards. In December 2009, the Department of Environment (D.O.E) Malaysia has revised the Environmental Quality (Sewage) Regulations 2009, which significantly reduced the discharge limit of ammonia-nitrogen from 50 mg/L to 5 mg/L. In addition, a new discharge limit was also set for nitrate at 10 mg/L (Kutty et al., 2011).

LITERATURE REVIEW

Constructed Wetland

Types of Constructed Wetland

It have been proven that, constructed wetlands are suitable wastewater treatment system for many developing countries in tropical areas where land is available at low cost and warm temperatures are suitable for biodegradation (Kantawanichkul et al., 2013). Simulating a natural wetland ecosystem in terms of constructed wetland, the treatment system can make use of the assimilation capacity of soil (living media) and aquatic plants (main up taker) to remove both pollutants (organic and inorganic compounds and microbial pathogens) and nutrients without additional energy demand (Inamori et al., 2007). Constructed wetlands are alternative wastewater treatment systems which are low-cost, easy operation and maintenance than conventional technologies (Stefanakis et al., 2014; Stefanakis et al., 2011). The correct functioning of a constructed wetland (CW) depends on the interaction between plants, substrate and microorganisms in relation to the type of structure and wastewater treatment being used (Stefanakis et al., 2011). In most of literatures pay attention vastly on engineering aspects of constructed wetland, however plant species make a significant contribution to natural decontamination of process of wastewater (Stottmeister et al., 2003). Removal of total nitrogen in different types of constructed wetlands varied between 40 and 55% with removed load ranging between 250 and 630 g $Nm^{-2}Yr^{1}$ depending on CWs type and inflow loading (Vymazal, 2007). Removal of phosphorus in all types of constructed wetlands is low unless Special substrates with high absorption capacity are used. Removal of total phosphorus varied between 40 and 60% in all types of constructed wetlands with removed load ranging between 45 and 75 g $Pm^{-2}y^{-1}$ depending on CWs type and inflow loading (Vymazal, 2007).

Horizontal Flow Constructed Wetland (HFCW)

By preparing suitable preplanning process both of HFCW and VFCW can effectively remove 90% of organic pollutants and all concentration of N and P from wastewater (Luederitz et al., 2001) meanwhile HFCW could be considered as an advanced performance for long time P removal system. CWs with free water surface (FWS) units remove substantially more total nitrogen as compared to other types of hybrid constructed wetlands (Vymazal, 2013). These systems are good for suspended solids removal and bacteria removal because of their ability to filter BOD removal up to a set oxygen transfer capacity. Poor for nitrification because of limited oxygen transfer capability (Cooper, 1999).

Vertical Flow Constructed Wetland (VFCW)

These systems are good for nitrification because of their high oxygen transfer capability which also leads to good removal of BODs and COD. VFCW can also remove some bacteria, less good for suspended solids removal, and can become clogged (Cooper, 1999). All types of hybrid CWs are comparable with single VFCWs in terms of NH_4-N removal rates (Vymazal, 2013). Advantages and disadvantages of vertical and horizontal constructed wetland summarized in Table 1.

Hybrid Constructed Wetland

Practical experiences shows that all kinds of hybrid constructed wetlands are more efficient in total nitrogen removal than single HF or VF constructed wetlands (Vymazal, 2013). Saeed and Sun (2011) reported high efficiency removal nitrogen (up to 100%) from artificial wastewater in vertical + horizontal constructed wetland. Additional structures, requiring more lands, electrical energy requirement for pomp and cost of operation are disadvantages of hybrid constructed wetland.

Mixed Constructed Wetland

New researchers are concentrating on mixing vertical and horizontal constructed wetland in specific single unit for decreasing operation cost and effective application of specific surface area for degradation process. The efficiency of a novel design (mixing two types of constructed wetland) for enhancing total nitrogen and phosphorus removal, consequently to design mixed vertical and horizontal wetland in a constructed system which not only have surface and subsurface water flow but also it have down to up and up to down water flow. An experience carried out recently by Tee et al. (2012) showed that jointed vertical and horizontal constructed wetland in one system could be considered as practical way

Table 1. Advantages and disadvantages of HF and VF constructed wetland

Type	Advantages	Disadvantages
VFCW	Smaller area Good oxygen supply " Nitrification " Organic and SS removal Simple hydraulics Long oxygen flowing distance	Short flow distance Poor denitrification Low nitrate removal High technical requirement
HFCW	Long flowing distance Efficient in organic and SS removal Possibility of denitrification Formation of humic acid for N & P removal	High area requirement Possibility of clogging problem Loss of P removal performance Sensitivity of nitrification by sulphur transformation Low ammonia oxidation Careful hydraulic calculation for O_2 supply Uniform passage of wastewater

(Saeed & Sun, 2012)

for enhancing efficiency of ammonia nitrogen removal. Figure 1 and 2 Shows two type of combined / mixed constructed wetland. Future of constructed wetland highly depends on specific designing of creative CW for improving decontamination efficiency. Another combining technique in CW is mixing physicochemical treatment reactors with constructed wetland. In such cases, two separated treatment method relocate in unique system of decontamination. Figure 3 shows the spectacular combining SBR and CW for wastewater treatment.

All aforementioned designs are looking for compacted effective treatment in the shortest time. These types of novel designing could be considered as capable method decreasing the required time for treatment and converting CW technique as non-time taker process.

Operating Techniques for Constructed Wetland

Aeration Effect

Artificial aeration is readily available to use at a low capital cost (Ouellet et al., 2006). The aerated and recirculated wetland resulted in a higher total nitrogen removal (8.6 gN m(-2)d(-1)) due to simultaneous nitrification/denitrification, (Foladori et al., 2013). Under higher organic load conditions, the benefit of artificial aeration may even be more pronounced (Ouellet et al., 2006). A recirculating vertical flow constructed wetland (RVFCW) produces high quality effluents (Sklarz et al., 2009). 1.8$m^3$of air per minute in 12 hours (6 Am to 6 Pm) used by Nivala et al. (2007) for treatment of landfill leachate. About 9 mmol of oxygen per liter of influent treated by Green et al. (1998). A system of compressor, air regu-

Figure 1.

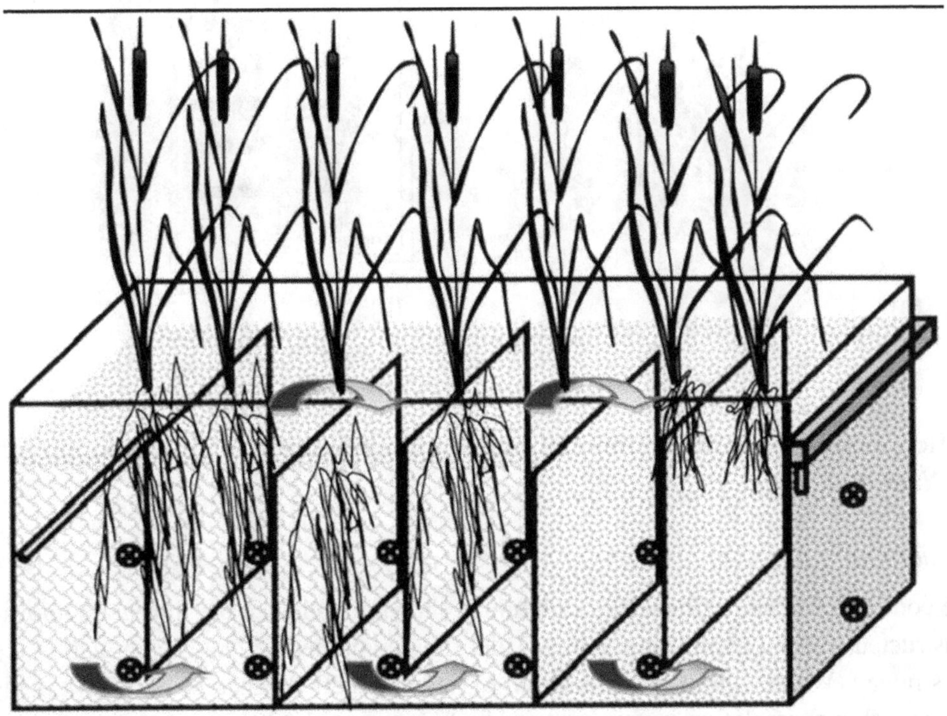

Figure 2.

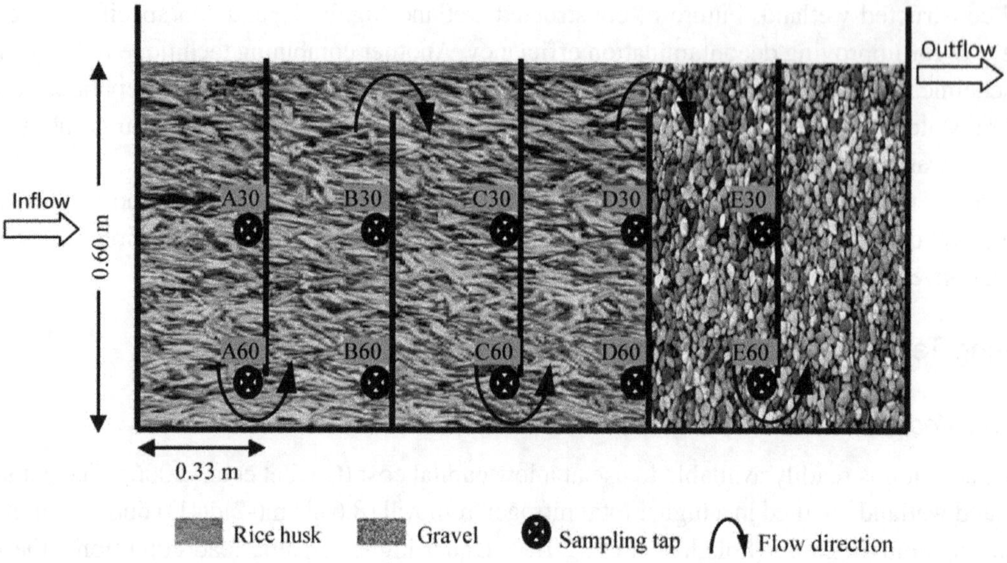

Figure 3. Combining sequencing batch reactor and constructed wetland (Guo et al., 2014)

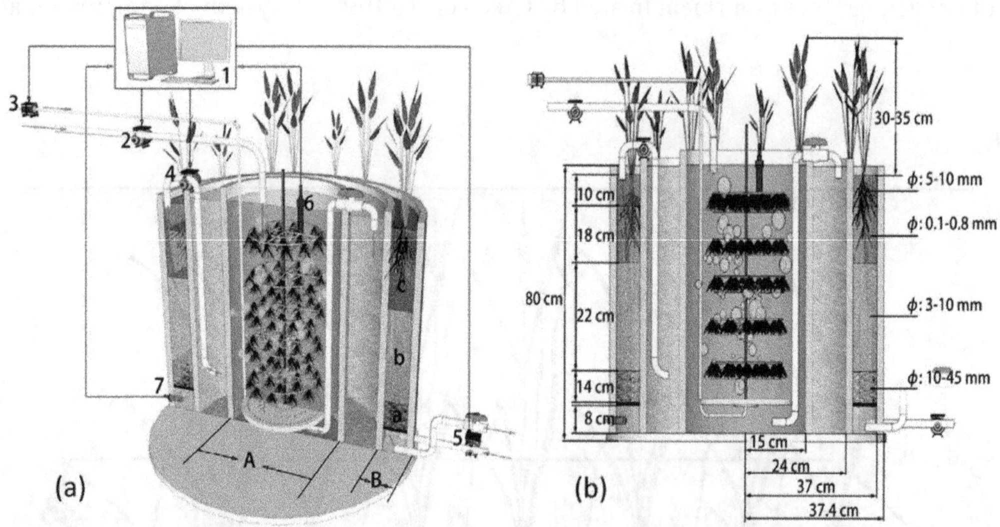

lator and flow meter will use for aeration and a timer will use to control the aeration and non-aeration schedule (Yang et al., 2003).

Recirculation

Inlet zone configuration and effluent recirculation is shown to enhance the nitrogen removal capability of constructed wetland treatment systems. Tidal flow combined effluent recirculation is favorable operation strategy (White, 1995;Kantawanichkul & Neamkam 2003; Zhao et al., 2004). A higher total nitrogen removal in recirculation system reported by Foladori et al. (2013).

Dry-to-Wet Duration Effect (Feeding Style)

Same as time sequence batch reactor rhythmical movement wastewater and aeration (Morgenroth & Wilderer 1998), to have prolonged and satisfying pollutant removal efficiencies, biological filters need to be rested periodically to allow break down of accumulated organic matter. For enhancing performance of system, aeration and reduces the likelihood of anoxia, (Okurut et al., 1999;Platzer 1999; Chan et al., 2008). Alternative to continuous flow system commonly used for removing nutrients and carbon from domestic wastewater (Artan et al., 2002). Chan et al. (2008) carried out study on treatment of domestic wastewater in constructed wetland by *Cyperus alternifolius* and following results reported:

- The "dry" period provides "resting" period of the pollutant retained inside the system to be completely degraded by microbial activities
- Wet and dry period allows higher oxygen level in domestic wastewater treatment constructed wetland beds
- The periodicities of wet" and "dry" periods were generally effective for ammonia nitrogen removal.
- Dry-to-wet duration have no effect on phosphorus and suspended solid removals.
- Efficiency of the pollutant removal from the conventional wetland treatment system could be modified by sophisticated control of sequential feed of wastewater

Another study carried out by Cui et al. (2009) in aspect of dry to wet period effect on domestic wastewater treatment in constructed wetland with use of *Cyperus alternifolius*, they did not give clear report on this part of study, it seems that dry and wet time operation management have critical role in positive efficiency of this technique. Chan et al. (2008) arranged 2 hours wet and followed it by 4 or 6 hours dry period and clear result obtained, meanwhile Cui et al. (2009) used 8 hours wet time and followed it by 24 or 36 hours dry period, on the other hand based on the Chan results, there is no significant different between wet to dry with ratio of 4:4 hours and ratio of 6:2 hours wet to dry period time, thus it seems short period of dry time (2-4 hours) will give higher efficiencies in domestic wastewater treatment process in constructed wetland system. More assessment required for confirming best dry period for earning maximum up taking or degradation of Nitrogen in constructed wetland especially in tropical weather of Malaysia.

Absorbents

Consideration of win-win plan in environmental problem solution helps us to use local waste which can use as absorbent. Domestic wastewater treated by local waste as absorbent by (Chan et al., 2008a), they used Coal slag, which is the waste residue from burning coal for electricity generation in China for filling up in the constructed wetland system as the supporting medium they also used *Cyperus alternifolius* in constructed wetland system. (Chan et al., 2008b) used Cinder for filling up constructed wetland as an waste absorbent which is a common Coal waste from electricity generation in six vertical constructed wetland, *Cyperus alternifolius* planted in half of their treatments. (Cui et al., 2009) applied yellow coal-burn-slag and black coal-burn- slag which are two local waste from student cafeteria, as absorbent for treatment of domestic wastewater by constructed wetland. The averaged percentages of TN contents in the biomass among Planted Yellow coal-burn-slag was higher than black coal-burn-slag (Cui et al., 2009). Coal bottom ash as by-product of power plants produce and available many millions of tons, free and causes main disposal problem for environment from 1920 that power generation began in Malaysia and fly ash which is the particulate material that remains after coal is burned, with annually production about 600 million tons (Ahmaruzzaman, 2010). Fly ash had been used for removing COD from domestic wastewater and efficiency of COD removal was lower than commercial activated carbon (Devi & Dahiya 2006). Another local waste material which is free and widely available is palm oil tree fly ash which could be used as an alternative low-cost sorption media/filter for removal of organic compounds from real industrial Wastewater (Laohaprapanon et al., 2010). There is a gap of knowledge on palm oil tree fly ash and coal bottom ash as local free widely available absorbent waste in constructed wetland for domestic wastewater treatment. The type of substrate and presence of the plant have major effect on the species richness and the structure of bacterial communities (Calheiros et al., 2009).Recently a case study carried out on rice husk and gravel for removing Nitrogen from domestic wastewater (Tee et al., 2012) meanwhile, based on our best knowledge there is no exprience on palm oil ash/rice husk ash treatment via *Cyperus alternifolius* planted constructed wetland.

Plant Effect

Plant species carry out a series of critical biological, chemical and physical processes within a constructed wetland wastewater treatment system and make a significant contribution to wastewater purification and decontamination (Leto et al., 2013). The selection of the plant species is a particularly critical aspect in the design and mange of all kind of constructed wetlands and various aspects such as geographical distribution, climate and habitat conditions, composition of wastewater, availability of the plant species in nurseries, long term maintenance and agronomic management costs, and objects of project must be considerate. There is no scientific research data available that is clearly able to explain and demonstrate a significant difference in purification ability between different macrophytes commonly used in constructed wetlands wastewater treatment systems (Brix, 1997). Effects of plants in constructed wetland summarized sentences:

- Plants reduced algal growth and lower system operating temperatures (Yeh et al., 2015).
- plants in a CW slows down the flow of the wastewater (Petticrew & Kalff, 1992).
- Plants provide large surface areas for the development of biofilm by aboveground and belowground parts (VanKempen-Fryling & Camper, 2017).
- Aquatic vegetation of machrophytes releases oxygen into the rhizosphere and sediment has on redox conditions (Moore et al., 1994).
- Plants increase contact time between substrate and the waste water (Brix 1997).
- Root growth of macrophytes influences a number of the hydraulic characteristics of the substrate (Stottmeister et al., 2003).
- Plants stabilize the sediment surface with reduce wind speed (Vymazal, 2011).
- Filtering, increased rate of sedimentation and reduced risk of re-suspension (Vymazal, 2011).
- Plants roots have effect on nitrogen removal and microbial activity by effect on redox potential (Białowiec et al., 2012).
- Plants can use as phytoindicator in water quality and salt phytoremediation (Shelef et al., 2013).

Although various researches (Brisson & Chazarenc 2009; Vymazal, 2011) confirm that plant uptake in a constructed wetland (CW) is not decisive while others (Weis & Weis, 2004; Lee & Scholz, 2007) maintain that it is significant. A study carried out in Australia by Langergraber (2005) and report that nutrient uptake by macrophytes has no a major effect on water treatment in comparison to the applied loads. Literature reviews on domestic wastewater treatment in constructed wetland by means of *Cyperus alternifolius* also reported various results like as Leto et al. (2013) that clearly confirmed positive enhancing effect of *Cyperus alternifolius* L. and *Typha latifolia* L in nutrient removal from domestic wastewater, meanwhile they did not used any special absorbent. Chan et al. (2008) and Chaz et al. (2008) reported no significant difference of pollutant removals in planted and unplanted constructed wetland with *Cyperus alternifolius* L treatment on nutrient removal from domestic wastewater. Cui et al. (2009) researches show that more TN was removed from the influent with the presence of *C.alternifolius* compared to without this plant species. Based on the researches of Kantawanichkul et al. (2013)the removal efficiency of nitrogen in terms of total Kjeldahl nitrogen (TKN) was 65% at 20 cm/d hydraulic loading rate (HLR) for *C.alternifolius*, furthermore they also did not used specific absorbent in their constructed wetland system.

- ***Cyperus alternifolius***

Cyperus alternifolius is a perennial herb which can grow well in humid areas or swampland It has deep roots and nutrient tolerant ability, Hocking (1985) reported that Cyperus genus are suitable plants species for constructed wetland. *Cyperus alternifolius* used for piggery wastewater treatment by Liao et al. (2003), Liao et al. (2005) and domestic wastewater by (Cui et al., 2009; Cu et al., 2011; Kantawanichkul et al., 2013; Leto et al. (2013). This plant have high ability to oxygen release by its root (Yao et al., 2011). There is a gap of knowledge in effect of plants (*C.alternifolius*) alone in treatment of domestic wastewater and in this study we will assess this issue as a phytoremediation project.

Dilution

There is an article on domestic wastewater treatment in constructed wetland which used dilution technique on different plant species ability to uptake nutrients from wastewater and based on result of this study the abilities of different wetland plants to directly take up nitrogen and phosphorus were species-specific, but within each species, uptake was similar between diluted and undiluted sewage. *Cyperus alternifolius* used via this study (Li et al., 2013).

Hydraulic Retention Time (HRT)

It is well known that critical pollution factors removal efficiencies in constructed wetlands are related to HRT (Persson et al., 1999; Belmont & Metcalfe, 2003; Chen et al., 2006; Kadlec & Wallace, 2008; Ghosh & Gopal, 2010). There is a positive relationship between microbial inactivation and HRT, but in general when the HRT is approximately 3 days the microbial inactivation reaches saturation values (Kröpfelová, 2008). When HRT is more than 4 days, concentrations of NO_2-N, NO_3-N, NH_4-N and TN change little (Xinshan et al., 2010). In a horizontal subsurface-flow(HSF) constructed wetland for domestic wastewater treatment which carried out in Malaysia, 2, 3, and 5-day HRT applied in planted wetland with cattails (*Typha latifolia*) and raw rice husk as media. Diameters of constructed wetland in their study wetland units of dimension 2.0 x 0.5 x 0.8m (Length x Width x Height) (Tee et al., 2012).

Carbone Resource

Domestic wastewater contains high concentration of Nitrogen and Phosphorus and organic material but concentration of Carbon should be available to reach reasonable C/N ratio amended by external addition of carbon source which can present suitable C/N ratio for efficiency of microbial growth and development as the fundamental mechanism of degradation. Summary of related studies presented in Table 2. Since of present Carbone in palm oil ash (5.4-12 percent) (Foo & Hameed, 2009) or rice husk ash (2.7-6.42 percent) (Nagrale et al., 2012), it seems no need to additional Carbone sources. So we use cost effective material as absorbent.

Summary of Literature Review

Results of 8 related studies on domestic wastewater treatment by *Cyperus alternifolius* in different type of constructed wetland indicate that: Removal efficiency for Nitrogen different from 20% to 89% and none of could pass Malaysian 2009 Nitrogen regular; consequently this is the main objectives of this study.

Unsolved Problems

- Normal treatment system, need a lot of fossil fuel energy and chemicals are typically necessary to meet treatment objectives (Kadlec & Wallace, 2008)
- Malaysia has new N regulation (10 times less than previous) which is very difficult to comply with usual using treatment system. Department of Environment Malaysia, (2009
- In the present scenario, in Malaysia low sewerage tariff is unable to support the high operation and maintenance costs, thus following cheap treatment system (such as constructed wetland) is compulsory.
- It had been reported frequently that vertically constructed wetlands do not need aeration and specific design of vertical wetland system present suitable oxygen for microbial degradation (Cooper, 1999; Vymazal, 2007; Saeed & Sun, 2012) but we need to use artificial aeration because of mixed type of designing constructed wetlands that used in our study.

Table 2. Nitrogen removal performances in wetland systems, when carbon had been provided externally or internally

Wetland Type	Wastewater Type	Influent Pollution	Concentration mg/L	Influent C/N Ratio	Source of C	Denitrification Rate
HF	Domestic	NO_3-N	36	60 ml/d	Methanol	58-89
HF	Runoff	TN	10.1	1.13-3	Methanol	40-97
HF	Leachate	NO_3-N	209	0.1-0.8	Acetic acid	10-75
HF	Synthetic	$NaNO_3$	30-40	3:1	Glucose	60
VF	Synthetic	TN	41.5-8.6	2.5-10	Glucose	25-62
VF	Synthetic	TN	40.19	2.5-10	Glucose	46-87
VF	Synthetic	TN	2.3-59.9	-	Wood mulch	63-98
VF	Domestic	TN	30.9-31.6	-	Rice husk	98-100

(Summarized from(Saeed & Sun, 2012).

Table 3. Domestic wastewater treatment by Cyperus alternifolius (Umbrella plant) in constructed wetland researches summary

Type of CW	Absorbent	Treatments	Objective	Analysis	Location	Reference
vertical flow six circular concrete tanks 0.8 m x1.23 m	Sand and gravel	1.two different plants 2. Intermittently times 3.HRT	Compare Cyperus alternifolius and Vetiveria zizanioides N and P removal ability	Two way ANOVA	Thailand	(Kantawanichkul et al. 2013)
batch-fed Three Identical Tanks 5 m X 3 m X 1.8 m 2 years	Gravel + Coal slag (Cinder)	Contact time, HLR	N and P removal mechanisms of the soil Plant microbial interactions of constructed wetlands	Prediction model "Black box"	China	(Chan et al. 2008)
Vertical flow 8 System Pilot study	1.Yellow coal-burn-slag 2. black coal-burn-slag	1. Two substrates 2. Wet-to-dry ratios 3. Planted-unplanted	Total nitrogen removal	SAS	China	(Cui et al. 2009)
vertical flow PVC 1.3 m X 0.2 m lab study 3 month	Coal slag (Cinder) Local absorbent	1.Three different wet-to-dry ratios 2.planted-unplanted	Effects of plant on the pollutant removal efficiencies (N, P, SS)	ANOVA	Hong Kong	(Chan et al. 2008)
12 plastic boxes 25cm × 40 cm × 30 cm Hydroponic	No Absorbent No Aeration	1.diluted and undiluted sewage 2-different plants	1.diffrent plant species (6) eco-physiological growth and characteristics 2.ROL effect on decontamination	two-way ANOVA SPSS	China	(Li et al. 2013)
Horizontal	Pebbles Gravel Soil	1.Twelve emergent plant species 2. different influent salinities concentrations	1. selecting the suitable salt-tolerant plant 2. the effect of influent salinity on The performance of CWs	SPSS	China	(Gao et al. 2013)
Full scale 1400m2 (Hybrid)	Gravel (8–10mm) Depth of 0.45m	Horizontal CW system	1. the oxidation-reduction potential 2. microbial population density 3. specific oxygen uptake rate	No Data	China	(Zhai et al. 2012)
horizontal	silica quartz river gravel	*Cyperus alternifolius* L. and *Typha latifolia* L	the use and the choice of plant species	one-way ANOVA	Italy	(Leto et al. 2013)

Note: *Cyperus alternifolius*, constructed wetland and domestic wastewater had been used in all of these studies

- Normal systems do not control nitrogen in domestic wastewater and in fact typically send oxidized nitrogen to groundwater (Kadlec & Wallace, 2008)

With the inception of the National Environmental Health Action Plan (NEHAP) for Malaysia, issues on safe use of wastewater for agriculture should be discussed and addressed accordingly, so constructed wetland could be considered as an applicable treatment method. Malaysian regular and assessment of related researches in ability to pass this standard. All of these researches are about Nitrogen removal from wastewater in constructed wetland with use of *Cyperus alternifolius*. Only continuous and intermittent aeration modes which were investigated in the study of Chiemchaisri et al. (1992) could achieve

Table 4. Assessment of nitrogen removal efficiency to achieve Malaysian new regular

N Concentration Before Treatment	N Removal Efficiency	N Concentration After Treatment	Malaysian Regular[1]	References
TN:35.56 mg/L Ammonia N: 32.71 mg/L	50%	9.59 mg /L 8.37 mg/L	> 5 mg/L > 5 mg/L	(Chan et al. 2008)
23.1-+2.08 mg/L	40%	13.86-15.11 mg/L	> 5mg/L	(Chan et al. 2008)
Not given	29.71%	12.88 mg/L	> 5 mg/L	(Cui et al. 2009)
74.01mg/L	87.61%	13.81mg/L	> 5mg/L	(Zhai et al. 2012)
------	Not Given	------		(Li et al. 2013)
27.4 mg/L	65%	9.59 mg/L	> 5 mg/L	(Kantawanichkul et al. 2013)
17.8±0.5 (14.7–21.7) mg/L	36.1%	11.2±0.2 (9.0–12.6) mg/L	> 5mg/L	(Leto et al. 2013)
30.9-31.6 mg/L	100%	0 mg/L	< 5 mg/L	(Tee et al. 2012)
95.mg/L	89%	10.2 mg/L	> 5 mg/L	(Gikas and Tsihrintzis 2012)
130 mg/L	20%	104 mg/L	> 5mg/L	(Liao et al. 2003)
48.11 mg/L	91.5%	4.09 mg/L	< 5 mg/L	(Guo et al. 2014)

1: Malaysian last (2009) regular for Nitrogen concentration in domestic wastewater (5mg/ l).

Nitrogen concentration lower than 5 mg/l and recently study occurred in China by Guo et al. (2014) that combined vertical constructed wetland with sequencing batch biofilm reactor could to reach lower than 5 mg/l TN in effluent. The summary of related studies gathered in Table 2.

Practical Elements in Constructed Wetlands

Sampling

Based on the reported of Bayley et al. (2003), there is no significant difference in BOD attenuation rate between the upper, middle and bottom layers within the 0.5 m deep pilot-scale constructed wetland planted with *Phragmites australis*. Meanwhile, resent researches indicate that in a single constructed wetland, spatial variable physicochemical and microbial community expected even in single treatment system (Button, Nivala, et al., 2015; Weber & Legge, 2013).

Diameters of Constructed Wetland System

Based on the literature dimension 1.5 x 0.5 x 0.58m (Length X width X height), Yong *C.alternifolius* plants with 25-30 cm roots long and depth of wetland for assessment of plant absorption and media effect on wastewater treatment also considered as same diameters and finally with 5 cm additional volume in bottom for aeration on the other hands 3 cm additional for optimal operation and wastewater safely flow considered in total 58 cm depth/height (Kantawanichkul et al., 2009; Tee et al., 2012; Leto et al., 2013). Total area for each plant in such system is around 0.05 m^2 (Cui et al., 2009; Cui et al., 2011). Influent and effluent could be chosen based on the (Persson et al., 1999) advise which considered as good hydraulic efficiency design.

Plant Density

In the literatures which related to domestic wastewater in constructed wetland, with assistance of plant species (*Cyperus alternifolius*) there are many different plant densities. Kantawanichkul et al. (2013) used one plant for each column. Chan et al. (2008) used 3-4 plants in squire meter, (Cui et al., 2009), five rooted seedling plants in squire meter used by Chan et al. (2008). 10 young plants in 30 liters plastic tank (Li et al., 2013), five rooted steam in squire meter plants whit 30 centimeters (seedling) greenhouse growth plants with 30 centimeters (Leto et al., 2013). A system of compressor, air regulator and flow meter will use for aeration and a timer will use to control the aeration and non-aeration schedule (Yang et al., 2003).

Running the Wetland

There are several methods for running the constructed wetlands. Kantawanichkul et al. (2013) used 4-5 months preparation time period for system and Cui et al. (2009) one month waited for system preparation meanwhile. Meanwhile, Li et al. (2013) used tap water one week and Hoogland nutrients for preparation plants in the second week. Some characteristics of wastewater such as Biological Oxygen Demand (BOD), Total Nitrogen (TN), Ammonia nitrogen, Nitrite, Nitrate and Total phosphorus (TP) often collected with standard of APHA (2005).

CONCLUSION

Constructed wetland in both types of horizontal and vertical have been recognized as qualitative and effective treatment method for municipal wastewater. After selection of a suitable plant species in a reasonable density and capable media for plantation, engineering responses such as artificial aeration, dry and wet duration, carbon source addition absorbent and recirculation of wastewater will influence the quality of treated wastewater. Wastewater engineering is a type of managing these responses based on the environmental circumstances, pollutants concentration environmental discharge regulations and cost efficiency of treatment. Results of this chapter clearly illustrate municipal wastewater could be effectively fully treated by constructed wetland even with limited engineering techniques.

REFERENCES

Ahmaruzzaman, M. (2010). A review on the utilization of fly ash. *Progress in Energy and Combustion Science*, *36*(3), 327–363. doi:10.1016/j.pecs.2009.11.003

APHA. (2005). *Standard methods for the examination of water and wastewater*. American Public Health Association.

Artan, N. D., Orhon, D., & Tasli, R. (2002). Design of SBR systems for nutrient removal from wastewaters subject to seasonal fluctuations. *Water Science and Technology*, *46*(8), 91–98. PMID:12420970

Bayley, M. L., Davison, L., & Headley, T. R. (2003). Nitrogen removal from domestic effluent using subsurface flow constructed wetlands: Influence of depth, hydraulic residence time and pre-nitrification. *Water Science and Technology*, *48*(5), 175–182. PMID:14621162

Białowiec, A. L., Davies, L., Albuquerque, A., & Randerson, P. F. (2012). The influence of plants on nitrogen removal from landfill leachate in discontinuous batch shallow constructed wetland with recirculating subsurface horizontal flow. *Ecological Engineering*, *40*, 44–52. doi:10.1016/j.ecoleng.2011.12.011

Brisson, J., & Chazarenc, F. (2009). Maximizing pollutant removal in constructed wetlands: Should we pay more attention to macrophyte species selection? *The Science of the Total Environment*, *407*(13), 3923–3930. doi:10.1016/j.scitotenv.2008.05.047 PMID:18625516

Brix, H. (1997). Do macrophytes play a role in constructed treatment wetlands? *Water Science and Technology*, *35*(5), 11–17.

Button, M., Nivala, J., Weber, K. P., Aubron, T., & Müller, R. A. (2015). Microbial community metabolic function in subsurface flow constructed wetlands of different designs. *Ecological Engineering*, *80*, 162–171. doi:10.1016/j.ecoleng.2014.09.073

Calheiros, C. S., Duque, A. F., Moura, A., Henriques, I. S., Correia, A., Rangel, A. O., & Castro, P. M. (2009). Substrate effect on bacterial communities from constructed wetlands planted with *Typha latifolia* treating industrial wastewater. *Ecological Engineering*, *35*(5), 744–753. doi:10.1016/j.ecoleng.2008.11.010

Chan, S. Y., Tsang, Y. F., & Chua, H. (2008a). Domestic wastewater treatment using tidal-flow cinder bed with Cyperus alternifolius. *Aquatic Ecosystem Health & Management*, *11*(2), 206–211. doi:10.1080/14634980802100022

Chan, S. Y., Tsang, Y. F., & Chua, H. (2008b). Domestic wastewater treatment using batch-fed constructed wetland and predictive model development for NH_3-N removal. *Process Biochemistry*, *43*(3), 297–305. doi:10.1016/j.procbio.2007.12.009

Chiemchaisri, C., Wong, Y. K., Urase, T., & Yamamoto, A. K. (1992). Organic stabilization and nitrogen removal in membrane separation bioreactor for domestic wastewater treatment. *Water Science and Technology*, *25*(10), 231–240.

Cooper, P. (1999). A review of the design and performance of vertical-flow and hybrid reed bed treatment systems. *Water Science and Technology*, *40*(3). PMID:17802831

Cui, L. H., Ouyang, Y., Chen, Y., Zhu, X. Z., & Zhu, W. L. (2009). Removal of total nitrogen by *Cyperus alternifolius* from wastewaters in simulated vertical-flow constructed wetlands. *Ecological Engineering*, *35*(8), 1271–1274. doi:10.1016/j.ecoleng.2009.05.017

Cui, L. H., Zhu, X. Z., Ouyang, Y., Chen, Y., & Yang, F. L. (2011). Total phosphorus removal from domestic wastewater with Cyperus Alternifolius in vertical-flow constructed wetlands at the microcosm level. *International Journal of Phytoremediation*, *13*(7), 692–701. doi:10.1080/15226514.2010.525552 PMID:21972496

Devi, R., & Dahiya, R. (2006). Chemical oxygen demand (COD) reduction in domestic wastewater by fly ash and brick kiln ash. *Water, Air, and Soil Pollution*, *174*(1-4), 33–46. doi:10.100711270-005-9022-9

Foladori, P., Ruaben, J., & Ortigara, A. R. (2013). Recirculation or artificial aeration in vertical flow constructed wetlands: A comparative study for treating high load wastewater. *Bioresource Technology*, *149*, 398–405. doi:10.1016/j.biortech.2013.09.099 PMID:24128403

Foo, K., & Hameed, B. (2009). Value-added utilization of oil palm ash: A superior recycling of the industrial agricultural waste. *Journal of Hazardous Materials*, *172*(2), 523–531. doi:10.1016/j.jhazmat.2009.07.091 PMID:19695771

Gao, F., Yang, Z. H., Li, C., & Jin, W. H. (2013). Saline domestic sewage treatment in constructed wetlands: study of plant selection and treatment characteristics. *Desalination and Water Treatment*.

Gikas, G. D., & Tsihrintzis, V. A. (2012). A small-size vertical flow constructed wetland for on-site treatment of household wastewater. *Ecological Engineering*, *44*, 337–343. doi:10.1016/j.ecoleng.2012.04.016

Green, M., Friedler, E., & Safrai, I. (1998). Enhancing nitrification in vertical flow constructed wetland utilizing a passive air pump. *Water Research*, *32*(12), 3513–3520. doi:10.1016/S0043-1354(98)00182-1

Guo, Y.-M., Liu, Y.-G., Zeng, G.-M., Hu, X.-J., Xu, W.-H., Liu, Y.-Q., ... Huang, H.-J. (2014). An integrated treatment of domestic wastewater using sequencing batch biofilm reactor combined with vertical flow constructed wetland and its artificial neural network simulation study. *Ecological Engineering*, *64*, 18–26. doi:10.1016/j.ecoleng.2013.12.040

Hocking, P. J. (1985). Effects of sodium and potassium chlorides on the growth and accumulation of mineral ions by Cyperus involucratus Rottb. *Aquatic Botany*, *21*(3), 201–217. doi:10.1016/0304-3770(85)90049-X

Inamori, R., Gui, P., Dass, P., Matsumura, M., Xu, K. Q., Kondo, T., ... Inamori, Y. (2007). Investigating CH_4 and N_2O emissions from eco-engineering wastewater treatment processes using constructed wetland microcosms. *Process Biochemistry*, *42*(3), 363–373. doi:10.1016/j.procbio.2006.09.007

Kadlec, R. H., & Wallace, S. (2008). *Treatment wetlands*. CRC press. doi:10.1201/9781420012514

Kantawanichkul, S., Kladprasert, S., & Brix, H. (2009). Treatment of high-strength wastewater in tropical vertical flow constructed wetlands planted with Typha angustifolia and Cyperus involucratus. *Ecological Engineering*, *35*(2), 238–247. doi:10.1016/j.ecoleng.2008.06.002

Kantawanichkul, S., & Neamkam, P. (2003). Optimum recirculation ratio for nitrogen removal in a combined system: vertical flow vegetated bed over horizontal flow sand bed.

Kantawanichkul, S., Sattayapanich, S., & Van Dien, F. (2013). Treatment of domestic wastewater by vertical flow constructed wetland planted with umbrella sedge and Vetiver grass. *Water Science and Technology*, *68*(6), 1345–1351. doi:10.2166/wst.2013.379 PMID:24056433

Kröpfelová, L. (2008). *Wastewater treatment in constructed wetlands with horizontal sub-surface flow*. Springer.

Kutty, S. R. M., Isa, M. H., & Leong, L. C. (2011). Removal of Ammonia-Nitrogen (NH_3-N) and Nitrate by Modified Conventional Activated-Sludge System to Meet New DOE Regulations.

Langergraber, G. (2005). The role of plant uptake on the removal of organic matter and nutrients in subsurface flow constructed wetlands: A simulation study. *Water Science and Technology*, *51*(9), 213–223. PMID:16042261

Laohaprapanon, S., Marques, M., & Hogland, W. (2010). Removal of Organic Pollutants from Wastewater Using Wood Fly Ash as a Low-Cost Sorbent. Clean–Soil, Air. *Water (Basel)*, *38*(11), 1055–1061.

Lee, B.-H., & Scholz, M. (2007). What is the role of *Phragmites australis* in experimental constructed wetland filters treating urban runoff? *Ecological Engineering*, *29*(1), 87–95. doi:10.1016/j.ecoleng.2006.08.001

Leto, C., Tuttolomondo, T., Bella, S. L., Leone, R., & Licata, M. (2013). Growth of Arundo donax L. and *Cyperus alternifolius* L. in a horizontal subsurface flow constructed wetland using pre-treated urban wastewater—a case study in Sicily (Italy). *Desalination and Water Treatment*, *51*(40-42), 7447–7459. doi:10.1080/19443994.2013.792134

Leto, C., Tuttolomondo, T., La Bella, S., Leone, R., & Licata, M. (2013). Effects of plant species in a horizontal subsurface flow constructed wetland–phytoremediation of treated urban wastewater with *Cyperus alternifolius* L. and *Typha latifolia* L. in the West of Sicily (Italy). *Ecological Engineering*, *61*, 282–291. doi:10.1016/j.ecoleng.2013.09.014

Li, L., Yang, Y., Tam, N. F., Yang, L., Mei, X. Q., & Yang, F. J. (2013). Growth characteristics of six wetland plants and their influences on domestic wastewater treatment efficiency. *Ecological Engineering*, *60*, 382–392. doi:10.1016/j.ecoleng.2013.09.044

Liao, X., Luo, S., Wu, Y., & Wang, Z. (2003). Studies on the abilities of *Vetiveria zizanioides* and *Cyperus alternifolius* for pig farm wastewater treatment. In *International conference on vetiver and exhibition*.

Liao, X., Luo, S., Wu, Y., & Wang, Z. (2005). Comparison of nutrient removal ability between *Cyperus alternifolius* and Vetiveria zizanioides in constructed wetlands. *Ying yong sheng tai xue bao [The journal of applied ecology]*, *16*(1), 156-160.

Luederitz, V., Eckert, E., Lange-Weber, M., Lange, A., & Gersberg, R. M. (2001). Nutrient removal efficiency and resource economics of vertical flow and horizontal flow constructed wetlands. *Ecological Engineering*, *18*(2), 157–171. doi:10.1016/S0925-8574(01)00075-1

Moore, B. C., Lafer, J. E., & Funk, W. H. (1994). Influence of aquatic macrophytes on phosphorus and sediment porewater chemistry in a freshwater wetland. *Aquatic Botany*, *49*(2), 137–148. doi:10.1016/0304-3770(94)90034-5

Morgenroth, E., & Wilderer, P. (1998). Sequencing Batch Reactor Technology: Concepts, Design and Experiences (Abridged). *Water and Environment Journal: the Journal / the Chartered Institution of Water and Environmental Management*, *12*(5), 314–320. doi:10.1111/j.1747-6593.1998.tb00192.x

Nagrale, S. D., Hajare, H., & Modak, P. R. (2012). Utilization of Rice Husk Ash. *Carbon*, *2*, 6=42.

Nivala, J., Hoos, M. B., Cross, C., Wallace, S., & Parkin, G. (2007). Treatment of landfill leachate using an aerated, horizontal subsurface-flow constructed wetland. *The Science of the Total Environment*, *380*(1), 19–27. doi:10.1016/j.scitotenv.2006.12.030 PMID:17241653

Okurut, T. O., Rijs, G. B. J., & Van Bruggen, J. J. A. (1999). Design and performance of experimental constructed wetlands in Uganda, planted with *Cyperus papyrus* and *Phragmites mauritianus*. *Water Science and Technology*, *40*(3), 265–271.

Ouellet-Plamondon, C., Chazarenc, F., Comeau, Y., & Brisson, J. (2006). Artificial aeration to increase pollutant removal efficiency of constructed wetlands in cold climate. *Ecological Engineering*, *27*(3), 258–264. doi:10.1016/j.ecoleng.2006.03.006

Persson, J., Somes, N. L. G., & Wong, T. H. F. (1999). Hydraulics efficiency of constructed wetlands and ponds. *Water Science and Technology*, *40*(3), 291–300.

Petticrew, E. L., & Kalff, J. (1992). Water flow and clay retention in submerged macrophyte beds. *Canadian Journal of Fisheries and Aquatic Sciences*, *49*(12), 2483–2489. doi:10.1139/f92-274

Platzer, C. (1999). Design recommendations for subsurface flow constructed wetlands for nitrification and denitrification. *Water Science and Technology*, *40*(3), 257–263.

Saeed, T., & Sun, G. (2012). A review on nitrogen and organics removal mechanisms in subsurface flow constructed wetlands: Dependency on environmental parameters, operating conditions and supporting media. *Journal of Environmental Management*, *112*, 429–448. doi:10.1016/j.jenvman.2012.08.011 PMID:23032989

Saeed, T., & Sun, G. (2011). Enhanced denitrification and organics removal in hybrid wetland columns: Comparative experiments. *Bioresource Technology*, *102*(2), 967–974. doi:10.1016/j.biortech.2010.09.056 PMID:20934326

Shelef, O., Gross, A., & Rachmilevitch, S. (2013). Role of Plants in a Constructed Wetland: Current and New Perspectives. *Water (Basel)*, *5*(2), 405–419. doi:10.3390/w5020405

Sklarz, M. Y., Gross, A., Yakirevich, A., & Soares, M. I. M. (2009). A recirculating vertical flow constructed wetland for the treatment of domestic wastewater. *Desalination*, *246*(1–3), 617–624. doi:10.1016/j.desal.2008.09.002

Stefanakis, A., Akratos, C. S., & Tsihrintzis, V. A. (2014). *Vertical flow constructed wetlands : eco-engineering systems for wastewater and sludge treatment*. Elsevier Science.

Stottmeister, U., Wießner, A., Kuschk, P., Kappelmeyer, U., Kästner, M., Bederski, O., ... Moormann, H. (2003). Effects of plants and microorganisms in constructed wetlands for wastewater treatment. *Biotechnology Advances*, *22*(1), 93–117. doi:10.1016/j.biotechadv.2003.08.010 PMID:14623046

Tee, H.-C., Lim, P.-E., Seng, C.-E., Mohd Nawi, M. A., & Adnan, R. (2015). Enhancement of azo dye Acid Orange 7 removal in newly developed horizontal subsurface-flow constructed wetland. *Journal of Environmental Management*, *147*, 349–355. doi:10.1016/j.jenvman.2014.09.025 PMID:25284799

Tee, H.-C., Lim, P.-E., Seng, C.-E., & Nawi, M.-A. M. (2012). Newly developed baffled subsurface-flow constructed wetland for the enhancement of nitrogen removal. *Bioresource Technology*, *104*, 235–242. doi:10.1016/j.biortech.2011.11.032 PMID:22130081

Van Kempen-Fryling, R. J., & Camper, A. K. (2017). Escherichia coli O157:H7 attachment and persistence within root biofilm of common treatment wetlands plants.

Vymazal, J. (2011) Plants used in constructed wetlands with horizontal subsurface flow: a review. *Hydrobiologia*, *674*(1), 133–156.

Vymazal, J. (2007). Removal of nutrients in various types of constructed wetlands. *The Science of the Total Environment*, *380*(1), 48–65. doi:10.1016/j.scitotenv.2006.09.014 PMID:17078997

Vymazal, J. (2013). The use of hybrid constructed wetlands for wastewater treatment with special attention to nitrogen removal: A review of a recent development. *Water Research*, *47*(14), 4795–4811. doi:10.1016/j.watres.2013.05.029 PMID:23866134

Weber, K. P., & Legge, R. L. (2013). Comparison of the catabolic activity and catabolic profiles of rhizospheric, gravel-associated and interstitial microbial communities in treatment wetlands. *Water Science and Technology*, *67*(4), 886–893. doi:10.2166/wst.2012.637 PMID:23306269

Yeh, T. Y., Wu, M. H., Cheng, C. Y., & Hsu, Y. H. (2015). *A Study and Analysis on the Physical Shading Effect of Water Quality Control in Constructed Wetlands in Research and Practices in Water Quality*. InTech.

Weis, J. S., & Weis, P. (2004). Metal uptake, transport and release by wetland plants: Implications for phytoremediation and restoration. *Environment International*, *30*(5), 685–700. doi:10.1016/j.envint.2003.11.002 PMID:15051245

White, K. D. (1995). Enhancement of nitrogen removal in subsurface flow constructed wetlands employing a 2-stage configuration, an unsaturated zone, and recirculation. *Water Science and Technology*, *32*(3), 59–67.

Xinshan, S., Qin, L., & Denghua, Y. (2010). Nutrient removal by hybrid subsurface flow constructed wetlands for high concentration ammonia nitrogen wastewater. *Procedia Environmental Sciences*, *2*, 1461–1468. doi:10.1016/j.proenv.2010.10.159

Yang, P. Y., Su, R., & Kim, S. J. (2003). EMMC process for combined removal of organics, nitrogen and an odor producing substance. *Journal of Environmental Management*, *69*(4), 381–389. doi:10.1016/j.jenvman.2003.09.015 PMID:14680899

Yao, F., Shen, G. X., Li, X. L., Li, H. Z., Hu, H., & Ni, W. Z. (2011). A comparative study on the potential of oxygen release by roots of selected wetland plants. *Physics and Chemistry of the Earth Parts A/B/C*, *36*(9), 475–478. doi:10.1016/j.pce.2010.11.001

Zeeman, G., Kujawa, K., De Mes, T., Hernandez, L., De Graaff, M., Abu-Ghunmi, L., ... Van Lier, J. (2008). Anaerobic treatment as a core technology for energy, nutrients and water recovery from source-separated domestic waste (water). *Water Science and Technology*, *57*(8), 1207. doi:10.2166/wst.2008.101 PMID:18469391

Zhai, J., Zou, J., He, Q., Ning, K., & Xiao, H. (2012). Variation of dissolved oxygen and redox potential and their correlation with microbial population along a novel horizontal subsurface flow wetland. *Environmental Technology*, *33*(17), 1999–2006. doi:10.1080/09593330.2012.655320 PMID:23240193

Zhao, Y. Q., Sun, G., & Allen, S. J. (2004). Purification capacity of a highly loaded laboratory scale tidal flow reed bed system with effluent recirculation. *The Science of the Total Environment*, *330*(1), 1–8. doi:10.1016/j.scitotenv.2004.03.002 PMID:15325153

Chapter 9
Role of Microorganisms in Bioremediation of Pesticides

Verinder Wahla
Gurukul Kangri University, India

Shruti Shukla
Gurukul Kangri University, India

ABSTRACT

This chapter describes how exposure to pesticides has become one of the most prominent components of current agriculture. Pesticides have high impact on farmer's economy as they prevent and minimize agricultural losses due to pests (insects, fungi, nematode), improve yield and quality food in terms of superficial appeal. However, these synthetic pesticides have serious environmental concerns and create a variety of toxic effects in living beings. In this regard, the bioremediation approach can be a suitable alternative over the conventional treatment methods for the treatment of pesticides-contaminated sites. Bioremediation is the use of microorganisms or their enzymes to degrade and detoxify the pesticide from the environment. Microbial interactions greatly influence natural bio remedial processes occurring in soil. Microorganisms involved various biochemical mechanisms for the degradation of pesticides, such as oxidation and reduction reactions. This chapter provides an overview on the nature, characteristics and toxicity of pesticides in environment and role of microbes in bioremediation of pesticides.

INTRODUCTION

A pesticide can be defined as any substance or mixture of substances that are used for preventing and destroying unwanted pests. Insecticides, herbicides and fungicides are various pesticide used to control different pests (EPA, 2015). Pesticides and agrochemicals, in general, became an important component of worldwide agriculture systems during the last century, allowing for a noticeable increase in crop yields and food production (Alexandratos & Bruinsma, 2012). In modern agriculture practices, the extensive use of pesticides is very frequent to fulfil higher yield requirements. Dispersion of pesticide residues in the environment and mass killing of nonhuman biota, such as bees, birds, amphibians, fish, and small mammals, were also reported (Paoli et al., 2015; WHO, 2017). The problem of environmental

DOI: 10.4018/978-1-5225-3540-9.ch009

contamination by pesticides goes beyond the locality where it is used. The agricultural pesticides that are exhaustively applied to the land surface travel long distances and can move downward until reaching the water table at detectable concentrations, reaching aquatic environments at significantly longer distances. Therefore, the fate of pesticides is often uncertain, they can contaminate other areas that are distant from where they were originally used. Thus, decontaminating pesticide-polluted areas is a very complex task (Gavrilescu, 2005). In addition to a high risk of contamination in ecosystem, pesticide also has chronic threats to human life, caused by long term exposure to pesticides. Pesticides can cause hormonal disruption, diminished intelligence, and reproductive abnormalities in humans (Gupta, 2004). Thus, the degradation of persistent pesticides is very essential for the environment (Hodaifa et al., 2009).

Earlier techniques or technologies which were used to remove pesticides from environment were landfills, recycling, pyrolysis etc., but these have opposing effects on the environment by formatting toxic intermediates (Debarati et al., 2005). All these methods have proved to be expensive and difficult in case of pesticides. One promising treatment method to remove or degrade toxic components from environment, is bioremediation. Bioremediation is the engineered process of toxic removal from environment. This is a promising alternative to physico-chemical methods of remediation, because it is reasonable and can selectively gain complete destruction of toxic components (Hussain et al., 2009). The use of microorganisms for the degradation and detoxification of numerous toxic pesticides, proved to be an efficient tool to decontaminate the polluted sites in the prevailing environment (Mervat et al., 2009). Pesticides in soil and water can be biodegraded and is the primary mechanism of pesticide breakdown and detoxification in many soils (Surekha et al., 2008). Thus, biological decontamination methods are preferable to conventional approaches because, in general, microorganisms degrade numerous environmental pollutants without producing toxic intermediates (Pieper & Reineke, 2000; Furukawa, 2003). Microbes which were commonly reported in pesticides bioremediation *include Pseudomonas sp., Bacillus sp, Klebsiella sp, Pandoraea sp, Phanerochaete Chrysosporium, Mycobacterium sp. Agrocybe semiorbicularis, Auricularia auricula, Coriolus versicolor, Dichomitus squalens, Flammulina velupites, Hypholoma fasciculare, Pleurotus ostreatus, Stereum hirsutum, and Avatha discolour.* (Colberg & Young, 1995). These microorganisms have shown their ability to degrade various pesticide groups like phenylamide, triazine, phenylurea, dicarboximide, chlorinated and organophosphorus compounds. In this chapter we are going to discuss about different toxic pesticide with their harmful effects and also mechanism of microorganism involved in biodegradation and bioremediation of Pesticide.

Status of Pesticides in India and World

Every year approximately two million tons of pesticides are consumed throughout the world and among which, 45%in Europe, 24% is consumed in USA, and 25% in rest of the world. Consumption of pesticide among the Asian countries, is highest in China followed by Korea, Japan and India. India is agriculture-based country so to fulfill large scale food demand of people many of the pesticides use commonly. Discussing the total consumption of different type of pesticides in India 80% are in the form of insecticides, 15% are herbicides, 2% are fungicides, and less than 3% are others. This is due to increased insect pest attack caused mainly by the prevailing warm humid climatic condition (Bhat & Padmaja, 2014). One more main cause of increasing demand of pesticides is the promotion of High Yielding Varieties that marked the green revolution has led to large scale use of chemicals as pesticides. At present, India is the largest producer of pesticides in Asia and ranks twelfth in the world, with an annual production of 90,000 tonnes (Abhilash & Singh, 2009; David et al., 2005).

Pesticide and Its Classification

The pesticides are classified depending upon target species and mode of action. According to the application the main classes of pesticides include herbicides (that kills weeds), fungicides (that kills fungi) and insecticides (that kills insects). Depending upon the chemical nature, the major classes of pesticides includes organochlorine, organophosphorus, thiocarbamates, pyrethroids, neonicotinoids, glycine derivatives and dipyridyl derivatives.

Organochlorines

The organochlorine pesticides are known to be highly persistent in the environment. it is an organic compound containing at least one covalently bonded chlorine atom. Their wide structural variety and diverse chemical properties lead to a broad range of applications. Many derivatives are controversial because of the effects of these compounds on the environment. These are insecticides composed primarily of carbon, hydrogen and chlorine. Most of them break down slowly and can remain in the environment long after application and in organisms long after exposure. The accumulation of organochlorine compounds is a result of their chemical structure and their physical properties such as polarity and solubility. These fat-soluble compounds persist in both the body and the environment (Barr & Needham, 2002). Organochlorine pesticide residues are capable of disrupting DNA (deoxyribonucleic acid) in unborn children, the endocrine system, as well as damaging nerves and brain cells (Oyeyiola et al., 2017). Some of the examples of organocholrine pesticides include alpha-BHC, beta-BHC, delta-BHC, gamma-BHC (Lindane), heptachlor, endosulfan, methoxychlor, aroclor and dichlorodiphenyltrichloroethane (DDT). DDT is banned from worldwide but its alternatives are used commonly in other names (Patnaik, 2003; Menone et al., 2001).

Organophosphates

The next large class developed was the organophosphates. Organophosphorus is the general name for organic derivatives of phosphorus. Among the different groups of pesticides, organophosphate are said to be major and probably the most widely used group of pesticides in the world, that nearly accounts for more than 36% of the total world market. Organophosphates pesticides are all esters of phosphoric acid which include aliphatic, phenyl and heterocyclic derivatives (Oga, 2003). They are the most commonly used insecticides because their unstable chemical structure leads to rapid hydrolysis and little long-term accumulation in the environment. They have characteristic of irreversibly inhibiting the enzyme acetylcholine esterase in the nervous system, which acts in both insects and in mammal (Griza et al., 2008). This results in disruption of nerve terminals resulting in agitation, hypersalivation, convulsion, and ultimately death of insects and mammals (King & Aaron, 2015; Rayu et al., 2017). Some of the organophosphorus compounds includes-Acephate, azinphos-methyl, bensulide, chlorethoxyfos, chlorpyrifos, diazinon, dicrotophos, dimethoate, disulfoton, ethoprop, fenamiphos, fenitrothion, fenthion, fosthiazate, methamidophos, monocrotophos and malathion (Ghosh et al., 2010). CP is a highly effective pesticide because it irreversibly inhibits acetylcholinesterase (AChE) enzymes leading to excess acetylcholine accumulation at nerve terminals resulting in agitation, hypersalivation, convulsion, and ultimately death of insects and mammals (King & Aaron, 2015).

Figure 1. General structures of Organochlorines pesticide

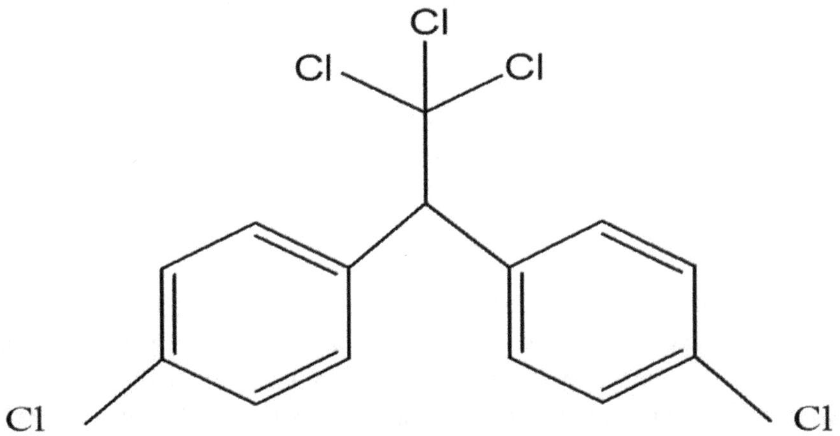

Figure 2. General structures of organophosphates pesticide

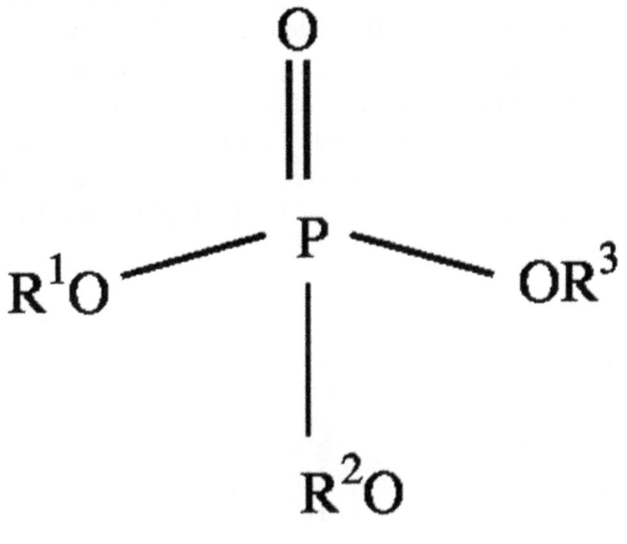

Carbamates

Carbamates pesticides were introduced around 1956 and are still being used extensively as pest control because of their effectiveness in broad spectrum of biological activities such as fungicides, insecticides, and herbicides. Chemically, the carbamate pesticides are esters of carbamates and organic compounds derived from carbamic acid. This group of pesticides can be divided into benzimidazole-, N-methyl-, N-phenyl-, and thiocarbamates. The compounds derived from carbamic acid are probably the insecticides with the widest range of biocide activities (Sogorb & Vilanova, 2002). Carbamates are closely related to organophosphorus compounds and kills insects in a similar fashion as organophosphate insecticides.

They are widely used in homes, gardens, and agriculture. Like the organophosphates, their mode of action is inhibition of cholinesterase enzymes, affecting nerve impulse transmission. High polarity and solubility in water and thermal instability are typical characteristics of carbamate pesticides, as well as high acute toxicity (Soriano et al., 2001). Symptoms of the carbamate poisoning are similar to that of organophosphorus poisoning. Bendiocarb, carbofuran, carbaryl, dioxacarb, fenobucarb, fenoxycarb, isoprocarb and methomyl are some of the examples of carbamate insecticides.

Pyrethrins and Pyrethroids

Pyrethroids are relatively low toxic in contrast to organophosphorus and carbamates pesticides, synthetic pyrethroids are one of the extensively used pesticides and account for more than one-third of the insecticides currently marketed in the world (Le et al., 2017). Pyrethrins and pyrethroids function mainly as insecticides. Pyrethrins are natural insecticides originating from the dried head of plant *Chrysanthemum cinerariaefolium.* which are cultivated as ornamentals for their showy flower. (Laskowski, 2002; Cycoń & Piotrowska, 2017.) They comprise active agents (pyrethrins I-VI), but pyrethrins I and II are the most active. These compounds decompose rapidly in the presence of light, but synthetic production of pyrethroids around 1950 overcame some disadvantages of natural pyrethrins. The chemical compound, pyrethroid pesticides, has been refined to enact the insecticidal activity of the natural compound pyrethrum (Ellenhorn & Barceloux, 1988). Compounds in this group are often applied against household pests. Cypermethrin, cyfluthrin, deltamethrin, etofenprox, fenvalerate, permethrin, phenothrin, prallethrin, resmethrin and tetramethrin are some of the pyrethroids. Cypermethrin (CM) is a neurotoxicant pyrethroid that is persistent in soil and sediment the compound is very toxic to aquatic organisms. Freshwater fish (Channa punctatus) exposed to cypermethrin has low levels of red blood cells and proteins in the blood (Saxena & Seth, 2002).

Figure 3. General structures of carbamate pesticide

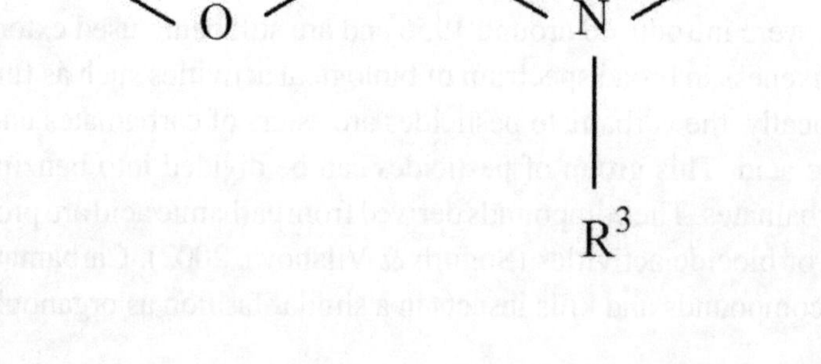

Figure 4. General structures of Pyrethrins pesticide

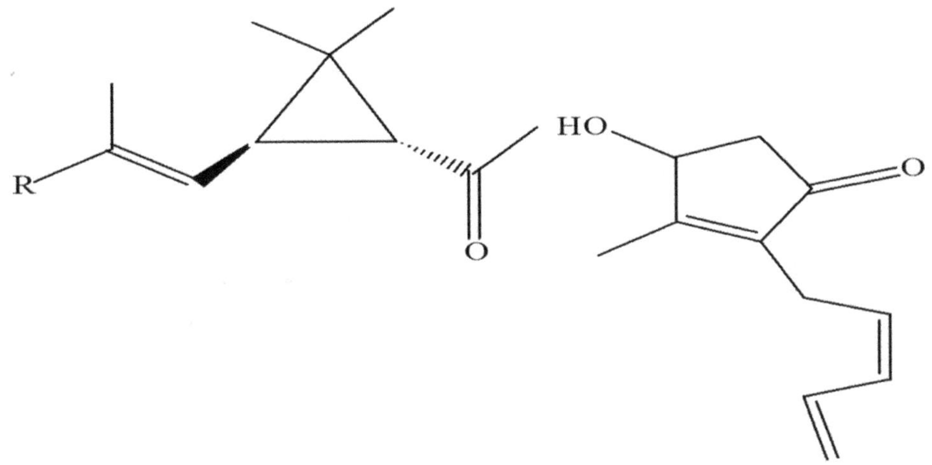

Neonicotinoids

Neonicotinoid pesticides were first introduced in the early 1990s and since then they are extensively applied and become the most widely used class of insecticide in the world. This increase in demand has broadly occurred from the early 2000s. Neonicotinoids are water-soluble, and broadly use in seed treatment. often as substitutes for organophosphates and carbamates (Douglas & Tooker, 2015). Neonicotinoids are synthetic mimic of the natural insecticide nicotine (very less toxic to mammals and have greater field persistence). These are wide spectrum insecticides with rapid action from minutes to hours. It includes acetamiprid, clothianidin, imidacloprid, nitenpyram, nithiazine, thiacloprid and thiamethoxam (Tomizawa & Casida, 2005). Imidacloprid is the most widely used insecticide in the world. Compared to organophosphate and carbamate insecticides, neonicotinoids cause less toxicity in birds and mammals than insects. Some breakdown products are also toxic to insects.

Triazines

The class of triazine is the largest proportion of herbicides used in agriculture and horticulture as they are inexpensive and very effective as herbicide. They kill broad spectrum of weeds present in crops and public gardens by inhibiting photosynthesis and electron transport chain in plants. (Stamatis et al., 2010; Kim et al., 2001). Physiological and molecular variations in organisms due to accumulation of these compounds is still very complex. The chemical structures of triazine herbicides correlate to the alkyl substituted 2,4-diamines of chlorotriazine). After absorption, these compounds undergo conjugation with glutathione or simply dealkylation. The chemical structures of triazine herbicides correspond to permutations of the alkyl substituted 2,4-diamines of chlorotriazine (Barr & Needham, 2002).After absorption, these compounds undergo conjugation with glutathione or simply dealkylation. Triazine herbicides intensive use indicates its potential as a pollutant of both ground waters and surface waters (Bowmer, 1991).

Figure 5. General structures of neonicotiniods pesticide

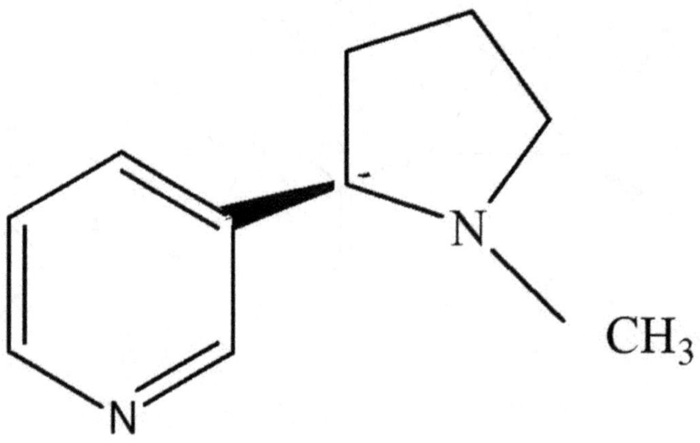

Figure 6. General structures of Triazines pesticide

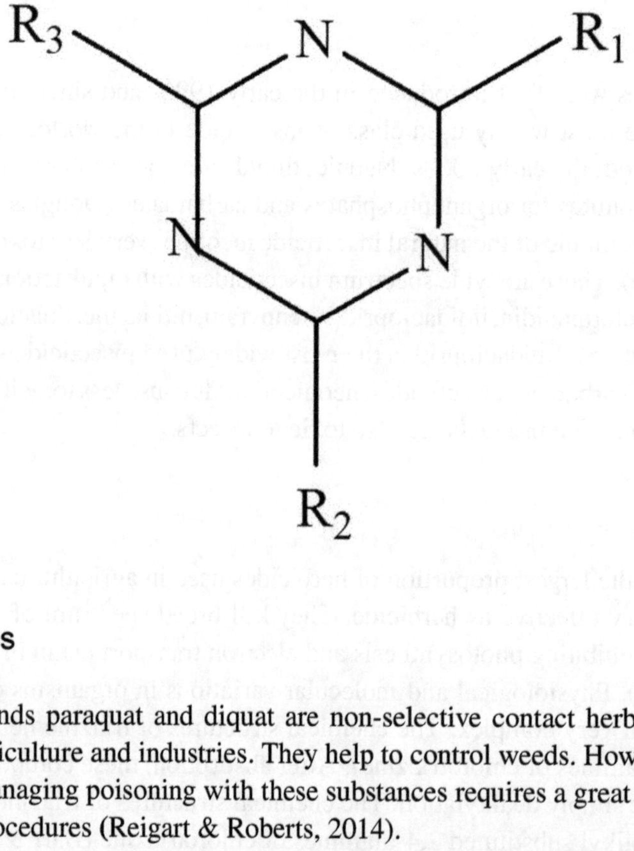

Dipyridyl Derivatives

The dipyridyl compounds paraquat and diquat are non-selective contact herbicides that have found wide application in agriculture and industries. They help to control weeds. However, these compounds are highly toxic and managing poisoning with these substances requires a great skill and knowledge of proper management procedures (Reigart & Roberts, 2014).

Glycine Derivatives

Two representatives of this class are glyphosate (N-phosphonomethyl glycine) and glufosinate (N-phosphonomethyl homoalanine), marketed primarily as the isopropylamine salt (glyphosate) or ammonium

Figure 7. General structures of Dipyridyl pesticide

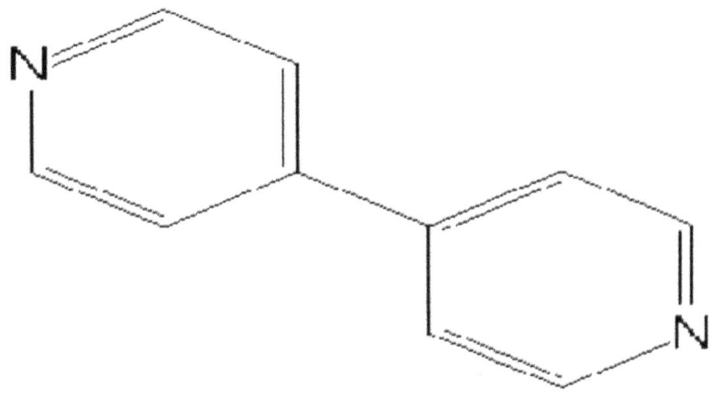

salt (glufosinate). Both substances are broad-spectrum nonselective systemic herbicides with application in for post-emergent control of annual and perennial plants. Glyphosate inhibits an enzyme in the biosynthesis of tryptophan, phenylalanine, and tyrosine, present in plants, fungi, and bacteria, but not in animals or humans (Campbell, 2014). However, according to literature reports, glyphosate can enter living organisms, including humans, where it exerts various toxic effects (De Liz Oliveira Cavalli et al., 2013).

Dithiocarbamates

According to WHO, yearly consumption of Dithiocarbamates is between 25,000 and 35,000 metric tonnes, and most of the DTCs applied as fungicides are classified as being hazardous (Kanchi et al., 2003). Dithiocarbamates are classified into two groups depending on which metal cation is present in the chemical structure, [I] dimethyldithiocarbamate and [II] ethylenebisdithiocarbamate. The nomenclature of various compounds of this class is related to the association of the metalcations; e. g., maneb (manganese), and zineb and ziram (zinc) (Alonzo & Corrêa, 2014). They are slow absorber of the means that they have low acute oral and dermal toxicity. On the other hand, chronic exposure to dithiocarbamates leads to adverse effects due to contact with dithiocarbamate acid or metal ligand (Costa, 2008). The metabolite that arises from dithiocarbamates biotransformation is ethylenethiourea, which induces thyroid cancer and modifies thyroid hormones. Although these products are little toxicity to humans, they are potential precursors of ethylenethiourea, which has carcinogenic and teratogenic action. There is no specific treatment for poisoning with this class of compounds, so only maintenance of vital functions and minimization of their absorption (activated charcoal) are necessary (Sampogna & Cunard, 2007).

Harmful Effects of Pesticides

Pesticides released into the environment may have several adverse ecological effects ranging from long-term effects to short-lived changes in the normal functioning of an ecosystem. Despite the good results of using pesticides in agriculture and public health, their use is usually accompanied with deleterious environmental and public health effects.

Figure 8. General structures of Dithiocarbamates pesticide

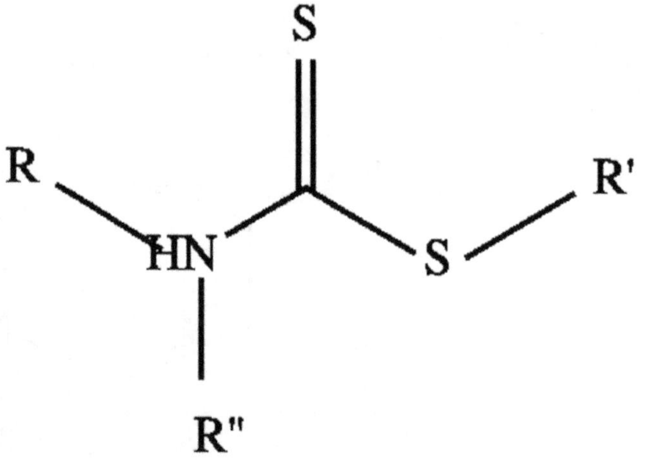

Effects of Pesticide on Human Health and Diseases

There is no doubt that pesticides have improved the standard of human health by controlling vector-borne diseases, however, their long-term use has resulted in dangerous health effects. Human beings, specifically children are highly prone to harmful effects of pesticides because of ist toxic nature and inadequate use. Due to increased use of pesticide over the past few decades, exposure of human beings to these chemicals has also increased considerably. The World Health Organisation estimates Worldwide pesticide production increased at a rate of about 11% per year, from 0.2 million tons in 1950s to more than 5 million tons by 2000 (FAO, 2017). About 2.2 million people, mainly belonging to developing countries are at increased risk of exposure to pesticides (Hicks, 2013). Pesticides enter the human body through ingestion, inhalation or penetration via skin (Spear, 1991). But the majority of people get affected via the intake of pesticide contaminated food. Through different pathways, they ultimately reach human tissues and organs. Toxic effects are produced when the concentration of pesticide in the body increases far more than its initial concentration in the environment (Hayo & Werf, 1996). The effects of pesticides on human health are highly variable. According to the Environmental Health Investigations Branch of the California Department of Health, "exposures to pesticides may cause a wide variety of both acute (short-term) and chronic (long-term) health effects, depending upon the type of pesticide and the amount of exposure. Signs of acute poisoning may include diarrhoea, pinpoint pupils, rashes, nausea and vomiting. Some pesticides may cause eye or throat irritation. Chronic exposure (greater than 1 year) to some types of pesticides may aggravate asthma symptoms; other types may increase the risk for certain types of cancers and birth defects, or cause damage to the genetic and immune systems." More severe pesticide poisoning can cause respiratory failure and convulsions, leading to coma and death. Exposure to organophosphates and carbamates causes, symptoms similar to those of increased neurotransmitter-acetylcholine. These pesticides interfere with the normal nerve signal transduction and exposure to them causes headaches, dizziness, confusion, nausea and vomiting, muscle and chest pain. Difficulty breathing, convulsions, coma and death may occur in severe cases. Pyrethroids can cause an allergic skin response, aggressiveness, hyper- excitation, reproductive or developmental effects in addition to causing tremors and seizures. It is observed that there is a relationship between pesticides and Parkinson's disease and Alzheimer's disease (Casida & Durkin, 2013). Recent reviews of exposure and health impact of pesticides on human health have underlined the burden on human health and re-evaluated the current toxicity of legacy pesticide residues (Mrema et al., 2013). The WHO and IARC, among other organizations, keep under close scrutiny and revision the advisories on toxicity of new and old chemicals. Many agrochemicals were related to prostate cancer and other types of cancer and are increasingly regulated (Singh et al., 2016; ECA, 2017).

Effects of Pesticide on Soil and Water

Pesticides enter the soil through spray drift during foliage treatment, wash-off from treated vegetation, or from treated seeds in soil. Some pesticides are applied directly into soil to control pests and plant diseases presented in soil. The transport, persistence or degradation of pesticides in soil depends on their chemical properties of the soil. All these factors affect sorption/ desorption, volatilisation, degradation, uptake by plants, run-off, and leaching of pesticides. Persistence of pesticides in soil can vary from few hours to many years in case of OC pesticides. Despite OC pesticides were banned or restricted in many countries, they are still detecting in soils (Shegunova et al., 2007; Toan et al., 2007; Li et al., 2008; Jiang et al., 2009). Pesticides can get into water by drift during pesticide spraying, leaching through the soil, by runoff from treated area. In some cases, pesticides can be applied directly onto water surface e.g. for control of mosquitoes.

In cases of water contamination, it depends mainly on nature of pesticides (water solubility, hydrophobicity), landscape, soil properties, weather conditions, and also on the distance from an application site to a water source. Heavy rainfall shortly after application of the pesticide to soils is the main reason of rapid transport to groundwater may. In 1990, herbicide Atrazine and Endosulphan were found most often in surface waters in the USA and Australia due to their widespread use. Other pesticides detected included Pronofos, Dimethoate, Chlordane, Diuron, Prometryn and Fluometuron. More recent studies also reported presence of pesticides in surface water and groundwater close to agriculture lands over the world (Cerejeira et al., 2003; Konstantinou et al., 2006; Gilliom, 2007; Woudneh et al., 2009; Añasco et al., 2010).

In general, the compounds most frequently detected were currently used pesticides (herbicides Atrazine, Simazine, Alachlor, Metolachlor and Trifluralin, insecticides Diazinon, Parathion methyl, and organochlorine compounds due to their long persistence (lindane, endosulfan, aldrin, and other organochlorine pesticides).

The geographic distribution of pesticide depends on patterns in land use and pesticide use. Rivers and streams were more polluted than ground waters and more near the areas with substantial agricultural and urban land use. Pesticides usually occurred in mixture of multiple compounds, even if individual pesticide were detected bellow limits. This potentially can lead to underestimation of toxicity when assessments are based on individual compounds. High levels of pesticides chlordecone were detected in coastline, rivers, sediments and groundwater in the Caribbean island of Martinique due to its massive application on bananas plantations (Bocquené & Franco, 2005).

Bioremediation of Pesticides by Different Microorganisms

Bioremediation can be defined as any process that uses microorganisms or their enzymes to return the environment altered by contaminants to its original condition. Early studies provided the evidence that microbial degradation played a vital role in enhancing the degradation of pesticide. (Shanahan, 2004; Pieper & Reineke, 2000; Furukawa, 2003). Pesticide degrading microorganisms can be found throughout the microbial world in the domains eubacteria, archae and eukarya. Every physiological type is represented in the pesticide –degrading microflora: aerobic, anaerobic (fermentive, methanogenic, sulphate-reducing), chemolithotropic and photosynthetic organisms. Pesticide residues can be mineralised by single microorganism or by assemblages of microorganism. Many different soil microorganisms catalysing the same or different types of transformation reactions may simultaneously attack a pesticide. Complete biodegradation of a pesticide involves the oxidation of parent compound to form carbon dioxide and water. This process provides both carbon and energy for the growth and reproduction of microbes. Each degradation step is catalyzed by specific enzyme produced by a degrading cell or enzyme found external to the cell. Degradation of pesticide by either external or internal enzyme will stop at any step if an appropriate enzyme is not present. Absence of an appropriate enzyme is one of the common reasons for persistence of any pesticide. If an appropriate microorganism is absent in soil or if biodegrading microbial population has been reduced due to toxicity of pesticide in that case a specific microorganism can be added or introduced in soil to enhance the activity of the existing population. The use of microorganisms for the degradation and detoxification of numerous toxic chemicals such as pesticides is an effective tool to decontaminate the polluted sites. Degradation by microbes depends not only on the presence of degradative enzymes, but also on a wide range of environmental parameters. Temperature, pH, water potential, nutrients and the amount of pesticide or metabolite in soil may also act as limiting factor for pesticide degrading microorganisms, which requires further exploration in relation to total microbial population and their biochemical activities (Singh, 2008). Some pesticides are readily degraded by microorganisms; others have proven to be recalcitrant (Mulchandani et al., 1999).

Microrganism degrade pesticides by introducing minor structural changes and render them into non-toxic substances and release them into soil where it is susceptible to further degradation (Uqab et al., 2016) changes to the pesticides rendering it non-toxic and are released to soil, where it is susceptible to further biodegradation by bacteria (Gianfreda et al., 2004). Several fungi such as *Agrocybe semiorbicularis, Auricularia auricula, Coriolus versicolor, Dichomitus squalens, Flammulina velupites, Hypholoma fasciculare, Pleurotus ostreatus, Stereum hirsutum,* and *Avatha discolor* have shown their ability to degrade various pesticide groups like phenylamide, triazine, phenylurea, dicarboximide, chlorinated and organophosphorus compounds (Bending et al., 2002). Several classes of pesticides such as lindane, atrazine, diuron, terbuthylazine, metalaxyl, DDT, gammahexachlorocyclohexane (g-HCH), dieldrin, aldrin, heptachlor, chlordane, lindane, mirex, etc. have been degraded to different extent by white-rot fungi. Fungi isolated from oil spill environment can reduce oil pollution (Das & Chandran, 2011). Bacteria have been used extensively for bioremediation purposes. These studies have focused on the employment of bacteria, consortia or on the search for biotransformation enzymes (Cycon & Piotrowska-Seget, 2009). The fast growth, easy handling and low cost make them suitable for bioremediation. Bacteria play role in degradation of many pesticides such as metamitron, chlorpyrifos, atrazine, iprodione pesticides

Most bacterial species which degrade pesticides belong to the genera Flavobacterium, *Arthrobacter, Azotobacter, Burkholderia and Pseudomonas.* The nature of degradation varies among species and the target compound. *Pseudomonas sp. and Klebsiella pneumoniae* have been shown to possess hydrolytic enzymes that are capable of breaking down s-triazine herbicides, such as atrazine (Mulbry & Kearney, 1991). Besides organophosphorus compounds, the Neonicotionoids are degraded by the *Pseudomonas species* (Uqab et al., 2016). Two more strains of *Pseudomonas putida* are reported to utilize diclofop methyl as a source of carbon and energy (Ralebits et al ., 2002; Karpouzas et al., 2000).

Microbial Transformation Reactions Applied for Pesticide Biodegradation

Given the great chemical diversity of pesticide and pesticide degrading microorganisms it is to be expected that a wide variety of transformation reaction catalysed by microorganisms should be involved in their degradation. A given pesticide is usually biodegraded by a limited number of consistent reactions in soils. Microbial pesticide transformation reaction can be grouped into those that are oxidative, reductive hydrolytic or synthetic in nature. A brief discussion of the major pesticide transformation reaction follows.

Table 1. Showing half-life of pesticides and microorganism degrading pesticides

Pesticides	Half- life	Degrading Microorganisms
Chlorinated hydrocarbons DDT	5-10 years	*Alcaligenes eutrophus* *Pseudomonas sp. Sphingobacterium sp*
EPTC	30 days	*Arthrobacter* *Rhodococcus* *Rhodococcus*
Carbamates carbofuran	8 days	*Achromobacter* *Pseudomonas* *Flavo bacterium* *Flavobacterium*
Phenoxy compounds 2,4-D	10-20 years	*Alcaligenes* *eutrophus* *Flavobacterium* *Arthrobacter* *Pseudomonas cepacia*
Organophosphates parathion	1-10 days	*Flavobacterium ATCC 27551* *Pseudomonas diminut*
Triazine atrazine	60 days	*Norcardia* *Pseudomonas* *Pseudomonas* *Rhodococcus* *Rhodococcus*
diazinon fenthion	37.4 days 2.9 to 21.1 days	*Flavobacterium ATCC 27551* *Bacillus*
Aromatic compounds	-	*Geobacter sp.*
Cyclohexylamine	-	*Brevibacterium oxydans DH35A*
Lindane	708 days	*Phaenerochaete chrysosporium*
Benzene	Only few days	*Arhodomonas rozel*
Naphthalene	2.8 hours	*Marinobacter vinifirmus, Stappia aggregate* *Pseudoalteromonas ganghwensis*

1. Oxidative Reaction

The major reactions are hydroxylation, dealkylation, epoxidation and sulfoxidation. Oxidation reaction are among the most important pesticide transforming reaction in aerobic environments. They are catalysed by monoxygenase, deoxygenase, and peroxidase enzymes. Aerobic condition are required for these reaction because in addition to functioning as a terminal electron acceptor for respiration, oxygen in an obligate reagent in the transformation reaction. The products of oxidative reaction frequently have anionic hydroxyl or carboxyl substituent and are more polar and water soluble than the parent pesticide, they are thus more biodegradable and more likely to be immobilised by covalent attachment to humic materials. Example of each of the major transformation reaction follows:

Hydroxylation Reaction

Hydroxylation reaction are those in which hydroxyl groups, the oxygen atoms of which originate from dioxygen gas, are inserted into target molecules. In contrast in hydrolytic reaction the oxygen atom orginates from water. Hydroxylation is a very frequent mechanism of initial attack on pesticide molecules under aerobic condition. Both alkyl and aromatic group can be hydroxylated. For example, an *actinomycetes* hydroxylates the alkyl groups of the herbicide metolachlor. (Krause et al., 1985; Cesare et al., 2005).

A *rhodococcus* strain hydroxylates the benzene nucleus of the insecticide carbofuran, producing 5-hydroxycarbonfuran as the end product (Figure 10 pathway B) (Stompor et al., 2013).

A *Flavobacterium* strain hydroxylates the biocides pentachlorophenol (pcp) producing tetrachlorobparahydroquinone (Figure 11 step A (Xun et al., 1992; Lifeng & Jian, 2008). The ligninolytic white rot fungus *Phanerochaete chrysosporium* oxidizes and ultimately mineralises numerous pesticide including the herbicide 2,4 D and 2,4,5 trichlorophenoxyacetic acid (2,4,5 –T) (Yadav & Reddy, 1993). The substrate specificity for these reaction is, in some cases, narrow and in other broad. For example, the *Flavobacterium* strain will hydroxylate only at the para position of a rather narrow range of halogenated phenols, whereas *Phanerochaete chrysosporium* will hydroxylate many aromatic and aliphatic organic pollutants that are extremely varied in structure.

Dealkylation Reactions

Many pesticides possess alkyl groups linked to the rest of the molecules via nitrogen or oxygen atoms. The biodegradation of these pesticide is frequently initiated by oxidative removal of these substituents. Dealkylation reaction by mono and deoxygenase require oxygen and reduction in the form of reduced nicotinamide adenine dinucleotide phosphate(NADPH). N – dealkylation is the primary route of degradation of the s-triazine herbicide in soil and by pure culture (see Figure 12) (Behki et al., 1993; Sawangjit, 2016).

The first reaction in the degradation of 2, 4-D by *Alcaligenes eutrophus* JMP 134 (pjp4) is removal of the alkylether group by means of α ketoglutarate dependent dioxygenase, formerly thought to be monooxygenase (Fukumori and Hausinger, 1993). Hydroxylation of the methoxy group on metolacholar by an *actinomycete* was accompanied by removal of the methyl group i.e. demethylation.

Figure 9. Hydroxylation of the herbicide metolachor by an actinomycete. Hydroxylation of the alkyl chain in pathway A result in demethylation

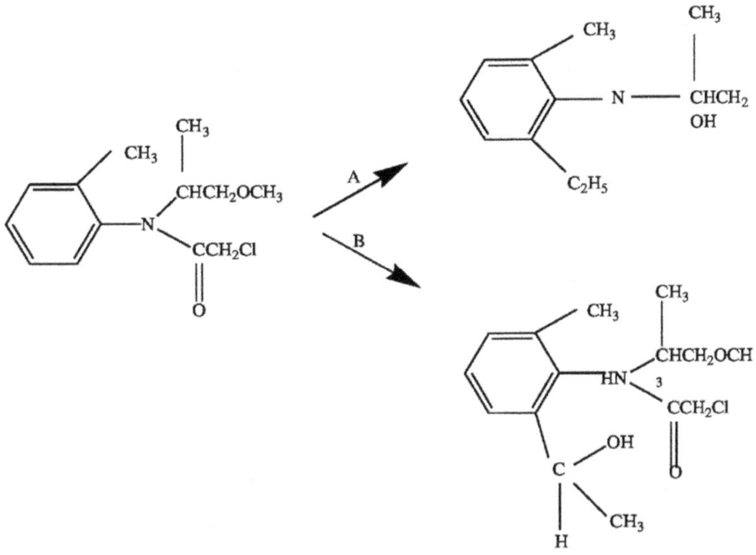

Figure 10. A. Metabolism of the insecticide carbofuran by gram negative methylotrophic bacterium, strain ER2. Hydrolysis of the alkyl group byb a carbofuran hydrolyase produces methyl amine, carbon dioxide and 7-phenolcarbofuran, which is excreted as an endproduct. Methylamine supports the growth of the organism as the sole source of carbon and nitrogen B. Cometabolic ring hydroxylation of carbofuran to form the endproduct 5-hydroxycarbofuran by Rhodoccous.

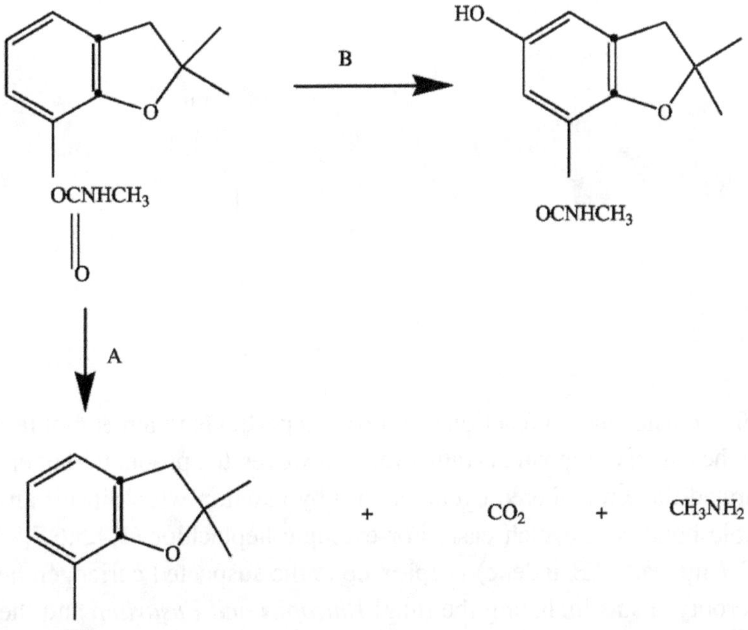

Figure 11. Transformation of pentachlorophenol: A, Metabolism by a Flavobacterium sp. An initial hydroxylation is followed by sequential glutathione-dependent reductive dehalogenation steps. The intermediate chlorohydroquinone is mineralised by as yet unknown reaction. B, Sequential reductive dechlorination in anerobic sludges. Various isomers of tetra, tri and dichlorophenol accumulate sequentially and transiently. C, Cometabolic methylation to produce volatile tetrachloroanisole.

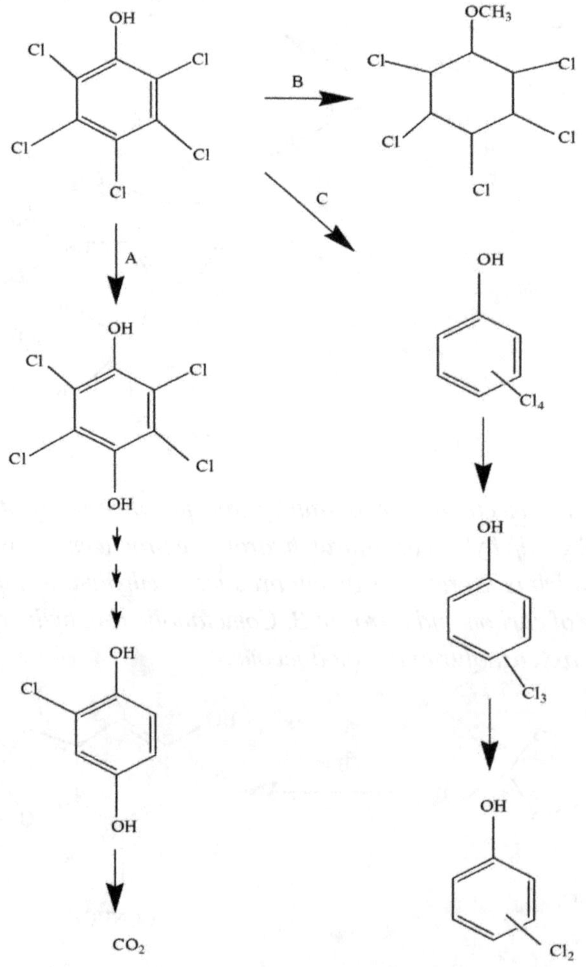

Epoxidation

Example of microbial transformation reaction converting a pesticide to a metabolite which is more of a health concern than the parent compound is rather rare. However, the production of epioxide (compound which contain saturated three membered cyclic ethers) by reaction which insert an oxygen atom into carbon-carbon double bond is one such case. For example heptachlor (1,4,5,6,7,8,8-heptachloro-3α, 4,7,7α-tetrahydro-4,7-methano-1H-indene) is epioxide to the suspected caricogen heptachlor epioxide by a number of microorganisms including the fungi *Rhizopus and Fusarium* and the bacteria *Bacillus* and *Streptomyces* (Xun and Orser, 1992).

Figure 12. Dealkylation of atrazine by rhodoccus TE1. Both the ethyl (pathway A) and the isopropyl (pathway B) groups can be removed.

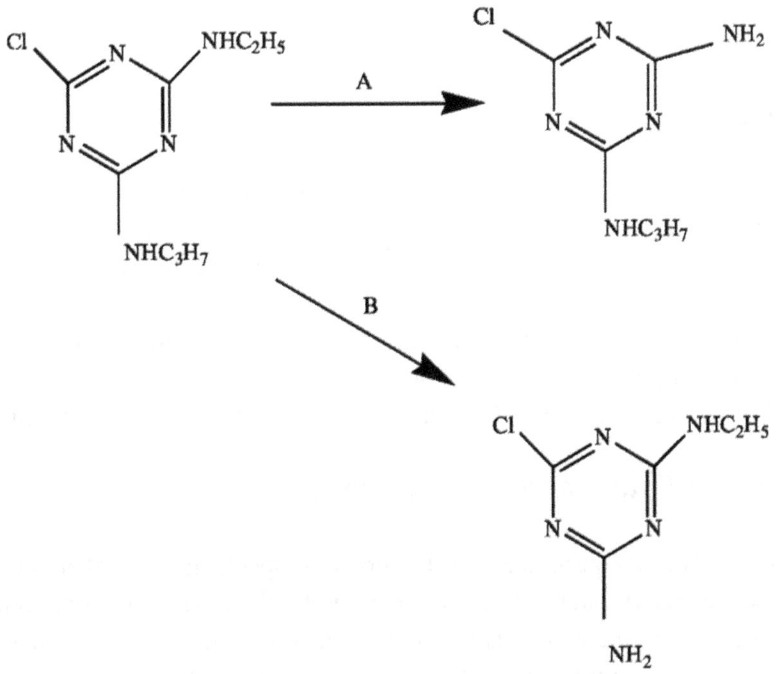

Sulfoxidation

Sulfoxidation reaction oxidizes a divalent sulphur atom to the sulfoxide and sometimes to the sulfone. Numerous pesticide such as the insecticide 2-methyl-2-(methylthio) propanol O-[(methylamino)carbonyl]oxime (adicarb) are sulphides having the structure RSR'. Sulfoxidation is a frequent mechanism of degradation of these compounds in soils. Sulfoxidation of some sulphides in soil probably occurs chemically as well as biologically because this reaction can also occur in sterile soils.

2. Reduction Reactions

Reaction which add electron to pesticide molecules are common, particularly in flooded soils. The saturation of double bonds is important in the anaerobic degradation of aromatic molecules. (Grbic´-Galic´ et al., 1990).

Reductive dehalogenation reactions are extremely important in the transformation of halogenated pesticide in anerobic environments. A number of mixed or pure bacterial culture reductively dehalogenate the organochlorine pesticide DDT, dieldrin, heptachlor, the γ-isomer of 1,2,3,4,5,6-hexachloro-cyclohexane (lindane), and 2,2-bis(4-methoxyphenyl)-1,1,1-trichloroethane (methoxychlor). The heterocyclic herbicide 5-bromo-6-methyl-3-(1-methyl-propyl)-2,4(1H,3H)-pyrimidinedione (bromocil) is dehalogenation in anaerobic slurries of aquifer material (Adrian & Suflita, 1990).

Reductive dehalogenation is favoured under extremely reducing methanogenic condition and is frequently inhibited by oxygen, nitrate or sulphate. However reductive dehalogenetaion reaction can occur at higher redox potential. For example, several reductive dehalogenation steps are involved in the aerobic mineralization of PCP by a *Flavobacterium* (Figure 11).

3. Hydrolytic Reaction

Hydrolytic reactions are very important in the degradation of numerous pesticide. The hydroxyl group (OH) originates from water, and therefore hydrolytic reaction can occur in the presence or absences of oxygen. Since no cofactors are required, hydrolytic reaction can be catalysed by intracellular enzymes or extracellulary hydrolyzed by soil microorganisms. Example of such pesticide include the organophosphorus and methylcarbamate insecticide such as parathion and carbofuran and the phenylurea herbicide such as linuron. A carbofuran by hydrolyzing the carbamate linkage producing 7-phenolcarbofuran, carbondioxide and methylamine.(Figure 10, pathway A) (Topp et al., 1993; Sorensen et al., 2007).

Criteria for Bioremediation and Biodegradation

For successful biodegradation of pesticide in soil, following aspects must be taken into consideration. i) Organisms must have necessary catabolic activity required for degradation of contaminant at fast rate to bring down the concentration of contaminant, ii) the target contaminant must be bioavailability, iii) soil conditions must be congenial for microbial /plant growth and enzymatic activity iv) cost of bioremediation must be less than other technologies of removal of contaminant

Types of Bioremediation Technologies

Bioremediation technologies have been broadly divided into two categories based on whether is stimulated in situ or carried out ex situ in compost heaps or bioreactors. Pollutants like pesticide can be bioremediated using either in situ or ex situ systems.

In Situ Bioremediation Techniques

In situ bioremediation techniques involve enhancement of the biodegradation rate of organic contaminants within the affected area (soil, sediment and surface water or groundwater environments). It is the clean-up approach that directly involves the contact between microorganisms and the dissolved and sorbed contaminants for biotransformation (Alcalde et al., 2006). When native microorganisms lack biodegradable capacity, genetically engineered microorganisms (GEMs) may be added to the surface during in situ bioremediation, although stimulation of for the removal of by-products of raw materials and wastes. Bacteria are the most commonly used agents in these biodegradation processes.In situ bioremediation techniques have been successfully used to treat chlorinated solvents, dyes, heavy metals, and hydrocarbons polluted sites (Folch et al., 2013; Kim et al., 2014; Frascari et al., 2015). In situ remediation is less costly due to the lack of excavation and transportation costs, but these remediation techniques are less controllable and less effective (Koning et al., 2000).

Types of In Situ Bioremediation

1. **Bioventing**

Bioventing is an *in situ* remediation technology that uses microorganisms to biodegrade organic constituents adsorbed on soils in the unsaturated zone. Bioventing enhances the activity of indigenous bacteria and simulates the natural in situ biodegradation of hydrocarbons in soil by inducing air or oxygen flow into the unsaturated zone and necessary, by adding nutrients (Van Deuren et al., 2002).

2. **Biosparging**

Biosparging involves the injection of air under pressure below the water level to increase groundwater oxygen concentrations and enhance the rate of biological degradation of contaminants by indigenous microorganisms. Biosparging increases the mixing in the saturated zone and thereby increases the contact between the soil and groundwater. The ease and low cost of installing small-diameter air injection points allows considerable flexibility in the design and construction of the system (Held and Dörr, 2000). The effectiveness of biosparging depends on two major factors namely: soil permeability, which determines pollutant bioavailability to microorganisms, and pollutant biodegradability (Philp & Atlas, 2005).

3. **Bioslurping**

Bioslurping is a unique in situ treatment technique in that it also treats free product phases floating on top of the groundwater. This technique applies a vacuum to extract, soil vapor, water, and free product from the subsurface. Each of those products is separated and then treated. This technique is cost effective because only a small amount of groundwater and soil vapor are pumped at a time, therefore the treatment plant used to treat the vapor and free product can be small (Held & Dörr, 2000). Following complete free products removal, thesystem can easily be made to operate as a conventional bioventing system to complete remediation process (Kim et al., 2014).

4. **Phytoremediation.**

Phytoremediation is an *in situ* technique that uses plants and their associated microorganisms in order to remove, degrade or isolate toxic substances from the environment Phytoremediation is most suited for sites where other remediation options are not costs effective, low-level contaminated sites, or in conjunction with other remediation techniques (Chaney et al., 1997; Ensley, 2000; Dickinson et al., 2009).

Ex Situ Bioremediation Techniques

Ex situ bioremediation techniques are usually aerobic and involve the treatment of contaminated soil or sediments using solid or slurry phase systems. It actually involves the removal of waste material from polluted area and their collection at a place distant from the contaminated site to facilitate microbial degradation (Koning et al., 2000). The use of this particular ex situ technique is increasingly being considered due to its constructive features including cost effectiveness, which enables effective biodegradation on the condition that nutrient, temperature and aeration are adequately controlled (Whelan et al., 2015).

Based on the physical nature of the contaminant under treatment, this technology is classified into two types

1. Solid phase system (including land treatment and soil piles)
2. Slurry phase systems (including solid-liquid suspensions in bioreactors)

1. **Solid Phase Treatment:** It includes organic wastes such as animal manures, leaves and agricultural wastes and waste to be treated eg: domestic and industrial watses, sewage sludge and municipal solid wastes. Solid-phase soil treatment processes include composting. landfarming, and soil biopiles (Koning et al., 2000).
2. **Slurry-Phase Bioremediation:** Slurry phase bioremediation is a relatively more rapid process compared to the other treatment processes. Contaminated soil is combined with water and other additives in a large tank called a bioreactor and mixed to keep the microorganisms, which are already present in the soil, in contact with the contaminants in the soil. Nutrients and oxygen are added, and conditions in the bioreactor are controlled to create the optimum environment for the microorganisms to degrade the contaminants. When the treatment is completed, the water is removed from the solids, which are disposed of or treated further if they still contain pollutants (Koning et al., 2000).

CONCLUSION

Pesticides are often considered a quick, easy, and inexpensive solution for controlling weeds and insect pests in urban landscapes. However, pesticides use comes at a significant cost. They have contaminated almost every part of our environment. Pesticides residues are found in soil, air, in surface and ground water across the countries. Bioremediation is a characteristic procedure for the remediation of polluted site and to expel the contamination from nature environment with collaborative use of biological agents (bacteria, fungi, plants). For bioremediation to be effective, the pesticides must be subject to microbial attack or microbial transformation, the metabolic product must be safe and the process must not cause adverse ecological side effects. The foremost step to a successful bioremediation is site characterization, which helps establish the most suitable and feasible bioremediation technique (ex situ or in situ). Environmental conditions must permit in situ growth of the microorganism capable of bioremediation or transformation of the pollutant so that it can be biodegraded ex situ in bioreactors. There is no doubt that bioremediation is in the process of providing a healthy environment. Regardless of which aspect of bioremediation that is used, this technology offers an efficient and cost-effective way to treat contaminated ground water and soil. Its advantages generally outweigh the disadvantages, leading to maximum use of this technology. We should be thankful to bioremediation technology for using beneficial microorganisms to clean environment from harmful pesticides and helping in maintaing healthy environment.

REFERENCES

Abhilash, P. C., & Singh, N. (2009). Pesticide use and application: An Indian scenario. *Journal of Hazardous Materials, 165*(1-3). doi:10.1016/j.jhazmat.2008.10.061 PMID:19081675

Adrian, N. R., & Suflita, J. M. (1990). Reductive dehalogenation of a nitrogen heterocyclic herbicide in anoxic aquifer slurries. *Applied and Environmental Microbiology, 56*(1), 292–294. PMID:16348103

Alcalde, M., Ferrer, M., Plou, F. J., & Ballesteros, A. (2006). Environmental biocatalysis: From remediation with enzymes to novel green processes. *Trends in Biotechnology, 24*(6), 281–287. doi:10.1016/j.tibtech.2006.04.002 PMID:16647150

Alexandratos, N., & Bruinsma, J. (2012). World Agriculture Towards 2030/2050 (The 2012 Revision) (ESA Working paper No. 12-03). FAO.

Alonzo, H. G. A., & Corrêa, C. L. (2014). Praguicidas. In S. Oga, M. M. A. Camargo, & J. A. Batistuzzo (Eds.), *Fundamentos de Toxicologia* (4th ed., pp. 323–341). Editora Atheneu, São Paulo.

Añasco, N., Uno, S., Koyama, J., Matsuoka, T., & Kuwahara, N. (2010). Assessment of pesticide residues in freshwater areas affected by rice paddy effluents in Southern Japan. *Environmental Monitoring and Assessment, 160*(1), 371–383. doi:10.100710661-008-0701-z PMID:19089593

Barr, D. B., & Needham, L. L. (2002). Analytical methods for biological monitoring of exposure to pesticides: A review. *Journal of Chromatography. A, 778*, 5–29. PMID:12376114

Behki, R., Topp, E., Dick, W., & Germon, P. (1993). Metabolism of the herbicide atrazine by *Rhodococcus* strains. *Applied and Environmental Microbiology, 59*(61), 955–1959. PMID:8328812

Bending, G. D., Turner, M. K., & Jones, J. E. (2002). Interactions between crop residue and soil organic matter quality and the functional diversity of soil microbial communities. *Soil Biology & Biochemistry, 34*(8), 1073–1082. doi:10.1016/S0038-0717(02)00040-8

Bhat, D., & Padmaja, P. (2014). Assessment of organic pesticides in Ground and surface water in Bhopal India. *IOSR journal of Environmental Science. Toxicology, and Food Technology, 8*(5), 51–52. doi:10.9790/2402-08535152

Bocquené, G., & Franco, A. (2005). Pesticide contamination of the coastline of Martinique. *Marine Pollution Bulletin, 51*(5-7), 612–619. doi:10.1016/j.marpolbul.2005.06.026 PMID:16045943

Bowmer, K. H. (1991). Atrazine persistence and toxicity in two irrigated soils in Australia. *Australian Journal of Soil Research, 29*(2), 339–350. doi:10.1071/SR9910339

Campbell, A. W. (2014). Glyphosate: Its effects on humans. *Alternative Therapies in Health and Medicine, 20*, 9–11. PMID:24755565

Casida, J. E., & Durkin, K. A. (2013). Neuroactive insecticides: Targets, selectivity, resistance, and secondary effects. *Annual Review of Entomology, 58*(1), 99–117. doi:10.1146/annurev-ento-120811-153645 PMID:23317040

Cerejeira, M. J., Viana, P., Batista, S., Pereira, T., Silva, E., Valério, M. J., ... Silva-Fernandes, A. M. (2003). Pesticides in Portuguese surface and ground waters. *Water Research*, *37*(5), 1055–1063. doi:10.1016/S0043-1354(01)00462-6 PMID:12553980

Cesare, A., Claudio, S., & Alberto, V. (2005). Influence of flooding on the degradation of linuron, isoproturon and metolachlor in soil. *Agronomy for Sustainable Development*, *25*(3), 401–406. doi:10.1051/agro:2005030

Chaney, R. L., Malik, M., Li, Y. M., Brown, S. L., Angle, J. S., & Baker, A. J. M. (1997). Phytoremediation of soil metals. *Current Opinion in Biotechnology*, *8*(3), 279–284. doi:10.1016/S0958-1669(97)80004-3 PMID:9206007

Colberg, P. J. S., & Young, L. Y. (1995). Anaerobic degradation of no halogenated homocyclic aromatic compounds coupled with nitrate, iron, or sulfate reduction. In Microbial Transformation and Degradation of Toxic Organic Chemicals (pp. 307–330). New York: Wiley-Liss.

Costa, L. G. (2008). Toxic Effects of Pesticides. In C. D. Klaassen (Ed.), *Cassarett and Doull's Toxicology The Basic Science of Poisons* (7th ed.). New York: McGraw-Hill.

Cycon, M., & Piotrowska-Seget, Z. (2009). Changes in bacterial diversity and community structure following pesticides addition to soil estimated by cultivation technique. *Ecotoxicology (London, England)*, *18*(5), 632–642. doi:10.100710646-009-0321-6 PMID:19444604

Das, N., and Chandran, P. (2011). Microbial degradation of petroleum hydrocarbon contaminants: An overview. *Biotechnology Research International.*

Debarati, P., Gunjan, P., Janmejay, P., & Rakesh, V. J. K. (2005). Accessing Microbial Diversity for Bioremediation and Environmental Restoration. *Trends in Biotechnology*, *23*(3), 135–142. doi:10.1016/j.tibtech.2005.01.001 PMID:15734556

Dickinson, N. M, Baker, A. J. M., Doronila, A., Laidlaw, S., and Reeves, R. D. (2009). Phytoremediation of inorganics: realism and synergies. *International Journal of Phytoremediation*, *11*(2), 97-114.

Douglas, M. R., & Tooker, J. F. (2015). Large-scale deployment of seed treatments has driven rapid increase in use of neonicotinoid insecticides and preemptive pest management in U.S. field crops. *Environmental Science & Technology*, *49*(8), 5088–5097. doi:10.1021/es506141g PMID:25793443

ECA. (2017). European Chemicals Agency. Retrieved from https://echa.europa.eu/regulations/reach/legislation

Ellenhorn, M. J., & Barceloux, D. G. (1988). Medical Toxicology. In Diagnosis and Treatment of Human Poisoning. New York: Elsevier.

Ensley, B. D. (2000). Rationale for use of phytoremediation. In I. Raskin & B. D. Ensley (Eds.), *Phytoremediation of toxic metals. Using plants to clean up the environment* (pp. 3–11). New York: John Wiley & Sons.

EPA. (April 2015). What is a Pesticide? Retrieved from https://www.epa.gov/minimum-risk-pesticides/what-pesticide

FAO. (2017). Retrieved from http://www.fao.org/faostat/en/#home

Folch, A., Vilaplana, M., Amado, L., Vicent, R., & Caminal, G. (2013). Fungal permeable reactive barrier to remediate groundwater in an artificial aquifer. *Journal of Hazardous Materials*, *262*, 554–560. doi:10.1016/j.jhazmat.2013.09.004 PMID:24095995

Frascari, D., Zanaroli, G., & Danko, A. S. (2015). In situ aerobic cometabolism of chlorinated solvents: A review. *Journal of Hazardous Materials*, *283*, 382–399. doi:10.1016/j.jhazmat.2014.09.041 PMID:25306537

Fukumori, F., & Hausinger, R. P. (1993). Alcaligenes eutrophus JMP134 "2,4-dichlorophenoxyacetate monooxygenase" is an alpha-ketoglutarate-dependent dioxygenase. *Journal of Bacteriology*, *175*(7), 2083–2086. doi:10.1128/jb.175.7.2083-2086.1993 PMID:8458850

Furukawa, K. (2003). 'Super bugs' for Bioremediation. *Trends in Biotechnology*, *21*(5), 187–190. doi:10.1016/S0167-7799(03)00054-4 PMID:12727376

Gavrilescu, M. (2005). Fate of Pesticides in the Environment and its Bioremediation. *Engineer in Life Science*, 5-6, 497-526.

Ghosh, P. G., Sawant, N. A., Patil, S. N., & Aglave, B. A. (2010). Microbial Biodegradation of Organophosphate Pesticides. *International Journal of Biotechnology and Biochemistry*, *6*(6), 871–876.

Gianfreda, L., & Rao, M. A. (2004). Potential of extra cellular enzymes in remediation of polluted soils: A review. *Enzyme and Microbial Technology*, *35*(4), 339–354. doi:10.1016/j.enzmictec.2004.05.006

Gilliom, R. J. (2007). Pesticides in U.S. Streams and Groundwater. *Environmental Science & Technology*, *41*(10), 3408–3414. doi:10.1021/es072531u PMID:17547156

Grbic´-Galic, ´. D., Churchman-Eisel, N., & Mrakovic´, I. (1990). Microbial transformation of styrene by anaerobic consortia. *The Journal of Applied Bacteriology*, *69*(2), 247–260. doi:10.1111/j.1365-2672.1990.tb01516.x PMID:2272946

Griza, F. T., Ortiz, K. S., Geremias, G., & Thiesen, F. V. (2008). Avaliação da Contaminação por Organofosforados em Águas Superficiais no Município de Rondinha- Rio Grandedo Sul. *Quimica Nova*, *31*(7), 1631–1635. doi:10.1590/S0100-40422008000700006

Gupta, P. K. (2004). Pesticide exposure—Indian scene. *Toxicology*, *198*(1–3), 83–90. doi:10.1016/j.tox.2004.01.021 PMID:15138033

Hayo, M. G., & Werf, V. D. (1996). Assessing the impact of pesticides on the environment. *Agriculture, Ecosystems & Environment*, *60*(2-3), 81–96. doi:10.1016/S0167-8809(96)01096-1

Held, T., & Dörr, H. (2000). In Situ Remediation. *Biotechnology*, *11b*, 350–370.

Hicks, B. (2013). Agricultural pesticides and human health. In: National Association of Geoscience Teachers. Retrieved from http://serc.carleton.edu/NAGTWorkshops/health/case_studies/pesticides.html

Hodaifa, G., Nieto, L. M., & Casanova, M. S. (2009). Elimination of pesticide residues from virgin olive oil by ultraviolet light: Preliminary results. *Journal of Hazardous Materials*, *168*(1), 555–559. doi:10.1016/j.jhazmat.2009.02.030 PMID:19272709

Hussain, S., Siddique, T., Arshad, M., & Saleem, M. (2009). Bioremediation and Phytoremediation of Pesticides: Recent Advances. *Critical Reviews in Environmental Science and Technology*, *39*(10), 843–907. doi:10.1080/10643380801910090

Jiang, Y. F., Wang, X. T., Jia, Y., Wang, F., Wu, M.-H., Sheng, G.-Y., & Fu, J.-M. (2009). Occurrence, distribution and possible sources of organochlorine pesticides in agricultural soil of Shanghai, China. *Journal of Hazardous Materials*, *170*(2-3), 989–997. doi:10.1016/j.jhazmat.2009.05.082 PMID:19535203

Karpouzas, D. G., Morgan, J. A., & Walker, A. (2000). Isolation and Characterization of 23 carbofuran degrading bacteria from soils from distant geographical areas. *Letters in Applied Microbiology*, *31*(5), 353–358. doi:10.1046/j.1472-765x.2000.00823.x PMID:11069636

Kim, S., Krajmalnik-Brown, R., Kim, J.-O., & Chung, J. (2014). Remediation of petroleum hydrocarbon-contaminated sites by DNA diagnosis-based bioslurping technology. *The Science of the Total Environment*, *497-498*, 250–259. doi:10.1016/j.scitotenv.2014.08.002 PMID:25129160

Kim, S. Y., Jo, Y. A., Choi, J., & Choi, M. J. (2001). Characterization of s-Triazine Antibodies and Comparison of Enzyme Immunoassay for the Determination of s-Triazine. *Microchemical Journal*, *68*, 163–172. doi:10.1016/S0026-265X(00)00143-0

King, A. M., & Aaron, C. K. (2015). Organophosphate and carbamate poisoning. Organophosphate and carbamate poisoning. *Emergency Medicine Clinics*, *33*(1), 133–151.

Koning, M., Hupe, K., & Stegmann, R. (2000). Thermal Processes, Scrubbing/Extraction, Bioremediation and Disposal. *Biotechnology*, *11b*, 306–317.

Konstantinou, I. K., Hela, D. G., & Albanis, T. A. (2006). The status of pesticide pollution in surface waters (rivers and lakes) of Greece. Part I. Review on occurrence and levels. *Environmental Pollution*, *141*(3), 555–570. doi:10.1016/j.envpol.2005.07.024 PMID:16226830

Krause, A., Hancock, W. G., Minard, R. D., Freyer, A. J., Honeycutt, R. C., LeBaron, H. M., ... Bollang, J. M. (1985). Microbial transformation of the herbicide metolachlor by a soil *actinomycete*. *Journal of Agricultural and Food Chemistry*, *33*(4), 584–589. doi:10.1021/jf00064a008

Laskowski, D. A. (2002). Physical and chemical properties of pyrethroids. *Reviews of Environmental Contamination and Toxicology*, *174*, 49–170. doi:10.1007/978-1-4757-4260-2_3 PMID:12132343

Lei, W., Wang, D. D., Dou, T. Y., Hou, J., Feng, L., Yin, H., ... Yang, L. (2017). Assessment of the inhibitory effects of pyrethroids against human carboxylesterases. *Toxicology and Applied Pharmacology*, *321*, 48–56. doi:10.1016/j.taap.2017.02.018 PMID:28242322

Li, X., Wang, W., Wang, J., Cao, X., Wang, X., Liu, J., ... Jiang, X. (2008). Contamination of soils with organochlorine pesticides in urban parks in Beijing, China. *Chemosphere*, *70*(9), 1660–1668. doi:10.1016/j.chemosphere.2007.07.078 PMID:17869322

Lifeng, C., & Jian, Y. (2008). Biochemical Characterization of the Tetrachlorobenzoquinone Reductase Involved in the Biodegradation of Pentachlorophenol. *International Journal of Molecular Sciences*, *9*(3), 198–212. doi:10.3390/ijms9030198 PMID:19325743

Martin-Reina, J., Duarte, J. A., Cerrillos, L., Bautista, J. D., & Moreno, I. (2017). Insecticide Reproductive Toxicity Profile: Organophosphate, Carbamate and Pyrethroids. *J. Toxins*, *4*(1), 7.

Mervat, S. M. (2009). Degradation of methomyl by the novelbacterial strain *Stenotrophomonas maltophilia* M1. *Electronic Journal of Biotechnology*, *12*(4).

Mrema, E. J., Rubino, F. M., Mandic-Rajcevic, S., Sturchio, E., Turci, R., Osculati, A., ... Colosio, C. (2013). Exposure to priority organochlorine contaminants in the Italian general population. Part 1. Eight priority organochlorinated pesticides in blood serum. *Human and Experimental Toxicology*, *32*(12), 1323–1339. doi:10.1177/0960327113485255 PMID:23857032

Mulchandani, A., Kaneva, I., & And Chen, W. (1999). Detoxification of organophosphate pesticides by immobilized Escherichia coli expressing organophosphorus hydrolase on cell surface. *Biotechnology and Bioengineering*, *63*, 216–223. doi:10.1002/(SICI)1097-0290(19990420)63:2<216::AID-BIT103.0.CO;2-0 PMID:10099598

Oga, S. (2003). Fundamentos de toxicologia (2nd ed.). São Paulo: Atheneu Editora.

Menone, M. L., De Moreno, J. A., Moreno, V. J., Lanfranchi, A. L., Metcalfe, T. L., & Metcalfe, C. D. (2001). Organochlorine Pesticides and PCBs in a Southern Atlantic Coastal Lagoon Watershed, Argentina. *Archives of Environmental Contamination and Toxicology*, *40*(3), 355–362. doi:10.1007002440010183 PMID:11443366

Oyeyiola, A. O., Fatunsin, O. T., Akanbi, L. M., Fadahunsi, D. E., & Moshood, O. M. (2017). Human Health Risk of Organochlorine Pesticides in Foods Grown in Nigeria. *Journal of Health & Pollution*, *7*(15), 63–70. doi:10.5696/2156-9614-7.15.63

Paoli, D., Giannandrea, F., Gallo, M., Turci, R., Cattaruzza, M. S., & Lombardo, F. (2015). Exposure to polychlorinated biphenyls and hexachlorobenzene, semen quality and testicular cancer risk. *The Journal of Endocrinology*. PMID:25770454

Patnaik, P. (2003). *A Comprehensive Guide to the Hazardous Properties of Chemical Substances*. New Jersey, USA: John Wiley & Sons.

Philp, J. C., & Atlas, R. M. (2005). Bioremediation of contaminated soils and aquifers. In Atlas RM, Philp JC, editors. Bioremediation: applied microbial solutions for real-world environmental cleanup (pp. 139–236). Washington: American Society for Microbiology (ASM) Press. doi:10.1128/9781555817596.ch5

Pieper, D. H., & Reineke, W. (2000). Engineering bacteria for bioremediation. *Current Opinion in Biotechnology*, *11*(3), 262–270. doi:10.1016/S0958-1669(00)00094-X PMID:10851148

Ralebits, T. K., Senior, E., & Van Versevell, H. W. (2002). Microbial aspects of atrazine degradation in natural environments. *Biodegradation*, *13*(1), 11–19. doi:10.1023/A:1016329628618 PMID:12222950

Rayu, S., Nielsen, U. N., Nazaries, L., & Singh, B. K. (2017). Isolation and Molecular Characterization of Novel Chlorpyrifos and 3,5,6- trichloro-2-pyridinol-degrading Bacteria from Sugarcane Farm Soils. *Frontiers in Microbiology*, *8*, 518. doi:10.3389/fmicb.2017.00518 PMID:28421040

Reigart, J. R., & Roberts, J. R. (2014). *Recognition and Management of Pesticide Poisonings* (5ª ed.). United States Environmental Protection Agency. Retrieved from http://www.epa.gov/oppfead1/safety/healthcare/handbook/handbook.pdf

Cavalli, V. L. D. L. O., Cattani, D., Rieg, C. E. H., Pierozan, P., Zanatta, L., Parisotto, E. B., ... & Zamoner, A. (2013). Roundup disrupted male reproductive functions by triggering calcium-mediated cell death in rat testis and Sertoli cells. *Free Radicals, Biology and Medicine*, *65*, 335–346.

Sampogna, R. V., & Cunard, R. (2007). Roundup intoxication and a rationale for treatment. *Clinical Nephrology*, *68*(3), 190–196. doi:10.5414/CNP68190 PMID:17915625

Sawangjit, S. (2016). Isolation and Characterization of Atrazine-Degrading Xanthomonas sp. ARB2 and Its Use in Bioremediation of Contaminated Soils. *International Journal of Environmental Sciences and Development*, *7*(5), 351–354. doi:10.7763/IJESD.2016.V7.798

Saxena, K. K., & Seth, N. (2002). Toxic effects of cypermethrin on certain hematological aspects of fresh water fish *Channa punctatus. Bulletin of Environmental Contamination and Toxicology*, *69*(3), 364–369. doi:10.100700128-002-0071-0 PMID:12177757

Shanahan, P. (2004). *Bioremediation. Waste Containment and RemediationTechnology*. Springer.

Shegunova, P., Klánová, J., & Holoubek, I. (2007). Residues of organochlorinated pesticides in soils from the Czech Republic. *Environmental Pollution*, *146*(1), 257–261. doi:10.1016/j.envpol.2006.03.057 PMID:17045372

Singh, D. K. (2008). Biodegradation and bioremediation of pesticide in soil:concept, method and recent developments. *Indian Journal of Medical Microbiology*, *48*(1), 35–40. doi:10.100712088-008-0004-7 PMID:23100698

Singh, Z. J., Kaur, R., & Hundal, S. S. (2016). Toxic effects of organochlorine pesticides: A review. *American Journal of Bioscience*, *4*(3), 11–18. doi:10.11648/j.ajbio.s.2016040301.13

Sogorb, M. A., & Vilanova, E. (2002). Enzymes Involved in the Detoxification of, Organophosphorus, Carbamate and Pyrethroid Insecticides Through Hydrolysis. *Toxicology Letters*, *128*(1-3), 215–228. doi:10.1016/S0378-4274(01)00543-4 PMID:11869832

Sorensen, S., Holtez, M. S., Simosen, A., & Aamand, J. (2007). Degradation and Mineralization of Nanomolar Concentrations of the Herbicide Dichlobenil and Its Persistent Metabolite 2,6-Dichlorobenzamide by *Aminobacter* spp. Isolated from Dichlobenil-Treated Soils. *Applied and Environmental Microbiology*, *73*(2), 399–406. doi:10.1128/AEM.01498-06 PMID:17114317

Soriano, J. M., Jiménez, B., Font, G., & Moltó, J. C. (2001). Analysis of Carbamate Pesticides and Their Metabolites in Water by Solid Phase Extraction and Liquid Chromatography: A Review. *Critical Reviews in Analytical Chemistry*, *31*(1), 19–5. doi:10.1080/20014091076677

Spear, R. (1991). Recognised and possible exposure to pesticides. In W. J. Hayes & E. R. Laws (Eds.), *Handbook of pesticide toxicology* (pp. 245–274). San Diego, CA: Academic.

Stamatis, N., Konstantinou, I., & Hela, D. (2010). Pesticide Inputs from the Sewage Treatment Plant of Agrinio to River Acheloos, Western Greece: Occurrence and Removal. *Water Science and Technology*, *62*(5), 1098–1105. doi:10.2166/wst.2010.932 PMID:20818051

Stompor, M., Potaniec, B., Szumny, A., Zielinski, P., & Anna Katarzyna Zołnierczyk, A. K. (2013). Microbial synthesis of dihydrochalcones using *Rhodococcus* and *Gordonia* species. *Journal of Molecular Catalysis. B, Enzymatic*, *97*, 283–288. doi:10.1016/j.molcatb.2013.09.009

Surekha, R. M., Lakshmi, P. K. L., Suvarnalatha, D., Jaya, M., Aruna, S., Jyothi, K., ... Venkateswarlu, K. (2008). Isolation and characterization of a chlorpyrifos degrading bacterium from agricultural soil and its growth response. *African Journal of Microbiological Research*, *2*, 26–31.

Toan, V., Thao, V., Walder, J., Schmutz, H.-R., & Ha, C. (2007). Contamination by Selected Organochlorine Pesticides (OCPs) in Surface Soils in Hanoi, Vietnam. *Bulletin of Environmental Contamination and Toxicology*, *78*(3), 195–200. doi:10.100700128-007-9149-z PMID:17476441

Tomizawa, M., & Casida, J. E. (2005). Neonicotinoid insecticide toxicology: Mechanisms of selective action. *Annual Review of Pharmacology and Toxicology*, 24. PMID:15822177

Topp, E., Hanson, R. S., Ringelberg, D. B., White, D. C., & Wheatcroft, R. (1993). Isolation and characterization of an N-methylcarbamate insecticide-degrading *methylotrophic* bacterium. *Applied and Environmental Microbiology*, *59*, 3339–3349. PMID:7504430

Uqab, B., Mudasir, S., & Nazir, R. (2016). Review on Bioremediation of Pesticides. *Journal of Bioremediation & Biodegradation*, *7*, 3.

Vaccari, D. A., Strom, P. F., & Alleman, J. E. (2005). Environmental Biology for Engineers and Scientists.

Van Deuren, J., Lloyd, T., Chhetry, S., Raycharn, L., & And Peck, J. (2002). Remediation Technologies Screening Matrix and Reference Guide. In *Federal Remediation Technologies Roundtable*.

Veiga, M. M., Silva, D. M., & Veiga, L. B. E., and De CastroFaria, M. V. (. (2006). Pesticide pollution in water systems in a small rural community in Southeast Brazil. *Cadernos de Saude Publica*, *22*(11), 2391–2399. doi:10.1590/S0102-311X2006001100013 PMID:17091176

Whelan, M. J., Coulon, F., Hince, G., Rayner, J., McWatters, R., Spedding, T., & Snape, I. (2015). Fate and transport of petroleum hydrocarbons in engineered biopiles in polar regions. *Chemosphere*, *131*, 232–240. doi:10.1016/j.chemosphere.2014.10.088 PMID:25563162

WHO. 9/2017). Agrochemicals, health and environment: directory of resources. Retrieved from http://www.who.int/ heli/risks/toxics/chemicalsdirectory/en/index1.html

Woudneh, M. B., Ou, Z., Sekela, M., Tuominen, T., & Gledhill, M. (2009). Pesticide Multiresidues in Waters of the Lower Fraser Valley, British Columbia, Canada. Part I. Surface Water. *Journal of Environmental Quality*, *38*(3), 940–947. doi:10.2134/jeq2007.0524 PMID:19329682

Xun, L., Topp, E., & Orser, C. S. (1992). Purification and characterization of a tetrachloro-*p*-hydroquinone reductive dehalogenase from a *Flavobacterium* sp. *Journal of Bacteriology*, *174*(24), 8003–8007. doi:10.1128/jb.174.24.8003-8007.1992 PMID:1459949

Yadav, J. S., & Reddy, C. A. (1993). Degradation of benzene, toluene, ethylbenzene, and xylenes (BTEX) by the lignin-degrading basidiomycete *Phanerochaete chrysosporium. Applied and Environmental Microbiology*, *59*, 756–762. PMID:8481002

Chapter 10
Sources of Soil Pollution

Abderrezak Khelfi
National Center of Toxicology, Algeria

ABSTRACT

Soil, though often overlooked, is a critical component of the earth and is essential for human existence. Soil pollution has deteriorated large areas of agricultural land around the globe. Human health is at risk due to the high concentration of pollutants found in soils and food produced in polluted lands. Soil pollution may arise from a wide range of sources that can be discrete point sources, or diffuse sources. It is usually due to domestic, municipal, industrial, mining, and agricultural wastes as well as agrochemicals such as fertilizers and pesticides. Each source has specific characteristics, briefly presented below. The most dangerous categories of soil pollutants are polycyclic aromatic hydrocarbons, polychlorinated biphenyl, pesticides, heavy metals, and radionuclides. New soil pollutants, such as veterinary medicines, have emerged with industrialization and globalization. This chapter deals with sources of soil pollution, pollutants generated from these sources and their possible adverse effects on the environment.

INTRODUCTION

Soil, though often overlooked, is a critical component of the earth and is essential for human existence. From an environmental fate perspective soil is clearly a well-defined, though highly variable, environmental compartment. It serves as the major source of food production for humans.

Compared to air or water pollution, soil pollution is much less apparent and its sources cannot be easily detected. These could be two of several reasons why it is much less frequently mentioned in the literature. However, concerns about soil pollution have increased in recent decades because of some famous cases in which soil pollution has caused detrimental impacts on populations and the environment. In fact, soil pollutants have decreased soil biodiversity and deteriorated large areas of agricultural land around the globe. Human health is also at risk due to the high concentration of pollutants found in soils and food produced in polluted lands. It is claimed that almost 25% of the soil worldwide is highly degraded, whereas 44% soil is moderately degraded (Tripathi et al., 2015).

DOI: 10.4018/978-1-5225-3540-9.ch010

Soil pollution is a broad term that refers to the accumulation of native or introduced substances in soil at a harmful level for the growth and health of organisms (micro-organisms, plants, animals and human beings). However, the term "soil contamination" describes only the introduction of a chemical substance that is not originally present in the soil (De Haan & Van Riemsdijk, 1986). However, many regulatory and non-regulatory programs and assessments use soil pollution and soil contamination as equivalent terms.

In this chapter we use the term soil pollution to mean the presence of a substance, either introduced or native, above the threshold level.

Some of the common types of wastes that lead to the development of soil pollution include municipal wastes, commercial wastes, ashes, non-biodegradable wastes, biodegradable wastes and animal wastes.

Soil pollution may arise from a wide range of sources that can be discrete point sources (commercial activities), or diffuse sources (sprayed pesticides and air deposition). The pollution process itself may be deliberate, as in the case of fertilization processes or following accidents such as oil spills in seawater. Each source has specific characteristics, briefly presented below.

Overall, pollutants find their way to the soil via the air path (dry deposition); the water path (sewage water, precipitation, irrigation water, surface water and wet deposition); and via organic solids, such as sewage sludges, composts and other agricultural inputs (fertilizers, manure or plant-protection agents) (Auerswald et al., 2000).

The most dangerous categories of soil pollutants are the persistent organic compounds such as polycyclic aromatic hydrocarbons (PAHs), polychlorinated biphenyl (PCBs) and pesticides as well as heavy metals such as lead, cadmium, arsenic, mercury, zinc, copper and radionuclides. New soil pollutants, such as veterinary medicines and surfactants, have emerged with industrialization and globalization.

This chapter deals with sources of soil pollution, pollutants generated from these sources and their possible adverse effects on the environment.

SOIL POLLUTION SOURCES

Mining Activities

Mining activities represent one of the major sources of soil pollution. The number of mining sites may be limited (almost 2% in Europe) but the extent of the contaminated areas is huge, as are the consequences of such pollution for human health and the environment (Tarazona, 2014). Chemical wastes are generated in every stage of mining activity, particularly during the extraction of minerals and the processing of huge quantities of ores. Other concerns during mining operation are related to the tailings.

Minerals are extracted by crushing and processing ores using massive volumes of water and several chemical and physical methods. The mineral content represents a small part of the ore (from 0.00005% to 5%). This means that the rest of the ore is a waste product known as tailings. Tailings usually contain a mixture of water, finely ground ore and any residual chemicals used at large quantities in all processing stages such as cyanide, xanthates and other flotation agents, surfactants and sulfuric acid (Tarazona, 2014). Tailings may also contain undesired metals (arsenic, mercury, copper, lead, cadmium, selenium, zinc and nickel) that can create environmental concerns as well as sulfides due to the natural composition of the ore.

Tailings may be very reactive wastes and can represent serious environmental risks especially when acid mine drainage occurs leading to the release of significant amounts of toxic metals like nickel, copper, cadmium and aluminum. Once dried, tailings can also be transported by wind causing widespread soil pollution. Acid mine drainage occurs when rocks with sulfur bearing minerals like pyrites are exposed to air, water and micro-organisms. The sulfur in the minerals turns into sulfuric acid and moves in drainage water (Osman, 2014). Also, the mining potential for soil pollution does not disappear at the end of the mining activities, therefore, abandoned mines may represent a huge environmental concern. In the end, all of the pollutants mentioned above contaminate surface soils and waters and find their way into the food chain resulting in dangerous impacts on flora and fauna.

The following are some examples of the involvement of mining activities in soil pollution. In April 1998, a dam break in Aznalcollar mine, Spain, sent acidic and toxic metal-rich sludge over large areas of agricultural land and threatened one an important nature reserves, Doñana National Park. Berkeley pit in Montana is a good example of the impact of acidic toxic runoff and drainage. The open pit of this former copper mine is filled with water at pH 2.5 rich in heavy metals like zinc, cadmium and arsenic. The impact of the tailings covers a large area; thus this location became one of the largest superfund sites (Tarazona, 2014).

Finally, uranium mining activities produce large volumes of residues from ore processing. The radionuclides released from these wastes may lead to human exposure by air, water or soil contamination.

Agricultural Activities

The term "Agrochemicals" refers to various chemical substances used to increase agricultural production. They include a broad range of pesticides. Three main chemical groups should be considered in this category: insecticides, herbicides and fungicides. Agrochemicals also include fertilizers, hormones and other growth agents, as well as animal manure. Some agrochemicals may be dangerous particularly in the event of accidental spills. Fertilizers and manures may cause soil pollution by introducing different compounds into the soil.

Fertilizers

During the nineteenth century, commercial mineral fertilizers were introduced and now a large variety is available to meet specific soil needs. Fertilizers are directly applied to soil in order to increase the level of nutrients and consequently increase crop yields. Nitrogen and phosphorus are the main nutrients used. Other elements are used such as potassium, calcium, sulfur, magnesium, iron, chlorine, copper, manganese, molybdenum and zinc. The main way in which fertilizers and manure cause soil pollution is by introducing heavy metals and their compounds into the soil. Fertilizers, particularly some phosphate fertilizers, may contain highly toxic metals such as arsenic, cadmium or lead. These metals can be related to the mineral used for the production of the fertilizer and therefore related to the origin of mineral, or may be byproducts of the production process. The maximum content of these metals is usually regulated (Osman, 2014).

Organic fertilizers such as manure have been traditionally used for providing nutrients to agricultural soil and may contaminate soil with some heavy metals. Poultry manure and pig manure may pollute the soil with zinc, copper and arsenic. Several jurisdictions have developed specific limits on metal content in manure.

Another group of organic fertilizers has been introduced: composted material produced from different type of organic wastes, including municipal wastes in particular, sludge from wastewater treatment plants and a variety of green residues.

The recurrent application of fertilizers may lead to soil pollution, particularly with nutrients with low mobility in soil. However, the main contamination problem is related to the use of fertilizers containing undesired and highly toxic components.

Veterinary Medicines

The use of veterinary pharmaceuticals is a result of growing animal production. The concern related to these chemical products is relatively recent, though regulatory measures have already been implemented in some jurisdictions. There are two groups of veterinary medicines of particular concern regarding soil pollution: antimicrobials and antiparasitic drugs. The concern regarding environmental contamination with antimicrobials is associated with the possibility of generating antimicrobial-resistant strains. The concerns regarding antiparasitic drugs are due to their very high toxicity to soil and dung invertebrates.

The occurrence of 13 veterinary drugs was studied by Wei et al. (2016) in soil fertilized with animal manures in Eastern China. Results showed that animal wastes, especially from poultry farms, were one of the pollution sources of veterinary drugs in soil. The detection frequency of veterinary drugs in soil was near 90% at different soil depths. The detection rates for the five classes of drugs in soil followed the rank order: cyromazine > tetracyclines > sulfonamides > fluoroquinolones > florfenicol. The authors pointed out that veterinary drugs accumulate easily and persist in the deeper soil. In addition, residue levels of veterinary drugs in soil were related to animal species from which the manure was produced. Overall, the predominance of tetracyclines in sampled soils under scored the need to regulate their veterinary use in order to improve the management and treatment of associated releases (Wei et al., 2016).

Zhang et al. (2016) have evaluated a total of 125 surface soils, collected throughout China for the presence of 17 antibiotics in order to identify antibiotic contamination caused by long-term manures application. The results indicated that the agricultural land has accumulated a higher concentration of antibiotics than conventional open lands used for agriculture activities. The maximum oxytetracycline concentration was 8,400 $\mu g\ kg^{-1}$, the highest level that has ever been reported for oxytetracycline in soils. The residual concentration was decided by both plant duration and manure type. Short-term (<5 years) planting showed the highest residues of tetracyclines and fluoroquinolones in the tested soils (Zhang et al., 2016).

Guo et al. (2016) have studied a total of 53 manure and 50 amended soil samples from 16 livestock farms in the Jiangsu province of China. In the manure samples, the highest detected frequencies and concentrations were those of tetracyclines (TCs, 54.1 $\mu g\ kg^{-1}$), followed by fluoroquinolones (8.4 $\mu g\ kg^{-1}$), sulfonamides (3.2 $\mu g\ kg^{-1}$) and macrolides (0.4 $\mu g\ kg^{-1}$). Statistical analysis was used to illuminate the pollution characteristics of 23 veterinary antibiotics for various animal types and different regions in the Jiangsu province. The results showed that the pollution level in cow manure was relatively low compared to levels in pig and chicken manure because of the relative restriction of medication (Guo et al., 2016).

The same observations were made in other countries concerning the involvement of the use of veterinary medicines in soil pollution. Leal et al. (2012) have studied the occurrence of four commonly used fluoroquinolones (norfloxacin, ciprofloxacin, danofloxacin and enrofloxacin) in soil samples from São Paulo State, Brazil. Fluoroquinolones' mean concentrations found in soil (22.93 $\mu g\ kg^{-1}$) were compatible to those found elsewhere. Enrofloxacin was the most often detected compound at the highest mean concentration (22.93 μg kg^{-1}) (Leal et al., 2012).

Kay et al. (2004) studied 1.55 ha arable field located on a working farm in the upper reaches of the River Soar catchment in Leicestershire, United Kingdom. They investigated the fate of three antibiotics from the sulfonamide, tetracycline and macrolide groups. Pig slurry was applied to a field in arable production during two consecutive years. Both sulfachloropyridazine and oxytetracycline were detected in soil at concentrations up to 365 and 1691 μg kg^{-1} respectively (Kay et al., 2004).

Pesticides

Pesticides are widely used in agriculture. They protect agricultural products from undesired plants, micro-organism and insects. Thereby, they ensure increased yield and the quality of production. They can be sprayed on vegetation or directly applied to soil.

Concerns about pesticide pollution have increased in previous decades. Their residues constitute the main pollutant group in agricultural soils, particularly for the more persistent. In fact, persistency is one of the major factors leading to the long-term toxicity of such chemical products.

At the present time, marketing and use of pesticides are regulated all over the world, with exhaustive ecological risk assessment in many jurisdictions.

Pesticides are usually classified into the three groups according to their target:

- **Insecticide:** These are chemical compounds used to eliminate insects
- **Herbicides:** Chemicals used to combat the growth of undesired plants.
- **Fungicides:** A group of compounds used to suppress the growth of fungus.

Insecticides

The worldwide use of insecticides has been greatly increased in agriculture since the end of the Second World War. Currently, several commercial formulations for these products have been introduced to the market. There are a great number of substances classified as insecticides, but they mainly belong to four groups of organic compounds: organochlorines, organophosphorus compounds, carbamates and pyrethroids.

- **Organochlorines**

During the Second World War, a group of organochlorine compounds were found to be highly effective in controlling pests responsible for diseases such as malaria and yellow fever. These compounds are cheap and readily available. They belonged to three chemical groups: the DDT (dichlorodiphenyl trichloroethane) group, the BHC (β-hexachlorocyclohexane) group and the cyclodiene group. The principal representative of the BHC family is called lindane. Aldrin, dieldrin and heptachlor are stereochemically related compounds and belong to the cyclodiene group, which were used effectively for controlling locusts.

Most organochlorines have a very high persistency and stability, with a half-life in soil of several decades. Due to their massive use in the past decades, they are still the main contaminants in several areas. Their stability results from the non-reactive nature of the C–C, C–H and C–Cl bonds forming these compounds, which makes them very persistent and hence dangerous for humans and animals. In addition, their partition coefficients promote their accumulation in biolipids leading to their accumulation in body lipids of exposed organisms. Because of these properties and their indiscriminate killing of beneficial as well as harmful insects, their use was banned in many developed countries. However, some organochlorines are the active ingredients of some garden and agricultural pest-control products (Tsai, 2010).

- **Organophosphorus Compounds**

Organophosphorus compounds are used in different ways in agriculture and animal hygiene. Some of them are used as fumigants, others as contact poisons, while others are used as systemic pesticides. The most important member of this group of insecticides is tetraethyl pyrophosphate (TEPP) and O-isopropyl methylphosphonofluoridate know also as sarin.

Malathion, dibrom, chlorpyrifos, temephos, diazinon and terbufos are other organophosphates largely used in farming activities. This group of pesticides inhibit irreversibly the enzymatic activity of acetylcholinesterase and may cause a severe form of cholinergic poisoning in humans (Čolović et al., 2013).

- **Carbamates**

Carbamates are derivatives of carbamic acid (NH_2–COOH), of which about 40 commercial (aldicarb, carbaryl, propoxur, oxamyl and terbucarb, etc.) compounds, used as insecticides, are on the market. When used in agriculture, they usually persist from few hours to several months in soil. Their toxic effect, like that of the organophosphates, arises from their disruption of the nervous system by inhibiting acetylcholinesterase. Unlike organophosphates, they kill insects by reversibly inactivating this enzyme. However, they have also been fatal to a large number of birds in agriculture lands and have negatively affected their breeding success.

- **Pyrethroid Insecticides**

Pyrethroids are originally quite effective natural pesticides, extracted from *Chrysanthemum cineraria folium*. They are comprised of six active ingredients: pyrethrum I and II, cinerins I and II and jasmolines I and II. The synthetic pyrethroids include allethrin, bioresmethrin, permethrin and fenvalerate groups.

Pyrethroids are preferred for their lower persistence and lower toxicity to mammals and birds. Currently, they are used in many household insecticides. They provide quick action on insects and rapidly degrade in heat and sunlight. This property reduces their usefulness in killing agricultural pests. However, they are highly toxic to aquatic organisms and bees.

Herbicides

After the Second World War, a considerable number of commercial organic compounds have replaced the traditional herbicides, such as copper sulfate solutions, dilute sulfuric acid and petroleum oil. Now, hundreds of organic and inorganic substances classified in several groups are used as herbicides. The main herbicides belong to the following groups: Organochlorine compounds, organophosphorus herbicides, derivatives of carbamic acid and triazine derivatives.

- **Organochlorine Herbicides**

The organochlorine herbicide group contains derivatives of phenoxyacetic acid such as 2,4-dichlorophenoxyacetic acid (2,4-D), 2,4,5-trichlorophenoxyacetic acid (2,4,5-T) or 2-methyl-4,6-dichlorophenoxyacetic acid (MPCA). The mixture of 2,4-D and 2,4,5-T is known as "Agent Orange". Besides derivatives of phenoxyacetic acid, there are derivatives of aniline among organochlorine herbicides such as propanil and alachlor. The US Army sprayed millions of hectares of woodlands with Agent Orange during the Vietnam war, causing persistent environmental damage. To this day, the Vietnamese still suffer from damages caused by this agent (Patterson et al., 2016).

- **Organophosphorus Herbicides**

Organophosphorus herbicide known also as glyphosate is widely used in agriculture because of its effectiveness against weeds and its non-carcinogenic character. It is the active ingredient in the herbicide Roundup, considered to be nearly non-toxic to humans. However, a recent paper argued that glyphosate may be a key contributor to several diseases and conditions (Samsel & Seneff, 2013).

- **Derivatives of Carbamic Acid**

In addition to being insect killers, derivatives of carbamic acid are used in order to be absorbed by the root systems of weeds, where they operate as systemic pesticides. Examples include several derivatives of urea such as diuron, fluometuron, linuron and chlorobromuron.

- **Triazine Derivatives**

Other herbicides which pollute the soil include triazine derivatives. They are used as systematic weed control agents of relatively low toxicity for mammals. This group of herbicides, which includes mainly atrazine, has the current distinction of being the greatest groundwater pollutants. Furthermore, atrazine can persist for several weeks after being applied to soils. Though banned in Europe, triazine derivatives are widely used in many countries to control weeds in forest plantations and in some crops such as canola, sugar cane and maize. They are degraded by soil, produce a large number of degradation products and affect soil biodiversity.

Fungicides

Fungicides are a large group of chemical compounds of inorganic and organic nature, that are used to control the growth of fungi. They include:

- Inorganic and organic compounds of heavy metals: copper derivatives (e.g., Bordeaux mixtures), organotins, which are represented by tributyltinacetate and triphenyltin acetate.
- Derivatives of phthalic acid with the familiar name "captan".
- Other fungicides: benzimidazoles and barbituric acid derivatives.

Persistence, Translocation and Accumulation of Pesticides in Soil

Many pesticides persist for a long time in soil. They can be divided into three groups based on their half-lives: non-persistent pesticides with a soil half-life of less than 30 days, moderately persistent pesticides with a soil half-life of 30 to 100 days, or persistent pesticides with a soil half-life of more than 100 days (Kerle et al., 2007).

Pesticide persistence and mobility are influenced by their properties. These properties are in turn, influenced by the soil environment such as site conditions and weather (Figure 1). Many other factors can influence the persistence of pesticides in soil like soil pH and photo/microbial degradation that can cause the breakdown of pesticides. These factors strongly interact with one another. On the other hand, soil temperature, climatic conditions, soil moisture, soil texture and the chemical nature may affect the translocation of pesticides in soil. Depending on these factors, the pesticide may have a short or long persistence in soil (Osman, 2014). Table 1 lists half-lives and soil adsorption of some commonly used pesticides.

Organochlorines are the most persistent pesticides due to their high stability and resistance to biodegradation and photodegradation in the environment. This ability of organochlorine pesticides made them highly effective and widely used in agriculture. DDT and dieldrin are two examples of organochlorines with low water solubility and slow rates of microbial degradation, all of which result in persistence. The properties that made these chemicals persistent also made them dangerous for humans and the environment where they accumulate in fat tissue.

Figure 1. Factors influencing pesticide fate in soil

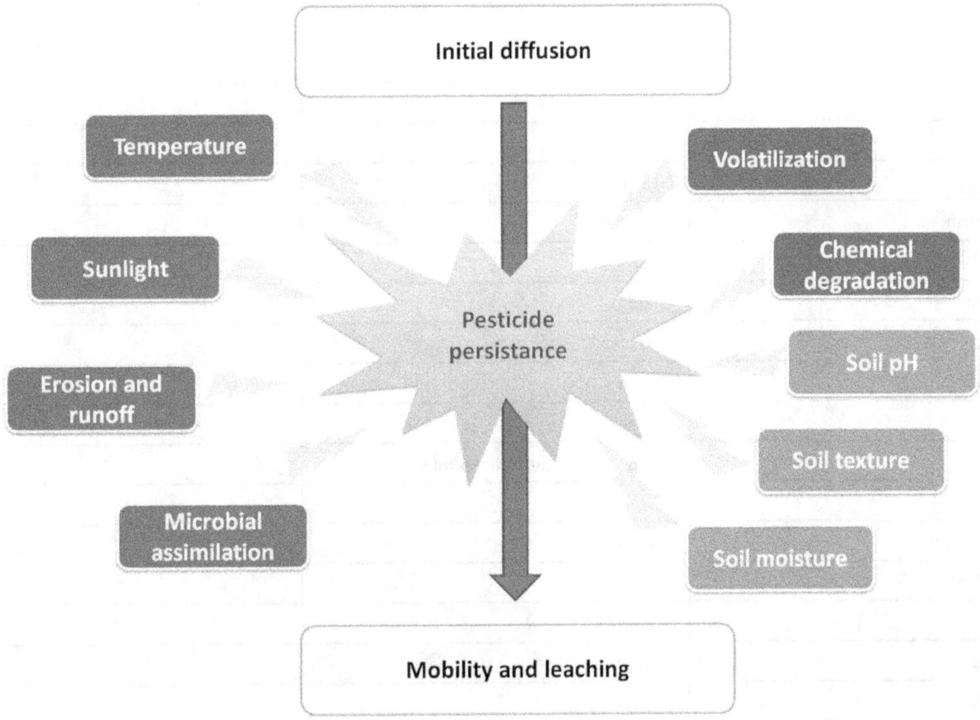

Organophosphorus, carbamate and pyrethroid insecticides are less persistent due to their high rates of degradation by ester hydrolysis.

Herbicides vary in their potential to persist in the soil. Some herbicide families that have persistent members include the triazines, uracils, phenylureas, sulfonylureas and some plant-growth regulators.

Apart from degradation, pesticides undergo several processes that define their movement in the environment. Among these processes are volatilization in the atmosphere, erosion and runoff, leaching to the groundwater and accumulation in soil (Figure 1).

The principal process of pesticide accumulation in soil is adsorption, which is managed by several interactions; covalent, electrostatic, hydrogen and Van der Waals interactions as well as ligand exchange and hydrophobic partitioning. Interactions between pesticides and their metabolites often lead to the formation of strong bonds (chemisorption), which increase pesticide persistence in the soil, but decrease their toxicity, leaching and transport properties (Gevao et al., 2000).

Table 1. Persistence of pesticides in soil

Pesticides	Half-Lives	Soil Sorption Coefficients (μg g^{-1}) Koc
	Organochlorines	
Endosulfan	60 days	-
Dicofol	57 days	-
Aldrin	1-4 years	-
chlordane	2-4 years	-
Chlorobenzilate	22 days	-
DDT	3-10 years	-
Dieidrin	1-7 years	-
a-Endosulfan	60 days	-
p-Endosulfan	800 days	-
HCH	2 years	-
Heptachlor	7-12 years	-
Isodrin/endrin	4-8 years	-
Lindane	35 days	-
Methoxychlor	120 days	-
Toxaphene	10 years	-
Methoxychlor	128 days	-
Lindane	423 days	-
	Organophosphates	
Acephate	3 days	2
Chlorfenvinphos	87 days	-
Chlorpyrifos	41 days	-
Fenthion	34 days	-
Chlorthion	36 days	-
Demeton S	54 days	-

continued on following page

Table 1. Continued

Pesticides	Half-Lives	Soil Sorption Coefficients (μg g^{-1}) Koc
Diazinon	16 days	-
Dichlorvos	17 days	-
Dimethoate	4 days	20
Disulfoton	290 days	-
Fenitrothion	16 days	-
Fenthion	34 days	-
Ethoprop	25 days	70
Terbufos	5 days	500
Methamidophos	6 days	5
Acephate	3 days	-
Fonofos	40 days	870
Azinphos-methyl	10 days	1000
Malathion	3 days	1800
Methyl demeton S	26 days	-
Mothyï parathion	45 days	-
Chlorpyrifos	36 days	6070
Diazinon	40 days	1000
Monocrotophos	3 days	-
Phorate	48 days	1000
Parathion	180 days	-
Phorate	2 days	-
Trichlorfon	140 days	-
Carbamates		
Aldicarb	30 days	30
Bendiocarb	5 days	-
Carbaryl	17 days	300
Carbofuran	68 days	22
Oxamyl	4 days	25
Methiocarb	15 days	-
Propoxur	51 days	-
Bifenthrîn	126 days	-
Cyfluthrin	84 days	-
Methomyl	30 days	72
Cyhalothrin	14 days	-
Cypcrmelhrin	8 days	-
Deltamethrin	15 days	-
Esfenvalerate	75 days	-

continued on following page

Table 1. Continued

Pesticides	Half-Lives	Soil Sorption Coefficients (μg g^{-1}) Koc
Permethrin	20 days	-
Tefluthrin	30 days	-
Tralomethrin	63 days	-
Benomyl	153 days	1900
Pyrethroids		
Esfenvalerate	42 days	-
Permethrin	42 days	-
Fenvalerate	46 days	5300
Herbicides		
Diuron	90 days	480
Prometon	500 days	150
Bromacil	60 days	32
Bentazon	20 days	34
Hexazinone	90 days	54
Terbacil	120 days	55
Bensulide	92 days	1000
Tebuthiuron	360 days	80
Atrazine	60 days	100
Acifluorfen	14 days	113
Simazine	60 days	130
Prometon	1-2 years	150
Alachlor	15 days	170
Cyanazine	14 days	190
S-ethyl dipropylcarbamothioate	6 days	200
Metolachlor	90 days	200
Linuron	60 days	400
Diuron	90 days	480
Oxadiazon	75 days	-
Norflurazon	90 days	600
Oryzalin	20 days	600
Fluazifop-p-butyl	15 days	5700
Trifluralin	60 days	8000
Diclofop-methyl	37 days	16000
Picloram	90 days	16
Dicamba	14 days	2
Metolachlor MAIS ORGANOCHLORES	15 days	-
Oxasulfuron	5 days	-

continued on following page

Table 1. Continued

Pesticides	Half-Lives	Soil Sorption Coefficients (μg g^{-1}) Koc
Primisulfuron	11 days	-
Prosulfuron	11 days	-
Terbuthylazine	36 days	-
Trifluralin	81 days	-
Ethalfluralin	50 days	4000
Triallate	74 days	-
Dacthal	50 days	-
Glyphosate	38 days	-
Pendimethalin	174 days	-
Benfluralin	80 days	-
Fungicides		
Hexachlorobenzene	4 years	-
Metalaxyl	27 days	50
Chlorothalonil	30 days	1380
Captan	3 days	200
Pentachlorophenol	48 days	-
Dichlone	82 days	-
Pentachloronitrobenzene	65 days	-

Municipal Wastes

Municipal wastes include various types of materials that may contribute to changes in the composition of soil. They comprise household wastes, market wastes, hospital wastes, slaughterhouse wastes, metals, glass and ceramic wastes as well as leakages of sewage water, which can contribute significantly to soil pollution.

Municipal wastes can be classified into different groups: biodegradable materials (food waste such as meat trimmings or vegetable peelings, yard waste, clothing, paper, newspapers and milk cartons), non-biodegradable/recyclable materials (glass, plastic (plastic plates, plastic wrap and tires) and metal (aluminum cans) materials and hazardous wastes (medicines, paint, batteries, fertilizer and pesticide containers). Non-biodegradable materials clog soil pores and hamper natural drainage, tillage and planting operations.

Disposal of municipal wastes by incineration may lead to soil contamination by organic matters and heavy metals, such as cadmium, copper, lead, tin and zinc, or by ash fallout from incinerating plants. Several studies have been conducted to assess the environmental effects of incinerators.

According to Xu et al. (2009), the overall soil PCDD/Fs (polychlorodibenzo-*p*-dioxins and polychlorodibenzofuran) levels increased significantly, 33% and 39% for total concentration and I-TEQ (median value) respectively in agricultural soils in the vicinity of a municipal solid waste incinerator in Eastern China during 2006–2007 (Xu et al., 2009).

Collet et al. (1998) found that atmospheric emissions of lead from the Baldovie municipal waste incinerator, Scotland, significantly determines the local distribution of lead in soils within the immediate vicinity of the incinerator (Collett et al., 1998).

Bretzel and Calderisi (2011) demonstrated that the increase in Pb and Zn emitted from the municipal solid waste incinerator plant Pisa, Italy represented the major soil pollutants (Bretzel & Calderisi, 2011).

Other studies have been carried out on the impact of dumping sites on soil properties. According to Ali et al. (2014), significant modifications were observed in the soil properties of the dumping sites in Pakistan. Soils at the disposal sites showed high pH in comparison to control sites. Various heavy metal concentrations (i.e., lead, copper, nickel, chromium and zinc) were also found to be higher at the dumping sites (Ali et al., 2014).

Ihedioha et al. (2016) assessed the levels of some heavy metals in soils in the vicinity of a municipal solid waste dumpsite in Uyo, Nigeria. They found that soil samples from the dumpsite and distances from 10 and 20 m East of the dumpsite were highly polluted with cadmium. Moreover, the ecological risk assessment results showed that cadmium was the only metal posing a potentially high risk to the local environment (Ihedioha et al., 2017).

A huge quantity of waste is generated in health-care facilities. Several types of waste can be considered to be pollutants such as sharps, chemicals, medicines, medical devices and radioactive materials. Moreover, potentially infectious wastes like used catheters and gloves, cultures and stocks of infectious agents, empty bottles of lab, diagnostic kits, waste materials from dialysis and dentistry units, waste from isolation units, bandages, dressings, swabs, gauze and disposable medical devices, if they don't undergo proper treatment, represent hazardous pollutants for humans and the environment. Radioactive hospital wastes that may pollute soil include contaminated glassware, packages, or absorbent paper. These types of waste are often disposed with municipal wastes in open dumps.

Soil naturally contains more than 50 radionuclides, among them plutonium, radon, strontium, tritium, thorium and uranium.

Artificial radioactivity in soil may occur during the normal operations of nuclear installations such as nuclear ore processing, uranium enrichment, fuel fabrication, reactor operations and application of radioisotopes in research field. Soils may receive some radionuclides from these activities if they are buried for disposal. Plutonium, uranium, americium, neptunium, curium and cesium are among the elements whose nuclides occur in radioactive wastes.

Landfills

Urban activities produce large quantities of wastes. These wastes are comprised of mainly paper, plastic, glass, aluminum, food materials and fecal micro-organisms. In developing countries, municipal wastes and sewage are dumped in open dumps, whereas in developed countries, despite increasing efforts to recycle these wastes, much of them ends up in landfills. However, severe environmental problems like soil pollution may occur by open dumping and landfilling. Landfills constructed prior to the regulations were usually not well constructed and therefore, many older landfills have problems associated with leachates. Moreover, many of them were located in lowlands near streams overlying porous sediments.

Municipal wastes potentially accepted by landfills include household waste, municipal sludge, municipal waste combustion ash, waste tires, some infectious waste, industrial non-hazardous waste, construction and demolition waste and agricultural wastes.

Landfills are dangerous places; there is continuous release of gases. Under anaerobic conditions, microorganisms produce two toxic gases methane and CO_2. However, the primary means by which landfills can serve as a source of soil pollution is through the generation and leakage of landfill leachate. It consists of water in which is dissolved hazardous chemicals that, as a result either of the concentrations involved or of their considerable toxicity, pose a risk to human health and the environment. As the water moves through the waste, components of the waste are dissolved. Leachate collected from landfills often contains heavy metals and synthetic organic compounds some of them are VOCs. In fact, the environmental threats from two landfill areas of the USA, Volney and Colesville, show the danger of landfills. The groundwater near these areas contains volatile organic compounds (VOCs) (i.e., benzene, toluene and chlorinated hydrocarbons) and heavy metals (i.e., arsenic, cadmium, chromium, mercury and nickel). Potential pathways of exposure to these pollutants are drinking contaminated water, particularly groundwater, as well as accidental ingestion of fruits and vegetables from contaminated soils. Current regulations require landfills to be constructed so that leachate production and escape are minimized.

Several other studies have been conducted on the impact of landfills on the environment. Kasassi et al. (2008) have studied soil samples of a closed unlined landfill at different depths (2.5-17.5 m) located northwest of Thessaloniki, North Greece. They have found that the metal values varied over a wide range: from 0.50 to 18.75 mg kg-1 for cadmium, 3.88–171.88 mg kg-1 for chromium, 8.13–356.25 mg kg^{-1} for Cu, 5.63–63.75 mg kg^{-1} for nickel, 2.50–92.50 mg kg^{-1} for lead and 6.38–343.75 mg kg^{-1} for zinc (Kasassi et al., 2008).

Melnyk et al. (2015) studied the concentration of PAHs in areas situated around the municipal solid waste landfill in Gdańsk in northern Poland. They found that the total concentration of ΣPAHs ranged from 892 to 3514 μg kg^{-1} with a mean of 1974 μg kg^{-1}. They demonstrated that the concentrations of PAHs in six samples were above the acceptable level. They found that the highest total concentration of PAHs was 3.5 mg kg^{-1} and in most samples, the most dominant were PAHs containing 2–3 rings in a molecule (Melnyk et al., 2015).

Chrysikou et al. (2008) have studied the concentration of PAHs, PCBs, as well as trace elements in soil samples collected from the surrounding area of the "Tagarades" landfill, in northern Greece. High concentrations of total PAHs, PCBs and heavy metals were found inside the landfill (1475 μg kg^{-1}, 399 μg kg^{-1} and 29.8 mg kg^{-1} respectively) (Chrysikou et al., 2008).

Sewage Sludges

Municipal, sanitary and storm sewages are a mixture of solids and dissolved materials. They are processed through wastewater treatment plants. In this process, the water is separated from the solid part by several treatments before its safe discharge into streams and lakes (treated waste-water effluent). The solid waste (sewage sludge) is often discharged in open dumps, landfills, incinerators, or is composted for agricultural use. In almost all developed countries, the disposal of this sludge by dumping it at sea is being phased out.

Sewage sludge contains organic and inorganic materials. Soil application of treated sewage sludge provides a large part of the nutrient requirements of many crops, but it has the risk of polluting the environment. In fact, the use of sludge is an advantageous process in itself. It adds essential organic matter as well as useful nutritive elements like nitrogen and phosphorus to the soil. Sewage sludge is also rich in organic substances, sulfur, calcium, magnesium and other microelements necessary for plants and soil fauna. Apart from the indispensable elements, sludge can contain hazardous compounds such as pesticides, heavy metals and microbial contaminants.

Heavy metals, which are usually concentrated in sewage sludge, may accumulate in the soil after repeated application and may eventually be taken up by food crops such as leafy vegetables, which are known to preferentially take up cadmium. In addition to being phytotoxic, these metals are toxic to humans. Once the waste has been disposed of in the soil, rain or irrigation water can dissolve chemical compounds from waste, thereby moving the contaminants deeper into the soil. Some of these compounds can enter the food chain by plant or animal uptake. Currently, recommendations for sludge application rates on soil are based on some nutritive elements (nitrogen, phosphor and potassium) and levels of some metals present in sludge. The metals of primary concern are copper, zinc, lead, nickel, cadmium, mercury and chromium. When applied to soils in excessive amounts, these metals may accumulate in plant tissues, impair the quality of food and can be hazardous to humans.

Apart from heavy metals, sewage sludge may contain organic toxic pollutants such as PAHs, PCDDs and PCDFs. One of the most common form of soil pollutant dioxins is the 2,3,7,8-tetrachlorodibenzo-*p*-dioxin. PCDFs such as 2,3,7,8-tetrachlorodibenzofuran are also highly toxic pollutants. Both of these chemical substances are considered to be the most lethal synthetic chemicals. Other substances like surfactants enter the soil environment via sewage water and sewage sludge. It is estimated, for example, that large quantities of linear alkyl benzene sulfonates remain in these materials, which can also contribute significantly to soil pollution.

Sewage sludge can contain microbial contaminants (bacteria, viruses and protozoa) along with other parasitic helminths which can give rise to potentially severe pathologies in humans, animals and plants.

Some studies have focused on soil pollution by polybrominated diphenyl ethers (PBDE) present in sewage sludges. Research conducted by Eljarrat et al. (2008) on sewage sludge sampled in different locations throughout Spain and sludge amended soil samples collected at different selected sites in the province of Barcelona has demonstrated that sewage-sludge amendment at the research stations increased concentrations of all PBDE congeners 1.2- to 45-fold, with the highest increases for BDE-209 (Eljarrat et al., 2008). According to Kim et al. (2017), BDE-209 is the most significant congener present in sludges, followed by BDE-99 and BDE-47. The highest concentrations, up to 23,500 ng g^{-1}, of PBDEs in sludge were found in North America until 2004-2007, whereas since then sludge PBDE concentrations, up to 6600 ng g^{-1} have been higher in Asia than on the other two continents. The estimated concentrations of BDE-47, -99 and -209 in soils receiving sludge applications during the past 15 years are 40-300 times higher than in soils after the initial sludge application (Kim et al., 2017).

Lourival et al. (2011) were interested in the PAHs content of sewage sludges obtained from the municipal Sanitation Plant of Franca, State of São Paulo, Brazil. The results showed that PAH concentration levels found in sewage sludge might raise potential contamination risks to the soil. Simulations of the concentrations of PAHs in soil, by mathematical modeling, allowed to indicate a priority order for monitoring soil quality in terms of these contaminants: phenanthrene > chrysene > benzo(k)fluoranthene > benzo(a)pyrene > pyrene > benzo(b)fluoranthene > benzo(g,h,i)perylene > indeno(1,2,3-c,d)pyrene > fluoranthene > benzo(a)-anthracene > dibenzo(a,h)anthracene > fluorene > anthracene > naphthalene > acenaphthene > acenaphthylene (Lourival et al., 2011).

On the other hand, repeated applications of compost from sewage sludge on agricultural soil have significant effects on the physical and chemical properties of soil. Several studies have been carried out in order to evaluate heavy metals in composts from sewage sludge. For example, Alvarenga et al. (2015)

have demonstrated the existence of heavy metal concentrations exceeding the proposed limit of the draft European legislation in some composts collected at different locations in Portugal (Alvarenga et al., 2015).

Finally, to control the quality of sewage sludge and prevent any possible hazardous impact of fertilisation, some mandatory limits of heavy metal content have been established by the European Commission (Sewage Sludge Directive 86/278/EEC) and independently by some countries (Table 2). Individual member nations could choose to adopt more stringent standards than those recommended in the directive. Some member states have set their limit values below the maximums allowed by the Directive.

Industrial Wastes

Industrial wastes are the major sources of severe environmental pollution by toxic chemical compounds and heavy metals. They are generated at every step in the production process and contribute largely to soil pollution. Industrial wastes include also oil, dirt and gravel, wood and scrap lumber. They may be toxic, corrosive, or reactive. If improperly disposed of, this waste can affect public health and can have environmental consequences.

Different kinds of industries add different types of waste materials to the soil. For example, metal producing industrial wastes and smelting industrial wastes are comprised of heavy metals including zinc, lead, arsenic, chromium, mercury, copper, cobalt, nickel, cadmium and iron. Similarly, the waste of the leather industry is comprised of chromium whereas that of the textile industry includes manganese and cadmium.

Many new industrial products are introduced every year in the market, but the human health and environmental impacts of many of these products are largely unknown (Table 3). For example, there are more than 70,000 different chemicals being manufactured in the USA with some 1,000 new chemicals being added each year. Some of them are hazardous to the environment, like plastics, drugs, textiles, paints, dyes and pesticides. Two types of chemical products of particular concern are dry cleaning fluids and embalming fluid (Osman, 2014).

One of the most important incidents related to industrial activities that caused extensive soil pollution is the Seveso disaster. The Seveso disaster is an industrial accident that occurred on July 10, 1976, in a chemical manufacturing plant in the Italian town of Seveso. The accident was due to the overheating of the reactor. This industrial accident exposed a large population to substantial amounts of relatively pure 2,3,7,8-tetrachlorodibenzo-p-dioxin (TCDD). Early health investigations including assessment of liver function, immune function, neurologic impairment and reproductive effects yielded inconclusive results. However, nearly 200 cases of chloracne, mostly in children have been identified with certainty. Soil toxicological analyses were repeatedly performed between 1976 and 1986 and thousands of soil

Table 2. Limit Values for Heavy Metals in sewage sludges (mg kg^{-1})

Country	Cadmium	Copper	Nickel	Lead	Zinc	Mercury	Chromium
Directive 86/278/EEC	20-40	1000-1750	300-400	750-1200	2500-4000	16-25	-
United States	39	1500	420	300	2800	17	-
China	20	1500	200	1000	3000	15	1000
France	20	1000	200	800	3000	10	1000
Germany	10	800	200	900	2500	8	900

Table 3. Soil pollutants generated by industries

Industrial Activity	Waste Generated
Oil refinery	Wastewater containing benzene and other hydrocarbons
Leather manufacturing	Chromium Toluene Benzene
Paper industry	Heavy metals Organic matters
Metal manufacturing	Cyanide waste Paint waste Sludges containing heavy metals
Printing industry	Ink sludges containing heavy metals Solvents
Chemical manufacturing	Acids and bases Solvents Reactive waste

samples were analyzed, which allowed territorial mapping of TCDD distribution and monitoring of its behavior over time (di Domenico et al., 1980). Based on these measurements and measurements of a limited number of human blood samples, the exposed population and areas were classified into three categories, A (highest exposure), B (median exposure) and R (lowest exposure). Zone A, the most heavily contaminated area, had mean soil levels of TCDD ranging from 15.5 to 580 pg/m^2. In zone B the levels of TCDD did not exceed 5 pg/m^2 on average. In the third zone with lower contamination levels, TCDD levels generally were below 1.5 pg/m^2. Toxicological analyses of contaminated soils also indicated that 2,3,7,8-TCDD accounted for more than 90% of total dioxins in the environment and that other isomers were present like 1,3,7,8-TCDD in zone A and inside the plant and 1,3,6,8-TCDD and 1,3,7,9-TCDD in zone R. Moreover, over 90% of detectable TCDD was found in the thick upper 15 cm layer in 1977 and no further significant level changes were observed after that time (di Domenico et al., 1980).

From these alarming investigations, the entire top soil layer (at least 25 cm) was removed in zone A and replaced by fresh uncontaminated soil. Rehabilitation of zone B resulted in soil removal in public and private gardens, however in the remaining territory, the upper soil layer underwent treatments such as harrowing, plowing and sowing to dilute surface TCDD and help in its dispersal. TCDD levels in sediment and in ground and surface water showed the low solubility of TCDD (Bertazzi & di Domenico, 1994).

This catastrophic accident prompted the adoption of legislation on the prevention and control of such accidents. The so-called Seveso-Directive (Directive 82/501/EEC) was later changed in view of other accidents such as Bhopal, Toulouse or Enschede resulting into Seveso-II (Directive 96/82/EC). In 2012 Seveso-III (Directive 2012/18/EU) was adopted taking into account the changes in the European Union legislation on the classification of chemicals and increased rights for citizens to access information and justice.

On other hand, several studies have shown the implication of industrial activities in soil contamination by heavy metals and organic compounds. The most important metals are mercury, zinc, arsenic, cadmium, iron, copper, lead, chrome, manganese and vanadium, which can even be found at high concentrations in vegetables growing in industrial-waste-water-irrigated-agricultural fields.

Wang et al. (2017) investigated soil and crop contamination and associated health risks by PAHs in industrial areas of the Yangtze River Delta region, China. Soil and crops were contaminated by PAHs with concentrations in the range of 103.8-1070.4 and 244.1-2234.6 μg kg^{-1} respectively. Based on both the results of a principal component analysis and the PAH ratios, the authors found that the main sources of the PAHs in soils and crops were from the combustion of coal and petroleum. The cancer risk analysis indicated that soil contaminated with PAHs posed a lower risk than crops, but the risk was still much higher than the baseline value (Wang et al., 2017).

A study performed by Liu et al. (2017) showed that an organic chemical plant operating for more than 50 years in Chongqing, Southwest China, caused severe soil PAHs contamination. The total concentrations of 16 priority PAHs in each borehole ranged from 0.7 to 4889.5 mg kg^{-1}, with an average of 130 mg kg^{-1} and the pollutants mainly accumulated in the first layer of the site (0-1 m) (Liu et al., 2017).

Turrio-Baldassarri et al. (2007) studied PCB, PCDD and PCDF contamination in Brescia, a city in the North-West of Italy. The area is close to an industrial plant that produced, in total, some 31000 tons of PCB. In several cases, the legal limit of PCDD/Fs was exceeded by more than one order of magnitude, with the highest contamination in some agricultural areas and in the surrounding zones. According to the authors, the source of the contamination is probably a heavy PCB mixture with a prevalence of hexa/hepta chlorinated congeners (such as an Aroclor 1262 or Aroclor 1264 type of mixture) with a minor component of lighter congeners (Turrio-Baldassarri et al., 2007).

Transport Activities

Heavy traffic constitutes one of the main sources of environmental pollution and significantly contributes to soil pollution. Soil pollution occurs along major highways due mostly to emission of particulate matter, as well as gaseous pollutants, which can settle in soil by both wet and dry deposition.

Many of the chemical substances released from vehicles after combustion of fossil fuels have high adsorption ability and tend to bind to inorganic and organic particles in soil. The contribution of transport activities to soil pollution is due also to the emissions from airplanes. Chemical substances emitted from all means of transportation that use internal combustion engines, include carbon, nitrogen and sulfur oxides as well as some heavy metals. In fact, roadside soils usually contain high concentrations of metallic elements like copper, zinc, iron cadmium and lead which is found in lead sulfate form. These metals are essential components of many alloys, wires and tires and could be released into the roadside soil as a result of mechanical abrasion and normal wear. In several studies conducted in France, China and Saudi Arabia, It was observed that roadside soils contain considerable concentrations of lead, cadmium, copper, zinc, arsenic, chromium and nickel (Chen et al., 2010; Kadi, 2009; Viard et al., 2004).

Combustion of fossil fuels in vehicle engines is also an important source of highly toxic components like PAHs. In fact, some studies conducted in India have demonstrated that total PAHs concentration in soil at traffic sites are higher than those at rural sites. The carcinogenic potency of PAHs load in traffic soil was nearly 6 to 21 times higher as compared to the rural soil. The observed PAHs pattern was dominated by four, five- and six-ring PAHs (contributing >50% to the total PAHs) at all the traffic sites examined (Agarwal, 2009; Suman et al., 2016).

Other possibilities for soil pollution due to highway activities may be encountered:

- Pollution due to accumulations of chemicals, oils and hazardous substances as a result of accidental spills.

- Pollution resulting from maintenance activities requiring the use of weed and insect control compounds, as well as salts used to control the formation of ice in winter.

In a study conducted in Mexico, it was observed that cesium, zinc, antimony and arsenic concentrations are significantly high in urban soils. These concentrations are much more than the standard guidelines of the US EPA and are not without harmful effects on the environment and human health (Mireles et al., 2012).

Power Generation Emissions

Energy-production activities such as electricity generation can be an important source of soil pollution. Emissions from power plants include NOx, SOx, COx and PAHs generated from coal-fired power stations and radionuclides emitted from nuclear power plants. The most abundant radionuclides found in soils and originating from nuclear power emissions are 137Cs and 134Cs.

Pollutants from power plants may be deposited on the ground either directly by dry deposition or in a wet form after being dissolved in water droplets during precipitation (see air deposition and soil pollution).

Large-scale production of coal ash from thermal power plants around the world is now causing a serious threat to the environment by polluting the soil. When coal is used for electricity generation, devices are often placed in the plant smoke stack to remove most pollutants so that they are not released into the atmosphere. The waste materials that are collected in this manner are often disposed of on site by application to soil. Residues of coal combustion are considered as hazardous solid waste around the globe. It has been observed that soils which are located in the vicinity of coal-fired thermal power stations are polluted not only by ashes but also by heavy metals (Bhattacharya et al., 2012; George et al., 2015). In fact, coal fired thermal power plants are a real source of soil pollution. Different studies carried out in different countries (India, China, South Africa, Turkey, Russia, Serbia, Brazil and Spain) have shown that these stations can lead to soil pH changes and can represent sources of sulfur and heavy metals like chromium, cadmium, nickel, lead, arsenic, nickel, strontium, chromium, copper, nickel, cobalt, manganese, strontium, zinc, iron, mercury, cadmium, barium, beryllium, titanium, tin, antimony and vanadium. Their concentrations are sometimes higher than the maximum allowable limits stipulated for most countries (Cicek & Koparal, 2004; Dragović et al., 2013; George et al., 2015; Li et al., 2017; Minnikova et al., 2017; Okedeyi et al., 2014; Ozkul, 2016; Raja et al., 2015; Rodriguez-Iruretagoiena et al., 2015; Rodríguez Martín & Nanos, 2016; Sengupta et al., 2010; Singh et al., 1995).

An experiment was conducted in China in which it was observed that the total yearly emissions of chromium, cadmium and lead have significantly increased in recent years, leading to soil pollution (Tian et al., 2012).

Nuclear power plants can be a source of soil pollution, particularly in the case of incidents endangering the integrity of the environment. On March 11, 2011, a magnitude 9.0 earthquake occurred 130 km off the Pacific coast of the Tohoku region of Japan. The quake struck a wide area of northeastern Japan and was followed by a large tsunami. This resulted in serious damage to the Fukushima Dai-ichi Nuclear Power Plant (FDNPP), operated by the Tokyo Electric Power Company (TEPCO). Three nuclear reactors experienced full nuclear meltdown and pressures inside the reactors mounted. Though TEPCO attempted to vent this pressure to the atmosphere, hydrogen explosions occurred.

The results of an analysis carried out by Fujiwara et al. (2012) on soil samples obtained from 50 locations in Fukushima prefecture on April 20, 2011, revealed the presence of a spectrum of radionuclides resulted from the accident of the FDNPP. The sum γ radioactivity concentration ranged in more than 3 orders of magnitude, depending on the sampling locations (Fujiwara et al., 2012).

Soil samples were also taken from areas around FDNPP, at four locations Northwest of FDNPP, at four schools and in four cities, including Fukushima City. Radioactive contaminants in soil samples were identified and included 129mTe, 129Te, 131I, 132Te, 132I, 134Cs, 136Cs, 137Cs, 140Ba and 140La. The highest soil depositions were measured to the northwest of FDNPP (Endo et al., 2012).

Harada and Nonaka (2012) investigated soil radiocesium contamination of rice fields in the area of Niigata, 130 to 200 km away from the FDNPP, Niigata being one of the largest rice growing regions in Japan. Soil samples were collected from the plow layer of five rice fields, 5-6 months after the FDNPP accident. Results showed that radiocesium concentrations (the sum of 134Cs and 137Cs) in the rice soil samples were ~300 Bq (kg dry soil)$^{-1}$ (Harada and Nonaka, 2012).

Military Activities

Soil Pollution Through Chemical Warfare

Agent Orange

The development and use of pesticides as chemical warfare agents showed its horrible face in the use of herbicides. Agent Orange, a defoliant chemical, is an illustrative example of such application and widely known for its use by the U.S. military during the Vietnam War. Agent Orange, as previously mentioned, is a mixture in equal amounts of two herbicides, 2,4-D and 2,4,5-T. Military operations involving the use of this agent were stopped in May 1970, after the appearance of an opposition in the United States. This war caused the pollution of immense forests and genetic damage to exposed people for many generations to come (Patterson et al., 2016).

Mustard Gas

Mustard gas, also known as sulfur mustard, was the most important vesicant poison gas used during the First World War. When used in impure form such as warfare agents, it appears usually as yellow-brown heavy liquid and has an odor resembling mustard plants, hence the name. Mustard gas was considered to be the most lethal of all poisonous chemicals used during this war. Its vapor has the ability to form extensive blisters and burns on exposed skin and in the lungs. Once in the soil, mustard gas has a dangerous effect, since they remain active for several decades in polluted soils even if it is covered by water. Mustard gas can significantly change the number and the specific composition of soil microbiota and inhibit the enzyme activity of soils (Medvedeva et al., 2008).

Mustard agents were regulated under the 1993 Chemical Weapons Convention. Sulfur and nitrogen mustard are categorized in Schedule 1, as substances with no use other than in chemical warfare (Organisation for the Prohibition of Chemical Weapons, 2017).

Soil Pollution Through Biological Warfare

The term of "biological warfare" refers to the use of toxins and/or infectious agents such as bacteria, viruses and fungi with the intent to kill humans, animals or plants as an act of war. In addition to their chemical counterparts, these agents are characterized by their ability to reproduce or replicate within their host victims, thereby increasing their toxicity and their ability to be transmitted from person to person leading to epidemics, which are difficult to control. Many of them can persist in a dormant form for a very long time in soil and under different environmental conditions:

Bacillus Anthracis

Anthrax is a disease caused by the bacterium *Bacillus anthracis* in humans and animals. This bacterium is a gram-positive sporulating bacillus commonly found in the soil of endemic areas. It exists in two forms, vegetative cells inside the host and persistent spores in the soil and the environment. The spores are resistant to extreme environmental conditions like heat, cold, drying, pH, chemicals, UV, deficiency of nutrients and chemical disinfectants. It can persist for decades (with a half life of about 100 years) in soil and remain viable in the upper 6 cm of soil. Experiments involving detonation of bombs containing *Bacillus anthracis* spores were conducted by US and their allies on Gruinard Island in 1943. After the experiments, live spores were detected in the soil at various depths and took 40 years and tons of formaldehyde before to disinfect the island and declare it free of anthrax spores. This bacterium persists for a long period in animal products and produces a horrible disease, which leads to death after a series of painful symptoms including respiratory and skin disorders. Exposed animals that die of anthrax release massive quantities of spores into the soil and this requires the burning of the infested animal carcasses to prevent long-term environmental contamination (Goel, 2015; Spencer, 2003).

Burkholderia Pseudomallei

Burkholderia pseudomallei is a free-living organism in soil and water. It's a gram-negative bacillus that produces an infectious disease, known as melioidosis. The infection occurs through inhalation, or skin abrasions that come in contact with contaminated soil or water. Melioidosis is endemic to Northern Australia and Southeast Asia especially Thailand. When used as a biological warfare agent, this bacterium causes disease that may appear as an acute pulmonary infection and an acute localized skin infection. This bacterium also has the capacity to remain latent in soil for long periods of time.

AIR DEPOSITION AND SOIL POLLUTION

Air deposition leading to soil pollution includes gas, ashes, smoke, particulate matter and even rain. Its occurrence is highest in the Northern hemisphere where industrial activity is concentrated. Oxides of nitrogen and sulfur, fluorides, chlorides and ammonium are the principal gas pollutants deposited from air. Air deposition may also contain toxic particles like persistent organic pollutants as PCBs, PAHs and heavy metals.

Usually, most of the heavy metals (lead, cadmium, etc.) in the air are in fine particulate form (<2.5 μm) with the exception of mercury.

Atmospheric deposition may occur in different ways:

1. Cloud deposition which means the deposition of droplets of clouds and fogs containing air pollutants.
2. Dry deposition, which corresponds to the deposition of suspended toxic materials by the gravitational field in gaseous or particle-bound form. For example, substances with low vapor pressure, like PCBs, PAHs and other chloro-organic compounds are present in the air mainly in the gas phase and are subsequently deposited on the ground.

Some elements produced during soil erosion (aluminum, Silicium, Magnesium, etc) can aggregate in suspended particles. Once the weight of these particles becomes large enough, they can reach the soil and change its composition.

3. Wet deposition which corresponds to deposition of toxic material through precipitation. It occurs when air pollutants, in gas or particle forms, are incorporated into cloud droplets and deposited by rain precipitation. This process takes place during the formation of water droplets and when the droplets fall to the ground. Generally, this kind of air deposition contains mainly acid forms of oxides of sulfur and nitrogen. Sulfur dioxide (SO_2) and nitrogen dioxide (NO_2), formed mainly from the burning of fossil fuels, react in the atmosphere to form sulfuric and nitric acids respectively. Sulfate deposits can also be formed from the burning of biomass and from volcanic eruption.

Upon oxidation by photochemical reactions in the atmosphere, sulfur and nitrogen oxides react with water droplets in the air to produce strong acids such as HNO_3 and $H2SO_4$. These acids dissolve in the rain, forming what is often referred to as acid rain (Figure 2).

The consequences of acid atmospheric deposition are principally environmental and include depletion of soil nutrients, damage to vegetation and acidification of stream water. The decrease in pH promotes leaching of nutrients from the upper soil and the release of aluminum ions from clay minerals, which can lead to depletion of fish populations in headwater streams.

Figure 2. Acid rain formation and soil effects

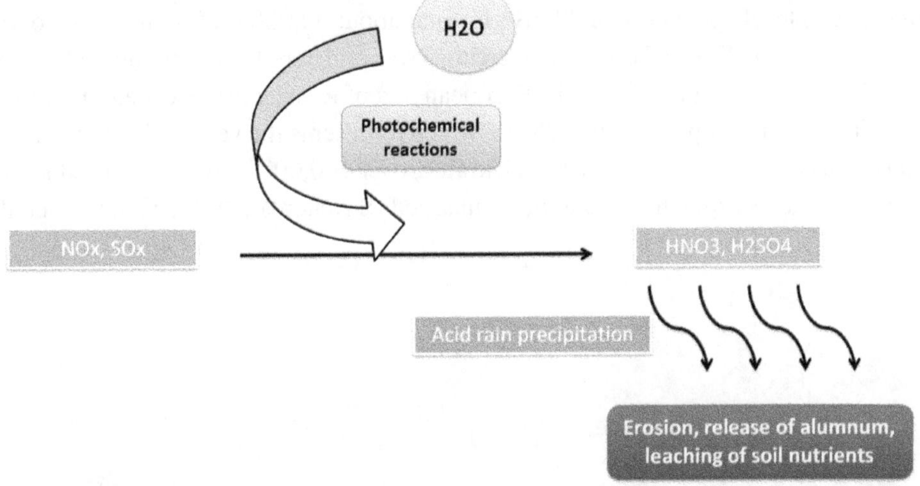

Some particles remain in the air longer and undergo chemical reactions there. Usually, they are transported over long distances and are deposited in other areas. In fact, air deposition can be classified according to the traveled distance in the atmosphere, into short-range, medium-range and long-range transported particles.

Short-range transported particles are deposited in the vicinity of their respective source and are transported according to wind conditions. Typical examples are particles comprised of pesticide following spray applications, dust spreading around industrial areas, or particles emitted in the vicinity of highways. The most important chemicals associated with highway emissions are lead, chromium, zinc, copper, iron, magnesium, manganese and cadmium.

Medium-range transported particles include all particles that undergo volatilization, transport and deposition processes at several kilometers from their sources. These particles contain mostly volatile pesticides from plants and soil that are deposited by a continuous process instead of a single event. Medium-range transported particles are also created from industrial and urban emissions of volatile chemicals that bind to larger particles or dissolve in water droplets and are subject to dry and wet deposition respectively.

Long-range transported particles are particularly relevant to some pollutants particularly persistent organic pollutants and mercury. The long transport distance is due to the persistence and the volatilization of the substances forming these particles. Their movement is related to global air circulation patterns, which lead to their long-range transport and deposition in the polar areas and high altitudes.

Radionuclides deposition can also be a significant pathway to soil pollution and human exposure to poisonous elements by ingestion of contaminated food and animal products. One of the most cited examples of such contamination is the catastrophic accident that occurred at the Chernobyl Nuclear Plant in Ukraine on April 26, 1986. An explosion occurred late at night in the N°4 light water graphite moderated reactor. It produced considerable updrafts for about 9 days that lofted plumes of fission products into the atmosphere. The prevailing winds sent the plume in a North to Northeasterly direction. The plume spread over Europe, resulting in significant fallout of radiation in Austria, Switzerland, Germany, Britain and Sweden. Moreover, this accident caused a radioactive contamination of a considerable surface area that included agricultural lands in Belarus, Russia and Ukraine, where about 5 million people were living. In fact, the releases contaminated about 125,000 km^2 of land in Belarus, Ukraine and Russia with radiocesium (137Cs) levels greater than 37 kBq m^{-2} and about 30,000 km^2 with radiostrontium greater than 10 kBq m^{-2}. A total of 23% of Belarussian territory was registered as contaminated by this measure, compared with 7% in Ukraine and 1.5% in the European part of Russia. In these areas, radionuclides have been measured in the soil to depths of up to 25 cm, which represents the vertical layer in which most crop cultivation takes place. As a result, in Northern Ukraine, over 100,000 ha of agricultural land containing some of the world's richest soils has had to be abandoned (Savchenko, 1995; Yablokov et al., 2009).

CONCLUSION

The problem of contaminated soil has been the object of intensive research in recent years. There are many anthropogenic sources, which cause important environmental risks.

Comprehensive assessment of soil pollution originating from mining activities, agricultural activities, municipalities, landfills, sewage sludges, industries, transport activities and power generation activities is required. This will help in identification of recommended policies to mitigate pollution and save the human health and environment.

Soil contamination can contribute to human toxic exposure via a number of routes. These mainly include plant uptakes of soil pollutants, including fertilizers and pesticides, which are directly eaten or passed up the food chain

It is clear that preventing pollution from occurring is a much more effective strategy that requires technical methodologies. Technical methodologies include methods of identification and characterization *in situ* and off-site, including the implementation of a number of techniques for identification of solid, liquid and gaseous pollution.

It is also clear that pollution prevention programs can be fully successful only when the public, industries, researchers, and governments work together.

Development of guidelines may offer the possibility of optimizing the investigation and identification of pollutant sources, and provide decision support at various stages of a comprehensive process that can ultimately lead to clean-up of the soil.

Finally, appropriate measures need to be taken to control levels of organic and inorganic pollutants and their sources particularly in cultivatable soil. This will in turn result in enhancement of life quality and human health.

REFERENCES

Agarwal, T. (2009). Concentration level, pattern and toxic potential of PAHs in traffic soil of Delhi, India. *Journal of Hazardous Materials, 171*(1-3), 894–900. doi:10.1016/j.jhazmat.2009.06.081 PMID:19615818

Ali, S. M., Pervaiz, A., Afzal, B., Hamid, N., & Yasmin, A. (2014). Open dumping of municipal solid waste and its hazardous impacts on soil and vegetation diversity at waste dumping sites of Islamabad city. *Journal of King Saud University - Science, 26*(1), 59-65. doi:10.1016/j.jksus.2013.08.003

Alvarenga, P., Mourinha, C., Farto, M., Santos, T., Palma, P., Sengo, J., ... Cunha-Queda, C. (2015). Sewage sludge, compost and other representative organic wastes as agricultural soil amendments: Benefits versus limiting factors. *Waste Management (New York, N.Y.), 40*, 44–52. doi:10.1016/j.wasman.2015.01.027 PMID:25708406

Auerswald, K., Litz, N., Mayer, R., Stoy, A., Alef, K., & Kerndorff, H. (2000). Soil, 2. Damage and Soil Pollution. In Ullmann's Encyclopedia of Industrial Chemistry. Wiley-VCH Verlag GmbH & Co. KGaA.

Bertazzi, P. A., & di Domenico, A. (1994). Chemical, Environmental, and Health Aspects of the Seveso, Italy, Accident. In A. Schecter (Ed.), *Dioxins and Health* (pp. 587–632). Boston, MA: Springer US. doi:10.1007/978-1-4899-1462-0_18

Bhattacharya, S. S., Iftikar, W., Sahariah, B., & Chattopadhyay, G. N. (2012). Vermicomposting converts fly ash to enrich soil fertility and sustain crop growth in red and lateritic soils. *Resources, Conservation and Recycling*, *65*, 100–106. doi:10.1016/j.resconrec.2012.05.008

Bretzel, F. C., & Calderisi, M. (2011). Contribution of a municipal solid waste incinerator to the trace metals in the surrounding soil. *Environmental Monitoring and Assessment*, *182*(1-4), 523–533. doi:10.100710661-011-1894-0 PMID:21327482

Chen, X., Xia, X., Zhao, Y., & Zhang, P. (2010). Heavy metal concentrations in roadside soils and correlation with urban traffic in Beijing, China. *Journal of Hazardous Materials*, *181*(1), 640–646. doi:10.1016/j.jhazmat.2010.05.060 PMID:20541319

Chrysikou, L., Gemenetzis, P., Kouras, A., Manoli, E., Terzi, E., & Samara, C. (2008). Distribution of persistent organic pollutants, polycyclic aromatic hydrocarbons and trace elements in soil and vegetation following a large scale landfill fire in northern Greece. *Environment International*, *34*(2), 210–225. doi:10.1016/j.envint.2007.08.007 PMID:17900688

Cicek, A., & Koparal, A. S. (2004). Accumulation of sulfur and heavy metals in soil and tree leaves sampled from the surroundings of Tunçbilek Thermal Power Plant. *Chemosphere*, *57*(8), 1031–1036. doi:10.1016/j.chemosphere.2004.07.038 PMID:15488593

Collett, R. S., Oduyemi, K., & Lill, D. E. (1998). An investigation of environmental levels of cadmium and lead in airborne matter and surface soils within the locality of a municipal waste incinerator. *The Science of the Total Environment*, *209*(2-3), 157–167. doi:10.1016/S0048-9697(98)80107-1 PMID:9514037

Čolović, M. B., Krstić, D. Z., Lazarević-Pašti, T. D., Bondžić, A. M., & Vasić, V. M. (2013). Acetylcholinesterase Inhibitors: Pharmacology and Toxicology. *Current Neuropharmacology*, *11*(3), 315–335. doi:10.2174/1570159X11311030006 PMID:24179466

De Haan, F. A. M., & Van Riemsdijk, W. H. (1986). Behaviour of Inorganic Contaminants in Soil. In J. W. Assink & W. J. Van Den Brink (Eds.), *Contaminated Soil: First International TNO Conference on Contaminated Soil, November 11–15, Utrecht, The Netherlands* (pp. 19-32). Dordrecht: Springer Netherlands. 10.1007/978-94-009-5181-5_3

di Domenico, A., Silano, V., Viviano, G., & Zapponi, G. (1980). Accidental release of 2,3,7,8-tetrachlorodibenzo-p-dioxin (TCDD) at Seveso, Italy. IV. Vertical distribution of TCDD in soil. *Ecotoxicology and Environmental Safety*, *4*(3), 327–338. doi:10.1016/0147-6513(80)90034-2 PMID:7439102

Dragović, S., Ćujić, M., Slavković-Beškoski, L., Gajić, B., Bajat, B., Kilibarda, M., & Onjia, A. (2013). Trace element distribution in surface soils from a coal burning power production area: A case study from the largest power plant site in Serbia. *Catena*, *104*, 288–296. doi:10.1016/j.catena.2012.12.004

Eljarrat, E., Marsh, G., Labandeira, A., & Barcelo, D. (2008). Effect of sewage sludges contaminated with polybrominated diphenylethers on agricultural soils. *Chemosphere*, *71*(6), 1079–1086. doi:10.1016/j.chemosphere.2007.10.047 PMID:18068210

Endo, S., Kimura, S., Takatsuji, T., Nanasawa, K., Imanaka, T., & Shizuma, K. (2012). Measurement of soil contamination by radionuclides due to the Fukushima Dai-ichi Nuclear Power Plant accident and associated estimated cumulative external dose estimation. *Journal of Environmental Radioactivity, 111*, 18–27. doi:10.1016/j.jenvrad.2011.11.006 PMID:22137554

Fujiwara, T., Saito, T., Muroya, Y., Sawahata, H., Yamashita, Y., Nagasaki, S., ... Tanaka, S. (2012). Isotopic ratio and vertical distribution of radionuclides in soil affected by the accident of Fukushima Dai-ichi nuclear power plants. *Journal of Environmental Radioactivity, 113*, 37–44. doi:10.1016/j.jenvrad.2012.04.007 PMID:22634028

George, J., Masto, R. E., Ram, L. C., Das, T. B., Rout, T. K., & Mohan, M. (2015). Human exposure risks for metals in soil near a coal-fired power-generating plant. *Archives of Environmental Contamination and Toxicology, 68*(3), 451–461. doi:10.100700244-014-0111-x PMID:25537921

Gevao, B., Semple, K. T., & Jones, K. C. (2000). Bound pesticide residues in soils: a review. *Environmental Pollution (Barking, Essex: 1987), 108*(1), 3-14.

Goel, A. K. (2015). Anthrax: A disease of biowarfare and public health importance. *World Journal of Clinical Cases: WJCC, 3*(1), 20–33. doi:10.12998/wjcc.v3.i1.20 PMID:25610847

Guo, X. Y., Hao, L. J., Qiu, P. Z., Chen, R., Xu, J., Kong, X. J., ... Wang, N. (2016). Pollution characteristics of 23 veterinary antibiotics in livestock manure and manure-amended soils in Jiangsu province, China. *Journal of Environmental Science and Health. Part. B, Pesticides, Food Contaminants, and Agricultural Wastes, 51*(6), 383–392. doi:10.1080/03601234.2016.1142743 PMID:26963628

Harada, N., & Nonaka, M. (2012). Soil radiocesium distribution in rice fields disturbed by farming process after the Fukushima Dai-ichi Nuclear Power Plant accident. *The Science of the Total Environment, 438*, 242–247. doi:10.1016/j.scitotenv.2012.05.032 PMID:23000550

Ihedioha, J. N., Ukoha, P. O., & Ekere, N. R. (2017). Ecological and human health risk assessment of heavy metal contamination in soil of a municipal solid waste dump in Uyo, Nigeria. *Environmental Geochemistry and Health, 39*(3), 497–515. doi:10.100710653-016-9830-4 PMID:27170184

Kadi, M. W. (2009). "Soil Pollution Hazardous to Environment": A case study on the chemical composition and correlation to automobile traffic of the roadside soil of Jeddah city, Saudi Arabia. *Journal of Hazardous Materials, 168*(2-3), 1280–1283. doi:10.1016/j.jhazmat.2009.03.015 PMID:19345483

Kasassi, A., Rakimbei, P., Karagiannidis, A., Zabaniotou, A., Tsiouvaras, K., Nastis, A., & Tzafeiropoulou, K. (2008). Soil contamination by heavy metals: Measurements from a closed unlined landfill. *Bioresource Technology, 99*(18), 8578–8584. doi:10.1016/j.biortech.2008.04.010 PMID:18508262

Kay, P., Blackwell, P. A., & Boxall, A. B. (2004). Fate of veterinary antibiotics in a macroporous tile drained clay soil. *Environmental Toxicology and Chemistry, 23*(5), 1136–1144. doi:10.1897/03-374 PMID:15180364

Kerle, E. A., Jenkins, J. J., & Vogue, P. A. (2007). Understanding pesticide persistence and mobility for groundwater and surface water protection. Oregon State University.

Kim, M., Li, L. Y., Gorgy, T., & Grace, J. R. (2017). Review of contamination of sewage sludge and amended soils by polybrominated diphenyl ethers based on meta-analysis. *Environmental Pollution, 220*(Pt B), 753-765. doi:10.1016/j.envpol.2016.10.053

Leal, R. M., Figueira, R. F., Tornisielo, V. L., & Regitano, J. B. (2012). Occurrence and sorption of fluoroquinolones in poultry litters and soils from Sao Paulo State, Brazil. *The Science of the Total Environment*, *432*, 344–349. doi:10.1016/j.scitotenv.2012.06.002 PMID:22750180

Li, R., Wu, H., Ding, J., Fu, W., Gan, L., & Li, Y. (2017). Mercury pollution in vegetables, grains and soils from areas surrounding coal-fired power plants. *Scientific Reports*, *7*, 46545. doi:10.1038rep46545 PMID:28484233

Liu, G., Niu, J., Guo, W., Zhao, L., Zhang, C., Wang, M., ... Guo, G. (2017). Assessment of terrain factors on the pattern and extent of soil contamination surrounding a chemical industry in Chongqing, Southwest China. *Catena*, *156*, 237–243. doi:10.1016/j.catena.2017.04.005

Lourival, C., Sonia, C. N., Debora, R. C., & Maria, L. (2011). Risk Simulation of Soil Contamination by Polycyclic Aromatic Hydrocarbons from Sewage Sludge used as Fertilizers. *Journal of the Brazilian Chemical Society*, *22*(6), 1156–1163. doi:10.1590/S0103-50532011000600022

Medvedeva, N., Polyak, Y., Kuzikova, I., Orlova, O., & Zharikov, G. (2008). The effect of mustard gas on the biological activity of soil. *Environmental Research*, *106*(3), 289–295. doi:10.1016/j.envres.2007.04.003 PMID:17537425

Melnyk, A., Dettlaff, A., Kuklińska, K., Namieśnik, J., & Wolska, L. (2015). Concentration and sources of polycyclic aromatic hydrocarbons (PAHs) and polychlorinated biphenyls (PCBs) in surface soil near a municipal solid waste (MSW) landfill. *The Science of the Total Environment*, *530*, 18–27. doi:10.1016/j.scitotenv.2015.05.092 PMID:26026405

Minnikova, T. V., Denisova, T. V., Mandzhieva, S. S., Kolesnikov, S. I., Minkina, T. M., Chaplygin, V. A., ... Bauer, T. V. (2017). Assessing the effect of heavy metals from the Novocherkassk power station emissions on the biological activity of soils in the adjacent areas. *Journal of Geochemical Exploration*, *174*, 70–78. doi:10.1016/j.gexplo.2016.06.007

Mireles, F., Davila, J. I., Pinedo, J. L., Reyes, E., Speakman, R. J., & Glascock, M. D. (2012). Assessing urban soil pollution in the cities of Zacatecas and Guadalupe, Mexico by instrumental neutron activation analysis. *Microchemical Journal*, *103*, 158–164. doi:10.1016/j.microc.2012.02.009

Okedeyi, O. O., Dube, S., Awofolu, O. R., & Nindi, M. M. (2014). Assessing the enrichment of heavy metals in surface soil and plant (Digitaria eriantha) around coal-fired power plants in South Africa. *Environmental Science and Pollution Research International*, *21*(6), 4686–4696. doi:10.100711356-013-2432-0 PMID:24352550

Organization for the Prohibition of Chemical Weapons. (2017, December 10). Genesis and Historical Development. Retrieved from https://www.opcw.org/chemical-weapons-convention/genesis-and-historical-development/

Osman, K. T. (2014). Soil Pollution. In K. T. Osman (Ed.), *Soil Degradation, Conservation and Remediation* (pp. 149–226). Dordrecht: Springer Netherlands. doi:10.1007/978-94-007-7590-9_6

Ozkul, C. (2016). Heavy metal contamination in soils around the Tuncbilek Thermal Power Plant (Kutahya, Turkey). *Environmental Monitoring and Assessment*, *188*(5), 284. doi:10.100710661-016-5295-2 PMID:27071662

Patterson, A. T., Kaffenberger, B. H., Keller, R. A., & Elston, D. M. (2016). Skin diseases associated with Agent Orange and other organochlorine exposures. *Journal of the American Academy of Dermatology*, *74*(1), 143–170. doi:10.1016/j.jaad.2015.05.006 PMID:26210237

Raja, R., Nayak, A. K., Shukla, A. K., Rao, K. S., Gautam, P., Lal, B., ... Patra, D. K. (2015). Impairment of soil health due to fly ash-fugitive dust deposition from coal-fired thermal power plants. *Environmental Monitoring and Assessment*, *187*(11), 679. doi:10.100710661-015-4902-y PMID:26450689

Rodriguez-Iruretagoiena, A., Fdez-Ortiz de Vallejuelo, S., Gredilla, A., Ramos, C. G., Oliveira, M. L. S., Arana, G., ... Silva, L. F. O. (2015). Fate of hazardous elements in agricultural soils surrounding a coal power plant complex from Santa Catarina (Brazil). *The Science of the Total Environment*, *508*, 374–382. doi:10.1016/j.scitotenv.2014.12.015 PMID:25497677

Rodríguez Martín, J. A., & Nanos, N. (2016). Soil as an archive of coal-fired power plant mercury deposition. *Journal of Hazardous Materials*, *308*, 131–138. doi:10.1016/j.jhazmat.2016.01.026 PMID:26808251

Samsel, A., & Seneff, S. (2013). Glyphosate, pathways to modern diseases II: Celiac sprue and gluten intolerance. *Interdisciplinary Toxicology*, *6*(4), 159–184. doi:10.2478/intox-2013-0026 PMID:24678255

Savchenko, V. K. (1995). *The ecology of the Chernobyl catastrophe Scientific outlines of an international programme of collaborative research*. UNESCO France.

Sengupta, S., Chatterjee, T., Ghosh, P. B., & Saha, T. (2010). Heavy metal accumulation in agricultural soils around a coal fired thermal power plant (Farakka) in India. *Journal of Environmental Science & Engineering*, *52*(4), 299–306. PMID:22312798

Singh, J., Agrawal, M., & Narayan, D. (1995). Changes in soil characteristics around coal-fired power plants. *Environment International*, *21*(1), 93–102. doi:10.1016/0160-4120(94)00035-6

Spencer, R. C. (2003). Bacillus anthracis. *Journal of Clinical Pathology*, *56*(3), 182–187. doi:10.1136/jcp.56.3.182 PMID:12610093

Suman, S., Sinha, A., & Tarafdar, A. (2016). Polycyclic aromatic hydrocarbons (PAHs) concentration levels, pattern, source identification and soil toxicity assessment in urban traffic soil of Dhanbad, India. *The Science of the Total Environment*, *545-546*, 353–360. doi:10.1016/j.scitotenv.2015.12.061 PMID:26747999

Tarazona, J. V. (2014). *Pollution, Soil A2 - Wexler, Philip Encyclopedia of Toxicology* (3rd ed., pp. 1019–1023). Oxford: Academic Press.

Tian, H., Cheng, K., Wang, Y., Zhao, D., Lu, L., Jia, W., & Hao, J. (2012). Temporal and spatial variation characteristics of atmospheric emissions of Cd, Cr, and Pb from coal in China. *Atmospheric Environment*, *50*, 157–163. doi:10.1016/j.atmosenv.2011.12.045

Tripathi, V., Fraceto, L. F., & Abhilash, P. C. (2015). Sustainable clean-up technologies for soils contaminated with multiple pollutants: Plant-microbe-pollutant and climate nexus. *Ecological Engineering*, *82*, 330–335. doi:10.1016/j.ecoleng.2015.05.027

Tsai, W.-T. (2010). Current Status and Regulatory Aspects of Pesticides Considered to be Persistent Organic Pollutants (POPs) in Taiwan. *International Journal of Environmental Research and Public Health*, *7*(10), 3615–3627. doi:10.3390/ijerph7103615 PMID:21139852

Turrio-Baldassarri, L., Abate, V., Alivernini, S., Battistelli, C. L., Carasi, S., Casella, M., ... La Rocca, C. (2007). A study on PCB, PCDD/PCDF industrial contamination in a mixed urban-agricultural area significantly affecting the food chain and the human exposure. Part I: Soil and feed. *Chemosphere*, *67*(9), 1822–1830. doi:10.1016/j.chemosphere.2006.05.124 PMID:17234238

Viard, B., Pihan, F., Promeyrat, S., & Pihan, J.-C. (2004). Integrated assessment of heavy metal (Pb, Zn, Cd) highway pollution: Bioaccumulation in soil, Graminaceae and land snails. *Chemosphere*, *55*(10), 1349–1359. doi:10.1016/j.chemosphere.2004.01.003 PMID:15081778

Wang, J., Zhang, X., Ling, W., Liu, R., Liu, J., Kang, F., & Gao, Y. (2017). Contamination and health risk assessment of PAHs in soils and crops in industrial areas of the Yangtze River Delta region, China. *Chemosphere*, *168*, 976–987. doi:10.1016/j.chemosphere.2016.10.113 PMID:27816287

Wei, R., Ge, F., Zhang, L., Hou, X., Cao, Y., Gong, L., ... Bao, E. (2016). Occurrence of 13 veterinary drugs in animal manure-amended soils in Eastern China. *Chemosphere*, *144*, 2377–2383. doi:10.1016/j.chemosphere.2015.10.126 PMID:26610297

Xu, M. X., Yan, J. H., Lu, S. Y., Li, X. D., Chen, T., Ni, M. J., ... Cen, K. F. (2009). Agricultural soil monitoring of PCDD/Fs in the vicinity of a municipal solid waste incinerator in Eastern China: Temporal variations and possible sources. *Journal of Hazardous Materials*, *166*(2-3), 628–634. doi:10.1016/j.jhazmat.2008.11.072 PMID:19135306

Yablokov, A. V., Nesterenko, V. B., & Nesterenko, A. V. (2009). 8. Atmospheric, water, and soil contamination after Chernobyl. *Annals of the New York Academy of Sciences*, *1181*, 223–236. doi:10.1111/j.1749-6632.2009.04831.x PMID:20002050

Zhang, H., Zhou, Y., Huang, Y., Wu, L., Liu, X., & Luo, Y. (2016). Residues and risks of veterinary antibiotics in protected vegetable soils following application of different manures. *Chemosphere*, *152*, 229–237. doi:10.1016/j.chemosphere.2016.02.111 PMID:26971176

Section 3
Removal of Toxic Product and Hazardous Waste

Chapter 11
Sources of Air Pollution

Abderrezak Khelfi
National Center of Toxicology, Algeria

ABSTRACT

This chapter describes how air is a complex natural gaseous system essential to support life on Earth. Air pollution comes from a wide variety of sources, which discharge of harmful substances into the atmosphere, causing adverse effects to humans and the environment. They can be natural or anthropogenic. Natural air pollution sources are multiple and include volcanic eruption, fire, ocean vapors, dust storms and fermentation of organic materials. However, the range and quantities of chemicals discharged into the atmosphere from industry, transport, agriculture, energy production, domestic heating, and many other human activities, have increased dramatically. Some pollutants are emitted directly into the atmosphere and are known as primary pollutants (NOx, SOx, particulate matter, etc.). Others are formed in the air as a result of chemical reactions with other pollutants and atmospheric gases; these are known as secondary pollutants like ozone. This chapter provides an overview on air pollution sources as well as the ways in which pollutants can affect human health and the environment.

INTRODUCTION

Air is a complex natural gaseous system essential to support life on Earth. It is a dynamic system, where the composition and concentration of air pollutants may change over a period of time.

According to the WHO "air pollution is contamination of the indoor or outdoor environment by any chemical, physical or biological agent that modifies the natural characteristics of the atmosphere". It occurs mainly in the troposphere, the region closest to the Earth's surface into which most of pollutant emissions are released. Air pollution comes from a wide variety of sources, which discharge of harmful substances into the atmosphere, causing adverse effects to man and the environment. They can be natural or anthropogenic. Natural air pollution sources are multiple and include volcanic eruption, fire, ocean vapors, dust storms and fermentation of organic materials. However, it has become increasingly apparent in recent years that human activities have a significant effect on air composition. During the past century, the range and quantities of chemicals discharged into the atmosphere from industry, transport, agriculture, energy production, domestic heating and many other human activities have increased

DOI: 10.4018/978-1-5225-3540-9.ch011

dramatically. Both stationary and mobile sources contribute to the release of air pollutants. Fossil fuel-powered electricity plants, heat generators and waste incineration sites represent the major stationary point sources of air pollution. However, the major portion of air pollution results from gasoline- and diesel-powered automobile and truck emissions. The mobile and ubiquitous nature of motor vehicles makes their pollution products widespread in several countries (Gordon et al., 2014).

Some pollutants are emitted directly into the atmosphere and are known as primary pollutants (NOx, SOx, hydrocarbons, volatile organic compounds (VOCs), carbon monoxide (CO) and particulate matter (PM)). Others are formed in the air because of chemical reactions with other pollutants and atmospheric gases; these are known as secondary pollutants like ozone. Air pollutants are usually removed from the atmosphere via wet or dry deposition, resulting in their transfer to other environmental compartments such as aquatic systems, soil or plants (Forbes & Garland, 2016).

Air pollution can be influenced by weather parameters such as ambient temperature, wind speed and direction, cloud cover, rain, snow or hail. It varies greatly in their spatial scales. Its local impacts on biodiversity can be dramatic close to large point sources of emissions, but significant impacts on biodiversity over much wider areas can also result from the long-range transport of pollutants. Similarly, air pollution impacts may vary on different temporal scales. Some impacts, for example, are the result of an accidental release of large pollutant concentrations, whereas others are the result of an accumulation of pollutant deposition over years or even decades.

Air pollution is known to contribute to mortality and morbidity and has been associated with health outcomes such as respiratory symptoms, reduced lung function and chronic bronchitis. This chapter provides an overview on air pollution sources as well as the ways in which pollutants can affect human health and ecological processes. This chapter discusses the major sources of the main primary air pollutants including the contribution of the different sources to total emissions of a range of pollutants. The chemistry associated with the majority of these pollutants is also discussed as well as the impacts of each pollutant on human health and the environment.

SOURCES OF AIR POLLUTANTS

Sources of Primary Air Pollutants

Carbon Monoxide

According to data of the US Environmental Protection Agency (EPA) and the European Environment Agency (EEA), one of the highest emissions of air pollutants for several years corresponds to emissions of CO (Tables 1-19) (EEA, 2010B; EPA, 2017a).

CO is a colorless and odorless gas, which is formed when carbon in fossil fuels or other materials such as wood is incompletely combusted. If the fuel contains only carbon, hydrogen and, eventually, oxygen atoms, like in the example of methane, CO will be formed instead of carbon dioxide (CO_2), as shown in the following reaction:

$$2\,CH_4 + 3\,O_2 \rightarrow 4\,H_2O + 2\,CO$$

Table 1. Sources of air pollution by CO in the Unites States (EPA)

General Sources	Detailed Sources	Emissions by Years		
		1996 (10^3 Tons/Year (%))	2006 (10^3 Tons/Year (%))	2016 (10^3 Tons/Year (%))
Stationary fuel combustion	Fuel comb. elec. util.	407.749 (0.32)	672.894 (0.78)	716.621 (1.21)
	Fuel comb. industrial	1188.116 (0.92)	1067.464 (1.24)	956.51 (1.62)
	Fuel comb. other	2740.534 (2.13)	3136.878 (3.65)	2452.055 (4.16)
	Total	4336.399 (3.37)	4877.236 (5.68)	4125.186 (6.99)
Industrial and other processes	Chemical and allied product MFG	1052.988 (0.82)	199.642 (0.23)	128.557 (0.22)
	Metals processing	1598.517 (1.24)	832.926 (0.97)	609.445 (1.03)
	Petroleum and related industries	353.756 (0.27)	322.178 (0.38)	859.393 (1.46)
	Other industrial processes	560.626 (0.44)	498.178 (0.58)	607.8 (1.03)
	Solvent utilization	1.472 (0.00)	3.997 (0.00)	1.635 (0.00)
	Storage and transport	69.771 (0.05)	77.612 (0.09)	10.044 (0.02)
	Waste disposal and recycling	2903.791 (2.25)	1495.445 (1.74)	2155.76 (3.65)
	Total	6540.921 (5.08)	3429.978 (4.00)	4372.634 (7.41)
Transportation	Highway vehicles	78605.995 (61.00)	40045.945 (46.65)	17871.689 (30.30)
	Off-highway	24358.497 (18.90)	20372.737 (23.73)	12989.497 (22.02)
	Total	102964.492 (79.91)	60418.682 (70.39)	30861.186 (52.32)
Wildfires		14502.104 (11.25)	7900.048 (9.20)	10326.712 (17.51)
Miscellaneous		514.224 (0.40)	9211.155 (10.73)	9296.82 (15.76)
Total		128858.14	85837.099	58982.538

Carbon monoxide is also released from the photochemical conversion of atmospheric CH_4 and other hydrocarbons (isoprene, terpenes, etc.).

The majority of CO in the atmosphere comes from anthropogenic sources. In urban areas, the majority of CO emissions into ambient air come from mobile sources (large parts of mobile sources are vehicles such as cars and trucks moving by the combustion of fossil derived fuels). Thus, higher levels of CO generally occur in areas with heavy traffic congestion, such as near entrances to tunnels. Off-road engines and vehicles (such as construction equipment and boats) contribute also in CO emissions. On average, mobile sources (on-road vehicles) account for about 52.32% of the total CO emissions in the USA. Another 6.99% come from fuel combustion processes in stationary sources and 7.41% from industrial production processes such as metals processing and chemical manufacturing (Table 1). In Europe, a total of 20.74% of all outdoor CO emissions nationwide result from on-road vehicles (Table 2).

The observed decreases in CO emissions from 1996 to the present are largely a result of controls, through the introduction of catalysts on gasoline engine vehicles, new vehicle technologies and clean fuels programs. Stationary sources of CO include residential wood burning, public power, waste disposal and natural sources such as forest fires and oceanic emissions. Biomass burning can increase the concentration of CO above background levels over large areas (Forbes & Garland, 2016).

Table 2. Sources of air pollution by CO in the Unites States (EEA)

General Sources	Detailed Sources	Emissions by Years	
		2006 (10^3 Tons/Year (%))	2014 (10^3 Tons/Year (%))
Stationary fuel combustion	Public electricity and heat production	483.219 (1.65)	473.880 (2.15)
	Fuel comb. other	3187.295 (10.88)	2436.101 (11.04)
	Total	3670.514 (12.53)	2909.981 (13.19)
Industrial and other processes	Chemical and allied product MFG	218.876 (0.75)	100.885 (0.46)
	Metals processing	2796.067 (9.54)	2231.980 (10.12)
	Petroleum and related industries	219.407 (0.75)	168.077 (0.76)
	Other industrial processes	107.566 (0.37)	61.308 (0.28)
	Total	3341.916 (11.41)	2562.250 (11.61)
Agricultural activities	Agricultural livestock	0.000 (0.00)	0.000 (0.00)
	Other agricultural activities	842.563 (2.88)	1034.202 (4.69)
	Total	842.563 (2.88)	1034.202 (4.69)
Traffic activities	Road transport	9910.955 (33.82)	4575.932 (20.74)
	Other transport means	616.128 (2.10)	464.951 (2.11)
	Total	10527.084 (35.93)	5040.883 (22.85)
Waste disposal and treatment		113.945 (0.39)	110.592 (0.50)
Solvents		27.813 (0.09)	19.741 (0.09)
Miscellaneous		10028.369 (34.22)	9763.257 (44.26)
Anthropogenic sources		28552.204 (97.44)	21440.906 (97.19)
Forest fires		571.029 (1.95)	446.897 (2.03)
Other natural		178.350 (0.61)	173.224 (0.79)
Natural sources		749.379 (2.56)	620.121 (2.81)
Total		29301.582	22061.027

CO levels vary by season. The highest levels of CO in the outside air typically occur during the colder months of the year, when atmospheric inversion conditions are more frequent, leading to higher localized pollution levels. Thereby, air pollutants become trapped near the ground beneath a layer of warm air. They are also highest during cold weather, because vehicles and other equipment do not burn fuels as efficiently in cold weather as in warm weather. Another reason is that homes heated by fuel combustion (wood and fuel oils) in the winter can generate high levels of CO.

CO is not of significant concern in terms of its impact on the environment, but it is highly toxic when inhaled. The negative effects of carbon monoxide are most often felt by persons with cardiovascular disease. When inhaled, CO can bind to hemoglobin, thus reducing the oxygen carrying capacity of the blood. The affinity between CO and hemoglobin is much stronger than that between hemoglobin and oxygen. Therefore, it prevents hemoglobin from transporting oxygen from the respiratory organs to other organs including the heart and brain and can cause harmful effects by reducing oxygen delivery to tissues. Prolonged exposure to concentrations above 50-100 ppm causes fatigue, nausea and headache, while several hours of exposure to concentrations exceeding 400-500 ppm will gradually lead to dizziness. At extremely high levels, CO can cause death (Kennes & Veiga, 2013).

While individuals with coronary artery disease and individuals with chronic obstructive pulmonary disease or anemia are the most susceptible to the effects of ambient CO exposure, very young children and fetuses (the gas crosses the placental barrier) may also be at risk from high CO levels (Martin & Dombrowski, 2008).

CO is also involved in the production of tropospheric ozone. The following reactions highlight the chain mechanism for the production of ozone from CO:

$$CO + OH\cdot \rightarrow CO_2 + H\cdot$$

$$H\cdot + O_2 + M \rightarrow HO_2\cdot + M$$

$$HO_2\cdot + NO \rightarrow OH\cdot + NO_2$$

$$NO2 + h\nu \rightarrow NO + O(^3P)$$

$$O(^3P) + O_2 + M \rightarrow O_3 + M$$

Therefore, CO has an impact on atmospheric chemistry through these linkages with HOx and is involved in the production of high levels of ozone that can affect human health and environment.

Carbon Dioxide

CO_2 is an important trace gas in Earth's atmosphere. Currently it constitutes about 0.04% of the atmosphere. Despite its relatively small concentration, CO_2 plays a vital role in regulating Earth's surface temperature through radiative protection and the greenhouse effect. Human activities have steadily increased the content of CO_2 in the atmosphere by 40% over the past 150 years at around 400 ppm. Such activities have pumped enough CO_2 into the atmosphere to raise its levels higher than they have been for hundreds of thousands of years (280 ppm) (NOAA, n.d.; Letcher, 2014).

This increase in CO_2 content in the air is mainly due to anthropogenic activities from a number of lines of evidence. For example, atmospheric oxygen is decreasing at approximately the same rate as the atmospheric CO_2 increase, which means that the source of increasing CO_2 is from a release of carbon combining with atmospheric oxygen rather than a natural release of CO_2.

There are some positive effects of global warming due to increased levels of CO_2 in the air like improvement of agricultural yield and increase in vegetation growth in some circumstances. However, the negatives will far outweigh the positives like melting ice threatens most of the available drinking water on Earth. In fact, the current episode of global warming is largely attributed to increasing emissions of CO_2 and other greenhouse gases into Earth's atmosphere. Of all the greenhouse gases, CO_2 has been undergoing the largest changes in concentration. The average annual rate of increase is about 1.2 ppm or 0.4% per year. Moreover, CO_2 is considered to contribute 70% of the enhanced greenhouse effect that has been experienced to date (EEA, 2010a).

Another impact of increasing atmospheric CO_2 emissions is ocean acidification as shown in the following reaction:

$$CO_2 + H_2O \rightarrow H_2CO_3 \rightarrow H^+ + HCO_3^-$$

At high concentrations, CO_2 can also have direct human health effects. If inhaled, it can be toxic in high concentrations, causing an increase in breathing rate, unconsciousness and death.

There are both natural and human sources of CO_2 emissions. Natural sources include decomposition, volcanic outgassing, hot springs, carbonate rocks, ocean release and respiration of living aerobic organisms. The turnover of carbon is large in scale, with an estimated 60 Gt yr^{-1} processed via photosynthesis and respiratory processes and around 90 Gt yr^{-1} via ocean-atmosphere exchange (Jackson, 2007).

Anthropogenic emissions of CO_2 are much smaller than natural emissions but they have upset the natural balance that existed for many thousands of years before the influence of humans. There are many sources of direct anthropogenic CO_2 emissions to the air like cement production, deforestation, power generation as well as the burning of fossil fuels like coal, oil and natural gas.

Fossil Fuels

About 87% of all human-produced CO_2 emissions come from fossil fuels. The three types of fossil fuels that are used the most are coal, natural gas and oil. Coal is responsible for 43% of CO_2 emissions from fuel combustion, 36% is produced by oil and 20% from natural gas (GEOELECT, n.d.).

CO_2 is emitted when carbon-containing fossil fuels are oxidized by combustion. Burning these fuels releases energy but CO_2 also gets produced as a byproduct. This is because almost all the carbon that is stored in fossil fuels gets transformed to CO_2 during this process.

Once fossil fuel has been recovered, processed and burned, the CO_2, which would normally have been released over tens of millions of years, is suddenly released within a period of a few hundred years, thus increasing the amount of CO_2 in the environment. Finally, CO_2 is widely considered to be a pollutant when associated with cars, planes, power plants and other human activities that involve the burning of fossil fuels such as gasoline, diesel and natural gas.

Power Plants

Electricity generation produced 35% of fossil fuel related CO_2 emissions in 2015 in the USA. Around the world, this sector relies heavily on coal, the most carbon-intensive of fossil fuels, explaining this sector giant carbon footprint.

Transportation

Road transport accounts for 32% of this sector's CO_2 emissions in 2015 in the USA. Cars, freight and light-duty trucks are the main sources of emissions for the whole transport sector and emissions from these three have steadily grown since 1990. Apart from road vehicles, the other important sources of emissions for this sector are marine shipping and global aviation.

Industry

Industry is one of the major consumer sector of fossil fuels because some manufacturing processes are very energy intensive. Specifically, the production of chemicals, iron/steel, cement, aluminum as well as pulp and paper account for the great majority of industrial CO_2 releases (15%).

Residential and Commercial Sectors

These sectors are also heavily reliant on electricity for meeting their energy needs, particularly for lighting, heating, air conditioning and appliances and account for 10% of the global emission budget of CO_2 (EPA, 2017b).

Deforestation

When trees die and decompose, CO_2 is released. This is part of the normal carbon cycle. When trees are cut down and used for fuel, the CO_2 is also released. Moreover, with the reduction in vegetation, there is a reduction in photosynthesis. As a result, the carbon cycle is interrupted and the CO_2 is not being converted into sugars and oxygen.

Nitrogen Oxides

Nitrogen dioxide (NO_2) is one of a group of highly reactive gases known as "oxides of nitrogen", or "nitrogen oxides (NOx)". Other nitrogen oxides of tropospheric significance are nitric oxide (NO), nitrate (NO_3), nitrous oxide (N_2O), dinitrogen trioxide (N_2O_3), dinitrogen tetroxide (N_2O_4) and dinitrogen pentoxide (N_2O_5).

Nitrogen oxides are released into the troposphere from a variety of biogenic and anthropogenic sources. The major sources of such pollutants are industrial and traffic activities, biomass burning, microbiological emission and lightning.

NOx From Fossil Fuel Combustion

Nitrogen oxides are formed mainly from the interaction between oxygen naturally present in combustion air and nitrogen compounds found in fuels. This interaction is significant only at high temperatures and favors NO formation. High-temperature combustion conditions result in less than 10% of the NOx in typical exhausts being in the form of NO_2. However, NO generated during combustion can eventually be converted to NO_2 in the atmosphere. NO_2 is formed quickly from emissions from different technologies using fuels like cars, trucks and buses, power plants and off-road equipment.

In the USA, transportation emissions account for over half of all NOx emissions. The other half of US outdoor NOx emissions is largely a result of industrial processes and stationary fuel combustion emissions, which are split between electric utilities and all other industrial, commercial and residential combustion (Table 3). In Europe, almost half of all annual NO_2 emissions are from road transport, as shown in Table 4. Almost similar contributions are seen between Europe and the USA (EEA, 2010b; EPA, 2017a).

The US NOx emissions declined between 1996 and 2006 as a result of the introduction of catalytic converters on gasoline vehicles. Total US NOx emissions have subsequently declined rapidly in recent years, largely due to the introduction of three-way catalysts to gasoline-engine cars, but also due to improved abatement in the energy and industry sectors.

Some studies based on satellite measurements have shown that NO_2 levels are decreasing in the USA and Europe, but are greatly increasing in the industrial regions of China where increased fossil fuel use is taking place particularly in industrial areas (Richter et al., 2005).

Table 3. Sources of air pollution by NOx in the Unites States (EPA)

General Sources	Detailed Sources	Emissions by Years		
		1996 (10^3 Tons/Year (%))	2006 (10^3 Tons/Year (%))	2016 (10^3 Tons/Year (%))
Stationary fuel combustion	Fuel comb. elec. util.	6164.219 (24.87)	3587.749 (18.66)	1052.514 (10.03)
	Fuel comb. industrial	3151.408 (12.71)	1681.336 (8.74)	1188.934 (11.34)
	Fuel comb. other	1196.955 (4.83)	681.834 (3.55)	534.129 (5.09)
	Total	10512.582 (42.41)	5950.919 (30.95)	2775.577 (26.46)
Industrial and other processes	Chemical and allied product MFG	124.778 (0.50)	58.818 (0.31)	46.598 (0.44)
	Metals processing	83.408 (0.34)	70.561 (0.37)	74.488 (0.71)
	Petroleum and related industries	139.083 (0.56)	379.462 (1.97)	821.689 (7.83)
	Other industrial processes	432.8 (1.75)	460.843 (2.40)	336.937 (3.21)
	Solvent utilization	2.394 (0.01)	4.963 (0.03)	1.219 (0.01)
	Storage and transport	15.416 (0.06)	14.161 (0.07)	6.028 (0.06)
	Waste disposal and recycling	152.588 (0.62)	129.512 (0.67)	113.836 (1.09)
	Total	950.467 (3.83)	1118.32 (5.82)	1400.795 (13.35)
Transportation	Highway vehicles	8732.744 (35.23)	7992.075 (41.57)	3605.533 (34.37)
	Off-highway	4179.209 (16.86)	3897.901 (20.27)	2415.402 (23.03)
	Total	12911.953 (52.09)	11889.976 (61.84)	6020.935 (57.40)
Wildfires		404.986 (1.63)	93.809 (0.49)	117.578 (1.12)
Miscellaneous		7.375 (0.03)	174 (0.90)	174.14 (1.66)
Total		24787.363	19227.024	10489.025

Lightning Production

Lightning generates relatively large but uncertain quantities of nitrogen oxides, critical precursors for ozone and hydroxyl radical (OH), the primary tropospheric oxidants. Lightning is a rapid large-scale discharge, with high voltage and current. NO is produced within the very hot lightning channel as O_2 and N_2 dissociation occurs.

Lightning NO accounts for small quantity to the global NOx sources, but is the most dominant source in the upper troposphere. Though the global source of NOx produced by lightning (2-8 Tg N yr^{-1}) is several times smaller than other sources, lightning strongly influences background ozone due to high ozone production efficiencies in the free troposphere, effecting small but non-negligible contributions to surface pollutant concentrations, owing to the longer lifetime of NOx (Lightning NOx lifetimes are a few days instead of hours as at the surface) (Fengxia et al., 2016).

Production of NOx by Micro-Organisms

Micro-organisms can produce some NOx species through nitrification and denitrification processes. Well aerated N fertilized arable soils can be a significant source of NO emission. Estimates of the soil contribution to the European total NOx budget range from 2 to 23%, with increased contributions on hot summer season.

Table 4. Sources of air pollution by NO_2 in the Unites States (EEA)

General Sources	Detailed Sources	Emissions by Years	
		2006 (10^3 Tons/Year (%))	2014 (10^3 Tons/Year (%))
Stationary fuel combustion	Public electricity and heat production	2163.144 (18.62)	1344.050 (17.10)
	Fuel comb. other	1231.348 (10.60)	819.220 (10.42)
	Total	3394.492 (29.21)	2163.270 (27.52)
Industrial and other processes	Chemical and allied product MFG	116.099 (1.00)	94.644 (1.20)
	Metals processing	55.160 (0.47)	39.958 (0.51)
	Petroleum and related industries	359.164 (3.09)	229.921 (2.92)
	Other industrial processes	94.129 (0.81)	74.610 (0.95)
	Total	624.552 (5.38)	439.132 (5.59)
Agricultural activities	Agricultural livestock	10.039 (0.09)	9.865 (0.13)
	Other agricultural activities	461.439 (3.97)	427.586 (5.44)
	Total	471.478 (4.06)	437.451 (5.56)
Traffic activities	Road transport	4762.251 (40.99)	3078.049 (39.16)
	Other transport means	814.681 (7.01)	613.444 (7.80)
	Total	5576.932 (48.00)	3691.493 (46.96)
Waste disposal and treatment		11.290 (0.10)	11.001 (0.14)
Solvents		3.305 (0.03)	2.321 (0.03)
Miscellaneous		1486.729 (12.80)	1074.848 (13.67)
Anthropogenic sources		11568.779 (99.56)	7819.516 (99.47)
Forest fires		13.534 (0.12)	8.556 (0.11)
Other natural		37.139 (0.32)	32.810 (0.42)
Natural sources		50.673 (0.44)	41.366 (0.53)
Total		11619.452	7860.882

Briefly, nitrification is the oxidation of ammonium ions in the soil that leads to the production of nitrate ions. This process is mediated by autotrophic bacteria such as *Nitrosomonas* spp. and *Nitrobacter* spp. under aerobic conditions (Figure 1). Denitrification processes occur under anaerobic conditions in the soil, by heterotrophic bacteria such as *Pseudomonas* and *Achromobacter*. Denitrification reaction is the reduction of nitrate to form nitrogen gas, a process involving several steps with the production of nitrite ion and nitric oxide intermediates as shown in Figure 1 (Ferguson et al., 2007).

NOx can also be produced in soils by purely chemical means. Sources include the photolysis of HONO and aqueous nitrite and the disproportionation of nitrous acid.

The most abundant nitrogen oxides in the atmosphere is nitrous oxide (N_2O), produced by natural microbiological processes in the soil. In fact, agriculture is a major source of anthropogenic N_2O emissions to the atmosphere, through nitrification of ammonium-containing fertilizers and animal waste or denitrification of NO_3 in soils. Though it is relatively unreactive in the troposphere, it is a major active greenhouse gas similarly to methane and CO_2, but with a global warming potential almost 300 times higher than that of CO_2. It has also a deleterious effect on the content of ozone in stratosphere.

Figure 1. Nitrification and denitrification processes

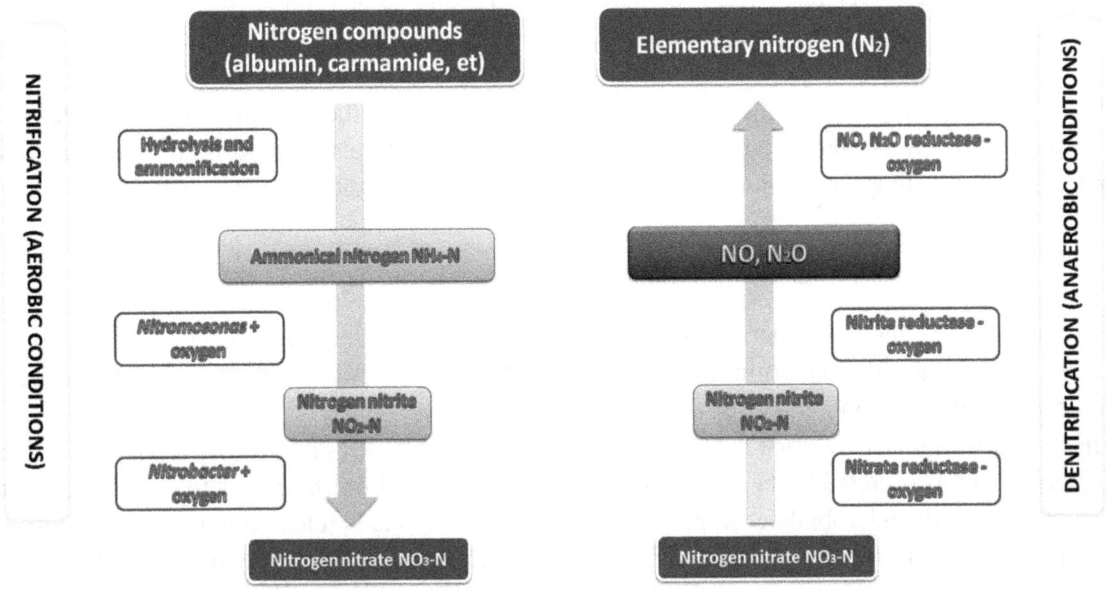

Other Sources

Other natural sources of NOx include volcanoes, oceans, biological decay and biomass burning. Indoor sources of NOx include kerosene heaters, unvented gas stoves and heaters and cigarette smoking.

In the atmosphere, NOx reacts to form nitrate particles and acidic aerosols, which are removed from the atmosphere by both wet and dry deposition, the overall process being termed acid deposition. The latter process contributes to the formation of acid rain, which results in increased levels of nitric acid deposits in soils and plants, as summarized in the following reactions:

Action of $OH^\bullet$ (slow reaction): $NO_2 + OH^\bullet + M \rightarrow HNO_3 + M$

Action of ozone: $NO_2 + O_3 \rightarrow NO_3^\bullet + O_2$

$+ \rightarrow N_2O_5$ (hemi-epoxide) $+ M$

$NO_2 + M$

Hemi-epoxyde is a generator of nitric acid: $N_2O_5 + H_2O$ (atmospheric vapor) $\rightarrow 2\ HNO_3$

This process is known to have detrimental effects on watercourses, vegetation and buildings. Nitrate is also involved in the formation of fine atmospheric particles that can cause visibility impairment.

NOx reacts with VOCs and oxygen in the presence of high temperature and ultraviolet (UV) radiation to produce photochemical smog, mainly in urban areas. Therefore, they play a key role in the photochemically induced catalytic production of ozone that has a number of known negative effects on human health. They can cause eye and skin irritation and have adverse effects on the respiratory system and plants.

In addition to contributing to the formation of ground-level ozone and fine particle pollution, NOx has several adverse effects on human health and is linked with a number of adverse effects on the respiratory system. Lin, *et al.* (2004) have shown that there is an immediate effect of NO_2, sulfur dioxide (SO_2) and other gaseous air pollutants like CO and ozone on hospital admissions in Vancouver, British Columbia, Canada. Studying hospitalization of children ages (6-12 years old), they found a direct dose-response association between the number of asthma hospitalizations and levels of nitrogen dioxide and sulfur dioxide after controlling for ambient temperatures and relative humidity (Lin *et al.*, 2004).

Ammonia

Ammonia (NH_3) is another important nitrogen-bearing pollutant. The major part of NH_3 emission is believed to originate from agricultural activities, mainly livestock wastes, poultry operations, with a small contribution from mineral fertilizer application.

In fact, farm animals are often fed with high protein content food. A non-negligible part of it is not metabolized and is excreted in the urine and feces, which are used for the production of manure. Furthermore, during the microbial decomposition of manure, ammonia is released in the air as pollutants. The other major source of NH_3 emission is the use of mineral fertilizers. Only about 40% of the nitrogen (mainly urea) present in high-nitrogen fertilizers is taken up by crops and forage species. The other part of nitrogen is released through gaseous emissions, leaching and erosion of soil as well as runoff. In the USA, minor sources of NH_3 are industrial activities and transportation (Table 5). It has been suggested that low-temperature combustion and industrial processes may generate a significant emission of NH_3.

In Europe in 2014 it was estimated that 93.57% of the NH_3 emissions were from agriculture, while road transport, waste treatment, production processes and minor sources contributed to the other 6.43% (Table 6). Within the agriculture sector livestock emission and the use of fertilizers are considered to be the major sources of production of NH_3 (EEA, 2010b; EPA, 2017b).

Sulfur Oxides

SOx are compounds of sulfur and oxygen molecules, sulfur dioxide (SO_2) being the predominant form found in the lower levels of the atmosphere. The main source of SOx in the air is the burning of sulfur-containing fuels. SO_2 appears in larger amounts than SO_3 in combustion gases. It is largely released during the combustion of fossil fuels, mainly coal, according to the following reactions:

$$S + O_2 \rightarrow SO_2$$

$$SO_2 + \frac{1}{2} O_2 \rightarrow SO_3$$

Natural gas and gasoline have a relatively low sulfur content. Until recently, diesel engine emissions led to small elevations in SO_2 alongside busy roads, but regulations have significantly reduced the sulfur content of this fuel. Its concentration in combustion gases will depend on the amount of sulfur present in the original fuel, which usually does not exceed 3-4% by weight but may occasionally reach 10%.

Stationary combustion processes represent by far the major source of SO_2, with around 73% and 50% of the total SOx emissions in the USA and Europe respectively (Table 7 and 8). Over 43% of SO_2 released to the air in the USA comes from electric utilities, primarily from units that burn coal. In fact,

Table 5. Sources of air pollution by NH_3 in the Unites States (EPA)

General Sources	Detailed Sources	Emissions by years		
		1996 (10^3 Tons/Year (%))	2006 (10^3 Tons/Year (%))	2016 (10^3 Tons/Year (%))
Stationary fuel combustion	Fuel comb. elec. util.	6	26.377 (0.65)	25.506 (0.66)
	Fuel comb. industrial	34	17.263 (0.42)	16.399 (0.43)
	Fuel comb. other	7	32.879 (0.81)	67.717 (1.76)
	Total	47	76.519 (1.88)	109.622 (2.84)
Industrial and other processes	Chemical and allied product MFG	123	18.071 (0.44)	22.004 (0.57)
	Metals processing	5	2.476 (0.06)	1.052 (0.03)
	Petroleum and related industries	16	1.442 (0.04)	1.992 (0.05)
	Other industrial processes	43	138.078 (3.39)	22.25 (0.58)
	Solvent utilization	0	0.406 (0.01)	0.55 (0.01)
	Storage and transport	1	2.101 (0.05)	5.426 (0.14)
	Waste disposal and recycling	84	41.015 (1.01)	23.22 (0.60)
	Total	272	203.589 (4.99)	76.494 (1.98)
Transportation	Highway vehicles	236	146.25 (3.59)	90.314 (2.34)
	Off-highway	34	2.644 (0.06)	3.267 (0.08)
	Total	270	148.894 (3.65)	93.581 (2.43)
Wildfires		-	129.925 (3.19)	168.969 (4.38)
Miscellaneous		4138	3517.963 (86.29)	3408.867 (88.37)
Total		-	4076.89	3857.533

the highest concentrations of SO_2 are usually recorded in the vicinity of facilities lacking modern emission control technologies, such as older coal-fired power plants (EEA, 2010b; EPA, 2017b).

SO_2 is also formed by metal smelting (iron and steel industry) and some industrial processes that derive their products from raw materials like metallic ore, coal and crude oil, or that burn coal or oil to produce process heat. Examples include petroleum refineries, cement manufacturing, smelting sulfur-containing ores and metal processing facilities.

The other anthropogenic sources of SO_2 in the USA and Europe are industry and transport. Locomotives such as large ships docked at deep ports still burn high sulfur fuels and release SO_2 to the air in large quantities.

Overall, anthropogenic sources of SO_2 are coal- and oil-fired plants and fossil fuel combustion, as sulfur is a natural contaminant of these fuels, though there are natural sources. Natural sources of SO_2 include releases from volcano eruptions, biological decay and forest fires. When volcanoes erupt and inject SO_2 into the stratosphere, the SO_2 aerosols can impact the troposphere and people living in volcanic regions may thus be exposed to high levels of SO_2 in ambient air.

In terms of environmental impact, SO_2 can be oxidized in the atmosphere to produce sulfuric acid, leading to acid precipitation commonly known as acid rain as shown in Figure 2. SO_2 can also interact with other gases and particles in the air to form secondary PM like PM2.5, which is a significant health concern as well as a main pollutant that impairs visibility. SO_3, another SOx, is either emitted directly

Table 6. Sources of air pollution by NH_3 in the Unites States (EEA)

General Sources	Detailed Sources	Emissions by Years	
		2006 (10^3 Tons/Year (%))	2014 (10^3 Tons/Year (%))
Stationary fuel combustion	Public electricity and heat production	3.688 (0.09)	4.276 (0.11)
	Fuel comb. other	7.691 (0.19)	7.330 (0.19)
	Total	11.378 (0.28)	11.606 (0.30)
Industrial and other processes	Chemical and allied product MFG	44.850 (1.10)	37.132 (0.94)
	Metals processing	0.754 (0.02)	0.327 (0.01)
	Petroleum and related industries	1.593 (0.04)	1.220 (0.03)
	Other industrial processes	9.733 (0.24)	8.252 (0.21)
	Total	56.929 (1.39)	46.931 (1.19)
Agricultural activities	Agricultural livestock	2392.540 (58.51)	2244.534 (57.12)
	Other agricultural activities	1384.986 (33.87)	1432.453 (36.45)
	Total	3777.526 (92.37)	3676.987 (93.57)
Traffic activities	Road transport	97.016 (2.37)	54.288 (1.38)
	Other transport means	0.277 (0.01)	0.277 (0.01)
	Total	97.293 (2.38)	54.565 (1.39)
Waste disposal and treatment		70.407 (1.72)	61.681 (1.57)
Solvents		5.904 (0.14)	5.621 (0.14)
Miscellaneous		57.156 (1.40)	60.772 (1.55)
Anthropogenic sources		4076.594 (99.69)	3918.162 (99.70)
Forest fires		3.996 (0.10)	2.607 (0.07)
Other natural		8.831 (0.22)	9.009 (0.23)
Natural sources		12.827 (0.31)	11.616 (0.30)
Total		4089.421	3929.778

into the atmosphere or produced from the oxidation of SO_2 and is rapidly converted to sulfuric acid, which is removed usually through wet deposition processes. It has a much higher corrosion potential than SO_2 and has been reported to be 10 times more toxic than SO_2, mainly for the respiratory system (Kikuchi, 2001).

Sulfuric acid can be transported to large distances before it is deposited to the surface. The consequence of this acid deposition on the environment include acidification of soils, lakes, rivers and streams, reduction in crop production and the accelerated corrosion of buildings, monuments and materials. Acid depositions can affect fish species and plant life, both, directly or through the dissolution of major mineral elements present in the sediments, such as aluminum. Sulfate aerosol particles can also have significant impacts on clouds, human health and climate change.

In terms of health effects, high exposures to sulfate aerosols or SO_2 adsorbed on PM result in breathing problems (constrict bronchi), alterations in the lung's defenses, irritation and aggravation of existing respiratory and cardiovascular disease. People most sensitive to SO_2 include asthmatics and individuals

Table 7. Sources of air pollution by SO_2 in the Unites States (EPA)

General Sources	Detailed Sources	Emissions by Years		
		1996 (10^3 Tons/Year (%))	2006 (10^3 Tons/Year (%))	2016 (10^3 Tons/Year (%))
Stationary fuel combustion	Fuel comb. elec. util.	12767.183 (69.44)	9545.56 (72.74)	1185.136 (43.76)
	Fuel comb. industrial	2848.732 (15.49)	1514.16 (11.54)	656.901 (24.25)
	Fuel comb. other	635.913 (3.46)	484.763 (3.69)	130.869 (4.83)
	Total	16251.828 (88.40)	11544.483 (87.97)	1972.906 (72.84)
Industrial and other processes	Chemical and allied product MFG	255.247 (1.39)	229.289 (1.75)	114.98 (4.25)
	Metals processing	388.806 (2.11)	175.472 (1.34)	107.485 (3.97)
	Petroleum and related industries	335.059 (1.82)	182.13 (1.39)	102.65 (3.79)
	Other industrial processes	385.994 (2.10)	325.026 (2.48)	172.406 (6.37)
	Solvent utilization	1.034 (0.01)	0.476 (0.00)	0.069 (0.00)
	Storage and transport	4.986 (0.03)	3.864 (0.03)	3.439 (0.13)
	Waste disposal and recycling	32.32 (0.18)	25.992 (0.20)	36.689 (1.35)
	Total	1403.446 (7.63)	942.249 (7.18)	537.718 (19.85)
Transportation	Highway vehicles	301.665 (1.64)	106.894 (0.81)	18.413 (0.68)
	Off-highway	413.121 (2.25)	398.748 (3.04)	30.63 (1.13)
	Total	714.786 (3.89)	505.642 (3.85)	49.043 (1.81)
Wildfires		14.794 (0.08)	55.916 (0.43)	70.433 (2.60)
Miscellaneous		0.415 (0.00)	74.264 (0.57)	78.465 (2.90)
Total		18385.269	13122.554	2708.565

with chronic lung diseases or cardiovascular diseases. Children and elderly are at higher risks compared to others.

The main effects of SOx on vegetation are related to loss of foliage, reduction of the productivity and premature death. Several adverse effects related to SOx were identified in pine trees, alfalfa and rye grass.

Volatile Organic Compounds

VOCs mean any vapor-phase atmospheric organics, excluding CO, CO_2, carbonic acid, metallic carbides or carbonates and ammonium carbonate. They include pure hydrocarbons containing C and H only (e.g. alkanes, alkenes, alkynes and aromatic compounds) as well as species that contain other elements, such as oxygen and chlorine (e.g. aldehydes, alcohols, esters, ketones, chlorinated alkenes, chloroflurocarbons (CFCs), hydrofluorocarbons, etc.).

Methane

Methane (CH_4) is the simplest VOC in terms of structure. It is also the most abundant VOC in the atmosphere (around 1800 ppb). Based on analyses of ice cores, the concentration of methane has more than doubled since preindustrial times.

Table 8. Sources of air pollution by SO_2 in the Unites States (EEA)

General Sources	Detailed Sources	Emissions by Years	
		2006 (10^3 Tons/Year (%))	2014 (10^3 Tons/Year (%))
Stationary fuel combustion	Public electricity and heat production	4607.590 (51.99)	1449.484 (35.97)
	Fuel comb. other	960.574 (10.84)	546.135 (13.55)
	Total	5568.164 (62.82)	1995.618 (49.53)
Industrial and other processes	Chemical and allied product MFG	100.474 (1.13)	70.630 (1.75)
	Metals processing	92.780 (1.05)	70.143 (1.74)
	Petroleum and related industries	722.756 (8.15)	299.973 (7.45)
	Other industrial processes	101.435 (1.14)	58.985 (1.46)
	Total	1017.446 (11.48)	499.731 (12.40)
Agricultural activities	Agricultural livestock	0.000 (0.00)	0.000 (0.00)
	Other agricultural activities	63.835 (0.72)	52.451 (1.30)
	Total	63.835 (0.72)	52.451 (1.30)
Traffic activities	Road transport	23.839 (0.27)	6.244 (0.15)
	Other transport means	36.653 (0.41)	6.384 (0.16)
	Total	60.492 (0.68)	12.628 (0.31)
Waste disposal and treatment		4.048 (0.05)	2.934 (0.07)
Solvents		2.222 (0.03)	1.637 (0.04)
Miscellaneous		834.411 (9.41)	517.893 (12.85)
Anthropogenic sources		7550.618 (85.19)	3082.893 (76.51)
Forest fires		3.567 (0.04)	2.329 (0.06)
Other natural		1308.938 (14.77)	943.923 (23.43)
Natural sources		1312.505 (14.81)	946.252 (23.49)
Total		8863.122	4029.145

Figure 2. Generation of sulfur trioxide and sulfuric acid from SO_2

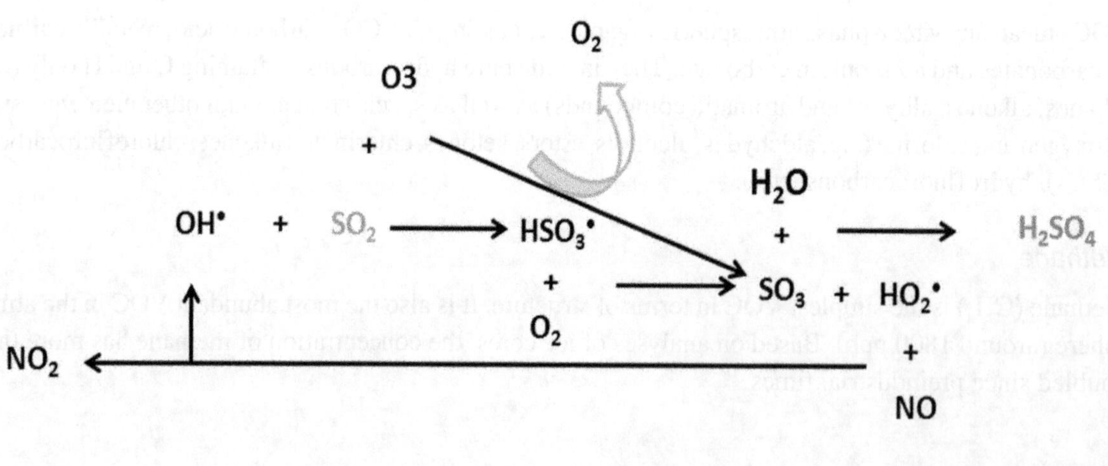

It is known for its impact on global warming, with a global-warming potential 50 times higher than that of CO_2. Methane emissions come from a number of different sources, both natural and anthropogenic. Some of the main sources of methane are agriculture, urban waste disposal, ruminants and enteric fermentation, wetlands, rice paddies, biomass burning and oceans.

Emissions also result from fossil fuel-related methane sources such as natural gas leakage and the petroleum industry. In fact, natural gas and petroleum systems are the largest source of CH_4 emissions in the USA (31% in 2015). Methane is the primary component of natural gas. It is emitted to the atmosphere during the production, processing, storage, transmission and distribution of natural gas. Because gas is often found alongside petroleum, the production, refinement, transportation and storage of crude oil is also a source of CH_4 emissions.

Ruminants represent an important source of CH_4 emissions, by enteric fermentation during the microbial digestion of feeds and from microbial degradation of excreted residues in manure. In fact, domestic livestock such as cattle, buffalo, sheep and goats produce large amounts of CH_4 as part of their normal digestive process and accounts with manure production and management for 35% of global emission in the USA.

Other major sources of CH_4 emission in the USA are landfills and coal mining, which account for 18% and 9% respectively (EPA, 2017c). In Europe the majority of CH_4 is also emitted from ruminant animals and fermentation reactions as well as from processes such as waste treatment and disposal, coal mining, landfills and leakage from natural gas distribution networks.

Wetlands represent one of the largest natural sources of CH_4 to the atmosphere, producing it by bacterial degradation of organic matter in anaerobic conditions. The role of rice paddies in the increase of atmospheric CH_4 is increasing with the global demand for rice, which is related to population growth in many countries.

Through chemical interactions CH_4 can indirectly influence climate. For example, oxidation of CH_4 eventually produces CO_2 from CO and significantly contributes to the elevation of tropospheric and stratospheric ozone levels by mean of high level production of NO_2 (Figure 3 and 4).

Nonmethane Volatile Organic Compounds (NMVOC)

The major classes of atmospheric NMVOCs are alkanes, alkenes, aromatic hydrocarbons and oxygenated compounds. Vast quantities of NMVOCs are emitted into the atmosphere from both biogenic and anthropogenic sources, where the latter includes emissions from vehicles, refineries, chemical solvents, landfills, petrochemical operations and solvent-based industries (painting, dry cleaning, printing, etc.).

Industries are one of the major sources of NMVOCs in the USA and Europe. They represent about 20 to 50% of the overall emission estimates. Industrial sources generally emit solvents like benzene, toluene, xylene and perchloroethylene (principal dry-cleaning solvent).

In the USA and Europe, vehicle emissions are also dominant sources of outdoor VOCs (Table 9 and 10), predominantly in the form of alkanes, isoprene compounds and aromatics, emitted mainly as a result of the incomplete combustion of fuel or its vaporization (EEA, 2010b; EPA, 2017b).

However, emissions from vehicles have declined as a result of the use of catalytic converters and the switch from gasoline to diesel cars.

NMVOCs are also released from solvent use and different burning fuel, such as gasoline, wood, coal and natural gas. Emissions from solvent use and industrial processes have also decreased as pollutant abatement technologies are introduced in the production processes.

Table 9. Sources of air pollution by VOC in the Unites States (EPA)

General Sources	Detailed Sources	Emissions by Years		
		1996 (10^3 Tons/Year (%))	2006 (10^3 Tons/Year (%))	2016 (10^3 Tons/Year (%))
Stationary fuel combustion	Fuel comb. elec. util.	49.74 (0.24)	46.811 (0.26)	37.358 (0.24)
	Fuel comb. industrial	179.145 (0.86)	124.092 (0.69)	110.064 (0.69)
	Fuel comb. other	893.317 (4.28)	527.827 (2.95)	384.159 (2.42)
	Total	1122.202 (5.38)	698.73 (3.90)	531.581 (3.35)
Industrial and other processes	Chemical and allied product MFG	388.259 (1.86)	186.092 (1.04)	74.573 (0.47)
	Metals processing	73.395 (0.35)	45.421 (0.25)	28.835 (0.18)
	Petroleum and related industries	476.949 (2.29)	956.231 (5.34)	3219.602 (20.30)
	Other industrial processes	434.733 (2.08)	425.483 (2.38)	352.846 (2.23)
	Solvent utilization	5476.63 (26.24)	3937.63 (22.00)	2941.728 (18.55)
	Storage and transport	1293.915 (6.20)	1359.665 (7.60)	680.705 (4.29)
	Waste disposal and recycling	508.956 (2.44)	323.16 (1.81)	197.027 (1.24)
	Total	8652.837 (41.46)	7233.682 (40.41)	7495.316 (47.27)
Transportation	Highway vehicles	6220.77 (29.81)	3463.536 (19.35)	1765.075 (11.13)
	Off-highway	2934.983 (14.06)	2746.318 (15.34)	1577.402 (9.95)
	Total	9155.753 (43.87)	6209.854 (34.69)	3342.477 (21.08)
Wildfires		1869.894 (8.96)	1867.67 (10.43)	2428.562 (15.31)
Miscellaneous		70.549 (0.34)	1891.802 (10.57)	2059.752 (12.99)
Total		20871.235	17901.738	15857.688

NMVOCs are also frequently found in indoor air, as they are present in most paints, varnishes, waxes, glues, cleansers and cosmetics, among other products used at home or work.

Apart from anthropogenic sources, organic compounds are released naturally to the atmosphere from many types of vegetation. Over 1,000 natural NMVOCs are known to enter the atmosphere and some of them can react with NOx via a complex series of atmospheric reactions to form ozone in the presence of high temperature and sunlight. In fact, in terms of environmental impact, NMVOCs play a key role in the formation of ozone in the lower atmosphere after reaction with NO, but they differ greatly in their potential to promote the production of ozone (Jackson, 2007). NMVOCs are also involved in the production of other tropospheric oxidants such as CO and aldehydes (Figure 3 and 4).

The main biogenic compounds (BVOCs) released to the air are isoprene and monoterpene families. Deciduous trees typically emit isoprene, while conifers typically emit terpenes (e.g. α- and β-pinene). Common terpenes include also sabinene, limonene, myrcene and α-phellandrene and β-caryophyllene. Isoprene and terpenes are involved in the production of air pollutants and greenhouse gases such as ozone, CO and CH_4. Other BVOCs of relevance are: alkanes, alkenes, alcohols (such as 2-methyl-3-buten-2-ol and methanol), esters, carbonyls (such as acetone) and acids.

The amount of isoprene emitted from traffic in urban areas compared to its biogenic emission has been the focus of several studies. A study performed by Borbon et al. in an urban centre site in Lille,

Table 10. Sources of air pollution by NMVOC in the Unites States (EEA)

General Sources	Detailed Sources	Emissions by Years	
		2006 (10^3 Tons/Year (%))	2014 (10^3 Tons/Year (%))
Stationary fuel combustion	Public electricity and heat production	77.016 (0.69)	68.140 (0.77)
	Fuel comb. other	112.754 (1.02)	99.188 (1.12)
	Total	189.771 (1.71)	167.329 (1.89)
Industrial and other processes	Chemical and allied product MFG	172.320 (1.55)	114.376 (1.29)
	Metals processing	28.196 (0.25)	22.629 (0.26)
	Petroleum and related industries	580.107 (5.23)	400.784 (4.53)
	Other industrial processes	1778.715 (16.03)	1284.662 (14.52)
	Total	2559.338 (23.07)	1822.451 (20.59)
Agricultural activities	Agricultural livestock	574.375 (5.18)	558.670 (6.31)
	Other agricultural activities	173.397 (1.56)	196.281 (2.22)
	Total	747.772 (6.74)	754.951 (8.53)
Traffic activities	Road transport	1600.259 (14.43)	715.233 (8.08)
	Other transport means	156.420 (1.41)	94.911 (1.07)
	Total	1756.678 (15.84)	810.144 (9.15)
Waste disposal and treatment		85.844 (0.77)	62.913 (0.71)
Solvents		2270.583 (20.47)	1898.061 (21.45)
Miscellaneous		1419.332 (12.79)	1206.789 (13.64)
Anthropogenic sources		9029.318 (81.40)	6722.639 (75.96)
Forest fires		22.617 (0.20)	12.605 (0.14)
Other natural		2041.165 (18.40)	2114.929 (23.90)
Natural sources		2063.781 (18.60)	2127.534 (24.04)
Total		11093.099	8850.173

France, from May 1997 to April 1999 showed that in urban areas isoprene of biogenic origin was most dominant in summer (traffic-related isoprene varied between 10 and 50%), while vehicle exhausts were mainly responsible for its high levels from late September to April (Borbon et al., 2001).

NMVOCs in ambient air may have human health effects including carcinogenicity and reproductive effects. Some NMVOCs like carbonyls (such as formaldehyde, acetaldehyde and acetone) contribute to eye, throat and lung irritations and BTEX (benzene, toluene, ethylbenzene and xylene) are known to affect the respiratory and central nervous system where they act as depressants (Zhang *et al.*, 2012).

Atmospheric halocarbons, known also as halogen-containing organic compounds, are considered as NMVOCs. They can be produced naturally by biological processes in the oceans from sea salt and from biomass burning. With the exception of the naturally occurring portions of CH_3Cl and CH_3Br emissions, all the halocarbons in the atmosphere are man-made. Their inertness and long lifetimes have made them attractive chemicals in chemical industry. For many of these species, industrial synthesis was performed for use as propellants, refrigerants and blowing agents. It is these man-made halocarbons that are of concern with regard to atmospheric composition and chemistry, in particular those called

Figure 3. Generation of aldehydes pollutants from VOC and mechanism of production of ozone by the excessive generation of NO_2

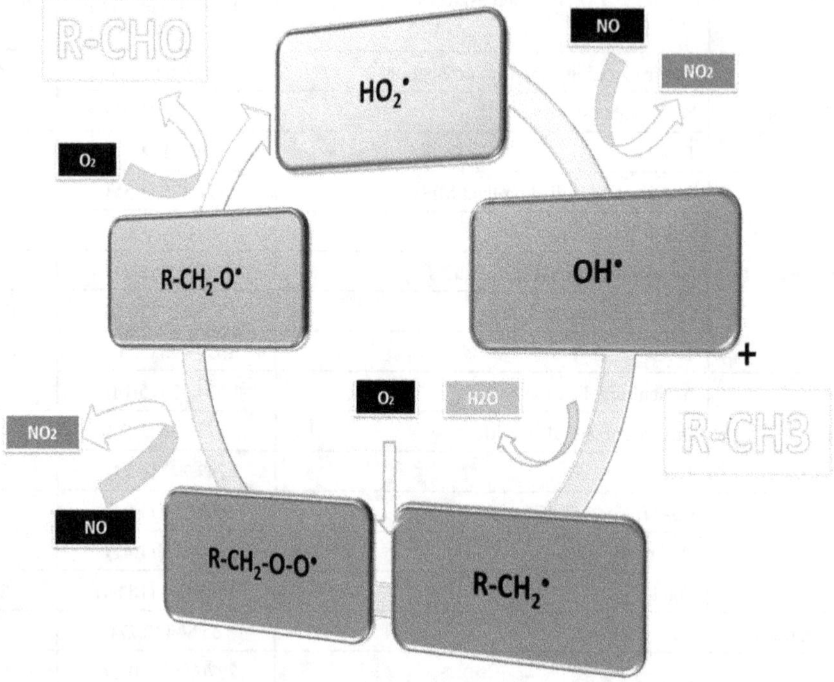

CFCs (chlorofluorocarbons). One of the unique properties of these compounds is their inertness in the troposphere, conditioning their very long lifetimes (Table 11). In fact, all of the fully halogenated CFCs have long atmospheric lifetimes particularly, CFC-11 and CFC-12 that have lifetimes of about 50 and 100 years respectively (Wuebbles, 2013).

CFC molecules have a greenhouse effect 5,000 to 15,000 thousand times greater than CO_2. Among the most potent greenhouse halocarbons in the current atmosphere are CFC-11 ($CFCl_3$) and CFC-12 (CF_2Cl_2). One molecule of CFC-11 or CFC-12 in the atmosphere is, respectively, 12,400 and 15,800 times more effective as a greenhouse gas than one molecule of CO_2.

CFCs play also an important role in the depletion of the stratospheric ozone layer and they are of particular concern with regard to destruction of stratospheric layer protecting living beings from solar radiation. These observations were explained by the catalytic depletion of ozone via chlorine, released from CFCs (Figure 5), which had entered the stratosphere from the troposphere as a result of their long lifetimes. When exposed to ultraviolet solar radiation, CFCs release chlorine atoms that react with ozone in the stratosphere, reducing the total amount of ozone available to intercept destructive ultraviolet radiation. Because of the reactional circle created by the CFCs, it is believed that the chlorine atoms in one CFC lead to the destruction of tens of thousands of ozone molecules.

The destruction of stratospheric ozone permits more of solar radiations to reach earth's surface, which could cause an increase in eye cataracts, skin cancer and disruption of animal and plant life particularly, phytoplankton the major source of food for several aquatic species.

Figure 4. Generation of formaldehydes pollutants and carbon monoxide from methane and mechanism of production of ozone by the excessive generation of NO_2

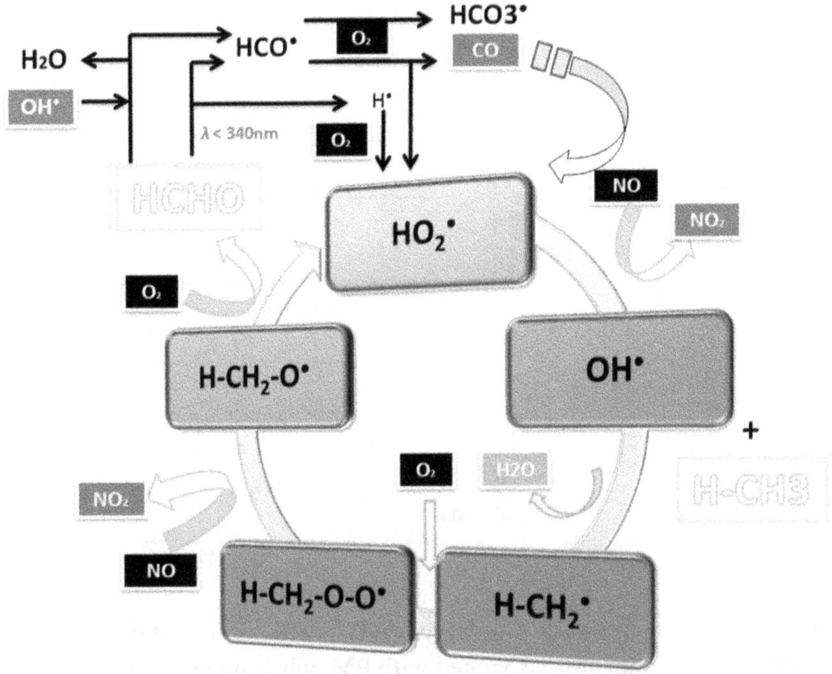

Table 11. Lifetimes of some CFC compounds

Name	Formula	Lifetime in Years
CFC-11 Trichlorofluoromethane	CCl_3F	45
CFC-12 Dichlorodifluoromethane	CCl_2F_2	100
CFC-113 1,1,2-Trichlorotrifluoroethane	$C2F_3Cl_3$	85
CFC-114 Dichlorotetrafluoroethane	$C_2F_4Cl_2$	190
CFC-115 Monochloropentafluoroethane	$C2F_5Cl$	1020

The use of CFCs has diminished greatly, with all but a few minor essential applications being banned as a result of the international agreement of the Montreal Protocol to protect the stratospheric ozone layer. In 27-29 June 1990, environment ministers from different countries met in London and agreed to phase out the production and use of CFCs and related chemicals by the end of the century (United Nations Environment Program: Ozone Secretariat, 2011).

Polycyclic Aromatic Hydrocarbons

Polycyclic aromatic hydrocarbons (PAHs) are chemicals which can cause adverse effects to human health and environment. They are a group of chemical hydrocarbons with two or more fused aromatic rings. More than 150 species have been identified in the atmosphere, ranging from naphthalene (two

Figure 5. Mechanism of depletion of stratospheric ozone by CFC compounds.

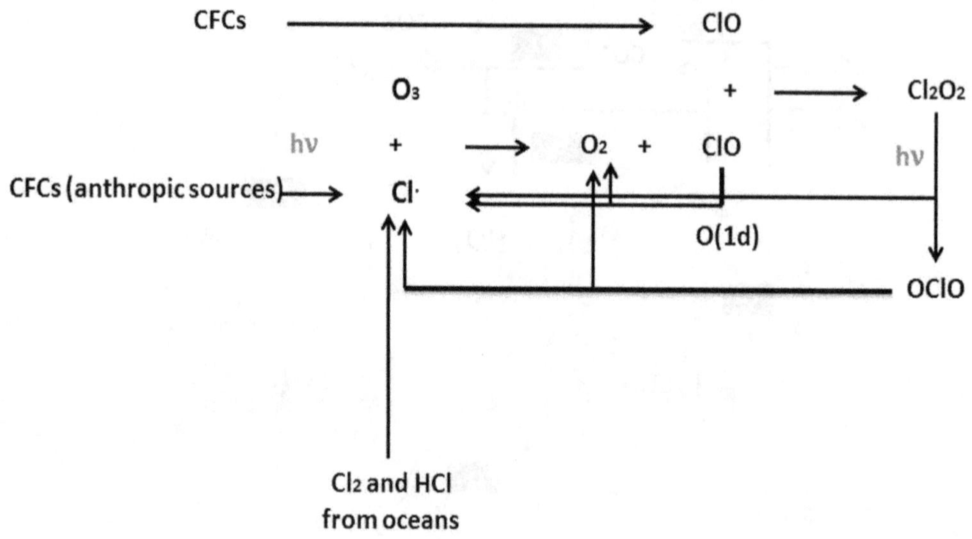

benzene rings) to coronene (seven rings). The larger PAHs, containing five or more aromatic rings, are found in the atmosphere predominantly associated with PM, while naphthalene exists exclusively in the gas phase. The primary PAHs found in both PM2.5 and PM10 included benzo[b]fluoranthene, pyrene, fluoranthene, indeno[1,2,3-cd]pyrene and chrysene.

The major sources of PAHs are the incomplete combustion of organic materials. Therefore, the predominant sources are anthropogenic and include stationary power plants (coal- and oil combustion), deliberate biomass burning, tyre burning, vehicular emissions (both diesel and gasoline) and even domestic sources (coal and wood burning for heating and cooking purposes and tobacco smoke). A small contribution comes from natural sources such as forest fires or volcanoes. Diagnostic ratio of PAH concentrations is used to identify emission sources of these pollutants, such as petroleum production, petroleum combustion, coal burning or biomass burning (Forbes & Garland, 2016).

Road transport represents one of the predominant sources of benzo[a]pyrene in Europe. Other major sources include field burning of agricultural residues and industrial activities (e.g. aluminum production), accounting for about 13 and 4% of the total emission respectively (Table 12). The relatively inefficient combustion conditions of domestic fires and boilers (coal and wood burning and residential heating) compared to power stations and industrial combustion accounts for the higher emissions from domestic sources (EEA, 2010b).

PAHs are very persistent organic pollutants in the environment, the importance of which resides in their impact upon human health. Wind can expand PAHs far away from their sources, though the concentration would be considerably diluted by mixing with the surrounding ambient air.

The list of the US EPA PAHs comprises more than 150 pollutants, including substances that can negatively impact the neurological, respiratory and reproductive systems. Exposure to particle-phase PAHs originating from traffic emissions has been found to increase the oxidative burden in human exposed to heavy traffic volume, which can potentially lead to DNA damage (Wei *et al.*, 2010). In fact,

Table 12. Sources of air pollution by benzo[a]pyrene in the Unites States (EEA)

General Sources	Detailed Sources	Emissions by Years	
		2006 (Tons/Year (%))	2014 (Tons/Year (%))
Stationary fuel combustion	Public electricity and heat production	0.896 (0.47)	1.358 (0.71)
	Fuel comb. other	5.144 (2.70)	2.916 (1.53)
	Total	6.041 (3.17)	4.274 (2.24)
Industrial and other processes	Chemical and allied product MFG	0.011 (0.01)	0.009 (0.00)
	Metals processing	11.568 (6.07)	6.860 (3.60)
	Petroleum and related industries	0.229 (0.12)	0.318 (0.17)
	Other industrial processes	0.069 (0.04)	0.055 (0.03)
	Total	11.877 (6.24)	7.242 (3.80)
Agricultural activities	Agricultural livestock	0.000 (0.00)	0.000 (0.00)
	Other agricultural activities	22.382 (11.75)	24.898 (13.07)
	Total	22.382 (11.75)	24.898 (13.07)
Traffic activities	Road transport	4.879 (2.56)	9.000 (4.73)
	Other transport means	0.946 (0.50)	0.857 (0.45)
	Total	5.826 (3.06)	9.857 (5.18)
Waste disposal and treatment		1.048 (0.55)	0.682 (0.36)
Solvents		0.249 (0.13)	0.155 (0.08)
Miscellaneous		137.204 (72.04)	138.904 (72.94)
Anthropogenic sources		184.627 (96.94)	186.013 (97.67)
Forest fires		2.956 (1.55)	1.551 (0.81)
Other natural		2.880 (1.51)	2.880 (1.51)
Natural sources		5.836 (3.06)	4.431 (2.33)
Total		190.463	190.444

some of PAHs are carcinogenic, the most potent including benzo[a]pyrene, benzofluoranthenes, benz[a] anthracene, dibenzo[ah]anthracene and indeo[1,2,3-cd]pyrene.

Polychlorinated Biphenyls and Polybrominated Diphenyl Ethers

PCBs are more than 200 possible congeners of PCBs based on the number and position of chlorine atoms present. They are a group of synthetic organic compounds that were originally manufactured for use in electrical transformers and capacitors and were largely exploited as plasticizers. They were also used in carbon-less papers and inks. Though PCBs are no longer manufactured, they continue to be emitted into the atmosphere from capacitors and PCB-filled transformers still in use. PCBs in old appliances are disposed of to landfills where large quantities are emitted to the air, but now such equipment are mainly disposed of by incineration, where high temperatures and adequate oxygen ensure the destruction of PCBs.

Electricity utilities and metal processing represent around 9% and 21% of the total PCB emissions in Europe respectively (Table 13). There are also small emissions from petroleum and related industries (EEA, 2010b).

PCB concentrations in air are usually higher in urban areas, which emerge from indoor air representing a predominant source of PCBs. Volatilization from soils is a less significant source. Regional differences in congener profile emissions indicate a difference in their respective sources.

PCBs are extremely persistent compounds in the environment and possess the ability to accumulate in the food chain because of their lipophilic properties.

Some PCBs are classified as probably carcinogenic to humans and have been linked with chronic effects such as infertility, immune and endocrine dysfunctions, as well as behavioral and learning impairment.

Polybrominated diphenyl ethers (PBDEs) are used as flame-retardants in electronics, furniture, textiles building materials, furnishings, motor vehicles, airplanes, plastics and polyurethane foams. They are structurally akin to the PCBs, consisting of two halogenated aromatic rings and are classified according to the average number of bromine atoms in their structure. Due to their high production volume,

Table 13. Sources of air pollution by PCBs in the Unites States (EEA)

General Sources	Detailed Sources	Emissions by Years	
		2006 (10^3 Tons/Year (%))	2014 (10^3 Tons/Year (%))
Stationary fuel combustion	Public electricity and heat production	368.976 (7.95)	352.661 (9.34)
	Fuel comb. other	111.404 (2.40)	90.391 (2.40)
	Total	480.380 (10.36)	443.052 (11.74)
Industrial and other processes	Chemical and allied product MFG	0.000 (0.00)	0.007 (0.00)
	Metals processing	1068.261 (23.03)	777.924 (20.61)
	Petroleum and related industries	21.767 (0.47)	17.636 (0.47)
	Other industrial processes	1469.710 (31.69)	1154.661 (30.60)
	Total	2559.738 (55.19)	1950.228 (51.68)
Agricultural activities	Agricultural livestock	0.000 (0.00)	0.000 (0.00)
	Other agricultural activities	25.271 (0.54)	5.762 (0.15)
	Total	25.271 (0.54)	5.762 (0.15)
Traffic activities	Road transport	145.398 (3.13)	109.837 (2.91)
	Other transport means	9.960 (0.21)	10.254 (0.27)
	Total	155.358 (3.35)	120.091 (3.18)
Waste disposal and treatment		756.292 (16.31)	601.409 (15.94)
Solvents		0.003 (0.00)	0.002 (0.00)
Miscellaneous		661.239 (14.26)	653.375 (17.31)
Anthropogenic sources		4638.281 (100.00)	3773.919 (100.0)
Forest fires		0.102 (0.00)	0.078 (0.00)
Other natural		0.000 (0.00)	0.000 (0.00)
Natural sources		0.102 (0.00)	0.078 (0.00)
Total		4638.383	3773.997

widespread usage and environmental persistence, PBDEs have become ubiquitous contaminants in environmental media. Their release to the atmosphere is via volatilization or formation of dusts during use, in addition to emissions during manufacture and disposal. They are present in ambient air in the gas phase or associated with particles. These compounds are lipophilic and are known to bioaccumulate and be globally transported. The potential toxicity of these compounds has attracted increasing scrutiny. Because of their toxicity and persistence, the industrial production of some PBDEs is restricted under the Stockholm Convention. Humans are exposed to low levels of PBDEs through ingestion of food and by inhalation. Toxicity data for PBDEs is limited; however, they are recognized as endocrine disruptors and their capability to interfere with the thyroid functioning and reduce fertility have been shown. The number and position of bromine atoms on the phenyl rings of these molecules result in different congeners with differing human toxicities, where the lower brominated congeners (mono- to pentabrominated) are considered to be more harmful (Besis & Samara, 2012).

Dioxins

Dioxins refer to polychlorinated dibenzo-p-dioxins (PCDD) and polychlorinated dibenzofurans (PCDF). They represent unwanted byproducts of many anthropogenic processes, including incineration of hazardous waste and sintering operations. PCDDs and PCDFs are two groups of tricyclic, planar, chlorinated aromatic compounds. Their chemical properties are similar (lipophilic, nonpolar and stable compounds), which favor their persistence in the environment.

The main source of dioxins is the incomplete combustion of chlorine-containing compounds, in particular, the incineration of municipal solid wastes at insufficient temperatures.

Combustion of coal and other solid fuels can also result in dioxin formation, but power stations contribute only to a small percentage of dioxins to the atmosphere in Europe (Table 14). Other potential sources include vehicles, chemical production and use of polychlorinated aromatic pesticides and herbicides. The largest sources of dioxins and furans in Europe are stationary fuel combustion and incineration of municipal solid waste (municipal, medical and other biological incineration) (EEA, 2010b).

The monitoring of ambient airborne PCDD/PCDFs revealed extremely variable content of these compounds, based on the area where sampling was conducted as well as the season of the year. They are subject to long-range atmospheric transport to remote regions rendering them ubiquitous contaminants and are two of the thirty persistent organic pollutants (POPs) included in the Stockholm Convention. (UNEP Stockholm Convention, 2008).

Human exposure pathways include air-deposition leading to food and water chain transfer-human exposure, which are both important for these bioaccumulating compounds. PCDD and PCDFs possess a number of toxicological properties, but the main concern is regarding their possible role in immunologic and reproductive effects. Some of the PCDD/PCDFs, specifically the 2,3,7,8-substituted congeners, are known to be carcinogenic and endocrine disruptors.

Benzene

Benzene is among the main pollutants commonly monitored in several countries due to its carcinogenic properties. Studies of industrial workers exposed to high levels of benzene have demonstrated an excess risk of leukemia, which increased with working exposure lifetime (Armenta & de la Guardia, 2016). Sources of this highly carcinogenic compound can be local, but they are ubiquitous. The combustion of

Table 14. Sources of air pollution by PCDD/PDCF in the Unites States (EEA)

General Sources	Detailed Sources	Emissions by Years	
		2006 (g I-TEQ/Year (%))	2014 (g I-TEQ/Year (%))
Stationary fuel combustion	Public electricity and heat production	81.703 (3.56)	89.785 (4.83)
	Fuel comb. other	423.352 (18.42)	293.963 (15.81)
	Total	505.055 (21.98)	383.748 (20.64)
Industrial and other processes	Chemical and allied product MFG	0.323 (0.01)	0.441 (0.02)
	Metals processing	329.068 (14.32)	218.321 (11.74)
	Petroleum and related industries	21.097 (0.92)	23.374 (1.26)
	Other industrial processes	25.567 (1.11)	24.817 (1.33)
	Total	376.054 (16.37)	266.954 (14.36)
Agricultural activities	Agricultural livestock	0.000 (0.00)	0.000 (0.00)
	Other agricultural activities	47.746 (2.08)	61.970 (3.33)
	Total	47.746 (2.08)	61.970 (3.33)
Traffic activities	Road transport	100.897 (4.39)	74.355 (4.00)
	Other transport means	4.439 (0.19)	3.137 (0.17)
	Total	105.336 (4.58)	77.492 (4.17)
Waste disposal and treatment		440.028 (19.15)	282.696 (15.21)
Solvents		19.311 (0.84)	14.640 (0.79)
Miscellaneous		738.942 (32.16)	723.126 (38.90)
Anthropogenic sources		2232.473 (97.15)	1810.626 (97.39)
Forest fires		9.446 (0.41)	8.644 (0.46)
Other natural		55.976 (2.44)	39.813 (2.14)
Natural sources		65.422 (2.85)	48.457 (2.61)
Total		2297.895	1859.083

gasoline, of which it is a minor constituent, is the most important source of benzene in the air. Benzene is also found in air as a result of emissions from burning coal, production of fuels, gasoline vapours at gasoline service stations, waste incineration, cigarette smoke and wood-burning fires. Benzene can also evaporate into the air from fuel storage containers and during refueling.

Vehicles working with gasoline emit more benzene in exhaust than diesel vehicles, even when catalytic converters are used to control emissions in these cars. Exhaust gases and evaporation from gasoline vehicles were high in 1997 and accounted for about 56 and 4% of total benzene emissions respectively.

Benzene emission occurs also during the combustion of aromatics in the gasoline. The aromatic content is not explicitly controlled in either USA or EU automotive fuel specifications, though the EU Fuel Quality Directive 2009/30/EC introduced a maximum total aromatic content of 35%. It also introduced a 1% maximum for benzene in automotive fuel (Eur-Lex, 2009). Currently in the USA there are significant concentrations of benzene in ambient air, largely due to vehicle emissions though there is considerable variation across this country. The control of hazardous air pollutants from mobile sources final rule (also known as the Mobile Source Air Toxics rule or MSAT2 was published in October 16,

2008 in the USA. This rule requires that refiners and importers produce gasoline that has an annual average benzene content of 0.62% or less at the of beginning of 2011 (EPA, 2008).

1,3-Butadiene

1,3-butaiene is used as an industrial chemical, particularly in the manufacture of synthetic rubber for tires and some liquefied petroleum gas which contains up to 8%. The major source of 1,3-butadiene in the air is the combustion of petrol, gasoline and diesel fuels, as well as the burning of other fossil fuels. In fact, the main source of 1,3-butadiene is traffic activities. Gasoline and diesel contains little 1,3-butadiene, but it is formed during combustion in the engine.

Other sources arising from extraction and distribution of fossil fuels as well as waste disposal and chemical industry through the manufacture and use of 1,3-butadiene. In some areas forest fires can be a significant source of 1,3-butadiene in the atmosphere.

Exposure to 1,3-butadiene is thought to induce cancers of the lymphoid system and blood-forming tissues, lymphomas and leukemia.

Lead

Lead is a toxic heavy metal that can have acute and chronic impacts on human health. It is found naturally in the environment and in the anthropogenic emissions. There are several natural sources of lead in the atmosphere including erosion of rocks and volcanic activity. Other natural sources include sea spray, wind-blown dust and forest fires.

The major sources of atmospheric lead emissions have historically been motor vehicles with leaded gasoline (such as cars and trucks) and industrial sources (for example lead smelters, ore and metal processing). As a result of regulatory efforts to remove the use of alkyl lead from gasoline, levels of lead in the air declined between 1980 and 2000. Road traffic source is rapidly declining as a result of the increased use of unleaded fuels (which is not completely lead-free) and a reduction in the maximum permitted level in leaded fuels. Though emissions have been largely declined, road traffic is still responsible for a big part of all emissions in several countries (USA 2014: 461 tons; Europe 2014: 300 tons). Apart from road traffic emission, metals processing has become one of the major source of lead emissions to the atmosphere, representing an important part of all emissions in the USA and Europe (USA 2014: 157 tons; Europe 2014: 348 tons (Table 15)). Other stationary sources include fuel combustion and disposal of lead-containing products accounting for 92 and 891 tons in the USA (EPA, 2014) and Europe (EEA, 2014) respectively (EEA, 2010b; EPA, 2017b).

Exposure to lead can have multiple health effects, where exposure to high levels can affect the function of almost all organs and systems in the body. Lead distributes throughout the body and is accumulated in the bones. This metal adversely affects the nervous system, kidney function, immune and reproductive systems, development and the cardiovascular system. Lead exposure also affects the oxygen-carrying capacity of the blood. Exposure of children to lead can cause a largely irreversible impact on their cognitive development. In fact, the most critical of these adverse health effects have occurred in children resulting from fetal, neonatal and later postnatal lead exposures and include deficits in neurocognitive and neurobehavioral development.

Table 15. Sources of air pollution by Lead in the Unites States (EEA)

General Sources	Detailed Sources	Emissions by Years	
		2006 (Tons/Year (%))	2014 (Tons/Year (%))
Stationary fuel combustion	Public electricity and heat production	130.307 (5.60)	113.606 (5.90)
	Fuel comb. other	775.019 (33.28)	777.523 (40.38)
	Total	905.326 (38.88)	891.129 (46.28)
Industrial and other processes	Chemical and allied product MFG	6.450 (0.28)	3.292 (0.17)
	Metals processing	684.924 (29.41)	348.243 (18.09)
	Petroleum and related industries	24.830 (1.07)	11.699 (0.61)
	Other industrial processes	51.239 (2.20)	47.781 (2.48)
	Total	767.444 (32.96)	411.016 (21.35)
Agricultural activities	Agricultural livestock	0.000 (0.00)	0.000 (0.00)
	Other agricultural activities	9.305 (0.40)	8.416 (0.44)
	Total	9.305 (0.40)	8.416 (0.44)
Traffic activities	Road transport	307.535 (13.21)	285.210 (14.81)
	Other transport means	20.870 (0.90)	15.340 (0.80)
	Total	328.406 (14.10)	300.550 (15.61)
Waste disposal and treatment		17.642 (0.76)	13.379 (0.69)
Solvents		39.830 (1.71)	38.235 (1.99)
Miscellaneous		260.788 (11.20)	262.727 (13.64)
Anthropogenic sources		2328.741 (100.00)	1925.451 (100.00)
Forest fires		NE	NE
Other natural		0.000 (0.00)	0.000 (0.00)
Natural sources		0.000 (0.00)	0.000 (0.00)
Total		2328.741	1925.451

Sources of Secondary Air Pollutants

Secondary pollutants are the result of the transformation of primary compounds (NO, CO, unburnt hydrocarbons (e.g. alkanes and alkenes) and partially oxidized fuel (e.g. aldehydes) usually increased by high temperatures and bright sunlight.

The most well-known secondary pollutants is ozone. Other secondary pollutants include peroxyacyl nitrates, aldehydes, alkyl nitrates, oxidized hydrocarbon and secondary aerosol.

Ozone

Ozone in the upper atmosphere (stratosphere) is essential to protect living beings from solar radiation by absorbing the sun's most harmful ultraviolet rays. Exposure to excess ultraviolet light is known to cause skin cancer and other related problems. However, in the ground level, ozone is a secondary gas that is not usually emitted directly into the air, but is formed in the atmosphere by photochemical reactions in the presence of NOx, VOCs and sunlight (Figure 3 and 4). Therefore, at the surface, the presence of ozone is largely not natural and can result in adverse human health and environmental troubles.

The phenomenon of photochemical smog was first recognized in Los Angeles in 1943, where the basin-like topography, high temperatures and high solar flux in this region are ideal for promoting the photochemical processes. The historical word "smog" implies that this pollution is formed from the condensation of smoke and fog. It caused plant damage, eye irritation and decrease in visibility. Because of the region where it was initially observed, this smog quickly became known as "Los Angeles smog" but was soon found to occur in big cities worldwide like Beijing, Delhi and Mexico (Folberth *et al.*, 2015).

Emissions from industrial facilities, electric utilities, motor vehicle exhausts, gasoline vapours and chemical solvents are some of the major sources of NOx and VOCs.

As shown in the following reactions, NO_2 is converted to unwanted ground ozone. As mentioned above, these reactions require the presence of ultraviolet light and VOC compounds. Temperature plays also an important role revealed by higher formation of ozone in the air at higher temperature (i.e. during the day and in the summer period).

$$NO_2 + \mathbf{h\nu} \rightarrow NO + O.$$

$$O\cdot + O_2 + M \rightarrow O_3 + M$$

Ozone has harmful effects on sensitive vegetation and ecosystems. Ozone is considered to be the most damaging air pollutant to vegetation, with significant effects on the growth of trees and important crops like wheat, soybeans and rice.

Ozone can trigger a variety of health problems, particularly for children, the elderly and people who have lung diseases like asthma. Ozone can have several effects, such as increasing susceptibility to respiratory infections, lung function decrements, inflammation and irritation of the lining of the lungs causing coughing and throat irritation, as well as epithelial damage and increases in airway responsiveness, a condition in which the conducting airways have an enhanced bronchoconstriction following exposure to variety of stimuli like allergens and cold air. Exacerbation of asthma increases in hospital admissions for respiratory infections and excess mortality has also been reported to be associated with oxidant air pollution episodes (Armenta & de la Guardia, 2016).

Particulate Matter

Particulate matter/aerosol, also known as particle pollution or PM, are the general term used for solid particles or liquid droplets that are suspended in the air. They are not a single pollutant, but are made up of particles of many different sizes and chemical composition, from a wide range of natural and anthropogenic sources. Some particles are large and dark enough to be seen as soot or smoke; others are much smaller and can only be detected with an electron microscope. Different terms can be used for PM (e.g. dust, smoke, mist) depending on their nature and characteristics. They are made up of a number of components, including oxides, acids (nitrates and sulfates), organic compounds, metals and soil. A water droplet of acid rain, carrying sulfuric acid or nitric acid produced from NOx or SOx, can be classified as PM.

The size of particles is directly linked to their potential risk on human health at pulmonary level. The effect of particles on health is more pronounced in the case of smaller particles. Particles greater than 10 μm are thought to be less harmful to health than smaller particles (less than 2.5 μm).

Apart from their size, PM can be characterized by their formation mechanism, origin and chemical composition.

PM air pollution can be viewed in two major components: primary PM (soot, carbonaceous particles (organic matter, black smock), industrial dust, soil dust and sea salt), emitted directly into the atmosphere from pollution sources such as industry, electric power plants, diesel buses and automobiles and secondary PM, formed in the atmosphere as a result of reactions or interactions between primary gaseous pollutants, such as SOx and SOx, VOC gases and compounds present in the atmosphere.

Anthropogenic primary particles originate from a wide variety of sources, including transportation, coal combustion, cement manufacturing, industry and metallurgy activities, waste incineration, biomass burning, oceans, soils, while primary biogenic particles consist of plant debris, humic matter and microbial particles.

Secondary PM consisting of sulfate are produced by chemical reactions in the atmosphere from gaseous precursors like SOx originating from different sources (see Sulfur oxides) and dimethyl sulfide (from marine plankton). Secondary nitrate PM are also closely tied to the relative abundance of NOx and NH_3 (see Nitrogen oxides and Ammonia).

Some sources can generate primary and secondary PM. The typical example is volcanic outgassing that are important contributors to global PM emissions through direct emission of primary volcanic dust, or through emission of gaseous sulfur species capable of reacting to form secondary sulfate particles.

Though natural sources represent around 90% of the total PM emissions, it is anthropogenic sources that result in over half of the direct adverse effects of PM, the major contribution arising from SO_2. In urban areas, natural sources are often dwarfed by particles arising from human activity.

PM exposure causes health problems when particulate becomes lodged in the smallest bronchi of lung tissue and, as a result, impedes the ability of this tissue to absorb oxygen. This produces increases of respiratory efforts and prompts the heart to work at higher rates to transport sufficient oxygen. Thus, persons particularly sensitive to high concentrations of PM include people with heart (cardiac arrhythmias and heart attacks) and lung disease (asthma), older adults and children.

Black Smoke

Black smoke is related to the blackness or soiling capacity of particles. Carbonaceous particles, representing one of several types of black smokes and consisting of organic substances and different forms of black carbon, make up a large, but highly variable, proportion of global PM. Road transport, particularly coal and diesel fuel burning are the main sources of black smoke emissions. Other sources are biomass and fossil fuel burning and the oxidation of biogenic and anthropogenic VOC compounds. During the London smog of December 1952, airborne concentrations of black smoke exceeded 1,500 μgm^{-3}, accompanied by massive concentrations of SO_2 (Jackson, 2007).

PM10

PM10 or coarse particles are defined as all particles equal to or less than 10 μm and lager than 2.5μm in diameter. Particles larger than 10 μm are not generally able to reach the lungs and are caught in the nose and throat. PM10 do not remain suspended in the air for significant periods of time, but they have been shown to be harmful to health (Martin & Dombrowski, 2008). They include both primary and secondary PM:

- Primary particles emitted directly by combustion and non-combustion processes. Non-combustion processes include natural events such as wind-blown dusts and soils and wildfires. They are also mechanically generated outdoors by crushing or grinding operations like the formation tire debris, but atmospheric concentrations are often dominated by dusts material resuspended by traffic on paved or unpaved roads. Other outdoor sources of coarse particles include those from construction, farming and mining activities.
- Secondary particles formed in the atmosphere from gas-to-particle conversion processes in the atmosphere, consequently composed substantially of sulfates and nitrates formed from the reactions of emissions of SOx and NOx, together with organic and elemental carbon and a range of trace metals.

1. Sources of Primary PM10

The estimated emissions of anthropogenic primary PM10 to the USA and Europe are given in Table 16 and 17. One of the main sources of primary PM10 is road transport, predominantly from diesel vehicles. Emissions also arise from brake and tire wear and from the reentrainment of dust on the road surface. While road transport accounts for about 2-13% of global emissions, it accounts for over 50% of primary emissions inside cities. Power stations are the largest users of coal, but all coal-fired stations are equipped with electrostatic precipitators, which remove on average 99.5% of the PM. The main stationary combustion source of PM10 is from domestic premises and this has traditionally been from the use of coal burnt in open fires. There are a large number of non-combustion industrial sources, including iron and steel, pesticide production, construction and quarrying. Agricultural sources such as land preparation, fertilizer application and harvesting also contribute; these tend to be very seasonal (EEA, 2010b; EPA, 2017b).

2. Sources of Secondary PM10

Sulfate produced by the photochemical oxidation of SO_2 emitted from combustion sources (see Sulfur oxides) is perhaps the most well-known secondary PM10, in the form of particulate sulfuric acid and ammonium sulfate. This is due to its role in health impacts, reduction in visibility, acid rain and environmental acidification. The contribution to urban PM10 from particulate nitrate (see Nitrogen oxides) is quite smaller than that of particulate sulfate, though it is likely also formed photochemically. Some organic compounds can also be precursors to secondary PM10. In the winter, periods of elevated PM10 concentrations are associated mainly with vehicle exhaust emissions, while in the summer there is a greater contribution from secondary PM.

a. PM2.5

PM2.5, also known as fine fraction particles (generally defined as those particles with a diameter of 2.5 μm or less but higher than 0.1 μm), are the most harmful particulates that can be breathed into the deepest areas of the lungs. They are so small that an electron microscope is required to detect their pres-

Table 16. Sources of air pollution by primary PM10 in the Unites States (EPA)

General Sources	Detailed Sources	Emissions by Years		
		1996 (10^3 Tons/ Year (%))	2006 (10^3 Tons/ Year (%))	2016 (10^3 Tons/ Year (%))
Stationary fuel combustion	Fuel comb. elec. util.	**288.798**	**555.521 (2.60)**	**228.58 (0.94)**
	Fuel comb. industrial	238.407	296.577 (1.39)	324.838 (1.33)
	Fuel comb. other	380.174	439.028 (2.05)	368.257 (1.51)
	Total	907.379	1291.126 (6.03)	921.675 (3.77)
Industrial and other processes	Chemical and allied product MFG	63.114	32.503 (0.15)	19.741 (0.08)
	Metals processing	144.051	82.469 (0.39)	58.471 (0.24)
	Petroleum and related industries	29.281	28.623 (0.13)	36.118 (0.15)
	Other industrial processes	325.167	1 032.05 (4.82)	776.983 (3.18)
	Solvent utilization	5.767	6.982 (0.03)	4.24 (0.02)
	Storage and transport	80.506	55.65 (0.26)	67.129 (0.27)
	Waste disposal and recycling	453.848	272.008 (1.27)	305.312 (1.25)
	Total	1101.734	1510.284 (7.06)	1267.994 (5.19)
Transportation	Highway vehicles	287.147	376.276 (1.76)	286.597 (1.17)
	Off-highway	341.425	283.353 (1.32)	162.626 (0.67)
	Total	628.572	659.629 (3.08)	449.223 (1.84)
Wildfires		-	805.007 (3.76)	1 029.99 (4.21)
Miscellaneous		20 219.31	17 135.09 (80.07)	20 772.66 (84.99)
Total		-	21401.132	24 441.54

ence. Some of them may carry heavy metals and carcinogenic molecules. They can also cause disorders of the respiratory system, asthma, bronchitis and even heart problems.

Fine particles are commonly directly emitted from combustion sources. In some conditions, they are also formed secondarily from gaseous precursors such as SOx, NOx and organic compounds. Such secondary particles are usually composed of different chemicals like nitrate, sulfate, chloride and ammonium compounds, organic and elemental carbon, as well as metals.

Stationary combustion of coal, oil and wood, as well as industrial process sources (such as smelters and steel mills), waste disposal, forest fires and agricultural burning can produce emissions that contribute to ambient fine particle concentrations. Mobile sources contribute largely in the total budget of fine particle like combustion of gasoline and diesel in motor vehicles (Table 18 and 19). Several studies based on population data have consistently reported associations between short-term exposures to PM2.5 and cardiovascular and respiratory hospital admissions in a number of regions throughout the USA, Europe and Asia. Long-term exposures to PM2.5 have been linked to cardiovascular diseases, accelerated development of atherosclerosis, decrements in lung function growth, increased respiratory symptoms and asthma development (EEA, 2010b; EPA, 2017b).

Table 17. Sources of air pollution by PM10 in the Unites States (EEA)

General Sources	Detailed Sources	Emissions by Years	
		2006 (10^3 Tons/Year (%))	2014 (10^3 Tons/Year (%))
Stationary fuel combustion	Public electricity and heat production	134.997 (5.88)	75.620 (3.96)
	Fuel comb. other	124.403 (5.42)	88.621 (4.64)
	Total	259.400 (11.30)	164.241 (8.61)
Industrial and other processes	Chemical and allied product MFG	12.242 (0.53)	11.784 (0.62)
	Metals processing	54.377 (2.37)	30.952 (1.62)
	Petroleum and related industries	29.499 (1.29)	10.924 (0.57)
	Other industrial processes	286.077 (12.47)	222.723 (11.67)
	Total	382.195 (16.66)	276.382 (14.48)
Agricultural activities	Agricultural livestock	185.188 (8.07)	188.058 (9.85)
	Other agricultural activities	157.114 (6.85)	164.923 (8.64)
	Total	342.302 (14.92)	352.980 (18.50)
Traffic activities	Road transport	308.964 (13.46)	214.527 (11.24)
	Other transport means	67.454 (2.94)	43.093 (2.26)
	Total	376.418 (16.40)	257.620 (13.50)
Waste disposal and treatment		27.635 (1.20)	25.021 (1.31)
Solvents		68.042 (2.97)	49.978 (2.62)
Miscellaneous		798.397 (34.79)	743.394 (38.95)
Anthropogenic sources		2254.388 (98.24)	1869.617 (97.97)
Forest fires		12.188 (0.53)	10.686 (0.56)
Other natural		28.150 (1.23)	28.090 (1.47)
Natural sources		40.338 (1.76)	38.776 (2.03)
Total		2294.726	1908.393

EMERGING AIR POLLUTANTS

Biological Pollutants

Airborne micro-organisms can be found in almost all environments, even in the air. They usually do not present a health hazard to the exposed individuals. Nevertheless, most airborne micro-organisms have the property of self-replication, leading to allergic reactions and even diseases in humans. They enter into the atmosphere from a variety of natural sources like soil, lakes, animals and humans as well as anthropogenic sources like sewage treatment, agricultural livestock and fermentation processes. Though some bioparticles are dominant (genus *Bacillus, Micrococcus, Staphylococcus* and *Pseudomonas*), their proportions differed from place to place (Armenta & de la Guardia, 2016).

Table 18. Sources of air pollution by primary PM2.5 in the Unites States (EPA)

General Sources	Detailed Sources	Emissions by Years		
		1996 (10^3 Tons/Year (%))	2006 (10^3 Tons/Year (%))	2016 (10^3 Tons/Year (%))
Stationary fuel combustion	Fuel comb. elec. util.	156.839	447.536 (7.80)	177.331 (2.88)
	Fuel comb. industrial	152.239	178.245 (3.11)	240.375 (3.90)
	Fuel comb. other	355.847	407.702 (7.11)	363.082 (5.89)
	Total	664.925	1033.483 (18.02)	780.788 (12.67)
Industrial and other processes	Chemical and allied product MFG	38.542	26.04 (0.45)	13.982 (0.23)
	Metals processing	100.665	61.095 (1.07)	45.052 (0.73)
	Petroleum and related industries	16.516	21.806 (0.38)	31.968 (0.52)
	Other industrial processes	179.593	372.199 (6.49)	307.227 (4.99)
	Solvent utilization	5.107	6.2 (0.11)	3.751 (0.06)
	Storage and transport	29.806	24.638 (0.43)	18.16 (0.29)
	Waste disposal and recycling	426.759	245.658 (4.28)	252.571 (4.10)
	Total	796.988	757.636 (13.21)	672.711 (10.92)
Transportation	Highway vehicles	228.579	295.216 (5.15)	132.622 (2.15)
	Off-highway	312.925	267.122 (4.66)	151.021 (2.45)
	Total	541.504	562.338 (9.80)	283.643 (4.60)
Wildfires		-	673.037 (11.73)	872.89 (14.16)
Miscellaneous		4 720.93	2 709.96 (47.24)	3 552.57 (57.65)
Total		-	5736.449	6162.601

Pesticides

In general, pesticide contents in the air vary seasonally (with higher levels in summer season) and geographically, where soil influences the volatility of the pesticides. Many (now typically banned) pesticides have been associated with endocrine disrupting effects, such as organochlorine pesticides, whilst some other in-use pesticides such as the pyrethroids and chlorpyrifos are also of concern in this regard (Forbes & Garland, 2016). Most of the as organochlorine pesticides belong to POPs, which have high toxicity, persistency and long-range migration property. Moreover, organochlorine pesticides have high vapor pressure, low water solubility and high lipid solubility, which undergo limited decomposition/degradation. Many of them, including DDTs, have been gradually phased out due to their high bioaccumulation, toxicity and persistence in the environment. However, these phased-out organochlorine pesticides are still widely detected in air, water and soil largely because organochlorine pesticide associated with soils or dissolved in surface water are continuously evaporated to air.

Atmospheric deposition including both dry and wet deposition is responsible for returning of the organochlorine pesticides from air to the ground surface, contaminating soil and water that may be far away from the source areas.

Table 19. Sources of air pollution by PM2.5 in the Unites States (EEA)

General Sources	Detailed Sources	Emissions by Years	
		2006 (10^3 Tons/Year (%))	2014 (10^3 Tons/Year (%))
Stationary fuel combustion	Public electricity and heat production	76.371 (5.12)	47.183 (3.82)
	Fuel comb. other	96.415 (6.46)	70.388 (5.70)
	Total	172.787 (11.58)	117.571 (9.52)
Industrial and other processes	Chemical and allied product MFG	9.010 (0.60)	8.877 (0.72)
	Metals processing	38.046 (2.55)	21.014 (1.70)
	Petroleum and related industries	21.123 (1.42)	7.812 (0.63)
	Other industrial processes	91.173 (6.11)	74.555 (6.04)
	Total	159.353 (10.68)	112.259 (9.09)
Agricultural activities	Agricultural livestock	38.515 (2.58)	38.835 (3.14)
	Other agricultural activities	36.061 (2.42)	39.128 (3.17)
	Total	74.576 (5.00)	77.963 (6.31)
Traffic activities	Road transport	252.136 (16.90)	158.651 (12.84)
	Other transport means	63.265 (4.24)	39.620 (3.21)
	Total	315.401 (21.14)	198.271 (16.05)
Waste disposal and treatment		20.443 (1.37)	19.380 (1.57)
Solvents		25.887 (1.74)	20.605 (1.67)
Miscellaneous		701.434 (47.02)	668.374 (54.11)
Anthropogenic sources		1469.881 (98.52)	1214.422 (98.32)
Forest fires		9.993 (0.67)	8.748 (0.71)
Other natural		12.038 (0.81)	11.978 (0.97)
Natural sources		22.031 (1.48)	20.726 (1.68)
Total		1491.911	1235.148

Atmospheric concentrations of organochlorine pesticides were measured in Nuuk, Greenland, where α-hexachlorocyclohexane was present at nearly 20 pg/m^3, and lindane, endosulfan, dieldrin and DDEs were detected in the low pg/m^3 range (Bossi *et al.*, 2008).

High values have been found in a typical pesticide contaminated site located in Southwestern of China. ΣHCHs (52.3 ng/m^3) and ΣDDTs (130 ng/m^3) were the dominant pollutants at the contaminated site. The authors showed that the volatilization of the pesticides from soil into air is the obvious pollution source for the atmosphere especially at the site. The volatilization tendency from soil into air at and around the contaminated site showed that the site is a serious pollution source for the local atmosphere (Fang *et al.*, 2017). Huang *et al.* (2013) assessed the potential of organochlorine pesticides pollution via dry and wet deposition over the fast-developing Pearl River Delta area, South China. They quantified both the dry and wet deposition fluxes of 19 organochlorine pesticides including DDTs, endosulfans and hexachlorocyclohexanes. The results showed that each year about 67.4, 42.0, 15.0 and 8.07 kg of total organochlorine pesticides, DDTs, endosulfans and hexachlorocyclohexanes were returned to the ground, among which 11.7, 10.4, 0.84 and 0.16 kg were in the dry deposition forms. The large spatial variations in organochlorine pesticide deposition fluxes indicated that organochlorine pesticides pollution in the air was mainly influenced on local scales because evaporation from local soil was likely the major source of the phased out OCPs (Huang *et al.*, 2014).

Nanomaterials

Nanomaterials or ultrafine materials are defined as engineered structures with at least one dimension of 100 nm or less. They are classified as nanoparticles, nanofibres and nanotubes, composite materials and nanostructured surfaces. They can be also subdivided into carbon-based nanoparticles, such as carbon nanotubes, nanohorns, nanowires or fullerenes and so on and metal based nanoparticles, such as noble metal nanoparticles, metal oxides or quantum dots. Since ancient times, humans have been exposed to nanosized materials originating from combustion processes, such as forest fires and volcanoes, as well as natural sources, such as sea spray. This exposure drastically increased with the industrial revolution, when anthropogenic sources of nanosized particles were introduced along with the rapid development of nanotechnology.

The widespread use of nanomaterials will inevitably raise concern about possible air pollution with materials that may have adverse biological effects. The current knowledge of the toxicology of nanomaterials is poor but suggests that they may be able to have adverse effects at their portal of entry, for example, the lungs, but some of them may also escape the normal defenses and translocate from their portal of entry to have diverse effects in other target organs.

Manufactured nanomaterials are likely to enter the environment for several reasons. Some of them will be produced by the ton and any material produced in such mass quantities is likely to reach with the environment from manufacturing effluent or from spillage during shipping and handling. They are being used in personal-care products such as cosmetics and sunscreens, in electronics, tires, fuel cells and many other products. They are also being used in disposable materials such as filters and electronics and may therefore reach the environment through landfills and other methods of disposal. Moreover, their extended use in motor catalyzers together with their incorporation in clothes and packaging has dramatically increased the presence of terribly small particles in the air.

Our understanding of the ecotoxicology of nanomaterials is very limited compared to other chemicals with a similar distribution in the environment and till now there are not too many available data on the real distribution mobility and potential toxic effects of these materials.

Health professionals do not yet understand the toxicity of nanomaterials in relation to environmental and occupational diseases. Some scientists have hypothesized that ultrafine particles, because of their small size and large surface area to mass ratio, may be the most toxic PM. Concerns over the lack of understanding of the potential environmental impact of nanotechnologies has led to some organizations such as the action group on erosion, Technology and Concentration (ETC Group) to call for a moratorium on the use and manufacture of nanomaterials (Hood, 2004).

CONCLUSION

Air pollution is a global concern that creates a need for international strategies and agreements-reflections of the social demands of the way we live. Numerous potential air pollutants may enter the environment from many natural and anthropogenic sources. They are known to have negative effects on pulmonary and cardiovascular functioning. Air pollutants cause serious health problems together with adverse environmental and ecological effects. The most common are carbon monoxide (the most abundant air pollutant), PM, ammonia, NOx, SOx and VOCs. They are emitted from either stationary sources or mobile sources. Hence, air pollutants must be controlled at the source or diluted before being emitted into the atmosphere.

It is well known that emissions from transportation sources are a major component of the total ambient air pollution concentration in large urban cities. Thus, development of alternative fuels that can reduce emissions of air pollutants as well as the use of public transport are necessary. From a policy perspective, the close association between traffic patterns and air pollution leads to the development of transport policies and regulations with population health improvements as a primary objective. Other important sources of air pollution are power stations. Emissions from transportation and power generation sources have begun to decline in recent years owing to the introduction of efficient abatement measures. Urban inventories are important for identifying these sources and their impact on the environment.

Remarkable advances have been made in reducing air pollution in different countries over the last half century. However, atmospheric pollution still occupies the central focus in global environmental issues, which is not surprising given population growth and associated demands for energy and transportation.

The results of the research presented here provide many opportunities and challenges for the development of policies for improvements to air quality and human health. Important decisions need to be taken by governments to reduce air pollution and its hazardous effects on the environment. The challenge remains, however, of how best to achieve these reductions. Health professionals must raise their voices in the political decision process to give strong support for clean air policies, on both a national and an international level.

Responsibility for the future is both national and global, with many actors, comprising national environmental agencies, international organizations, academics, etc.

Communication between the scientific and regulatory communities with regard to ambient air quality standards and regulatory approaches is critical to developing policies to reduce the human health risks due to air pollution. New studies on the long-term effects of low doses of air pollutants are needed, including important partnerships between epidemiologists, toxicologists, metrologists, chemists, physicists and biologists.

Related pollutants must be monitored across multiple spatial and temporal scales to integrate studies of their effects ranging from microenvironmental to global and from less than an hour to decades.

Fostering the growth of exposure assessment sciences requires improvement of current methods and development of new and innovative methods that can be used for better risk assessment, better environmental epidemiology, better environmental toxicology and better source/effects modeling. In addition, we certainly need more sophisticated and integrated approaches to effectively deal with air pollution.

The information related to air pollution presented in this chapter is intended to increase awareness of the problem and of the need for prevention and control. More drastic solutions will be needed so that air quality can continue to improve. Economic attempts to reduce energy use and to advance and expand the use of pollutant-reducing technologies must be increased. Improved scientific understanding of the links between air pollution and health effects will help both to sustain this progress and to improve it by better targeting the most toxic emissions.

REFERENCES

Armenta, S., & de la Guardia, M. (2016). Pollutants and Air Pollution. *Comprehensive Analytical Chemistry, 73*, 27–44. doi:10.1016/bs.coac.2016.03.002

Besis, A., & Samara, C. (2012). Polybrominated diphenyl ethers (PBDEs) in the indoor and outdoor environments--a review on occurrence and human exposure. *Environmental Pollution, 169*, 217-229. doi:10.1016/j.envpol.2012.04.009

Borbon, A., Fontaine, H., Veillerot, M., Locoge, N., Galloo, J. C., & Guillermo, R. (2001). An investigation into the traffic-related fraction of isoprene at an urban location. *Atmospheric Environment, 35*(22), 3749–3760. doi:10.1016/S1352-2310(01)00170-4

Bossi, R., Skov, H., Vorkamp, K., Christensen, J., Rastogi, S. C., Egeløv, A., & Petersen, D. (2008). Atmospheric concentrations of organochlorine pesticides, polybrominated diphenyl ethers and polychloronaphthalenes in Nuuk, South-West Greenland. *Atmospheric Environment, 42*(31), 7293–7303. doi:10.1016/j.atmosenv.2008.06.049

Environmental Protection Agency. (2008, October 16). Control of Hazardous Air Pollutants From Mobile Sources: Early Credit Technology Requirement Revision. *Federal Register, 73*(201), 61358–61363. Retrieved from https://www.gpo.gov/fdsys/pkg/FR-2008-10-16/pdf/E8-24591.pdf

Eur-Lex. (2009). Title and reference. Retrieved from http://eur-lex.europa.eu/legal-content/EN/TXT/?uri=CELEX:32009L0030

European Environment Agency (EEA). (2010, November 8). Atmoshperic Greenhouse Gas Concentrations. Retrieved July 6, 2017 from https://www.eea.europa.eu/data-and-maps/indicators/atmospheric-greenhouse-gas-concentrations/atmospheric-greenhouse-gas-concentrations-assessment-3

European Environment Agency (EEA). (2010, July 9). European Union Emmision Inventory report 1990-2008 under the UNECE Convention on Long-Range Transboundary Air Pollution (LRTAP). Retrieved June 1, 2017 from https://www.eea.europa.eu/publications/european-union-emission-inventory-report/

Fang, Y., Nie, Z., Die, Q., Tian, Y., Liu, F., He, J., & Huang, Q. (2017). Organochlorine pesticides in soil, air, and vegetation at and around a contaminated site in southwestern China: Concentration, transmission, and risk evaluation. *Chemosphere, 178*, 340–349. doi:10.1016/j.chemosphere.2017.02.151 PMID:28334674

Fengxia, G., Min, B., Yijun, M., Zupei, L., Yawen, L., & Haifeng, S. (2016). Temporal and spatial characteristics of lightning-produced nitrogen oxides in China. *Journal of Atmospheric and Solar-Terrestrial Physics*, *149*, 100–107. doi:10.1016/j.jastp.2016.10.007

Ferguson, S. J., Richardson, D. J., & van Spanning, R. J. M. (2007). *Biochemistry and Molecular Biology of Nitrification Biology of the Nitrogen Cycle* (pp. 209–222). Amsterdam: Elsevier. doi:10.1016/B978-044452857-5.50015-1

Folberth, G. A., Butler, T. M., Collins, W. J., & Rumbold, S. T. (2015). Megacities and climate change – A brief overview. *Environmental Pollution*, *203*(Suppl. C), 235–242. doi:10.1016/j.envpol.2014.09.004 PMID:25300966

Forbes, P. B. C., & Garland, R. M. (2016). Outdoor Air Pollution. *Comprehensive Analytical Chemistry*, *73*, 73–96. doi:10.1016/bs.coac.2016.02.004

GEOELEC. (n.d.). Geothermal Electricity: Potential for CO2 Mitigation. Retrieved on July 7, 2017 from http://www.geoelec.eu/wp-content/uploads/2013/11/Geoelec-CO2-mitigation.pdf

Gordon, T., Stanek, L. W., & Brown, J. (2014). *Pollution. In Air in Encyclopedia of Toxicology A2 - Wexler, Philip Encyclopedia of Toxicology* (3rd ed., pp. 995–1002). Oxford: Academic Press.

Hood, E. (2004). Nanotechnology: Looking As We Leap. *Environmental Health Perspectives*, *112*(13), A740–A749. doi:10.1289/ehp.112-a740 PMID:15345364

Huang, Q., Song, J., Zhong, Y., Peng, P., & Huang, W. (2014). Atmospheric depositional fluxes and sources apportionment of organochlorine pesticides in the Pearl River Delta region, South China. *Environmental Monitoring and Assessment*, *186*(1), 247–256. doi:10.100710661-013-3370-5 PMID:23943243

Jackson, A. V. (2007). *Sources of Air Pollution Handbook of Atmospheric Science* (pp. 124–155). Blackwell Science Ltd.

Kennes, C., & Veiga, M. C. (2013). Introduction to Air Pollution. *Air Pollution Prevention and Control; the Official Journal of the World Heart Federation.*

Kikuchi, R. (2001). Environmental management of sulfur trioxide emission: Impact of SO3 on human health. *Environmental Management*, *27*(6), 837–844. doi:10.1007002670010192 PMID:11393318

Letcher, T. M. (2014). *Introduction with a Focus on Atmospheric Carbon Dioxide and Climate Change Future Energy* (2nd ed., pp. 3–16). Boston: Elsevier.

Lin, M., Chen, Y., Villeneuve, P. J., Burnett, R. T., Lemyre, L., Hertzman, C., ... Krewski, D. (2004). Gaseous air pollutants and asthma hospitalization of children with low household income in Vancouver, British Columbia, Canada. *American Journal of Epidemiology*, *159*(3), 294–303. doi:10.1093/aje/kwh043 PMID:14742290

Martin, P., & Dombrowski, C. (2008). Air and Water Pollution. In R. P. Martin & S. C. Dombrowski (Eds.), *Prenatal Exposures: Psychological and Educational Consequences for Children* (pp. 185–199). Boston, MA: Springer US.

National Oceanic & Atmospheric Administration (NOAA). (n.d.). Trends in Atmospheric Carbon Dioxide). Retrieved June 30, 2017 from https://www.esrl.noaa.gov/gmd/ccgg/trends/weekly.html

Richter, A., Burrows, J. P., Nusz, H., Granier, C., & Niemeier, U. (2005). Increase in tropospheric nitrogen dioxide over China observed from space. *Nature, 437*(7055), 129-132. Retrieved from http://www.nature.com/nature/journal/v437/n7055/suppinfo/nature04092_S1.html

UNEP Stockholm Convention. (2008). All POPs listed in the Stockholm Convention. Retrieved on June 30, 2017 from http://chm.pops.int/TheConvention/ThePOPs/AllPOPs/tabid/2509/Default.aspx

United Nations Environment Program, Ozone Secretariat. (2011). The London Amendment (1990): The amendment to the Montreal Protocol agreed by the Second Meeting of the Parties (London, 27-29 June 1990) Retrieved June 30, 2017 from https://ozone.unep.org/new_site/en/Treaties/treaties_decisions-hb.php?dec_id_anx_auto=780

United States Environmental Protection Agency (EPA). (2017a). Air Emissions Inventories. Retrieved June 1, 2017 from https://www.epa.gov/air-emissions-inventories/air-emissions-sources

United States Environmental Protection Agency (EPA). (2017b). Greenhouse Gas Emissions: Carbon Dioxide Emissions. Retrieved June 30, 2017 from https://www.epa.gov/ghgemissions/overview-greenhouse-gases#carbon-dioxide

United States Environmental Protection Agency (EPA). (2017c). Greenhouse Gas Emissions: Methane Emissions. Retrieved June 30, 2017 from https://www.epa.gov/ghgemissions/overview-greenhouse-gases#methane

Wei, Y., Han, I.-K., Hu, M., Shao, M., Zhang, J., & Tang, X. (2010). Personal exposure to particulate PAHs and anthraquinone and oxidative DNA damages in humans. *Chemosphere, 81*(10), 1280–1285. doi:10.1016/j.chemosphere.2010.08.055 PMID:20869742

Wuebbles, D. J. (2013). *Atmospheric Gases A2. In A Encyclopedia of Biodiversity* (2nd ed., pp. 281–290). Waltham: Academic Press. doi:10.1016/B978-0-12-384719-5.00275-6

Zhang, Y., Mu, Y., Liu, J., & Mellouki, A. (2012). Levels, sources and health risks of carbonyls and BTEX in the ambient air of Beijing, China. *Journal of Environmental Sciences (China), 24*(1), 124–130. doi:10.1016/S1001-0742(11)60735-3 PMID:22783623

Chapter 12
Anaerobic Digestion Enhancement With Microbial Electrolysis Cells:
Is Biomethane Production the Direction to Go for Commercialization?

Ellie Vipond
Teesside University, UK

Pattanathu K.S.M. Rahman
Teesside University, UK

ABSTRACT

The engineering of replacements for crude oil is a priority within industrial biotechnology. Biogas, produced by anaerobic digestion (AD) during organic waste degradation, has been used for electricity generation and heating. Microbial electrolysis cells (MECs) are an emerging technology which when combined with AD can produce higher yields of such energy whilst simultaneously treating waste water and sludge. MECs are bioelectrochemical systems which utilize the metabolism of microbes to oxidize organics. The majority of the research has been focused on biohydrogen production, despite associated issues, which has resulted in poor commercialization prospects. Consequently, scientists are now suggesting that methane production should be the focus of MEC technology. This chapter presents lab research on the bioprocessing of biomethane using AD and MECs and addresses important issues, namely the lack of pilot-scale studies. Downstream processing techniques are discussed, as well as a novel suggestion of further utilising MECs in the purification process.

DOI: 10.4018/978-1-5225-3540-9.ch012

INTRODUCTION

The heavy use of fossil fuels has resulted in a global energy crisis which has provided an incentive for researchers to find sustainable alternatives to meet modern-day society's demands. It has been estimated that current treatment of municipal waste water accounts for approximately 3% of global electricity consumption (Li et al., 2014). Waste water contains significant amounts of energy e.g. domestic waste water contains 7.6 $kJ\ L^{-1}$ (Heidrich & Dolfing, 2011). Harnessing the energy contained in these dissolved organics would mitigate the burdens associated with treatment, whilst simultaneously producing renewable fuels. The use of novel bio-technologies is allowing for this energy capture, which has caused a shift in what was traditionally classified as waste. Anaerobic digestion (AD) is a popular technology used to extract bio-energy in the form of biogas during the microbial conversion of organic substrates (Bharathiraja et al., 2016) e.g. high strength waste water and sludges. Unfortunately, there are several bottlenecks to AD technology application such as slow metabolism of methanogens and accumulation of inhibitory volatile fatty acids (VFAs) (Zhang et al., 2009). This combined with the fact that AD can be energy expensive, makes it apparent that more efficient bio-catalytic routes need to be developed which convert organic waste into green fuels.

Over the past decade, bioelectrochemical systems (BSEs) have been gaining speed as a new generation of technologies which show great potential for waste treatment whilst generating hydrogen or methane. A form of these advancements called Microbial Electrolysis Cells (MECs) has been studied as a hopeful candidate to improve gas energy production. The objective of this chapter is to discuss MEC assisted methane production as well as what needs further investigation if MECs are to achieve commercialization.

Background

MECs utilize exoelectrogenic microbes to oxidize organic waste substrates which transfer the electrons to the anode. The electrons are then transferred to the cathode where they can reduce protons for hydrogen production, or produce methane via several pathways (Lu & Ren, 2016). A small external voltage is required between the electrodes to overcome the thermodynamic barrier, which can be supplied by microbial fuel cells. It is apparent that research into hydrogen production using MECs has been prioritised. Hydrogen makes an ideal fuel due to being carbon free and having high energy content (Zhang & Angelidaki, 2014). However, the production of hydrogen presents numerous challenges that prohibit upscaling and commercialization. This was demonstrated in the largest MEC reactor reported to date. After 43 days, methanogenisis led to H_2 consumption, and consequently methane became the dominant product (Cusick et al., 2011). Methane is relatively safer than hydrogen for storage, and most importantly, its production has been demonstrated to be more robust and consistent (Clauwaert & Verstraete, 2008). Moreover, the incorporation of MECs into traditional AD systems is proving to be an extremely hopeful technology. All of this suggests that methane production should be at the forefront of MEC research.

BIOMETHANE BIOPROCESSING

Microorganisms

Commonly in bioprocessing, analysis and optimization of microbes occurs prior to production. However in this case, molecular techniques are employed afterwards to determine the bacterial dynamics involved. The main species responsible for methane formation belong to *Geobacter* and methanogenic archaea. Figure 1 demonstrates the current consensus on the various catalytic pathways. *Geobacter* sp. adapt to grow with electrodes and have a key role in stimulating electron transfer from organic matters to the anode, giving a higher current density (Yin et al., 2016; Zhao et al., 2015). Generally, CO_2 reduction to methane is processed via hydrogenotrophic methanogens in sequential pathways which first requires the formation of hydrogen (Figure 1) (Bo et al., 2014). It was then discovered that exoelectrogenic methanogens have the ability to produce methane from CO_2 reduction by using the cathode as a direct electron donor (electromethanogenesis) (Cheng et al., 2009). Albeit, current density readings of a MEC reactor indicated that the methane production via this cathodic reduction only accounted for 6.3% of the increment (Zhao et al., 2016). The enhancement of sludge decomposition and methane production in this work was shown to be a result of direct interspecies electron transfer (DIET). This is a syntrophic interaction (Figure 1) recently observed between *Geobacter* sp. (*G. metallireducens*) and methanogens (*Methanosaeta* and *Methanosarcina* sp.), which are highly abundant in AD systems. These interactions provide efficiency in methane production as *Methanosarcina* can utilize two pathways in AD-MEC reactors: hydrogen interspecies transfer (HIT) and DIET (Yin et al., 2016). In addition, DIET in the presence of conductive material did not require electrically conductive pili or outer surface C-type cytochromes (Zhao et al., 2016). This offers an ecological advantage for users as investments in DIET can be reduced.

Figure 1. The steps involved in the sequential break-down of organic waste by bacteria in anaerobic digestion. Acetate (acetic acid) is central to the process as it directly feeds into methanogenisis. Abbreviations in Figure 1: VFA (Volatile fatty acids) DIET (Direct interspecies electron transfer)

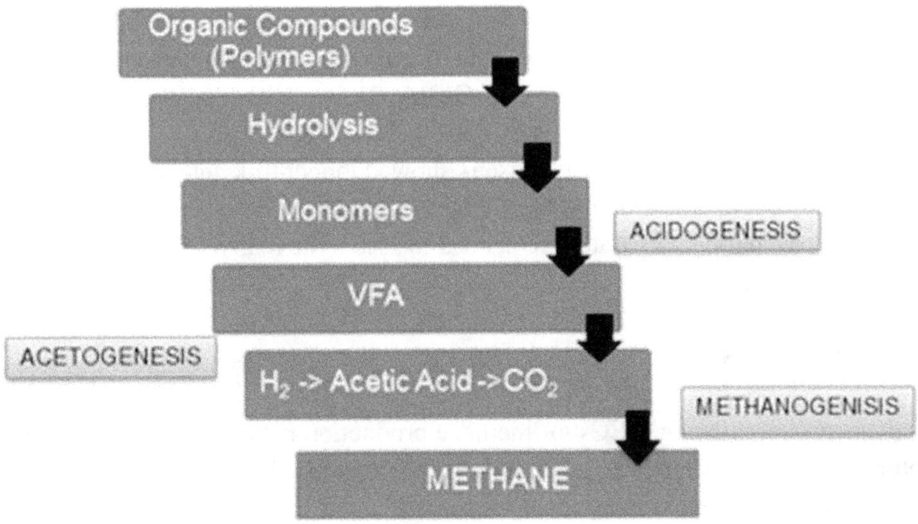

Figure 2. A typical MEC cell with the various pathways of bacterial methane production presented.

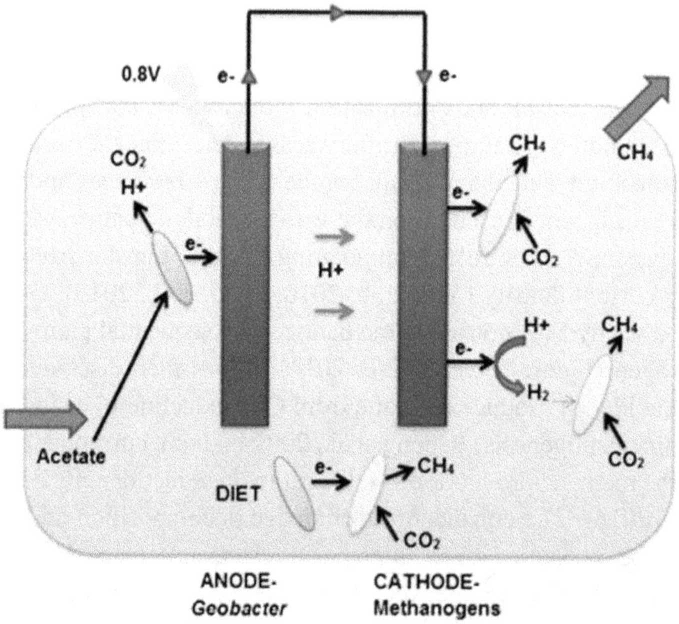

Substrates

Due to its application niche, MEC studies are predominantly based on the use of WASTE WATER or sludge as the substrate. Table 1 presents some typical substrates that have been tested in different reactors under varying parameters. The biodegradability of sludge and waste water can be characterised using the following measurements- total and volatile suspended solids, chemical oxygen demand (COD) and total: carbohydrate, protein and short-chain fatty acids (mg COD/L) (Zhao et al., 2016). Additionally, Figure 1 demonstrates how these are eventually deduced to methane by bacteria. A lot of studies use specified amounts of simple substrate e.g. acetate which allows the researchers to control substrate composition (Xafenias & Mapelli, 2014). Yin et al. (2016) achieved one of the highest methane yields (360.2 mL/g-COD), which was able to compete with the maximum yield in an anaerobic digester (370.0 mL/g-COD), by feeding with acetate for 3 months. Gajaraj, Huang et al. (2017) compared glucose (1000 mg/L) and waste activated sludge (WAS) as substrates in batch-digestion experiments. They found that when sludge was used as the substrate, the digesters showed longer time intervals for volatile fatty acid (VFA) production and degradation, due to the sludge complexity. These studies highlight the fact that studies utilising unconventional substrates are needed to assess the real-world potential of this technology and improve MEC performance.

Operating Parameters

One of the advantages of utilising MECs for methane production is the process which occurs at ambient temperatures (Table 1), thereby saving energy. Heidrich et al. (2014) found their large-scale MEC functioned to produce hydrogen in North East England at water temperature 1-5 °C. However, metha-

Table 1. Production rates of methane achieved with various microbial electrolysis cell (MEC) reactor designs and operating parameters

MEC Configuration	Substrate	Voltage (V)	Temperature (°C)	Production Rate	Source
Single chamber	Synthetic Media Waste water	1.0 1.0	21 ± 2 21 ± 2	8.0 ± 0.2 mL-CH_4 Lr^{-1} d^{-1} 1.4 ± 0.3 mL-CH_4 Lr^{-1} d^{-1}	(Moreno et al., 2016)
Two chambers	Acetate	0.2	24 ± 2	0.28 L/L d	(Villano et al., 2013)
Single Chamber	Waste activated sludge	0.6	35 ± 1	0.020 ± 0.002 L CH4 L^{-1} day^{-1}	(Gajaraj et al., 2017)
Two chambers	Acetate	0.5	24	0.018 L L^{-1} d^{-1}	(Villano et al., 2011)
Single chamber	Acetate	1.0	25 ± 2	360.2 mL/g-COD	(Yin et al., 2016)
Single chamber	Raw sludge	0.3	35	170.2 L/kg-VSS	(Feng et al., 2015)
Single Chamber	Acetate	0.8	22	0.17-0.75 L L^{-1} d^{-1}	See (Villano et al., 2011)
Single Chamber	Acetate	0.9	30	0.12 L L^{-1} d^{-1}	(Rader, & Logan, 2010)

nogenic wastewater treatment systems are limited by low temperatures (Bowen et al., 2014). One theory to overcome this is to use a seed from low temperature environments or use the autochthonous bacteria present (Heidrich et al., 2014; Jadhav, & Ghangrekar, 2009).

Changes in pH can occur throughout the AD procedure, typically associated with the accumulation of VFAs in acetogenesis. There are no major reports of pH controls being set in place for MEC reactors as small fluctuations are usually short-lived if the reactor design is functional. Gajaraj et al. (2017) found there was a decrease in pH (8.0 to 7.1) in all digesters during the first 24 h of AD, which recovered due to the rapid consumption of VFAs.

Applied voltage affects the activities of microorganisms in MEC and hence impacts methane generation. The right voltage application improves MEC performance compared to the presence of conductive material alone. Zhao et al. (2016) showed that carbon felt could improve methanogenisis as compared to control however, adding a small voltage (0.6 V) improved this further (32%). Ding et al. (2016) suggested the optimum voltage to operate a MEC to treat waste water is 0.8 V. They reported excessive voltages (1.0 and 2.0 V) decreased COD removal efficiency and methane yield due to higher plasmatorrhexis and lower growth and metabolism.

Electrodes, Membranes and Reactor Configurations

For upscaling the above should allow high conductivity leading to efficiency and controlled internal resistance. They should possess biological and physio/chemical stability e.g. with stand corrosion. In addition they must be economically viable which involves using less expensive materials (Escapa et al., 2016). Carbon-based materials e.g. carbon felt and graphite granules meet many of these requirements and this has made them a popular electrode material (Zhao et al., 2016; Villano et al., 2011). Microbial biocathodes have received attention as cheaper, self-regenerating and sustainable alternative to abiotic cathodes. The bacteria here act as catalytic agents, functioning via aforementioned pathways. Using a hydrogenotrophic methanogenic culture as the biocathode has proven to reduce CO_2 to methane, at coulombic efficiencies exceeding 80% (Villano et al., 2010).

Downstream Processing

Most lab scale studies report collection of methane through gas bags. This can then be analysed by using gas chromatography techniques e.g. two-channel gas chromatographer (Xafenias, & Mapelli, 2014). Although not always needed, the techniques for biomethane purification produced in this way will be homologous to those used for biogas purification. Removal of water can be achieved through physical (condensation) or chemical (adsorption, absorption) drying. For CO_2 removal absorption with water or polyethylene glycol, PSA (e.g. carbon molecular sieves) and membrane technology are in operation on a large-scale (Ryckebosch et al., 2011). Recently, swollen polyamide thin-film composite membranes were proposed for effective CO_2/CH_4 separation, based on the higher solubility of CO_2 in water as compared to that of methane (Simcik et al., 2016). The type of technique that is implemented must be an economically relevant choice that maintains methane yield. In Sweden, water scrubbers are used mostly where as in the Netherlands they also use PSA-units and membrane technology (Ryckebosch et al., 2011). Figure 3 depicts a process flow diagram of the upstream and downstream processes involved in the production of biomethane.

Other researchers have used continuous modes of operation. A higher HRT with continuous mode resulted in a better chemical oxygen demand removal. Despite this, a significant amount of the acetate consumed was not converted to electricity. This was most likely due to the presence of acetoclastic microorganisms that consume acetate for non-electrogenic methane production (Moreno et al., 2016). Other researchers have used semi-batch mode in a two chamber MEC where retention times can be up to 72 days (Villano et al., 2013).

Figure 3. A process flow diagram for the production of biomethane in batch mode using microbial electrolysis cell (MEC) and anaerobic digestion (AD) technology. Abbreviations:HRT (Hydraulic retention time), WAS (Waste activated sludge, AD (Anaerobic Digestion), MEC (Microbial Electrolysis Cell).

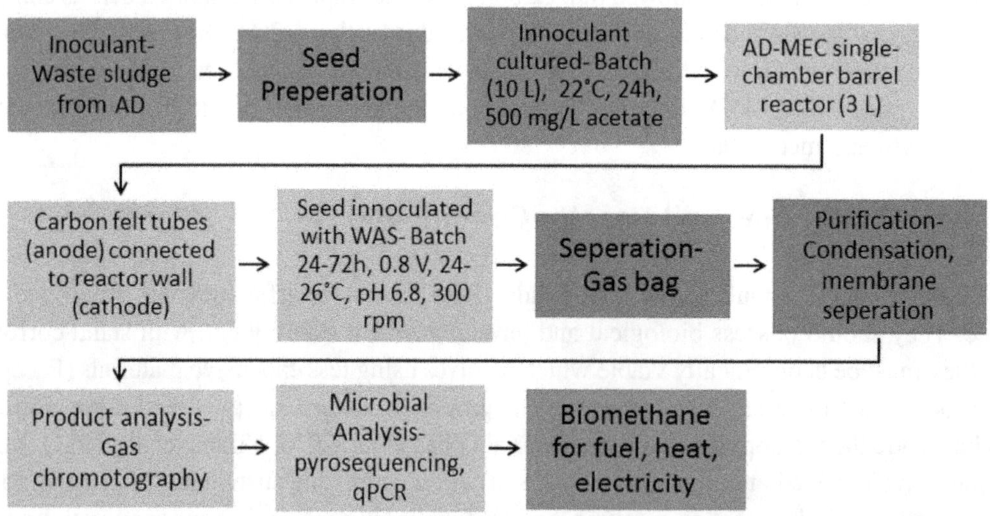

SOLUTIONS AND RECOMMENDATIONS

Several reactor designs have been put forward each bearing strengths and limitations, as previously discussed. Another design, the MEC coupled baffled reactor, represents an innovative configuration which can produce both hydrogen and methane. Ran et al., (2014) showed the level of efficiency this system can achieve. In the first compartment 20.7% proportion of hydrogen was generated. The anaerobic intermediate metabolites formed the substrates for the last three continuous MECs, which achieved up to 98.8% proportion of methane and a 98.8% COD removal rate.

Furthermore, originally in MEC design the anode and cathode were separated usually using an ion exchange membrane (IEM). This offered advantages such as protecting the methanogenic consortia against inhibitory compounds (Villano et al., 2010). However, IEMs can also increase internal resistance by reducing charge circulation (Escapa et al, 2016) and DIET. More recent studies are moving towards designs that couple AD and MEC in single-chamber (Yin et al, 2016). These membraneless reactors are more economically feasible and have shown successful results e.g. CH_4 content >98% was achieved when this design was used (Bo et al, 2014).

The biggest limitation of this technology is the lack of pilot-scale studies or research into "real-world" scenarios. At present, most lab studies are conducted using acetate in a 10–100 mL reactor with many papers not stating the duration of experiments. Clearly research has a long way to go before the MEC can meet its potential as a full-scale waste treatment/methane producing technology. This will involve treating complex waste with indigenous populations, operating throughout yearly temperatures and working at industrial scales (Heidrich et al, 2014). To also bridge the gap between lab-scale and commercial development, improved data reporting and standardised methods of characterising MECs are required (Escapa et al., 2016).

FUTURE RESEARCH DIRECTIONS

One of the main challenges involves gaining a better understanding of the syntrophic and competitive interactions amongst specified microbial groups. This will allow methods for promoting beneficial interactions or minimizing energy losses to be developed for system scale up. Research is expanding in the area of synthetic microbial consortia (SMC) for bioprocessing, which is particularly interested in biofilms (Shong et al., 2012). The authors therefore suggest that MEC and SMC technologies should be brought together. The reasons being, MECs operate on the basis of biofilms and synthetic biology research involves developing metabolic models that can predict community behaviour and synergy between organisms.

In terms of downstream processing, a method of biogas purification known as biological methane enrichment (BME) is not yet operated on a large scale. Strevett et al. (1995) investigated the mechanism and kinetics of this 'chemo-autotrophic biogas upgrading'. Different methanogens, using only CO_2 as a carbon source and H_2 as an energy source, were examined. When a synthetic biogas of ~50% CH_4, ~40% CO_2 and 1-2% H_2S was mixed with H_2 and fed to the organisms, it was found the biological system could effectively remove CO_2, while approximately doubling the original CH_4 mass (Strevett et al., 1995).

This was due to methanogens exhibiting methanogenisis and conversion of CO_2 by pathways already detailed. There are noticeable similarities between this, as a purification technique, and the production of methane in MECs. It is therefore justifiable to suggest that MEC technology could also be used as a purification strategy to purify methane produced in AD-MEC reactors. As usual, the voltage would be used to promote oxidation of the electrode by exoelectrogens. This would then promote the formation of methane by hydrogenotrophic methanogens, which would utilise the H_2 energy source, as well as the direct reduction of the CO_2 carbon source. Hence in theory with a real biogas mixture this would remove the CO_2 contaminants whilst increasing the CH_4 yield further, as observed in BME.

CONCLUSION

BESs have shown promise as being a robust, economically viable and environmentally friendly way to produce renewable energy. Methane-producing MECs should be recognised as suitable alternatives to hydrogen-producing MECs. Designs which combine both methane and hydrogen production are also an option as seen in the baffled reactor and BME. Notwithstanding, there is much more research to be done if this technology is to be taken seriously as a suitable way of producing replacements for fossil fuel derivatives. Before commercialization can occur, large scale pilot studies must be carried out to confirm real-world applications as well as further molecular analysis of the complex microbial interactions involved. MECs are not likely to replace conventional AD, due to being more favorable for low-strength waste streams. Despite this AD and MECs, which were once considered competing technologies, are showing great promise when combined. In addition, this is a relatively new technology and in future years it is likely MECs will be an additional tool for scientists faced with tackling the energy crisis.

REFERENCES

Bharathiraja, B., Sudharsanaa, T., Bharghavi, A., Jayamuthunagai, J., & Praveenkumar, R. (2016). Biohydrogen and Biogas – An overview on feedstocks and enhancement process. *Fuel*, *185*, 810–828. doi:10.1016/j.fuel.2016.08.030

Bo, T., Zhu, X., Zhang, L., Tao, Y., He, X., Li, D., & Yan, Z. (2014). A new upgraded biogas production process: Coupling microbial electrolysis cell and anaerobic digestion in single-chamber, barrel-shape stainless steel reactor. *Electrochemistry Communications*, *45*, 67–70. doi:10.1016/j.elecom.2014.05.026

Bowen, E., Dolfing, J., Davenport, R., Read, F., & Curtis, T. (2014). Low-temperature limitation of bioreactor sludge in anaerobic treatment of domestic wastewater. *Water Science and Technology*, *69*(5), 1004. doi:10.2166/wst.2013.821 PMID:24622549

Cheng, S., Xing, D., Call, D., & Logan, B. (2009). Direct Biological Conversion of Electrical Current into Methane by Electromethanogenesis. *Environmental Science & Technology*, *43*(10), 3953–3958. doi:10.1021/es803531g PMID:19544913

Clauwaert, P., & Verstraete, W. (2008). Methanogenesis in membraneless microbial electrolysis cells. *Applied Microbiology and Biotechnology*, *82*(5), 829–836. doi:10.100700253-008-1796-4 PMID:19050859

Cusick, R., Bryan, B., Parker, D., Merrill, M., Mehanna, M., Kiely, P., ... Logan, B. E. (2011). Performance of a pilot-scale continuous flow microbial electrolysis cell fed winery wastewater. *Applied Microbiology and Biotechnology*, *89*(6), 2053–2063. doi:10.100700253-011-3130-9 PMID:21305277

Ding, A., Yang, Y., Sun, G., & Wu, D. (2016). Impact of applied voltage on methane generation and microbial activities in an anaerobic microbial electrolysis cell (MEC). *Chemical Engineering Journal*, *283*, 260–265. doi:10.1016/j.cej.2015.07.054

Du, Z., Li, H., & Gu, T. (2007). A state of the art review on microbial fuel cells: A promising technology for wastewater treatment and bioenergy. *Biotechnology Advances*, *25*(5), 464–482. doi:10.1016/j.biotechadv.2007.05.004 PMID:17582720

Escapa, A., Mateos, R., Martínez, E., & Blanes, J. (2016). Microbial electrolysis cells: An emerging technology for wastewater treatment and energy recovery. From laboratory to pilot plant and beyond. *Renewable & Sustainable Energy Reviews*, *55*, 942–956. doi:10.1016/j.rser.2015.11.029

Feng, Y., Zhang, Y., Chen, S., & Quan, X. (2015). Enhanced production of methane from waste activated sludge by the combination of high-solid anaerobic digestion and microbial electrolysis cell with iron–graphite electrode. *Chemical Engineering Journal*, *259*, 787–794. doi:10.1016/j.cej.2014.08.048

Gajaraj, S., Huang, Y., Zheng, P., & Hu, Z. (2017). Methane production improvement and associated methanogenic assemblages in bioelectrochemically assisted anaerobic digestion. *Biochemical Engineering Journal*, *117*, 105–112. doi:10.1016/j.bej.2016.11.003

Heidrich, E., Curtis, T., & Dolfing, J. (2011). Determination of the Internal Chemical Energy of Wastewater. *Environmental Science & Technology*, *45*(2), 827–832. doi:10.1021/es103058w PMID:21142001

Heidrich, E., Edwards, S., Dolfing, J., Cotterill, S., & Curtis, T. (2014). Performance of a pilot scale microbial electrolysis cell fed on domestic wastewater at ambient temperatures for a 12month period. *Bioresource Technology*, *173*, 87–95. doi:10.1016/j.biortech.2014.09.083 PMID:25285764

Jadhav, G., & Ghangrekar, M. (2009). Performance of microbial fuel cell subjected to variation in pH, temperature, external load and substrate concentration. *Bioresource Technology*, *100*(2), 717–723. doi:10.1016/j.biortech.2008.07.041 PMID:18768312

Li, X., Liang, D., Bai, Y., Fan, Y., & Hou, H. (2014). Enhanced H2 production from corn stalk by integrating dark fermentation and single chamber microbial electrolysis cells with double anode arrangement. *International Journal of Hydrogen Energy*, *39*(17), 8977–8982. doi:10.1016/j.ijhydene.2014.03.065

Lu, L., & Ren, Z. (2016). Microbial electrolysis cells for waste biorefinery: A state of the art review. *Bioresource Technology*, *215*, 254–264. doi:10.1016/j.biortech.2016.03.034 PMID:27020129

Moreno, R., San-Martín, M., Escapa, A., & Morán, A. (2016). Domestic wastewater treatment in parallel with methane production in a microbial electrolysis cell. *Renewable Energy*, *93*, 442–448. doi:10.1016/j.renene.2016.02.083

Rader, G., & Logan, B. (2010). Multi-electrode continuous flow microbial electrolysis cell for biogas production from acetate. *International Journal of Hydrogen Energy*, *35*(17), 8848–8854. doi:10.1016/j.ijhydene.2010.06.033

Ran, Z., Gefu, Z., Kumar, J., Chaoxiang, L., Xu, H., & Lin, L. (2014). Hydrogen and methane production in a bio-electrochemical system assisted anaerobic baffled reactor. *International Journal of Hydrogen Energy, 39*(25), 13498–13504. doi:10.1016/j.ijhydene.2014.02.086

Ryckebosch, E., Drouillon, M., & Vervaeren, H. (2011). Techniques for transformation of biogas to biomethane. *Biomass and Bioenergy, 35*(5), 1633–1645. doi:10.1016/j.biombioe.2011.02.033

Shong, J., Jimenez Diaz, M., & Collins, C. (2012). Towards synthetic microbial consortia for bioprocessing. *Current Opinion in Biotechnology, 23*(5), 798–802. doi:10.1016/j.copbio.2012.02.001 PMID:22387100

Simcik, M., Ruzicka, M., Karaszova, M., Sedlakova, Z., Vejrazka, J., Vesely, M., ... Izak, P. (2016). Polyamide thin-film composite membranes for potential raw biogas purification: Experiments and modeling. *Separation and Purification Technology, 167*, 163–173. doi:10.1016/j.seppur.2016.05.008

Strevett, K., Vieth, R., & Grasso, D. (1995). Chemo-autotrophic biogas purification for methane enrichment: Mechanism and kinetics. *The Chemical Engineering Journal and the Biochemical Engineering Journal, 58*(1), 71–79. doi:10.1016/0923-0467(95)06095-2

Villano, M., Aulenta, F., Beccari, M., & Majone, M. (2010). Microbial generation of H_2 or CH_4 coupled to wastewater treatment in bioelectrochemical systems. *Chemical Engineering Transactions, 20*, 163–168. doi:10.3303/CET1020028

Villano, M., Monaco, G., Aulenta, F., & Majone, M. (2011). Electrochemically assisted methane production in a biofilm reactor. *Journal of Power Sources, 196*(22), 9467–9472. doi:10.1016/j.jpowsour.2011.07.016

Villano, M., Scardala, S., Aulenta, F., & Majone, M. (2013). Carbon and nitrogen removal and enhanced methane production in a microbial electrolysis cell. *Bioresource Technology, 130*, 366–371. doi:10.1016/j.biortech.2012.11.080 PMID:23313682

Xafenias, N., & Mapelli, V. (2014). Performance and bacterial enrichment of bioelectrochemical systems during methane and acetate production. *International Journal of Hydrogen Energy, 39*(36), 21864–21875. doi:10.1016/j.ijhydene.2014.05.038

Yin, Q., Zhu, X., Zhan, G., Bo, T., Yang, Y., Tao, Y., ... Yan, Z. (2016). Enhanced methane production in an anaerobic digestion and microbial electrolysis cell coupled system with co-cultivation of *Geobacte*r and *Methanosarcina. Journal of Environmental Sciences (China), 42*, 210–214. doi:10.1016/j.jes.2015.07.006 PMID:27090713

Zhang, P., Chen, Y., & Zhou, Q. (2009). Waste activated sludge hydrolysis and short-chain fatty acids accumulation under mesophilic and thermophilic conditions: Effect of pH. *Water Research, 43*(15), 3735–3742. doi:10.1016/j.watres.2009.05.036 PMID:19555988

Zhang, Y., & Angelidaki, I. (2014). Microbial electrolysis cells turning to be versatile technology: Recent advances and future challenges. *Water Research, 56*, 11–25. doi:10.1016/j.watres.2014.02.031 PMID:24631941

Zhao, Z., Zhang, Y., Quan, X., & Zhao, H. (2016). Evaluation on direct interspecies electron transfer in anaerobic sludge digestion of microbial electrolysis cell. *Bioresource Technology, 200*, 235–244. doi:10.1016/j.biortech.2015.10.021 PMID:26492177

Zhao, Z., Zhang, Y., Wang, L., & Quan, X. (2015). Potential for direct interspecies electron transfer in an electric-anaerobic system to increase methane production from sludge digestion. *Scientific Reports*, *5*(1), 11094. doi:10.1038rep11094 PMID:26057581

KEY TERMS AND DEFINITIONS

Anaerobic Digestion: A sequence of biological processes in which microorganisms break down organic material under hypoxic conditions. One of the metabolic end products is biogas.

Biocathode: In microbial electrolysis cells the cathode is the 'electron sink', meaning it accepts electrons and protons produced from the anodic oxidation of the substrate. A bio-cathode utilises microorganisms for this catalytic purpose.

Bioelectrochemical Systems (BESs): Systems which harness the metabolisms of microbes to convert chemical energy into electrical energy e.g. Microbial Electrolysis Cell and Microbial Fuel Cell.

Direct Interspecies Electron Transfer (DIET): DIET is a syntrophic interaction that occurs in methanogenic communities. Free electrons are transferred directly from one microbial cell to another rather than being shuttled by reduced molecules such as hydrogen.

Downstream Processing: The later stages of bioprocessing namely recovery and purification of biosynthetic products.

Energy Crisis: This is an ongoing global issue that is concerned with the fact that natural resources (fossil fuels), used to power industrial society, are diminishing and the demand for them continues to rise. Suitable renewable replacements must be sought which is the aim of industrial biotechnology.

Exoelectrogen: Bacteria which can transfer electrons extracellularly e.g. *Geobacter* species.

Methanogen: Microorganisms that produce methane as a metabolic by product under anaerobic conditions.

Sludge: Sludge is semi-solid slurry and can be produced as part of numerous industrial processes. For example, sewage sludge refers to the residual material that is produced as a by-product during sewage treatment of wastewater. The term activated sludge refers to suspended sludge containing many active bacteria, which remove biodegradable substances from wastewater or waste activated sludge (WAS). WAS refers to excess sludge produced in an activated sludge system which must be processed in a further treatment chain.

Chapter 13
Biosorption of Heavy Metals:
Biological Approach to Control the Industrial Waste

Neelesh Babu
Gurukul Kangri University, India

Vinay Mohan Pathak
Gurukul Kangri University, India

Akash
Gurukul Kangri University, India

Navneet
Gurukul Kangri University, India

ABSTRACT

Large-scale production of commodities for mankind by industries did huge damage to the environment. Industrial waste contains lots of toxic materials including heavy metals were drained to water bodies like river, lakes, ponds, etc. These effluents drastically ruin water quality as well as the soil fertility. Type of industry and its raw material decides quantity and quality of the emerged wastes including both biodegradable as well as non-biodegradable. Among non-biodegradable wastes, copper, chromium, nickel, cadmium, etc. are widespread contaminants of soil, water, and these are most common heavy metals. Several heavy metals such as cadmium, mercury, and lead are highly poisonous and fatal to human as well as animals. Several plants as well as microbes respond to heavy metals by diverse biological processes like biosorption to their cell wall and entrapment in their capsule, oxidation and reduction, precipitation, complexation, etc. These responses may help significantly in the remediation of heavy metals from the contaminated sites.

DOI: 10.4018/978-1-5225-3540-9.ch013

INTRODUCTION

All over the world ceaseless growth of mankind has resulted into the more demand of commodities or goods which are meant to be life supportive for them. As a result industrialization occurs on large scale to fulfill their requirements by introducing new alternatives (Ahemad & Khan, 2012; Ahemad & Malik, 2011). Large scale production of such commodities by industries did huge damage to the environment. Industrial effluent contains several toxic materials including heavy metals were drained into several water bodies like river, lakes, ponds etc. during their various operations. These effluents drastically ruin not only water quality but also damage the soil fertility as well (Singh *et al.*, 2010; Ahemad & Malik, 2011). The type of industries and their raw materials decides quantity and quality of the emerged wastes including both biodegradable as well as non-biodegradable. Among non-biodegradable wastes copper, chromium, nickel, cadmium etc. are widespread contaminants of soil, water and these are most common heavy metals (Ahemad & Malik, 2011). Several heavy metals such as cadmium, mercury and lead are highly poisonous and fatal to human as well as animals. Ordinarily, some heavy metals are either essential nutrients (typically iron, cobalt and zinc) or relativity harmless such as ruthenium, silver and indium. They serve as micronutrients for prokaryotes as well as eukaryotes in their life cycle, but high concentration of these metals could be toxic (Ahemad, 2012). Several plants as well as microbes respond to heavy metals by diverse biological processes like biosorption to their cell wall and entrapment in their capsule, oxidation and reduction, precipitation, complexation etc. (Lawrence *et al.,* 2010). These responses may help significantly in the remediation of heavy metals from the contaminated sites.

BIOSORPTION

Before the starting we should have to know about the term biosorption. What does it means? What does this term defines? One cannot easily define biosorption because of its complex mechanism. Bio and

Table 1. Some of the heavy metal effluents from industries

S.No.	Metals	Industries
1	Copper	Metal refining, electroplating industry, plastic industry etc.
2	Chromium	Electroplating industry, leather, chrome plating, petroleum refining, tanning, textile.
3	Lead	Automobile batteries, Petrol based materials, pesticides, Paints
4	Nickel	Metal refining, galvanization, paint and powder, batteries processing units, super phosphate fertilizers.
5	Zinc	Rubber industries, paints, dyes, wood preservatives and ointments.
6	Mercury	Thermometers, adhesives, paints, light bulb industry, wood preservatives, leather tanning, ointments producing industry
7	Iron	Metal refining, galvanization engine parts.
8	Arsenic	Semiconductors, petroleum refining, wood preservatives, animal feed additives, coal power plants Automobile exhaust, industrial dust and dyes.

(Lone *et al.,* 2008)

sorption collectively form the term biosorption in which prefix bio used for the biological entities which are been involved in the process. It could be either plants, fungus, bacteria or their products. Suffix sorption denotes adsorption as well as absorption. Adsorption means physical adherence of the ions on the surface of any substance whereas absorption is the state of being absorbed in some material. Adsorption results the formation of stable bonding between the adsorbate and adsorbent which can be described as surface complex. Reason behind the surface complexation is the presence of functional groups on to the adsorbent (fungus, bacteria etc.) and adsorbate. Most of the time, these groups are sulfhydryl, carboxylic, hydroxyl etc. These groups are having tendency to donate an electron to accepter compounds (metallic ions which acts as Lewis acid) and form stable bonding between them (Gadd, 2009). Therefore, one can conclude that biological entities like bacteria, fungi, algae, plants which are having these tendencies are able detoxify heavy metal impurities. Basically biosorption is the removal of materials from the solution by non- living biological entities. Materials could be organic as well as inorganic. Although, living biomass also able to remove such kinds of impurities by the process of bioaccumulation. Biosorption and bioaccumulation are actually different, over bioaccumulation, biosorption is more advantageous. Unlike bioaccumulation there is no need of growth medium for the biosorption as the materials are available in the form of byproduct or waste (Farooq *et al.,* 2011).

Biosorbents

Biological substances have affinity towards the metals, among them selection of appropriate biosorbent is more important. These sorbents are judged according to their ability to attract sorbate and retain in immobilized form. Availability and economy of sorbent material is another important factor for sorbent material selection, whether they can be recovered or not. Biosorbents recovery enhances the economy because of their reuse in several biosorption cycles. Due to this, heavy metals can be easily recovered and effluent will be metal free (Kratochivi & Volesky, 1998). There are number of biosorbent materials have been investigated for their ability to bind metals. Kratochivi and Volesky (1998) demonstrated about the better sorbent material by comparison of their sorption performance using sorption isotherms which includes two classical models Langmuir model and Freundlich model.

$$q = q_{max}\frac{bCf}{1+bCf} \text{ (Langmuir equation)}$$

$$q = KC_f^{1/n} \text{ (Freundlich equation)}$$

Considering low cost biosorbents, natural materials could be the best option because these are available in large quantities and are environmentally friendly (Deans & Dixon, 1998). Different biomaterials as biosorbents used by several researchers like *Azadirachata* biomass, bark, potato peel waste, husk, Bengal grams, *Cicer arientium* citrus pectin (Ahalya *et al.,* 2005; King *et al.,* 2007; Arshad *et al.,* 2008; Balaria & Schiewer, 2008). Beside the plant material several microbes are also have potential of biosorption. *Bacillus, Pseudomonas* and *Streptomyces* are the potent genera under the class bacteria for biosorption as several functional groups are present on the cell wall of the bacteria. These functional groups are helpful in the binding of metals by chelation, electrostatic interactions etc (Kang *et al.,* 2007; Vijayaraghavan & Yun, 2008). Similarly fungi and algae are also known to be economic because they are efficient and cheap.

Microorganisms as Biosorbents

Microbes have the ability of transformation of metals and metalloids by means of oxidation, reduction, methylation and dealkylation. Immobilized biomass in the form of biofilm may provide additional capacity for the removal of several pollutants such as nitrates, sulphates, hydrocarbons and pesticides. Other than biofilm living or dead microbial biomass can be immobilized by the support of ethyle glycol, agar, cellulose, silica gel and can be used in their natural state (Gadd & White, 1993). Several fungi potentially assimilate the complex carbohydrates by their extracellular enzymes for former hydrolysis which makes capable degradation of several degree of pollutants and their biomass can be easily separated (Leitao, 2009; Li *et al.,* 2015).

Bacteria

As we know that bacteria is a major group of microbes and are ubiquitous. They could be excellent biosorbent because of their high surface to volume ratio and potentially active functional groups on their cell wall (Beveredge, 1981). Due to ubiquity, small size, resistance against several environmental conditions and their growth under controlled conditions they were used as a potent biosorbents in past (Urrutia, 1997; Eman, 2012; Kinoshita *et al.,*2013). Later their biomass which is industrial waste product extensively utilized as biosorbent (Sag & Kutsal, 2001). Many bacterial species includes *Pseudomonas, Streptomyces, Bacillus, Micrococcus* etc. were tested for the uptake of metal and their metal uptake capacity ranges from 568 to .070 mg g^{-1} (Wang and Chen, 2009). Among several functional groups (phosphate, carboxyl, amine and hydroxyl) present on the cell wall of bacteria, carboxyl groups are most abundantly available which binds to the metals cations because of their anionic nature. Beside carboxylic groups, amine groups are also available on to the cell wall and potentially participate in adsorption of anionic metal species by electrostatic interaction and also chelate cationic metal ions (Kang *et al.,* 2007; Vijayaraghavan & Yun, 2008). Bacteria are classified into two groups Gram positive and Gram negative according to their cell wall composition. Gram positive bacteria constitute of peptidoglycan surface and have techoic acid which appears too extended to the peptidoglycan surface. It also contains phosphoryl group of secondary polymers and the carboxyl groups of the peptide chain. Because of these Gram positive bacterial cell wall is anionic as they are negatively charged species which helps to bind cationic metal species (Prescott *et al.,* 2002). Other than Gram positive bacteria, Gram negative bacteria have lipopolysaccharide which also provides anionic character. Therefore bacterial cell walls are potent chelating agent. Heavy metal binding on to the surface of bacterial cell wall is multiple step process. First step involves interaction between the metals and ions and the second one is deposition of heavy metal species onto the cell wall (Vasudevan *et al.,* 2001). Carboxyl groups which are present on the cell wall are main site of metal deposition (Wang & Chen, 2009). Seki *et al.,* 1998, reported that heavy metals (Cd and Pb) can be recovered by *Rhodobacter sphaeroides* and *Alkaligenes eutropus* H16 from aqueous environment. According to Ilhem *et al.,* (2004), biosoroption of Cr, Pb and Cu can be done by *Sthaphylococcus saprophytics* at pH 2.0, 4.5 and 3.5. In another study different isolates of bacteria were able to tolerate 50 to 200 mg/l of Cd, Cr, Pb and Cu (Mustapha & Halimoon, 2015). Syed and Chinthala, (2015) concluded that the isolates, *B. licheniformis* NSPA5, *B. cereus* NSPA8, and *B. subtilis* NSPA13, exhibited maximumbiosorption of lead fromtested heavy metals. Among all the isolated strains in their study, *B. cereus* NSPA8 has showed maximum biosorption of lead (87%), followed by *B. subtilis* NSPA13 (85%) and *B. cereus* NSPA8 (78%).

Fungi

Biosorption by fungi has many advantages as it can be easily cultivated via simple fermentation techniques in cheap media. Media could be the industrial waste product which provides economic benefit to the process. Use of fungi is safe because fungi which are being used are non pathogenic and can be easily grown on large scale as it has sort multiplication cycle which leads to give high yield biomass. 30% of dry weight of fungi is its cell wall constitutes of mostly polysaccharides which could be excellent biosorbents (Horikoshi *et al.,* 1981). These are actually large amount of metal binding polymers including chitin, chitoson, mannan and glucan having several metal binding ligands which includes carboxyl, phosphate, hydroxyl and amine groups providing huge potential for metal binding (Tobin *et al.,* 1984). Many fungal strains were used as biosorbents for different heavy metals like Cd, Cu, Cr, Pb, Hg etc. Stokes and Lindsay, 1979 found that *Penicillium ochrochloron* can uptake copper up to 4×10^5µg/dry weight of biomass at pH 2.8. Similarly, cultural experiment on *Penicillium chrysogenum* was done by using radium (Tsezos & Keller, 1983). *Rhizopus arrhizus* can uptake uranium and thorium (Tsezos & Volesky, 1981) where as *Agaricus macroporus, Termitomyces alypeates* and *Aspergillus parasiticus* can uptake Hg, Cd, Cu, Cr, and Pb (Gracia *et al.,* 2005; Das & Guha, 2007; Akar *et al.,* 2007). Fungi have the capability to assimilate complex carbohydrates by its extracellular enzymes which makes easy to degrade various pollutants biomass. These studies show that fungi could be safe, economic and reliable biosorbent for the biosorption process. Fawzy *et al.,* (2017) studied Biosorption of heavy metals onto different eco-friendly substrates. In their study they isolated 5 metal tolerated fungal species from different contaminated soils. Among these isolates *Rhizopus stolonifier* and *Macrophomina phaseolina* recorded significant roles as good biosorbent agents for Pb, Cd, Cu and Zn and showed better uptake capacity for Pb, Zn and Cd as compared to Cu. This uptake capacity increased in *M. phaseolina* as compared to *R. stolonifier. Spirulina platensis* has also been reported as an impressive biosorbent for removal of several heavy metals from industrial effluents. *S. Platensis* is a good accumulator and phytoremediation of zinc because biosorption of zinc by this microalgae was estimated as maximum in effluent containing 100 mg of dry biomass for one-liter (0.1 g/L) at pH 8 at 35°C with the optimal contact time of 60 min. (Palaniswamy & Veluchamy, 2017).

Algae

Algae are another promising type of biosorbents which requires high metal uptake (Trinelli *et al.,* 2013; Sulaymon *et al.,* 2012; Flouty, & Estephane, 2012; Khani, 2011). These are available in huge amount in sea and oceans (Kuyicak & Volesky, 1990; Rincon *et al.,* 2005). Although they have not been used as bacteria and fungi but they have significant economic benefits over them because they require low amount nutrients and produce large biomass as compare to bacteria and fungi (Romera *et al.,* 2006). Among three groups of algae namely Green algae (micro or fresh water algae), Brown algae (macro or marine water algae) and Red algae, Brown algae have high capacity of metal uptake (Wang *et al.,*1998). More specifically, its chemical composition of cell wall includes carboxyl, amino, sulfate and sulphonate groups are responsible for high metal uptake (Wilke *et al.,* 2006). Binding of metal ions on the surface of algae is depends upon ionic charge of metal ion, algal species and the composition of metal ions in the solution (Crist *et al.,* 1981). There are number of marine and macro algae being used as biosorbents which includes *Cyclotella cryptica, Lyngbya taylorii, Phaeodactylum tricornutum, Porphyridium purpureum, Scenedesmus quadricauda, Spirulina platensis, Chlamydomonas reinhardtii, Chlorella salina, Chlorococcum* sp, *Stigeoclonium tenue, Ascophyllum nodosum, Cladophora fascicularis, Codium fragile,*

Corallina officinalis, Ecklonia sp., *Padina pavonia, Porphyra columbina, Sargassum asperifolium, Turbinaria conoides* etc.(Brinza *et al.,* 2007). In a study (Raheman & Sathasivan, 2015) reported that the dried biomass of *Kappaphylus* sp. may be used as low cost biosorbent for the removal of heavy metal ions for aquous solution by performing activity on four heavy metals Pb, Cu, Fe and Zn. Marine macroalga *Ulva lactuca* is another important biosorbent.KOH-activated carbon based *U. lactuca* can be used as an effective technology for removal of metal ions from contaminated environment (Ibrahim *et al.*, 2016). *Phaeodactylum tricornutum, Nitzschia* sp., *Skeletonema* sp., and *Chlorella vulgaris* were investigated as good biosorbents Cu and Zn. Sudy reveals that phytoremediation using LEDs and SPM tube containing appropriate microalgae could be utilized as a useful technique for remediating contaminated sediment (Kwon *et al.,* 2016). In another study algal biomass from the stable growth phase works as a potential biosorbent for Pb(II), dry biomass of *Chlorella* Sp. QB-102 was investigated for this study. Results reveals that each growth phase represented highly efficient removal of Pb(II) from water with the highest Pb(II) biosorption capacity of 298.17 mg g_1 at stable phase(Li *et al.,*2017).

Mechanism of Biosorption

Various concept of mechanism of biosorption are being assumed, many of them are brought to industrial level. Mechanism of biosorption may be classified according to the metabolism of cell known as metabolism dependent and according to the location of recovered metal whether metals are accumulated extracellularly or intracellularly precipitated (Davis *et al.,* 2003; Neetu *et al.,* 2015). Biosorption mainly occurs through several interactions such as physical adsorption, ion exchange, complexation, precipitation and entrapment in inner space (Lawrence *et al.,* 2010). Just like metabolically important ions are transported across the cell membrane of microbes in the same way heavy metals are also transported. Aksu *et al.,* 1992 demonstrated that biosorption of Cu by bacterium *Zoogloea ramigera* and algae *Chlorella vulgaris* due to electrostatic interactions.

Complexation

Complexation is the formation of a species by the association of two or more species which involves ligand centers in the organic species i.e. the presence of an atom which donate lone pairs of electron. When one of the species is a metal ion, the resulting entity is known as a metal complex. For example Cadmium will form the following complexes with chloride ions: $CdCl^+$, $CdCl_2^0$, $CdCl_3^-$, $CdCl_4^{2-}$. Hard metals such as Na, K, Ca, Mg, often essential nutrients for microbial growth, they bind preferentially to oxygen-containing "hard" ligands such as OH^-, HPO_4^{2-}, CO_3^{2-}, R-COO^-, and =C=O. (Akshu *et al.,* 1992; Naja & Volesky 2011).

Coordination

A compound is said to be a coordination compound if the central metal atom of a complex is bound to its neighbors by covalent bonds by accepting an electron pair from each non-metal atom, the latter is called the donor and the former the acceptor atom. Chemical coordination of metal species and their stereochemical characteristics play very important role for their attachment on the available ligands by the microbial cell (Weber & Walt, 1979; Naja & Volesky 2011).

Ion Exchange

Gram-positive bacteria have great capacity of metal binding because of their negative charge density. This negative charge density is due to the components of their cell wall teichoic and teichuronic acid which are attached to the peptidoglycan network The phosphodiesters of the teichoic acid and the carboxylate groups of the teichuronic acid contribute ion-exchange capacity to the cell wall. Accumulation evidences of metals shows that the biosorption involves ion exchange to a high degree, some biosorbents may be more selective toward cations (e.g. Cd, Cu, Ni, Pb, Cr3+, etc.) or anions (e.g. As, Se, V, Cr6+, etc.) (Davis *et al.,* 2003).

Chelation

When more than one atom bonded to a central metal ion to form a ring structure known as chelation. If a chelating agent may bond to a metal ion in more than one place simultaneously, chelates are more stable than complexes involving unidentate ligands. Therefore stability tends to increase with the number of chelating sites available on the ligand. A molecule is said to be a chelating agent if possess at least two appropriate functional groups, the donor atoms which are capable of combining with a metal atom by donating a pair of electrons. These electrons may be contributed by basic coordinating groups such as NH or groups functioning as acids by losing a proton (Naja & Volesky 2011; Lawrence *et al.,* 2010).

Metal Recovery

Metal recovery is an important step for the economy of the process. Desorption of heavy metals should be done in such an way that the recovery of metal is maximum and also restore the biosorbent quite similar to original state so that it can be reuse for another biosorption cycle (Alluri *et al.,*2007) For the desorption assay dilute mineral acids as well as organic acids are used with complexing agents like EDTA (Alluri *et al.,* 2007; Kiran *et al.,* 2005). HNO_3, H_2SO_4, citric acid, lactic acid, and acitic acid are frequently used for the process.

FUTURE RESEARCH DIRECTIONS

As lots of difficulties are during the biosorption mechanism people are considering new type of technologies for the better results. These include hybrid technology for metal removal of heavy metals and the synthesis of ion exchange resins (Tsezos, 2001; Volesky, 2007). The researchers from various field are working together to give promising results. A variety of alternatives are demonstrated by several researchers such as application of living cells in spite of dead cells (Mallik, 2004). Researchers are paying attention on the economy of the biosorbents, source and the type of biosorbents plays an important role in the economy of the process. Most economic biosorbents includes several fungi, algae which can be successfully grown for large scale because of their minimal requirement of nutrients.

REFERENCES

Ahalya, N., Kanamadi, R. D., & Ramachandra, T. V. (2005). Biosorption of chromium (VI) from aqueous solutions by the husk of Bengal gram (Cicer arientinum). *Electronic Journal of Biotechnology, 8*(3).

Ahemad, M. (2012). Implications of bacterial resistance against heavy metals in bioremediation: A review. *Journal of Institute of Integrative Omics and Applied Biotechnology*, *3*(3).

Ahemad, M., & Khan, M. S. (2012). A Comparative analysis of the growth parameters of different legume crops grown in quizalafop-p-ethyl applied soils. *Insight Botany*, *2*(1), 1–6. doi:10.5567/BOTANY-IK.2012.1.6

Ahemad, M., & Malik, A. (2011). Bioaccumulation of heavy metals by zinc resistant bacteria isolated from agricultural soils irrigated with wastewater. *Journal of Bacteriology*, *2*(1), 12–21. doi:10.3923/bj.2012.12.21

Ahemad, M., & Malik, A. (2011). *Bioremediation Potentials of Heavy Metal Resistant Bacteria: A Practical Approach*. VDM Publishing.

Akar, T., Tunali, S., & Çabuk, A. (2007). Study on the characterization of lead (II) biosorption by fungus Aspergillus parasiticus. *Applied Biochemistry and Biotechnology*, *136*(3), 389–405. doi:10.100712010-007-9032-8 PMID:17625240

Aksu, Z., Sag, Y., & Kutsal, T. (1992). The biosorpnon of copperod by C. vulgaris and Z. ramigera. *Environmental Technology*, *13*(6), 579–586. doi:10.1080/09593339209385186

Alluri, H. K., Ronda, S. R., Settalluri, V. S., Bondili, J. S., Suryanarayana, V., & Venkateshwar, P. (2007). Biosorption: An eco-friendly alternative for heavy metal removal. *African Journal of Biotechnology*, *6*(25).

Arshad, M., Zafar, M. N., Younis, S., & Nadeem, R. (2008). The use of Neem biomass for the biosorption of zinc from aqueous solutions. *Journal of Hazardous Materials*, *157*(2), 534–540. doi:10.1016/j.jhazmat.2008.01.017 PMID:18289783

Balaria, A., & Schiewer, S. (2008). Assessment of biosorption mechanism for Pb binding by citrus pectin. *Separation and Purification Technology*, *63*(3), 577–581. doi:10.1016/j.seppur.2008.06.023

Beveridge, T. J. (1981). Ultrastructure, chemistry, and function of the bacterial wall. *International Review of Cytology*, *72*, 229–317. doi:10.1016/S0074-7696(08)61198-5 PMID:6166584

Brinza, L., Dring, M. J., & Gavrilescu, M. (2007). Marine micro-and macro-algal species as biosorbents for heavy metals. *Environmental Engineering and Management Journal*, *6*(3), 237–251.

Crist, R. H., Oberholser, K., Shank, N., & Nguyen, M. (1981). Nature of bonding between metallic ions and algal cell walls. *Environmental Science & Technology*, *15*(10), 1212–1217. doi:10.1021/es00092a010

Das, S. K., & Guha, A. K. (2007). Biosorption of chromium by Termitomyces clypeatus. *Colloids and Surfaces. B, Biointerfaces*, *60*(1), 46–54. doi:10.1016/j.colsurfb.2007.05.021 PMID:17618091

Davis, T. A., Volesky, B., & Mucci, A. (2003). A review of the biochemistry of heavy metal biosorption by brown algae. *Water Research*, *37*(18), 4311–4330. doi:10.1016/S0043-1354(03)00293-8 PMID:14511701

Deans, J. R., & Dixon, B. G. (1992). Uptake of Pb2+ and Cu2+ by novel biopolymers. *Water Research, 26*(4), 469–472. doi:10.1016/0043-1354(92)90047-8

Farooq, U., Khan, M. A., Athar, M., & Kozinski, J. A. (2011). Effect of modification of environmentally friendly biosorbent wheat (Triticum aestivum) on the biosorptive removal of cadmium (II) ions from aqueous solution. *Chemical Engineering Journal, 171*(2), 400–410. doi:10.1016/j.cej.2011.03.094

Fawzy, E. M., Abdel-Motaal, F. F., & El-zayat, S. A. (2017). Biosorption of heavy metals onto different eco-friendly substrates. *Journal of Toxicology and Environmental Health Sciences, 9*(5), 35–44.

Flouty, R., & Estephane, G. (2012). Bioaccumulation and biosorption of copper and lead by a unicellular algae Chlamydomonas reinhardtii in single and binary metal systems: A comparative study. *Journal of Environmental Management, 111*, 106–114. doi:10.1016/j.jenvman.2012.06.042 PMID:22835654

Gadd, G. M. (2009). Biosorption: Critical review of scientific rationale, environmental importance and significance for pollution treatment. *Journal of Chemical Technology and Biotechnology (Oxford, Oxfordshire), 84*(1), 13–28. doi:10.1002/jctb.1999

Gadd, G. M., & White, C. (1993). Microbial treatment of metal pollution—a working biotechnology? *Trends in Biotechnology, 11*(8), 353–359. doi:10.1016/0167-7799(93)90158-6 PMID:7764182

Garcia, M. A., Alonso, J., & Melgar, M. J. (2005). Agaricus macrosporus as a potential bioremediation agent for substrates contaminated with heavy metals. *Journal of Chemical Technology and Biotechnology (Oxford, Oxfordshire), 80*(3), 325–330. doi:10.1002/jctb.1203

Gomaa, E. Z. (2012). Production and characteristics of a heavy metals removing bioflocculant produced by Pseudomonas aeruginosa. *Polish Journal of Microbiology, 61*(4), 281–289. PMID:23484411

Ibrahim, W. M., Hassan, A. F., & Azab, Y. A. (2016). Biosorption of toxic heavy metals from aqueous solution by Ulva lactuca activated carbon. *Egyptian Journal of Basic and Applied Sciences, 3*(3), 241–249. doi:10.1016/j.ejbas.2016.07.005

Ilhan, S., Nourbakhsh, M. N., Kiliçarslan, S., & Ozdag, H. (2004). Removal of chromium, lead and copper ions from industrial waste waters by Staphylococcus saprophyticus. *Turkish Electronic Journal of Biotechnology, 2*(2), 50–57.

Kang, S. Y., Lee, J. U., & Kim, K. W. (2007). Biosorption of Cr (III) and Cr (VI) onto the cell surface of Pseudomonas aeruginosa. *Biochemical Engineering Journal, 36*(1), 54–58. doi:10.1016/j.bej.2006.06.005

Khani, M. H. (2011). Uranium biosorption by Padina sp. algae biomass: Kinetics and thermodynamics. *Environmental Science and Pollution Research International, 18*(9), 1593–1605. doi:10.100711356-011-0518-0 PMID:21607729

King, P., Anuradha, K., Lahari, S. B., Kumar, Y. P., & Prasad, V. S. R. K. (2008). Biosorption of zinc from aqueous solution using Azadirachta indica bark: Equilibrium and kinetic studies. *Journal of Hazardous Materials, 152*(1), 324–329. doi:10.1016/j.jhazmat.2007.06.101 PMID:17681426

Kinoshita, H., Sohma, Y., Ohtake, F., Ishida, M., Kawai, Y., Kitazawa, H., ... Kimura, K. (2013). Biosorption of heavy metals by lactic acid bacteria and identification of mercury binding protein. *Research in Microbiology, 164*(7), 701–709. doi:10.1016/j.resmic.2013.04.004 PMID:23603782

Kiran, I., Akar, T., & Tunali, S. (2005). Biosorption of Pb (II) and Cu (II) from aqueous solutions by pretreated biomass of Neurospora crassa. *Process Biochemistry*, *40*(11), 3550–3558. doi:10.1016/j.procbio.2005.03.051

Kratochvil, D., & Volesky, B. (1998). Advances in the biosorption of heavy metals. *Trends in Biotechnology*, *16*(7), 291–300. doi:10.1016/S0167-7799(98)01218-9

Kuyicak, N., & Volesky, B. (1990). Biosorption by fungal biomass. In B. Volesky (Ed.), *Biosorption of heavy metals* (pp. 173–198). CRC Press.

Kwon, H. K., Jeon, J. Y., & Oh, S. J. (2017). Potential for heavy metal (copper and zinc) removal from contaminated marine sediments using microalgae and light emitting diodes. *Ocean Science Journal*, *52*(1), 57–66. doi:10.100712601-017-0001-z

Leitão, A. L. (2009). Potential of Penicillium species in the bioremediation field. *International Journal of Environmental Research and Public Health*, *6*(4), 1393–1417. doi:10.3390/ijerph6041393 PMID:19440525

Li, X., Li, A., Long, M., & Tian, X. (2015). Equilibrium and kinetic studies of copper biosorption by dead Ceriporia lacerata biomass isolated from the litter of an invasive plant in China. *Journal of Environmental Health Science & Engineering*, *13*(1), 37. doi:10.118640201-015-0191-1 PMID:25954509

Li, Y., Xia, L., Huang, R., Xia, C., & Song, S. (2017). Algal biomass from the stable growth phase as a potential biosorbent for Pb (II) removal from water. *RSC Advances*, *7*(55), 34600–34608. doi:10.1039/C7RA06749F

Lone, M. I., He, Z. L., Stoffella, P. J., & Yang, X. E. (2008). Phytoremediation of heavy metal polluted soils and water: Progresses and perspectives. *Journal of Zhejiang University. Science. B.*, *9*(3), 210–220. doi:10.1631/jzus.B0710633 PMID:18357623

Malik, A. (2004). Metal bioremediation through growing cells. *Environment International*, *30*(2), 261–278. doi:10.1016/j.envint.2003.08.001 PMID:14749114

Mustapha, M. U., & Halimoon, N. (2015). Screening and isolation of heavy metal tolerant bacteria in industrial effluent. *Procedia Environmental Sciences*, *30*, 33–37. doi:10.1016/j.proenv.2015.10.006

Naja, G., & Volesky, B. (2011). The mechanism of metal cation and anion biosorption. In *Microbial biosorption of metals* (pp. 19–58). Springer Netherlands. doi:10.1007/978-94-007-0443-5_3

Neethu, C. S., Mujeeb Rahiman, K. M., Saramma, A. V., & Mohamed Hatha, A. A. (2015). Heavy-metal resistance in Gram-negative bacteria isolated from Kongsfjord, Arctic. *Canadian Journal of Microbiology*, *61*(6), 429–435. doi:10.1139/cjm-2014-0803 PMID:25942102

Neu, T. R., & Lawrence, J. R. (2009). Extracellular polymeric substances in microbial biofilms. In Microbial Glycobiology (pp. 733-758). Academic Press.

Palaniswamy, R., & Veluchamy, C. (2017). *Removal of chromium from effluent by biosorption using spirulina platensis*. Academic Press.

Prescott, M., Harley, J. P., & Klein, D. A. (2002). *Industrial Microbiology and Biotechnology in General Microbiology*. Academic Press.

Rahman, M. S., & Sathasivam, K. V. (2015). Heavy metal adsorption onto kappaphycus sp. from aqueous solutions: The use of error functions for validation of isotherm and kinetics models. *BioMed Research International*. PMID:26295032

Rincon, J., Gonzalez, F., Ballester, A., Blazquez, M. L., & Munoz, J. A. (2005). Biosorption of heavy metals by chemically-activated alga Fucus vesiculosus. *Journal of Chemical Technology and Biotechnology (Oxford, Oxfordshire)*, *80*(12), 1403–1407. doi:10.1002/jctb.1342

Romera, E., Gonzalez, F., Ballester, A., Blazquez, M. L., & Munoz, J. A. (2006). Biosorption with algae: A statistical review. *Critical Reviews in Biotechnology*, *26*(4), 223–235. doi:10.1080/07388550600972153 PMID:17095433

Sag, Y., & Kutsal, T. (2001). Recent trends in the biosorption of heavy metals: A review. *Biotechnology and Bioprocess Engineering; BBE*, *6*(6), 376–385. doi:10.1007/BF02932318

Sakaguchi, T., Nakajima, A., & Horikoshi, T. (1981). Studies on the accumulation of heavy metal elements in biological systems. *European Journal of Applied Microbiology and Biotechnology, 12*(2), 84-89.

Seki, H., Suzuki, A., & Mitsueda, S. I. (1998). Biosorption of heavy metal ions onRhodobacter sphaeroidesandAlcaligenes eutrophusH16. *Journal of Colloid and Interface Science*, *197*(2), 185–190. doi:10.1006/jcis.1997.5284 PMID:9466859

Singh, V., Chauhan, P. K., Kanta, R., Dhewa, T., & Kumar, V. (2010). Isolation and characterization of pseudomonas resistant to heavy metals contaminants. *International Journal of Pharmaceutical Sciences Review and Research*, *3*, 164–167.

Stokes, P. M., & Lindsay, J. E. (1979). Copper tolerance and accumulation in Penicillium ochro-chloron isolated from copper-plating solution. *Mycologia*, *71*(4), 796–806. doi:10.2307/3759190 PMID:530290

Sulaymon, A. H., Mohammed, A. A., & Al-Musawi, T. J. (2013). Competitive biosorption of lead, cadmium, copper, and arsenic ions using algae. *Environmental Science and Pollution Research International*, *20*(5), 3011–3023. doi:10.100711356-012-1208-2 PMID:23054774

Syed, S., & Chinthala, P. (2015). Heavy metal detoxification by different Bacillus species isolated from solar salterns. *Scientifica*. PMID:26525498

Tobin, J. M., Cooper, D. G., & Neufeld, R. J. (1984). Uptake of metal ions by Rhizopus arrhizus biomass. *Applied and Environmental Microbiology*, *47*(4), 821–824. PMID:16346521

Trinelli, M. A., Areco, M. M., & dos Santos Afonso, M. (2013). Co-biosorption of copper and glyphosate by Ulva lactuca. *Colloids and Surfaces. B, Biointerfaces*, *105*, 251–258. doi:10.1016/j.colsurfb.2012.12.047 PMID:23376752

Tsezos, M. (2001). Biosorption of metals. The experience accumulated and the outlook for technology development. *Hydrometallurgy*, *59*(2), 241–243. doi:10.1016/S0304-386X(99)00056-0

Tsezos, M., & Keller, D. M. (1983). Adsorption of radium-226 by biological origin absorbents. *Biotechnology and Bioengineering*, *25*(1), 201–215. doi:10.1002/bit.260250116 PMID:18548548

Tsezos, M., & Volesky, B. (1981). Biosorption of uranium and thorium. *Biotechnology and Bioengineering, 23*(3), 583–604. doi:10.1002/bit.260230309

Urrutia, M. M. (1997). General bacterial sorption processes. *Biosorbents for metal ions*, 39-66.

Vasudevan, P., Padmavathy, V., Tewari, N., & Dhingra, S. C. (2001). Biosorption of heavy metal ions. *Journal of Scientific and Industrial Research (New Delhi, India), 60*, 112–120.

Vijayaraghavan, K., & Yun, Y. S. (2008). Bacterial biosorbents and biosorption. *Biotechnology Advances, 26*(3), 266–291. doi:10.1016/j.biotechadv.2008.02.002 PMID:18353595

Volesky, B. (2007). Biosorption and me. *Water Research, 41*(18), 4017–4029. doi:10.1016/j.watres.2007.05.062 PMID:17632204

Wang, J., & Chen, C. (2009). Biosorbents for heavy metals removal and their future. *Biotechnology Advances, 27*(2), 195–226. doi:10.1016/j.biotechadv.2008.11.002 PMID:19103274

Wang, L. K., Tay, J. H., Tay, S. T. L., & Hung, Y. T. (Eds.). (2010). *Environmental bioengineering* (Vol. 11). Springer Science & Business Media. doi:10.1007/978-1-60327-031-1

Wang, T. C., Weissman, J. C., Ramesh, G., Varadarajan, R., & Benemann, J. R. (1998). Heavy metal binding and removal by Phormidium. *Bulletin of Environmental Contamination and Toxicology, 60*(5), 739–744. doi:10.1007001289900688 PMID:9595189

Weber, W. J. (1972). *Physicochemical processes for water quality control.* Wiley Interscience.

Wilke, A., Buchholz, R., & Bunke, G. (2006). Selective biosorption of heavy metals by algae. *Environmental Biotechnology, 2*, 47–56.

Chapter 14
Biodegradation of Low Density Polyethylene Films

Soumita Dutta Laha
University of Calcutta, India

Kingshuk Dutta
University of Calcutta, India & Cornell University, USA

Patit Paban Kundu
Indian Institute of Technology Roorkee, India

ABSTRACT

With its ever-increasing demand, the enormous production of low density polyethylene (LDPE) is leading to its huge accumulation in the environment. LDPE remains durable and inert to natural degradation and deterioration. This chapter focuses on various microbial effects and approaches on biodegradation of LDPE. Biodegradation takes place through several different strategies, such as assimilation, biodeterioration, biofragmentation, etc. Few microorganisms have so far been isolated which can grow on LDPE. Degradation by microbial enzyme is considered to be one of the most powerful tools to study the biodegradation of LDPE. Some blends and composites of LDPE with natural polymers have been found to be biodegradable; however, their manufacturing is costly. Thorough analysis of the microbial degradation of LDPE helps us realize the overall mechanism involved. In essence, this chapter aims to objectify the in situ biodegradation of LDPE via development of microbial biofilm on the polymer surface.

INTRODUCTION

Polyethylene (PE) represents a versatile synthetic hydrocarbon resin composed of long polymeric chains synthesized via combination of the ethylene (C_2H_4) monomer. As per the reports, PE is the most widely utilized non-degradable solid waste. The most important grades of PE consist of high density polyethylene (HDPE), low density polyethylene (LDPE) and linear low density polyethylene (LLDPE). LDPE, made from petroleum, is semicrystalline in nature. Materials and commodities made from versatile and effective LDPE are found to be strong, durable, light-weight and are thus widely utilized for the production

DOI: 10.4018/978-1-5225-3540-9.ch014

and fabrication of dispensing bottles, various containers, various molded laboratory wares and plastic bags. It has been estimated that annually 500 billion to 1 trillion bags made from LDPE get consumed worldwide. After a very short service life, these materials are dumped in landfills; leading to pollution. PE, especially LDPE, amounts to ~64% of the total synthetic plastics which gets discarded (Lee, Pometto, Fratzke, & Bailey, 1991). This accumulated LDPE wastes in landfills and natural habitats demand strategizing efficient waste management techniques (Thompson, Swan, Moore, & vom Saal, 2009). In addition, the fact that the bio-degradation of PE under normal environmental conditions is extremely slow aggravates the situation (Gu, Ford, Mitton, & Mitchel, 2000; Hakkarainen & Albertsson, 2004; Koutny, Lemaire, & Delort, 2006a; Arutchelvi *et al.*, 2008; Eubeler, Bernhard, & Knepper, 2010). The bio-degradation rate of PE has been reported to be <0.5% in 100 years and ~1% if pre-exposed to sunlight for 2 years (Khabbaz & Albertsson, 2001).

LDPE is an odorless and non-toxic material. LDPE thin films possess low production cost, excellent electrical insulation, easy to process, transparency, flexibility, and toughness (Brydson, 1999). In addition, it is resistant to acids, alcohols, bases, esters and is biologically inactive. This inertness is due to its high molecular weight, hydrophobicity and lack of functional groups that can be recognized by microbial enzymatic systems (Harshvardhan & Jha, 2013). The recalcitrant behavior of LDPE towards attack by biological microorganisms is believed to be owing to the contribution of the hydrophobic backbone, 3-dimensional structure and high molecular weight. As a result, the biodegradation aspect of LDPE films has served as a subject of concern over years among waste management researchers and activists.

Several studies have revealed that PE disposed in moist soil exhibited no signs of weight loss even after 12 years, owing to its inert nature (Potts, 1978). In a separate study, the authors have observed partial degradation of LDPE films after 32 years of long incubation in soil (Otake, Kobayashi, Asabe, Murakami, & Ono, 1995). There are reports that state blockage in the intestines of marine mammals, fish and birds due to intake of PE as food. Ingestion of this waste, thereby, results in endangering hundreds of different species (Secchi & Zarzur, 1999; Spear, Ainley, & Ribic, 1995). Finally, it leads to an ever-increasing ecological threat to marine and terrestrial wildlife (Shimao, 2001; Barnes, Galgani, Thompson, & Barlaz, 2009). Thus, a cheap, effective, less time consuming and completely environmental friendly process to degrade LDPE films is required to be developed in order to save the environment from this enormous pollution. A safe and eco-friendly method of recycling of PE was conducted by Sivan *et al.* (2006); however, they failed to provide a safe disposal for all PE materials. Therefore, microbial degradation remains as one of the best options for PE degradation.

Definition and Categorization of Biodegradation of LDPE

Degradation is defined as any chemical or physical alteration in LDPE occurring as a result of any environmental factor, including heat, light, moisture, biological activity, chemical conditions or wind. By the help of physical means, LDPE films can be subjected to specific temperatures varying from 50°C to 80°C under UV rays for a particular time period to achieve highest level of oxidation (Figure 1).

Biodegradation is a subcategory of degradation, which involves degradation via biological activities. As mentioned above, biodegradation of LDPE is a slow process owing to its inert attribute and resistance to attack by microorganisms. This essentially leads to its environmental accumulation and pollution. In recent times, the prospect of biodegradation of LDPE and the employment of microbes to degrade them have earned significant importance owing to the lack of efficiency of the physical and chemical methods of its disposal. Biodegradation of LDPE films can be broadly classified into abiotic and biotic measures.

Figure 1. The mechanism of LDPE oxidation by UV and heat

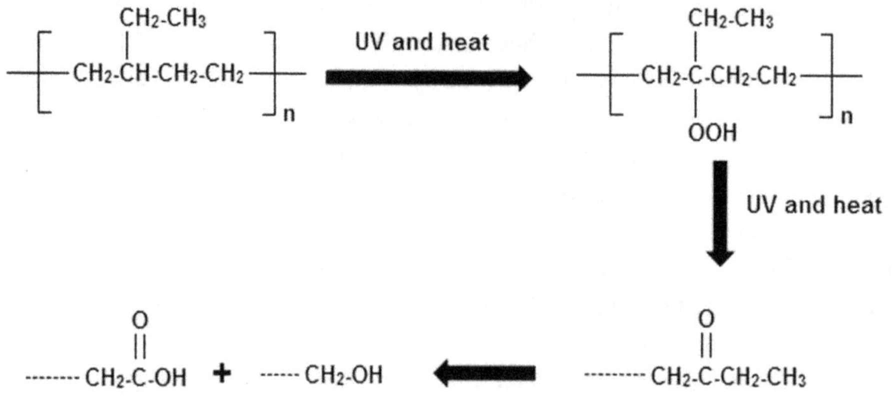

'Abiotic' measure refers to deterioration of LDPE due to environmental factors, such as temperature and UV irradiation, in absence of oxygen (Equation 1). Certain anaerobic bacteria are known to use sulfate, nitrate, carbon dioxide, manganese and iron as acceptors of electrons, and breakdown organic compounds into smaller entities. The abiotic mechanisms of deterioration of PE have been described extensively by Hakkarainen and Albertsson (2004) in their work.

$$C_{PLASTIC} \rightarrow CH_4 + CO_2 + H_2O + C_{RESIDUAL} + BIOMASS \quad (1)$$

On the other hand, 'biotic' measure refers to biodegradation caused by the action of microorganisms that modify and consume the polymer leading to changes in its properties. It is an important measure of naturally attenuating hazardous contaminants at many waste sites. Aerobic bacteria utilize oxygen as an acceptor of electron, and break down organic compounds into smaller organic entities and release CO_2 and water as the final products (Equation 2).

$$C_{PLASTIC} + O_2 \rightarrow CO_2 + H_2O + C_{RESIDUAL} + Biomass \quad (2)$$

Santhoskumar *et al.* (2010) clearly showed that the alteration of LDPE is found to be the same in case of both abiotic (photo-degradation) and biotic samples (Figure 2). Absence of double bond peak in IR spectra was prominent when carbonyl groups were formed from abiotically degraded LDPE. Results revealed that carbonyl index got increased upon prolonged incubation under abiotic atmosphere. The peak at 1636 cm^{-1} confirmed the formation of alkene in biologically affected films.

Based on the factors responsible, polymer degradation can be categorized into three methods: photo-degradation, thermo-oxidative degradation and biodegradation (Zenteno *et al.*, 2017; Antunes, Agnelli, Babetto, Bonse, & Bettini, 2017; Abraham, Ghosh, Mukherjee, & Gajendiran, 2017) Pathak & Navneet, 2017a; Pathak & Navneet, 2017b; Pathak & Navneet, 2017c; Pathak & Navneet, 2017d; Pathak et al., 2018). Among these methods, biodegradation accounts for the major loss of most compounds that get released in the environment. It refers to dissolution of materials by fungi, bacteria and other biological processes. This degradation process occurs via degradation and assimilation of polymers by living

Figure 2. Proposed mechanisms for the biodegradation of PE (Santhoskumar, Palanivelu, Sharma, & Nayak, 2010)

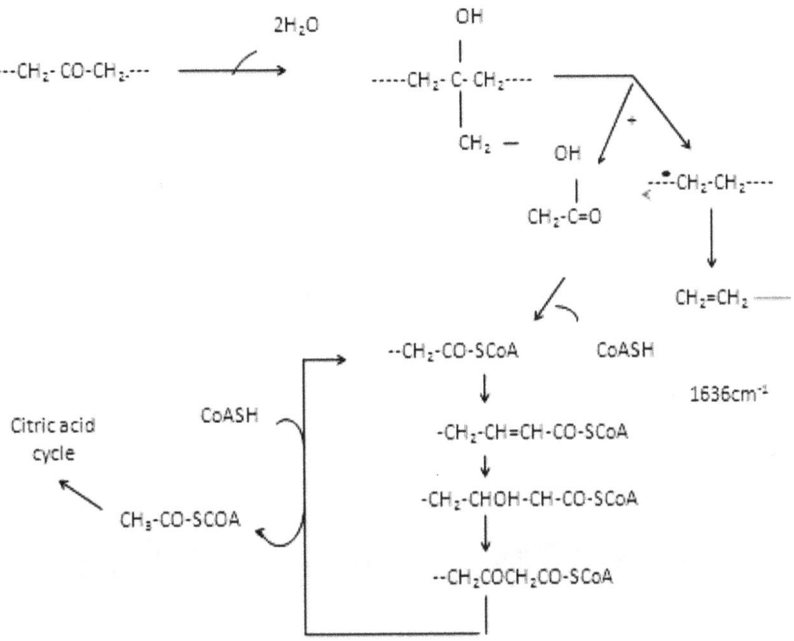

microorganisms, leading to formation of degraded products (Figure 3) (Mueller, 2006). According to Arutchelvi *et al.* (2008), the utilization of PE by microorganisms is physically limited by its insolubility in aqueous media, lack of functional groups and high molecular weight. Certain works on biodegradation of PE have revealed complete metabolic pathways involved in the process, structural changes and identity of all the enzymes involved (Wasserbauer, Beranová, Vancurová, & Doležel, 1990; Pometto, Lee, & Johnson, 1992; Santo, Weitsman, & Sivan, 2013; Yoon, Jeon, & Kim, 2012). The percentage of degradation can be mathematically summarized as:

$$\% \text{ degradation} = \frac{1-\text{final weight}}{\text{initial weight}} \times 100 \quad (3)$$

Generation of biodegradable LDPE requires modification of its degradation resistant properties. Photo, chemical and/or thermal oxidation of LDPE, before exposing it to a biotic environment, increase the rate of its biodegradation (Potts, 1978). These pretreatments enhance the hydrophilic nature of the surface of LDPE via generation of functional moieties, such as carbonyls, that can be later utilized by the microbes. The microorganisms secrete extracellular enzymes, which adhere to the plastic surface. This adherence of enzymes leads to cleavage of the polymer chains. Further cleavage results in the erosion of LDPE surface, producing CO_2, H_2O and CH_4 as end products.

Figure 3. Mechanism of microbial degradation of LDPE (Mueller, 2006)

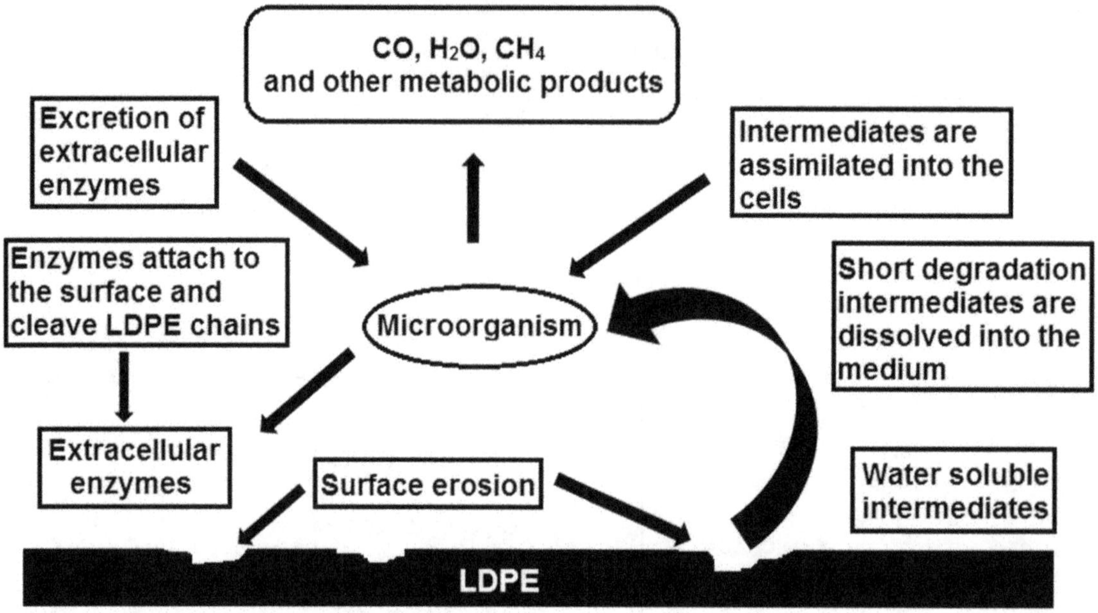

Components Affecting Biodegradation of LDPE Films

Microorganisms can degrade LDPE films based on their metabolic system and capability to accustom to inhospitable environments. The efficiency of microorganisms to biodegrade is dependent upon many parameters that include the concentration of the pollutants, the availability of the pollutants to microbes, the chemical nature of the pollutants and the physical and chemical attributes of the surroundings (Fantroussi & Agathos, 2005). Therefore, parameters that enhance the rate of LDPE degradation by microbes are related to the biological factors (i.e. nature of the microbes involved and their nutritional needs) and the environmental factors (i.e. physical and chemical nature of the environment).

Biological Components

A biotic component represents the metabolic ability that microorganisms possess. The biotic components that causes the microbial degradation of organic substrates include: (a) direct inhibition of enzymatic activities and (b) proliferation procedures of degrading microbes (Riser-Roberts, 1998). Other components that affect the biodegradation rate (via modulating the reaction rates that are catalyzed by enzymes) are moisture, pH and temperature. Bio-enzymes that are involved in the degradation route possess an optimum temperature as well as the same metabolic output for each temperature value (Van der Heul, 2011). It has been observed that the biodegradation rate gets lowered by ~1/2 for every 10°C reduction in temperature. Although biodegradation can take place over a wide pH range, a pH range of 6.5 to 8.5 is usually considered to be optimum for most aquatic and terrestrial surfaces biodegradation. Moisture can also influence the rate of contaminant metabolism; since it affects the amount and kind of available soluble materials as well as the pH and osmotic pressure of aquatic and terrestrial systems (Cases & de Lorenzo, 2005).

Environmental Components

The nature and the amount of organic matter present in the soil influence the potential towards biodegradation of LDPE. The rate of biodegradation gets affected mostly due to the transmission of certain gases like oxygen, methane and carbon dioxide as well as when the water saturation of the soil becomes high. These factors affect the type and rate of biodegradation. The redox potential of the soil also affects the rate of biodegradation.

Genetically Engineered Microorganisms

A genetically engineered microorganism (GEM) or a genetically modified microorganism (GMM) can be defined as a microorganism with altered genetic material. Molecular biology provides the means for modification and modulation of the biodegradative capacities of microbes, acceleration in the evolution of novel biodegradation activities and construction of totally novel biodegradation routes via the congregation of catabolic segments from various microorganisms (Ramos *et al.*, 1994). These alterations are usually performed by utilizing genetic engineering approaches, which are influenced by the phenomenon of natural genetic exchange between microbes. These approaches are commonly identified as recombinant DNA technology. GEMs have demonstrated potentiality towards biodegradation of LDPE. However, the practical prospect of the release of GMMs for LDPE biodegradation has led to concerns regarding biosafety and risk assessment.

Several microbial strains have been created via genetic manipulations, which have exhibited the ability to biodegrade various types of hydrocarbons. A multiplasmid-containing *Pseudomonas* strain, capable to oxidizing terpenic, aromatic, polyaromatic and aliphatic hydrocarbons, has been successfully produced. Research works based on the degradation of LDPE by GEMs have focused on the usage of different genetic engineering approaches, such as modification of the biodegradation pathway and the substrate-specific nature of the microbes (Joutey, Bahafid, Sayel, & Ghachtouli, 2013).

Biodegradability of LDPE can be very well determined by chemical and physical attributes, like (a) availability of functional moieties that accelerates hydrophilicity, (b) molecular composition, i.e. availability of easily degradable bonds (such as ester and amide bonds as against carbon-carbon bonds), (c) complex nature of the molecular structure such as linear or branched structure), (d) degree of crystallinity, (e) nature of the physical forms, such as pellets, films, powders, etc. and (f) density, molecular weight and molecular size of LDPE. Therefore, in summary, the biodegradation of LDPE films gets determined by the following factors:

1. Availability of enzymes and microorganisms:
    a. Induction in the synthesis of the degradative enzyme of the polymer.
    b. Optionally, a consecutive secretive enzyme.
2. Biotic availability of the polymeric structure:
    a. Crystallinity of the polymer.
    b. Accessibility of the cleavable bond to the enzyme.
    c. Formation of microbially metabolizable products due to the enzymatic cleavage
3. Abiotic Factors:
    a. Convenient environment, like the availability of nutrients, the presence or absence of oxygen, temperature and suitable pH.

Microorganisms Involved in the Biodegradation of LDPE Films

Biodegradation of LDPE can be elucidated by two different potential mechanisms. In the first mechanism, pure strains are used to degrade LDPE films to study the rate of degradation. Several studies have been performed to study the rate of degradation by using pure cultures (Sivan *et al.*, 2006; Pometto *et al.*, 1992; Santo *et al.*, 2013; Yoon *et al.*, 2012; Markandey & Rajvaidya, 2004; Albertsson, Barenstedt, Karlsson, & Lindberg, 1995; Yamada-Onodera, Mukumoto, Katsuyaya, Saiganji, & Tani, 2001; Volke-Sepúlveda, Saucedo-Castañeda, Gutiérrez-Rojas, Manzur, & Favela-Torres, 2002; Gilan (Orr), Hadar, & Sivan, 2004; Hadad, Geresh, & Sivan, 2005; Koutny *et al.*, 2006b; Balasubramanian *et al.*, 2010; Fontanella *et al.*, 2010; Rajandas, Parimannan, Sathasivam, Ravichandran, & Yin, 2012; Tribedi & Sil, 2013; Pathak & Navneet, 2017a). Researchers have isolated and identified a fungal strain of *Penicillium simplicissimum* YK for biodegradation of LDPE in absence of additives. Oxidizing agents or UV light have been used to activate the inert material, i.e. LDPE (Yamada-Onodera *et al.*, 2001). It is also reacted with nitric acid at a temperature of 80°C for a period of 6 days before insertion of microbe-susceptive functional moieties. The obtained results showed that LDPE, possessing molecular weight range of 4000 to 28000, got degraded to smaller units of molecular weight of ~500 after 3 months of liquid cultivation; thus, showing biological degradation of LDPE to fungal activity (Wasserbauer *et al.*, 1990). Physico-chemically treated LDPE films are also found to be effectively degraded by the fungal isolates as compared to the untreated films. The hypothesis is that the physicochemical treatment of the LDPE results in the oxidation and simultaneous breakdown, helping in easy assimilation by the fungus; and thus, can be effectively utilized as a pretreatment strategy before subjecting the LDPE to biological degradation (Arkatkar, Arutchelvi, Bhaduri, Uppara, & Doble, 2009). The oxidized LDPE on the other hand, helps in adhesion of microorganisms and serves as a prerequisite for biodegradation (Tribedi & Sil, 2013).

In the second approach, the biodegradation of LDPE is done in presence of complex environments and microbial communities (Albertsson, 1980; Albertsson, Andersson, & Karlsson, 1987; Karlsson, Ljungquist, & Albertsson, 1988; Pegram & Andrady, 1989; Albertsson & Karlsson, 1990; Orhan & Büyükgüngör, 2000; Chiellini, Corti, & Swift, 2003; Artham *et al.*, 2009; Mumtaz, Khan, & Hassan, 2010; Lobelle & Cunliffe, 2011; Nowak, Pająk, Drozd-Bratkowicz, & Rymarz, 2011). The substrate i.e. the type of polymer used depends on the structure of the microbial colony isolated on the surface of LDPE during biodegradation assays. In various studies, it has been proved that the physiological condition of a surface determines the ability of the microorganism to build biofilm structures (Pegram & Andrady, 1989; Albertsson & Karlsson, 1990; Orhan & Büyükgüngör, 2000).

LDPE degrading microbes are basically found in several sites, like rhizosphere soil of mangroves, LDPE buried in the soil, plastic and soil at the dumping sites and also in marine water as given in Table 1.

Table 1. Several microenvironments used to study LDPE biodegradation.

Microenvironment	Reference Number
Marine exposure conditions	Pegram & Andrady, 1989; Chiellini *et al.*, 2003; Lobelle & Cunliffe, 2011
Soil burial conditions	Albertsson *et al.*, 1987; Karlsson *et al.*, 1988; Albertsson & Karlsson, 1990; Orhan & Büyükgüngör, 2000; Chiellini *et al.*, 2003; Mumtaz *et al.*, 2010; Nowak *et al.*, 2011
Composting conditions	Chiellini *et al.*, 2003
Mangrove soil conditions	Kumar, Hatha, & Christi, 2007; Kathiresan, 2003

Based on several microenvironments, research works have either used pure strains of microorganisms or microbial community for over 50 years. It has been reported that a number of strains have the ability to interact with LDPE causing some kind of deterioration and degradation. These microorganisms play an effective role in degrading LDPEs, which are collected from sewage sludge or landfill by different techniques. Two categories of microorganisms are particularly prospective in the biodegradation of LDPE: bacteria (eubacteria, actinomycetes, etc.) and fungi. The richness of microorganisms that are able to degrade LDPE is represented by 19 genera in case of bacteria and 12 genera in case of fungi. These numbers can, however, get increased based on more sensitive and careful isolation and characterization techniques, which include sequencing of rDNA. This technology also allows an approach to assess the composition of a microbial community, non-culturable fraction of microorganisms which constitutes up to 90% of the real biodiversity in an ecosystem according to Hugenholtz *et al.* (1998).

An expansion in the rate of biodegradation of LDPE takes place with an extension in exposure time to UV, as observed by Hadad *et al.* (2005). After 126 days, biodegradation in the uninoculated treatments has been found to be ~7.6% and ~8.6% after mineralization with the non-UV irradiated and UV irradiated respectively, indicating towards a slow process. Gilan *et al.* (2004) states that a new strain *Rhodococcus ruber* is found to colonize and biodegrade LDPE. This bacterium has the ability to form biofilm due to the hydrophobicity of the cell surface.

Tables 2 and 3 present exhaustive lists of microorganisms that are extensively connected with LDPE colonization or biodegradation. Each of these microorganisms consists of several characteristics, so the degradation rate varies accordingly.

Effect of Microbial Activity on LDPE Films

The microorganisms start colonizing on the surface of LDPE by using it as the sole carbon source. In primary degradation, low-molecular weight fragments, like oligomers, dimers and monomers are formed by cleaving the main chain. Further enzymatic decomposition of the macromolecules progress from the polymer chain ends due to metabolism of the split portions. This phenomenon eventually results in shortening of the chain fragments up to the extent that they get consumed by the microbes (Barnes *et al.*, 2009). The microorganisms further utilize these low molecular weight compounds as sources of carbon and energy. The complete utilization of the fragmented components by the microbes prevents creation of hazard to the environment and health (Sivan *et al.*, 2006). This degradation is basically due to the secretion of extracellular enzymes by the microorganisms. The enzyme gets bound to the LDPE substrate and catalyzes the hydrolytic cleavage (Table 4).

Fungi have also been examined for the potentiality toward biodegradation of LDPE (Awasthi, Srivastava, Singh, Tiwary, & Mishra, 2017). It is known that fungi produce extracellular polysaccharides and degrading enzymes, which possess the ability to assist colonization of the LDPE surface. The fact that fungal hyphae possess the ability of penetration and distribution has been considered to be an added advantage. LDPE degradation studies have been carried out by utilizing fungal isolates, namely *Phanerochaete chrysosporium*, *Aspergillus fumigatus*, *Aspergillus flavus*, *Aspergillus niger* and *Aspergillus terreus* (Lee *et al.*, 1991; Singh *et al.*, 2012; Volke-Sepúlveda *et al.*, 2002; Raaman, Rajitha, Jayshree, & Jegadeesh, 2012).

Table 2. Bacterial strains associated with biodegradation of LDPE films

Genus	Species	Reference Number
Acinetobacter, Arthrobacter	*baumanni* sp.	Balasubramanian *et al.*, 2010; Nowak *et al.*, 2011; Alok *et al.*, 2008
	paraffineus	Albertsson *et al.*, 1995; Albertsson *et al.*, 1998
	viscosus	Nowak *et al.*, 2011
Bacillus	*amyloliquefaciens*	Nowak *et al.*, 2011
	brevies	Watanabe *et al.*, 2009
	cereus	Nowak *et al.*, 2011; Alok *et al.*, 2008; Roy *et al.*, 2008; Sudhakar *et al.*, 2008
	circulans	Watanabe *et al.*, 2009
	halodenitrificans	Nowak *et al.*, 2011; Roy *et al.*, 2008
Bacillus	*mycoides*	Seneviratne *et al.*, 2006
	pumilus	Nowak *et al.*, 2011; Alok *et al.*, 2008; Roy *et al.*, 2008
	sphericus	Sudhakar *et al.*, 2008; Kawai *et al.*, 2004
	thuringiensis	Nowak *et al.*, 2011
Brevibacillus	*borstelensis*	Hadad *et al.*, 2005
Delftia	*acidovorans*	Koutny *et al.*, 2009
Flavobacterium	sp.	Koutny *et al.*, 2009
Micrococcus	*luteus*	Nowak *et al.*, 2011
	lylae	Nowak *et al.*, 2011
Microbacterium	*paraoxydans*	Rajandas *et al.*, 2012
Nocardia	*asteroides*	Koutny *et al.*, 2006b; Bonhomme *et al.*, 2003
Paenibacillus	*macerans*	Nowak *et al.*, 2011
Pseudomonas	sp.	Yoon *et al.*, 2012; Balasubramanian *et al.*, 2010; Tribedi & Sil, 2013
	aeruginosa	Rajandas *et al.*, 2012; Koutny *et al.*, 2009
	fluorescens	Nowak *et al.*, 2011
Rahnella	*aquatilis*	Nowak *et al.*, 2011
Ralstonia	sp.	Koutny *et al.*, 2009
Rhodococcus	*ruber*	Sivan *et al.*, 2006; Santo *et al.*, 2013; Gilan *et al.*, 2004
	rhodocrous	Koutny *et al.*, 2006b; Bonhomme *et al.*, 2003; Fontanella *et al.*, 2010
	erythropolis	Koutny *et al.*, 2009
Staphylococcus	*epidermidis*	Chatterjee *et al.*, 2010
	cohnii	Nowak *et al.*, 2011
	xylosus	Nowak *et al.*, 2011
Stenotrophomonas	sp.	Koutny *et al.*, 2009
Streptomyces	*badius*	Pometto *et al.*, 1992
	setonii	Pometto *et al.*, 1992
	viridosporus	Pometto *et al.*, 1992

Table 3. Fungal strains associated with biodegradation of LDPE films

Genus	Species	Reference Number
Acremonium	*kiliense*	Karlsson *et al.*, 1988
Aspergillus	*niger*	Kawai *et al.*, 2004; Volke-Sepúlveda *et al.*, 2002; Raghavan & Torma, 1992
	versicolor	Fontanella *et al.*, 2010; Koutny *et al.*, 2009
	flavus	Pramila & Ramesh, 2011a; Manzur *et al.*, 1997
Chaetomium	sp.	Sowmya *et al.*, 2012
Cladosporium	*cladosporioides*	Bonhomme *et al.*, 2003; Manzur *et al.*, 1997
Fusarium	*redolans*	Albertsson, 1980; Karlsson *et al.*, 1988; Albertsson & Karlsson, 1990
Gliocladium	*virens*	Manzur *et al.*, 2004
Mortierella	*alpina*	Albertsson *et al.*, 1995
Mucor	*circinelloides*	Pramila & Ramesh, 2011a
Penicillium	*frequentans*	Seneviratne *et al.*, 2006
	pinophilum	Volke-Sepúlveda *et al.*, 2002; Manzur *et al.*, 2004
	simplicissimum	Yamada-Onodera *et al.*, 2001
Phanerochaete	*chrysosporium*	Arkatkar *et al.*, 2009; Kawai *et al.*, 2004; Chatterjee *et al.*, 2010
Verticillium	*lecanii*	Fontanella *et al.*, 2010

Table 4. Useful bacterial and fungal enzymes for degradation of different types of plastic

Source	Enzyme	Microorganism	Plasticsas a Substrate	Reference Number
Bacteria	Lipase	*Rhizopus delemer*	PCL (Polycaprolactone)	Spear *et al.*, 1995
		Rhizopus arrizus	PCL, PBS (Polybutylene succinate), PEA (Polyethylene adipate)	Spear *et al.*, 1995
	Unknown	*Firmicutes* sp.	PHB (Polyhydroxybutyrate), PBS, PCL	Spear *et al.*, 1995
		Protobacteria sp.	PHB	Spear *et al.*, 1995
	Serine Hydrolase	*Pseudomonas stutzeri*	PHA (Polyhydroxyalkanoate)	El-Rehim *et al.*, 2004
Fungi	Glucosidase	*Aspergillus flavus*	PCL	Spear *et al.*, 1995
	Unknown	*Penicillium funiculosum*	PHB	Spear *et al.*, 1995
		Amycolaptosis sp.	PLA (Polylactic acid)	El-Rehim *et al.*, 2004
	Unknown	*Streptomyces* sp.	PHB, PCL	Spear *et al.*, 1995
	Cutinase	*Aspergillus oryzae*	PBS	Danjaji *et al.*, 2002
		Fusarium sp.	PCL	El-Rehim *et al.*, 2004
	Catalase, Protease	*Aspergillus niger*	PCL	Spear *et al.*, 1995
	Urease	*Trichoderma* sp.	Polyurethane	Pometto *et al.*, 1992
	Lipase Peroxidase, Manganese Peroxidase	*Phanerochaete chrysosporium*	PE	Santo *et al.*, 2013
	Serine Hydrolase	*Pestalotiopsis microspora*	Polyurethane	Pometto *et al.*, 1992

Different Characteristics to Establish the Extent of Biodegradation of LDPE Films

In order to establish the extent of biodegradation of LDPE, six different characteristics are usually monitored, like functional groups of the surface, wettability, crystallinity, surface topography, mechanical properties, molecular weight distribution and consumption of the LDPE.

Functional Groups on the Surface

An enhancement in the fungal growth and associated structural changes have been observed by FT-IR for treated LDPE, which indicate the fragmentation of LDPE chains and the existence of oxidatively produced fragments of LDPE. It is to be noted that non-degraded LDPE exhibits almost no absorbance values corresponding to those wave numbers. Absorbance peaks appearing at 1710 cm^{-1} to 1715 cm^{-1} (representing carbonyl compound), 1640 cm^{-1} and 830 cm^{-1} to 880 cm^{-1} (representing -C=C) after treatment with UV and nitric acid were found to decrease after cultivation with microbial consortia (Volke-Sepúlveda, Saucedo-Castañeda, Gutiérrez-Rojas, Manzur, & Favela-Torres, 2002). Certain reports showing typical biodegradation of modified LDPE-starch mixture as well as formation of bands at 1620 cm^{-1} to 1640 cm^{-1} and 840 cm^{-1} to 880 cm^{-1} also attribute to the oxidative decomposition of LDPE (Riser-Roberts, 1998).

LDPE degradation can be summarized as a combination of photo-degradation and biodegradation processes. At first, required abiotic precursors are acquired either by treatment with heat or by exposure to UV light (i.e. abiotic oxidation). This is followed by complete biodegradation via degradation of the low molecular weight oxidation products by selected microorganisms. Certain reports demonstrate that chain segments ruptured upon biological treatment can result in the formation of vinyl groups and asimultaneous increment in the double bond index (DBI) (Manzur, Cuamatzi, & Favela, 1997). Again, biotic dehydrogenation can also add to this enhancement in the DBI. LDPE biodegradation further confirms an enhancement in the vinyl bond index, the ester carbonyl bond index and the keto carbonyl bond index, which can be quantitatively calculated from the FT-IR spectra (Shah, Hasan, Hameed, & Ahmed, 2008). FT-IR findings reveal that the degradation of LDPE is a complex process that can differ for different microorganisms and different communities. Incubation with microorganisms generates changes in the concentrations of functional groups at the surface of LDPE either because of their consumption or production. In a complex microbial community, in which abiotic factors are also affecting the chemistry of the LDPE, the net effect depends on the balance of rates of oxidation and degradation, which in turn depends on the nature of the microorganisms present. The study of the chemistry of LDPE surface turns out to be very important because oxidized groups are more easily degraded by microorganisms. Oxidized groups present on the surface modulate microbial attachment by increasing the hydrophilicity of the surface, which implies LDPE degradation is enhanced if a more oxidized surface is used as the substrate.

Hydrophobicity/Hydrophilicity

Measurement of microbial hydrophobicity can be of importance in many research areas and biodegradation studies e.g. biofouling, oral microbiology, phagocytosis, soil microbiology etc. Therefore, a good measure for microbial hydrophobicity is required. However, for this purpose a wide variety of experimental analyses is employed, which makes it difficult to compare the results reported in different studies. The relation between surface hydrophobicity and microorganisms determines the extent of colonization on the LDPE substrate. It is accepted that more hydrophilic surfaces are more easily colonized by microorganisms. Oxidation due to abiotic factors, like activity of enzymes or UV light, which is higher than the extent of consumption of functional groups can lead to an increase in the hydrophilicity. 2,3,5-triphenyltetrazolium chloride (TTC) is used as a redox probe, facilitating direct monitoring of actively respiring microbes. Thus, TTC reduction test is done to confirm the viability of the isolates growing on LDPE surface. This viability can be correlated with the simultaneous enhancement in the biomass protein density. Hydrophobicity can be determined by the contact angle of the LDPE surface with a probe liquid, such as water. The contact angle can be defined as the angle made by the probe liquid on a solid surface, measured at a point where a liquid/vapor interface meets a solid surface. It helps in determining the surface hydrophobicity of a solid by a liquid. The Young-Dupré equation estimates the energy of adhesion to the solid as well as acid, base and Van der Waals components.

The Young-Dupré equation is designed as follows:

$$Y\,(1 + \cos \theta C) = \Delta WSLV \quad (4)$$

where, ΔWSLV represents the energy of adhesion per unit area of the solid-liquid interface when measured in a medium V and θC is the equilibrium contact angle.

Crystallinity

Crystallinity can be defined as a state of order of greater degree, which implies a long range of periodic geometric pattern of atomic spacing. In polymers which are semicrystalline, such as LDPE, the crystalline degree affects the degree of hardness, stiffness and resistance to heat. Calculation of crystallinity can be performed from a colorimetry curve by using the following equation:

$$\text{Crystallinity (\%)} = \text{Measured heat of fusion/heat of fusion of 100\% crystalline material} \quad (5)$$

LDPE resins generally exhibit crystallinity from 35% to 55%. LDPE possess low crystallinity and has less tightly packed molecules as it has side branches. It is experimentally stated that at first the amorphous regions get consumed since they are more accessible to microbes. Prior consumption of amorphous surfaces leads to an initial increase in the percentage of crystallinity (Figure 4). Once the accessible amorphous regions get depleted, microorganisms start consuming the smaller crystals present. This leads to an increase in the proportion of larger crystals. Insufficient data till date prevents us from gaining precise understanding of events that occur after the amorphous regions get consumed.

Surface Topography

Colonization of LDPE surfaces by microorganisms usually generates changes in the surface topography as has been proven extensively by different researchers. Penetration of hyphal structures has also been reported as common features after microbial attack. Several works prove that some superficial damage can be observed after LDPE surfaces get exposed to biodegradation. It has been evident that surface topography gets modified by microbial colonization, but it is still unknown that how topography gets modified if microorganisms are removed. Surface damage of LDPE in the form of pits and erosion with *Fusarium* species AF4 has been observed with the help of scanning electron microscopy (Karlsson *et*

Figure 4. Crystallinity characteristic to establish the extent of biodegradation of PE

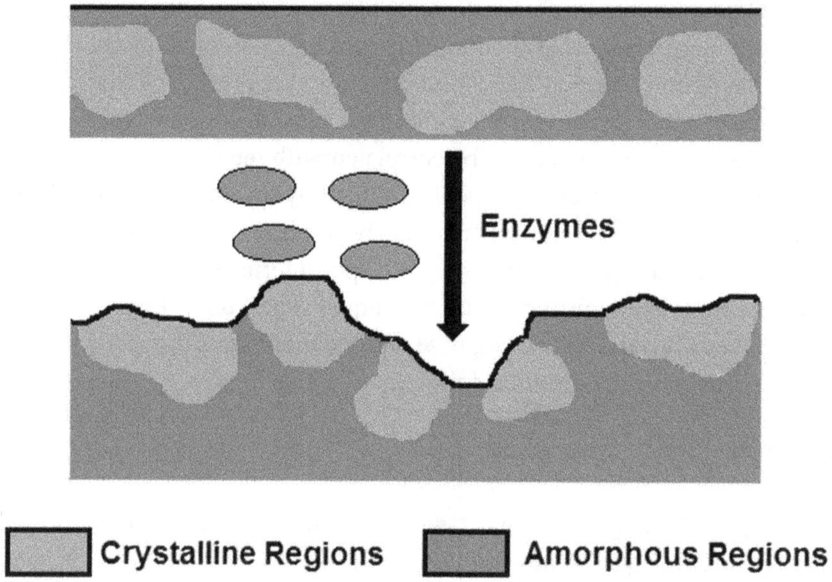

al., 1988; Albertsson & Karlsson, 1990). The *Fusarium* species AF4 can stick to the LDPE surface and can lead to surface damage. Several studies also confirm that after removal of the microorganisms, grown on the surface of LDPE, the surface becomes physically pitted and eroded. Otake *et al.* (1995) have clearly reported the changes that occurred to LDPE films after burial in the soil for 32 years. The authors observed formation of small holes on the surface and whitening of the degraded area.

Mechanical Properties

Most of the studies on LDPE biodegradation show breaking load is a common mechanical property that leads to substrate deterioration. Mechanical properties can be modified by oxidation which induces changes in crystallinity and also by average molecular weight. These modified LDPE films can be further modified biochemically. Biodegradation of the further modified films in soil show significant changes, like elongation at brake of 98%. It is also studied that these films can undergo rapid degradation in soils which are rich in organic matter than those in waste coal soil. Nowak *et al.* (2011) states that certain fungi like *Gliocardium viride*, *Aspergillus awamori* and *Mortieralla subtilissima* can easily colonize polypropylene (PP) and PE. The storage and loss modulus of the LDPE can be determined by rheological analysis (Nowak *et al.*, 2011). Universal mechanical testing system (UMTS) have been utilized to determine the mechanical properties of any polymer sample. So far, the effects of biological activity on LDPE samples have been basically studied on thin films; however thick walls have become common application of this polymer in the manufacture of certain tanks (Pegram & Andrady, 1989; Nowak *et al.*, 2011; Vasile, 1993). Therefore, the changes in the mechanical properties due to microbial activity are still an active area of research. Microorganism's effects are often found to be superficial. The resolutions of bulk testing methods are commonly employed for the abovementioned measurements which diminish the local surface related damage and influence less on the overall sample (Table 5) (Pegram & Andrady, 1989; Nowak *et al.*, 2011; Vasile, 1993).

Table 5. Changes in mechanical properties due to microbial activity in different PE samples

Substrate	Environment	Time (days)	%Δ Elongation	%ΔTensile	Reference Number
LDPE	Crater soil	225	-1.5	-19.5	Nowak *et al.*, 2011
	Forest soil	225	-4	-16.4	Nowak *et al.*, 2011
	Sea water	365	-12	-15	Pegram & Andrady, 1989
	Waste coal	225	+4	-16.4	Nowak *et al.*, 2011
	Sterile sea water + *B. sphericus*	365	+2.7	-3.8	Sudhakar *et al.*, 2008
	Mineral media + *Pseudomonas* sp.	45	NR	-30	Sudhakar *et al.*, 2008
HDPE	Sterile sea water + *B. sphericus*	365	+8.9	-9.7	Sudhakar *et al.*, 2008

Molecular Weight Distribution

The basic limitation in LDPE biodegradation is its high molecular weight. After microbial attack, the average molecular weight gets enhanced owing to the consumption of the fragments of lower molar mass (Pometto *et al.*, 1992; Santo *et al.*, 2013; Riser-Roberts, 1998; Ramos *et al.*, 1994). Some researchers have also observed slight changes in the molecular weight distribution (Rajandas *et al.*, 2012; Kumar, Hatha, & Christi, 2007). Fontanella *et al.* (2010) have pointed out another fact that the major factor influencing the molecular weight is the effect caused by abiotic factors, such as UV irradiation, rather than direct microbial attack. On the other hand, Ohtaki *et al.* (1998) have tested bottles of LDPE which were kept for over 30 years in exposure to aerobic soil and have realized loss in molar masses revealed by time of flight mass spectrometry (TOF-MS). For the determination of molecular weight distribution, two different approaches have been used. Firstly, size exclusion chromatographic techniques have been used at high temperatures (Pometto *et al.*, 1992; Fantroussi & Agathos, 2005; Riser-Roberts, 1998; Albertsson *et al.*, 1995; Volke-Sepúlveda *et al.*, 2002; Tribedi & Sil, 2013; Albertsson, 1980; Watanabe *et al.*, 2009). The second approach has been the use of rheological measurements that correlate indirectly with the molecular weight distribution, as reported by Hadad *et al.* (2005). On the other hand, some results have shown the extent of reduction on the basis of number average molecular weight of the LDPE samples. This has been clearly depicted in Tables 6 and 7.

Table 6. Changes in the molecular number due to microbial activity in different LDPE samples

Sample	%Δ Molecular Number	Reference Number
LDPE	-15	Santo *et al.*, 2013
LDPE + Starch	-17	Pometto *et al.*, 1992
LDPE + UV irradiated	-34	Ramos *et al.*, 1994

Table 7. Percentage of loss of mass of LDPE films at different intervals of time (Burlein & Rocha, 2014)

Blends	Loss of Mass (%)	Standard Deviation	Loss of Mass (%)	Standard Deviation	Loss of Mass (%)	Standard Deviation
	For 30 days		For 80 days		For 120 days	
LDPE/PHB (90/10)	0.18	0.06	-0.01	0.15	0.18	0.02
LDPE/PHB (80/ 20)	0.09	0.03	0.75	0.38	1.30	0.26
LDPE/PHB (70/30)	0.22	0.04	2.23	0.82	2.48	0.30
LDPE/CC (90/10)	0.04	0.07	-0.14	0.06	-0.11	0.26
LDPE/CC (80/20)	-0.07	0.04	-0.14	0.05	-0.33	0.10
LDPE/CC (70/30)	-0.19	0.01	-0.69	0.16	-0.56	0.20
LDPE/PHB/CC (80/10/10)	0.17	0.07	-0.13	0.18	0.25	0.13
LDPE/PHB/CC (70/20/10)	-0.39	0.05	4.25	0.92	7.95	2.33
LDPE/PHB/CC (70/10/20)	-0.31	0.06	0.81	0.47	3.05	0.55

Biodegradable LDPE Films

Degradation may be defined as degeneration of substances, especially complex compounds like proteins and polymers, and demonstrating well-defined intermediate products. However, biodegradation can be defined as chemical dissolution of substrates by a variety of microbes and other biological means. Biodegradation of several categories of PEs, like LDPE, HDPE and medium density polyethylene, can take place through three different strategies, namely biodeterioration, assimilation and bio-fragmentation without neglecting the involvement of abiotic factors.

Biodegradation of LDPE provides evidence of its assimilation but the slowness of this process makes it very difficult to detect. Several studies have shown a reduction in the weight of samples determined by gravimetric measurements (Ramos *et al.*, 1994; Markandey & Rajvaidya, 2004; Hadad *et al.*, 2005; Albertsson, 1987; Albertsson & Karlsson, 1990; Hasan *et al.*, 2007). Some studies have also suggested the reduction in the weight of the samples by CO_2 evolution (Koutny *et al.*, 2006b; Fontanella *et al.*, 2010; Tribedi & Sil, 2013; Alok *et al.*, 2008; Bonhomme *et al.*, 2003). Albertsson (1980) has reported that biodegradation of LDPE films has taken place from 0.2% to 8.4% by irradiating the samples with UV light before a biotic treatment. Here the microorganisms use LDPE as the only source of carbon, finally converting it to carbon dioxide during respiration. This technique is used as an indirect measurement of the amount of LDPE that has been used by the microorganisms (Karlsson *et al.*, 1988). Albertsson *et al.* (1987) and Karlsson *et al.* (1988) have reported that CO_2 evolution of any system can be maintained by determining the total consumption of the polymer and also the rate of degradation continuously. This technique can successfully degrade LDPE of low molar mass in order to check the ability of some strains of microorganisms (Yoon *et al.*, 2012). Initial reduction in weight can be found when LDPE is mixed with starch, and the initial reduction in weight can be due to the starch consumption rather than LDPE usage (Markandey & Rajvaidya, 2004).

Abiotic factors can promote oxidation and can extensively influence the rate and extent of polymer consumption, which has been elucidated in Table 8. The table demonstrates the extent of biodegradation found in different PE types prepared without any oxidative treatment.

Table 8. Various studies on weight loss percentage due to biological action for different types of PE in various environments

Substrate	Environment	Time (days)	% of Weight Loss	Reference Number
LDPE	Waste coal	225	0.26	Nowak *et al.*, 2011
	Forest soil	225	0.13	Nowak *et al.*, 2011
	Crater soil	225	0.28	Nowak *et al.*, 2011
	Sea water	365	1.9	Artham *et al.*, 2009
	Soil + *Fusarium redolens*	3650	0.2	Albertsson & Karlsson, 1990
	Soil	800	0.1	Albertsson, 1980
	Mineral media + *Rhodococcus ruber*	56	7.5	Markandey & Rajvaidya, 2004
	Mineral media + *Rhodococcus ruber*	30	2.5	Santo *et al.*, 2013
	Mineral media + *Brevibacillus borstelensis*	30	2.5	Ramos *et al.*, 1994
	Mineral media + Pseudomonas sp.	45	5	Tribedi & Sil, 2013
HDPE	Sea water	365	1.6	Artham *et al.*, 2009
	Soil	800	0.4	Albertsson, 1980

Microbial degradation gets facilitated by chemical and physical degradation. Recent works on biodegradation of LDPE suggest that biodegradation can take place via two approaches i.e. oxo-biodegradation and hydro-biodegradation.

Hydro-Biodegradable LDPE Films

Hydro-biodegradation is the term coined from hydrolysis. This class of LDPE possesses high starch content and are made from renewable sources. Crops which are genetically modified have also been utilized in the production of hydro-biodegradable LDPE. Many of the hydro-biodegradable LDPEs contain up to 50% of the synthetic plastics produced from oil and certain other aliphatic polyesters which are totally based on intermediates that have been obtained from oil. Hydro-biodegradable LDPEs have not been precisely termed as "renewable", as the approach involved in producing them from crops significantly utilizes energy from fossil-fuel and therefore, generates greenhouse gases (Kyrikou & Briassoulis, 2007; Luckachan & Pillai, 2011)

Oxo-Biodegradable LDPE Films

Oxo-biodegradable LDPEs are prepared from polymers, like PE, PP and polystyrene (PS) possessing additional ingredients (other than heavy metals). These LDPEs are tested following the AFNOR AC T51-808 or BS8472 or ASTM D6954 standards in order to analyze their biodegradation in open environments. CEN (the European Standards Organization) defines oxo-biodegradation as "degradation resulting from oxidative and cell-mediated phenomena, either simultaneously or successively". This category of plastics is prepared from polymers which are petroleum based (usually PE). They also contain certain additional ingredients that cause LDPE to degrade easily. Small amounts of metal salts are added, which catalyze and speed up the process of degradation. As a result, the oxo-biodegradable LDPEs decompose abiotically in the presence of oxygen at the termination of its service life much faster compared to ordinary LDPEs. LDPEs are then converted to small chain organic chemicals via carboxylation, hydroxylation and get ready to biodegrade. The biodegradation rate varies according to the type of microorganisms present in the environment. Major differences are however found between oxo- and hydro-biodegradable LDPEs as shown in Table 9.

Degradation Process for Oxo-Biodegradable LDPE Films

The first process of degradation is chain scission via oxidation. Metal salts catalyzes this reaction step and leads to the formation of oxygenated molecules of shorter chain lengths. If the oxo-biodegradable LDPEs get discarded in the open environment, they decompose to oxygenated chains of lower molecular weights of ~5.000 amu to 10.000 amu within 2 months to 18 months. This phenomenon, in turn, depends on the type of substrate, temperature and other environmental parameters. This very chemical process helps in enabling further substrate breakdown by microorganisms that are present in nature.

Benefits and Challenges of Oxo-Biodegradable LDPE Films

Oxo-biodegradable LDPEs serve as a remedy to littering issues. However, they fail to satisfy the requirement criteria of home and industrial compostability set up as per various established standards. Nowadays, oxo-biodegradable LDPE bags are available to substitute traditional LDPE bags. The oxo-degradable LDPE bags are not biodegradable but they breakdown into small fragments after exposure to oxygen (Wrona, Salafranca, & Nerin, 2017). The small pieces are often consumed by the animals or are scattered over the ground, which may lead to environmental problems. Many researchers, like Natarajan *et al.*, have proved that oxo-degradable PE fragments concentrate pesticide residues in the soil, while other researchers have reported PE and PP fragments in the marine environment (Koutny *et al.*, 2006b; Balasubramanian *et al.*, 2010; Fontanella *et al.*, 2010). Therefore, during conversion to final products, oxo-biodegradable ingredients are often introduced to conventional plastics like PE, PP, polyvinyl chloride (PVC), polyethylene terephthalate (PET) and polystyrene (PS).

These modifications are done with two additives i.e. pro-oxidant and starch, which are added during the production of biodegradable LDPE. The overall mechanical steps involve four different stages i.e. microbe attachment to the LDPE surface, microbe growth utilizing the LDPE as the source of carbon, primary degradation of LDPE and finally the ultimate degradation.

Table 9. Comparisons and similarities between hydro- and oxo-biodegradable LDPE films

Oxo-Biodegradable LDPE	Hydro-Biodegradable LDPE
Generally synthesized from oil-refining by-products	Generally synthesized from starch
Possesses the ability to be recycled from a normal stream of plastic waste	Causes damage to recyclate, if not extracted from feedstock
Possesses the ability to be synthesized from recyclate	Lacks the ability to be synthesized from recyclate
Exhibits slow emission of carbon dioxide during degradation	Exhibits rapid emission of carbon dioxide during degradation
Exhibits inertness when present deep inside landfill	Exhibits emission of methane when present deep inside landfill
Uses workforce and machinery same as those used for conventional plastic	Requires special workforce and machinery
Exhibits suitability towards utilization in high speed machinery	Exhibits unsuitability towards utilization in high speed machinery
Exhibits compostability in-vessel	Exhibits compostability
Demonstrates minor or zero on-cost	Demonstrates 4 to 5 times higher cost compared to conventional plastic
Exhibits strength same as conventional plastic	Exhibits strength lower than conventional plastic
Exhibits weight same as conventional plastic	Exhibits weight higher and width thicker than conventional plastic
Demonstrates Leak-proof nature	Demonstrates proneness to leakage
Demonstrates degradation anywhere on sea or land	Demonstrates degradation only in environment which are highly microbial
Degradation time can be determined and set during manufacture	Control of degradation time is not possible
Does not contain ingredients that are genetically modified	There is possibility of containing ingredients that are genetically modified
Demonstrates safety towards food contact	Demonstrates safety towards food contact
Does not utilize water, pesticides or fertilizers for production	Utilizes water, pesticides and fertilizers for production
Devoid of PCB's heavy-metals or organo-chlorines	Devoid of PCB's heavy-metals or organo-chlorines
Demonstrates ability to be incinerated with recovery of high energy	Demonstrates ability to be incinerated, but possesses lower value of calorie
Exhibits limitless feedstock availability	Exhibits limited feedstock availability
Its demand does not involve increase in vehicle fuel cost	Its demand involves increase in cost of foodstuffs for animals and humans

LDPE has only $-CH_2$ groups in their main chain; thus, the surfaces are found to be hydrophobic. However, microorganisms can only attach to the LDPE surface if it is hydrophilic. In this regard, initial chemical and physical degradations result in introduction of hydrophilic moieties on the surface of LDPE. Microorganisms require carbon source for their growth and after achievement of the required hydrophilicity of the LDPE surface, they start growing by getting attached to the surface (Sen & Raut, 2015).

During primary degradation of LDPE, cleavage of the main chain occurs which leads to the generation of fragments of lower molar mass i.e. monomers or dimers. This initial breakdown of LDPE can result in physical changes like wetting or drying, freezing or thawing and heating or cooling. These physical changes can even result in cracking of the polymeric materials and other mechanical damages. The microorganisms secrete extracellular enzymes and the degradation occurs due to the secretion of these extracellular enzymes. LDPE gets depolymerized by these enzymes, followed by absorption of the monomers within the microbial cells and getting further biodegraded. The enzymatic degradation can be further divided into two steps by hydrolysis. Degradation of LDPE occurs either by the extracellular or the intracellular depolymerases. The enzyme first gets bound to the LDPE, after which they catalyze hydrolytic cleavages. During the process of biodegradation, the microbial extracellular enzymes rupture complex polymeric structures to chains of shorter lengths, like oligomers, dimers or monomers, which can pass easily through the semi-permeable membranes. This process is known as depolymerization. After depolymerization the short chains get mineralized to form CH_4, H_2O or CO_2; and this process is known as mineralization. The resulting compounds of lower molar mass get further utilized by the microorganisms as the sources of energy and carbon. Oligomeric species get diffused into the micro-organisms and results in the formation of final products, like biomass, H_2O and CO_2, after degradation under aerobic conditions (Figure 5).

Various studies have reported the generation of carbonyl moieties by oxidation, accompanied by a reduction of the polymeric molecular weight, under the treatment of UV light (Albertsson, 1980; Albertsson *et al.*, 1987). The process of degradation of LDPE, caused by microbes, gets started upon secretion of extracellular enzymes. Metabolism of the fragmented portions lead to further progress in enzymatic dissimilation of the polymers from the ends of the polymeric chains. Finally, the fragmented chains shorten and get consumed by the microbes (Karlsson *et al.*, 1988).

Strategies Used to Characterize Biodegradability of LDPE Films

At first, the polymers having high molar mass get fragmented into oligomers. A fraction of these oligomers is found to be soluble in water; while others get further fragmented into intermediate organic compounds, like acids, alcohols, ketones etc. Biomass accumulation, which determines the rate of growth of the microbes (using LDPE as the only source of carbon) is known as the major strategy to assess and monitor the rate of biodegradation. Several other strategies are also found, which include surface changes, CO_2 evolution rate, modulations in the physical and mechanical properties of the LDPE, O_2 uptake and also products of reaction using chemical analysis.

Figure 5. Biodegradation of LDPE

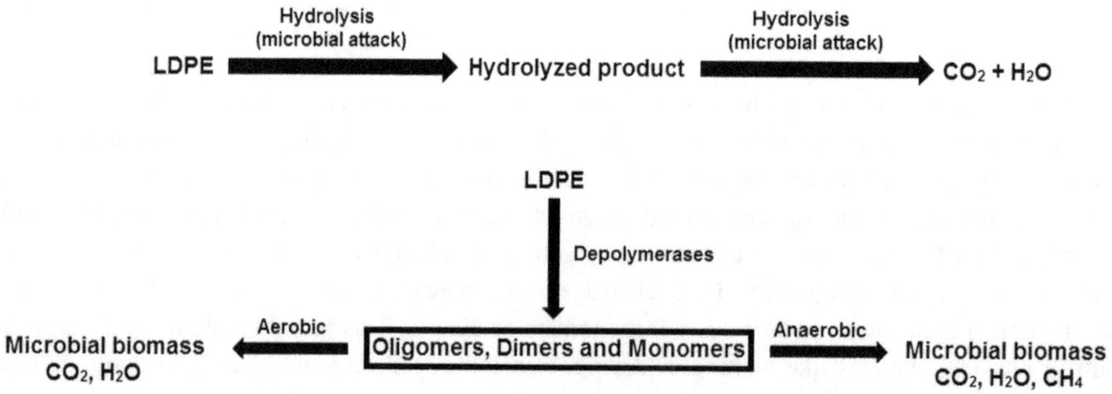

Experimental Methodology of Biodegradation of LDPE Films

Different methodologies are known for the LDPE biodegradation studies (Ali *et al.*, 2009). These methodologies have been listed and briefly described in this section.

Soil Burial Treatment

Otake *et al.* (1995) have analyzed different polymers that have been kept buried for a period of greater than 32 years under soil. They observed significant degradation for LDPE thin films that were kept in direct contact with sewage sludge. Biodegradation by burial into soil has also been previously reported for polymers like poly(vinyl chloride) (PVC) and plasticized PVC (pPVC).

Plate Assay

For the plate assay technique, cellulose blended polymer films are incubated on a medium of mineral salt agar, which are inoculated with isolated microbial strains. Only the growth of strain PV1 has been observed on the films. The visibility of clear zones surrounding the colonies provides an indication of the polymer getting enzymatically hydrolyzed into products that are water soluble. After six months of burial into soil, the polymer-cellulose blended film surface gets completely changed. This observation has been found missing in the control samples. The polymer-cellulose blended film samples get discolored as well as some hyphal growth on the film surface gets clearly visible to the bare eye. Visibility of film erosion and cracks has also been reported (Ali *et al.*, 2009).

Shake Flask Analysis

Polymer films blended with cellulose have been incubated under shaking condition with the microorganisms isolated from soil burial analyses. Mineral salt media contains KH_2PO_4 (1 g), K_2HPO_4 (0.2 g), NaCl (1 g), $CaCl_2.2H_2O$ (0.002 g), boric acid (0.005 g), $(NH_4)_2SO_4$ (1 g), $MgSO_4.7H_2O$ (0.5 g), $MnSO_4.H_2O$ (0.001 g), $ZnSO_{4.}7H_2O$ (0.001 g), $CuSO_4.5H_2O$ (0.001 g), $FeSO_{4.}7H_2O$ (0.01 g) per 1000 mL of distilled water. Three pieces of cellulose blended polymer films in 90 mL of mineral salt media have been inoculated with spore suspension (10 mL), which contains $10\pm2.1\times10^6$ spores mL^{-1} and have been incubated for a period of 3 months at a temperature of 30°C. The polymer samples have been retrieved after every 4 weeks and evaluated visually and also with the help of infrared spectroscopy (El-Shafei *et al.*, 1998).

Sturm Test

Biodegradation of cellulose blended polymer results in CO_2 evolution, which can be measured by Sturm Test. Several polymer pieces have been introduced into the culture bottles, which contains 285 mL of mineral salt media devoid of any source of carbon. *Phanerochaete chyrosporium* spore suspension is employed as inoculum 5% (v/v) in control and test bottles (devoid of plastic). Supply of sterilized air is done to maintain aerobic conditions and continuous stirring of reaction bottles are done by magnetic stirrer. Gravimetric tests to analyze the production of CO_2 have been performed after a period of 30 days by trapping the gaseous CO_2 in KOH-containing adsorption bottle. 1 M concentration of KOH has been used for this purpose. The precipitates, thus formed of the control and the test samples, after titration with a 1 M solution of barium chloride have been filtered and weighed; followed by calculating the production of CO_2 per liter.

Chiellini, Corti, aand Swift (2003) have used another sophisticated method, known as respirometric biodegradation test. Here biodegradation analyses are performed in cylindrical glass vessels of 500 mL to 600 mL capacity, known as biometer flasks. These flasks contain a multilayer substrate, a specified amount of forest sandy soil (10 g to 15 g) or mature compost (5 g). LDPE test samples can be positioned as the middle layer between 35 mg to 70 mg per g of soil and 70 mg per g of mature compost concentrations. The flasks can then be placed in dark, followed by incubation at 55°C in case of mature compost being employed as the incubation medium or at room temperature for soil burial tests.

In order to trap the evolved CO_2 from the control and the test samples, the test vessels have to be fitted with a beaker containing 40 mL to 50 mL of KOH solution (0.05 N) and back titrated with 0.1 N HCl. This solution has to be substituted after every 3 days to 14 days. The extent of biodegradation of the test samples can be measured as a fraction of the total theoretical CO_2 produced. This total CO_2 produced can be calculated from the measured carbon content of the sample (Pramila & Ramesh, 2011a, 2011b).

Composting Technique

Composting is basically done under controlled conditions. A heat insulated composting chamber is used for this purpose. A standard mixture of raw materials utilized for the purpose of cultivating white button mushroom (*Agaricus bisporus*) is utilized. Polymer foil specimens placed in small containers made from stainless steel can be buried into the pile of compost for durations of 180 days, 100 days or 42 days. The temperature of the pile is maintained at ~60°C during the initial weeks.

Fungal Treatment

Certain fungal strains can be acquired from the *Basidiomycetes* culture collection. These cultures are pretreated using nutrient and glucose rich extracted agar in petridishes. Surface sterilization of polymer filaments, using 3% H_2O_2 for 2 min, are then placed on agar medium surface, and simultaneously inoculated with the fungus. Other degradation experiments are carried out in liquid media, where specified strains of fungus are statically cultivated in Erlenmeyer flask (250 mL), containing 15 mL of culture media within which polymer film sample pieces are to be immersed. Prior to insertion within nutrient medium, the LDPE pieces are surface sterilized using 3% H_2O_2 for 2 min. For this purpose, two culture media can be employed, namely malt extract and nitrogen limited media.

LDPE Blends and Composites

LDPE can be blended with PP in order to enhance its physical properties, like its impact resistance at low temperatures. The constituent detail of this blend is critical for attainment of a balanced performance. Judicious mixing of the pristine homopolymers essentially eliminates the requirement for expensive chemical synthesis of new block copolymers. Researchers have clearly reported the influence of the compatibilizer poly(acrylic acid) (PAA) on X-ray diffraction, mechanical properties, water absorption and morphology properties of composites of LDPE/nanoclay, fabricated by melt intercalation. The obtained enhancements in modulus of elasticity and tensile strength have been attributed to the homogenous dispersion of nanoclay within the LDPE matrix which got induced by the presence of PAA. Another important methodology of preparing blends is by mixing the melts of LDPE and PP, using a single screw extruder.

Obasi and Onuegbu (2013) reported that the mechanical properties of LDPE/waste maize cob flour (WMCF) composites with different filler loadings were found to be insignificant. Further compatibilization of LDPE/WMCF with maleated polyethylene exhibited lower water uptake compared to the uncompatibilized LDPE/WMCF. Blending of LDPE with wheat maize cob flour were performed in an injection machine, using a 50 rpm screw speed and a temperature range of 150°C to 170°C. Biodegradability was analyzed from weight loss of the composite within a certain time interval. Soil burial test was used to biodegrade the composite, which continued for a week. The percentage of weight loss was measured using the formula:

$$\text{Weight loss } (\%) = [(M_0 - M_d)/M_0] \times 100 \quad (6)$$

where, M_0 and M_d represent respectively the initial mass and the degradation mass at each designated week.

The loss in weight of WMCF indicated that the biodegradation degree increased as the amount of filler content increased. Here, the composites which had 60% of the filler content, exhibited rapid decrease within six weeks. Thus, it was confirmed that the presence of soil microorganisms, like bacteria and fungi, created pores within the LDPE matrix by consuming the maize filler.

Constantin, Răut, Vasilescu, Arsene, and Jecu (2012) have prepared blend films with 75% LDPE, 15% wood flour and 10% glycerol. They have used isolated strains of *Fusarium* sp. (strain 2, 6 and 9), *Aspergillus* sp. (strain 5) and *Penicillium* sp. (strain 7) for the biodegradation process. Here, the cultures along with the LDPE blends were kept in an agitating condition at 160 rpm for 3 months on a rotary shaker by maintaining a temperature of 28°C (Constantin *et al.*, 2012). The morphological analysis of the biodegraded film surfaces was performed using SEM, where the micrographs at different magnifications of the samples were analyzed in order to find out the changes that occurred during the degradation process on the sample surfaces. Certain observations were made, including creation of holes, peeling and exfoliation within the structure of the films. The weight loss percentage of the inoculated sample was measured before and after the microbial treatment. The weight loss percentage of the inoculated samples was calculated using the method reported by Obasi and Onuegbu (2013). The highest rate of biodegradation was observed in case *Aspergillus niger* 105, where the rate of degradation was found to be 5.95%; followed by *Phanerochaete chrysosporium* with a rate of 4.25%, *Fusarium* species strains (2, 6 and 9) with a degradation rate of 4.05%, 4.00% and 3.90% respectively. The lowest rate of degradation was shown by *Penicillium* species strain 7 with a degradation rate of 2.70%. The results obtained in this work was later confirmed by several other works (Mathur, Mathur, & Prasad, 2011; Konduri, Anupam, Vivek, & Kumar, 2010).

Zan, Fa, and Wang (2006) have characterized the prepared LDPE-TiO_2 nanocomposite films by using SEM, HT-GPC, FT-IR spectroscopy, XPS and weight loss by photo-induction. They have reported that the nanocomposite films were efficiently degradable under sun or UV light. The composite films exhibited a 68.38% loss of the original weight and 93.75% loss of the initial average molecular weight. FT-IR and XPS analyses indicated that LDPE was oxidized during UV light irradiation.

El-Rehim, Hegazy, Ali, and Rabie (2004) aimed on accelerating the biodegradation of LDPE by blending it with starch, followed by exposing the LDPE/starch blends to UV and soil burial treatments. Alterations that occurred in the morphological, structural and thermal properties of the blends after the UV and microbial treatments were observed with DSC, FT-IR and SEM. These techniques clearly showed the synergistic influence on the biodegradation process of the blends when they were exposed to light followed by soil burial. The obtained results indicated that photo-induced oxidation accelerate the rate of biodegradation of the blends; since oxidized LDPE/starch is very susceptible to biotic reactions. Another implication of the treatments (i.e. photo and soil burial) was the reduction in the blends' tensile strength of up to 20% to 66%. However, blends treated with only soil burial exhibited a reduction in tensile strength of up to 10% to 26%.

LDPE/starch composites were found to absorb moisture in accordance to the amount of added starch. This moisture absorption resulted in reduction of their mechanical properties and enhancement in their susceptibility to fungal attack. Hydrolysis of the starch granules, in presence of α-amylase solution, was found to be restricted only on the surface. This, in turn, led to a notable reduction in the mechanical properties. Soil burial resulted in surface tearing, creation of holes, reduction in mechanical properties and weight loss. Embrittlement, total matrix collapse and exposure of the embedded granules of starch represent some of the alterations that occur in the composites.

Danjaji, Nawang, Ishiaku, Ismail, and Ishak (2002) have reported a 15% weight loss in linear LDPE (LLDPE)/starch composite after immersion for a period of 14 days in α-amylase solution. They have also observed decrease in elongation at break and tensile strength with immersion duration. Similar to this work, Strantz and Zottola (1992) have also reported that the elongation at break and the tensile strength got decreased in LDPE-corn starch composites.

Borghei, Karbassi, Khoramnejadian, Oromiehie, and Javid (2010) performed degradation of LDPE/potato starch and pristine LDPE strips in soil. The soil was composed of garden soil, municipal waste, compost and an extra decomposer (*P. aeruginosa*). Weight losses of the polymer strips indicated the rate of biodegradation. The microorganisms, at first, got attracted to the potato starch component of the blend. This was followed by consumption of the starch present within the polymer matrix by the microbes. This resulted in causing fragmentations of the LDPE chains, owing to the presence of maleic anhydride. FT-IR analyses was done in order to confirm the biodegradation rate before and after the burial of the polymer strips. The results obtained revealed a little loss in weight for pure LDPE. This loss in weight was owing to the biodegradation caused by *P. aeruginosa*; since media free from carbon can cause degradation of LDPE (Agamuthu & Faizura, 2005). On the other hand, LDPE/potato starch showed a significant loss in weight. In this work, potato starches were the only carbon source for the microbes. The weight change of LDPE/potato starch composite after exposure to *P. aeruginosa* for 84 days has been explained in Table 10 (Borghei *et al.*, 2010).

Works have also been carried out using unpretreated and thermally pretreated LDPEs and HDPEs, and unpretreated LDPE/starch blends, subjected to *in vitro* biodegradation. Specifically, two microorganisms have been selected, namely *Bacillus cereus* subgroup A and *Bacillus sphericus* GC subgroup IV (Danjaji *et al.*, 2002). The PEs have been used as the only source of carbon. They have been kept for 1 year at 30°C under a pH of 7.5. The loss in weight of the heat-treated HDPE and LDPE samples were found to be 9% and 19%, respectively; whereas, the untreated samples exhibited 3.5% and 10% weight loss, respectively, with *Bacillus sphericus*. On the other hand, the loss in weight of the LDPE/starch blends with *Bacillus cereus* was found to be 25%. The thermally pretreated HDPE, unpretreated LDPE/starch blends and thermally pretreated LDPE exhibited decrease in tensile strengths by 14.8%, 30.5% and 27%,

Table 10. Weight change of LDPE/potato starch composite after exposure to P. aeruginosa for 84 days (Borghei et al., 2010)

Samples	Weight Loss (%)
LDPE	0.43
LDPE/Potato starches	2.19

respectively. Similarly, the crystallinity was also found to decrease to 2.2%, 8.5% and 8%, respectively. The polymer surfaces were found to become more hydrophilic after exposure.

Several reports demonstrated the biodegradation rate of LDPE/starch composites by sampling the LDPE films for over 28 months. Continued fragmentation resulted in weight losses of 49% to 95% for the LDPE/starch composite films. This magnitude of degradation of the films within the site of exposure was analyzed from the toughness of the plastic sheet samples. It was found that the plastic sheets were able to retain only 7% of the initial toughness (Kaur & Gautam, 2010).

Kaur and Gautam (2010) have reported that the percentage of weight loss for samples treated for a duration of 4 months varied under two different soil conditions. In the first case, an 88% weight loss was found when the LDPE-starch composite was buried under simple soil. In the second case, the weight loss was found to be 90% when buried under urea enriched soil.

Several reports showed that 30% starch blended with LDPE exhibited maximum biodegradation, leading to a loss of 6.3% weight. Biodegradation of the same film was found to accelerate upon incubation with an acclimatized culture in a shake flask, showing 11.2% loss of weight after only 6 months. FT-IR and C-NMR analyses have revealed loss of starch in the biodegradable films. This was accompanied by alterations in the LDPE structure, such as formation of short LDPE chain branches and decrease in the LDPE backbone peak area (Ruiz-Dueñas & Martínez, 2009). It is a well-known fact that UV light possesses a decaying influence on LDPE substrates. When exposed to open environment, LDPE undergoes significant changes, like photo-degradation, which leads to loss of mechanical properties. This photo-degradation depends on the intensity of sunlight and other environmental parameters. Studies have demonstrated no significant alterations in the tensile strengths of LDPE/starch blends, containing different weight percentages of incorporated starch, and pristine LDPE with variable duration of exposure to sunlight. The photo-oxidation of LDPE/starch blend and pristine LDPE can be distinguished by a period of induction in which uptake of oxygen, which is responsible for the formation of intermediate products including alcohols, peroxides, ketones and hydroperoxides, takes place at a very slow rate. The effect of light irradiation on LDPE/starch blend and pristine LDPE gets mainly induced by slight scission of chains, resulting in reorientation of chains to shorter and more readily crystallizable forms. Several studies indicated that the rate of biodegradation of LDPE/starch blends is much higher when exposed to UV light/soil burial treatment as compared to similarly treated pristine LDPE. This concludes that starch plays a pivotal role in increasing the rate of LDPE biodegradation. On the other hand, studies revealed that LDPE samples containing pro-oxidant demonstrated greatest susceptibility towards thermal degradation. In this case, starch acts as an inhibitor and reduces the rate of thermal degradation of LDPE.

Mechanisms of Biodegradation of LDPE Films

The mechanisms of biodegradation of LDPE can be studied from three different perspectives: colonization on LDPE by microorganisms, chemical or biochemical reactive pathways and the impact of macromolecular structure of the polymer on microbial usage.

LDPE is a hydrophobic molecule; and therefore, it has been usually accepted that colonization of biofilm is the introductory step for its degradation (Cases & de Lorenzo, 2005). Various bacteria, such as *Streptomyces setonii* 75Vi2, *Streptomyces badius* 252, *Streptomyces viridosporus* T7A and wood degrading fungi produce some extracellular enzymes which results in degradation of LDPE (Hadad *et al.*, 2005). In wood degrading fungi, the extracellular enzymatic complex (ligninolytic system) contains oxidases, laccases and peroxidases, which leads to formation of extracellular H_2O_2 (Ruiz-Dueñas & Martínez, 2009). Depending upon the type of the organism or strain and culture condition, the characteristics of this enzyme system varies (Kirk & Farrell, 1987). For degradation of lignin, three enzymes such as manganese peroxidase, lignin peroxidase and phenoloxidase containing copper (commonly called laccase) are used (Maciel, e Silva, & Ribeiro, 2010; Wariishi, Valli, & Gold, 1989; Kim, Yeo, Kum, Song, & Choi, 2005). Lingolytic enzymes are used in various industries, such as agricultural, chemical, cosmetics, food, fuel, paper, textile and more interestingly they are also reported to induce the degradation process of xenobiotic compounds and dyes (Maciel *et al.*, 2010). During lignin degradation, phenolic compounds are oxidized in the presence of H_2O_2 and manganese by manganese peroxidase. Manganese peroxidase further oxidizes Mn^{2+} to Mn^{3+}; and monomeric phenols, phenolic lignin dimmers and synthetic lignin are in turn oxidized by Mn^{3+} via formation of phenoxy radicals as intermediates (Koutny *et al.*, 2006b; Albertsson, 1980; Wariishi, Valli, & Gold, 1989; Kim, Yeo, Kum, Song, & Choi, 2005). The LDPE by-products vary depending upon the conditions of degradation. Under aerobic conditions, CO_2, H_2O and microbial biomass are the final degradation products; whereas, in case of anaerobic or methanogenic conditions, CO_2, H_2O, CH_4 and microbial biomass are the end products. Again, under sulfidogenic conditions, H_2S, CO_2, H_2O and microbial biomass are reported to be the end products (Arutchelvi *et al.*, 2008).

Biofilms are sessile communities of microorganisms developed on a surface that can be composed of individuals from the same or different species (Cunliffe, Smart, Alexander, & Vulfson, 1999). Complex biofilm communities comprising of different microorganisms have been detected on LDPE surfaces once they are exposed to different biotic environments (Gilan *et al.*, 2004; Hadad *et al.*, 2005; Koutny *et al.*, 2006b; Fontanella *et al.*, 2010; Arkatkar *et al.*, 2009; Orhan & Büyükgüngör, 2000; Chiellini *et al.*, 2003; Nowak *et al.*, 2011; Alok *et al.*, 2008; Albertsson *et al.*, 1995; Koutny *et al.*, 2009; Fontanella *et al.*, 2010; Sudhakar *et al.*, 2008; Cunliffe *et al.*, 1999; Lobelle & Cunliffe, 2011). Studies on microorganism attachment to LDPE have identified that the main limitation of the colonization process is the relatively high hydrophobicity of the LDPE in contrast to the regularly hydrophilic surfaces of most microorganisms (Hadad *et al.*, 2005; Arkatkar *et al.*, 2009). It has been proposed that strains with more hydrophobic surfaces can potentially play a major role in the initial colonization of LDPE. Other metabolic adaptations that are important in LDPE colonization are the production of surfactants and molecules that can mediate the attachment process of microorganisms to the hydrophobic surface (Arkatkar *et al.*, 2009; Karlsson *et al.*, 1988). Theoretically, LDPE can be potentially utilized as a source of carbon for microorganisms that are similar to many other hydrocarbons; however, its high molecular weight limits its use as a substrate for enzymatic reactions to take place. In terms of the chemical or biochemical processes involved in LDPE biodegradation, it can be stated that there are two key reactions: the first one being the reduction of its molecular weight and the second being the oxidation of the molecules. Reduction of molecular weight is required for two reasons, firstly to enable transport of molecules through the cell membrane and secondly because enzymatic systems present in the microorganisms are only able to attack certain molecular weights, usually in the range of 10 to 50 carbons, though there has been a report of enzymatic activity up to 2000 carbons (Yoon *et al.*, 2012).

Once the size of the molecule is reduced, oxidation is required in order to transform the hydrocarbon into a carboxylic acid that is metabolized by means of beta oxidation and Krebs cycle (Albertsson *et al.*, 1995). Figure 1 shows the proposed mechanisms of biodegradation for PE. Both oxidation and molecular weight reduction during the biodegradation process are a result of synergistic effects between biotic and abiotic factors (photo-oxidation or heat treatment). There are several papers reporting both the formation of carbonyl groups (oxidation) and reduction of molecular weight after treatment with UV

Table 11. Brief summary for the biodegradation of PE and LDPE samples and composites

Material	Condition	Time/Duration of Degradation	Organism	Reported Degradation (Observation)	Reference Number
LDPE	Culture medium at 29°C	30 days	*Aspergillus fumigatus* and *Penicillium* sp.	*Penicillium* sp. – 6.58% *Aspergillus fumigatus* – 4.65%	Singh *et al.*, 2012
LDPE	Petriplates containing synthetic medium without yeast extract at room temperature	1 week to 28 days.	*Aspergillus flavus* and *Mucor cicinelloides*	After 7 days the fresh weight of *Aspergillus flavus*was 0.015 whereas after 28 days the fresh weight was 0.34. In case of *Mucor cicinelloides* the fresh weight was 0.007 after 7 days and after 28 days the fresh weight was 0.029	Pramila & Ramesh, 2011a
LDPE	LDPE sheets were kept in growth medium with 0.2% glucose and 0.2% LDPE	*In vitro* assay for 4 days	Different strains of *Pseudomonas citronellois*	Four strains had different degradation rates 5%, 17.8%, 0.9% and 0.6%. *Pseudomonas citronellois* EMBS027 strain showed maximum degradation rate of 17.8% based on weight loss	Bhatia *et al.*, 2014
HDPE, LDPE	Minimal broth with LDPE and HDPE	2 weeks	*Microbacterium* sp., *Pseudomonas putida,* Bacterium Te68R.	Consortium treated samples of HDPE had a greater extent of weight loss of 22.41% as compared to LDPE carrying out 21.7% weight loss at 400°C	Alok *et al.*, 2008
PE	(i) Site of disposal of domestic waste, which are dumped with vegetable waste and household garbage (ii) soil from drainage site of effluents from textile industries (iii) soil dumped with sewage sludge	Incubation in shake flask for a period of 8 weeks	*Pseudomonas* sp.	(i) *Pseudomonas* sp. from sewage sludge dump degraded 29.1% for synthetic and 46.2% for natural PE (ii) biodegradability rate was 16.3% for synthetic and 31.4% for natural PE for household garbage dump (iii) from drainage sites of effluents from textile industries, biodegradability was 19.7% and 39.7% for natural and synthetic PE, respectively	Nanda *et al.*, 2010
PE	PE films, preheated at 70°C for 10 days, were disinfected chemically and incubated at 125 rpm in 0.6% yeast extract medium at pH 7.5 and 3% yeast extract medium at pH 5.5.	1, 2 and 4 weeks	*Mucor rouxii* NRRL 1835 and *Aspergilus flavus* as fungi and eight different *Streptomyces* sp.	Heat treatment for 10 days at 70°C resulted in exhibition of 60% elongation in *Streptomyces* sp. and 46.5% in fungi *Mucor rouxii*	El-Shafei *et al.*, 1998
LDPE (20 µm thick)	Bacterial isolates were grown in nutrient broth with PE film as the only source of carbon	30 days at 37°C	bacterial isolates (namely 1and 2)	Incubated untreated LDPE with bacterial isolate 2 and 1 causes reduction in mass by 1.3% and 1.29%, respectively	Sudhakar *et al.*, 2008
PE	Sterilized PE samples were at UV exposure at 254 nm; controls and assays were incubated for a period of 6 months	6 months at 27°C	*Cladosporium cladosporoides* ATCC 20251 and *Rhodococcus rhodochrous* ATCC 29672 (bacteria)	Surface erosion of the film was visualized in the vicinity of the microbes and decay of oxidation products in the polymer film surface	Bonhomme *et al.*, 2003

continued on following page

Table 11. Continued

Material	Condition	Time/Duration of Degradation	Organism	Reported Degradation (Observation)	Reference Number
PE	Spores are added to media for fungal growth, spores were spread on 3 kinds of plates: i.) medium containing no PE ii.) medium containing PE (0.1%) iii.) medium containing PE (0.5%)	28°C for 1 week	*Penicillium simplicissimum* YK	PE, possessing a molar mass of 1000 to10000, cultivated for 1 month, or 4000 to 28000, cultivated for 3 months, with fungal hyphae	Yamada-Onodera *et al.*, 2001
LDPE	30 mL of basal mineral medium with 600 µL of bacterial stock solution were mixed with 9 mg of LDPE films	120 days at 37°C.	*Pseudomonas aeruginosa* (ATCC 15692), *Pseudomonas syringae* (DC3000 ATCC 10862), *Pseudomonas putida* (KT2440, ATCC 47054) and *Pseudomonas aeruginosa* PA01 (ATCC 15729)	*Pseudomonas* strain causes weight loss of upto 20% in the tested PE within 120 days	Kyaw *et al.*, 2012
PE	PE pieces (100 mg) and bacterial suspensions (10 mL) were added with 40 mL of liquid carbon-free basal medium (LCFBM)	60 days at 30°C on a rotary shaker	*Enterobacter asburiae* and *Bacillus* sp.	*Enterobacter asburiae* degrade approximately 6.1 ± 0.3% of PE films (100 mg) and *Bacillus* sp. degrade approximately 10.7 ± 0.2% of the PE films (100 mg)	Yang *et al.*, 2014
PE carry bags	Regular shaking of the PE bags with culture media	3 months at shaking condition	*Bacillus cereus* and *Pseudomonas* sp.	After 3 months of regular shaking the PE discs were corroded on the surface and tensile strength decreases and maximum 12.5% weight loss was recorded	Yabannavar & Bartha, 1994
LPDE in powder form	Sea water	1, 2 and 3 weeks	*Aspergillus versicolor* and *Aspergillus* sp.	Per week maximum 4.1594 gL^{-1} of CO2 was released after degradation of the PE	Pramila & Ramesh, 2011b
LDPE and PE	Plastics and soil from the plastic dumping site	15, 30 and 45 days' time interval	*Pseudomonas stutzeri*	After 45 days maximum change in percent extension (73.38% reduction), tensile strength (0.01 Ncm^{-2} and it was similar even after 15 and 30 days) and elongation (1.8 cm) of the PE was recorded	Sharma & Sharma, 2004
HDPE and LDPE	Mangrove soil sample from Suva, Fiji Islands	8 weeks of incubation	*Bacillus, Micrococcus, Listeria* and *Vibrio*	Nearly 5% of weight loss after a period of eight weeks	Kathiresan, 2003

light (Albertsson *et al.*, 1995; Volke-Sepúlveda *et al.*, 2002; Balasubramanian *et al.*, 2010; Fontanella *et al.*, 2010; Tribedi & Sil, 2013).

The biotic factor is determined by groups of enzymes which are able to degrade oxidized or reduced LDPE molecules. However, there are very few works devoted to studying the enzymes involved in these processes. Breaking down large LDPE molecules can be accomplished by enzymatic action, as proven by Santo *et al.* (2013). They have found that by incubation with the enzyme laccase, the molecular weight of LDPE got reduced and its keto-carbonyl index got increased. These two factors indicate that both scission and oxidation reactions are catalyzed by the same enzyme. In regards to the oxidation process, there is another important work performed by Yoon *et al.* (2012). They have isolated an alkane hydroxylase from the AlkB family that was active to LDPE samples with molecular weights up to 27,000 Da. It is interesting to note that enzymes of this family have been described as microbes that have the ability to degrade several hydrocarbons. Thus, it has been concluded that alkane hydroxylase performs the first oxidation that leads to the subsequent degradation of a hydrocarbon (Ali *et al.*, 2009). A brief summary of the various studies reported in the literature are hereby presented to estimate the rate of biodegradation of PE samples and various PE composites (Table 11).

CONCLUSION

The ever-increasing use of LDPE has resulted in a great challenge regarding its degradation. It has been reported in a survey that ~500 billion to 1 trillion LDPE carry bags get consumed worldwide per year, which has been found to be durable. This huge quantity will require up to 1000 years to get naturally degraded in the environment. The LDPE composites (i.e. LDPE mixed with starch, cellulose, pro-oxidant additives, etc.) have been developed to make LDPE biodegradable; however, the problem with these composites lie in their processing. New materials need new process requirements, such as extruders; and this causes escalation in cost. As a result, the composite films are costly and there are few takers in their daily applications. Since pure LDPE films possess a micron level (~10 μm to 20 μm) thickness, therefore the requirement is that they should be biodegradable after their use; otherwise they will cause environmental problems by littering. The present chapter also described the *in situ* biodegradation of LDPE, which depends on several factors like hydrophilicity or hydrophobicity, thickness of the film, types of microbes engaged for biodegradation, etc. It appeared that when hydrophilicity is more, the rate of biodegradation is also more. Again, the microbes which are oxidative produce more effective biodegradation rate. Therefore, *in situ* biodegradation process renders the film hydrophilic; thereby, forming biofilm on the surface by the microbes. This results in a faster rate of degradation.

REFERENCES

Abraham, J., Ghosh, E., Mukherjee, P., & Gajendiran, A. (2017). Microbial degradation of low density polyethylene. *Environmental Progress & Sustainable Energy*, *36*(1), 147–154. doi:10.1177/0734242X05051045 PMID:15864950

Agamuthu, P., & Faizura, P. N. (2005). Biodegradability of degradable plastic waste. *Waste Management & Research*, *23*(2), 95–100. doi:10.1177/0734242X05051045 PMID:15864950

Albertsson, A.-C. (1980). The shape of the biodegradation curve for low and high density polyethenes in prolonged series of experiments. *European Polymer Journal*, *16*(7), 623–630. doi:10.1016/0014-3057(80)90100-7

Albertsson, A.-C., Andersson, S. O., & Karlsson, S. (1987). The mechanism of biodegradation of polyethylene. *Polymer Degradation & Stability*, *18*(1), 73–87. doi:10.1016/0141-3910(87)90084-X

Albertsson, A.-C., Barenstedt, C., Karlsson, S., & Lindberg, T. (1995). Degradation product pattern and morphology changes as means to differentiate abiotically and biotically aged degradable polyethylene. *Polymer*, *36*(16), 3075–3083. doi:10.1016/0032-3861(95)97868-G

Albertsson, A.-C., Erlandsson, B., Hakkarainen, M., & Karlsson, S. (1998). Molecular weight changes and polymeric matrix changes correlated with the formation of degradation products in biodegraded polyethylene. *Journal of Environmental Polymer Degradation*, *6*(4), 187–195. doi:10.1023/A:1021873631162

Albertsson, A.-C., & Karlsson, S. (1990). The influence of biotic and abiotic environments on the degradation of polyethylene. *Progress in Polymer Science*, *15*(2), 177–192. doi:10.1016/0079-6700(90)90027-X

Ali, M. I., Perveen, Q., Ahmad, B., Javed, I., Razi-Ul-Hussnain, R., Andleeb, S., ... Hameed, A. (2009). Studies on biodegradation of cellulose blended polyvinyl chloride films. *International Journal of Agriculture and Biology*, *11*, 577–580.

Alok, S., Soni, R., Zaidi, M., Shouche, Y., & Goel, R. (2008). Comparative biodegradation of HDPE and LDPE using an indigenously developed microbial consortium. *Journal of Microbiology and Biotechnology*, *18*, 477–482. PMID:18388465

Antunes, M. C., Agnelli, J. A. M., Babetto, A. S., Bonse, B. C., & Bettini, S. H. P. (2017). Abiotic thermos-oxidative degradation of high density polyethylene: Effect of manganese stearate concentration. *Polymer Degradation & Stability*, *143*, 95–103. doi:10.1016/j.polymdegradstab.2017.06.012

Arkatkar, A., Arutchelvi, J., Bhaduri, S., Uppara, P. V., & Doble, M. (2009). Degradation of unpretreated and thermally pretreated polypropylene by soil consortia. *International Biodeterioration & Biodegradation*, *63*(1), 106–111. doi:10.1016/j.ibiod.2008.06.005

Artham, T., Sudhakar, M., Venkatesan, R., Nair, C. M., Murty, K. V. G. K., & Doble, M. (2009). Biofouling and stability of synthetic polymers in sea water. *International Biodeterioration & Biodegradation*, *63*(7), 884–890. doi:10.1016/j.ibiod.2009.03.003

Arutchelvi, J., Sudhakar, M., Arkatkar, A., Doble, M., Bhaduri, S., & Uppara, P. V. (2008). Biodegradation of polyethylene and polypropylene. *Indian Journal of Biotechnology*, *7*, 9–22.

Awasthi, S., Srivastava, N., Singh, T., Tiwary, D., Mishra, P. K. (2017). Biodegradation of thermally treated low density polyethylene by fungus *Rhizopus oryzae* NS5. *3 Biotech, 7*, 73.

Balasubramanian, V., Natarajan, K., Hemambika, B., Ramesh, N., Sumathi, C. S., Kottaimuthu, R., & Kannan, V. R. (2010). High-density polyethylene (HDPE)-degrading potential bacteria from marine ecosystem of Gulf of Mannar, India. *Letters in Applied Microbiology*, *51*, 205–211. PMID:20586938

Barnes, D. K. A., Galgani, F., Thompson, R. C., & Barlaz, M. (2009). Accumulation and fragmentation of plastic debris in global environments. *Philosophical Transactions of the Royal Society of London. Series B, Biological Sciences, 364*(1526), 1985–1998. doi:10.1098/rstb.2008.0205 PMID:19528051

Bhatia, M., Girdhar, A., Tiwari, A., & Nayarisseri, A. (2014). Implications of a novel *Pseudomonas* species on low density polyethylene biodegradation: An *in vitro* to *in silico* approach. *SpringerPlus, 3*, 497.

Bonhomme, S., Cuer, A., Delort, A.-M., Lemaire, J., Sancelme, M., & Scott, G. (2003). Environmental biodegradation of polyethylene. *Polymer Degradation & Stability, 81*(3), 441–452. doi:10.1016/S0141-3910(03)00129-0

Borghei, M., Karbassi, A., Khoramnejadian, S., Oromiehie, A., & Javid, A. H. (2010). Microbial biodegradable potato starch based low density polyethylene. *African Journal of Biotechnology, 9*, 4075–4080.

Breslin, V. T., & Swanson, R. L. (1993). Deterioration of starch-plastic composites in the environment. *Journal of the Air & Waste Management Association, 43*(3), 325–335. doi:10.1080/1073161X.1993.10467137

Brydson, J. A. (1999). *Plastic materials*. Oxford, UK: Butterworth-Heinemann.

Burlein, G. A. D., & Rocha, M. C. G. (2014). LDPE/PHB blends filled with castor oil pressed cake. *Materials Research, 17*(1), 203–212. doi:10.1590/S1516-14392013005000166

Cases, I., & de Lorenzo, V. (2005). Genetically modified organisms for the environment: Stories of success and failure and what we have learned from them. *International Microbiology, 8*, 213–222. PMID:16200500

Chatterjee, S., Roy, B., Roy, D., & Banerjee, R. (2010). Enzyme-mediated biodegradation of heat treated commercial polyethylene by *Staphylococcal* species. *Polymer Degradation & Stability, 95*(2), 195–200. doi:10.1016/j.polymdegradstab.2009.11.025

Chiellini, E., Corti, A., & Swift, G. (2003). Biodegradation of thermally-oxidized, fragmented low-density polyethylenes. *Polymer Degradation & Stability, 81*(2), 341–351. doi:10.1016/S0141-3910(03)00105-8

Constantin, M., Răut, I., Vasilescu, G., Arsene, M. L., & Jecu, L. (2012). Colonization and degradation of polyethylene composites by fungal strains isolated. *Scientific Bulletin. Series F. Biotechnologies, 16*, 109–112.

Cunliffe, D., Smart, C. A., Alexander, C., & Vulfson, E. N. (1999). Bacterial adhesion at synthetic surfaces. *Applied and Environmental Microbiology, 65*, 4995–5002. PMID:10543814

Danjaji, I. D., Nawang, R., Ishiaku, U. S., Ismail, H., & Ishak, Z. A. M. M. (2002). Degradation studies and moisture uptake of sago-starch-filled linear low-density polyethylene composites. *Polymer Testing, 21*(1), 75–81. doi:10.1016/S0142-9418(01)00051-4

Dashtban, M., Schraft, H., Syed, T. A., & Qin, W. (2010). Fungal biodegradation and enzymatic modification of lignin. *International Journal of Biochemistry and Molecular Biology, 1*, 36–50. PMID:21968746

Dave, H., Rao, P. V. C., & Desai, J. D. (1997). Biodegradation of starch-polyethylene films in soil and by microbial cultures. *World Journal of Microbiology & Biotechnology, 13*(6), 655–658. doi:10.1023/A:1018566820659

Donlan, R. M. (2002). Biofilms: Microbial life on surfaces. *Emerging Infectious Diseases, 8*(9), 881–890. doi:10.3201/eid0809.020063 PMID:12194761

El-Rehim, H. A. A., Hegazy, E.-S. A., Ali, A. M., & Rabie, A. M. (2004). Synergistic effect of combining UV-sunlight-soil burial treatment on the biodegradation rate of LDPE/starch blends. *Journal of Photochemistry and Photobiology A Chemistry, 163*(3), 547–556. doi:10.1016/j.jphotochem.2004.02.003

El-Shafei, H. A., El-Nasser, N. H. A., Kansoh, A. L., & Ali, A. M. (1998). Biodegradation of disposable polyethylene by fungi and *Streptomyces* species. *Polymer Degradation & Stability, 62*(2), 361–365. doi:10.1016/S0141-3910(98)00019-6

Eubeler, J. P., Bernhard, M., & Knepper, T. P. (2010). Environmental biodegradation of synthetic polymers II. Biodegradation of different polymer groups. *Trends in Analytical Chemistry, 29*(1), 84–100. doi:10.1016/j.trac.2009.09.005

Fantroussi, S. E., & Agathos, S. N. (2005). Is bioaugmentation a feasible strategy for pollutant removal and site remediation? *Current Opinion in Microbiology, 8*(3), 268–275. doi:10.1016/j.mib.2005.04.011 PMID:15939349

Fontanella, S., Bonhomme, S., Koutny, M., Husarova, L., Brusson, J.-M., Courdavault, J.-P., ... Delort, A.-M. (2010). Comparison of the biodegradability of various polyethylene films containing pro-oxidant additives. *Polymer Degradation & Stability, 95*(6), 1011–1021. doi:10.1016/j.polymdegradstab.2010.03.009

Gilan (Orr), I., Hadar, Y., & Sivan, A. (2004). Colonization, biofilm formation and biodegradation of polyethylene by a strain of *Rhodococcus ruber. Applied Microbiology and Biotechnology, 65*, 97–104. PMID:15221232

Gu, J. D., Ford, T. E., Mitton, D. B., & Mitchel, R. (2000). Microbial degradation of polymeric materials. In R. W. Revie (Ed.), *The Uhlig's corrosion handbook* (pp. 421–438). New York, NY: Wiley.

Hadad, D., Geresh, S., & Sivan, A. (2005). Biodegradation of polyethylene by the thermophilic bacterium *Brevibacillus borstelensis. Journal of Applied Microbiology, 98*(5), 1093–1100. doi:10.1111/j.1365-2672.2005.02553.x PMID:15836478

Hakkarainen, M., & Albertsson, A.-C. (2004). Environmental degradation of polyethylene. *Advances in Polymer Science, 169*, 177–199. doi:10.1007/b13523

Harshvardhan, K., & Jha, B. (2013). Biodegradation of low-density polyethylene by marine bacteria from pelagic waters, Arabian sea, India. *Marine Pollution Bulletin, 77*(1-2), 100–106. doi:10.1016/j.marpolbul.2013.10.025 PMID:24210946

Hasan, F., Shah, A. A., Hameed, A., & Ahmed, S. (2007). Synergistic effect of photo- and chemical treatment on the rate of biodegradation of low density polyethylene by Fusarium sp. AF4. *Journal of Applied Polymer Science, 105*(3), 1466–1470. doi:10.1002/app.26328

Howard, G. T. (2002). Biodegradation of polyurethane: A review. *International Biodeterioration & Biodegradation*, *49*(4), 245–252. doi:10.1016/S0964-8305(02)00051-3

Hugenholtz, P., Goebel, B. M., & Pace, N. R. (1998). Impact of culture-independent studies on the emerging phylogenetic view of bacterial diversity. *Journal of Bacteriology*, *180*, 4765–4774. PMID:9733676

Joutey, N. T., Bahafid, W., Sayel, H., & Ghachtouli, N. E. (2013). Biodegradation: Involved microorganisms and genetically engineered microorganisms. In R. Chamy & F. Rosenkranz (Eds.), *Biodegradation – Life of science* (pp. 289–320). InTech.

Karlsson, S., Ljungquist, O., & Albertsson, A.-C. (1988). Biodegradation of polyethylene and the influence of surfactants. *Polymer Degradation & Stability*, *21*(3), 237–250. doi:10.1016/0141-3910(88)90030-4

Kathiresan, K. (2003). Polythene and plastic degrading microbes from mangrove soil. *Revista de Biología Tropical*, *51*, 629–633. PMID:15162769

Kaur, I., & Gautam, N. (2010). Starch grafted polyethylene evincing biodegradation behaviour. *Malaysian Polymer Journal*, *5*, 26–38.

Kawai, F., Watanabe, M., Shibata, M., Yokoyama, S., Sudate, Y., & Hayashi, S. (2004). Comparative study on biodegradability of polyethylene wax by bacteria and fungi. *Polymer Degradation & Stability*, *86*(1), 105–114. doi:10.1016/j.polymdegradstab.2004.03.015

Khabbaz, F., & Albertsson, A. C. (2001). Rapid test methods for analyzing degradable polyolefins with a pro-oxidant system. *Journal of Applied Polymer Science*, *79*(12), 2309–2316. doi:10.1002/1097-4628(20010321)79:12<2309::AID-APP10383.0.CO;2-E

Kim, Y., Yeo, S., Kum, J., Song, H. G., & Choi, H. T. (2005). Cloning of a manganese peroxidase cDNA gene repressed by manganese in *Trametes versicolor. Journal of Microbiology (Seoul, Korea)*, *43*, 569–571. PMID:16410775

Kirk, T. K., & Farrell, R. L. (1987). Enzymatic "combustion": The microbial degradation of lignin. *Annual Review of Microbiology*, *41*(1), 465–505. doi:10.1146/annurev.mi.41.100187.002341 PMID:3318677

Konduri, M. K. R., Anupam, K. S., Vivek, J. S., & Kumar, D. B. R. (2010). Synergistic effect of chemical and photo treatment on the rate of biodegradation of high density polyethylene by indigenous fungal isolates. *International Journal of Biotechnology and Biochemistry*, *6*, 157–174.

Koutny, M., Amato, P., Muchova, M., Ruzicka, J., & Delort, A.-M. (2009). Soil bacterial strains able to grow on the surface of oxidized polyethylene film containing prooxidant additives. *International Biodeterioration & Biodegradation*, *63*(3), 354–357. doi:10.1016/j.ibiod.2008.11.003

Koutny, M., Lemaire, J., & Delort, A.-M. (2006a). Biodegradation of polyethylene films with prooxidant additives. *Chemosphere*, *64*(8), 1243–1252. doi:10.1016/j.chemosphere.2005.12.060 PMID:16487569

Koutny, M., Sancelme, M., Dabin, C., Pichon, N., Delort, A.-M., & Lemaire, J. (2006b). Acquired biodegradability of polyethylenes containing pro-oxidant additives. *Polymer Degradation & Stability*, *91*(7), 1495–1503. doi:10.1016/j.polymdegradstab.2005.10.007

Kumar, S., Hatha, A. A. M., & Christi, K. S. (2007). Diversity and effectiveness of tropical mangrove soil microflora on the degradation of polythene carry bags. *Revista de Biología Tropical, 55*(3-4), 777–786. doi:10.15517/rbt.v55i3-4.5954 PMID:19086383

Kyaw, B. M., Champakalakshmi, R., Sakharkar, M. K., Lim, C. S., & Sakharkar, K. R. (2012). Biodegradation of low density polythene (LDPE) by *Pseudomonas* species. *Indian Journal of Microbiology, 52*(3), 411–419. doi:10.100712088-012-0250-6 PMID:23997333

Kyrikou, I., & Briassoulis, D. (2007). Biodegradation of agricultural plastic films: A critical review. *Journal of Polymers and the Environment, 15*(2), 125–150. doi:10.100710924-007-0053-8

Lee, B., Pometto, A. L. III, Fratzke, A., & Bailey, T. B. Jr. (1991). Biodegradation of degradable plastic polyethylene by *Phanerochaete* and *Streptomyces* species. *Applied and Environmental Microbiology, 57*, 678–685. PMID:16348434

Lobelle, D., & Cunliffe, M. (2011). Early microbial biofilm formation on marine plastic debris. *Marine Pollution Bulletin, 62*(1), 197–200. doi:10.1016/j.marpolbul.2010.10.013 PMID:21093883

Luckachan, G. E., & Pillai, C. K. S. (2011). Biodegradable polymers – A review on recent trends and emerging perspectives. *Journal of Polymers and the Environment, 19*(3), 637–676. doi:10.100710924-011-0317-1

Maciel, M. J. M., Silva, A. C., & Ribeiro, H. C. T. (2010). Industrial and biotechnological applications of ligninolytic enzymes of the basidiomycota: A review. *Electronic Journal of Biotechnology, 13*. doi:10.2225/vol13-issue6-fulltext-2

Manzur, A., Cuamatzi, F., & Favela, E. (1997). Effect of the growth of *Phanerochaete chrysosporium* in a blend of low density polyethylene and sugar cane bagasse. *Journal of Applied Polymer Science, 66*(1), 105–111. doi:10.1002/(SICI)1097-4628(19971003)66:1<105::AID-APP123.0.CO;2-0

Manzur, A., Limón-González, M., & Favela-Torres, E. (2004). Biodegradation of physicochemically treated LDPE by a consortium of filamentous fungi. *Journal of Applied Polymer Science, 92*(1), 265–271. doi:10.1002/app.13644

Markandey, D. K., & Rajvaidya, N. (2004). *Environmental Biotechnology*. New Delhi: APH Publishing Corporation.

Mathur, G., Mathur, A., & Prasad, R. (2011). Colonization and degradation of thermally oxidized high-density polyethylene by Aspergillus niger (ITCC No. 6052) isolated from plastic waste dumpsite. *Bioremediation Journal, 15*(2), 69–76. doi:10.1080/10889868.2011.570281

Mueller, R.-J. (2006). Biological degradation of synthetic polyesters – Enzymes as potential catalysts for polyester recycling. *Process Biochemistry, 41*(10), 2124–2128. doi:10.1016/j.procbio.2006.05.018

Mumtaz, T., Khan, M. R., & Hassan, M. A. (2010). Study of environmental biodegradation of LDPE films in soil using optical and scanning electron microscopy. *Micron (Oxford, England), 41*(5), 430–438. doi:10.1016/j.micron.2010.02.008 PMID:20207547

Nanda, S., Sahu, S. S., & Abraham, J. (2010). Studies on the biodegradation of natural and synthetic polyethylene by Pseudomonas spp. *Journal of Applied Science & Environmental Management, 14*(2), 57–60. doi:10.4314/jasem.v14i2.57839

Nowak, B., Pająk, J., Drozd-Bratkowicz, M., & Rymarz, G. (2011). Microorganisms participating in the biodegradation of modified polyethylene films in different soils under laboratory conditions. *International Biodeterioration & Biodegradation, 65*(6), 757–767. doi:10.1016/j.ibiod.2011.04.007

Obasi, H. C., & Onuegbu, G. C. (2013). Biodegradability and mechanical properties of low density polyethylene/waste maize cob flour blends. *International Journal of Applied Sciences and Engineering Research, 2*, 2277–9442.

Ohtaki, A., Sato, N., & Nakasaki, K. (1998). Biodegradation of poly-ε-caprolactone under controlled composting conditions. *Polymer Degradation & Stability, 61*(3), 499–505. doi:10.1016/S0141-3910(97)00238-3

Orhan, Y., & Büyükgüngör, H. (2000). Enhancement of biodegradability of disposable polyethylene in controlled biological soil. *International Biodeterioration & Biodegradation, 45*(1-2), 49–55. doi:10.1016/S0964-8305(00)00048-2

Otake, Y., Kobayashi, T., Asabe, H., Murakami, N., & Ono, K. (1995). Biodegradation of low-density polyethylene, polystyrene, polyvinyl chloride, and urea formaldehyde resin buried under soil for over 32 years. *Journal of Applied Polymer Science, 56*(13), 1789–1796. doi:10.1002/app.1995.070561309

Pathak, V, M. & Navneet. (2017a). Review on the current status of polymer degradation: A microbial approach. *Bioresour Bioprocess. Springer Berlin Heidelberg, 4*, 15.

Pathak, V. M., & Navneet. (2017b). Dataset on the impact of UV, nitric acid and surfactant treatments on low-density polyethylene biodegradation. *Data in Brief, 14*, 393.

Pathak, V. M., & Navneet. (2017c). Biodegradation of Polyurethane. *Life Science Journal, 14*(1), 60–62.

Pathak, V. M., & Navneet. (2017d). Implications of SiO 2 nanoparticles for in vitro biodegradation of low-density polyethylene with potential isolates of Bacillus, Pseudomonas, and their synergistic effect on Vigna mungo growth. Energy. *Ecology & Environment, 2*(6), 418–427.

Pathak, V. M., Navneet, Bisht, T. S., & Pandey, A. (2018). Biodegradation of Natural and Synthetic Polymer. Microbial Biotechnology in Environmental Monitoring and Cleanup, 106.

Pegram, J. E., & Andrady, A. L. (1989). Outdoor weathering of selected polymeric materials under marine exposure conditions. *Polymer Degradation & Stability, 26*(4), 333–345. doi:10.1016/0141-3910(89)90112-2

Pometto, A. L. III, Lee, B., & Johnson, K. E. (1992). Production of an extracellular polyethylene-degrading enzyme(s) by *Streptomyces* species. *Applied and Environmental Microbiology, 58*, 731–733. PMID:1610196

Potts, J. E. (1978). Biodegradation. In H. H. G. Jelink (Ed.), *Aspects of degradation and stabilization of polymers* (pp. 617–658). New York, NY: Elsevier.

Pramila, R., & Ramesh, K. V. (2011a). Biodegradation of low density polyethylene (LDPE) by fungi isolated from municipal landfill area. *Journal of Microbiology and Biotechnology Research, 1*, 131–136.

Pramila, R., & Ramesh, K. V. (2011b). Biodegradation of low density polyethylene (LDPE) by fungi isolated from marine water – a SEM analysis. *African Journal of Microbiological Research, 5*, 5013–5018. doi:10.5897/AJMR11.670

Raaman, N., Rajitha, N., Jayshree, A., & Jegadeesh, R. (2012). Biodegradation of plastic by *Aspergillus* spp. isolated from polythene polluted sites around Chennai. *Journal of Academic and Industrial Research, 1*, 313–316.

Raghavan, D., & Torma, A. E. (1992). DSC and FTIR characterization of biodegradation of polyethylene. *Polymer Engineering and Science, 32*(6), 438–442. doi:10.1002/pen.760320609

Rajandas, H., Parimannan, S., Sathasivam, K., Ravichandran, M., & Yin, L. S. (2012). A novel FTIR-ATR spectroscopy based technique for the estimation of low-density polyethylene biodegradation. *Polymer Testing, 31*(8), 1094–1099. doi:10.1016/j.polymertesting.2012.07.015

Ramos, J. L., Díaz, E., Dowling, D., de Lorenzo, V., Molin, S., O'Gara, F., ... Timmis, K. N. (1994). The behavior of bacteria designed for biodegradation. *Nature Biotechnology, 12*(12), 1349–1356. doi:10.1038/nbt1294-1349 PMID:7765565

Riser-Roberts, E. (1998). *Remediation of petroleum contaminated soils. Biological, physical and chemical process*. USA: Lewis Publishers Inc, CRC Press LLC. doi:10.1201/9781420050578

Rojo, F. (2010). Enzymes for aerobic degradation of alkanes. In Handbook of Hydrocarbon and Lipid Microbiology (pp. 781-798). Berlin: Springer. doi:10.1007/978-3-540-77587-4_59

Roy, P. K., Titus, S., Surekha, P., Tulsi, E., Deshmukh, C., & Rajagopal, C. (2008). Degradation of abiotically aged LDPE films containing pro-oxidant by bacterial consortium. *Polymer Degradation & Stability, 93*(10), 1917–1922. doi:10.1016/j.polymdegradstab.2008.07.016

Ruiz-Dueñas, F. J., & Martínez, A. T. (2009). Microbial degradation of lignin: How a bulky recalcitrant polymer is efficiently recycled in nature and how we can take advantage of this. *Microbial Biotechnology, 2*(2), 164–177. doi:10.1111/j.1751-7915.2008.00078.x PMID:21261911

Santhoskumar, A. U., Palanivelu, K., Sharma, S. K., & Nayak, S. K. (2010). A new synthesis of nickel 12-hydroxy oleate formulation to improve polyolefin's degradation. *Journal of Bioremediation and Biodegradation, 1*.

Santo, M., Weitsman, R., & Sivan, A. (2013). The role of the copper-binding enzyme – laccase – in the biodegradation of polyethylene by the actinomycete *Rhodococcus ruber. International Biodeterioration & Biodegradation, 84*, 204–210. doi:10.1016/j.ibiod.2012.03.001

Secchi, E. R., & Zarzur, S. (1999). Plastic debris ingested by a Blainville's beaked whale, *Mesoplodon densirostris*, washed ashore in Brazil. *Aquatic Mammals, 25*, 21–24.

Sen, S. K., & Raut, S. (2015). Microbial degradation of low density polyethylene (LDPE): A review. *Journal of Environmental Chemical Engineering, 3*(1), 462–473. doi:10.1016/j.jece.2015.01.003

Seneviratne, G., Tennalkoon, N., & Nandasena, K. (2006). Polyethylene biodegradation by a developed Penicillium-Bacillus biofilm. *Current Science*, *90*, 20–21.

Shah, A. A., Hasan, F., Hameed, A., & Ahmed, S. (2008). Biological degradation of plastics: A comprehensive review. *Biotechnology Advances*, *26*(3), 246–265. doi:10.1016/j.biotechadv.2007.12.005 PMID:18337047

Sharma, A., & Sharma, A. (2004). Degradation assessment of low density polythene (LDP) and polythene (PP) by an indigenous isolate of *Pseudomonas stutzeri*. *Journal of Scientific and Industrial Research*, *63*, 293–296.

Shimao, M. (2001). Biodegradation of plastics. *Current Opinion in Biotechnology*, *12*(3), 242–247. doi:10.1016/S0958-1669(00)00206-8 PMID:11404101

Singh, V., Dubey, M., & Bhadauria, S. (2012). Microbial degradation of polyethylene (low density) by *Aspergillus fumigatus* and *Penicillium* sp. *Asian Journal of Experimental Biological Sciences*, *3*, 498–501.

Sivan, A., Szanto, M., & Pavlov, V. (2006). Biofilm development of the polyethylene-degrading bacterium *Rhodococcus ruber*. *Applied Microbiology and Biotechnology*, *72*(2), 346–352. doi:10.100700253-005-0259-4 PMID:16534612

Sowmya, H. V., Ramalingappa, M., & Krishnappa, M. (2012). Degradation of polyethylene by *Chaetomium* sp. and *Aspergillus flavus*. *International Journal of Recent Science Research*, *3*, 513–517.

Spear, L. B., Ainley, D. G., & Ribic, C. A. (1995). Incidence of plastic in seabirds from the tropical pacific, 1984-91: Relation with distribution of species, sex, age, season, year and body weight. *Marine Environmental Research*, *40*(2), 123–146. doi:10.1016/0141-1136(94)00140-K

Strantz, A. A., & Zottola, E. A. (1992). Stability of cornstarch-containing polyethylene films to starch-degrading enzymes. *Journal of Food Protection*, *55*(9), 736–738. doi:10.4315/0362-028X-55.9.736

Sudhakar, M., Doble, M., Murthy, P. S., & Venkatesan, R. (2008). Marine microbe-mediated biodegradation of low- and high-density polyethylenes. *International Biodeterioration & Biodegradation*, *61*(3), 203–213. doi:10.1016/j.ibiod.2007.07.011

Thompson, R. C., Swan, S. H., Moore, C. J., & vom Saal, F. S. (2009). Our plastic age. *Philosophical Transactions of the Royal Society of London. Series B, Biological Sciences*, *364*(1526), 1973–1976. doi:10.1098/rstb.2009.0054 PMID:19528049

Tribedi, P., & Sil, A. K. (2013). Low-density polyethylene degradation by *Pseudomonas* sp. AKS2 biofilm. *Environmental Science and Pollution Research International*, *20*(6), 4146–4153. doi:10.100711356-012-1378-y PMID:23242625

Van der Heul, R. M. (2011). *Environmental degradation of petroleum hydrocarbons* (Unpublished master's dissertation). Utrecht University, Netherlands.

Vasile, C. (1993). Degradation and decomposition. In C. Vasile & R. B. Seymour (Eds.), *Handbook of polyolefins synthesis and properties* (pp. 479–509). New York, NY: Marcel Dekker Inc.

Volke-Sepúlveda, T., Saucedo-Castañeda, G., Gutiérrez-Rojas, M., Manzur, A., & Favela-Torres, E. (2002). Thermally treated low density polyethylene biodegradation by *Penicillium pinophilum* and *Aspergillus niger. Journal of Applied Polymer Science*, *83*(2), 305–314. doi:10.1002/app.2245

Wariishi, H., Valli, K., & Gold, M. H. (1989). Oxidative cleavage of a phenolic diarylpropane lignin model dimer by manganese peroxidase from *Phanerochaete chrysosporium. Biochemistry*, *28*(14), 6017–6023. doi:10.1021/bi00440a044

Wasserbauer, R., Beranová, M., Vancurová, D., & Doležel, B. (1990). Biodegradation of polyethylene foils by bacterial and liver homogenates. *Biomaterials*, *11*(1), 36–40. doi:10.1016/0142-9612(90)90049-V PMID:2302448

Watanabe, T., Ohtake, Y., Asabe, H., Murakami, N., & Furukawa, M. (2009). Biodegradability and degrading microbes of low-density polyethylene. *Journal of Applied Polymer Science*, *111*(1), 551–559. doi:10.1002/app.29102

Wrona, M., Salafranca, J., & Nerin, C. (2017). Fast assessment of oxo-biodegradable polyethylene film oxidation by surface-enhanced Raman scattering with *in situ* formation of a silver nanoparticle substrate. *Journal of Materials Chemistry. C, Materials for Optical and Electronic Devices*, *5*(2), 463–469. doi:10.1039/C6TC04401H

Yabannavar, A. V., & Bartha, R. (1994). Methods for assessment of biodegradability of plastic films in soil. *Applied and Environmental Microbiology*, *60*, 3608–3614. PMID:16349408

Yamada-Onodera, K., Mukumoto, H., Katsuyaya, Y., Saiganji, A., & Tani, Y. (2001). Degradation of polyethylene by a fungus, *Penicillium simplicissimum* YK. *Polymer Degradation & Stability*, *72*(2), 323–327. doi:10.1016/S0141-3910(01)00027-1

Yang, J., Yang, Y., Wu, W.-M., Zhao, J., & Jiang, L. (2014). Evidence of polyethylene biodegradation by bacterial strains from the guts of plastic-eating waxworms. *Environmental Science & Technology*, *48*(23), 13776–13784. doi:10.1021/es504038a PMID:25384056

Yoon, M. G., Jeon, H. J., & Kim, M. N. (2012). Biodegradation of polyethylene by a soil bacterium and AlkB cloned recombinant cell. *Journal of Bioremediation & Biodegradation*, *3*, 145–151.

Zan, L., Fa, W., & Wang, S. (2006). Novel photodegradable low-density polyethylene-TiO_2 nanocomposite film. *Environmental Science & Technology*, *40*(5), 1681–1685. doi:10.1021/es051173x PMID:16568787

Zenteno, A., Lieberwirth, I., Catalina, F., Corrales, T., Guerrero, S., Vasco, D. A., & Zapata, P. A. (2017). Study of the effect of the incorporation of TiO2 nanotubes on the mechanical and photodegradation properties of polyethylenes. *Composites. Part B, Engineering*, *112*, 66–73. doi:10.1016/j.compositesb.2016.12.007

Chapter 15
Microbial Bioreactor Systems for Dehalogenation of Organic Pollutants

Raghunath Satpathy
MITS Engineering College, India

ABSTRACT

Halogenated organic compounds having many beneficial applications, both in industries and agriculture sectors. Basically, the uses are as pesticides, solvents, surfactants, and plastics. However, their large, widespread uses throughout the world have resulted the negative impact on the environment. Considering their treatment process are widely accepted by using the bioreactor systems. The large variety of micro-organisms present in the bioreactor and their interaction is the key to the effective treatment and removal of these compounds. Usually the microbes produce the enzymes known as dehalogenase to remove the halogen form the compounds to make it non-toxic. Many of the different steps and about the microbial groups in degradation process of halogenated compounds are well understood, but more details concerning the microbial community are yet to be discovered. This chapter describes about the different dehalogenation systems available in microbes and their ultimate application in different bioreactor systems for the degradation analysis of several harmful halogenated compounds.

INTRODUCTION

Halogenated organic compounds are the product from the industries and agriculture sectors and extensively used in many ways (Gribble, 1994; Song *et al.*, 2000). Due to massive nature and continuous entry into the environment, the probability for the accumulation of these compounds in the habitats indicates a global threating. For example, some of these compounds like dichloro-diphenyl trichloroethane (DDT), polychlorinated biphenyl (PCB), dioxins and halogenated flame retardants have a destructive impact on human health also these are persistent in nature (Perocco *et al.*, 1983; Muir *et al.*, 2006; Rees *et al.*, 2007). An important step in the degradation of harmful organohalide compounds is the cleavage of the carbon-halogen bond and the process is known as dehalogenation. This process is mediated by a diverse

DOI: 10.4018/978-1-5225-3540-9.ch015

group of microorganisms (Schultz *et al.,* 2001; Lin *et al.,* 2011).The basis of microbial dehalogenation is the induction of an enzyme of dehalogenases that basically catalyzes the cleavage of carbon–halogen (C-X) bonds (Janssen *et al.,* 1994; Chaudhry & Chapalamadugu 2001; Van Pee & Unversucht, 2003). Microbiological dehalogenation reactions have immense particular practical interests because of their potential biotechnological applications in the bioremediation of halogenated environmental pollutants (Fishbein, 1979). Many types of dehalogenase enzymes have been discovered in the microorganisms and that corresponds to the types of the halogenated pollutant to which the microorganism is exposed. The global halogen metabolism that occurs in the environment consists of both biotic and abiotic modes of halo-dehalogenation mechanism. For the last few years the number of natural organo-halogen compounds have been discovered more than 4000 today (Gribble, 2003; Velazquez *et al.,* 2011). Examples of the halogenated hydrocarbons which are produced from biological sources are basically chlorinated phenols and phenolic ethers, halogenated terpenes, chlorinated amino acids and peptides, halogenated alkaloids, bromo- and chloro- substituted pyrroles, chlorinated insole, halogenated thiophenes, chlorinated prostaglandins, and various antibiotics (Gribble, 2004). However the toxic effect exhibited by different organohalidic compounds and their capacity for bioaccumulation in the food chain, food web and the creation of environmental contamination is a greater concern now (Table 1). So the major efforts have been taken to degrade the halogenated substances by using the microorganisms that play a major role the major mediators for the re-cycling of these halogenated organic compounds. Therefore, it is essential to study and understand the diverse microbial effect on the biodegradation and biotransformation processes (Hardman, 1991).

Mechanism of Microbial Dehalogenation Reaction

This mechanism by which the toxic organic compound is converted to non-toxic metabolite by a biological organism is referred to as bioremediation. One amongst the potential bioremediation technique is dehalogenation, where the microbe has the ability to degrade halogenated toxic organic compounds (Satpathy *et al.,* 2015). The organo-halide compounds are among the largest group of environmental chemicals and also important intermediate substances of the global halogen cycle occurring in nature. The degradation mechanisms for the halogenated substances are available in databases. Many attempts are made to replace halogenated chemicals by using microorganisms that play a major role as mediators for recycling of halogenated compounds. Thus, it is all important to study and understand the diverse microbial impact on the biodegradation and biotransformation processes (Hardman,1991).

The study of halogenated substances in the environment and their presence in life forms open the door to understanding the carbon–halogen bond formation and cleaving process. The interdisciplinary methodology is truly essential to determine the basic nature, metabolic potential of these biochemical reactions takes place in different microbiological systems (Erable *et al.,* 2006; Krupp *et al.,* 2006). Different types of dehalogenase enzymes are evolved in the microorganisms to catalyze different types of the halogenated pollutant to which the microorganism is exposed. The dehalogenase enzymes from diverse microbial sources have been classified as many categories, depending upon the type of halogenated substances it degrades (Hamid *et al.,* 2013). The eight categories of dehalogenase enzyme and the reaction mechanism of the substrate also has been described by Janssen (Janssen *et al.,* 1994) and tabulated on the Table 2.

Table 1. Showing the potential use of various organo halide compounds and their effect

Halogenated Compounds	Properties of the Compound	Harmful Effect	Major (Industrial) Uses	References
Tetrachloroethene	Colourless liquid and sweet odour at 1ppm	soil contaminant	Good solvent, industrial degreasing, washing in dry cleaning	(Ryoo *et al.*, 2001)
Pentachlorophenol	Colourless crystals and highly soluble in water	Affects kidneys, blood, lungs, nervous system and immune system,	Pesticide, disinfectant	(Yang *et al.*, 2005)
2,4,6 trichlorophenol	Yellow solid with a strong and sweet odour.	carcinogenic in animals, causing lymphomas, leukaemia and liver cancer	Used as a fungicide, herbicide, insecticide, antiseptic, defoliant, and glue preservative .	(Sánchez and Gonzále 2007)
PCB	Odourless, tasteless, clear to pale-yellow and viscous liquids	classified as a persistent organic pollutant, Toxic effects such as endocrine disruption and neurotoxicity	Dielectric and coolant fluids, in transformers, capacitors, and electric motors	(Williams *et al.*, 1997)
Atrazine	white, crystalline solid organic compound	contamination in drinking water, birth defects and menstrual problems	Widely used herbicide	(Muller *et al.*, 2011)
Dichloromethane	Colourless, volatile liquid with a moderately sweet aroma is widely used as a solvent. Although it is not miscible with water	carbon monoxide poisoning (after metabolised) optic neuropathy and hepatitis, skin irritation and may cause cancer	Used as a solvent, manufacture of steroids, antibiotics, vitamins, and tablet coatings, paint removers, adhesives and as a flame retardant	(Topp 2011)
Lindane (Hexachlorocyclohexane (HCH))	White crystalline solid that is volatile in air and insoluble in water.	is a neurotoxin, affects the liver and kidneys, and harmful to insects and fish	agricultural insecticide and pharmaceutical treatment for lice and scabies	(Singh, and Lal 2009)
DDT (dichlorodiphenyltrichloroethane)	Colourless,crystalline solid, tasteless and almost odourless chemical compound.	Toxic to marine animals, human effects are genotoxicity and endocrine disruption and may cause cancer	Insecticide	(Nadeau *et al.*, 1994)
1,2,3trichloro propane (TCP)	Colourless liquid with a strong acid odour. Denser than water and slightly soluble in water.	highly toxic, contaminant of groundwater and may cause cancers in human	Industrial solvent, production of chemicals such as dichloropropene, Glycerol, a cleaning and degreasing agent	(Samin and Janssen 2012)
Dichloroethane-1,2 (DCE)	It is a colourless liquid with a chloroform-like odour.	May cause cancer and genetic damage, Large exposure may result in harm to wildlife	Manufacture of vinyl chloride and PVC and good solvent	(Van den Wijngaard *et al.*, 1993)
Chloro-alkanes (C10-13)	(50-70% w/w chlorine) yellowish oily liquids, runny or thick and without a distinct melting point	May affect the kidney, liver and thyroid gland, and there is some evidence that these substances may cause cancer	industrial metal working fluids used in cutting, drilling, machining, fire retarding properties and also used in paints and other coatings	(Allpress & Gowland 1999)
Hexachloro-butadiene (HCBD)	Colourless liquid at room temperature that has an odour similar to that of turpentine.	May be carcinogenic, genetic damage, toxic effect on aquatic organisms	In the recovery of chlorine containing gas in chlorine plants, manufacture of lubricants and rubber compounds	(Li *et al.*, 2008)
Dioxins	Melting point > 593° K, a low vapour pressure, a low water solubility	Adverse effects on the heart, immune system, liver, skin and thyroid gland	For research and analytical purposes	(Nam *et al.*, 2016)
Tetrachloroethylene	colourless liquid	May cause health effects on the brain, eye, kidney, liver, skin, throat, and there is some evidence that the substance may cause cancer	Dry cleaning, lubricating and degreasing agent, regeneration of oil refinery catalysts, Manufacture of printing inks, adhesives, dye carriers	(Yamasaki *et al.*, 2012)
Chlorothalonil	Grayish to colourless crystalline solid that is odourless to slightly pungent	Long-term exposure results, kidney damage, tumours, highly toxic to fishes	A fungicide, also used as a wood protectant	(Putnam *et al.*, 2003)
Trichlorobenzenes (TCBs)	colourless liquid	High amounts of trichlorobenzenes exposure may develop liver problems	Chemical intermediates for the production of herbicides, pigments and dyes, solvents, dielectric fluids	(Wang *et al.*, 2007)
Trichloroethane-1,1,1 (TCE)	colourless, sweet-smelling liquid	May cause cancer and genetic damage. Excessive exposure may affect the brain (central nervous system effects), heart, kidney, and liver, and cause skin and eye irritation	Degreasing and cleaning of metal parts, adhesives, for synthesis in the chemical industry as a feedstock for the CFC substitutes	(Grostern & Edwards 2006)
Trichloromethane (Chloroform)	colourless, sweet-smelling, dense liquid	Excessive exposure to trichloromethane may affect the brain, digestive system, eye, kidney, liver and skin, and may cause cancer	Solvent, production of hydrochlorofluorocarbon-22 (HCFC 22) and pesticide formulations	(Wahman *et al.*, 2005)

Table 2. Showing categories of dehalogenation reactions and mechanisms

Categories	Name Reaction	General Mechanism	Examples of Enzymes	Examples of the Pollutants Degradation
A	Hydrolytic dehalogenation	**R-X + H_2O → R-OH + H^+ + X^-**	Haloalkane dehalogenase, 2-haloacid dehalogenase	4- chlorobenzoate,Chl oroacetate,1,2 dichloro ethane,chloroacetate,
B	Glutathione substitution	**R-X + GSH → [GS-R-X] + HOH → GSH + R-COH + X**	Dichloromethane dehalogenase, Tetrachloro-p-hydroquinone dehalogenase	Dichloromethane, Pentachlorophenol
C	Hydration	**C2X2HCOOH + H_2O →CH2(CHO)COOH**	4-chlorobenzozte dehalogenase	D-chloropropionicacid, chloroacrylic acid
D	Intramolecular substitution	**R-X^1 + XH → R-X + H^+ + X^{1-}**	3-chlorobenzoate hydroxylase	3,chlorobenzozte,2,4 dichlorophenoxy acetate
E	Dehydrohalogenation	**R-CH_2-CHX-R → R-CH=CH-R + H^+ + X**	g-HCH dehydrochlorinase	Gamma hexachlorocyclohexane
F	Reduction	**R-X + 2 [H] → R-H + H^+ + X^-**	Chlorobenzene Dioxygenase	2,4 dichlorophenoxy acetate,1,2,4 trichlorobenzene
G	Oxygenation	**R-CH_2-X + O_2 + 2[H^+] → R-CHO + H^+ + X^-**	Pentachlorophenol 4 monoxygenase	Pentachlorophenol

Bioreactor Systems

The bioreactor is an apparatus for production being used from the ancient times. At that time, the microbial cultures were used for the production of preparation food products. However, recently the process is successfully implemented in industrial scale for food production as well as solving the environmental problems by using bioremediation techniques (Mandenius, 2016). Bio-reactor is generally defined as a system in which the biological processes are developed under the closely monitored and controlled environmental with optimized operating systems such as pH, temp, pressure, aeration and agitation, nutrient supply and so on. (Garcia-Ochoa & Gomez, 2009; Galliher *et al.*, 2011). This bioreactor system usually applied for degradation of contaminants in the sample along with suitable microorganisms (Figure 1 and Figure 2). The reactors may be closed or open systems, such as a constructed wetland or an enclosed chamber. In comparison to natural attenuation and *in-situ* bioremediation, the bioreactors based degradation can control the problems of ineffective indigenous microorganisms. The high degree of reproducibility, efficiency in degradation, low risk for contamination and automatization process for a suitable microbe has made the bioreactor system as potential devices in microbial biotechnology. Recently, several categories of bioreactors have been used widely for biodegradation of several halogenated compounds. The commonly available bioreactors are fixed film bioreactor, static flask culture, two-step sequential bioreactor and so on (Aiken & Logan, 1989; Alleman *et al.*, 1995; Togna *et al.*, 1995). Many attempts are made to replace toxic halogenated chemicals by using microorganisms that play a major role in reducing environmental contamination. Thus, the bioreactor based study is important to understand impact of the microorganisms on the biodegradation and biotransformation processes (Hardman, 1991; Erable *et al.*, 2006; Hamid *et al.*, 2013).

Therefore, review of this literature is needed that specifically focus on different bioreactor system applicable for microbe-based dehalogenation of harmful compounds. This chapter discuss about the different types of microbe-based bioreactor systems, the formulation and design strategies that are implemented in dehalogenation processes.

Factors Affecting Dehalogenation of Halo Organic Compounds

Bioreactors are one of the most crucial part to study the degradation of the pollutant molecule by microbes. However, the important part is to select a proper reactor system and to set the optimal parameters associated with that. Several requirements must be satisfied for the biotransformation of organic pollutants to proceed. Generally the halogenated hydrocarbons undergo microbial degradation by the microorganisms if the organism utilizes the substances as the substrates for their metabolism. The factors that are responsible for the effective dehalogenation system is summarized in the Figure 3.A wide varieties of bioreactor systems has been developed to improve the bioremediation process. However, monitoring the internal effect is important. Several factors are responsible with in the bioreactor for the effective dehalogenation systems as given in the Table 2.

Figure 1. Showing a typical bioreactor system and its components

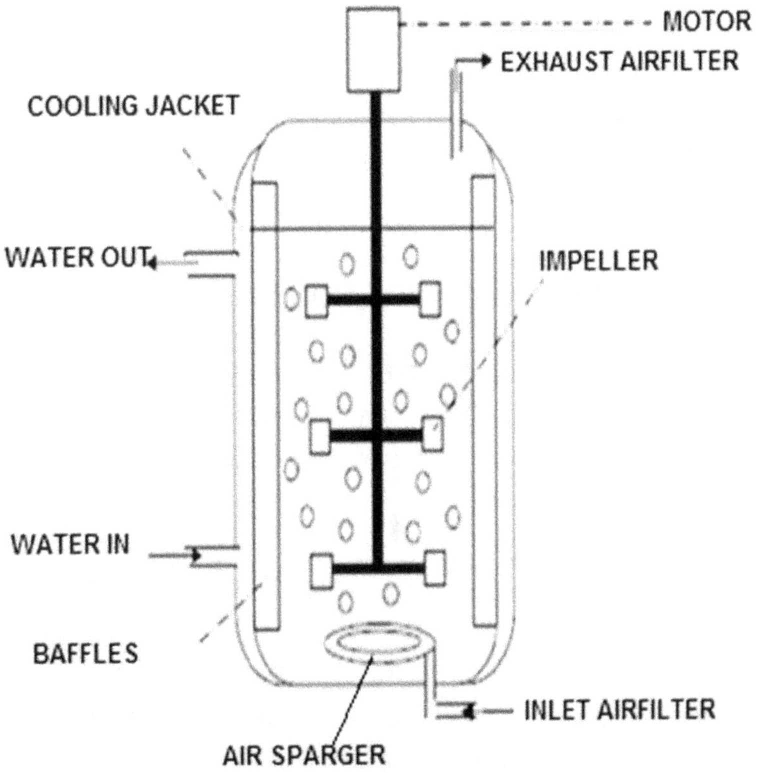

Figure 2. Showing the general outline for microbe-based bioreactor design and analysis and factors of consideration for dehalogenation study

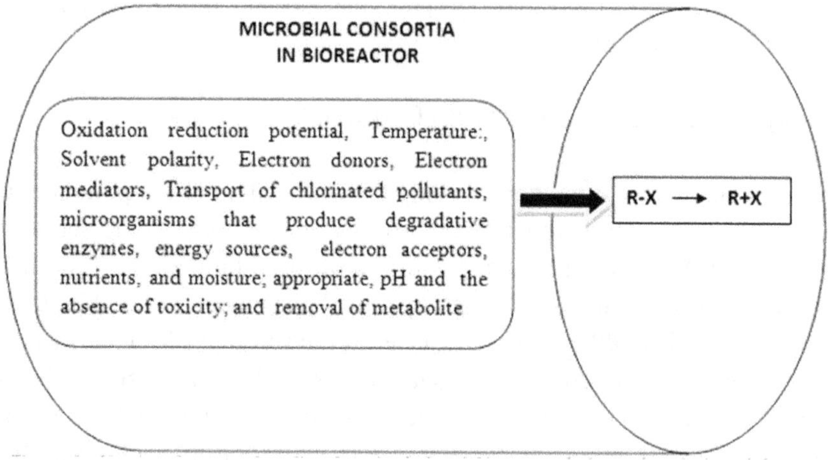

Figure 3. Describing the common factors for dehalogenation and in a bioreactor system

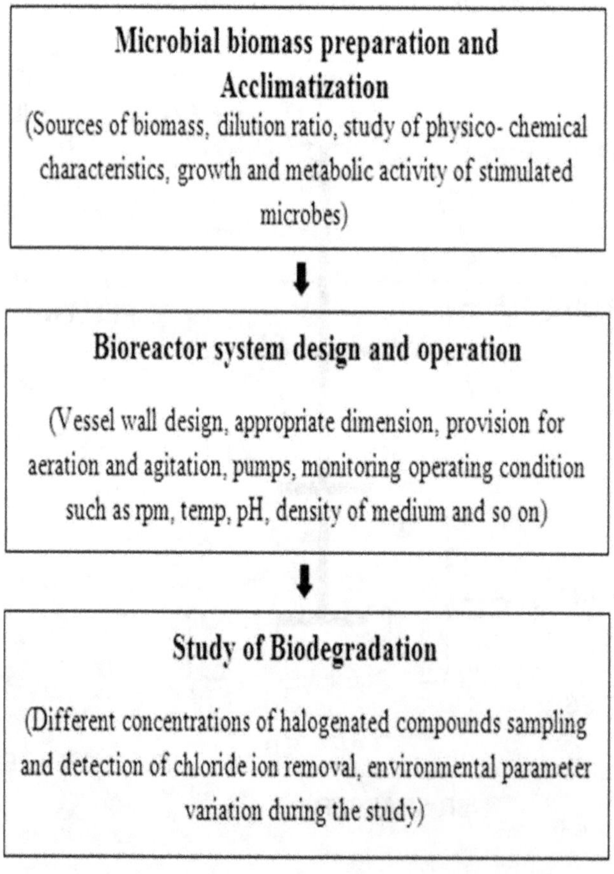

Table 2. General considerations involving in the operation of different bioreactor systems

Sl No.	Factors	Explanation
1	Flow and load variations	The variations in influent flow and organic loads can upset the balance in the dehalogenation processes.
2	Organic concentrations and temperatures	The preferred temperature of the bioreactor is to be maintained 25-35 °C to support optimal biological reaction rates.
3	Fraction of non-dissolved organic material	Pollutant samples (such as wastewaters) with high solids concentrations are to be treated preliminarily
4	Mixing	Thorough mixing is particularly important because it prevents the development of microenvironments unfavorable to the microbial population
5	Alkalinity	The alkalinity of the reactor is be indicative of its buffering capacity and the higher the alkalinity the more stable the pH is likely to be.
6	Nutrients	The amounts of nitrogen and phosphorus required per unit mass of cells formed are the same
7	Treatment efficiency needed	Anaerobic treatment processes are capable of high (Chemical Oxygen Demand) COD conversion efficiency with minimal biomass production.
8	Light metals ions (Na, K, Mg, Ca and Al)	High salt levels cause microbial cells to dehydrate due to osmotic pressure. So the cations of salts in solution must always be associated with the anions.
9	Heavy metals (Cr, Fe, Co, Cu, Zn, Cd and Ni)	The heavy metals identified to be of particular concern include chromium, iron, cobalt, copper, zinc, cadmium and nickel.
10	Toxicity caused by organics	These compound serving as a growth substrate with fermentation and the examples are non-halogenated aromatic and aliphatic compounds such as phenol, toluene, alcohols, and ketones.
11	Ammonia content	Ammonia is produced by the biological degradation of the nitrogenous matter, mostly in the form of proteins and urea.

Principles of Major Types of Bioreactors Involved in Dehalogenation Process

The general working principles of some major types of bioreactor systems has been given as below.

1. Rotating Biological Contactor (RBC)

The RBC type of bioreactor are basically used for waste water treatment purposes. This reactor involves allowing the wastewater to come in contact with a biological medium in order to remove pollutants in the wastewater before discharge of the treated wastewater to the environment, usually a body of water (river, lake or ocean). A rotating biological contractor is a type of the secondary treatment process. It consists of a series of closely spaced, parallel discs mounted on a rotating shaft which is supported just above the surface of the waste water (Figure 4). The microorganisms usually grow on the surface of the discs where biological degradation of the pollutants takes place (Grady *et al.*, 2011).

2. Anaerobic Packed Bed Bioreactor

Figure 4. Showing the Rotating Biological Contractor (RBC)

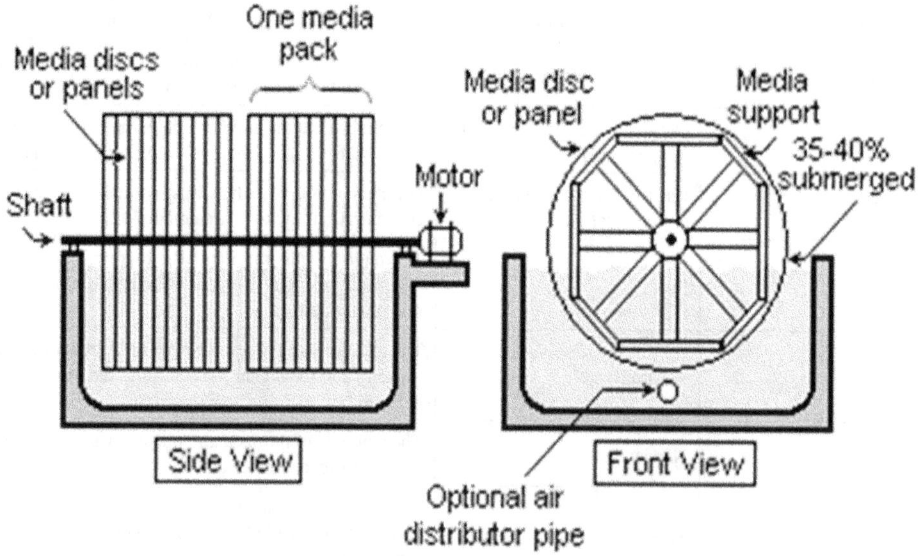

Anaerobic filter is a fixed-film biological wastewater treatment process in which a fixed matrix (support medium) provides an attachment surface that supports the anaerobic microorganisms in the form of a biofilm. The actual cause of remediation is during the flow of polluted wastewater through the bed and the dissolved pollutants are absorbed by the biofilm. Anaerobic filters were the first anaerobic systems that eliminated the need for solids separation and recycle (Singh & Prema, 2009). Various types of support material can be used, such as plastics, granular activated carbon (GAC), sand, reticulated foam polymers, granite, quartz and stone (Figure 5).

3. Expanded Granular Sludge Bed-Anaerobic Filter (EGSB-AF) Bioreactors

An expanded granular sludge bed (EGSB) reactor is a concept for anaerobic wastewater treatment. The unique characteristics of this bioreactor is the upward-flow velocity is monitored for the influent passing through the sludge bed (Figure 6). Also recycling is also accomplished by considering the flow velocity (Bajpai, 2017).

4. Up-Flow Anaerobic Sludge Bed (UASB) Reactor

Up flow anaerobic sludge blanket technology also known as UASB reactor is a form of anaerobic digester which used in wastewater treatment. UASB reactor is a methane-producing digester, which uses an anaerobic process and forming a blanket of granular sludge and is processed by the anaerobic microorganisms (figure 7) (Lettinga, 1995).

5. Fluidized Bed Bioreactors (FBBRs)

Figure 5. Showing the anaerobic packed-bed bioreactor systems

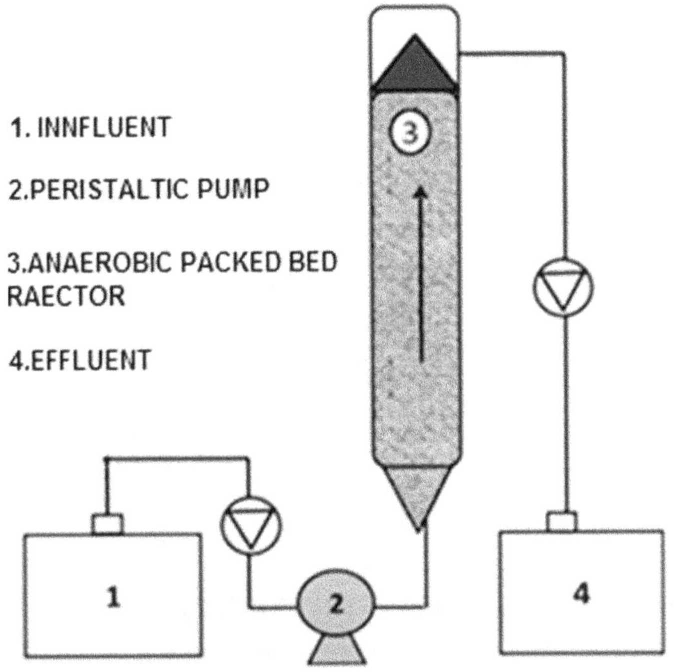

Figure 6. Showing the Expanded Granular Sludge Bed-Anaerobic Filter (ESGB-AF) bioreactor

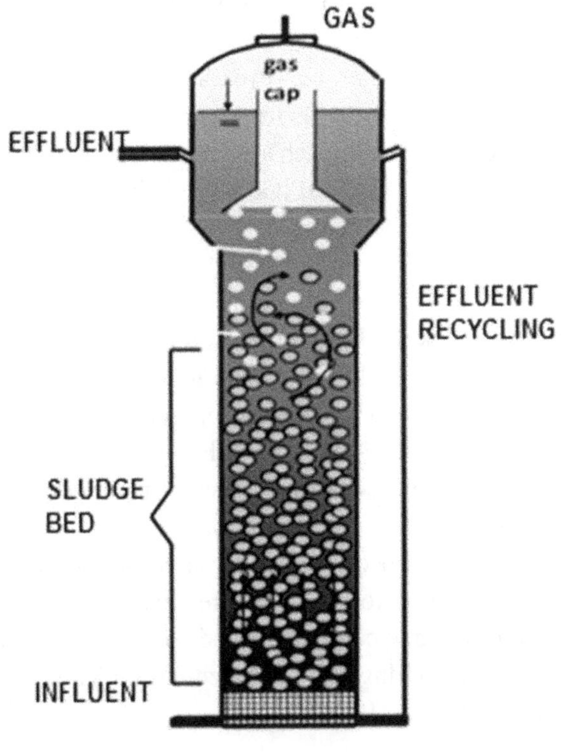

Figure 7. Showing the Up-Flow Anaerobic Sludge Bed (UASB) bioreactor

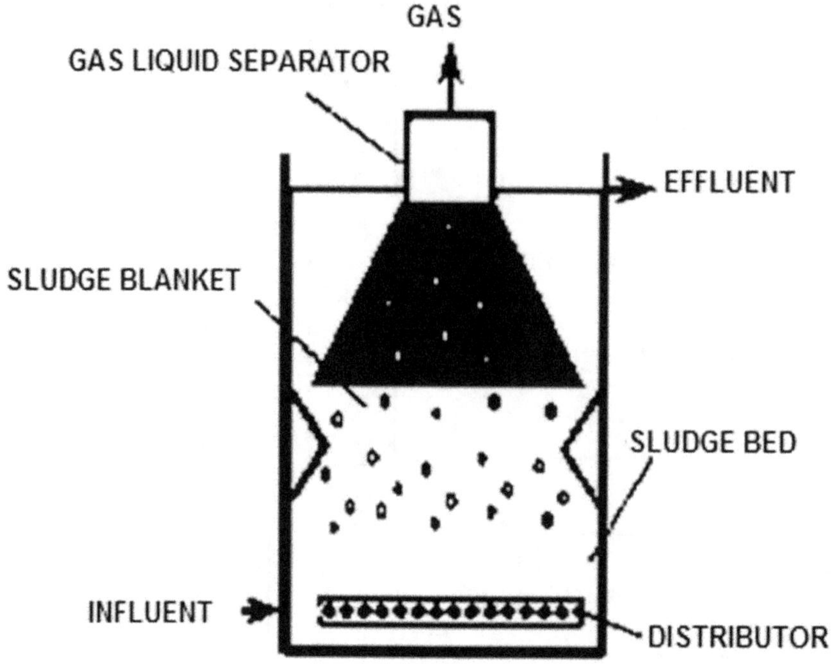

A fluidized bed bioreactor (FBBR) is a type of reactor device that can be used to carry out a variety of multiphase bio- chemical reactions. In this type of reactor (figure 8), a fluid (gas or liquid) is passed through a solid material at a high velocity to suspend the solid and the process is known as fluidization, imparts many important advantages to this bioreactors also used in many industrial applications. The fluid is then forced through the solid material. However, at a lower fluid velocities, the solids remain in place as the fluid passes through the voids in the material. This is known as a packed bed reactor. As the fluid velocity is increased, the reactor will reach a stage where the force of the fluid on the solids is enough to balance the weight of the solid material (Godia & Sola, 1995).

6. Biotrickling Filter

The terms such as trickle filter, trickling biofilter, biofilter, biological filter and biological trickling filter are often used to refer to a trickling filter. A trickling filter is a type of pollutant (usually waste water) treatment system that contains a fixed bed consists of rocks, lava, coke, gravel, slag, polyurethane foam ceramic and plastic media. Over which sewage or other wastewater flows downward and causes a layer of microbial slime (biofilm) to grow, covering the bed of media (Figure 9). Aerobic or anaerobic conditions can be maintained in this bioreactor. These systems have also been described as roughing filters, intermittent filters, packed media bed filters, alternative septic systems, percolating filters, attached growth processes, and fixed film processes. The removal of pollutants from the feed involves both absorption and adsorption of organic compounds and some inorganic species such as nitrite and nitrate ions by the layer of microbial bio film. The filter media is typically chosen to provide a very high surface area to volume (Cox & Deshusses, 1998).

Figure 8. Showing the Fluidized Bed bioreactor (FBR)

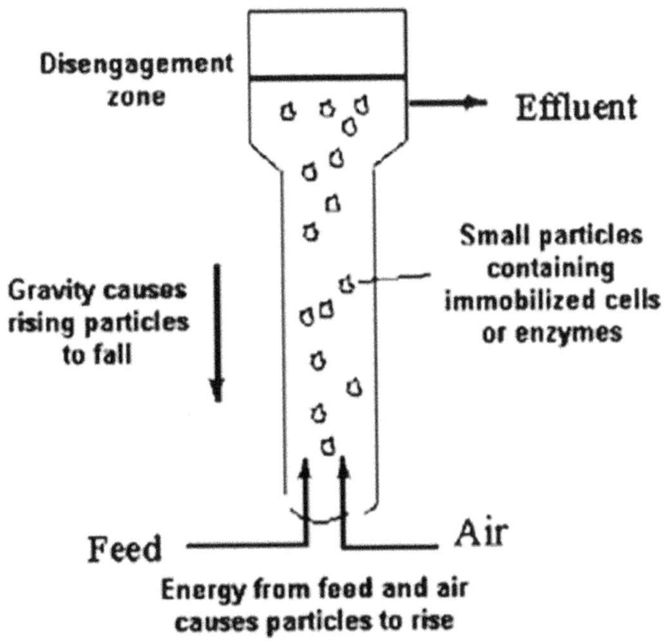

7. Membrane Biofilm Reactors (MBfRs)

The design of this reactor constitutes a membrane module that is connected to a recirculation loop. Due to high re-circulation rate, the system behaved as a completely mixed biofilm reactor. Usually, the membrane module consisted of a bundle of hydrophobic hollow-fiber membranes inside a PVC pipe shell (Figure 10). The free end of each fiber was sealed so that the total of the supplied hydrogen was transferred through the fiber walls to the surrounding biofilm. Also, from the bottom region about 10 cm of each fiber was deadened to minimize gas transfer and avoid excessive growth of biofilm that could cause the fibers to stick together (Casey *et al.,* 1999).

8. Batch Type Bioreactor

A batch reactor contains all the materials needed by the cell culture are introduced at the beginning of the process and there is no introduction or removal of substances or fresh medium (Figure 11). The processing in this bioreactor consists of two phases: The first phase of operating this bioreactor is maintaining the sterilization, after which the sterile culture medium is inoculated with microorganisms that have been cultivated to achieve a specific result. Next phase consists of monitoring of the dynamic reaction period. In which the concentration of the cells, substrates as well as the products vary with time. Therefore, proper mixing having an impact to maintain the equilibrium between them to an acceptable level. Also, to promote aerobic cultivation, in this bioreactor, the medium is aerated to provide a continuous flow of oxygen. Gaseous byproducts formed, such as CO_2 are removed and aeration along with gas-removal processes take place semi-continuously. Again to maintain the pH, an acid or alkali is

Figure 9. Showing the Biotrickling Filter bioreactor

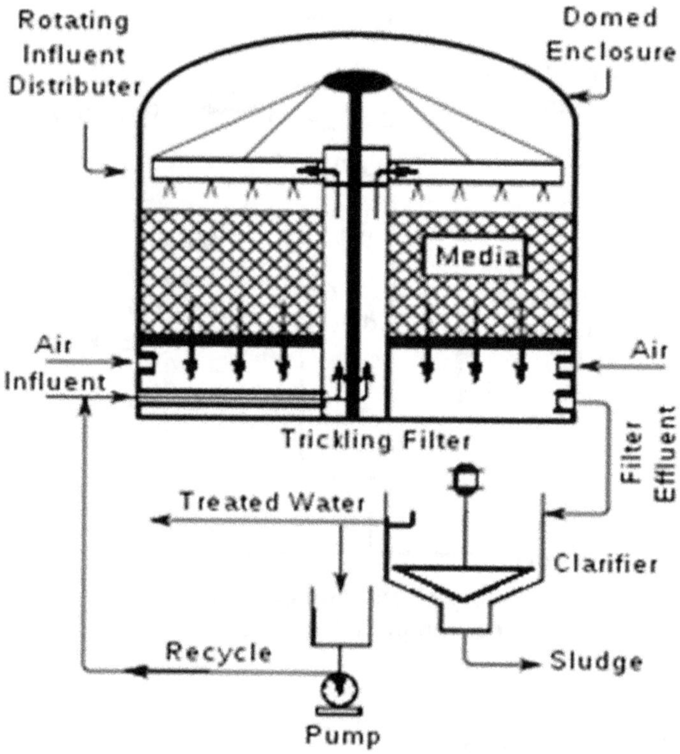

Figure 10. Showing the Membrane Film bioreactor (MBR)

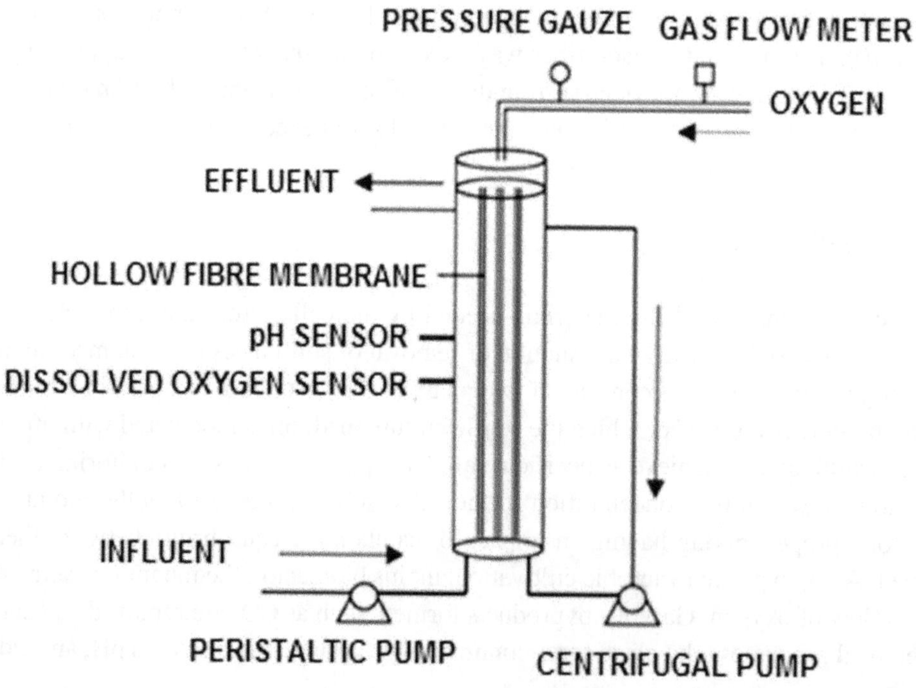

Figure 11. Showing the batch type bioreactor

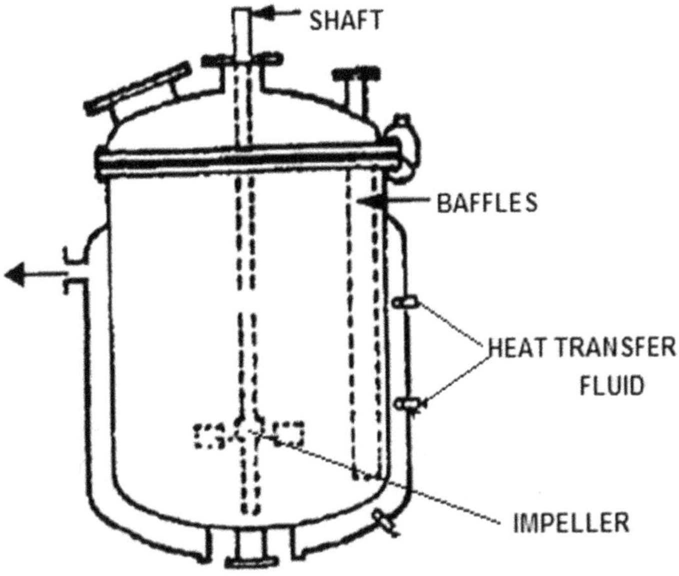

added if it is necessary. To keep foaming to acceptable levels, antifoaming agents may be added when indicated by a foam sensor (Bonvin, 1998).

Different Types of Bioreactor Systems Used for Microbial Dehalogenation

Many categories of halo organic compounds are dehalogenated by using them in a suitable bioreactor systems few of the case studies has been explained below. The investigation about degradation of different halogenated compounds such as trihalome thanes (THMs) in wastewater has been explained by Fitch in 1998 (Fitch *et al.*, 1998). They used various types of treatment plants including activated sludge, trickling filters and rotating biological contactors and evaluated the efficiency of selected techniques for THM removal. Also, it was observed that, particularly, use of a rotating biological contactor (RBC) composed of concentric Plexiglas cylinders was used to degrade cis-1, 2-dichoroethylene (c-DCE) in the presence of mixed culture of methanotrophs. De Best et al., in 1999 described about the degradation of a carbon tetrachloride (CT) that is commonly used as an industrial solvent and known as a common water pollutant. As the CT resists aerobic biotransformation, but by use of anaerobic packed-bed reactor without addition of an external electron donor it was biodegraded for about > 72% (de Best *et al.*, 1999). Application of an autotrophic packed-bed biofilm reactor inoculated with a perchlorate-reducing, hydrogen-oxidizing autotrophic bacterial consortium has shown its potential to degrade the perchlorate explained by Miller (Miller *et al.*, 2000). Similarly a continuously-fed up-flow, anaerobic column reactor was developed to degrade the PCE in a 12 hour of resident time interval in the column contains microbial concertia (Sato *et al.*, 2002). A sediment column study was carried out to demonstrate the bioremediation of tetra- and trichloroethenes (PCE and TCE) to ethene and ethane in a single anaerobic step under sulfate reducing conditions has been explained by Drzyzga (Drzyzga *et al.*, 2002). The immobilized form of isolated strain of *Penicillium camemberti* has been studied to evaluate its potential to degrade chlorinated pulping wastes and are found to be successful by using the strain in an upflow glass wool packed column reactor (Taşeli *et al.,* 2004). Similarly, the expanded granular sludge bed-anaerobic filter (EGSB-AF) bioreactors system containing 2, 4, 6-trichlo-rophenol (TCP) were operated in the presence of bacterial communities like *Campylobacter*, *Acidimicrobium* and *Heliophilum* like organisms (Collins *et al.,* 2005). The sludge from different sources that contains the pentachlorophenol (PCP) was evaluated for degradation by using of bioreactors in anaerobic conditions by (Bolaños *et al.,* 2005). To understand the microbial ecology underlying trichloethene (TCE) degradation by using an upflow anaerobic sludge bed (UASB) reactor has been explained by Tresse (Tresse *et al.,* 2005). In this experiment, TCE had shown a toxic effect on two uncultured bacterial populations, whereas aerobic conditions favored the growth of aerobic species belonging to Rhizobiaceae and Dechloromonas. Pratibha Singh in 2007 studied the sequential anaerobic and aerobic treatment in two step bioreactor was performed for removal of colour in the pulp and paper mill effluent. An interesting experiment was conducted by using the treated effluent in an aerobic bioreactor in presence of fungal strain, *Paecilomyces* sp. and bacterial strain *Microbrevis luteum.* It was observed that, the reduction of AOX occurs at the rate of 74%, and that was much higher in comparison to anaerobic treatment (~25%) (Singh, 2007). Similarly, the anaerobic dechlorination study of the technical grade hexachlorocyclohexane (THCH) was conducted in a continuous upflow anaerobic sludge blanket (UASB) reactor with methanol as a supplementary substrate and electron donor by Bhatt (Bhatt *et al.,* 2008). A H_2-based, denitrifying membrane-biofilm reactor (MBfR) was observed effective for removing trichloroethene (TCE) by reductive dechlorination. After addition of trichloroethene to the MBfR, due to the reductive dechlorination reaction took place and TCE was observed to be completely dechlorinated to ethene in presence of the bacteria *Dehalococcoides* (Chung *et al.,* 2007). A lab scale anaerobic bio trickling filter inoculated with a mixed culture containing multiple *Dehalococcoides* strains was used to convert to ethene, to cis 1,2dichloroethene (Popat & Deshusses, 2009). Similarly, the removal of dichloromethane (DCM) from polluted air was studied both in biotrickling filters and in continuous stirred tank bioreactors, using either a single-liquid aqueous phase or a combination of an aqueous–organic liquid phase by inoculation of *Hyphomicrobium* strains (Bailón *et al.,* 2009). The anaerobic bacterium *Clostridium bifermentans* DPH-1 was has shown its potential to dechlorinate PCE to cis-1, 2- dichloroethylene (cDCE) via trichloroethylene (TCE) at high rates in the monoculture biofilm of an up flow fixed-bed column reactor (Chang *et al.,* 2010). The perchloroethylene (PCE) degradation has been studied successfully in presence of coupled zero valent iron (ZVI) on a fluidized bed bioreactors (FBBRs) containing the anchored microbial consortia (Ozdemir *et al.,* 2010). Reductive dehalogenation was observed to be facilitated in the presence of an electron donor, hydrogen, to a biofilm in the liquid phase of a reactor (Zhang *et al.,* 2010). In a specific study, by using an anaerobic biotrickling filter inoculated with a mixed culture containing *Dehalococcoides* sp. show significant accumulation of *cis*-DCE and VC was observed in the bioreactor effluent (Popat *et al.,* 2012). Perchloroethylene (PCE) can be removed via a coupling biological processes like reductive dehalogenation and halorespiration, and by abiotic mechanisms such as zero-valent iron (ZVI) and zinc in an methanogenic and hybrid fluidized bed bioreactors have been studied by Breton-Deval (Breton-Deval *et al.,* 2012). A study with H_2-based membrane biofilm reactors (MBfRs) was considered for ex-situ reductive dechlorination of chlorinated ethenes. By examining the effectiveness of direct H_2 delivery during the dechlorination, it was observed the process efficiency goes up-to 95% in presence of the dominant bacterial strain such as *Dehalococcoides*, *Geobacter* and *Homoacetogens* (Ziv-El *et al.,* 2012). A novel aerobic to anaerobic to aerobic mode of treatment was implemented in batch reactors for successful degradation of tetrachloroethylene (TCE), trichloroethylene (TCE), vinyl chloride (VC), 1,1,2-trichloroethane (1,1,2-TCA) and chloroform (CF) (Frascari *et al.,* 2013). A hydrogen (H_2)-based biofilm to treat a groundwater contaminated with perchlorate (ClO_4^-) for the presence of perchlorate-reducing bacteria (PRB), denitrifying bacteria (DB), and sulfate-reducing bacteria (SRB) (Ontiveros-Valencia *et al.,* 2013). Chlorinated ethenes in drinking water could be reductively dechlorinated to non-toxic ethene by using a hydrogen based membrane biofilm reactor (H_2-MBfR) for dechlorinating bacteria in biofilm communities. *Dehalococcoides* were present in autotrophic biofilm community de-chlorinating PCE to ethane (Karataş *et al.,* 2014). A novel biodegradation pathways that avoids the production of dechlorination (1, 1, 2-TCA to 1, 1, 2-trichloroethane) product vinyl chloride, was established by Zhao et al in 2015. The bacterial strain like *Desulfitobacterium* sp. strain PR, which was isolated from an anaerobic bioreactor (Zhao *et al.,* 2015). Recently, the water contaminated with PCE to bioreactors seeded with inoculum was used implemented in four lab-scale fluidized bed bioreactors. Two bioreactors were operated as simultaneous methanogenic-denitrifying (MD) units, whereas the other two were operated in partially aerated methanogenic (PAM) mode. PCE removals in the range 86 to 97% were generally observed; the highest removal corresponded to PAM bioreactors along with the highest dehalogenation efficiency (94%) (Moreno-Medina *et al.,* 2016). The PCE degrading bacterium also has been used to degradation of the non-ionic surfactant Tween 80 by a batch lab-scale bioreactors (Bretón-Deval *et al.,* 2016). A recent report states that, 1, 1, 2-Trichloroethane (1, 1, 2-TCA) is a non-flammable organic solvent and common environmental contaminant in groundwater. Co-culture of organohalide-respiring bacteria such as *Dehalogenimonas and Dehalococcoides mccartyi* strains are key microorganisms to remediate 1, 1, 2-TCA under anaerobic conditions in a bioreactor (Mortan *et al.,* 2017).

CONCLUSION

The reduction of toxic halogenated compounds in the contaminated material be of economic interest. In spite of several chemical based and other methods, the biological (microorganism mediated) is preferred due to its environmental friendly nature. The work reported in this chapter represents about the harmful aspects of several halogenated compounds. Also the importance of and utilization of bioreactor systems has been explained. The microbial dehalogenation process that occurs in the different bioreactor system has been described. This kind of study is may be further utilized in elucidation of novel biodegradation pathways of toxic halogenated compounds by microorganisms. Also, the environmental condition required to enhance the process. In this chapter the author reviewed about the different bioreactor systems used exclusively for microbial dehalogenation of some toxic organic halogenated substances.

REFERENCES

Aiken, B. S., & Logan, B. E. (1989). Degradation of pentachlorophenol by the white rot fungus Phanerochaetechrysosporium grown in ammonium lignosulphonate media. *Biodegradation, 7*(3), 175–182. doi:10.1007/BF00058177 PMID:8782389

Alleman, B. C., Logan, B. E., & Gilbertson, R. L. (1995). Degradation of pentachlorophenol by fixed films of white rot fungi in rotating tube bioreactors. *Water Research, 29*(1), 61–67. doi:10.1016/0043-1354(94)E0106-G

Allpress, J. D., & Gowland, P. C. (1999). Biodegradation of chlorinated paraffins and long-chain chloro-alkanes by Rhodococcus sp. S45-1. *International Biodeterioration & Biodegradation*, *43*(4), 173–179. doi:10.1016/S0964-8305(99)00050-5

Bailón, L., Nikolausz, M., Kästner, M., Veiga, M. C., & Kennes, C. (2009). Removal of dichloromethane from waste gases in one-and two-liquid-phase stirred tank bioreactors and biotrickling filters. *Water Research*, *43*(1), 11–20. doi:10.1016/j.watres.2008.09.031 PMID:18945466

Bajpai, P. (2017). Anaerobic Reactors Used for Waste Water Treatment. In *Anaerobic Technology in Pulp and Paper Industry* (pp. 37–53). Springer Singapore. doi:10.1007/978-981-10-4130-3_5

Bhatt, P., Kumar, M. S., Mudliar, S., & Chakrabarti, T. (2008). Enhanced biodegradation of hexachloro-cyclohexane in upflow anaerobic sludge blanket reactor using methanol as an electron donor. *Bioresource Technology*, *99*(7), 2594–2602. doi:10.1016/j.biortech.2007.04.057 PMID:17574842

Bolaños, R. M. L., Damianovic, M. H. R. Z., Zaiat, M., & Foresti, E. (2005). Assessment of the ability of sludge to degrade PCP under anaerobic conditions. *Brazilian Journal of Chemical Engineering*, *22*(4), 611–617. doi:10.1590/S0104-66322005000400014

Bonvin, D. (1998). Optimal operation of batch reactors—a personal view. *Journal of Process Control*, *8*(5), 355–368. doi:10.1016/S0959-1524(98)00010-9

Breton-Deval, L., Galindez-Mayer, J., Moreno-Medina, C. U., Ríos-Leal, E., Rinderknecht-Seijas, N., & Poggi-Varaldo, H. M. (2012). Operation of hybrid fluidized bioreactors for on site bioremediation of water polluted with high concentration of PCE. *Environmental Engineering and Management Journal*, *11*(3), S25.

Bretón-Deval, L., Rios-Leal, E., Poggi-Varaldo, H. M., & Ponce-Noyola, T. (2016). Biodegradability of Nonionic Surfactant Used in the Remediation of Groundwaters Polluted with PCE. *Water Environment Research*, *88*(11), 2159–2168. doi:10.2175/106143016X14733681695564 PMID:28661332

Casey, E., Glennon, B., & Hamer, G. (1999). Review of membrane aerated biofilm reactors. *Resources, Conservation and Recycling*, *27*(1), 203–215. doi:10.1016/S0921-3449(99)00007-5

Chang, Y. C., Park, C. K., Jung, K., & Kikuchi, S. (2010). Dechlorination of High Concentrations of Tetrachloroethylene Using a Fixed-bed Reactor. *Korean Journal of Environmental Health Sciences*, *36*(4), 323–336. doi:10.5668/JEHS.2010.36.4.323

Chaudhry, G. R., & Chapalamadugu, S. (1991). Biodegradation of halogenated organic compounds. *Microbiological Reviews*, *55*(1), 59–79. PMID:2030673

Chung, J., Krajmalnik-Brown, R., & Rittmann, B. E. (2007). Bioreduction of trichloroethene using a hydrogen-based membrane biofilm reactor. *Environmental Science & Technology*, *42*(2), 477–483. doi:10.1021/es702422d PMID:18284150

Collins, G., Foy, C., McHugh, S., & O'Flaherty, V. (2005). Anaerobic treatment of 2, 4, 6-trichlorophenol in an expanded granular sludge bed-anaerobic filter (EGSB-AF) bioreactor at 15 C. *FEMS Microbiology Ecology*, *53*(1), 167–178. doi:10.1016/j.femsec.2004.10.008 PMID:16329938

Cox, H. H., & Deshusses, M. A. (1998). Biological waste air treatment in biotrickling filters. *Current Opinion in Biotechnology*, *9*(3), 256–262. doi:10.1016/S0958-1669(98)80056-6 PMID:9690989

de Best, J. H., Hunneman, P., Doddema, H. J., Janssen, D. B., & Harder, W. (1999). Transformation of carbon tetrachloride in an anaerobic packed-bed reactor without addition of another electron donor. *Biodegradation*, *10*(4), 287–295. doi:10.1023/A:1008309003059 PMID:10633544

Drzyzga, O., El Mamouni, R., Agathos, S. N., & Gottschal, J. C. (2002). Dehalogenation of chlorinated ethenes and immobilization of nickel in anaerobic sediment columns under sulfidogenic conditions. *Environmental Science & Technology*, *36*(12), 2630–2635. doi:10.1021/es010184x PMID:12099458

Erable, B., Goubet, I., Lamare, S., Legoy, M. D., & Maugard, T. (2006). Bioremediation of halogenated compounds: Comparison of dehalogenating bacteria and improvement of catalyst stability. *Chemosphere*, *65*(7), 1146–1152. doi:10.1016/j.chemosphere.2006.04.007 PMID:16723151

Erable, B., Goubet, I., Lamare, S., Legoy, M. D., & Maugard, T. (2006). Bioremediation of halogenated compounds: Comparison of dehalogenating bacteria and improvement of catalyst stability. *Chemosphere*, *65*(7), 1146–1152. doi:10.1016/j.chemosphere.2006.04.007 PMID:16723151

Fishbein, L. (1979). Potential halogenated industrial carcinogenic and mutagenic chemicals II. Halogenated saturated hydrocarbons. *The Science of the Total Environment*, *11*(2), 163–195. doi:10.1016/0048-9697(79)90026-3 PMID:373115

Fitch, M. W., Pearson, N., Richards, G., & Burken, J. G. (1998). Biological fixed-film systems. *Water Environment Research*, *70*(4), 495–518. doi:10.2175/106143098X134226

Frascari, D., Fraraccio, S., Nocentini, M., & Pinelli, D. (2013). Aerobic/anaerobic/aerobic sequenced biodegradation of a mixture of chlorinated ethenes, ethanes and methanes in batch bioreactors. *Bioresource Technology*, *128*, 479–486. doi:10.1016/j.biortech.2012.10.026 PMID:23201903

Galliher, P. M., Hodge, G., Guertin, P., Chew, L., & Deloggio, T. (2011). Single-Use Bioreactor Platform for Microbial Fermentation. *Single-Use Technology in Biopharmaceutical Manufacture*, 241-250.

Garcia-Ochoa, F., & Gomez, E. (2009). Bioreactor scale-up and oxygen transfer rate in microbial processes: An overview. *Biotechnology Advances*, *27*(2), 153–176. doi:10.1016/j.biotechadv.2008.10.006 PMID:19041387

Godia, F., & Sola, C. (1995). Fluidized-bed bioreactors. *Biotechnology Progress*, *11*(5), 479–497. doi:10.1021/bp00035a001

Grady, C. L. Jr, Daigger, G. T., Love, N. G., & Filipe, C. D. (2011). *Biological wastewater treatment*. CRC press.

Gribble, G. W. (1994). The natural production of chlorinated compounds. *Environmental Science & Technology*, *28*(7), 310–319. doi:10.1021/es00056a712 PMID:22662801

Gribble, G. W. (2003). The diversity of naturally produced organohalogens. *Chemosphere*, *52*(2), 289–297. doi:10.1016/S0045-6535(03)00207-8 PMID:12738253

Gribble, G. W. (2004). Natural organohalogens: A new frontier for medicinal agents? *Journal of Chemical Education*, *81*(10), 1441. doi:10.1021/ed081p1441

Grostern, A., & Edwards, E. A. (2006). A 1, 1, 1-trichloroethane-degrading anaerobic mixed microbial culture enhances biotransformation of mixtures of chlorinated ethenes and ethanes. *Applied and Environmental Microbiology*, *72*(12), 7849–7856. doi:10.1128/AEM.01269-06 PMID:17056695

Hamid, T. H. T. A., Hamid, A. A. A., & Huyop, F. (2013). A review on non-stereospecific haloalkanoic acid dehalogenases. *African Journal of Biotechnology*, *10*(48), 9725–9736. doi:10.5897/AJB11.934

Hardman, D. J. (1991). Biotransformation of halogenated compounds. *Critical Reviews in Biotechnology*, *11*(1), 1–40. doi:10.3109/07388559109069182 PMID:1863980

Janssen, D. B., Pries, F., & Van der Ploeg, J. R. (1994). Genetics and biochemistry of dehalogenating enzymes. *Annual Review of Microbiology*, *48*(1), 163–191. doi:10.1146/annurev.mi.48.100194.001115 PMID:7826004

Karataş, S., Hasar, H., Taşkan, E., Özkaya, B., & Şahinkaya, E. (2014). Bio-reduction of tetrachloroethen using a H2-based membrane biofilm reactor and community fingerprinting. *Water Research*, *58*, 21–28. doi:10.1016/j.watres.2014.03.053 PMID:24731873

Krupp, M., Weinmann, A., Galle, P. R., & Teufel, A. (2006). Actin binding LIM protein 3 (abLIM3). *International Journal of Molecular Medicine*, *17*(1), 129–133. PMID:16328021

Lettinga, G. (1995). Anaerobic digestion and wastewater treatment systems. *Antonie van Leeuwenhoek*, *67*(1), 3–28. doi:10.1007/BF00872193 PMID:7741528

Li, M. T., Hao, L. L., Sheng, L. X., & Xu, J. B. (2008). Identification and degradation characterization of hexachlorobutadiene degrading strain Serratia marcescens HL1. *Bioresource Technology*, *99*(15), 6878–6884. doi:10.1016/j.biortech.2008.01.048 PMID:18337093

Lin, C., Yang, L., Xu, G., & Wu, J. (2011). Biodegradation and metabolic pathway of β-chlorinated aliphatic acid in Bacillus sp. CGMCC no. 4196. *Applied Microbiology and Biotechnology*, *90*(2), 689–696. doi:10.100700253-010-3081-6 PMID:21210101

Lorah, M. M., Walker, C., & Graves, D. (2015). Performance of an anaerobic, static bed, fixed film bioreactor for chlorinated solvent treatment. *Biodegradation*, *26*(5), 341–357. doi:10.100710532-015-9738-1 PMID:26154697

Mandenius, C. F. (2016). *Bioreactors: Design, Operation and Novel Applications*. John Wiley & Sons. doi:10.1002/9783527683369

Miller, J. P., & Logan, B. E. (2000). Sustained perchlorate degradation in an autotrophic, gas-phase, packed-bed bioreactor. *Environmental Science & Technology*, *34*(14), 3018–3022. doi:10.1021/es991155d

Moreno-Medina, C. U., Poggi-Varaldo, H. M., Breton-Deval, L., & Rinderknecht-Seijas, N. (2016). Effect of sudden addition of PCE and bioreactor coupling to ZVI filters on performance of fluidized bed bioreactors operated in simultaneous electron acceptor modes. *Environmental Science and Pollution Research International*, 1–16. PMID:27498752

Mortan, S. H., Martín-González, L., Vicent, T., Caminal, G., Nijenhuis, I., Adrian, L., & Marco-Urrea, E. (2017). Detoxification of 1, 1, 2-trichloroethane to ethene in a bioreactor co-culture of Dehalogenimonas and Dehalococcoides mccartyi strains. *Journal of Hazardous Materials, 331*, 218–225. doi:10.1016/j.jhazmat.2017.02.043 PMID:28273571

Muir, D. C., & Howard, P. H. (2006). Are there other persistent organic pollutants? A challenge for environmental chemists. *Environmental Science & Technology, 40*(23), 7157–7166. doi:10.1021/es061677a PMID:17180962

Muller, E. E., Bringel, F., & Vuilleumier, S. (2011). Dichloromethane-degrading bacteria in the genomic age. *Research in Microbiology, 162*(9), 869–876. doi:10.1016/j.resmic.2011.01.008 PMID:21288483

Nadeau, L. J., Menn, F. M., Breen, A., & Sayler, G. S. (1994). Aerobic degradation of 1, 1, 1-trichloro-2, 2-bis (4-chlorophenyl) ethane (DDT) by Alcaligenes eutrophus A5. *Applied and Environmental Microbiology, 60*(1), 51–55. PMID:8117093

Nam, I. H., Kim, Y. M., Schmidt, S., & Chang, Y. S. (2006). Biotransformation of 1, 2, 3-tri-and 1, 2, 3, 4, 7, 8-hexachlorodibenzo-p-dioxin by *Sphingomonas wittichii* strain RW1. *Applied and Environmental Microbiology, 72*(1), 112–116. doi:10.1128/AEM.72.1.112-116.2006 PMID:16391032

Ontiveros-Valencia, A., Tang, Y., Krajmalnik-Brown, R., & Rittmann, B. E. (2013). Perchlorate reduction from a highly contaminated groundwater in the presence of sulfate-reducing bacteria in a hydrogen-fed biofilm. *Biotechnology and Bioengineering, 110*(12), 3139–3147. doi:10.1002/bit.24987 PMID:23797735

Ozdemir, C., Sen, N., Dursun, S., & Kalipci, E. (2010). Removal of dichloromethane in up-flow anaerobic sludge bed reactors and methane production. *Asian Journal of Chemistry, 22*(8), 6423.

Perocco, P., Bolognesi, S., & Alberghini, W. (1983). Toxic activity of seventeen industrial solvents and halogenated compounds on human lymphocytes cultured in vitro. *Toxicology Letters, 16*(1), 69–75. doi:10.1016/0378-4274(83)90012-7 PMID:6836616

Popat, S. C., & Deshusses, M. A. (2009). Reductive dehalogenation of trichloroethene vapors in an anaerobic biotrickling filter. *Environmental Science & Technology, 43*(20), 7856–7861. doi:10.1021/es901305x PMID:19921905

Popat, S. C., Zhao, K., & Deshusses, M. A. (2012). Bioaugmentation of an anaerobic biotrickling filter for enhanced conversion of trichloroethene to ethene. *Chemical Engineering Journal, 183*, 98–103. doi:10.1016/j.cej.2011.12.026

Putnam, R. A., Nelson, J. O., & Clark, J. M. (2003). The persistence and degradation of chlorothalonil and chlorpyrifos in a cranberry bog. *Journal of Agricultural and Food Chemistry, 51*(1), 170–176. doi:10.1021/jf020744r PMID:12502403

Rees, H. C., Oswald, S. E., Banwart, S. A., Pickup, R. W., & Lerner, D. N. (2007). Biodegradation processes in a laboratory-scale groundwater contaminant plume assessed by fluorescence imaging and microbial analysis. *Applied and Environmental Microbiology, 73*(12), 3865–3876. doi:10.1128/AEM.02933-06 PMID:17468279

Ryoo, D., Shim, H., Arenghi, F., Barbieri, P., & Wood, T. (2001). Tetrachloroethylene, trichloroethylene, and chlorinated phenols induce toluene-o-xylene monooxygenase activity in Pseudomonas stutzeri OX1. *Applied Microbiology and Biotechnology*, *56*(3-4), 545–549. doi:10.1007002530100675 PMID:11549035

Samin, G., & Janssen, D. B. (2012). Transformation and biodegradation of 1, 2, 3-trichloropropane (TCP). *Environmental Science and Pollution Research International*, *19*(8), 3067–3078. doi:10.100711356-012-0859-3 PMID:22875418

Sánchez, M. A., & González, B. (2007). Genetic characterization of 2, 4, 6-trichlorophenol degradation in Cupriavidus necator JMP134. *Applied and Environmental Microbiology*, *73*(9), 2769–2776. doi:10.1128/AEM.02584-06 PMID:17322325

Sato, T., Kimura, Y., & Takamizawa, K. (2002). Modeling transport and biodegradation of PCE in sandy soil. *WIT Transactions on Ecology and the Environment, 55.*

Satpathy, R., Konkimalla, V. B., & Ratha, J. (2015). Application of bioinformatics tools and databases in microbial dehalogenation research: A review. *Applied Biochemistry and Microbiology*, *51*(1), 11–20. doi:10.1134/S0003683815010147 PMID:25842899

Schultz, A., Jonas, U., Hammer, E., & Schauer, F. (2001). Dehalogenation of chlorinated hydroxybiphenyls by fungal laccase. *Applied and Environmental Microbiology*, *67*(9), 4377–4381. doi:10.1128/AEM.67.9.4377-4381.2001 PMID:11526052

Singh, A., & Lal, R. (2009). Sphingobium ummariense sp. nov., a hexachlorocyclohexane (HCH)-degrading bacterium, isolated from HCH-contaminated soil. *International Journal of Systematic and Evolutionary Microbiology*, *59*(1), 162–166. doi:10.1099/ijs.0.65712-0 PMID:19126742

Singh, P. (2007). Sequential anaerobic and aerobic treatment of pulp and paper mill effluent in pilot scale bioreactor. *Journal of Environmental Biology*, *28*(1), 77–82. PMID:17717990

Singh, S. P., & Prerna, P. (2009). Review of recent advances in anaerobic packed-bed biogas reactors. *Renewable & Sustainable Energy Reviews*, *13*(6), 1569–1575. doi:10.1016/j.rser.2008.08.006

Song, B., Palleroni, N. J., & Häggblom, M. M. (2000). Isolation and characterization of diverse halobenzoate-degrading denitrifying bacteria from soils and sediments. *Applied and Environmental Microbiology*, *66*(8), 3446–3453. doi:10.1128/AEM.66.8.3446-3453.2000 PMID:10919805

Taşeli, B. K., Gökçay, C. F., & Taşeli, H. (2004). Up flow column reactor design for dechlorination of chlorinated pulping wastes by Penicillium camemberti. *Journal of Environmental Management*, *72*(3), 175–179. doi:10.1016/j.jenvman.2004.04.008 PMID:15251223

Togna, M. T., Kafkewitz, D., & Armenante, P. M. (1995). Rapid dehalogenation of 2, 4, 6-trichlorophenol at alkaline pH by an anaerobic enrichment culture. *Letters in Applied Microbiology*, *20*(2), 113–116. doi:10.1111/j.1472-765X.1995.tb01299.x

Topp, E. (2001). A comparison of three atrazine-degrading bacteria for soil bioremediation. *Biology and Fertility of Soils*, *33*(6), 529–534. doi:10.1007003740100371

Tresse, O., Mounien, F., Lévesque, M. J., & Guiot, S. (2005). Comparison of the microbial population dynamics and phylogenetic characterization of a CANOXIS reactor and a UASB reactor degrading trichloroethene. *Journal of Applied Microbiology*, *98*(2), 440–449. doi:10.1111/j.1365-2672.2004.02475.x PMID:15659198

Van den Wijngaard, A. J., Van Der Kleij, R. G., Doornweerd, R. E., & Janssen, D. B. (1993). Influence of organic nutrients and cocultures on the competitive behavior of 1, 2-dichloroethane-degrading bacteria. *Applied and Environmental Microbiology*, *59*(10), 3400–3405. PMID:8250561

Van Pee, K. H., & Unversucht, S. (2003). Biological dehalogenation and halogenation reactions. *Chemosphere*, *52*(2), 299–312. doi:10.1016/S0045-6535(03)00204-2 PMID:12738254

Velazquez, F., Peak-Chew, S. Y., Fernández, I. S., Neumann, C. S., & Kay, R. R. (2011). Identification of a eukaryotic reductive dechlorinase and characterization of its mechanism of action on its natural substrate. *Chemistry & Biology*, *18*(10), 1252–1260. doi:10.1016/j.chembiol.2011.08.003 PMID:22035794

Wahman, D. G., Katz, L. E., & Speitel, G. E. (2005). Cometabolism of trihalomethanes by Nitrosomonas europaea. *Applied and Environmental Microbiology*, *71*(12), 7980–7986. doi:10.1128/AEM.71.12.7980-7986.2005 PMID:16332776

Wang, F., Grundmann, S., Schmid, M., Dörfler, U., Roherer, S., Munch, J. C., & Schroll, R. (2007). Isolation and characterization of 1, 2, 4-trichlorobenzene mineralizing Bordetella sp. and its bioremediation potential in soil. *Chemosphere*, *67*(5), 896–902. doi:10.1016/j.chemosphere.2006.11.019 PMID:17204305

Williams, W. A., Lobos, J. H., & Cheetham, W. E. (1997). Notes; A Phylogenetic Analysis of Aerobic Polychlorinated Biphenyl-Degrading Bacteria. *International Journal of Systematic and Evolutionary Microbiology*, *47*(1), 207–210. PMID:8995824

Yamasaki, S., Nomura, N., Nakajima, T., & Uchiyama, H. (2012). Cultivation-independent identification of candidate dehalorespiring bacteria in tetrachloroethylene degradation. *Environmental Science & Technology*, *46*(14), 7709–7716. doi:10.1021/es301288y PMID:22708499

Yang, C. F., Lee, C. M., & Wang, C. C. (2005). Degradation of chlorophenols using pentachlorophenol-degrading bacteria Sphingomonas chlorophenolica in a batch reactor. *Current Microbiology*, *51*(3), 156–160. doi:10.100700284-005-4481-1 PMID:16086107

Zhang, H., Ziv-El, M., Rittmann, B. E., & Krajmalnik-Brown, R. (2010). Effect of dechlorination and sulfate reduction on the microbial community structure in denitrifying membrane-biofilm reactors. *Environmental Science & Technology*, *44*(13), 5159–5164. doi:10.1021/es100695n PMID:20524654

Zhao, S., Ding, C., & He, J. (2015). Detoxification of 1, 1, 2-trichloroethane to ethene by Desulfitobacterium and identification of its functional reductase gene. *PLoS One*, *10*(4), e0119507. doi:10.1371/journal.pone.0119507 PMID:25835017

Ziv-El, M., Popat, S. C., Cai, K., Halden, R. U., Krajmalnik-Brown, R., & Rittmann, B. E. (2012). Managing methanogens and homoacetogens to promote reductive dechlorination of trichloroethene with direct delivery of H2 in a membrane biofilm reactor. *Biotechnology and Bioengineering*, *109*(9), 2200–2210. doi:10.1002/bit.24487 PMID:22392141

KEY TERMS AND DEFINITIONS

Bioreactor: A bioreactor is defined as a vessel system (may be flask, roller bottle, tank, or other container) used for bioprocessing that supports the growth of cells by formulating the medium and optimizing the conditions.

Bioremediation: Bioremediation is a treatment process that uses naturally occurring organisms to convert environmentally hazardous substances into its less toxic or non-toxic form.

Chemical Oxygen Demand (COD): It is measure of the amount of oxygen that can oxidize the soluble and particulate organic matter in water.

Dehalogenase: The dehalogenases are a group of enzymes produced especially in microbes that catalyzes the removal of a halogen atom from a halogenated substrate.

Microbial Consortia: A microbial consortium is the interaction (preferably symbiotic type) between two or more microbial groups in a given environment.

Organic Pollutants: Organic compounds that are resistant to environmental degradation through chemical, biological, and photolytic processes.

Toxic Substances: A toxic substance is a substance that can be lethal or cause health hazards when uptake by the body.

Chapter 16
Microbial Degradation of Azo Dyes:
The Role of Azoreductase to Initiate Degradation

Dirk Tischler
TU Bergakademie Freiberg, Germany

Jingxian Qi
TU Bergakademie Freiberg, Germany

Anna Christina R. Ngo
Far Eastern University, Philippines

Michael Schlömann
TU Bergakademie Freiberg, Germany

ABSTRACT

Azo dyes are considered as xenobiotic compounds, which are often recalcitrant to be biodegraded. Many dyes are relevant for staining in industries, especially textile and food related. Among other environmental problems, one obvious issue is the coloring effect on effluents from industrial sites and thus the release of xenobiotics into nature. Microorganisms (algae, fungi, yeast, and bacteria) have been found to decolorize a number of azo dyes. Decolorization of azo dyes by microbial cultures is summarized and this is mostly linked to initial activation or even cleavage of the azo bond (e.g., by azoreductases). However, it does not necessarily mean that the compound is degraded to non-toxic products. Various mechanisms of microbial decolorization processes were discovered, including adsorption, enzymatic degradation, or a combination of both. Oxidases and reductases were found to be involved, which contain azoreductase, lignin peroxidase, Mn peroxidase, laccase, tyrosinase, and so on. A focus is on the azoreductases including classification, activity, and applicability.

DOI: 10.4018/978-1-5225-3540-9.ch016

INTRODUCTION

Dyes and Their Application

Dyes are natural or chemical substances, which can impart colors to various materials. Mankind has a long history to use dyes. The earliest written record about the dyeing with plants, barks and insects was found in China 2600 BC (Bafana *et al.,* 2011). Also, tyrian purple was described as early as about 1500 BC, and is still known as the color of nobility (McGovern & Michel, 1985). Dyes were initially made from natural plants or animal products. However, William Henry Perkin introduced the world's first synthetic dye in 1856, marking the start of dye manufacturing industry (Bafana *et al,*. 2011). Compared to natural dyes, synthetic dyes are cheaper to produce, easier to apply, and could provide more abundant colors. This is why these synthetic dyes are now widely used in textile, print, food, cosmetics and pharmaceutical industries.

In 1876, Otto Witt developed a theory that a colored dye should have a chromophore group and an auxochrome group. Chromophores are capable of absorbing light in the visible region (e.g. nitro, azo, quinoid groups), while auxochromes deepen the color when introduced into a colored molecule (e.g. -COOH, -OH, $-SO_3H$ and $-NH_2$). However, Witt's theory was later replaced by the modern electronic theory. According to this theory, color is from the excitation of valance π electrons by visible light (Murrell, 1973). Generally, dyes are classified according to two aspects: chemical composition and application class (Garfield, 2002). In terms of chemical composition, it can be divided into azo, nitro, nitroso, diarylmethane, triarylmethane, xanthene, anthraquinoid, acridine, cyanine, quinoneimine, pthalocyanine and thiazole dyes. Alternatively, based on the application method, dyes can also be divided into acid, azoic, basic, direct, disperse, mordant, reactive, sulfur and vat dyes (Booth *et al.,* 2000). Some typical dyes applied in industries and research are presented in Figure. 1.

Textile Industry: An Overview of the Problem

The textile industry contributes a part in world economy and to a country's growth as it generates more income and wages for people. It leaves a significant trail, with each step of the process causing more environmental problems. It primarily utilizes large volumes of water and chemicals for desizing, scouring, bleaching, dyeing, printing, and finishing. These chemicals vary from inorganic materials to polymers and organic products (Banat *et al.,* 1996). Among these chemicals, a large number can be attributed to dyes as 10,000 different dyes, with a yearly production of more than $7x10^5$ metric tons, are just used for dyeing and printing (McMullan *et al.,* 2001) and 5 to 10% are released with wastewater (Alalewi & Jiang, 2012).

One of the most significant environmental problems is the presence of color in water courses. Though some of the colors in the aquatic ecosystem are product of biological activities, a considerable amount can be pointed out to industrial effluents (Alalewi & Jiang, 2012). A trace concentration of it is highly undesirable as it leads to aesthetic pollution, eutrophication and perturbation in the aquatic ecosystem. Their presence and by-products are not removed easily by conventional treatments (Kurade *et al.,* 2011).

The synthetic dyes are already recognized as toxic around the globe and more natural dyes enter the market (Kant, 2012).

Figure 1. Research and industrially relevant dyes (a selection)

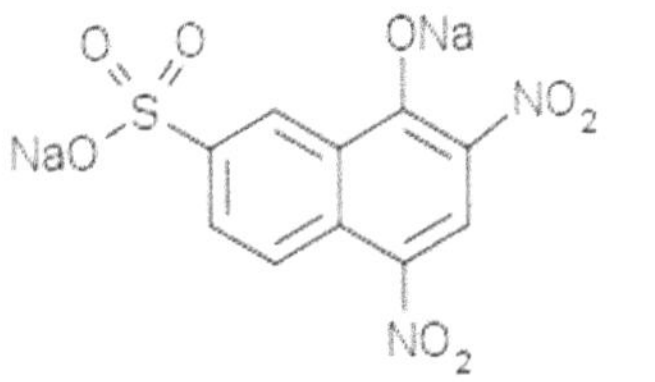

Naphthol Yellow S
(It was used in food industry and later to stain nucleic acids and proteins for microscopy)

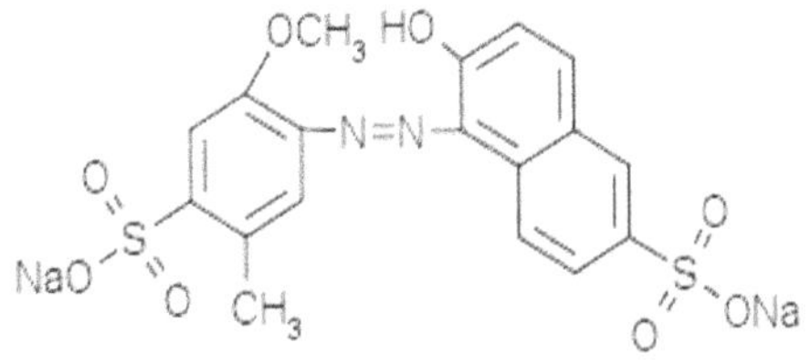

Allura Red AC
(It is applied in drug, cosmetic and food industries, but, only in a few countries)

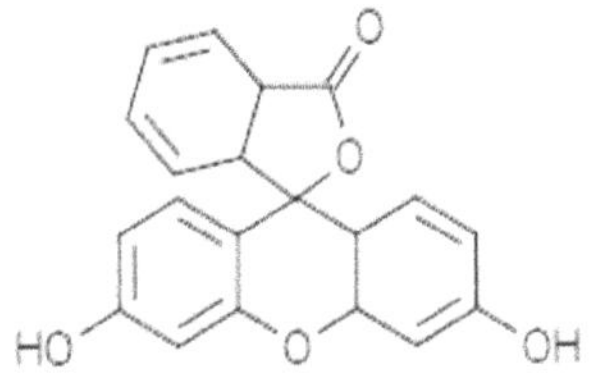

Fluorescein
(It is mostly used as a fluorescent marker compound for various applications.)

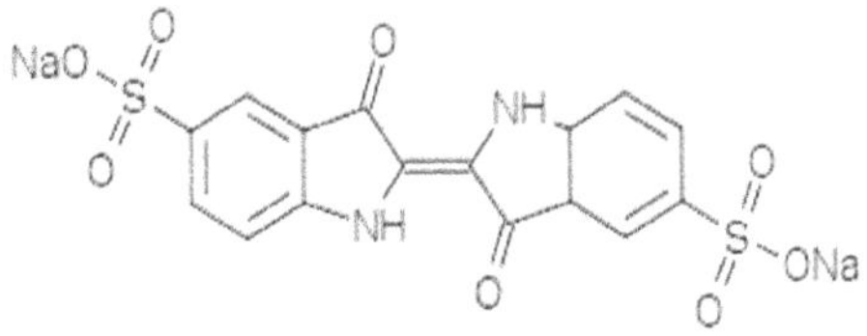

Indigo Carmine
(The dye is widely used to colorize foods with the respective number E132)

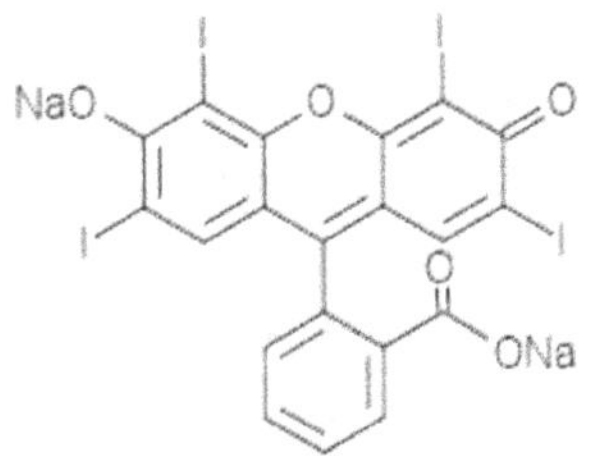

Acridine Orange
(It interacts with nucleic acids and thus is employed as a fluorescent stain/marker)

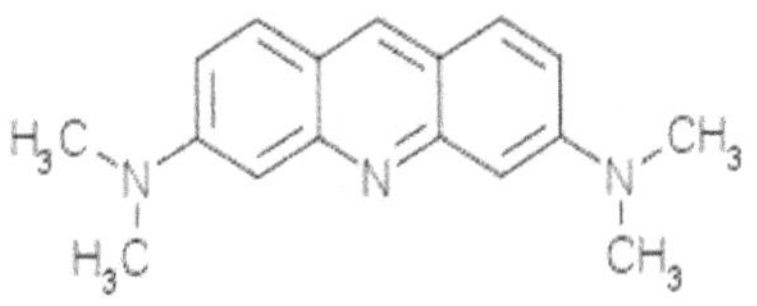

Erythrosine
(It is applied in various industries such as ink production, food, research among others.)

Environmental Concern of Textile Dyes

Most dyes used for textiles are synthetic in nature and are structurally diverse as they may comprise of benzidine and aromatic compounds (Banat *et al.*, 1996). They have chromophores that are responsible for the color and auxochromes that intensify it (dos Santos *et al.*, 2007). The most common dyes often employed in the industry are azo, anthraquinone, sulfur, indigoid, triphenylmethyl, and phtalocyanine (Forgacs *et al.*, 2004). Synthetic dyes prompted the interest of researchers because most of these are toxic, carcinogenic, and mutagenic (Erkurt *et al.*, 2007). They are difficult to treat primarily because of their synthetic origin and complex structures which make them stable and less susceptible to biodegradation (Banat *et al.*, 1996). Their improper disposal in the effluents raised so much concern as it leads to poor sunlight penetration, decrease of photosynthetic activity, reduction of dissolved oxygen concentration, poor water quality, and stimulates acute effects on the aquatic community (Vandevivire *et al.*, 1998; Saratale *et al.*, 2009). In addition, these dyes and the contaminated effluents affect germination rates and growth of various plants which have significant role in ecology such as serving as a habitat for different organisms, giving organic matters which contribute to soil fertility, and keeping the soil from erosion (Saratale *et al.,* 2011).

Azo dyes are one of the largest groups of dyes used in the industry for they are cost-efficient, stable, and easier to synthesize, and have variety of colors (Saratale *et al.,* 2009). They are identified by their azo chromophores (-N=N-) and the bonds between two or more aromatic rings (Jamal *et al.,* 2011). They contain xenobiotic components which make them very resistant to treatment (Solis *et al.,* 2012). They also increase carcinogenic potential due to the electron-donating substituents in *ortho-* and *para*-position (Carmen & Daniela, 2012). These dyes have raised concerns in terms of human health as they may increase the risk of human bladder cancer, splenic sarcomas, hepatocarcinomas, and nuclear anomalies based from conducted studies (Puvaneswari *et al,.* 2006). They can also cause allergies, dermatitis, and DNA damage that result to the formation of malignant tumors (Carmen & Daniela, 2012).

Azo Dyes

Azo dyes, as briefly introduced above, are characterized by containing one or more azo bonds (R_1-N=N-R_2; see Figures 1 and 2). Because of the simple synthetic process, great structural diversity, and medium-to-high fastness against light and wetness, azo dyes are popular, belonging to one of the largest groups of colorants. In general, azo dyes are released to the environment along with the effluent from factories. According to estimations, 280,000 metric tons of textile dyes are discharged globally into industrial effluents every year (Maas & Chaudhari, 2005). Furthermore, aquatic ecosystems including algae, plants and fishes may be contaminated and affected, if the drainage was not appropriately dealt with. One major concern is color, which causes an aesthetic pollution in aquatic systems (Kurade *et al.,* 2011). Chemical Oxygen Demand (COD) and turbidity increase, and then less sunlight penetrates. Eco-processes like photosynthesis of algae, plants growth (Puvaneswari *et al.,* 2006) and respiratory activities of microorganisms may be inhibited due to the presence of dyes. Azo dyes are generally considered as xenobiotic. Because they possess electron withdrawing groups like sulfonate groups which cause electron deficiency, resulting in the recalcitrance of biodegradation (Pandey *et al.,* 2007; Singh *et al.,* 2014). High durability is one major feature of synthetic dyes and appreciable kinds of azo dyes are considered to be toxic, even mutagenic or carcinogenic. The LD_{50} of majority of azo dyes is between 250–2000 mg kg^{-1} (Clarke & Anliker, 1980). Reactive Brilliant Red was found to cause inhibition of the normal function of human serum albumin. Aniline Yellow was classified into Group 2B (possibly carcinogenic) by the International Agency for Research on Cancer (IARC). Carcinogenicity of Direct Black 38 was demonstrated experimentally (Bafana *et al.,* 2011). Therefore, it is necessary to treat the wastewater before discharge.

A model compound is represented by methyl red (2-(N,N-dimethyl-4-aminophenyl) azobenzenecarboxylic acid; abbreviation: MR), also called C.I. Acid Red 2, which is a pH indicator dye widely used in laboratory (Table 1). Being one of the simplest azo dyes, it has been proved to be the best choice to study the mechanism of dye degradation (Kumar *et al.,* 2006) by means of microbial studies or even the enzymatic activation/conversion. Methyl Red is mutagenic in nature and thus also relevant from an environmental as well as social point of view (Ayed *et al.,* 2011; Jadhav *et al.,* 2012).

Conventionally, physical and chemical technologies have been applied in water treatment for removal of azo dyes, such as coagulation-flocculation, filtration, electrolysis, oxidation and adsorption and so on (Kant, 2012; Robinson *et al.,* 2001; Saratale *et al.,* 2011). The potential risks of generating by-products

Figure 2. Chemical structures of azo dyes: some examples

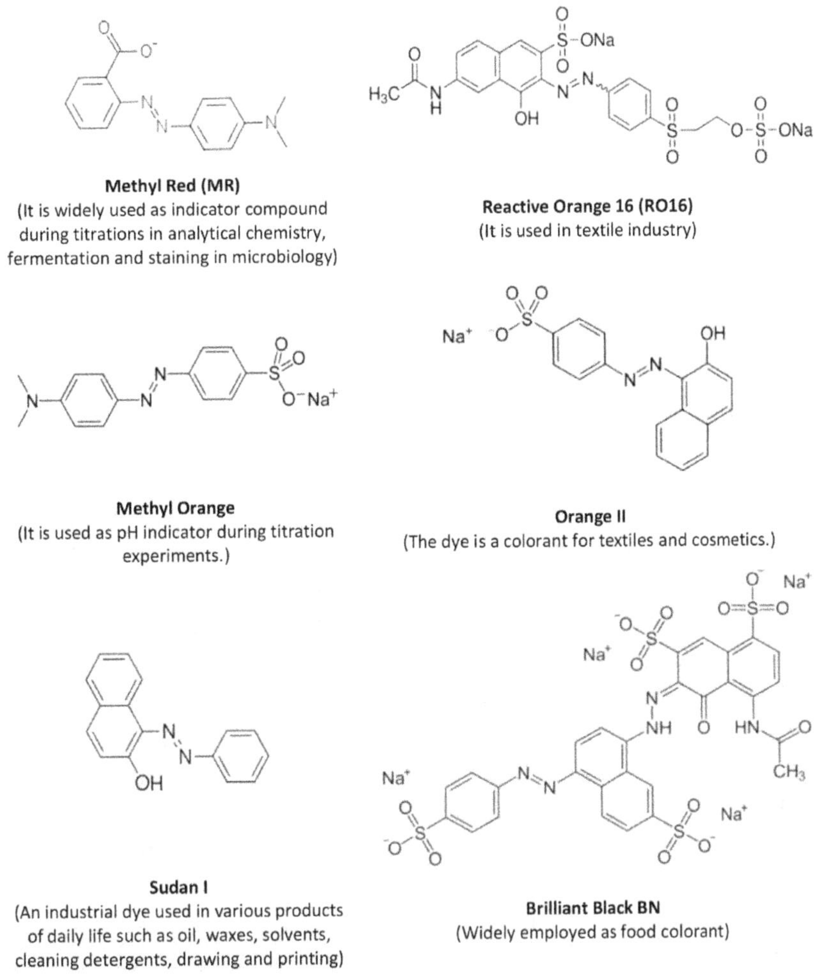

and causing secondary contamination, as well as economical unfeasibility are major concerns when applying physicochemical methods. Meanwhile, bioremediation technologies provide other solution possibilities to deal with the existing dyes in wastewater. The biodegradation of azo dyes raised attention due to following advantages: potential for complete mineralization, environmental friendliness, competitive cost and less sludge production (Saratale *et al.*, 2011).

Decolorization Techniques

An overview on treatment methods regarding dyes and related waste streams is provided by several research articles and reviews such as: (Carmen & Daniela, 2012; Forgacs *et al.*, 2004; Kant, 2012; Saratale *et al.*, 2011; Vandevivere *et al.*, 1998; Wang *et al.*, 2011) (Table 2). It has been estimated that 10% of the dyes used are released into the sewage treatment systems (Mane *et al.*, 2008). A number of methods have been developed for their removal so as to decrease the damage on the environment. However, physical and chemical methods such as coagulation-flocculation, membrane filtration, adsorption (using

Table 1. Physiochemical properties of some azo dyes

Dye	methyl red	brilliant black BN	reactive orange 16	sudan I
CAS No.	493-52-7	2519-30-4	12225-88-6	842-07-9
Chemical formula	$C_{15}H_{15}N_3O_2$	$C_{28}H_{17}N_5Na_4O_{14}S_4$	$C_{20}H_{17}N_3Na_2O_{11}S_3$	$C_{16}H_{12}N_2O$
Molecular weight [g mol⁻¹]	269.30	867.68	616.98	248.28
Solubility	ethanol, hardly in water	little in ethanol, water	water	organic solvents
λ_{max}	410 nm	407 nm 570 nm	388 nm 494 nm	418 nm 476 nm
Transition range	4.2 – 6.2, red to yellow	-	-	-
Melting point	179 – 182 °C	-	> 300 °C	~ 134 °C

(Ullmann's Encyclopedia of Industrial Chemistry as cited in Booth et al., 2000)

activated carbon, silica and clay) and chemical oxidation have disadvantages such as the expensive output, inefficiency to remove the dyes completely, and the generation of secondary pollution. Since most dyes are recalcitrant, these xenobiotics are less susceptible to these physical and chemical treatments. Furthermore, due to some treatments intermediates of degradation appear and those might be even more toxic or recalcitrant and therefore the problem might only be migrated to non-colorant compounds. This is the reason why the need for alternative strategies to degrade these components has increased.

Bioremediation is an interesting area in science as it uses microorganisms to solve pollution and other environmental problems. In this field, microbes adjust to the toxic wastes and develop resistant genes which then degrade or transform toxic chemicals into less harmful ones (Saratale *et al.,* 2011). There are already a number of approaches wherein microbes are used to solve different environmental problems. Biodegradation of dyes has called an emergent attention to try this approach mainly because of the recalcitrant nature of the dyes and their increasing usage over time. The use of microbial methods for the degradation of dyes has specific advantages such as it is environmental-friendly, it is cost-efficient, produces less secondary pollutants, and entails less water for the process (Banat *et al.,* 1996; Rai *et al.,* 2005; Saratale *et al.,* 2011). The efficacy of the biodegradation method entirely depends on the activity of the selected microorganism and its adaptability to the process. Subsequently, there are several organisms that have already been tested for their ability to biodegrade different dyes. A number of various microorganisms has been proven to decolorize a wide range of dyes such as: bacteria (Wang *et al.,* 2009;

Table 2. Methods to treat effluents loaded with dyes

Biological Methods	Chemical Methods	Physical Methods
Aeration: tanks, vessels, lagoons	Electrochemical treatment	(Membrane) filtration
Trickling filters, reactors	(Metal) catalysis	Solvent extraction
Activated sludge	Oxidative or reductive agents	Adsorption
Anaerobic fermentation	Activated carbon	Distillation
Aerobic fermentation		Flotation, foam fractionation
Bio-flocculation		Coagulation

Jadhav *et al.*, 2010; Gurav *et al.*, 2011), fungi (Erkurt *et al.*, 2007; Revankar & Lele 2007; Mishra *et al.*, 2011), actinomycetes (Mane *et al.*, 2008), and yeasts (Yang *et al.*, 2008; Saratale *et al.*, 2009).

Interestingly, actinomycetes are now being recognized in the field of microbial biotechnology as they are capable of releasing different enzymes that can be used in different industrial applications (Qi *et al.*, 2016, 2017). The genus *Rhodococcus* comprises actinomycetes known for their metabolic activities such as amino acid production and microbial transformation of steroids (Vobis & Miyadoh. 1997) as well as many other catabolic activities (Martínková *et al.*, 2009). However, azo reductase from this genus is rarely described to transform azo dyes (Qi *et al.*, 2016). Despite the number of genes from *Rhodococcus* sp. being sequenced and studied, only a single gene of an azo reductase had been discovered and described (Qi *et al.*, 2016). Within this study also the original host strain was investigated for its capability to degrade azo dyes (Figure 3).

Microbial Degradation (Decolorization) of Azo Dyes

As mentioned above, various microorganisms have demonstrated during laboratory investigations but also during field studies that they could be involved in decolorization or even degradation of azo dyes including: algae, fungi, yeast and bacteria which we summarized in Table 3. Algae could grow in the effluents from paper mill and pharmaceutical industries (Dubey *et al.*, 2011), which provides potential for treating dye-containing wastewater. Furthermore, the trait that algae derive carbon source from the air saves cultivation expenses. Lignin-degrading fungi have become popular for a long time because of their wild diversity and rapid adaption to various environments. White-rot fungi were quite often used to develop processes of mineralization of azo dyes (Machado *et al.*, 2006), especially *Phanaerochaete*. In comparison, fewer reports about utilizing yeast for decolorization of dyes were discovered. However, due to the high tolerance of dyes and heavy metals, some yeast strains were found to be effective (Ertuğrul *et al.*, 2008). Bacteria such as *Pseudomonas* (Kalyani *et al.*, 2009; Lin *et al.*, 2010), *Enterobacter* (Wang *et al.*, 2009), *Bacillus* (Dawkar *et al.*, 2009) and *Escherichia coli* (Jin *et al.*, 2009) have already been demonstrated possessing the capability of azo dyes degradation.

Two examples of agar plate-screening for azo dye decolorization by R. opacus 1CP have been performed and are highlighted. The assays were conducted in mineral medium agar plates of pH 6.0 and 7.0 respectively. Benzoate was used as a carbon source to led the strain initially form biomass at 30°C. *Left panel:* Methyl red was degraded while a light halo appeared within weeks and its size increased until the whole plate turned to be light (independent of pH). *Right panel:* In contrast, amido black seemed to be absorbed by strain 1CP, but not degraded as the color was still present even after one year.

A variety of azo dyes could decompose under anaerobic conditions, like Reactive Blue 13 (Lin *et al.*, 2010), Congo Red (Gopinath *et al.*, 2009), Reactive Black 5, Reactive Red 180 (Wang *et al.*, 2009) and so on. This reduction appears to be nonspecific (Stolz, 2001) but microbes decolorize azo dyes via the cleavage of azo bonds to subsequent hazardous aromatic amines (Pandey *et al.*, 2007), which hardly participate into further degradation process and accumulate if the conditions would not change (Bafana *et al.*, 2009). Under aerobic conditions, most azo dyes are resistant to degradation due to the competition of reduced electron carriers between oxygen and azo compounds (Stolz, 2001). However, still some aerobic strains such as *Geobacillus stearothermophilus* were found to be able to decolorize azo dyes (Evangelista-Barreto *et al.*, 2009). *Vibrio logei* and *P. nitroreducens* were isolated and identified using MR as the sole carbon source (Adedayo *et al.*, 2004). One essential advantage of aerobic systems is that aromatic amines could be further mineralized to less toxic products via oxidation. Therefore, a system

Figure 3. Methyl red degradation and amido black adsorption by Rhodococcus opacus 1CP (unpublished)

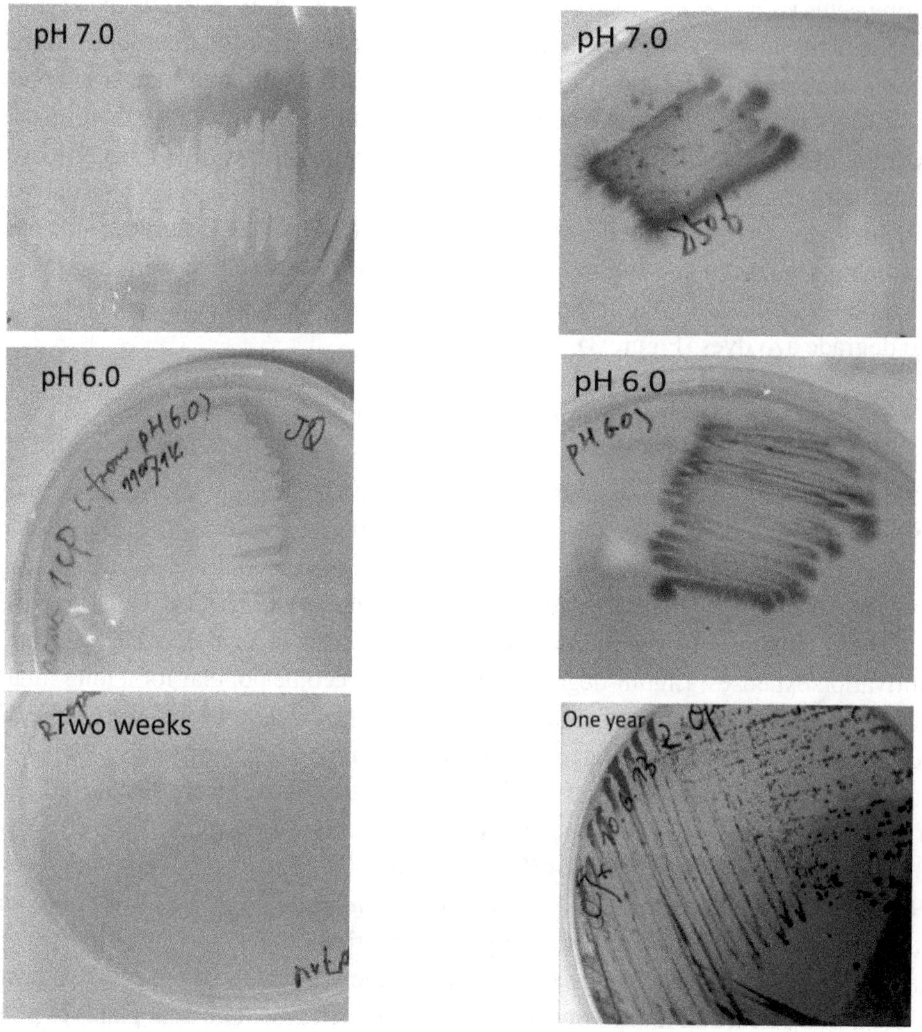

composing of an anaerobic treatment process followed by an aerobic process was proposed to achieve higher biodegradation efficiency of azo dyes (Khehra *et al.,* 2006; You & Teng 2009). A pathway for degradation of Reactive blue 13 was proposed, which adopted a two-stage anaerobic/aerobic fluidized beds system with a *Pseudomonas* sp. isolate (Lin *et al.,* 2010).

Besides to oxygen, microorganisms are also quite sensitive to other ambient conditions; the decolorization performance was highly influenced by factors including pH, temperature, carbon source, nitrogen source, the dye concentration and structure and so on. Generally, the presence of carbon and nitrogen source facilitates decolorization efficiency (Elisangela *et al.,* 2009; Moosvi *et al.,* 2007). A more inspiring finding shows that some microorganisms could utilize the azo dye as the sole carbon source, like the growth of *Saccharomyces cerevisiae* with Methyl Red (Jadhav *et al.,* 2007); the consortium composed of *Pseudomonas, Arthrobacter,* and *Rhizobium* with Acid Orange 7 (Ruiz-Arias *et al.,* 2010). The environmental temperature directly determines organismal temperature, influencing biochemical or enzymatic processes. The effects of temperature on the growth rate, biomass yield, and reaction

Table 3. Decolorization of various azo dyes by microorganisms

	Name	Conditions	Azo Dyes, Decolorization (%)	Reference
Algae	*Chlorella vulgaris, Lyngbya lagerlerimi, Nostoc lincki, Elkatothrix viridis, Volvox aureus*	pH 7, 25 °C, 7 days	MR, 82%; Orange II, 47%; G-Red (FN-3G), 59%	El-Sheekh *et al.*, 2009
	Oscillatoria curviceps BDU92191	pH 7, 4 °C, 100 rpm, 8 days	Acid Black, 84%	Priya *et al.*, 2011
Filamentous Fungi	*Phanerochaete chrysosporium* MTCC 787	pH 4.5, 30 °C, 150 rpm, 72 h	Direct Red 80, 100%	Sen *et al.*, 2012
	Trametes versicolor 1	pH 6.5, 26 °C, 240 rpm, 384 h	Reactive Blue 4, 90%	Yemendzhiev *et al.*, 2009
	Aspergillus niger HM11	pH 11, 20 °C, 72 h	Congo Red, 71%	Karthikeyan *et al.*, 2010
Yeast	*Pseudozyma rugulosa* Y-48	pH 5–6, 28 °C, 24 h	Reactive Brilliant Red -2BP, 99%; Weak Acid Brilliant Red B, 98%; Reactive Black KN-B, 96%; Acid Mordant Yellow, 94%; Reactive Brilliant Blue X-BR, 85%; Acid Mordant Red S-80, 22%; Acid Mordant light Blue B, 89%; Reactive Turquoise Blue KN-G, 67%	Yu & Wen 2005
	Issatchenkia orientalis JKS6	pH 7, 32 °C, 150 rpm, 24 h	Reactive Black 5, 99%; Reactive Orange 16, 99%; Reactive Red 198, 96%; Direct Yellow 12, 96%; Direct Blue 71, 95%; Direct Black 22, 97%	Jafari *et al.*, 2014
Bacteria	*Pseudomonas* sp.	pH 7, 35 °C, 70 h	Reactive Blue 13, 83%	Lin *et al.*, 2010
	Enterobacter sp. EC3	pH 7, 37 °C, anaerobic, 36 h	Reactive Black 5, 93%	Wang *et al.*, 2009
	Bacillus sp. VUS	pH 7, 40 °C, 18 h	Navy Blue 2GL, 94%	Dawkar *et al.*, 2009
	Escherichia coli JM109	pH 9, 30 °C, anaerobic, 12 h	Direct Blue 31, 100%	Jin *et al.*, 2009
	Rhodococcus opacus 1CP	pH 7, 30°C, aerobic 28 h	Methyl Red, 89%; Brilliant Black, 16%	Qi *et al.*, 2016

mechanism have been reported (Angelova *et al.*, 2008). Optimal pH values for bacterial decolorization are often between 6.0 and 10.0 (Kılıç *et al.*, 2007; Chen *et al.*, 2003; Guo *et al.*, 2007) and the effect of pH was proposed to be related to the transport of dye molecules through the cell membrane, which is the limiting step for the decolorization (Chang *et al.*, 2001; Kodam *et al.*, 2005). With the increase in the dye concentration, the decolorization rate gradually decreases, probably due to the toxic effect of dyes or the block of active sites of azoreductase by dye molecules (Sani & Banerjee 1999; Jadhav *et al.*, 2008; Tony *et al.*, 2009). Azo dyes with electron-withdrawing groups like $-SO_3H$ group would be easier to be decolorized than azo dyes with the electron-releasing groups like –NH-triazine. Moreover, azo dyes with more electron-withdrawing groups have shown faster decolorization rate. The ranking of decolorization rate influenced by the position of electron-withdrawing substituents in azo dyes is *para* > *ortho* > *meta* (Hsueh *et al.*, 2009).

Degradation Pathways

Mechanisms of microbial decolorization processes were discovered, including adsorption, enzymatic degradation or a combination of both. It has been demonstrated that biomass of algae, yeast, fungi and bacteria could be used as adsorbents for dye removal (Bhatnagar & Sillanpää 2010). The heteropolysaccharides and lipids in cell walls contribute to the adsorption through the various functional groups like amino, carboxyl, hydroxyl and other charged groups which are attractive to azo dyes (Srinivasan & Viraraghavan, 2010). Detailed pathways of the *in vivo* biodegradation may be different among various microorganisms and with respect to the structure of dyes. In general, the initial step is the reduction of azo bonds, which could result from enzymes, redox mediators or chemical reductants (Sandhya 2010). Oxidases and reductases were both found to be involved. It was proposed that peroxidases from *S. chromofuscus* A11 and *P. chrysosporium* could simultaneously split the azo linkage of sulfonated dyes both symmetrically and asymmetrically (Goszczynski *et al.,* 1994). However, normally azoreductase is responsible for the reduction of –N=N– bond into two $-NH_2$ groups (Figure 4), which involves a transfer of four electrons (Pearce *et al.,* 2006). Previous researches show azoreductase has been found to exist in several microorganisms.

Figure 4. Proposed biogradation pathway for Reactive blue 13 (Lin et al., 2010)

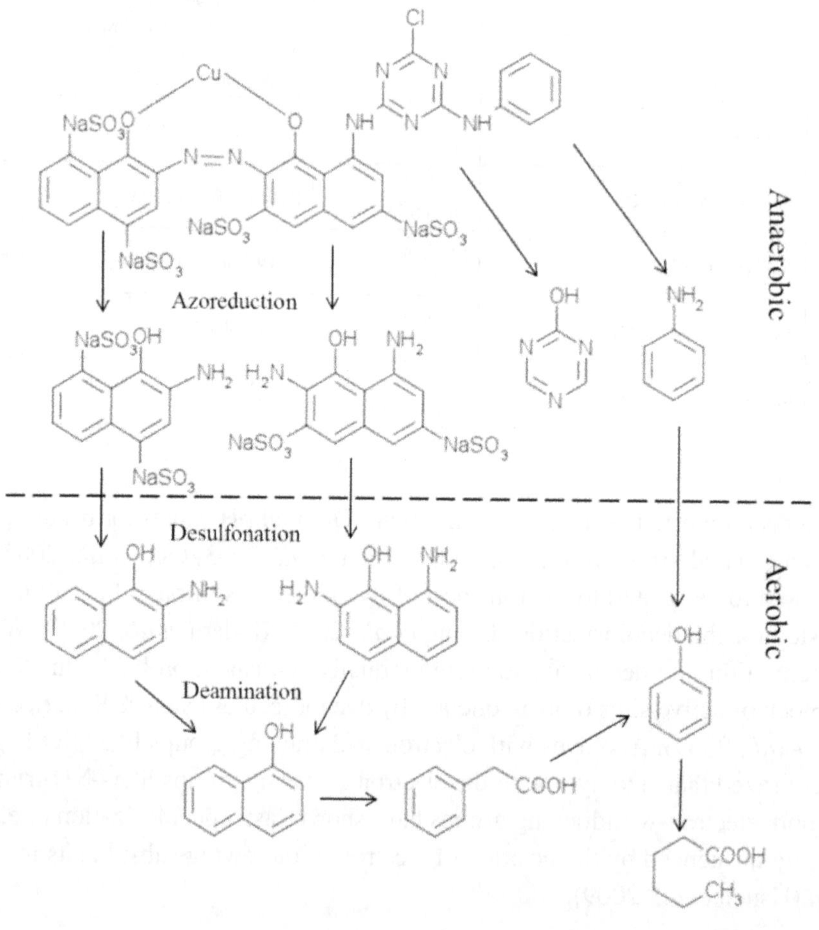

Enzymes Involved in Azo Dye Activation and Degradation

Enzymes from microorganisms can destroy the original structure of the dyes and split pollutants into fragments, achieving complete mineralization such as converting xenobiotic compounds into carbon dioxide, biomass, and other inorganic products (Khan *et al.,* 2013). A variety of enzymes have already been studied from different types of microorganisms such as laccases, manganese peroxidases, lignin peroxidase and azo reductases to name a few (Saratale *et al.,* 2011; Khan *et al.,* 2013). These enzymes are known to generate highly reactive free radicals to undergo a series of cleavage reactions (Khan *et al.,* 2013).

Laccases

Laccases are copper-containing enzymes which can oxidize a wide range of compounds such as polyaromatic hydrocarbons and synthetic dyes (Saratale *et al.,* 2011; Ashrafi *et al.,* 2013). It is widely used in biotechnological processes due to its non-specific oxidation capacity, no requirement for cofactors, and do not use readily available oxygen as an electron-acceptor (Telke *et al.,* 2011; Kalyani *et al.,* 2012; Singh *et al.,* 2015). Its monomeric, dimeric or tetrameric glycoproteins have four copper atoms, which are distributed among different binding sites, and are classified into three types: Copper I, II, and III, which play a role in the oxidation of the substrate (Khan *et al.,* 2013).

The use of laccases for decolorization have been mainly studied in lignolytic fungi (Abadulla *et al.,* 2000; Forootanfar *et al.,* 2012; Ashrafi *et al.,* 2013). In the study of Wong *et al.,* (2012), two laccase genes were cloned from *Lentinula edudes* and found that the efficient degradation of dyes take place by two allelic forms of *L. edodes* laccase as expressed by *Pichia pastoris.* In addition to this, laccase isolated from *Paraconiothyrium variabile* was able to decolorize 97% of Acid Red 18 after 15 minutes of incubation (Ashrafi *et al.,* 2013). Laccases are known for its ability to decolorize without the need to cleave azo bonds through a highly non-specific free radical mechanism. This avoids the formation of toxic aromatic amines (Kalme *et al.,* 2009; Saratale *et al.,* 2011).

Manganese Peroxidase

Manganese peroxidase (MnP) is an oxidative enzyme that is efficient in the decolorization of azo dyes and phthalocyanine complexes. It has been initially discovered in *Phanaerochaete chrysosporium* where in its first decolorization activity was first reported through a ligninolytic enzyme system (Chagas & Durran, 2001). Its catalytic mechanism starts with the initial oxidation by H_2O_2 to an intermediate that in turn promote the oxidation of Mn^{2+} to Mn^{3+} (Gold *et al.,* 2000; Khan *et al.,* 2013). This Mn^{3+} is further chelated with organic acids that in turn, serve as an active oxidant (Heinfling *et al.,* 1998; Mester & Field, 1998). MnP is known to not only catalyze the destruction of chromophores but also the breakdown of the chemical structure of the dye. Apart from white rot fungi, MnP activity has also been observed on bacteria such as in *Serratia marcescens* where in it efficiently decolorized azo and anthraquinone dyes such as Ranocid Fast Blue and Procion Brilliant Blue-H-GR (Verma & Madamwar, 2002).

Lignin Peroxidase

Lignin peroxidase has been widely used to mineralize a variety of recalcitrant compounds such as polychlorinated biphenyls and synthetic dyes (Duran & Esposito, 2000). Its decolorization mechanism involves the oxidation of the phenolic group to yield a carbon radical with the azo linkages. Moreover, the phenolic carbon is further cleaved producing phenyldiazene, which in turn can be also oxidized generating N_2 (Saratale *et al.,* 2011).

This enzyme system has been first discovered in white-rot fungi. In the study of Heinfling *et al.,* (1998), lignin peroxidase from *Bjerkandera adusta* was able to decolorize industrial azo and phthalocyanine dyes. Meanwhile, the study of Verma and Madamwar (2003) described the ability of lignin peroxidase from *Phanaerochaete chrysosporium* to decolorize dyes namely, Procion Brilliant Blue HGR, Ranocid Fast Blue, Acid Red 119, and Navidol Fast Black. Though this enzyme is rarely reported from bacteria, few studies have already been conducted to detect its presence in the said organism. Lignin peroxidase activity has already been observed from *Bacillus* sp. strain VUS and *Acinetobacter calcoaceticus* NCIM 2890 wherein the purified enzyme is efficient to degrade different azo dyes (Dawkar *et al.,* 2009; Ghodake *et al.,* 2011).

Azo Reductase

Azo reductases are enzymes that can cleave azo bonds (-N=N-) to produce colorless aromatic amine products (Khan *et al.,* 2013). Azo reductases can be differentiated as flavin-dependent azo reductases and flavin-independent azo reductases. Flavin-dependent azo reductases are further categorized based on their cofactors, NADH, NADPH or both. These coenzymes serve as electron donors. Azo reductases from microorganisms represent novel enzymes with little similarity to other reductases. A *Rhodococcus opacus* 1CP azoreductase (AzoRo) from the study of Qi *et al.,* (2016) shares the best identity with *Pseudomonas aeruginosa* azo reductase PaAzo1 with 33% similarity. Moreover, the expression of *Bacillus subtilis* azo reductase AzoR1 and AzoR2 shares around 30% amino acid sequence identity with *E. coli* Azo R (Leelakriangsak & Borisut, 2012; Leelakriangsak, 2013). Azo reductases have been purified and characterized from several microorganisms such as *Bacillus* sp. OY1-2, *Staphylococcus aureus*, and *Xenophilus azovorans* KF46F (Suzuki *et al.,* 2001; Chen *et al.,* 2005; Blümel *et al.,* 2002).

The required cofactor (NADH or NADPH) is quite expensive. Alternatively, a less costly cofactor, a synthetic 1-benzyl-1,4-dihydronicotinamide (BNAH) was employed as an electron donor with AzoRo for the reduction of azo dyes to show potential applications (Qi *et al.,* 2017B). One obstacle discovered about AzoRo was that it reached its rapidest degradation activity at the low pH (pH 4) where it is rather unstable. Thus AzoRo immobilization was studied, trying to improve its performance (Qi *et al.,* 2017A).

Azo dyes are xenobiotic chemical compounds which are recalcitrant to be degraded. Azoreductases catalyze the cleavage of the azo bond (-N=N-) which is an essential part of azo dyes. Most azoreductases already found were demonstrated to reduce azo compounds via Ping Pong Bi Bi mechanism (Nakanishi *et al.,* 2001; Liu *et al.,* 2008A, 2008B; Wang *et al.,* 2010), which requires two cycles of NAD(P)H-dependent reduction. In the first cycle, an azo bond is reduced to be hydrazine (-NH-NH-) which will be further reduced in the second cycle. In the end two amines ($-NH_2$) are generated (Figure 5) (Ito *et al.,* 2006A, 2006B; Wang *et al.,* 2010).

Figure 5. The proposed catalytic reaction of AzoR from E. coli reducing Methyl Red

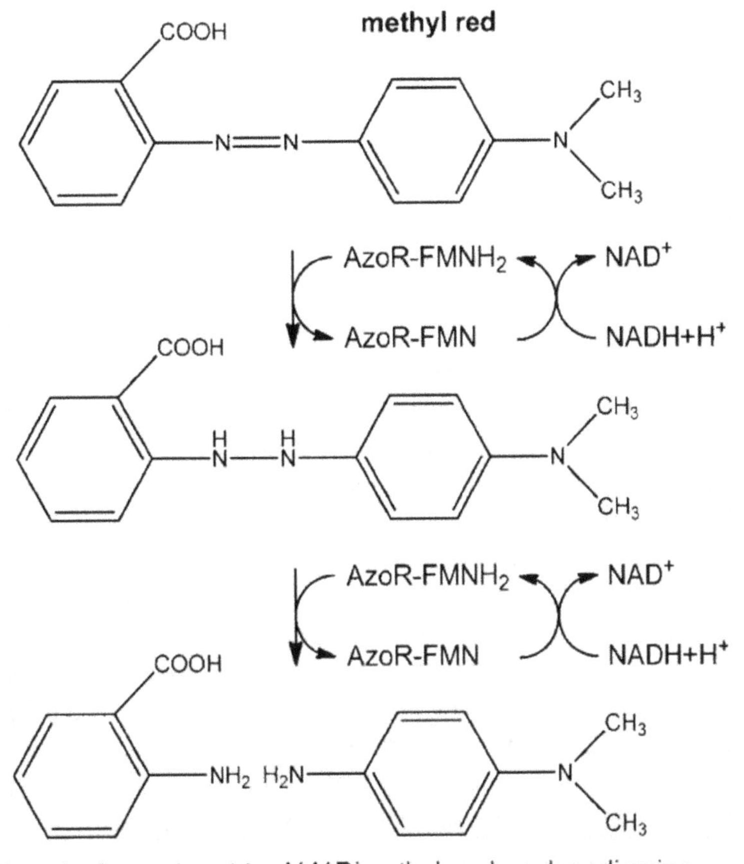

At first, azoreductases were classified depending on their oxygen requirement, being active in the presence or absence of oxygen (Bafana & Chakrabarti, 2008). Based on the structure and presence of cofactor, these enzymes could be classified into two groups: flavin-containing and flavin-free azoreductases. They could be further characterized based on their requirement for NADH, NADPH or both proposed a three-group classification system based on cofactor preference for aerobic azoreductases: 1) azoreductases closely related to AzoR from *E. coli* which contain FMN and usually show a preference of NADH as the reductant; 2) azoreductases containing FMN which have significant sequence similarities to AZR from *Bacillus* sp. OY1-2 and prefer NADPH; 3) flavin-free azoreductases from *P. kullae* and *X. azovorans* which do not show any significant sequence homologies to the two previous groups (Bürger & Stolz, 2010).

Mechanistic Aspects of Azoreductases

Majority of azoreductases were discovered due to the display of decolorization ability by source bacteria. Assays conducted in dyed plates or medium were usually utilized for observing the decolorization ability. So far, 20 of them have been expressed and studied (Table 4). These azoreductases expressed from different organisms are diverse and vary greatly with respect to sequence and functionality. The genes of azoreductases were discovered to exist in various bacteria, which have been cloned from *Escherichia coli* (Cui *et al.*, 2012; Liu *et al.*, 2009; Ito *et al.*, 2006; Langer *et al.*, 2013), *Bacillus* sp. OY1-2 (Suzuki *et al.*, 2001), *Bacillus subtilis* (Sugiura *et al.*, 2006; Deller *et al.*, 2006; Leelakriangsak & Borisut, 2012), *Enterococcus faecalis* (Feng *et al.*, 2012; Chen *et al.*, 2004; Chen *et al.*, 2008), *Rhodobacter sphaeroides* (Liu *et al.*, 2007; Bin *et al.*, 2004; Liu *et al.*, 2008), *Xenophilus azovorans* (Bürger & Stolz, 2010), *Pigmentiphaga kullae* (Chen *et al.*, 2010), *Pseudomonas aeruginosa* (Ryan *et al.*, 2014; Wang *et al.*, 2007; Nachiyar & Rajakumar, 2005; Ryan *et al.*, 2010; Wang *et al.*, 2010), *Pseudomonas putida* (Correia *et al.*, 2011; Gonçalves *et al.*, 2013), *Geobacillus stearothermophilus* (Matsumoto *et al.*, 2010; Sugiura *et al.*, 2006), *Rhodococcus opacus* (Qi *et al.*, 2016) and so on, a gene encoding NADPH-flavin Azo1 even from the skin bacterium *Staphylococcus aureus* ATCC 25923 (Chen *et al.*, 2005). Most of these enzymes contain 180 to 281 amino acids and have on average a molecular mass of around 18.7 – 30 kDa.

Biochemical properties, enzymatic characterization and kinetic studies were conducted utilizing purified azoreductases from whole cell extract from the source organism or from the recombinant host strain. The latter is usually represented by an *E. coli* derivative, but not limited to it. In most cases researchers would like to use gene-cloning methods to produce recombinant proteins to characterize azoreductases (Ito *et al.*, 2008; Chen *et al.*, 2010; Ryan *et al.*, 2010). Because the expression and purification of recombinant proteins facilitate the production and detailed characterization of proteins (Leelakriangsak, 2013). Once the gene of interest was identified, *E. coli* is always a preferred choice as the host to express azoreductases (Ooi *et al.*, 2007; Morrison *et al.*, 2012). However, due to the fact that *E. coli* strains harbor also an azoreductase gene and may produce it at same time during expression accidently it is necessary to consider this. Especially, during our work we have encountered this problem while trying to establish a whole cell biodegradation assay. In many cases therefore recombinant target proteins have been produced with a tag in order to facilitate subsequent purification. Purification and characterization of azoredutases were investigated and the results indicated that the enzyme activity differs in ambient conditions, substrate specificity, and also preferential electron-donors.

The optimal pH for listed bacterial azoreductases seems to vary within a wide range of 4 – 10, and with a temperature range between 25 and 86.5 °C. The recombinant azoreductase AzrA from *Bacillus* sp. strain B29 exhibited broad pH stability between 6 and 10 with an optimal temperature from 60 to 80 °C (Ooi *et al.*, 2007). The optimal temperature of AzrG from thermophilic *Geobacillus stearothermophilus* can achieve as high as 85 °C for MR degradation (Matsumoto *et al.*, 2010) and YhdA from the mesophilic bacterium *Bacillus subtilis* strain 168 also has unusually high thermostability with a melting temperature of 86.5 °C (Deller *et al.*, 2006).

In some cases, the optimum pH and temperature were different for each substrate. BTI1 from *Bacillus* sp. OY1-2 which is not listed in Table 4 is an azoreductase with pH-dependent substrate specificity. Optimal reduction was unexpectedly dye specific, a pH below 5.5 for Janus green B, pH 6.0 for MR, Methyl Orange, and Black Hole Quencher, pH above 8.3 for flame orange (Johansson *et al.*, 2011). However, this is usually not explored with other azoreductases but should be considered. The differences may be due to the protonation state of the various dyes under certain conditions which may change the azoreductase potential to reduce actually the dyes under those conditions. Moreover, when MR was used as the substrate for BSUa from *Bacillus subtilis* ATCC6633, GSTi from *Geobacillus stearothermophilus* IFO13737 and BSPo from *Bacillus* sp. OY1-2, the optimal temperatures were 25, 25, and 45

°C respectively. In contrast, the optimal temperatures for BSUa, GSTi, and BSPo were 45, 45, and 65 °C respectively when using Reactive Red 22 (Sugiura *et al.,* 2006).

Unlike many azoreductases, the PpAzoR could use either NADH or NADPH as the electron donor with approximate catalytic efficiencies while most azoreductases have a clear tendency to NADH or NADPH. AzoRo also could utilize both without significant activity difference (Qi *et al.,* 2017B). The dye degradation activity with NADPH was 90% of activity with NADH. Most azoredutases non-covenlantly bind FMN as the cofactor. The rate of NADPH oxidation was enhanced by the addition of external FMN for YhdA, therefore the dye reduction was accelerated (Deller *et al.,* 2006), but the additional FMN did not affect the reduction rate by paAzoR1 (Wang *et al.,* 2007). Until now, two flavin-free azoreductases were found: NADPH preferred azoB from *Pigmentiphaga kullae* K24 and AzoB from *Xenophilus azovorans* KF46F. Bürger and Stolz (2010) proposed an ordered reaction mechanism through the kinetic analysis of the turn-over of Orange II by AzoB, which was quite different from the Ping Pong mechanism suggested for most flavin-containing azoreductases.

Plenty of azo dyes were tested whether they could be qualified as substrates for azoreductases. A few azoreductases showed broad range of substrate specificity. Azo1 is able to decolorize a wide range of sulfonated azo dyes like Orange II, Amaranth, Ponceau BS and Ponceau S (Chen *et al.,* 2005). AZR from *Rhodobacter sphaeroides* AS1.1737 can reduce structurally complex azo dyes like Reactive Brilliant Red X-3B, Reactive Brilliant Red K-2BP and Acid Red G (Bin *et al.,* 2004). Certain specific activities were observed in BSUa, BSUi, and GSTi for Acid Red 88, Methyl Red, Mordant Orange 1, Reactive Black 5, Reactive Orange 16, Reactive Red 180, and Reactive Red 22, and faint activities were observed for Acid Orange 7, Acid Orange 52, Acid Yellow 23, Congo Red, Direct Brown 44, Direct Red 80, Reactive Red 120 and Reactive Yellow 17 (Sugiura *et al.,* 2006).

Nevertheless, azoredutases do not only degrade azo dyes but are also capable of reducing other substrates. For paAzoR1 from *Pseudomonas aeruginosa*, the catalytic efficiency with the azo pro-drug balsalazide ($V_{max(balsalazide)}$ = 81 μmol•min-1•mg-1) was almost twofold greater than with MR ($V_{max(MR)}$ = 28 μmol•min-1•mg-1) (Wang *et al.,* 2007). Interestingly, more evidences have shown recently that azoreductases possess quinone reductase and nitroreductase activity (Liu *et al.,* 2007; Liu *et al.,* 2008; Cui *et al.,* 2012; Gonçalves *et al.,* 2013). AZR from *Rhodobacter sphaeroides* functions as nitroreductase (Liu *et al.,* 2007). PpAzoR from *Pseudomonas putida* (Gonçalves *et al.,* 2013) and AzoR from *E. coli* respectively, overexpressed in *E. coli,* possess quinone reduction activity. Moreover, the quinone compounds were better substrates for AzoR than the model azo dye substrate MR (Liu *et al.,* 2009). Quinone compounds may be present the evolutionary background of those enzymes, which is still an ongoing debate.

Structural Properties of Azoreductases

Under primary amino acid level, azoreductases show low similarity between each other. With the development of X-ray crystallography, 3D-structure was analyzed to reveal the function mechanism from a molecular point of view. Structures of AzoR from *Escherichia coli* (2Z98), paAzoR1 from *Pseudomonas aeruginosa* (3KEG), AzoA from *Enterococcus faecalis* (2HPV), AzrA (3W77) and AzrC (3W79) from *Bacillus* sp. B29, PpAzoR from *Pseudomonas putida* (4C0W), YhdA from *Bacillus subtilis* (3GFQ), and azoreductases from *Yersinia pestis* CO92 (4ESE) and *Bacillus anthracis* (4M0C) have been published. It has turned out that similarities do exist on the tertiary structural level.

Figure 6. X-ray crystallographic structure of a Y131F mutant of Pseudomonas aeruginosa azoreductase in complex with Methyl Red (PDB: 3KEG)

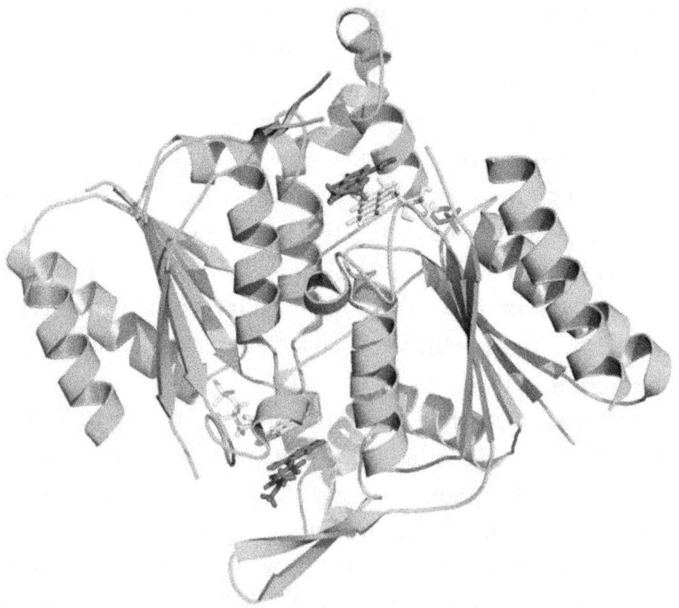

The protein structure of AzoR was firstly reported among FMN-dependent NADH-azoreductases. AzoR exists as a homodimer. Both its oxidized and reduced forms were crystalized (Ito *et al.,* 2006A, 2006B; Ito *et al.,* 2008). The isoalloxazine ring of FMN showed different conformation under the oxidized and reduced status. In oxidized AzoR, the isoalloxazine ring is nearly planar, having a slightly twisted conformation. But upon reduction, this ring employs a butterfly bend conformation along the N5–N10 axis by the dimethyl and pyrimidine rings turning toward the re face by an angle of ~15°.

In general, residues which are affinitive to the isoalloxazine ring are involved in binding of substrates (Ito *et al.,* 2006; Liu *et al.,* 2007). Several trials of site-directed mutagenesis were conducted, trying to verify the prediction from protein structure. Trough mutation, Tyr120 and Phe162 of AzoR on top of the isoalloxazine ring were found to be mainly responsible for NADH binding than MR binding and the interaction between Arg59 and the carboxyl group of azo compounds determines substrate specificity (Ito *et al.,* 2006). Tyr74, His75 and Lys109 in two conserved loops around the FMN of AZR from *Rhodobacter sphaeroides* are involved in the binding of substrates and crucial for enzyme activities (Liu *et al.,* 2008).

AzoA is also an asymmetrical homodimer. The FMN is stabilized by 22 amino acid residues, eight of which, Trp105, Asn106, Leu107, Gly150, Gly151, Tyr153, Asn121 and Tyr129 are involved in binding the isoalloxazine ring. Above all, Trp105 was intriguing, because of the presentational interaction between its indole group with all moieties of FMN. Complete loss of flavin of mutants which substituted Trp105 by short side-chain amino acids Ala and Gly was observed, resulting in no activity. However, substitution of Trp105 by the aromatic amino acids Tyr and Phe did not influence the bound FMN content. But the activity of mutants with bulky side-chains decreased. Therefore, Trp-105 is essential to both the FMN binding and catalytic activity of AzoA. NADH- and MR-binding sites were also probed through computational prediction and experimental confirmation (Feng *et al.,* 2012). A sandwiching system containing residues Tyr129 and Asp184 was proposed to play a critical role in MR and NADH binding.

Table 4. Biochemical properties of different azoreductases

Designation of Enzyme	Bacteria	pH	T (°C)	Main Substrates	Electron Donor(s)	Kinetic	Reference
Orange II azoreductase	*Pseudomonas* KF46	5.0–6.5	45	Orange II, carboxy-Orange II	NAD(P)H	$K_{m(Orange\ II)} = 1.5\ \mu M$, $K_{m(NADPH)} = 5\ \mu M$, $K_{m(NADH)} = 180\ \mu M$	Zimmermann *et al.*, 1982
Orange I azoreductase	*Pseudomonas* K24	6.2, 6.8[a]	41	Orange I, carboxy-Orange I	NAD(P)H	$K_{m(NADPH)} = 1.1\ \mu M$, $K_{m(NADH)} = 54\ \mu M$	Zimmermann *et al.*, 1984
p-Amino azobenzene reductase	*Pseudomonas cepacia*	6.8–7.0	37	*p*-aminoazobenzene	NAD(P)H	–	Idaka *et al.*, 1987
Azo1	*Staphylococcus aureus* ATCC 25923	6.0–6.6	35–40	MR, Orange II, Amaranth, Ponceau BS, Ponceau S	NADPH	$K_{m(NADPH)} = 74\ \mu M$, $K_{m(MR)} = 57\ \mu M$, $V_{max} = 0.4\ \mu M{\bullet}min^{-1}{\bullet}mg^{-1}$	Chen *et al.*, 2005
AzrA	*Bacillus* sp. B29	6.0–10.0	65–80	MR, 1-(2-pyridylazo)-2-naphthol, Orange I, Acid Red 88, Orange II	NADH	ping-pong mechanism	Ooi *et al.*, 2007
Azr	*Bacillus* sp. OY1-2	–	50/ 70[b]	Rocceline, Sumifix Black B, Solar Orange	NADPH	$V_{max}/K_m = 3.03$	Suzuki *et al.*, 2001
AzoR1	*Bacillus subtilis* JH642	5–9	25–45	Azobenzene, MR, Orange G, Congo Red	NADH	–	Leelakriangsak & Borisut 2012
YhdA	*Bacillus subtilis* 168	–	86.5[c]	Cibacron Marine	NAD(P)H	$K_{m(NADPH)} = 0.55\ mM$, $K_{m(NADH)} = 0.65\ mM$, $V_{max(NADPH)} = 57$, $V_{max(NADH)} = 0.7$	Deller *et al.*, 2006
AzoR	*Escherichia coli* JM109	6.3	–	MR, anthraquinone-2-sulfonic acid	NADH	$K_{m(MR)} = 27\ \mu M$, $K_{m(NADH)} = 260\ \mu M$, $V_{max} = 175 \pm 6\ \mu mol{\bullet}min^{-1}{\bullet}mg^{-1}$	Langer *et al.*, 2013
		–	–	MR, Ethyl red, Menadione	NADH	ping-pong Bi–Bi mechanism; $K_{m(MR)} = 17.9\ \mu M$, $K_{m(NADH)} = 31.6\ \mu M$	Nakanishi *et al.*, 2001
AzrG	*Geobacillus stearothermophilus*	–	85	MR, Ethyl Red, Acid Red 88, Orange I, Congo Red	NADH	–	Matsumoto *et al.*, 2010
AzoA	*Enterococcus faecalis* ATCC 19433	–	–	MR, Orange II, Orange G, Amaranth, Ponceau BS, Ponceau S	NADH	$K_{m(MR)} = 24\ \mu M$, $K_{m(NADH)} = 140\ \mu M$, $V_{max(MR)} = 89.2 U/mg$, $V_{max(NADH)} = 86.2$	Chen *et al.*, 2004
AZR	*Rhodobacter sphaeroides* AS1.1737	8.0	50	Reactive Brilliant Red X-3B, Orange II, Methyl Orange, Reactive Brilliant Red K-2BP, Acid Red G, MR	NADH	ping-pong mechanism; non-fusion enzyme: $K_{m(MR)} = 0.42\ mM$, $K_{m(NADH)} = 2.3\ mM$, $V_{max} = 65.2\ \mu mol{\bullet}min^{-1}{\bullet}mg^{-1}$; fusion enzyme: $K_{m(MR)} = 0.28\ mM$, $K_{m(NADH)} = 2.1\ mM$. $V_{max} = 87.3\ \mu mol{\bullet}min^{-1}{\bullet}mg^{-1}$	Bin *et al.*, 2004
		7.0	50	nitrofurazone	NAD(P)H	$K_{m(NADH)} = 0.5\ mM$, $K_{m(FMN)} = 14.2\ mM$, $V_{max} = 172.4\ \mu mol{\bullet}min^{-1}{\bullet}mg^{-1}$	Liu *et al.*, 2007
BSUa	*Bacillus subtilis* ATCC6633	–	25/ 45[d]	Acid Red 88, Methyl Red, Mordant Orange 1, Reactive Black 5, Reactive Orange 16, Reactive Red 180, Reactive Red 22	NADPH	$K_{m(MR)} = 91.7\ \mu M$, $V_{max} = 12\ \mu mol{\bullet}min^{-1}{\bullet}mg^{-1}$; $K_{m(Reactive\ Red\ 22)} = 25.4\ \mu M$, $V_{max} = 5.7\ \mu mol{\bullet}min^{-1}{\bullet}mg^{-1}$	Sugiura *et al.*, 2006
GSTi	*Geobacillus stearothermophilus* ATCC 7953	–				$K_{m(MR)} = 134\ \mu M$, $V_{max} = 13.1\ \mu mol{\bullet}min^{-1}{\bullet}mg^{-1}$; $K_{m(Reactive\ Red\ 22)} = 17.5\ \mu M$, $V_{max} = 3.3\ \mu mol{\bullet}min^{-1}{\bullet}mg^{-1}$	
AzoC	*Clostridium perfringens* ATCC 3626	9	30–37	Direct Blue 15	NADH	$K_{m(Direct\ Blue\ 15)} = 0.005\ \mu M$, $V_{max} = 0.42\ \mu M{\bullet}min^{-1}{\bullet}mg^{-1}$	Morrison *et al.*, 2012
PpAzoR	*Pseudomonas putida* MET94	–	–	MR, Reactive Black 5 (RB5), anthraquinone-2-sulfonate, 1,4-benzoquinone,	NAD(P)H	ping-pong Bi–Bi mechanism; $K_{m(MR)} = 0.4\ mM$, $V_{max} = 1.0\ \mu mol{\bullet}min^{-1}{\bullet}mg^{-1}$; $K_{m(RB5)} = 1.4\ mM$, $V_{max} = 2.2\ \mu mol{\bullet}min^{-1}{\bullet}mg^{-1}$	Mendes *et al.*, 2011; Gonçalves *et al.*, 2013
azoB	*Pigmentiphaga kullae* K24	6.0	37–45	Orange I	NAD(P)H	$K_{m(NADPH)} = 1.0\ \mu M$, $K_{m(NADH)} = 8.6\ \mu M$, $K_{m(Orange\ I)} = 170\ \mu M$	Chen *et al.*, 2010
PaAzoR1	*Pseudomonas aeruginosa* PAO1	–	50	MR, Sulfasalazine, Balsalazide, *p*-aminoazobenzene sulfonamide, Olsalazine,	NAD(P)H	$K_{m(NADH)} = 464\ \mu M$, $V_{max(NADH)} = 22\ \mu M{\bullet}s^{-1}{\bullet}mg^{-1}$; $K_{m(NADPH)} = 1100\ \mu M$, $V_{max(NADPH)} = 74\ \mu M{\bullet}s^{-1}{\bullet}mg^{-1}$; $K_{m(MR)} = 76\ \mu M$, $V_{max(MR)} = 28\ \mu mol{\bullet}min^{-1}{\bullet}mg^{-1}$	Wang *et al.*, 2007

continued on following page

Table 4. Continued

Designation of Enzyme	Bacteria	pH	T (°C)	Main Substrates	Electron Donor(s)	Kinetic	Reference
AzoB	*Xenophilus azovorans* KF46F	–	–	Orange II, MR	NAD(P)H	an ordered bireactant reaction mechanism; $K_{m(Orange\ II)} = 6.2$ μM and $V_{max(Orange\ II)} = 17.0$ μmol•min^{-1}•mg^{-1}, $K_{m(NADH)} = 180$ μM, $K_{m(NADPH)} = 10$ μM; $K_{m(MR)} = 49$ μM, $V_{max(MR)} = 0.40$ μmol•min^{-1}•mg^{-1}	Bürger & Stolz 2010
AzoRo	*Rhodococcus opacus* 1CP	4.0	53	MR	NAD(P)H	$K_{m(NADH)} = 10.07$ μM, $V_{max} = 51.48$ μmol•min^{-1}•mg^{-1}	Qi *et al.*, 2016; Qi *et al.*, 2017

[a]For Orange I is 6.2 and for carboxy-Orange I is 6.8.
[b]The optimal temperature for recombinant azoreductase is 50 °C whereas native enzyme is 70 °C.
[c]Melting temperature
[d]Optimum temperatures differ from substrates: 25 °C for MR and 45 °C for Reactive Red 22.
MR: methyl red
– Unknown

The crystal structures of PpAzoR, of PpAzoR in complex with substrates (anthraquinone-2-sulfonate and Reactive Black 5) were reported (Gonçalves *et al.*, 2013). Asn14 and Arg184 are involved in binding sulfonate groups of Reactive Black 5, and Arg184 also may participate in the NAD(P)H recognition.

To obtain a full understanding of substrates recognition, crystal structures of AzrA and of AzrC respectively complexed with the inhibitor Cibacron Blue of NAD(P)H and Acid Red 88 (high hydrophobicity) and Orange I were solved. Hydrophobic groups of a substrate interact with several hydrophobic residues of the substrate-binding pocket, and hydrophilic groups of the substrate are fixed by hydrogen bonds to a conserved asparagine residue (Asn101 in AzrA and Asn104 in AzrC). The azo bond is sandwiched and stabilized by π – stacking interaction with Tyr127 and FMN (Yu *et al.*, 2014). This π – stacking interaction has been seen in AzoR, AzoA, PpAzoR, paAzoR1 and also several other flavoproteins. The different manners of substrates binding observed from two structures of AzrC in complex with two different substrates indicate azoreductase can bind substrates in different orientations depending on substrates' characteristics.

Unlike AzoR, AzoA, PpAzoR, AzrA and AzrC are dimers, Azo1, YhdA and paAzoR1 are tetramers (Chen *et al.*, 2005; Deller *et al.*, 2006; Wang *et al.*, 2007). As mentioned before, paAzoR1 could use either NADH or NADPH, may due to conformational flexibility of the active site. Through the structural analysis of paAzoR1 in complex with MR, the FMN binding of paAzoR1 shows a planar isoalloxazine ring, bound in a hydrophobic pocket, where the substrate MR also is bound and oriented to the si face of FMN. And the re face of FMN is tightly packed against protein, so either MR or NAD(P)H could only bind to the si face. This directly correlates with the previously adopted kinetic mechanism of the azoreductase, because it would not be possible to bind the MR and NAD(P)H at the same time. The amino acids Gly147 and Phe173 in paAzoR1 were predicted to be key residues for binding MR. And Arg146 could stabilize FMN. Tyr131 was highly suggested as the key potential electron transferring residue due to its location during the reduction. After site-directed mutagenesis, Tyr131 was found to be important in the architecture of the active site but not essential for enzymatic activity (Wang *et al.*, 2010).

Although a majority of azoreductases show less genetic affinity, some conserved residues were still found. A Tyr residue existing in AzoA, AzoR and paAzoR1 is responsible for the stacking which facilitates the electron transfer from NAD(P)H to FMN, which is the first stage of the reduction reaction (Chen *et al.,* 2008). The NADH binding motif (GXGXXG) has been identified to be conserved in azoreductases from *Geobacillus* sp. OY1-2, *Rhodobacter sphaeroides*, *Pigmentiphaga kullae* K24 and *Bacillus* sp. OY1-2 (Suzuki *et al.,* 2001; Bin *et al.,* 2004; Liu *et al.,* 2008; Chen *et al.,* 2010). However, a key fingerprint motif ((T/S)XTGXT) for FMN binding (Drennan *et al.,* 1999) was just found in PpAzoR.

ACKNOWLEDGMENT

Jingxian Qi was supported by the CSC (China Scholarship Council) scholarship (201206400019), the "Landesstipendiums aus den Mitteln des Freistaates Sachsen" from GRAFA and the "Stipendium zur Förderung des Abschlusses von Promotionen" from TU Bergakademie Freiberg. Anna Christina Ngo applied in 2016 for a pre-doctoral scholarship (KAAD). Dirk Tischler was granted by the European Social Fund (ESF) and the Saxonian Government (GETGEOWEB: 100101363).

REFERENCES

Abadulla, E., Tzanov, T., Costa, S., Robra, K. H., Cavaco-Paulo, A., & Gubitz, G. M. (2000). Decolorization and detoxification of textile dyes with a laccase from *Trametes hirsute. Applied and Environmental Microbiology, 66*(8), 3357–3362. doi:10.1128/AEM.66.8.3357-3362.2000 PMID:10919791

Adedayo, O., Javadpour, S., Taylor, C., Anderson, W. A., & Moo-Young, M. (2004). Decolourization and detoxification of Methyl Red by aerobic bacteria from a wastewater treatment plant. *World Journal of Microbiology & Biotechnology, 20*(6), 545–550. doi:10.1023/B:WIBI.0000043150.37318.5f

Alalewi, A., & Jiang, C. (2012). Bacterial influence on textile wastewater decolorization. *Journal of Environmental Protection, 3*(08), 889–901. doi:10.4236/jep.2012.328104

Angelova, B., Avramova, T., Stefanova, L., & Mutafov, S. (2008). Temperature effect on bacterial azo bond reduction kinetics: An Arrhenius plot analysis. *Biodegradation, 19*(3), 387–393. doi:10.100710532-007-9144-4 PMID:17653820

Ashrafi, S. D., Rezaei, S., Forootanfar, H., Mahvi, A. H., & Faramarzi, M. A. (2013). The enzymatic decolorization and detoxification of synthetic dyes by the laccase from a soil-isolated ascomycete, *Paraconiothyrium variabile. International Biodeterioration & Biodegradation, 85*, 173–181. doi:10.1016/j.ibiod.2013.07.006

Ayed, L., Mahdhi, A., Cheref, A., & Bakhrouf, A. (2011). Decolorization and degradation of azo dye Methyl Red by an isolated *Sphingomonas paucimobilis*: Biotoxicity and metabolites characterization. *Desalination, 274*(1-3), 272–277. doi:10.1016/j.desal.2011.02.024

Bafana, A., & Chakrabarti, T. (2008). Lateral gene transfer in phylogeny of azoreductase enzyme. *Computational Biology and Chemistry*, *32*(3), 191–197. doi:10.1016/j.compbiolchem.2008.03.003 PMID:18440871

Bafana, A., Chakrabarti, T., Muthal, P., & Kanade, G. (2009). Detoxification of benzidine-based azo dye by *E. gallinarum*: Time-course study. *Ecotoxicology and Environmental Safety*, *72*(3), 960–964. doi:10.1016/j.ecoenv.2007.11.013 PMID:18206233

Bafana, A., Devi, S. S., & Chakrabarti, T. (2011). Azo dyes: Past, present and the future. *Environmental Reviews*, *19*(NA), 350–371. doi:10.1139/a11-018

Banat, I., Nigam, P., Singh, D., & Marchant, R. (1996). Microbial decolorization of textile dye containing effluents: A Review. *Bioresource Technology*, *58*(3), 217–227. doi:10.1016/S0960-8524(96)00113-7

Bhatnagar, A., & Sillanpää, M. (2010). Utilization of agro-industrial and municipal waste materials as potential adsorbents for water treatment — A review. *Chemical Engineering Journal*, *157*(2-3), 277–296. doi:10.1016/j.cej.2010.01.007

Bin, Y., Jiti, Z., Jing, W., Cuihong, D., Hongman, H., Zhiyong, S., & Yongming, B. (2004). Expression and characteristics of the gene encoding azoreductase from *Rhodobacter sphaeroides* AS1.1737. *FEMS Microbiology Letters*, *236*(1), 129–136. doi:10.1111/j.1574-6968.2004.tb09638.x PMID:15212802

Blümel, S., Knackmuss, H. J., & Stolz, A. (2002). Molecular cloning and characterization of the gene coding for the aerobic azoreductase from *Xenophilus azovorans* KF46F. *Applied and Environmental Microbiology*, *68*(8), 3948–3955. doi:10.1128/AEM.68.8.3948-3955.2002 PMID:12147495

Booth, G., Zollinger, H., McLaren, K., Sharpless, W. G., & Westwell, A. (2000). Dyes, General Survey. In *Ullmann's Encyclopedia of Industrial Chemistry*. Weinheim, Germany: Wiley-VCH. doi:10.1002/14356007.a09_073

Bürger, S., & Stolz, A. (2010). Characterisation of the flavin-free oxygen-tolerant azoreductase from *Xenophilus azovorans* KF46F in comparison to flavin-containing azoreductases. *Applied Microbiology and Biotechnology*, *87*(6), 2067–2076. doi:10.100700253-010-2669-1 PMID:20508929

Carmen, Z., & Daniela, Z. (2012). Textile organic dyes - characteristics, polluting effects and separation/elimination procedures from industrial effluents – A Critical Overview. In Organic Pollutants Ten Years After the Stockholm Convention - Environmental and Analytical Update. InTech.

Chagas, E. P., & Durrant, L. R. (2001). Decolorization of azo dyes by *Phanerochaete chrysosporium* and *Pleurotus sajorcaju*. *Enzyme and Microbial Technology*, *29*(8-9), 473–477. doi:10.1016/S0141-0229(01)00405-7

Chang, J.-S., Chou, C., Lin, Y. C., Lin, P. J., Ho, J. Y., & Hu, T. L. (2001). Kinetic characteristics of bacterial azo-dye decolorization by *Pseudomonas luteola*. *Water Research*, *35*(12), 2841–2850. doi:10.1016/S0043-1354(00)00581-9 PMID:11471684

Chen, H., Feng, J., Kweon, O., Xu, H., & Cerniglia, C. E. (2010). Identification and molecular characterization of a novel flavin-free NADPH preferred azoreductase encoded by *azoB* in *Pigmentiphaga kullae* K24. *BMC Biochemistry*, *11*(1), 13. doi:10.1186/1471-2091-11-13 PMID:20233432

Chen, H., Hopper, S. L., & Cerniglia, C. E. (2005). Biochemical and molecular characterization of an azoreductase from *Staphylococcus aureus*, a tetrameric NADPH-dependent flavoprotein. *Microbiology, 151*(5), 1433–1441. doi:10.1099/mic.0.27805-0 PMID:15870453

Chen, H., Wang, R., & Cerniglia, C. E. (2004). Molecular cloning, overexpression, purification, and characterization of an aerobic FMN-dependent azoreductase from *Enterococcus faecalis. Protein Expression and Purification, 34*(2), 302–310. doi:10.1016/j.pep.2003.12.016 PMID:15003265

Chen, H., Xu, H., Kweon, O., Chen, S., & Cerniglia, C. E. (2008). Functional role of Trp-105 of *Enterococcus faecalis* azoreductase (AzoA) as resolved by structural and mutational analysis. *Microbiology, 154*(9), 2659–2667. doi:10.1099/mic.0.2008/019877-0 PMID:18757799

Chen, K. C., Wu, J. Y., Liou, D. J., & Hwang, S. C. (2003). Decolorization of the textile dyes by newly isolated bacterial strains. *Journal of Biotechnology, 101*(1), 57–68. doi:10.1016/S0168-1656(02)00303-6 PMID:12523970

Chen, V. B., Arendall, W. B. III, Headd, J. J., Keedy, D. A., Immormino, R. M., Kapral, G. J., ... Richardson, D. C. (2010). MolProbity: All-atom structure validation for macromolecular crystallography. *Acta Crystallographica. Section D, Biological Crystallography, 66*(1), 12–21. doi:10.1107/S0907444909042073 PMID:20057044

Clarke, E. A., & Anliker, R. (1980). Organic dyes and pigments. In Anthropogenic Compounds. Springer.

Correia, B., Chen, Z., Mendes, S., Martins, L. O., & Bento, I. (2011). Crystallization and preliminary X-ray diffraction analysis of the azoreductase PpAzoR from *Pseudomonas putida* MET94. *Acta Crystallographica. Section F, Structural Biology and Crystallization Communications, 67*(1), 121–123. doi:10.1107/S1744309110048220 PMID:21206041

Cui, D., Li, G., Zhao, D., Gu, X., Wang, C., & Zhao, M. (2012). Purification and characterization of an azoreductase from *Escherichia coli* CD-2 possessing quinone reductase activity. *Process Biochemistry, 47*(3), 544–549. doi:10.1016/j.procbio.2011.12.013

Dawkar, V. V., Jadhav, U. U., Ghodake, G. S., & Govindwar, S. P. (2009). Effect of inducers on the decolorization and biodegradation of textile azo dye Navy blue 2GL by *Bacillus* sp. VUS. *Biodegradation, 20*(6), 777–787. doi:10.100710532-009-9266-y PMID:19468842

Deller, S., Sollner, S., Trenker-El-Toukhy, R., Jelesarov, I., Gübitz, G. M., & Macheroux, P. (2006). Characterization of a thermostable NADPH:FMN oxidoreductase from the mesophilic bacterium *Bacillus subtilis. Biochemistry, 45*(23), 7083–7091. doi:10.1021/bi052478r PMID:16752898

Dos Santos, A. B., Cervantes, F. J., & van Lier, J. B. (2007). Review paper on current technologies for decolourisation of textile wastewaters: Perspectives for anaerobic biotechnology. *Bioresource Technology, 98*(12), 2369–2385. doi:10.1016/j.biortech.2006.11.013 PMID:17204423

Drennan, C. L., Pattridge, K. A., Weber, C. H., Metzger, A. L., Hoover, D. M., & Ludwig, M. L. (1999). Refined structures of oxidized flavodoxin from *Anacystis nidulans. Journal of Molecular Biology, 294*(3), 711–724. doi:10.1006/jmbi.1999.3151 PMID:10610791

Dubey, S. K., Dubey, J., Viswas, A. J., & Tiwari, P. (2011). Studies on cyanobacterial biodiversity in paper mill and pharmaceutical industrial effluents. *British Biotechnology Journal*, *1*(3), 61–67. doi:10.9734/BBJ/2011/395

Duran, N., & Esposito, E. (2000). Potential applications of oxidative enzymes and phenoloxidase-like compounds in wastewater and soil treatment: A review. *Applied Catalysis B: Environmental*, *28*(2), 83–99. doi:10.1016/S0926-3373(00)00168-5

El-Sheekh, M. M., Gharieb, M. M., & Abou-El-Souod, G. W. (2009). Biodegradation of dyes by some green algae and cyanobacteria. *International Biodeterioration & Biodegradation*, *63*(6), 699–704. doi:10.1016/j.ibiod.2009.04.010

Elisangela, F., Andrea, Z., Fabio, D. G., de Menezes Cristiano, R., Regina, D. L., & Artur, C.-P. (2009). Biodegradation of textile azo dyes by a facultative *Staphylococcus arlettae* strain VN-11 using a sequential microaerophilic/aerobic process. *International Biodeterioration & Biodegradation*, *63*(3), 280–288. doi:10.1016/j.ibiod.2008.10.003

Erkurt, E., Unyayar, A., & Kumbur, H. (2007). Decolorization of synthetic dyes by White Rot Fungi, involving laccase enzyme in the process. *Process Biochemistry*, *42*(10), 1429–1435. doi:10.1016/j.procbio.2007.07.011

Ertuğrul, S., Bakır, M., & Dönmez, G. (2008). Treatment of dye-rich wastewater by an immobilized thermophilic cyanobacterial strain: *Phormidium* sp. *Ecological Engineering*, *32*(3), 244–248. doi:10.1016/j.ecoleng.2007.11.011

Evangelista-Barreto, N. S., Albuquerque, C. D., Vieira, R. H. S. F., & Campos-Takaki, G. M. (2009). Cometabolic decolorization of the reactive azo dye orange II by *Geobacillus stearothermophilus* UCP 986. *Textile Research Journal*, *79*(14), 1266–1273. doi:10.1177/0040517508087858

Feng, J., Kweon, O., Xu, H., Cerniglia, C. E., & Chen, H. (2012). Probing the NADH - and methyl red-binding site of a FMN-dependent azoreductase (AzoA) from *Enterococcus faecalis*. *Archives of Biochemistry and Biophysics*, *520*(2), 99–107. doi:10.1016/j.abb.2012.02.010 PMID:22387379

Forgacs, E., Cserhati, T., & Oros, G. (2004). Removal of synthetic dyes from wastewater: A Review. *Environment International*, *30*(7), 953–971. doi:10.1016/j.envint.2004.02.001 PMID:15196844

Forootanfar, H., Moezzi, A., Aghaie-Khozani, M., Mahmoudjanlou, Y., Ameri, A., Niknejad, F., & Faramarzi, M. A. (2012). Synthetic dye decolorization by three sources of fungal laccase. *Iranian Journal of Environmental Health Sciences & Engineering*, *9*(1), 27. doi:10.1186/1735-2746-9-27 PMID:23369690

Garfield, S. (2002). *Mauve: how one man invented a color that changed the world*. W. W. Norton & Company.

Ghodake, G., Jadhav, U., Tamboli, D., Kagalkar, A., & Govindwar, S. (2011). Decolorization of textile dyes and degradation of mono-azo dye amaranth by *Acinetobacter calcoaceticus* NCIM 2890. *Indian Journal of Microbiology*, *51*(4), 501–508. doi:10.100712088-011-0131-4 PMID:23024414

Gold, M. H., Youngs, H. L., & Gelpke, M. D. S. (2000). Manganese peroxidase. In A. Sigel & H. Sigel (Eds.), *Metal ions in biological systems* (pp. 559–586). New York: Marcel Dekker Inc.

Gonçalves, A. M. D., Mendes, S., de Sanctis, D., Martins, L. O., & Bento, I. (2013). The crystal structure of *Pseudomonas putida* azoreductase — The active site revisited. *The FEBS Journal*, *280*(24), 6643–6657. doi:10.1111/febs.12568 PMID:24127652

Gopinath, K. P., Murugesan, S., Abraham, J., & Muthukumar, K. (2009). *Bacillus* sp. mutant for improved biodegradation of Congo red: Random mutagenesis approach. *Bioresource Technology*, *100*(24), 6295–6300. doi:10.1016/j.biortech.2009.07.043 PMID:19692233

Goszczynski, S., Paszczynski, A., Pasti-Grigsby, M. B., Crawford, R. L., & Crawford, D. L. (1994). New pathway for degradation of sulfonated azo dyes by microbial peroxidases of *Phanerochaete chrysosporium* and *Streptomyces chromofuscus*. *Journal of Bacteriology*, *176*(5), 1339–1347. doi:10.1128/jb.176.5.1339-1347.1994 PMID:8113173

Guo, J., Zhou, J., Wang, D., Tian, C., Wang, P., Salah Uddin, M., & Yu, H. (2007). Biocalalyst effects of immobilized anthraquinone on the anaerobic reduction of azo dyes by the salt-tolerant bacteria. *Water Research*, *41*(2), 426–432. doi:10.1016/j.watres.2006.10.022 PMID:17129594

Gurav, A., Ghosh, J., & Kulkarni, G. (2011). Decolorization of anthroquinone based dye Vat Red 10 by *Pseudomonas desmolyticum* NCIM 2112 and *Galactomyces geotrichum* MTCC 1360. *International Journal for Biotechnology and Molecular Biology Research*, *2*, 93–97.

Heinfling, A., Martinez, M. J., Martínez, A. T., Bergbauer, M., & Szewzyk, U. (1998). Transformation of industrial dyes by manganese peroxidases from *Bjerkandera adusta* and *Pleurotus eryngii* in a manganese-independent reaction. *Applied and Environmental Microbiology*, *64*, 2788–2793. PMID:9687431

Hsueh, C.-C., Chen, B.-Y., & Yen, C.-Y. (2009). Understanding effects of chemical structure on azo dye decolorization characteristics by *Aeromonas hydrophila*. *Journal of Hazardous Materials*, *167*(1-3), 995–1001. doi:10.1016/j.jhazmat.2009.01.077 PMID:19237244

Idaka, E., Horitsu, H., & Ogawa, T. (1987). Some properties of azoreductase produced by *Pseudomonas cepacia*. *Bulletin of Environmental Contamination and Toxicology*, *39*(6), 982–989. doi:10.1007/BF01689588 PMID:3440155

Ito, K., Nakanishi, M., Lee, W. C., Sasaki, H., Zenno, S., Saigo, K., ... Tanokura, M. (2006). Three-dimensional structure of AzoR from *Escherichia coli*: An oxidereductase conserved in microorganisms. *The Journal of Biological Chemistry*, *281*(29), 20567–20576. doi:10.1074/jbc.M513345200 PMID:16684776

Ito, K., Nakanishi, M., Lee, W. C., Zhi, Y., Sasaki, H., Zenno, S., ... Tanokura, M. (2008). Expansion of substrate specificity and catalytic mechanism of azoreductase by X-ray crystallography and site-directed mutagenesis. *The Journal of Biological Chemistry*, *283*(20), 13889–13896. doi:10.1074/jbc.M710070200 PMID:18337254

Jadhav, J. P., Parshetti, G. K., Kalme, S. D., & Govindwar, S. P. (2007). Decolourization of azo dye methyl red by *Saccharomyces cerevisiae* MTCC 463. *Chemosphere*, *68*(2), 394–400. doi:10.1016/j.chemosphere.2006.12.087 PMID:17292452

Jadhav, J. P., Phugare, S. S., Dhanve, R. S., & Jadhav, S. B. (2010). Rapid biodegradation and decolorization of Direct Orange 39 (Orange TGLL) by an isolated bacterium *Pseudomonas aeruginosa* strain BCH. *Biodegradation*, *21*(3), 453–463. doi:10.100710532-009-9315-6 PMID:19937265

Jadhav, S. B., Surwase, S. N., Kalyani, D., Gurav, R. G., & Jadhav, J. P. (2012). Biodecolorization of azo dye remazol orange by *Pseudomonas aeruginosa* BCH and toxicity (Oxidative stress) reduction in allium cepa root cells. *Applied Biochemistry and Biotechnology*, *168*(5), 1319–1334. doi:10.100712010-012-9860-z PMID:22948606

Jadhav, S. U., Jadhav, M. U., Kagalkar, A. N., & Govindwar, S. P. (2008). Decolorization of Brilliant Blue G dye mediated by degradation of the microbial consortium of *Galactomyces geotrichum* and *Bacillus* sp. *Journal of the Chinese Institute of Chemical Engineers*, *39*(6), 563–570. doi:10.1016/j.jcice.2008.06.003

Jadhav, S. U., Kalme, S. D., & Govindwar, S. P. (2008). Biodegradation of methyl red by *Galactomyces geotrichum* MTCC 1360. *International Biodeterioration & Biodegradation*, *62*(2), 135–142. doi:10.1016/j.ibiod.2007.12.010

Jafari, N., Soudi, M. R., & Kasra-Kermanshahi, R. (2014). Biodecolorization of textile azo dyes by isolated yeast from activated sludge: *Issatchenkia orientalis* JKS6. *Annals of Microbiology*, *64*(2), 475–482. doi:10.100713213-013-0677-y

Jamal, F., Qidwai, T., Pandey, P., Singh, R., & Singh, S. (2011). Azo and anthraquinone dye decolorization in relation to its molecular structure using soluble *Trichosanthes dioica* peroxidase supplemented with redox mediator. *Catalysis Communications*, *12*(13), 1218–1233. doi:10.1016/j.catcom.2011.04.012

Jiang, W., & Fang, B. S. (2016). Construction and evaluation of a novel bifunctional phenylalanine–formate dehydrogenase fusion protein for bienzyme system with cofactor regeneration. *Journal of Industrial Microbiology & Biotechnology*, *43*(5), 577–584. doi:10.100710295-016-1738-6 PMID:26819086

Jin, R., Yang, H., Zhang, A., Wang, J., & Liu, G. (2009). Bioaugmentation on decolorization of C.I. Direct Blue 71 by using genetically engineered strain *Escherichia coli* JM109 (pGEX-AZR). *Journal of Hazardous Materials*, *163*(2-3), 1123–1128. doi:10.1016/j.jhazmat.2008.07.067 PMID:18755538

Jin, X., Liu, G., Xu, Z., & Tao, W. (2007). Decolourisation of a dye industry effluent by *Aspergillus fumigatus* XC6. *Applied Microbiology and Biotechnology*, *74*(1), 239–243. doi:10.100700253-006-0658-1 PMID:17086413

Johansson, H. E., Johansson, M. K., Wong, A. C., Armstrong, E. S., Peterson, E. J., Grant, R. E., ... Cook, R. M. (2011). BTI1, an azoreductase with pH-dependent substrate specificity. *Applied and Environmental Microbiology*, *77*(12), 4223–4225. doi:10.1128/AEM.02289-10 PMID:21531830

Kalme, S., Jadhav, S., Jadhav, M., & Govindwar, S. (2009). Textile dye degrading laccase from *Pseudomonas desmolyticum* NCIM 2112. *Enzyme and Microbial Technology*, *44*(2), 65–71. doi:10.1016/j.enzmictec.2008.10.005

Kalyani, D., Dhiman, S. S., Kim, H., Jeya, M., Kim, I. W., & Lee, J. K. (2012). Characterization of a novel laccase from the isolated *Coltricia perennis* and its application to detoxification of biomass. *Process Biochemistry*, *47*(4), 671–678. doi:10.1016/j.procbio.2012.01.013

Kalyani, D. C., Telke, A. A., Dhanve, R. S., & Jadhav, J. P. (2009). Ecofriendly biodegradation and detoxification of Reactive Red 2 textile dye by newly isolated *Pseudomonas* sp. SUK1. *Journal of Hazardous Materials*, *163*(2-3), 735–742. doi:10.1016/j.jhazmat.2008.07.020 PMID:18718713

Kant, R. (2012). Textile dyeing industry an environmental hazard. *Nature and Science*, *4*(01), 22–26. doi:10.4236/ns.2012.41004

Karthikeyan, K., Nanthakumar, K., Shanthi, K., & Lakshmanaperumalsamy, P. (2010). Response surface methodology for optimization of culture conditions for dye decolorization by a fungus, *Aspergillus niger* HM11 isolated from dye affected soil. *Iranian Journal of Microbiology*, *2*, 213–222. PMID:22347575

Khan, R., Bhawana, P., & Fulekar, M. H. (2013). Microbial decolorization and degradation of synthetic dyes: A review. *Reviews in Environmental Science and Biotechnology*, *12*(1), 75–97. doi:10.100711157-012-9287-6

Khehra, M. S., Saini, H. S., Sharma, D. K., Chadha, B. S., & Chimni, S. S. (2006). Biodegradation of azo dye C.I. Acid Red 88 by an anoxic–aerobic sequential bioreactor. *Dyes and Pigments*, *70*(1), 1–7. doi:10.1016/j.dyepig.2004.12.021

Kılıç, N. K., Nielsen, J. L., Yüce, M., & Dönmez, G. (2007). Characterization of a simple bacterial consortium for effective treatment of wastewaters with reactive dyes and Cr(VI). *Chemosphere*, *67*(4), 826–831. doi:10.1016/j.chemosphere.2006.08.041 PMID:17217991

Kodam, K. M., Soojhawon, I., Lokhande, P. D., & Gawai, K. R. (2005). Microbial decolorization of reactive azo dyes under aerobic conditions. *World Journal of Microbiology & Biotechnology*, *21*(3), 367–370. doi:10.100711274-004-5957-z

Kumar, S., Sharma, K. P., Sharma, S., Grover, R., Kumar, P., Soni, P., & Sharma, S. (2006). Optimization of microbial degradation of an azo dye (methyl red) in fixed film bioreactors. *Indian Journal of Biotechnology*, *5*, 68–75.

Kurade, M., Waghmode, T., Kagalkar, A., & Govindwar, S. (2012). Decolorization of textile industry effluent containing disperse dye Scarlet RR by a newly developed bacterial-yeast consortium BL-GG. *Chemical Engineering Journal*, *184*, 33–41. doi:10.1016/j.cej.2011.12.058

Kurade, M. B., Waghmode, T. R., & Govindwar, S. P. (2011). Preferential biodegradation of structurally dissimilar dyes from a mixture by *Brevibacillus laterosporus*. *Journal of Hazardous Materials*, *192*(3), 1746–1755. doi:10.1016/j.jhazmat.2011.07.004 PMID:21803494

Langer, S., Nakanishi, S., Mathes, T., Knaus, T., Binter, A., Macheroux, P., ... Mack, M. (2013). The flavoenzyme azobenzene reductase AzoR from Escherichia coli binds roseoflavin mononucleotide (RoFMN) with high affinity and is less active in its RoFMN form. *Biochemistry*, *52*(25), 4288–4295. doi:10.1021/bi400348d PMID:23713585

Leelakriangsak, M. (2013). Molecular approaches for bacterial azoreductases. *Songklanakarin Journal of Science and Technology*, *35*, 647–657.

Leelakriangsak, M., & Borisut, S. (2012). Characterization of the decolorizing activity of azo dyes by *Bacillus subtilis* azoreductase AzoR1. *Songklanakarin Journal of Science and Technology*, *34*, 509–516.

Lin, J., Zhang, X., Li, Z., & Lei, L. (2010). Biodegradation of Reactive blue 13 in a two-stage anaerobic/aerobic fluidized beds system with a *Pseudomonas* sp. isolate. *Bioresource Technology*, *101*(1), 34–40. doi:10.1016/j.biortech.2009.07.037 PMID:19713103

Liu, G., Zhou, J., Fu, Q. S., & Wang, J. (2009). The *Escherichia coli* azoreductase AzoR is involved in resistance to thiol-specific stress caused by electrophilic quinones. *Journal of Bacteriology, 191*(20), 6394–6400. doi:10.1128/JB.00552-09 PMID:19666717

Liu, G., Zhou, J., Jin, R., Zhou, M., Wang, J., Lu, H., & Qu, Y. (2008A). Enhancing survival of *Escherichia coli* by expression of azoreductase AZR possessing quinone reductase activity. *Applied Microbiology and Biotechnology, 80*(3), 409–416. doi:10.100700253-008-1555-6 PMID:18548247

Liu, G., Zhou, J., Lv, H., Xiang, X., Wang, J., Zhou, M., & Qv, Y. (2007). Azoreductase from *Rhodobacter sphaeroides* AS1.1737 is a flavodoxin that also functions as nitroreductase and flavin mononucleotide reductase. *Applied Microbiology and Biotechnology, 76*(6), 1271–1279. doi:10.100700253-007-1087-5 PMID:17846764

Liu, G., Zhou, J., Wang, J., Yan, B., Li, J., Lu, H., ... Jin, R. (2008B). Site-directed mutagenesis of substrate binding sites of azoreductase from *Rhodobacter sphaeroides. Biotechnology Letters, 30*(5), 869–875. doi:10.100710529-007-9627-8 PMID:18165868

Maas, R., & Chaudhari, S. (2005). Adsorption and biological decolourization of azo dye Reactive Red 2 in semicontinuous anaerobic reactors. *Process Biochemistry, 40*(2), 699–705. doi:10.1016/j.procbio.2004.01.038

Machado, K. M. G., Luciana, C. A., Compart, L. C. A., Morais, R. O., Luiz, H., Rosa, L. H., & Santos, M. H. (2006). Biodegradation of reactive textile dyes by basidiomycetous fungi from Brazilian ecosystems. *Brazilian Journal of Microbiology, 37*(4), 481–487. doi:10.1590/S1517-83822006000400015

Mane, U. V., Gurav, P. N., Deshmukh, A. M., & Govindwar, S. P. (2008). Degradation of Textile Dye Reactive Navy-Blue Rx (Reactive Blue 59) by an Isolated Actinomycete *Streptomyces krainskii* SUK-5. *Malaysian Journal of Microbiology, 4*, 1–5.

Martínková, L., Uhnáková, B., Pátek, M., Nesvera, J., & Kren, V. (2009). Biodegradation potential of the genus *Rhodococcus. Environment International, 35*(1), 162–177. doi:10.1016/j.envint.2008.07.018 PMID:18789530

Matsumoto, K., Mukai, Y., Ogata, D., Shozui, F., Nduko, J. M., Taguchi, S., & Ooi, T. (2010). Characterization of thermostable FMN-dependent NADH azoreductase from the moderate thermophile *Geobacillus stearothermophilus. Applied Microbiology and Biotechnology, 86*(5), 1431–1438. doi:10.100700253-009-2351-7 PMID:19997911

McGovern, P. E., & Michel, R. H. (1985). Royal Purple Dye: Tracing chemical origins of the industry. *Analytical Chemistry, 57*, 1514A–1522A.

McMullan, G., Meehan, C., Conneely, A., Kirby, N., Robinson, T., Nigam, P., ... Smyth, W. F. (2001). Microbial decolourisation and degradation of textile dyes. *Applied Microbiology and Biotechnology, 56*(1-2), 81–87. doi:10.1007002530000587 PMID:11499950

Mendes, S., Pereira, L., Batista, C., & Martins, L. O. (2011). Molecular determinants of azo reduction activity in the strain *Pseudomonas putida* MET94. *Applied Microbiology and Biotechnology, 92*(2), 393–405. doi:10.100700253-011-3366-4 PMID:21655981

Mester, T., & Field, J. A. (1998). Characterization of a novel manganese peroxidase-lignin peroxidase hybrid isozyme produced by *Bjerkandera* species strain BOS55 in the absence of manganese. *The Journal of Biological Chemistry, 273*(25), 15412–15417. doi:10.1074/jbc.273.25.15412 PMID:9624124

Mishra, A., Kumar, S., & Pandey, A. K. (2011). Laccase production and simultaneous decolorization of synthetic dyes in unique inexpensive medium by new isolates of White Rot Fungus. *International Biodeterioration & Biodegradation, 65*(3), 487–493. doi:10.1016/j.ibiod.2011.01.011

Moosvi, S., Kher, X., & Madamwar, D. (2007). Isolation, characterization and decolorization of textile dyes by a mixed bacterial consortium JW-2. *Dyes and Pigments, 74*(3), 723–729. doi:10.1016/j.dyepig.2006.05.005

Morrison, J. M., Wright, C. M., & John, G. H. (2012). Identification, isolation and characterization of a novel azoreductase from *Clostridium perfringens*. *Anaerobe, 18*(2), 229–234. doi:10.1016/j.anaerobe.2011.12.006 PMID:22182443

Murrell, J. N. (1973). *The theory of the electronic Spectra of Organic Molecules*. John Wiley & Sons.

Nachiyar, C. V., & Rajakumar, G. S. (2005). Purification and characterization of an oxygen insensitive azoreductase from *Pseudomonas aeruginosa*. *Enzyme and Microbial Technology, 36*(4), 503–509. doi:10.1016/j.enzmictec.2004.11.015

Ooi, T., Shibata, T., Sato, R., Ohno, H., Kinoshita, S., Thuoc, T. L., & Taguchi, S. (2007). An azoreductase, aerobic NADH-dependent flavoprotein discovered from *Bacillus* sp.: Functional expression and enzymatic characterization. *Applied Microbiology and Biotechnology, 75*(2), 377–386. doi:10.100700253-006-0836-1 PMID:17546472

Pandey, A., Singh, P., & Iyengar, L. (2007). Bacterial decolorization and degradation of azo dyes. *International Biodeterioration & Biodegradation, 59*(2), 73–84. doi:10.1016/j.ibiod.2006.08.006

Pearce, C. I., Christie, R., Boothman, C., von Canstein, H., Guthrie, J. T., & Lloyd, J. R. (2006). Reactive azo dye reduction by Shewanella strain J18 143. *Biotechnology and Bioengineering, 95*(4), 692–703. doi:10.1002/bit.21021 PMID:16804943

Priya, B., Uma, L., Ahamed, A. K., Subramanian, G., & Prabaharan, D. (2011). Ability to use the diazo dye, C.I. Acid Black 1 as a nitrogen source by the marine cyanobacterium *Oscillatoria curviceps* BDU92191. *Bioresource Technology, 102*(14), 7218–7223. doi:10.1016/j.biortech.2011.02.117 PMID:21571528

Puvaneswari, N., Muthukrishnan, J., & Gunasekaran, P. (2006). Toxicity assesment and microbial degradation of azo dyes. *Indian Journal of Experimental Biology, 44*, 618–626. PMID:16924831

Qi, J., Anke, M. K., Szymańska, K., & Tischler, D. (2017A). Immobilization of *Rhodococcus opacus* 1CP azoreductase to obtain azo dye degrading biocatalysts operative at acidic pH. *International Biodeterioration & Biodegradation, 118*, 89–94. doi:10.1016/j.ibiod.2017.01.027

Qi, J., Paul, C. E., Hollmann, F., & Tischler, D. (2017B). Changing the electron donor improves azoreductase dye degrading activity at neutral pH. *Enzyme and Microbial Technology, 100*, 17–19. doi:10.1016/j.enzmictec.2017.02.003 PMID:28284307

Qi, J., Schlömann, M., & Tischler, D. (2016). Biochemical characterization of an azoreductase from *Rhodococcus opacus* 1CP possessing methyl red degradation ability. *Journal of Molecular Catalysis. B, Enzymatic*, *130*, 9–17. doi:10.1016/j.molcatb.2016.04.012

Rai, J., Bhattacharya, M., Singh, J., Bansal, T. K., Vats, P., & Banerjee, U. C. (2005). Removal of dyes from the effluent of textile and dyestuff manufacturing industry: A review of emerging techniques with reference to biological treatment. *Critical Reviews in Environmental Science and Technology*, *35*(3), 219–238. doi:10.1080/10643380590917932

Revankar, M., & Lele, S. (2007). Synthetic dye decolorization by White Rot Fungus, *Ganoderma* sp. WR-1. *Bioresource Technology*, *98*(4), 775–780. doi:10.1016/j.biortech.2006.03.020 PMID:16730976

Robinson, T., McMullan, G., Marchant, R., & Nigam, P. (2001). Remediation of dyes in textile effluent: A critical review on current treatment technologies with a proposed alternative. *Bioresource Technology*, *77*(3), 247–255. doi:10.1016/S0960-8524(00)00080-8 PMID:11272011

Ruiz-Arias, A., Juárez-Ramírez, C., de los Cobos-Vasconcelos, D., Ruiz-Ordaz, N., Salmerón-Alcocer, A., Ahuatzi-Chacón, D., & Galíndez-Mayer, J. (2010). Aerobic biodegradation of a sulfonated phenyl-azonaphthol dye by a bacterial community immobilized in a multistage packed-bed BAC reactor. *Applied Biochemistry and Biotechnology*, *162*(6), 1689–1707. doi:10.100712010-010-8950-z PMID:20376575

Ryan, A., Kaplan, E., Nebel, J. C., Polycarpoum, E., Crescente, V., Lowe, E., ... Sim, E. (2014). Identification of NAD(P)H quinone oxidoreductase activity in azoreductases from *P. aeruginosa*: Azoreductases and NAD(P)H quinone oxidoreductases belong to the same FMN-dependent superfamily of enzymes. *PLoS One*, *9*(6), e98551. doi:10.1371/journal.pone.0098551 PMID:24915188

Ryan, A., Laurieri, N., Westwood, I., Wang, C. J., Lowe, E., & Sim, E. (2010). A novel mechanism for azo-reduction. *Journal of Molecular Biology*, *400*(1), 24–37. doi:10.1016/j.jmb.2010.04.023 PMID:20417637

Sandhya, S. (2010). Biodegradation of azo dyes under anaerobic condition: role of azoreductase. Handbook of Environmental Chemistry – Biodegradation of Azo Dyes Eds Erkurt, 39-57. doi:10.1007/698_2009_43

Sani, R. K., & Banerjee, U. C. (1999). Decolorization of triphenylmethane dyes and textile and dye-stuff effluent by *Kurthia* sp. *Enzyme and Microbial Technology*, *24*(7), 433–437. doi:10.1016/S0141-0229(98)00159-8

Saratale, R. G., Saratale, G. D., Chang, J. S., & Govindwar, S. P. (2009). Decolorization and biodegradation of textile dye Navy blue HER by *Trichosporon beigelii* NCIM-3326. *Journal of Hazardous Materials*, *166*(2-3), 1421–1428. doi:10.1016/j.jhazmat.2008.12.068 PMID:19157708

Saratale, R. G., Saratale, G. D., Chang, J. S., & Govindwar, S. P. (2011). Bacterial decolorization and degradation of azo dyes: A Review. *Journal of the Taiwan Institute of Chemical Engineers*, *42*(1), 138–157. doi:10.1016/j.jtice.2010.06.006

Sen, K., Pakshirajan, K., & Santra, S. B. (2012). Modelling the biomass growth and enzyme secretion by the White Rot Fungus *Phanerochaete chrysosporium* in presence of a toxic pollutant. *Journal of Environmental Protection*, *3*(01), 114–119. doi:10.4236/jep.2012.31014

Shah, V., Garg, N., & Madamwar, D. (1999). Exopolysaccharide production by a marine cyanobacterium *Cyanothece* sp. application in dye removal by its gelatin phenomenon. *Applied Biochemistry and Biotechnology*, *82*(2), 81–90. doi:10.1385/ABAB:82:2:81

Singh, R., Singh, R., & Singh, P. (2014). Bacterial decolorization of textile azo dye acid orange by *Staphylococcus hominis* RMLRT03. *Toxicology International*, *21*(2), 160–166. doi:10.4103/0971-6580.139797 PMID:25253925

Singh, R. L., Singh, P. K., & Singh, R. P. (2015). Enzymatic decolorization and degradation of azo dyes - A review. *International Biodeterioration & Biodegradation*, *104*, 21–31. doi:10.1016/j.ibiod.2015.04.027

Solis, M., Solis, A., Perez, H. I., Manjarrez, N., & Flores, M. (2012). Microbial decolouration of azo dyes: A Review. *Process Biochemistry*, *47*(12), 1723–1748. doi:10.1016/j.procbio.2012.08.014

Srinivasan, A., & Viraraghavan, T. (2010). Decolorization of dye wastewaters by biosorbents: A review. *Journal of Environmental Management*, *91*(10), 1915–1929. doi:10.1016/j.jenvman.2010.05.003 PMID:20627542

Stolz, A. (2001). Basic and applied aspects in the microbial degradation of azo dyes. *Applied Microbiology and Biotechnology*, *56*(1-2), 69–80. doi:10.1007002530100686 PMID:11499949

Sugiura, W., Yoda, T., Matsuba, T., Tanaka, Y., & Suzuki, Y. (2006). Expression and characterization of the genes encoding azoreductases from *Bacillus subtilis* and *Geobacillus stearothermophilus*. *Bioscience, Biotechnology, and Biochemistry*, *70*(7), 1655–1665. doi:10.1271/bbb.60014 PMID:16861800

Suzuki, Y., Yoda, T., Ruhul, A., & Sugiura, W. (2001). Molecular cloning and characterization of the gene coding for azoreductase from *Bacillus* sp. OY1-2 isolated from soil. *The Journal of Biological Chemistry*, *276*(12), 9059–9065. doi:10.1074/jbc.M008083200 PMID:11134015

Telke, A. A., Ghodake, G. S., Kalyani, D. C., Dhanve, R. S., & Govindwar, S. P. (2011). Biochemical characteristics of a textile dye degrading extracellular laccase from a *Bacillus* sp. ADR. *Bioresource Technology*, *102*(2), 1752–1756. doi:10.1016/j.biortech.2010.08.086 PMID:20855194

Tony, B. D., Goyal, D., & Khanna, S. (2009). Decolorization of textile azo dyes by aerobic bacterial consortium. *International Biodeterioration & Biodegradation*, *63*(4), 462–469. doi:10.1016/j.ibiod.2009.01.003

Vandevivere, P. C., Bianchi, R., & Verstraete, W. (1998). Treatment and reuse of wastewater from the textile wet-processing industry: Review of Emerging Technologies. *Journal of Chemical Technology and Biotechnology (Oxford, Oxfordshire)*, *72*(4), 289–302. doi:10.1002/(SICI)1097-4660(199808)72:4<289::AID-JCTB9053.0.CO;2-#

Verma, P., & Madamwar, D. (2002). Production of ligninolytic enzymes for dye decolorization by co-cultivation of white-rot fungi *Pleurotus ostreatus* and *Phanerochaete chrysosporium* under solid-state fermentation. *Applied Biochemistry and Biotechnology, 102*(1-6), 109–118. doi:10.1385/ABAB:102-103:1-6:109 PMID:12396115

Verma, P., & Madamwar, D. (2003). Decolourization of synthetic dyes by a newly isolated strain of *Serratia marcescens. World Journal of Microbiology & Biotechnology, 19*(6), 615–618. doi:10.1023/A:1025115801331

Vobis, G., & Miyadoh, S. (1997). Atlas of actinomycetes. Academic Press.

Wang, C. J., Hagemeier, C., Rahman, N., Lowe, E., Noble, M., Coughtrie, M., ... Westwood, I. (2007). Molecular cloning, characterisation and ligand-bound structure of an azoreductase from *Pseudomonas aeruginosa. Journal of Molecular Biology, 373*(5), 1213–1228. doi:10.1016/j.jmb.2007.08.048 PMID:17904577

Wang, C.-J., Laurieri, N., Abuhammad, A., Lowe, E., Westwood, I., Ryan, A., & Sim, E. (2010). Role of tyrosine 131 in the active site of paAzoR1, an azoreductase with specificity for the inflammatory bowel disease prodrug balsalazide. *Acta Crystallographica. Section F, Structural Biology and Crystallization Communications, 66*(1), 2–7. doi:10.1107/S1744309109044741 PMID:20057057

Wang, H., Su, J., Zheng, X., Tian, Y., Xiong, X., & Zheng, T. L. (2009). Bacterial decolorization and degradation of the reactive dye Reactive Red 180 by *Citrobacter* sp. SK3. *International Biodeterioration & Biodegradation, 63*(4), 395–399. doi:10.1016/j.ibiod.2008.11.006

Wang, H., Zheng, X. W., Su, J. Q., Tian, Y., Xiong, X. J., & Zheng, T. L. (2009). Biological decolorization of the reactive dyes Reactive Black 5 by a novel isolated bacterial strain *Enterobacter* sp. EC3. *Journal of Hazardous Materials, 171*(1-3), 654–659. doi:10.1016/j.jhazmat.2009.06.050 PMID:19577842

Wang, J., Qiao, M., Wei, K., Ding, J., Liu, Z., Zhang, K. Q., & Huang, X. (2011). Decolorizing activity of malachite green and its mechanism involved in dye biodegradation by *Achromobacter xylosoxidans* MG1. *Journal of Molecular Microbiology and Biotechnology, 20*(4), 220–227. doi:10.1159/000330669 PMID:21865764

Wong, K. S., Huang, Q., Au, C. H., Wang, J., & Kwan, H. S. (2012). Biodegradation of dyes and polyaromatic hydrocarbons by two allelic forms of *Lentinula edodes* laccase expressed from *Pichia pastoris. Bioresource Technology, 104*, 157–164. doi:10.1016/j.biortech.2011.10.097 PMID:22130082

Yang, Q. X., Tao, L. X., Yang, M., & Zhang, H. (2008). Effects of glucose on the decolorization of Reactive Black 5 by yeast isolates. *Journal of Environmental Sciences (China), 20*(1), 105–108. doi:10.1016/S1001-0742(08)60016-9 PMID:18572531

Yemendzhiev, H., Alexieva, Z., & Krastanov, A. (2009). Decolorization of synthetic dye reactive blue 4 by mycelial culture of white-rot fungi *Trametes versicolor* 1. *Biotechnology, Biotechnological Equipment, 23*(3), 1337–1339. doi:10.1080/13102818.2009.10817665

You, S.-J., & Teng, J.-Y. (2009). Anaerobic decolorization bacteria for the treatment of azo dye in a sequential anaerobic and aerobic membrane bioreactor. *Journal of the Taiwan Institute of Chemical Engineers*, *40*(5), 500–504. doi:10.1016/j.jtice.2009.01.007

Yu, J., Ogata, D., Gai, Z., Taguchi, S., Tanaka, I., Ooi, T., & Yao, M. (2014). Structures of AzrA and of AzrC complexed with substrate or inhibitor: Insight into substrate specificity and catalytic mechanism. *Acta Crystallographica. Section D, Biological Crystallography*, *70*(2), 553–564. doi:10.1107/S1399004713030988 PMID:24531489

Yu, Z., & Wen, X. (2005). Screening and identification of yeasts for decolorizing synthetic dyes in industrial wastewater. *International Biodeterioration & Biodegradation*, *56*(2), 109–114. doi:10.1016/j.ibiod.2005.05.006

Zimmermann, T., Gasser, F., Kulla, H. G., & Leisinger, T. (1984). Comparison of two bacterial azoreductases acquired during adaptation to growth on azo dyes. *Archives of Microbiology*, *138*(1), 37–43. doi:10.1007/BF00425404 PMID:6742955

Zimmermann, T., Kulla, H. G., & Leisinger, T. (1982). Properties of purified Orange II azoreductase, the enzyme initiating azo dye degradation by *Pseudomonas* KF46. *European Journal of Biochemistry*, *129*(1), 197–203. doi:10.1111/j.1432-1033.1982.tb07040.x PMID:7160382

Section 4

Advanced Approach for Waste Management

Chapter 17
Design and Development of a Sanitary Landfill for Low Income Countries for Optimal Waste Management

Musaida Mercy Manyuchi
University of Johannesburg, South Africa & Manicaland University of Applied Sciences, Zimbabwe

Edison Muzenda
University of Johannesburg, South Africa & Botswana International University of Science and Technology, Botswana

Charles Mbohwa
University of Johannesburg, South Africa

ABSTRACT

Proper waste management in developing countries is increasingly becoming popular, especially the use of engineered sanitary landfills. In this study, the classification, design, and operation of sanitary landfills are stated and suggestions made. The landfills were classified in accordance to waste type and population size. Leachate control mechanisms were put in place in a bid to avoid surface and underground water pollution. Several liner materials such as clay and geotextile material are recommended to avoid leachate perforating into the ground. Waste management is encouraged to be done through the cell system which must be compacted and covered on a daily basis. Resource recovery of biogas for power generation is recommended to recovery value from the waste as well as to lower landfill operation costs. Post-closure monitoring of the landfill must be done and recreation centers can be developed on the landfill as a rehabilitation strategy. Land filling is essential for proper waste management and is also governed by local acts.

DOI: 10.4018/978-1-5225-3540-9.ch017

INTRODUCTION

Proper waste management in developing countries is becoming critical in developing countries hence the need for properly engineered sanitary landfills. Land filling is defined as placing solid and semi-solid wastes on the ground, compacting and covering the waste with suitable materials to isolate it from the environment. Sanitary landfills have to be properly engineered and be able to accommodate a community waste quantity for a certain time period without compromising the resource expenditure and environmental health (Environment Agency, 2004). Various factors come into play in planning a sanitary landfill site construction and location. A sanitary landfill site should be planned for a long period of use in view of necessary investment in infrastructure and environmental protection measures (Ministry of Environment, 1999). Experience shows that it is economic to design a disposal site to operate for a minimum of 25 years. In order to meet these requirements sanitary landfills must both be affordable and environmentally acceptable, they must suit different needs and situations in a scientifically developed way. A village with a population of 100 people, for example, cannot be expected to have the same waste disposal needs, or meet the same set minimum requirements, as a city with a population of 1,000,000. A system for classifying sanitary landfills is therefore needed, as a basis for setting and applying the appropriately graded minimum requirements. This system has to recognize the inherent qualities and differences which characterize any sanitary landfill operation, i.e. the types of waste involved, the size of the waste stream and the potential for significant leachate generation. The sanitary landfill classification system will include waste type, size of operation and leachate generation capacity. In particular, the potential for significant leachate generation, the main cause of water pollution from sanitary landfills has to be preferentially considered.

This study focused on sanitary landfill classification systems are to be considered for waste disposal situations and needs in terms of combinations of waste type, size of waste stream and potential for significant leachate generation Furthermore, the sanitary landfill classification as the basis for setting standard requirements for the cost-effective selection, investigation, design, operation and closure of sanitary landfills.

Classification of Sanitary Landfills According to Waste

Using the classification system, sanitary landfills are grouped according to the type of waste involved, the size of the waste stream, and the potential for significant leachate generation (EPA Victoria, 2009). Note that the sanitary landfill classification system cannot address factors specific to a particular site, such as the sensitivity of the receiving environment. Such factors are addressed during site selection, investigation and environmental impact assessment, where any critical factor would be identified (Shah, 2000). Before a sanitary landfill can be developed or permitted for continued operation, it must to be demonstrated that any adverse environmental impacts can be addressed in the sanitary landfill design and operating plan.

Sanitary landfills can further be divided by the toxicity of waste to be accommodated. Conveniently, waste types can be grouped into two classes which are general and hazardous waste (EPA Victoria, 2009). Based on this, sanitary landfills can therefore also grouped into sanitary landfills suitable only for general waste, such as domestic refuse, and sanitary landfills for hazardous waste, generated by households but still toxic. This study mainly focused on municipal solid waste (MSW) which is general non-hazardous waste stream.

Classification According to Waste Toxicity

General Waste

General waste is a generic term for waste that, because of its composition and characteristics, does not pose a significant threat to public health or the environment if properly managed (EPA Victoria, 2009). Examples include domestic, commercial, certain industrial wastes and builders' rubble. General waste may have insignificant quantities of hazardous substances dispersed within it, for example, batteries, insecticides, weed-killers and medical waste discarded on domestic and commercial premises. General waste may be disposed of on any permitted sanitary landfill.

General waste can produce leachate with an unacceptably high pollution potential (Environment Agency, 2004). This may result from waste decomposition, together with the infiltration and/or percolation of water. Therefore, under certain conditions general sanitary landfills must have leachate management systems.

Therefore, in addition to being subdivided in terms of size of operation, general waste sanitary landfills are subdivided in terms of their potential to generate significant leachate.

Hazardous Waste

Hazardous waste is waste which can, even in low concentrations, have a significant adverse effect on public health and/or the environment (Hassan, 1996). This would be because of its inherent chemical and physical characteristics, such as toxic, ignitable, corrosive, carcinogenic or other properties. The following types of waste should be regarded as potentially hazardous:

- **Inorganic Waste:** This include acids and alkalis, cyanide waste, heavy metals, sludge and solutions, waste containing appreciable proportions of fibrous asbestos, oily waste *and* wastes primarily from the processing, storage and use of mineral oils.
- **Organic Waste:** This include halogenated solvent residues, *n*on-halogenated solvent residues, phenolic waste, polychlorinated biphenyl waste, paint and resin waste, Biocide waste and organic chemical residues.
- **Putrescible Organic Waste:** This include waste from the production of edible animal and vegetable oils, slaughter houses, tanneries and other animal and vegetable-based products.
- **High Volume/Low Hazard Waste:** This waste that contains small quantities of highly dispersed hazardous substances. This waste presents a relatively low hazard. Examples are harbor dredge spoils, sewage sludge, soils and builders' rubble, which are contaminated by heavy metals, oils and other pollutants.
- **Miscellaneous Hazardous Waste:** This include infectious waste such as diseased human/animal tissues, soiled bandages and syringes, commonly referred to as 'medical waste, redundant chemicals or medicines, laboratory waste, explosives waste from manufacturing operations or redundant munitions.

Hazardous wastes are grouped into nine classes, based on international danger groups*. They are also allocated a hazard rating *(Ministry of Environment, 1999). The hazard rating is based on acute mammalian toxicity, eco toxicity, environmental fate, chronic toxicity and other criteria. Hazardous waste is thus classified into: Hazard Rating 1: Extreme Hazard, Hazard Rating 2: High Hazard, Hazard Rating 3: Moderate Hazard and Hazard Rating 4: Low Hazard.

Hazardous waste sanitary landfills are therefore divided into two types according to the hazard rating of the waste that they are designed to handle. H: H sanitary landfills can accept all hazard ratings of waste, while H: h sanitary landfills can only accept Hazard Ratings 3 and 4 and general wastes.

Certain hazardous wastes may be delisted for disposal at an H:h sanitary landfill or an appropriately lined general waste site (EPA Victoria, 2009). This would be because the hazardous substance in the waste is of low mobility or concentration, or because the substance has been successfully treated to make it less hazardous. It must, however, be demonstrated to the satisfaction of the Department that the waste does not pose a risk to man or the environment. This would involve additional investigative testing. Once the waste class has been established, the class of sanitary landfill and the applicable Minimum Requirements for disposal can be determined.

SIZE OF A SANITARY LANDFILL

Sizing Based on Waste Generated

The ultimate physical size of a sanitary landfill will depend on the amount of waste it receives over its lifetime (U.S. Environmental Protection Agency, 2004). The size classification focuses on the size of the waste stream and the consequent size of the operation. This is because the immediate impacts of a sanitary landfill, the resources required to control them and, consequently, the minimum requirements applicable to the site will be dictated by the size of the operation. The size of the disposal site finally dictates the construction costs.

The size of a sanitary landfill operation depends on the daily rate of waste deposition. This in turn relates to, amongst other things, the size of the population served. To take time and growth into account sanitary landfills are classified using the maximum rate of deposition (MRD). This is simply the projected maximum average annual rate of waste deposition, expressed in tones per day, during the expected life of a sanitary landfill.

Determining the Maximum Rate of Waste Deposition

The initial rate of deposition (IRD) which is the measurement of the existing waste stream in tones per day must first be determined. The IRD is then escalated at a rate which is usually based on the projected population growth for the estimated or design life of a sanitary landfill.

The maximum average daily rate of deposition, which usually occurs in the final years of the operation, then represents the MRD. For consistency, the IRD (and hence the MRD) is based on a five day week. If waste is disposed of on weekends, this must be quantified and clearly indicated, but for purposes of standardization the total must still be presented as though for a five day week (i.e. a 260 day year).

Determining the Initial Rate of Deposition

There are several ways in which the IRD or existing waste stream can be calculated. These are as follows: If the waste stream already crosses a weighbridge at an existing sanitary landfill, an average daily tonnage for the latest 260 day year can often be calculated from historical data. This single figure is then recorded as the IRD.

If there is no weighbridge available, the average daily tonnage must be calculated on the basis of incoming volumes. This involves counting incoming vehicles and estimating the volumes carried in cubic meters. Thereafter, to convert cubic meters to tones, an appropriate density factor must be applied. These factors may vary from 0.15 tons per cubic meter (T/m^3) to more than 0.60 T/m^3, depending on waste constituents and compaction. The responsible person must therefore apply these factors with discretion, taking waste properties into account. In this way, an average daily tonnage or IRD can be arrived at for a given year.

If there is no weighbridge available, daily tonnages may also be obtained by applying per capita waste generation rates to the figures for the population served. In general, these rates vary with the socio-economic standing of the population, from 0.5 kg per capita per day in the poor areas, to 3.5 kg per capita per day in the affluent areas. Again, the responsible person must use discretion in applying these factors to arrive at a single figure for daily tonnage or IRD. Since waste frequently comes from different sources, it may be necessary to use a combination of the above methods in order to calculate the IRD and the MRD. It may also be useful to use more than one method for cross-checking purposes.

Determining the MRD From the IRD

The following formula provides the basis for calculating the MRD from the IRD.

$$MRD = (IRD)\,(1+d)^t$$

where:

IRD = initial rate of deposition of refuse on site in tons per day (T/day)
d= expected annual development rate, based on expected population growth rate in the area served by the sanitary landfill
t = years since deposition started at IRD
MRD = maximum rate of deposition after t years

Selection of Sanitary Landfill Size Based on Population

As a start, the Table 1 presents the sanitary landfill size classification based on the population sizes. General waste sanitary landfills are divided into four size categories depending on population. This will also determine the type of liner to be used.

Generally all sanitary landfills should be separated from the in situ soil by various materials for example compacted clay, geotextile layer and base layer. Each class of sanitary landfill should have a clay composition layer of around 450mm thickness for B layer and 150 mm for another B layer of compacted clay. Communal communities with less than 50 000 population size may have only clay liners while disposal sites serving population of more than 50 000 people should have a geotextile layer as one of the layers.

Table 1. Classification of sanitary landfills according to population

Type	Population Size	Sanitary Landfill Liner Requirement
Communal	<50 000	A clay as a liner (at 1m thick) and well compacted to hydraulic conductivity of < 10^{-7} m/s.
Small	50 000 - 200 000	HDPE or PVC liner (30mm thickness)
Medium	200 001 – 3000 000	HDPE or PVC liner(30mm thickness)
Large	300 000>	HDPE or PVC liner(30mm thickness)

LEACHATE GENERATION AND MANAGEMENT

Leachate refers to all the liquid waste generated in the landfill. To avoid surface water and ground water pollution, it is essential that significant leachate generation from sanitary landfills be managed by means of leachate collection and treatment systems (U.S. Environmental Protection Agency, 2004). All sanitary landfills are assumed to require leachate management systems. General waste sanitary landfills are classified in terms of their potential to generate leachate. This ensures that the risk of water pollution from leachate is identified at the earliest opportunity, even before a sanitary landfill site has been selected.

Any sanitary landfill has the capacity to generate random leachate in excessively wet weather conditions. It is only necessary, however, to install leachate management systems (under liners, drains and removal systems) when leachate generation could impact adversely on the environment.

A distinction is therefore drawn between general waste sanitary landfills that generate significant leachate and those that only generate sporadic leachate. Significant leachate requires to be managed by means of a proper leachate management system. Sporadic leachate, on the other hand, while requiring some management, does not warrant a costly leachate management system (Environment Agency, 2004).

Leachate Generation Potential

Leachate generation can be either seasonal or continuous throughout the year. It results mainly from climate and/or waste with a high moisture content. In the case of existing sanitary landfills that do not meet the minimum requirements, other factors may also exist. These include fundamental problems with the sanitary landfill sitting and drainage which result in significant entrance of ground or surface water into the waste body, and hence significant leachate generation. When constructing the landfill, determining whether significant leachate will be generated and if leachate management is required. It is a minimum requirement that, even before a specific sanitary landfill site is considered, the potential for significant leachate generation be assessed and any need for leachate management identified. The potential for leachate to be generated by a sanitary landfill depends on the water balance associated with the site, i.e. the Site Water Balance.

This is affected by such factors as rainfall, evaporation, moisture content of incoming waste and water entrance into the waste body on account of poor sanitary landfill site selection, design and operation. Of these, however, the relationship between rainfall and evaporation will, as a general rule, determine the Site Water Balance. Climate is the most common cause of leachate generation. As ambient climate is the major uncontrollable cause of significant leachate generation at a sanitary landfill, a Climatic Water Balance is used as the first step in determining the potential for significant leachate generation. The

Climatic Water Balance indicates whether the climate in which a sanitary landfill is located will cause it to generate significant leachate or not. It is thus a tool to alert the developer, as early as possible, to the need to address leachate management in the sanitary landfill design and costing. In many instances, this may be applied even before the site for the sanitary landfill is selected. Thereafter, Site Specific Factors, such as waste moisture content, and entrance of runoff and ground water into the waste body, must be taken into account.

Calculating the Climatic Water Balance

The Climatic Water Balance is not a detailed classical water balance, such as one that would be used to determine ground water recharge. It is a simple calculation that assists in deciding whether leachate management is required or not. It therefore provides a conservative means of determining whether or not significant leachate generation will occur.

The Climatic Water Balance (B) is calculated using only the two climatic components of the full water balance, namely Rainfall (R) and Evaporation (E). The data used are the precipitation and A-pan evaporation or S-pan evaporation, easily obtainable from the latest edition of the weather center's evaporation and precipitation records. The responsible person must identify the most representative weather station, or stations, on which to base the calculations. The Climatic Water Balance is defined by Equation 1:

$$B = R - E \qquad (1)$$

where:

B is the Climatic Water Balance in mm of water
R is the rainfall in mm of water
E is the evaporation from a soil surface in mm of water.

The value of B is calculated for the wet season of the wettest year on record.

B is then recalculated for successively drier years, because the wettest year on record may only be so on account of unseasonal rainfall, i.e. the wettest wet season does not always occur in the wettest year. This calculation is repeated until it is established whether:

B is positive for less than one year in five for the years for which data is available. If so: There should be no significant leachate generation on account of the climate. The site is classified B– If the Minimum Requirements for the sitting, design and operation are met and only dry waste is disposed of, no leachate management system should be necessary, or, B is positive for more than one year in five for the years for which data is available. If so: There should be significant leachate generation. The site is classified B +. As such leachate requires management, leachate management systems are a Minimum Requirement.

The Climatic Water Balance indicates where a specific sanitary landfill would plot on a hypothetical climatic index that ranges from arid to humid conditions. The cut-off point between sporadic leachate generation (B-) and significant leachate generation (B+) is where B is positive for more than one in five years or for 20% of the time for which data is available. This calibration is based on long term studies and observation of numerous sanitary landfills in Southern Africa. It is considered to be conservative, as a site which is classified as B+ is, in fact, subject to B – Climatic conditions for 80% of the time. The calculation is also conservative because it ignores run-off and thus assumes that: All precipitation falling on the sanitary landfill will infiltrate. It also ignores any moisture storage capacity of the waste body or the cover.

CONSTRUCTION OF AN ENGINEERED SANITARY LANDFILL

Various factors come into play in planning a sanitary landfill site construction. A sanitary landfill site should be planned for a long period of use in view of necessary investment in infrastructure and environmental protection measures. At least for the next 25 years, the sanitary landfill should operate. This means factoring the rate of population growth in a community, waste quantities per individual and estimated recycling rate, the final waste volume generated in 25 years should be able to be accommodated on the planned disposal site. Therefore the size of the disposal site is influenced by the population being served and this is related to the disposal site construction cost and time.

To develop an estimate of how long it should take to construct a composite liner and leachate collection system to standards, it is necessary to have an idea of how much production a contractor can realistically achieve. Contractors have varying capacity. Various contractors give varying rates of placement of composite liner system. A reasonable estimate for placement and compaction of clay liner is approximately about 5 days per hectare for a 600mm thick compacted clay liner.

A reasonable estimate for the installation of geo-membrane would be approximately 3 to 4 days per hectare, which includes deployment and production seaming of the material as well as typical repairs that are an inevitable part of the installation (US Environmental Protection Agency, 1991; Acosta et al., 2012).

Geo-synthetics installers will often estimate 2 hectares per day, but this may be too optimistic under the best of circumstances (US Environmental Protection Agency, 1994). The final operations are the installation of the drains for the leachate collection system and placement of the protective soil cover, which are typically performed simultaneously, and experience has shown that these operations typically require approximately 4 days per hectare for completion.

However earthworks contractors are capable of higher production rates and high production rates are generally achieved on large earth moving projects where the contractor has both a large equipment fleet and a large area over which to operate that fleet. Sanitary landfill cells are not large by earth moving project standards.

Delays are an inevitable part of any construction project; therefore, provision should be made in the schedule for reasonable delays. The causes for many delays are avoidable, or at least predictable, and an understanding of these causes can be helpful to an owner. There are basically three major reasons for delays during construction, and these are: delays due to weather, delays due to equipment problems, and delays due to material problems.

Construction Equipment

To construct a compacted clay liner it is essential that the contractor have the proper equipment to do the job. The following pieces of construction equipment are deemed appropriate and to be an absolute minimum: One (1) CAT 815 compactor, or equivalent; One (1) track mounted excavator; One (1) bulldozer (D-7, minimum); One (1) harrow disc or roan plow; Three (3) dump trucks; One (1) water truck, with water canon; One (1) motor grader; One (1) smooth drum roller; Diesel fuel storage tank sufficient to maintain the equipment; Pumps (typically 15.cm minimum) and hose sufficient to dewater the cell and borrow area(s).

The list of equipment above is an absolute minimum; however, it is adequate to achieve a production rate. You should seriously question the experience and abilities of a contractor who proposes to do the work with less equipment (World Bank, n.d.). Obviously, a contractor is potentially at risk if he provides the minimum amount of equipment without provision for backup. The contractor may be able to operate without one of his dump trucks; however, his operation will terminate if his excavator breaks down. Consequently, the age of the equipment is a factor if the contractor proposes to supply only the minimum number of pieces necessary to do the job. Conversely, if the contractor brings in more dump trucks, but only has a single compactor or a single excavator to load the trucks, his rate of production will remain the same.

Material of Construction

Two of the most important considerations in the project are the materials for construction, which are basically soil and geosynthetics (U.S. Environmental Protection Agency, 2004). Of the two, the geosynthetic component is the straighter forward to deal with because geosynthetics are man-made and as such, some level of quality control can be exercised by the manufacturer. The soil, however, is a naturally occurring material that is a bit more complicated than most people realize, and a basic understanding of the nature of clay is valuable in recognizing the potential effects that the material can have on the construction schedule.

Clay is the term given to the soil group made up of particles smaller than two (2) microns. The unique characteristics of clay are attributable to the size and chemical properties of the particles that make up the soil (US Environmental Protection Agency, 1994). The particular property of clay soils that make them desirable for application to sanitary landfill construction is their hydraulic conductivity, which is essentially the velocity of fluid passing through the soil and is typically expressed as centimetres per second (cm/s). As a measure to ensure that the desired hydraulic conductivity is achieved during construction, the stipulation is made that the clay be compacted to a certain standard. Achieving that standard is where problems arise. There exists a specific relationship between the density that can be achieved during compaction and the amount of moisture present in the soil. The density is commonly expressed in terms of the dry weight of soil in a unit volume. The moisture is expressed as the ratio of the weight of water in a unit volume of soil to the weight of dry soil in that same unit volume, typically expressed as a percentage. For a given compactive effort, say six passes with a CAT 815 compactor, the maximum dry density will be achieved at a specific moisture content, which is referred to as the optimum moisture content. The lowest hydraulic conductivity will normally be achieved when the soil is compacted slightly above the optimum moisture content. This relationship is very sensitive to the type of soil; therefore, the moisture/density relationship must be established for each soil type proposed for use in construction of the clay liner. Once this relationship is established it is very important to determine the moisture content of the clay at the proposed material source. If the soil is too dry, the moisture content will have to be increased, and if it is too wet it will have to be dried back.

Regrettably, the characteristic of clay soils that makes clay desirable as a natural liner material is the rate at which fluids pass through clay also makes them difficult to work with from a construction standpoint. Clays do not easily take on or give up water, which makes it difficult to increase or decrease their moisture content. What are typically required are time and a considerable amount of effort in repeatedly mixing the material, which is why the water truck and the harrow disc are essential pieces of equipment. Consequently, the closer the material is to the desired moisture content, the less time required for clay liner construction. Therefore, it is very important to know the characteristics of the clay material source so that the time necessary for moisture conditioning can be factored into the schedule. It is relevant to note that repeatedly running over the soil with a compactor will not achieve the desired product; however, this is a suggestion that is frequently offered by contractors as an alternative to the use of a harrow disc.

There are several points to keep in mind regarding the matter of moisture conditioning. As a general rule it is not practical to moisture condition material in the cell when the change in moisture content is greater than about 3% because of the time and amount of effort necessary. Rushing the process can result in an unacceptable product in terms of the engineering properties of the soil. Therefore, the soil should either be moisture conditioned in the borrow area or an area on site specifically designated for this purpose. This may require an additional bulldozer to tow the harrow disc, which may elongate operation period. If the material is extremely dry it is best to moisture condition the soil in the borrow area. One method is to open the ground with a ripper, water the area, then cover it back up and allow some time for the soil to absorb the moisture. A significant improvement in the material can be achieved over a period of days, but the process is slow and the effective depth is limited to the size of the ripper. Another extremely effective method is the use of a pulva-mixer, which simultaneously breaks up dry soil clods and mixes the material. The combined use of a water truck and pulva-mixer typically produces an excellent product, but the process translates into time and expense. Consequently, the owner must be aware of what will be required to achieve the required product in place.

Cost of Sanitary Landfill Construction

The costs of sanitary landfill construction vary depending on the size of the sanitary landfill, type of waste taken, and management measures in place. The costs include costs for sanitary landfill establishment, operation and end of life management. These costs are also sometimes termed private costs of sanitary landfill. Basically each sanitary landfill is unique and the complexity of construction has been worsened by ever increasing waste volumes and complexity. Sanitary landfills are developed or constructed with different future plans or downstream benefits gains: entrepreneurship, revenue collection, power generation etc. The bottom line is that they should be value creating infrastructures.

Generally the operation costs of a sanitary landfill are about 60 - 80% of the sanitary landfill (operation and infrastructure inclusive) total cost. Analysis has shown that 5 – 10% of the disposal site set up and operation is engineering costs while about 18 – 25% are cost associated with equipment purchase (US Environmental Protection Agency, 1991). Examples of the private costs of sanitary landfill include: Costs of land purchase; cost of environmental impact assessment (EIA) approval process; capital cost of equipment and buildings; cost of lining sanitary landfill bases to prevent leaching; cost of on-site gas recovery and flaring; cost of fencing and other measure to prevent waste from being blown into adjoining properties; operational costs including labour, fuel and materials; cost of capping sanitary landfills and landscaping; and cost of rehabilitation and aftercare.

The above costs can be divided into capital costs and operation costs. The two types of costs are capital and operating costs. Capital costs include building, land, equipment and large upfront costs. Whereas, operating costs include labour and equipment, daily, intermediate and final cover, maintenance, litter control, training and safety, administration, leachate and sanitary landfill gas management. Table 2 shows average costs for constructing a landfill in Zimbabwe.

Table 2. Estimated costs for large best practice sanitary landfill in Zimbabwe

Type of Cost	Cost Per Tonne of Waste
Land purchase including airspace	$2/m^3$
EMA approvals / site development	$20 000/5ha
Landfill cells development	$6.5/ton
Daily operations	$10/ton
Capping and rehabilitation	$2.5/ton
Landfill aftercare	$2/ton
Total	**$25/ton**

Methodology Used in This Study to Calculate the Private Costs of Sanitary Landfill

This section outlines our approach to estimating the private costs of waste disposal to sanitary landfill in Zimbabwe. The costs of waste disposal to sanitary landfill include: Costs of land purchase, cost of approvals process, capital cost of equipment and buildings, cost of lining sanitary landfill bases to prevent leaching, cost of on-site gas recovery and flaring, cost of fencing and other measures to prevent waste from being blown into adjoining properties, operational costs including labour, fuel and materials, cost of capping sanitary landfills and landscaping and the cost of rehabilitation and aftercare.

SANITARY LANDFILL DESIGN PARAMETERS AND DESIGN CONSIDERATIONS

Sanitary Landfill Design Considerations

A properly engineered sanitary landfill should have the following aspects: Fence which is 2 meters high, weigh bridge with a maximum weight 20 tons, bottom liner, leachate collection, leachate treatment, sanitary landfill gas collection, final cover layer, groundwater monitoring system, methane gas to energy production facility, sanitary landfill site selection, compaction then daily cell cover- 15cm of soil, final sanitary landfill cell cover-60cm of soil, slope of the sanitary landfill cell - 2: 1 i.e. 30o angle, leachate generation between 30-300 L/day and geology of area critical. From an economic perspective, a sanitary landfill site should be planned for a long period of at least for next 20 - 25 years in view of necessary investment in infrastructure and environmental protection measures. The design considerations are well tabulated in Table 3.

Auxiliary equipment is required for use at the sanitary landfill and the types of equipment required are shown in Table 4.

Sanitary Landfill Site Selection Factors

The selection of the sanitary landfill site is done through an environmental impact assessment process (Cointreau, 1996). There are key aspects to be considered. These are site must be 60m from lakes and

Table 3. Sanitary landfill design considerations

Landfill Design Factor	Material Used
Bottom liner	Geosythentic clay liner Geomembrane Geonet Geotextile
Leachate collection system	Gravel drainage bed HDPE pipes
Leachate treatment	Aeration ponds Release in receiving body of water
Landfill gas collection	Vertical wells
Landfill gas treatment	Flare Energy recovery
Final cover layer	Sand Geomembrane Organic soil
Groundwater monitoring	Boreholes sunk around the landfill
Energy production	Biogas electrical power station Natural gas industrial boiler

(US Environmental Protection Agency, 1991)

Table 4. Sanitary landfill auxiliary equipment and their use

Equipment	Use
Pumps	Leachate collection Collection and disposal of leachate sludge
Compressors	Landfill gas collection
Dehydrators and pipeline	Treatment and transport of landfill gas
Weighbridge	Quantifying incoming waste and waste types
Bowser	Water springing during waste spreading
Forklift	Waste lifting and spreading
Compactor	Waste compacting
Front end loader	Waste moving

streams, site must be 1 500m from human habitation, site must be 2 000m from an airport and the EIA process will also determine other site particular aspects which require appropriate mitigation measures.

SANITARY LANDFILL OPERATION

Daily operations on a sanitary landfill site involve the application of manpower, plant and materials (US Environmental Protection Agency, 1991). Site management is required to have a much better understanding and control of the site and, particularly, over its biodegradation process and associated impacts. This is a consequence of the increased sophistication of the technology used in sanitary landfill operations and the highly biodegradable nature of many of the wastes being accepted.

These developments indicate a need for comprehensive forward planning, so that the life span of the site is managed in an organized, structured manner. Such forward planning will also go hand-in-hand with the best operation practices which meets requirements for compliance with waste (the Environmental Management Act (EMA) Waste Management Act, 2007).

The Sanitary Landfill Environmental Management Plan

All sites should be subject to a detailed Environmental Management Plan (EMP). The Plan is not only necessary for competent site management, but will provide essential information for any application for a waste license under the Waste Management Act.

The EMP can be broken down into a number of sections. In certain cases, information on older, and already filled, parts of the site may not be readily available to the operator. It is essential that a detailed topographical survey of each sanitary landfill site is undertaken. This should be based upon identified fixed datum points located in areas of the site which are not likely to be disturbed and which relate to ordnance datum. These will provide a benchmark for subsequent site surveys, so that surveys are comparable and the resultant plans can be used to overlay each other.

The site survey and associated void space calculation should be repeated annually, so that the rate of fill can be assessed. The up-to-date survey can be used as the basis for other plans and for developments on the site. The survey should include filled areas so that the effects of settlement can be assessed. All leachate pumping chambers and other leachate and sanitary landfill gas monitoring points should be surveyed and accurately recorded. Benchmarks on all leachate level monitoring points should be accurately surveyed so that leachate levels can be assessed in respect of the fixed reference monitoring points at the sanitary landfill periphery. The accuracy of these benchmarks should be checked during the annual survey to ensure that factors such as settlement and lateral movement within the fill are taken into account. Adequate sanitary landfill human resources are critical and should include operations manager, security officers, machine operators and drivers.

Sanitary Landfill Records Keeping

A comprehensive series of site records should be maintained for a sanitary landfill. A full set of drawings of the site and structures on the site should be retained. As-built drawings of all lining, leachate drainage and collection systems should be included, along with drawings of all capping works. Plans should be accompanied by written descriptions of development works undertaken, as well as photographs of the site at various stages and details of particular capital projects. Site input records should be maintained for all wastes entering the site, including inert materials for restoration and cover.

The results and an interpretation of the results of environmental monitoring should always be included in the site records. Good record keeping is an essential component of the effective management of a sanitary landfill. Monitoring records are needed in order to assess and manage the biodegradation process. The necessity for an organized system of record keeping should be borne in mind and introduced at the earliest possible stage in the development of a sanitary landfill facility. Although older sites may have scant data, the introduction of a record keeping system is still necessary and should be considered a high priority for the operator.

Records should be retained throughout the life of the facility, and for the closure and aftercare periods. Documents should be organized, legible, dated and signed by the appropriate personnel. At least one complete duplicate set of records should be held at a location other than the site itself. The records should include the following: Waste analysis and quantification, manage hazardous materials, control release of gases and odors, quality of gas monitoring, borehole water monitoring, maintain equipment to minimize wear and tear as well as human resources adequacy.

Sanitary Landfill Inspection

During the operational phase, the sanitary landfill should receive thorough and regular inspections by the site manager. A written record of each inspection should be kept. This inspection should be carried out at least weekly and should cover the area of current operations, as well as completed areas. This inspection should also include the site perimeter and site security arrangements.

Sanitary Landfill Post Closure Management

At the end of sanitary landfill operation or when it has reached full capacity and a cover layer is laid; continuous operations should be implemented to ensure no pollution of the environment. There is need to do a post closure environmental monitoring system of up to 30 years as well as the sanitary landfill site remediation. The sanitary landfill site can be used for parks and golf course after rehabilitation or a power plant establishment (Pohland & Kim, 1999).

SANITARY LANDFILL LEGISLATION

Environmental legislation is statutory law that has been enacted by an Environmental Management Agency of Zimbabwe. The governing law is the Waste and Solid Waste Disposal Regulations, Statutory Instrument No. 6 of 2007. In order to meet local permit requirements, EIA Sanitary landfill construction should be done by a consultancy registered with EMA. EMA enforces the expected construction standards. Besides the EMA legislation the following are some of the legislation which governs the disposals site sitting and construction: The Environmental Management Act Chapter 20:27, the Urban Councils Act: Chapter 29:15, the Rural Councils Act: Chapter 29:13, the Regional Town and Country Planning Act: Chapter 29:12, the Water Act: Chapter 20:22 and the Public Health Act: 15:09.

FUTURE RESEARCH DIRECTIONS

Although sanitary landfills promote proper waste management in developing countries, it is important to encourage waste separation at source, to encourage recycling of waste to increase lifespan of landfill by adopting technologies like vermicomposting, as well as to recover energy form landfill and generate electricity or raw biogas to reduce landfill operation costs.

CONCLUSION

It is essential for developing counties to adopt the engineered sanitary landfill technology. This is essential for proper waste management and also proper leachate management. The environmental regulatory authorities must also ensure that the landfills are being constructed to specification according to the landfill design guidelines.

REFERENCES

Acosta, V., Paul, J, & Hanuschke Hanuschke, K. (2012). *Ecocenter integrated solid waste management facility with sanitary landfill and resource recovery technologies*. Academic Press.

Cointreau, S. (1996). *Sanitary landfill design and sitting criteria*. World Bank .

Environment Agency. (2004). *Guidance on the management of landfill gas* (LFTGN03). Author.

Hasan, S. E. (1996). *Geology and hazardous waste management*. Upper Saddle River, NJ: Prentice Hall.

Ministry of Environment. (1999). *Generation and discharge of municipal waste*. Author.

Pohland, F., & Kim, J. (1999). In situ anaerobic treatment of landfills for optimum Stabilization and biogas production. *Water Science and Technology*, *40*(8), 203–210.

Shah, K. L. (2000). *Basics of solid and hazardous waste management technology*. Upper Saddle River, NJ: Prentice Hall.

US Environmental Protection Agency. (1994). Review of liner and cap regulations for landfills. EPA/600/A-94/246., Risk Reduction Engineering Laboratory, Office of Research and Development, U.S. EPA.

U.S. Environmental Protection Agency. (2004). *Monitoring approaches for landfill bioreactors*. EPA/600/R-04/301.

EPA Victoria. (2009). *Solid industrial waste hazard categorization and management* (EPA publication IWRG631). Author.

World Bank. (n.d.). *Guidance Gact Sheet: Landfill Design*. Retrieved from http://siteresources.worldbank.org/INTCARBONFINANCE/Resources/MSW_SCS_FactSheet_Landfill_Design_Final.pdf

Yelda, S. (2005). Modified landfill design for sustainable waste management. *International Journal of Global Energy Issues*, *23*(1), 93–106. doi:10.1504/IJGEI.2005.006412

KEY TERMS AND DEFINITIONS

EIA: Environmental impact assessment.
EMA: Environmental management authority.
EMP: Environmental management plan.
HDPE: High density polyethylene.
IRD: The initial rate of deposition.
MRD: The maximum rate of deposition.
MSW: Municipal solid waste.
PVC: Poly vinyl chloride.

Chapter 18
Insilico Tools to Study the Bioremediation in Microorganisms

Pankaj Bhatt
Dolphin (P.G.) Institute of Biomedical and Natural Sciences, India

ABSTRACT

Bioinformatics is the new area of science to study various living organisms, plants and animals. Bioremediation is the removal of toxic pollutants from the environment using microorganisms. Many of the databases are available online and can be used for the study of the microorganism genomics, transcriptomics, proteomics. This chapter mainly focuses the uses of recent databases and tools for the exploration of microorganisms.

INTRODUCTION

Removal of toxic pollutant from environment using living microorganisms or plants called bioremediation. Toxic compounds and their residues are coming in environment through different methods of use. These toxic chemicals mainly include, pesticides, certain dyes, industrial waste, some other compounds synthesized in laboratory xenobiotics. Their nature is recalcitrant. Their effect in wide aspect, in animals, humans, plants, and also for microbial communities in environment. The residue of pesticides i.e. cypermethrin, fipronil and sulfosulfuron was found in different agriculture fields. The degrading microorganisms were isolated and characterized (Pankaj et al., 2015, 2016). Laboratory based experiments proof the presence of pollutants in environment and their degradation by using bacteria, fungi (Gangola et al., 2014, 2015; Negi et al., 2014, 2016). It was observed that microbial communities changed in response to environmental conditions, these toxic hazardous chemicals also affect the environmental microbial communities, and use of nanoparticles (Khati et al., 2017).

DOI: 10.4018/978-1-5225-3540-9.ch018

Up to today through wet laboratory, researchers already reported many microbes which is responsible for biodegradation. So they're in need to explore this data for more accurate interpretation of the microbial bioremediation by bioinformatic tools. Large number of proteins secreted by microorganisms during pesticide stress, their analysis by the online tools helped the researcher to find out the differential expression pattern (Pankaj et al., 2016).

Due to the development of next generation sequencing methods, microbial genome data have been accumulated at an accelerating pace, and now approximately 3000 complete sequences covering species across a wide taxonomic range are available. These complete microbial genomes are useful for the comparative analysis (Loman et al., 2015). Many of the genomic sequences are now released as draft genome sequences, for which some complete genome sequences of the same or closely related species are available. RefSeq microbial genome database has been recently reorganized to accommodate this situation, where genomes of the same species are collected into a directory and representative genomes are selected for each species after several quality checks (Uchiyama et al., 2015). The data from the databases is collectively used for the comparison between two genomes of bacteria under stress condition of toxic pollutants.

National Center for Biotechnology Information NCBI is a part of United states national library of Medicine NLM, a branch of the national Institute of Health. It acts as the combined houses of many of the databases. It includes Genbank,Pubmed, Pubchem, Entrez, and BLAST Basic local alignment search tools. These databases are used for different aspect of bioremediation, biodegradation studies under insilico conditions.

BIOINFORMATIC TOOLS FOR THE STUDY OF MICROORGANISMS

Identification and characterization of microorganisms is based on the molecular based methods. Interpretation of the results is based on the bioinformatics tools. Different databases are used for the microbial communities' identification and omics study. Table 1 shows the list of databases that can be explored for the bioremediation study associated with microorganisms.

Microbial Genome Database

The microbial genome database for comparative analysis (MBGD) (available at http://mbgd.genome.ad.jp/) is a comprehensive ortholog database for flexible comparative analysis of microbial genomes, where the users are allowed to create an ortholog table among any specified set of organisms. Due to rapid increase in microbial genome data owing to the next-generation sequencing technology, it becomes increasingly challenging to maintain high-quality orthology relationships while allowing the users to incorporate the latest genomic data available into an analysis. Because many of the recently accumulating genomic data are draft genome sequences for which some complete genome sequences of the same or closely related species are available, MBGD now stores draft genome data and allows the users to incorporate them into a user-specific ortholog database using the MyMBGD functionality. In this function, draft genome data are incorporated into an existing ortholog table created only from the complete genome data in an incremental manner to prevent low-quality draft data from affecting clustering results. In addition, to provide high-quality orthology relationships, the standard ortholog table containing all the representative genomes, which is first created by the rapid classification program DomClust, is now

Table 1. Tools for identification of microorganism involved in bioremediation

S.No	Database/Software	URL	Description	Type
1	Microbial Genome Database	http://mbgd.genome.ad.jp/)	Comparative analysis among genomes	Comprehensive ortholog database
2	Assembly	http://www.ncbi.nlm.nih.gov/assembly	Status of whole-genome shotgun assemblies	
3	BacWGSTdb	http://bacdb.org/BacWGSTdb/	Bacterial whole genome sequence typing database	
4	BIGNASim	http://mmb.irbbarcelona.org/BIGNASim/	Molecular dynamics simulations of nucleic acids	
5	CauloBrowser	http://www.caulobrowser.org	Biology of *Caulobacter crescentus*	
6	CSDB	http://csdb.glycoscience.ru/database	Carbohydrate structure database	**Bacteria and Archae, Plants, Fungi**
7	DESM	http://www.cbrc.kaust.edu.sa/desm	Microbial knowledge exploration systems	
8	EK3D	http://www.iith.ac.in/EK3D/	Capsular polysaccharide (K antigen) structures of various *E. coli* serotypes	
9	SigMol	http://bioinfo.imtech.res.in/manojk/sigmol	Quorum sensing signalling molecules	
10	SureChEMBL	https://www.surechembl.org/	Chemical compounds in patent documents	
11	BiGG Models	http://bigg.ucsd.edu	Biochemically, genetically and genomically structured metabolic network models	
12	CPPsite	http://crdd.osdd.net/raghava/cppsite/	Validated cell penetrating peptides	
13	DBAASP	http://dbaasp.org	Database of antimicrobial activity and structure of peptides	
14	MG-RAST	http://metagenomics.anl.gov/	Data portal for processing, analyzing, sharing and disseminating metagenomic data sets	
15	MNXref/MetaNetX	http://www.metanetx.org	Genome-scale metabolic networks	
16	sRNATarBase[a]	http://ccb1.bmi.ac.cn/srnatarbase/	sRNA-target interactions in bacteria	
17	**NCBI**	https://www.ncbi.nlm.nih.gov/	**Sequence of various organisms, Plants, Animals**	

(Rigden et al., 2016.)

refined using DomRefine, a recently developed program for improving domain-level clustering using multiple sequence alignment information (Uchiyama et al., 2015).

Genome Wide Identification

Genome wide identification mainly deals with the whole genome of organisms. Through genomics we can identify novel genes and their comparisons with previously reported genomes. Identification of gene cluster associated with triclosan catabolism reported in Sphingomonas sp. RD1 (Kagle et al., 2015). The Ribosomal Database Project (RDP) has become a reference point for almost all practical purposes in microbiology. The use of this gene is limited by the fact that it can be used to identify only to the extent to what has been known and is available in the RDP. In order to identify an organism whose *rrs* is not present in the RDP database, we need to generate novel markers to place the unknown on the evolutionary map. Kekre et al. (2015) have retrieved sequenced genomes of 27 *Clostridium* strains belonging to 9 species used to identify two sets of genes: (1) common to most of the species, and (2) unique to a species. Combinations of genes (*recN*, *dnaJ*, *secA*, *mutS*, and/or *grpE*) and their unique restriction endonuclease digestion (*AluI*, *BfaI* and/or *Tru9I*) patterns have been established to rapidly identify *Clostridium* species. Their strategy for identifying novel markers can be explored to find other organisms involved in bioremediation of toxic compounds.

The advent of functional genomics has enabled the molecular biosciences to come a long way towards characterizing the molecular constituents of life. Yet, the challenge for biology overall is to understand how organisms function. By discovering how function arises in dynamic interactions, systems biology addresses the missing links between molecules and physiology. Top-down systems biology identifies molecular interaction networks on the basis of correlated molecular behavior observed in genome-wide 'omics' studies. Bottom-up systems biology examines the mechanisms through which functional properties arise in the interactions of known components (Bruggeman et al., 2007).

System Biology

The concept of system biology is emerging very fast worldwide. Its main aim is to investigate the genomics, transcriptomics, metabolomics, proteomics, glycomics and other data to create a model how to cell work in presence of toxic compounds. For microbial sciences, the term Systems Microbiology was coined by the American Academy of Microbiology due to the importance of microbial world (Bruggemann & Westerhoff, 2006).

Systems biology is increasingly popular, but to many biologists it remains unclear what this new discipline actually encompasses. This brief personal perspective starts by outlining the aesthetic qualities that motivate systems biologists, discusses which activities do not belong to the core of systems biology, and finally explores the crucial link with synthetic biology. It concludes by attempting to define systems biology as the research endeavor that aims at providing the scientific foundation for successful synthetic biology.

Phylogenetic Analysis

The advancement in the field of microbial ecology was the discovery of 16S rRNA sequences with highly conserved gene and involved similar phylogentic characterstics of microorganisms with their communities (Amann et al., 1995). The advantage of 16S rRNA sequence analysis in bioremediation is to analyzed the sequence in contaminated environment and determine phylogenic characteristics of microorganisms and bioremediation process (Rogers & McClure, 2003). Previously Many researchers determined the phylogenetic relationship among microorganisms with MEGA software (Pankaj et al., 2015; Negi et al., 2014. Pankaj et al., 2016). For identification of microorganisms involved in bioremediation, 6S rRNA based phylogenetic tree by MEGA is important in research aspects. This software makes the tree on the basis of different desired algorithms.

MetaLab

It is a fully automatic software platform enabling an integrated data-processing pipeline for metaproteomics. The function of sample-specific database generation can be very advantageous for searching peptides against huge protein databases. It provides a seamless connection between peptide determination and taxonomic profiling; therefore, the peptide abundance is readily used for measuring the microbial variations. MetaLab is designed as a versatile, efficient, and easy-to-use tool which can greatly simplify the procedure of metaproteomic data analysis for researchers in microbiome studies during bioremediation (Cheng et al., 2017)

Design Expert

Design expert is statistical software Package of state ease incorporation mainly used for design of experiments. Design expert used for optimization, screening, characterization of bioremediation process. The optimization of biodegradation by bacteria and fungi is possible with design expert using response surface methodology. Pesticides biodegradation, enzymes of biowaste degradation is earlier optimized by the design expert (Pankaj et al., 2015; Chen et al., 2017). This optimization process mainly includes Box-Behnken and central composite design.

CONCLUSION

Omics technologies are critical value-added tools in microbial sciences. On the basis of known sequences from the database, we can analyse the sequences with respect to bioremediation. In laboratory from the isolation of microbes Bacteria, fungi to their final omics we need to know about the Bioinformatic tools used according to suitability and perfection.

ACKNOWLEDGMENT

The authors is thankful to the all researchers for their contribution, cited in the text and reference section.

REFERENCES

Amann, R. I., Ludwig, W., & Schleifer, K. H. (1995). Phylogenetic identification and in situ detection of individual microbial cells without cultivation. *Microbiological Reviews*, *59*(1), 143–169. PMID:7535888

Bhatt, P., Bhatt, R., & Bhatt, A. (2017). Soil Microbial Community affected by Climate change. *Scientific India Magazine*, *5*(3), 38–39.

Bhatt, P., & Nailwal, T. K. (2018). Crop Improvement Through Microbial Technology:a Step Towards Sustainable Agriculture. Elsevier. doi:/10.1016/B978-0-444-63987-5.00011-6

Breitling, R. (2010). What is Systems Biology? *Frontiers in Physiology, 1*. doi:10.3389/fphys.2010.00009 PMID:21423352

Bruggemann, F. J., & Westerhoff, H. V. (2006). The nature of systems biology. *Trends in Microbiology, 15*(1), 45–50. doi:10.1016/j.tim.2006.11.003 PMID:17113776

Cheng, K., Ning, Z., Zhang, X., Li, L., Liao, B., Mayne, J., ... Figeys, D. (2017). MetaLab: An automated pipeline for metaproteomic data analysis. *Microbiome, 5*(1), 157. doi:10.118640168-017-0375-2 PMID:29197424

Gangola, S., Pankaj, Negi, G., Srivastava, A., & Sharma, A. (2014). Enhanced Biodegradation of Endosulfan using a consortium of Aspergillus and Trichoderma spp. isolated from agriculture field of Kumaun region of Uttarakhand. *Pesticide Research Journal, 27*(2), 223–230.

Gangola, S., Pankaj, Khati, P., & Sharma, A. (2015). Mycoremediation of Imidaclopid in the presence of different soil amendment using trichoderma longibrachiatum and Aspergillus oryzae isolated from pesticide contaminated agricultural field of Uttarakhand. *Journal of Bioremediation & Biodegradation, 6*(5), 2–5.

Joshi, V., & Srivastava, A., Pankaj, Sharma, A., & Srivastava, P, C. (2015). Efficacy of cereal straw and its conjoint use with microbial consortium in reducing the leaching of chlorpyrifos: A soil column study. *Research Journal of Chemical Sciences., 5*(6), 9–14.

Kekre, A., Bhushan, A., Kumar, P., & Kalia, V. C. (2015). Genome Wide Analysis for Searching Novel Markers to Rapidly Identify Clostridium Strains. *Indian Journal of Microbiology, 55*(3), 250–257. doi:10.100712088-015-0535-7 PMID:26063934

Khati, P., Parul, Gangola, S., Bhatt, P., & Sharma, A. (2017). *Nanochitosan induced growth of Zea Mays with soil health maintenance.* Biotech. DOI 10.1007/s13205-017-0668-y

Khati, P., Sharma, A., Gangola, S., Kumar, R., Bhatt, P., & Kumar, G. (2017). Impact of agriusable nanocompounds on soil microbial activity: an indicator of soil health. *Soil, Air, Water, 45*(5). DOI: 10.1002/clen.201600458

Kumar, G., Arya, K., Verma, A., Pankaj, P., Khati, P., Gangola, S., ... Singh, H. (2017). Bioremediation of Petrol Engine Oil Polluted Soil Using Microbial Consortium and Wheat Crop. *Journal of Pure & Applied Microbiology, 11*(3), 1583–1588. doi:10.22207/JPAM.11.3.45

Loman, N. J., Constantinidou, C., Chan, J. Z., Halachev, M., Sergeant, M., Penn, C. W., ... Pallen, M. J. (2012). High-throughput bacterial genome sequencing: An embarrassment of choice, a world of opportunity. *Nature Reviews. Microbiology, 10*(9), 599–606. doi:10.1038/nrmicro2850 PMID:22864262

Negi, G., Pankaj, Srivastava, A., & Sharma, A. (2014). In situ Biodegradation of Endosulfan, Imidacloprid, and Carbendazim Using Indigenous Bacterial Cultures of Agriculture Fields of Uttarakhand, India. *Int. J. of Biological biomolecular agriculture food & biotech. Engineering, 8*(9), 953-961.

Negi, G., Pankaj, Srivastava, A., & Sharma, A. (2016). Optimization of Endosulfan biodegradation using indigenous bacterial isolate *Bacillus aryabhatti* through Response surface methodology. *Journal of Industrial Pollution Control, 32*(2), 638-644.

Pankaj, Bisht, T.S., Pathak, V.M., Barh, A., & Chandra, D. (2015). Optimization Of Amylase Production From The Fungal Isolates of Himalayan Region Uttarakhand. *Ecology. Environmental Conservation, 21*(3). 1517-1521.

Pankaj, Nailwal, T.K., Singh, L., & Amit, P. (2014). Isolation and characterization of Rhizobial isolates from the Rhizospheric soil of an Endangered plant Meizotropis pellita. *Asian Journal of Microbiology, Biotechnology & Environmental Science, 16*(2), 301–306.

Pankaj, Negi, G., Gangola, S., Khati, P., Srivastava, A., & Sharma, A. (2015). Optimization of Sulfo-sulfuron Biodegradation through Response Surface Methodology using Indigenous Bacterial Strain Isolated from Contaminated Agriculture Field. *International Journal of Current Microbiology and Applied Sciences, 4*(8), 105–112.

Pankaj, Sharma, A., Gangola, S., Khati, P., Kumar, G., & Srivastava, A. (2016). Novel pathway of cypermethrin biodegradation in a Bacillus sp. strain SG2 isolated from cypermethrin-contaminated agriculture field. *Biotech, 6*(1), 45. doi: 10.1007/s13205-016-0372-3

Pankaj & Sharma, A. (2018). *Microbial Biotechnology in Environmental Monitoring and Cleanup.* Hershey, PA: IGI Global. Doi:10.4018/978-1-5225-3126-5

Pankaj, Negi, G., Gangola, S., Khati, P., Kumar, G., Srivastava, A., & Sharma, A. (2016). Differential expression and characterization of cypermethrindegrading potential proteins in Bacillus thuringiensis strain, SG4. *Biotech, 6*, 225. DOI: 10.1007/s13205-016-0541-4

Pankaj, Tiwari, S., Gangola, S., Khati, P., Kumar, G., & Sharma, A. (2017). Removal of Xenobiotics from environment using microbial metabolism. *Scientific India Magzine, 5*(1), 33-34.

Rigden, D. J., Fernández-Suárez, X. M., & Galperin, M. Y. (2016). The 2016 database issue of *Nucleic Acids Research* and an updated molecular biology database collection. *Nucleic Acids Research, 44*(D1), D1–D6. doi:10.1093/nar/gkv1356 PMID:26740669

Sharma, A., Pankaj, Khati, P., Gangola, S., & Kumar, G. (2016). Microbial degradation of pesticides for environmental cleanup. Bioremediation of Industrial Pollutants. edited by Ram Naresh Bharagava and Gaurav Saxena. Write and Print Publication. Chapter-6.pp178-205.

Sharma, A., & Bhatt, P. (2016). Bioremediation: A microbial technology for improvising wildlife. Wildlife Management: Concept Analysis and Conservation. edited by Amita Saxena. Daya Publisher. Chapter-3, pp29-39.

Uchiyama, I., Mihara, M., Nishide, H., & Chiba, H. (2015). MBGD update: Microbial genome database for flexible ortholog analysis utilizing a diverse set of genomic data. *Nucleic Acids Research, 43*(D1), D270–D276. doi:10.1093/nar/gku1152 PMID:25398900

Chapter 19
Enzymatic Treatment of Petroleum-Based Hydrocarbons

Pankaj Kumar Chaurasia
Motilal Nehru National Institute of Technology Allahabad, India

Shashi Lata Bharati
North Eastern Regional Institute of Science and Technology, India

Ashutosh Mani
Motilal Nehru National Institute of Technology Allahabad, India

ABSTRACT

Petroleum and petroleum-based products are highly beneficial for our daily life system. Functioning of several industries and machineries directly depends on different forms of petroleum. There are strong possibilities of the release of such petroleum and petroleum-based compounds like aliphatic as well as aromatic hydrocarbons during their refining process, usage, storage, transport, and other activities due to the accidents, leakages, or just a little lack of awareness. Their exposure to soils or water causes serious problems for aquatic as well as human beings. The efficient removal or detoxification of such pollutants is the demand of the present time. The use of microbial sources for the bioremediation of such petroleum wastes may be promising technique because it does not require any drastic conditions for detoxification process and by-products produced by them are also harmless unlike chemical and other techniques. In this regard, this chapter discusses the enzymatic role in detoxification or bioremediation of petroleum-based hydrocarbons.

INTRODUCTION

In the present time, our environment is being contaminated with various types of pollutants released from pharmaceutical, leather, food, paper-pulp, synthetic industries etc. The chemicals released from the different types of industries are very hazardous for human beings as well as animals. Aquatic animals are strongly and adversely affected by sewage pollutants released from industries. Water pollution is being very dangerous for all types of livings depend directly or indirectly on water. It is well known that the

DOI: 10.4018/978-1-5225-3540-9.ch019

water is the most important part of our life and it is the valuable living place for huge aquatic animals whose life directly depends on the water medium. Pollution of water has been found the major cause of death and diseases (Pink, 2006; West, 2006). There is not a single factor responsible for the water pollution; however, the chemical pollution is one of the most important factors behind such death and diseases. Chemical pollution may be of any types that directly or indirectly based on chemical compounds. Organic pollutants may include detergents, oxygen demanding materials, grease, fats, volatile organic compounds, insecticides, herbicides, food processing wastes, chlorinated compounds, pharmaceutically active compounds, cosmetic compounds, synthetic compounds, petroleum hydrocarbons etc (Wikipedia, 2018a). Many research works are going on to solve the problems of contamination or pollutions worldwide. In order to solve the problem of pollution, many techniques either physical or chemicals are being applied. They may also create some types of pollutions like noise pollution, by-products pollutions, chemical pollution etc. This chapter focussed the role of enzymes in bioremediation of petroleum wastes/ hydrocarbons. Petroleum is a natural occurring liquid generally found in yellow to black in colour. It is found in geological formation beneath the surface of earth which undergoes refinement to various forms of fuels. This petroleum consists of different molecular weight hydrocarbons that are present in given compositions in crude oil: alkanes or paraffins (15-60%), naphthenes (30-60%), aromatics (3-30%) and asphaltics (remainder) (Wikipedia, 2018b). Petrochemicals are petroleum derived chemicals or products that can be divided in two major classes like olefins and aromatics (Wikipedia, 2018c). Olefins include ethylene, propylene and butadiene. Ethylene and propylene are the basis of industrial chemicals and plastic products while aromatics contain benzene, toluene and xylenes. Benzene is used as raw material for dye and synthetic detergents and xylenes are used in plastic and synthetic fibre industries (Wikipedia, 2018c). Petroleum contaminants released in soil and water are also hazardous for living organisms due to the carcinogenic properties of hydrocarbons. They become released in environment either from ruptured boats or automobiles or from oil plants, storage tanks, and from the accident of oil tankers and cause the sever effects on our systems (Wikipedia, 2018d). Bioremediation of pollutants are the highly preferable methods for the detoxification or removal of pollutants. In this method, either enzymes, microbes or other green sources are used for the bio-removal of such toxic materials. Different types of petroleum based organic compounds are shown in Figure 1 (Bahl & Bahl, 2003).

Crude Oil

Petroleum in its unprocessed form is called as crude oil that includes brent crude oil, heavy crude oil (HCO), light crude oil (LCO), sweet crude oil (SCO) and synthetic crude oil (Wikipedia, 2017a). Brent crude oil contains 0.37% sulphure and suitable for petrol and middle distillate productions' productions (Wikipedia, 2017b). Petrol (British english) also called as gasoline (American english) is a petroleum derived transparent liquid (Wikipedia, 2018e). It is mostly used in the internal combustion engines for the purpose of fuel. Organic compounds obtained from the fractional distillation are the main source of the gasoline (Wikipedia, 2018e). There is approximately 19 US gallons (72 L) of gasoline (petrol) production after processing in oil refinery from 42 gallon barrel of crude oil (159 L). This can be vary with crude oil source assay (Wikipedia, 2018e). Petroleum products can be divided into four different categories like light distillates, middle distillates, heavy distillates and residuum out of which light distillates contain LPG, naphtha, gasoline; middle distillates contain kerosene, jet fuel, diesel; while heavy distillates and residuum contain heavy fuel oil, lubricating oils, wax and asphalt (Wikipedia, 2018f). Hydrogen, light hydrocarbons, re-formate and pyrolysis gasoline become produce as intermediates from

Figure 1. Different types of petroleum based organic compounds (Bahl & Bahl, 2003)

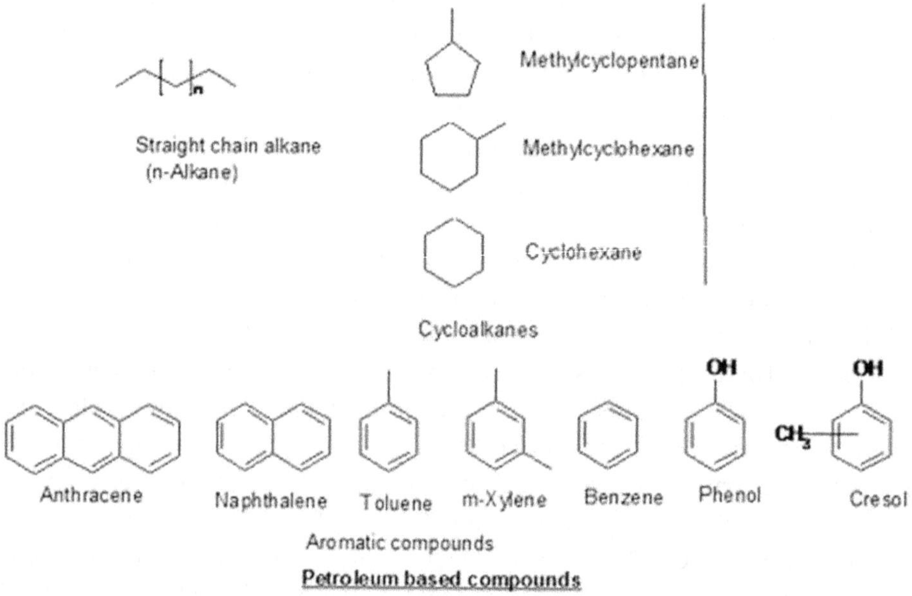

oil refineries. Pyrolysis gasoline also called as PyGas is a naphtha range product with high aromatics content (Wikipedia, 2016) It is found as by-product during the cracking of naphtha at high temperature. It is the mixture of compounds with high octane number contains aromatics, olefins and parafins ranging from carbon-5 to carbon-12 and has high potential for the use as gasoline blending mixture and or as source of aromatics. PyGas is generally used as a gasoline blending mixture due to its high octane number (Jayed, 2012).

Heavy crude oil does not flow under normal reservoir conditions due to their high viscous nature. Its density or specific gravity is more than that of light crude oil due to which its presented as heavy crude oil (Wikipedia, 2018g; Mai et al (2006). Light crude oil is of much more important than heavy crude oil and it is light due to low density and can free flow. It has high proportion of light petroleum fractions and has low wax content. Price of LCO is greater than that of HCO due to their high significance in the production of gasoline and diesel fuel (Wikipedia, 2017c). Sweet crude oil, a type of petroleum is known as sweet crude oil that contain less than 0.42% sulfure while petroleum containing high level of sulfure is known as sour crude oil. Low sulfure containing high quality petroleum is used for processing in gasoline and due to their high demand used in industrialized nations (Wikipedia, 2018h).

Enzymatic Bioremediation of Petroleum and Petrochemicals

Crude oil is a very complex mixture of different hydrocarbons. Polycyclic aromatic hydrocarbons are important part of crude oil which contains two to eight conjugated rings (Pampanin & Sydnes., 2013). A range of polyaromatic hydrocarbons are found in crude oil (Laughlin & Neff, 1979; Harvey, 1996; Feng et al., 2009). The precursors of polyaromatic hydrocarbons are the natural products that have been converted into aromatic hydrocarbons over time (Feng et al., 2009). Petroleum pollution is one of the most serious pollution in those countries which are under development (Ijah & Antai, 2003). Biodegradation of such molecules by using microbial methodologies is valuable at the contaminated site. Fungi, bacteria and enzymes are actively involved in such types of bioremediation process. Fungi are more efficient over bacteria due to the presence of their fungal hyphae and hydrolytic enzymes. These enzymes have capability to degrade hydrocarbons contaminated soil (Balaji & Ebenzer, 2008; Messias et al., 2009; Venkatesagowda et al., 2012). Balaji et al. (2014) studied the bio-treatment of polyaromatic hydrocarbons. They isolated the potential fungi and screened for the secretion of different enzymes like lipase, peroxidase, laccase and protease and studied the degradation of polyaromatic hydrocarbons. They isolated the twenty-one potential fungi that have potential capability of polyaromatic hydrocarbon's degradation. Their study showed that fungi and lipase from fungal consortia associated with polyaromatic hydrocarbons contaminated soil can be utilized in the bio-treatment of polyaromatic hydrocarbons contaminated habitat like motor oil, petrol, diesel, kerosene and grease (Balaji et al., 2014).

Thapa et al (2012) also summarized the information based on bioremediation of petroleum hydrocarbon contaminants in soil. The cytochrome P450 alkane hydroxylases enzymes have significant role in degradation of oil and chlorinated hydrocarbons (Van Beilen & Funhoff, 2007). According to the chain length of hydrocarbons, enzyme systems introduce oxygen in the substrate and start the degradation of hydrocarbons (Das & Chandran, 2011). For the degradation of petroleum hydrocarbons, there was active involvement of cytochrome P450 enzyme system which had been isolated from yeast species such as *Candida maltosa*, *Candida tropicalis* and *Candida apicola* (Scheuer et al., 1998). According to the chain length, soluble methane monooxygenase is suitable for alkanes, alkenes and cycloalkanes having one to eight carbons (McDonald et al., 2006). Particulate methane monooxygenases are suitable for halogenated alkanes and cycloalkanes of C_1-C_5 (McDonald et al., 2006). Similarly, alkanes of C_{10}-C_{16} are act as substrates for eukaryotic P450 (Iida et al., 2000), alkanes containing C_5-C_{16} system are substrates for bacterial P450 oxygenase system (Van Beilen et al., 2006) and alkane hydrocarbons with C_{10}-C_{30} carbons are act as substrate for dioxygenase (Maeng et al., 1996). The role of oxygenase enzymes in degradation of hydrocarbons are shown in Figure 2 (Fritsche & Hofrichter, 2000; Das & Chandran, 2011)

Monohydroxylation, desaturation, oxygen and nitrogen dealkylation, sulfoxidation like oxidation reactions are catalysed by naphthalene dioxygenase (Kolb et al., 1994; Van Hamme et al., 2003). It has broad specificity towards a large range of hydrocarbons due to which it can produce chiral petrochemical-based precursors that may be applicable for the synthesis of valuable chemicals (Boyd et al., 2001; Gibson & Parales, 2000; Resnick & Gibson, 1996a; Resnick et al., 1996b). Another enzyme namely alkane hydroxylase obtained from *Pseudomonas oleovorans* has wide range of specificity that can convert a variety of compounds like alkanes, alkanols, alkanals, alkenes etc in promising products (van Beilen et al., 1994; Schmid et al., 1998). Hamme et al (2003) and other discussed the role of enzymes for enhancing the bioremediation of petroleum contaminated soils (Fan & Krishnamurthy, 1995).

Laccase is an important enzyme of oxidoreductase group that have nice tendency to oxidize or bio-remediate the polyaromatic hydrocarbons either in presence or absence of mediator molecules. In some cases, it oxidizes in the absence of mediators while in many cases it oxidizes in the presence of mediators. Mediators generally enhance the range of substrates for laccase enzymes due to which applicability of such enzymes to oxidize or bio-remediate the hydrocarbons have been enhanced to wider range (Karigar & Rao, 2011; Sack et al., 1997; Salleh et al., 2003). In-vitro oxidation of anthracene and benzo[a]pyrene by laccases of *Trametes versicolor* had been studied by Collins et al. (1996). They found the enhanced oxidation of anthracene in the presence of ABTS (2,2'-azino-bis (3-ethylbenzothiazoline-6-sulphonic acid), while ABTS was essential for the oxidation of benzo[a]pyrene (Collins et al., 1996).

Figure 2. Oxygenase based degradation reaction of hydrocarbons (taken from Das & Chandran, 2011; Fritsche & Hofrichter, 2000)

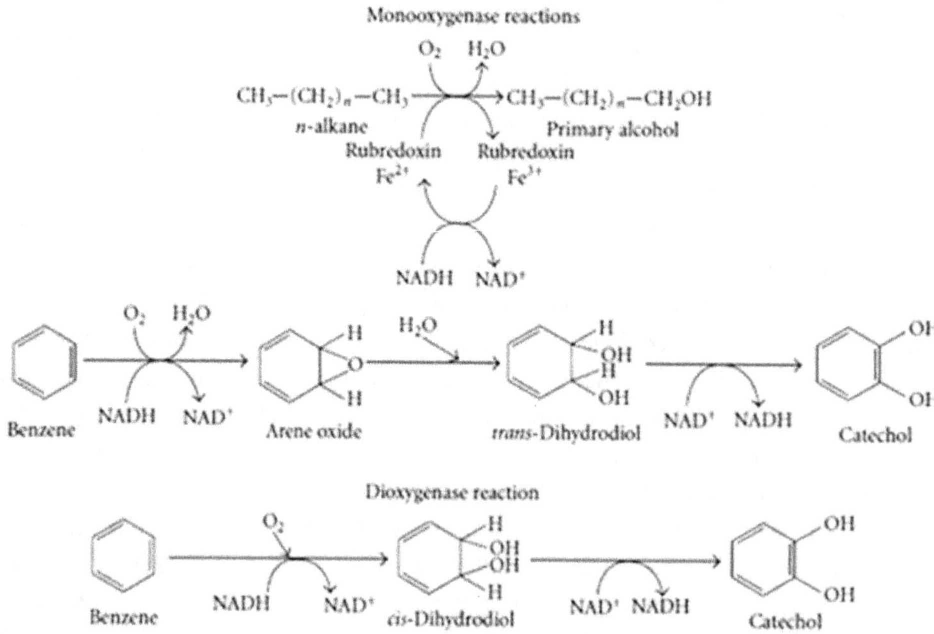

In order to cleaning up the petroleum contaminated soils, Fan and Krishnamurthy (1995) examined the biological treatment technology for the rate of degradation of the petroleum hydrocarbons by microbial enzymes. Johannes et al. (1998) performed the oxidation of acenaphthene and acenaphthylene by using laccase enzyme obtained from *Trametes versicolor* and mediator 1-hydroxybenzotriazole (HBT) and reported the oxidation of acenaphthylene and acenaphthene by 35% and 3% respectively. Zumarraga et al. (2007) also studies the oxidation of anthracene by mutant laccases. Zeng et al (2011) performed the oxidation of benzo[a]pyrene by bacterial laccase and findings proved the applicability of prokaryotic laccases in enhancing the mineralization of benzo[a]pyrene by PAH-degrading bacteria. Thongkred et al. (2011) was applied the isolate of *Pycnoporus coccineus* and its laccase for the oxidation of polycyclic aromatic hydrocarbons. They used five polycyclic aromatic hydrocarbons like anthracene, pyrene, fluoranthene, benzo[a]pyrene and phenanthrene and find good results in the form of their oxidation percentage (Thongkred et al., 2011). Liu and Hua (2014) studied the oxidation of two aromatic hydrocarbons namely acenaphthylene and anthracene by maleic anhydride modified laccase-2,2′-azino-bis-(3-ethylbenzothiazoline-6-sulfonic acid) (ABTS) system and found that oxidation of such polyaromatic hydrocarbons was more than 70%.

Moghimi et al. (2017) isolated the twenty fungal strains and selected the isolate G-05 with 70% degradation ability of petroleum hydrocarbons as a premium isolate after 15 days. G-05 has ability to degrade 90 and 100% aliphatic as well as some polycyclic aromatic hydrocarbons respectively. Such isolate has capability to produce laccase with high activity for oil removal. They identified strain as the member of the *Trematophoma* genus and found that this has strong potency in bio-treatment of crude oil contaminated soil (Moghimi et al., 2017).

Ameen et al. (2016) collected mangrove sediments. They purified the 45 isolated that were belonging to twelve genera. They also found that five isolates along with their consortium that were capable to grow in association with petroleum oil as sole carbon source under *in vitro* conditions. *Alternaria alternata*, *Aspergillus terreus*, *Cladosporium sphaerospermum*, *Eupenicillium hirayamae* and *Paecilomyces variotii* were the fungal strains that have the most activity to degrade diesel oil without showing antagonistic activity. They compared them with control and found that these fungi accumulated significantly as higher biomass. These strains produced extracellular enzymes and released greater quantity of carbon dioxide. They also concluded the potential role of laccase, manganese peroxidase, lignin peroxidase (LiP) and catalase in bioremoval of diesal fuel and alkane by mangrove fungi (Ameen et al., 2016).

Microorganisms play an important role in degradation of petroleum based polymers. Petroleum based polymer have various valuable applications in every area of human life. But due to their huge applications and their persistence in environment for long time, they create many global problems. Due to which their degradation study is become very valuable. Different microbial enzymes like laccase, manganese peroxidise and lignin peroxidase have been reported to be involved in the miodegrdation of polyethylene, the most common type plastic having C-C backbone (Restrepo-Flórez *et al.*, 2014; Krueger *et al.*, 2015; Wei & Zimmermann, 2017; Abimanyu et al., 2016; Pathak & Navneet, 2017).

Rhodococcus rubber C208 secretes a thermally stable laccase enzyme that has capability to degrade UV-irradiated polyethylene films in culture supernatants as well as in cell-free extracts. This takes place in the presence of copper (Santo et al., 2013). Due to oxidation and degradation in amorphous part of polyethylene films, there is carbonyl group enhancement and molecular weight lowering after the two weeks incubation with enzyme at 37oC observed. A *Trametes versicolor* based laccase also induced the reduction in molecular weight of polyethylene membrane in the presence of 1-hydroxybenzotriazole (Fujisawa et al., 2001). Panganese peroxidase obtained from *Phanerochaete chrysosporium* ME-446, a white rot fungi has strong capability to degrade high molecular weight polyethylene membrane (Liyoshi et al., 1998). Degradation of hydrocarbon oligomers can be catalysed by using alkane hydroxylase. It occurs by terminal or subterminal oxidations (Rojo, 2010). Figure 3 shows the different types of petroleum based polymers while Figure 4 shows the biodegradation of hydrocarbon based plastic.

Role of different enzymes in cleavage of polyurethane is shown in Figure 5 (Phua et al., 1987; Loredo-Trevino et al., 2012). It is clear from this figure that esterase produce carboxylic acid derivative and alcohol from the cleavage of polyurethane linkage using water molecule. In second reaction, protease

Figure 3. Different petroleum based plastics (redrawn from Wei & Zimmerman, 2008; Plastics Europe, 2016)

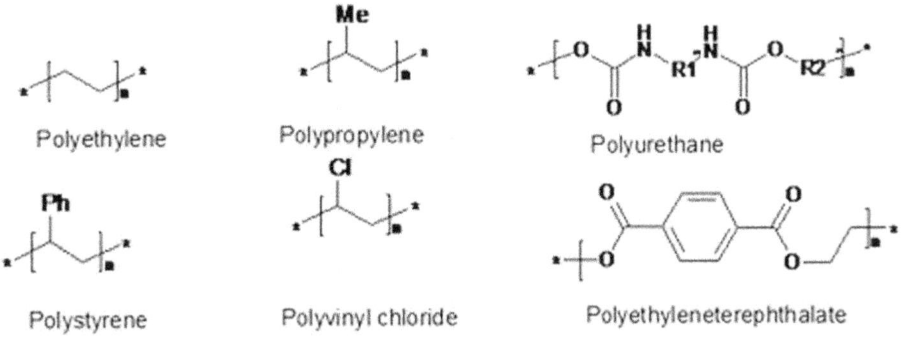

Figure 4. Biodegradation of plastic (Lucas et al., 2008; taken from Wei & Zimmermann, 2017)

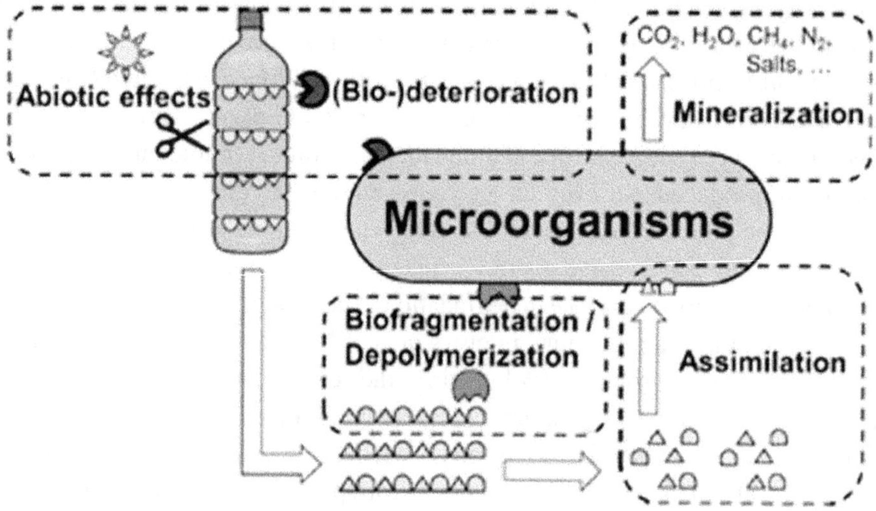

causes the addition of –OH group by removing the amine group (-NH-R_2) in the form of amine molecule (R_2NH_2). This reaction step also use water molecule. In the third reaction, urease cleaves the linkage using water molecule and produces two molecules of primary amine (R_1NH_2 and R_2NH_2) and carbon dioxide.

There are some studies on oxidation of aromatic methyl groups from substituted toluene using different laccase enzymes obtained from different sources that have been recently performed by Sharma et al (2016), Chaurasia et al (2015a, 2015b, 2014a, 2014b, 2014c, 2014d, 2014e, 2013a, 2013b). The enzymes used for oxidation of such compounds were very active and thermally stable due to which they can be potentially utilized for the oxidation or remediation of various petrochemicals and petroleum based hydrocarbons or substituted hydrocarbons (Sharma et al., 2016; Chaurasia et al., 2015a, 2015b, 2014a, 2014b, 2014c, 2014d, 2014e, 2013a, 2013b).

Figure 5. Role of Esterase, protease and urease in linkage cleavage in polyurethane (Taken from Wei & Zimmermann, 2017)

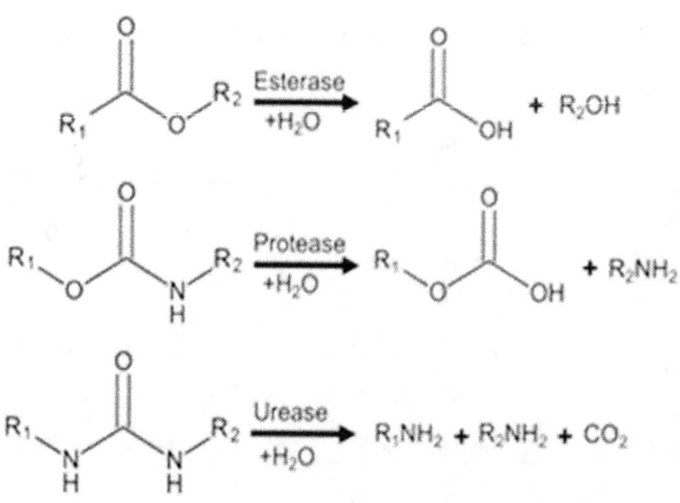

CONCLUSION

Above discussions show that in the modern time when there are more and more different types of pollutions being generated from human activities and technology developments, involvements of biological methods for their solution will be highly enthusiastic. In this context, study of the role of enzymes for removal of petroleum hydrocarbons may be promising due to their potential applicability in bioremediation of such hydrocarbons.

ACKNOWLEDGMENT

Dr PK Chaurasia is thankful to CSIR, New Delhi for the award of Research Associate-ship (RA). File No. 09/1032(0011)2K17. Department of Biotechnology, MNNIT, Allahabad, Uttar Pradesh and Department of Chemistry, NERIST, Arunachal Pradesh are also thankfully acknowledged.

REFERENCES

Abimanyu, N., & Andriana, N. (2016). Lisnawati, Rhokmatulloh HN,Shah DYF. Biochemestrical and environmental applications: Utilization of laccase enzyme as ioremediation in petroleum industry by reducing PAH (poly aromatic hydrocarbons). *International Journal of Advances in Science Engineering and Technology.*, *4*(2), 124.

Ameen, F., Moslem, M., Hadi, S., & Al-Sabri, A. E. (2016). Biodegradation of diesel fuel hydrocarbons by mangrove fungi from Red Sea Coast of Saudi Arabia. *Saudi Journal of Biological Sciences*, *23*(2), 211–218. doi:10.1016/j.sjbs.2015.04.005 PMID:26981002

Bahl, A., & Bahl, B. S. (2003). A text book of organic chemistry. S. Chand & Company Ltd.

Balaji, V., Arulazhagan, P., & Ebenezer, P. (2014). Enzymatic bioremediation of polyaromatic hydrocarbons by fungal consortia enriched from petroleum contaminated soil and oil seeds. *Journal of Environmental Biology*, *35*(3), 521–529. PMID:24813008

Balaji, V., & Ebenzer, P. (2008). Optimization of extracellular lipase production in Colletotrichum gloeosporioides by solid state fermentation. *Indian Journal of Science and Technology*, *1*(7), 1–8.

Boyd, D. R., Sharma, N. D., & Allen, C. C. R. (2001). Aromatic dioxygenases: Molecular biocatalysis and applications. *Current Opinion in Biotechnology*, *12*(6), 564–573. doi:10.1016/S0958-1669(01)00264-6 PMID:11849939

Chaurasia, P. K., Bharati, S. L., & Singh, S. K. (2014e). Role of Laccase from *Coriolus versicolor* MTCC-138 in Selective Oxidation of Aromatic Methyl Group. *Russian Journal of Bioorganic Chemistry*, *40*(3), 288–292. doi:10.1134/S1068162014020034 PMID:25898738

Chaurasia, P. K., Bharati, S. L., Singh, S. K., & Yadava, S. (2015b). Synthetic Applications of Purified Laccase from *Pleurotus sajor caju* MTCC-141. *Russian Journal of General Chemistry*, *85*(1), 173–175. doi:10.1134/S1070363215010302

Chaurasia, P. K., Bharati, S. L., Yadava, S., & Yadav, R. S. S. (2015a). Purification, characterization and synthetic application of a thermally stable laccase from *Hexagonia tenuis* MTCC-1119. *Indian Journal of Biochemistry & Biophysics*, *52*(1), 60–67. PMID:26040112

Chaurasia, P. K., Yadav, A., Yadav, R. S. S., & Yadava, S. (2013a). Purification and Characterization of Laccase Secreted by *Phellinus linteus* MTCC-1175 and its Role in the selective Oxidation of aromatic Methyl group. *Applied Biochemistry and Microbiology*, *49*(6), 592–599. doi:10.1134/S000368381306006 PMID:25434183

Chaurasia, P. K., Yadav, R. S. S., & Yadava, S. (2013b). Purification and Characterization of Laccase from *Coriolopsis floccosa* MTCC-1177 and its use in the selective oxidation of aromatic methyl group to aldehyde without mediators. *Journal of Chemical Sciences*, *125*(6), 1395–1403. doi:10.100712039-013-0525-4

Chaurasia, P. K., Yadav, R. S. S., & Yadava, S. (2014a). Purification and characterization of yellow laccase from *Trametes hirsuta* MTCC-1171 and its application in synthesis of aromatic aldehydes. *Process Biochemistry*, *49*(10), 1647–1655. doi:10.1016/j.procbio.2014.06.016

Chaurasia, P. K., Yadava, S., Bharati, S. L., & Singh, S. K. (2014b). Syntheses of aromatic aldehydes by laccase of *Pleurotus ostreatus* MTCC-1801. *Synthetic Communications*, *44*(17), 2535–2544. doi:10.1080/00397911.2014.904879

Chaurasia, P. K., Yadava, S., Bharati, S. L., & Singh, S. K. (2014c). Selective Oxidation and N-coupling of Amines by Purified Laccase of *Xylaria polymorpha* MTCC-1100. *Russian Journal of Bioorganic Chemistry*, *40*(4), 455–460. doi:10.1134/S1068162014040025

Chaurasia, P. K., Yadava, S., Bharati, S. L., & Singh, S. K. (2014d). Syntheses of aromatic aldehydes by laccase without the help of mediators. *Green Chemistry Letters and Reviews*, *7*(1), 100–104. doi:10.1080/17518253.2014.895869

Collins, P. J., Kotterman, M., Field, J. A., & Dobson, A. (1996). Oxidation of Anthracene and Benzo[a]pyrene by Laccases from *Trametes versicolor. Applied and Environmental Microbiology*, *62*(12), 4563–4567. PMID:16535468

Das N, Chandran P (2011) Microbial Degradation of Petroleum Hydrocarbon Contaminants: An Overview. *Biotechnology Research International*. doi:10.4061/2011/941810

Fan, C. Y., & Krishnamurthy, S. (1995). Enzymes for Enhancing Bioremediation of Petroleum-Contaminated Soils: A Brief Review. *Journal of the Air & Waste Management Association*, *45*(6), 453–460. doi:10.1080/10473289.1995.10467375 PMID:7788508

Feng, X., Pisula, W., & Mullen, K. (2009). Large polycyclic aromatic hydrocarbons: Synthesis and discotic organisation. *Pure and Applied Chemistry*, *81*(12), 2203–2224. doi:10.1351/PAC-CON-09-07-07

Fritsche, W., & Hofrichter, M. (2000). In J. Klein (Ed.), *Aerobic degradation by microorganisms in Environmental Processes- Soil Decontamination*. Weinheim, Germany: Wiley-VCH.

Fujisawa, M., Hiral, H., & Nishida, T. (2001). Degrdation of polyethylene and nylon-66 by the laccase-mediator system. *Journal of Polymers and the Environment*, *9*(3), 103–108. doi:10.1023/A:1020472426516

Gibson, D. T., & Parales, R. E. (2000). Aromatic hydrocarbon dioxygenases in environmental biotechnology. *Current Opinion in Biotechnology*, *11*(3), 236–243. doi:10.1016/S0958-1669(00)00090-2 PMID:10851146

Harvey, R. G. (1996). *Polycyclic aromatic hydrocarbons*. New York: Wiley-VCH.

Iida, T., Sumita, T., Ohta, A., & Takagi, M. (2000). The cytochrome P450ALKmultigene family of an n-alkane-assimilating yeast, *Yarrowia lipolytica*: Cloning and characterization of genes coding for new CYP52 family members. *Yeast (Chichester, England)*, *16*(12), 1077–1087. doi:10.1002/1097-0061(20000915)16:12<1077::AID-YEA6013.0.CO;2-K PMID:10953079

Iiyoshi, Y., Tsutsumi, Y., & Nishida, T. (1998). Polyethylene degradation by lignin-degrading fungi and manganese peroxidaes. *Journal of Wood Science*, *44*(3), 222–229. doi:10.1007/BF00521967

Ijah, U. J. J., & Antai, S. P. (2003). The potential use of chichen-drop micro-organisms for oil spill remediation. *The Environmentalist*, *23*(1), 89–95. doi:10.1023/A:1022947727324

Javed, A. (2012). *The Hydrogenation of Pyrolosis Gasoline (PyGas) over Nickel and Palladium catalysts* (PhD Thesis). University of Glasgow. Retrieved from http://theses.gla.ac.uk/3542/]

Johannes, C., Majcherczyk, A., & Huttermann, A. (1998). Oxidation of acenaphthene and acenaphthylene by laccase of Trametes versicolor in a laccase-mediator system. *Journal of Biotechnology*, *61*(2), 151–156. doi:10.1016/S0168-1656(98)00030-3

Karigar, C.S., & Rao, S.S. (2011). Role of Microbial Enzymes in the Bioremediation of Pollutants: A Review. *Enzyme Research*. doi:10.4061/2011/805187

Kolb, H. C., Van Nieuwenhze, M. S., & Sharpless, K. B. (1994). Catalytic asymmetric dihydroxylation. *Chemical Reviews*, *94*(8), 2483–2547. doi:10.1021/cr00032a009

Krueger, M. C., Harms, H., & Schlosser, D. (2015). Prospects for microbiological solutions to environmental pollution with plastics. *Applied Microbiology and Biotechnology*, *99*(21), 8857–8874. doi:10.100700253-015-6879-4 PMID:26318446

Laughlin, R. B. Jr, & Neff, J. M. (1979). Interactive effects of salinity, temperature and polycyclic aromatic hydrocarbons on the survival and development rate of larvae of the mud crab *Rhithropanopeus harrisii*. *Marine Biology*, *53*(3), 281–291. doi:10.1007/BF00952437

Liu, Y., & Hua, X. (2014). Degradation of acenaphthylene and anthracene by chemically modified laccase from *Trametes versicolor*. *RSC Advances*, *4*(59), 31120–31122. doi:10.1039/C4RA02807D

Loredo-Trevino, A., Gutierrez-Sanchez, G., Rodriguea-Herrera, R., & Aguilar, C. N. (2012). Microbial enzymes involved in polyurethane biodegradation: A review. *Journal of Polymers and the Environment*, *20*(1), 258–265. doi:10.100710924-011-0390-5

Lucas, N., Bienaime, C., Belloy, C., Quenedec, M., Silvestre, F., & Nava-Saucedo, J.-E. (2008). Polymer biodegradation: Mechanism and estimation techniques-a review. *Chemosphere*, *73*(4), 429–442. doi:10.1016/j.chemosphere.2008.06.064 PMID:18723204

Maeng, J. H. O., Sakai, Y., Tani, Y., & Kato, N. (1996). Isolation and characterization of a novel oxygenase that catalyzes the first step of n-alkane oxidation in *Acinetobacter* sp. strain M-1. *Journal of Bacteriology, 178*(13), 3695–3700. doi:10.1128/jb.178.13.3695-3700.1996 PMID:8682768

Mai, A., Bryan, J., Goodarzi, N., & Kantas, A. (2006). *Insights Into Non-Thermal Recovery of Heavy Oil. World Heavy Oil Conference (WHOC)*, Calgary, Alberta.

McDonald, I. R., Miguez, C. B., Rogge, G., Bourque, D., Wendlandt, K. D., Groleau, D., & Murrell, J. C. (2006). Diversity of solublemethane monooxygenase-containing methanotrophs isolated from polluted environments. *FEMS Microbiology Letters, 255*(2), 225–232. doi:10.1111/j.1574-6968.2005.00090.x PMID:16448499

Messias, J. M., da Costa, B. Z., de Lima, V. M. G., Dekker, R. F. G., Rezende, M. I., Krieger, N., & Barbosa, A. M. (2009). Screening Botryosphaeria species for lipase: Production of lipase by Botryosphaeria ribis EC-01 grown on soyabean oil and other carbon sources. *Enzyme and Microbial Technology, 45*(6-7), 426–431. doi:10.1016/j.enzmictec.2009.08.013

Moghimi, H., Tabar, R. H., & Hamedi, J. (2017). Assessing the biodegradation of polycyclic aromatic hydrocarbons and laccase production by new fungus *Trematophoma* sp. UTMC 5003. *World Journal of Microbiology & Biotechnology, 33*(7), 136. doi:10.100711274-017-2304-8 PMID:28585171

Pampanin, D. M., & Sydnes, M. O. (2013). Polycyclic Aromatic Hydrocarbons a Constituent of Petroleum: Presence and Influence in the Aquatic Environment. Academic Press. Doi:10.5772/48176

Pathak, V. M., & Navneet. (2017). Navneet. Review on the current status of polymer degradation: A microbial approach. *Bioresources and Bioprocessing, 4*(1), 15. doi:10.118640643-017-0145-9

Phua, S. K., Castillo, E., Anderson, J. M., & Hiltner, A. (1987). Biodegradation of polyurethane in vitro. *Journal of Biomedical Materials Research, 21*(2), 231–246. doi:10.1002/jbm.820210207 PMID:3818683

Pink, D.H. (2006). *Investing in Tomorrow's liquid gold.* Yahoo.

PlasticsEuorope. (2016). *Plastics-The facts 2016.* Brussels, Belgium: PlasticsEurope. Retrieved from http://www.plasticseurope.org/Document/plastics-the-facts-2016-15787.aspx?Page=DOCUMENT&FolID=2

Resnick, S. M., & Gibson, D. T. (1996). Regio- and stereospecific oxidation of fluorene, dibenzofuran, and dibenzothiophene by naphthalene dioxygenase from *Pseudomonas* sp. strain NCIB 9816–4. *Applied and Environmental Microbiology, 62*(11), 4073–4080. PMID:8899998

Resnick, S. M., Lee, K., & Gibson, D. T. (1996). Diverse reactions catalyzed by naphthalene dioxygenase from *Pseudomonas* sp strain NCIB 9816. *Journal of Industrial Microbiology & Biotechnology, 17*(5), 438–457. doi:10.1007/BF01574775

Rojo, F. (2010). *Enzymes for aerobic degradation of alkanes.* In K. N. Timmis (Ed.), *Handbook of Hydrocarbon and Lipid Microbiology* (pp. 781–797). Berlin: Springer-Verloa. doi:10.1007/978-3-540-77587-4_59

Rstrepo-Florez, J.-M., Bassi, A., & Thompson, M. R. (2014). Microbial degradation and deterioration of polyethylene – a review. *International Biodeterioration & Biodegradation*, *88*, 83–90. doi:10.1016/j.ibiod.2013.12.014

Sack, U., Hofrichter, M., & Fritsche, W. (1997). Degradation of phenanthrene and pyrene by *Nematoloma frowardii*. *Journal of Basic Microbiology*, *37*(4), 287–293. doi:10.1002/jobm.3620370408 PMID:9379367

Salleh, A. B., Ghazali, F. M., Rahman, R. N. Z. A., & Basri, M. (2003). Bioremediation of Petroleum hydrocarbon pollution. *Indian Journal of Biotechnology*, *2*, 411–425.

Scheuer, U., Zimmer, T., Becher, D., Schauer, F., & Schunck, W.-H. (1998). Oxygenation cascade in conversion of n-alkanes to α,ω-dioic acids catalyzed by cytochrome P450 52A3. *The Journal of Biological Chemistry*, *273*(49), 32528–32534. doi:10.1074/jbc.273.49.32528 PMID:9829987

Schmid, A., Sonnleitner, B., & Witholt, B. (1998). Medium chain length (mcl) alkane mass transfer rates in two-liquid phase *Pseudomonas oleovorans* cultivations. *Biotechnology and Bioengineering*, *60*(1), 10–23. doi:10.1002/(SICI)1097-0290(19981005)60:1<10::AID-BIT23.0.CO;2-3 PMID:10099401

Sharma, M., Chaurasia, P. K., Yadav, A., Yadav, R. S. S., Yadava, S., & Yadav, K. D. S. (2016). Purification and characterization of a thermally stable yellow laccase from *Daedalea flavida* MTCC-145 with higher catalytic performance towards selective synthesis of substituted benzaldehydes. *Russian Journal of Bioorganic Chemistry*, *42*(1), 59–68. doi:10.1134/S1068162016010143

Thapa, B., Ajay Kumar, K. C., & Ghimire, A. (2012). A review on bioremediation of petroleum hydrocarbon contaminants in soil. Kathmandu University Journal of Science. *Engineering and Technology.*, *8*(1), 164–170.

Thongkred, P., Lotrakul, P., Prasongsuk, S., Imai, T., & Punnapayak, H. (2011). Oxidation of polycyclic aromatic hydrocarbons by a tropical isolate of Pycnoporus coccineus and its laccase. *ScienceAsia*, *37*(3), 225–233. doi:10.2306cienceasia1513-1874.2011.37.225

Van Beilen, J. B., & Funhoff, E. G. (2007). Alkane hydroxylases involved in microbial alkane degradation. *Applied Microbiology and Biotechnology*, *74*(1), 13–21. doi:10.100700253-006-0748-0 PMID:17216462

Van Beilen, J. B., Funhoff, E. G., Van Loon, A., Just, A., Kaysser, L., Bouza, M., ... Witholt, B. (2006). Cytochrome P450 alkane hydroxylases of the CYP153 family are common in alkane-degrading eubacteria lacking integral membrane alkane hydroxylases. *Applied and Environmental Microbiology*, *72*(1), 59–65. doi:10.1128/AEM.72.1.59-65.2006 PMID:16391025

Van Beilen, J. B., Kingma, J., & Witholt, B. (1994). Substrate specificity of the alkane hydroxylase system of *Pseudomonas oleovorans* GPo1. *Enzyme and Microbial Technology*, *16*(10), 904–911. doi:10.1016/0141-0229(94)90066-3

Van Hamme, J. D., Singh, A., & Ward, O. P. (2003). Recent Advances in Petroleum Microbiology. *Microbiology and Molecular Biology Reviews*, *67*(4), 503–549. doi:10.1128/MMBR.67.4.503-549.2003 PMID:14665675

Venkatesagowda, B., Ponugupaty, E., Barbisa, A. M., & Dekker, R. F. (2012). Diversity of plant oil seed-associated fungi isolated from seven oil bearing seeds and their potential for the production of lipolytic enzymes. *World Journal of Microbiology & Biotechnology*, *28*(1), 71–80. doi:10.100711274-011-0793-4 PMID:22806781

Wei, R., & Zimmermann, W. (2017). *Microbial enzymes for the recycling of recalcitrant petroleum-based plastics: how far are we?* Academic Press. DOI: 10.1111/1751-7915.12710

West, L. (2006). *World water day: A billion people worldwide lack safe drinking water.* About.com.

Wikipedia. (2016). *Pyrolosis Gasoline*. Retrieved from https://en.wikipedia.org/wiki/Pyrolysis_gasoline#cite_note-1

Wikipedia. (2017a). *Crude*. Retrieved from https://en.wikipedia.org/wiki/Crude

Wikipedia. (2017b). *Brent Crude*. Retrieved from https://en.wikipedia.org/wiki/Brent_Crude

Wikipedia. (2017c). *Light Crude Oil*. Retrieved from https://en.wikipedia.org/wiki/Light_crude_oil

Wikipedia. (2018a). *Water Pollution*. Retrieved from https://en.wikipedia.org/wiki/Water_pollution

Wikipedia. (2018b). *Petroleum*. Retrieved from https://en.wikipedia.org/wiki/Petroleum

Wikipedia. (2018c). *Petrochemical*. Retrieved from https://en.wikipedia.org/wiki/Petrochemical

Wikipedia. (2018d). *Pollution*. Retrieved from https://en.wikipedia.org/wiki/Pollution#Pollutants

Wikipedia. (2018e). *Gasoline*. Retrieved from https://en.wikipedia.org/wiki/Gasoline

Wikipedia. (2018f). *Oil Refinery*. Retrieved from https://en.wikipedia.org/wiki/Oil_refinery#Major_products

Wikipedia. (2018g). *Heavy Crude Oil*. Retrieved from https://en.wikipedia.org/wiki/Heavy_crude_oil

Wikipedia. (2018h). *Sweet Crude Oil*. Retrieved from https://en.wikipedia.org/wiki/Sweet_crude_oil

Zeng, J., Lin, X., Zhang, J., Li, X., & Wong, M. H. (2011). Oxidation of polycyclic aromatic hydrocarbons by the bacterial laccase CueO from E. coli. *Applied Microbiology and Biotechnology*, *89*(6), 1841–1849. doi:10.100700253-010-3009-1 PMID:21120471

Zumarraga, M., Plou, F.J., Garcia-Arellano, H., Ballesteros, A., & Alcalde, M. (2007). *Bioremediation of polycyclic aromatic hydrocarbons by fungal laccases engineered by directed evolution*. Academic Press.

Compilation of References

Abadulla, E., Tzanov, T., Costa, S., Robra, K. H., Cavaco-Paulo, A., & Gubitz, G. M. (2000). Decolorization and detoxification of textile dyes with a laccase from *Trametes hirsute. Applied and Environmental Microbiology, 66*(8), 3357–3362. doi:10.1128/AEM.66.8.3357-3362.2000 PMID:10919791

Abd El Haleem, D., Von Wintzingerode, F., Moter, A., Moawad, H., & Göbel, U. B. (2000). Phylogenetic analysis of rhizosphere-associated β-subclass proteobacterial ammonia oxidizers in a municipal wastewater treatment plant based on rhizoremediation technology. *Letters in Applied Microbiology, 31*(1), 34–38. doi:10.1046/j.1472-765x.2000.00760.x PMID:10886611

Abdel-Ghani, N. T., Hefny, M., & El-Chaghaby, G. A. (2007). Removal of lead from aqueous solution using low cost abundantly available adsorbents. *International Journal of Environmental Science and Technology, 4*(1), 67–73. doi:10.1007/BF03325963

Abhilash, P. C., & Singh, N. (2009). Pesticide use and application: An Indian scenario. *Journal of Hazardous Materials, 165*(1-3). doi:10.1016/j.jhazmat.2008.10.061 PMID:19081675

Abimanyu, N., & Andriana, N. (2016). Lisnawati, Rhokmatulloh HN,Shah DYF. Biochemestrical and environmental applications: Utilization of laccase enzyme as ioremediation in petroleum industry by reducing PAH (poly aromatic hydrocarbons). *International Journal of Advances in Science Engineering and Technology., 4*(2), 124.

Abraham, J., Ghosh, E., Mukherjee, P., & Gajendiran, A. (2017). Microbial degradation of low density polyethylene. *Environmental Progress & Sustainable Energy, 36*(1), 147–154. doi:10.1177/0734242X05051045 PMID:15864950

Acosta, V., Paul, J, & Hanuschke Hanuschke, K. (2012). *Ecocenter integrated solid waste management facility with sanitary landfill and resource recovery technologies.* Academic Press.

Adedayo, M. R., Ajiboye, E. A., Akintunde, J. K., & Odaibo, A. (2011). Single cell proteins: As nutritional enhancer. *Advances in Applied Science Research, 2*(5), 396–409.

Adedayo, O., Javadpour, S., Taylor, C., Anderson, W. A., & Moo-Young, M. (2004). Decolourization and detoxification of Methyl Red by aerobic bacteria from a wastewater treatment plant. *World Journal of Microbiology & Biotechnology, 20*(6), 545–550. doi:10.1023/B:WIBI.0000043150.37318.5f

Adler, T. (1996). Botanical cleanup crews. *Science News, 150*(3), 42–43. doi:10.2307/3980349

Adrian, N. R., & Suflita, J. M. (1990). Reductive dehalogenation of a nitrogen heterocyclic herbicide in anoxic aquifer slurries. *Applied and Environmental Microbiology, 56*(1), 292–294. PMID:16348103

Aftab, M. N., & Hameed, A., & Ikram-ul-Haq, C. H. E. N. (2006). Biodegradation of leather waste by enzymatic treatment. *The Chinese Journal of Process Engineering, 6*(3), 462–465.

Agarwal, T. (2009). Concentration level, pattern and toxic potential of PAHs in traffic soil of Delhi, India. *Journal of Hazardous Materials*, *171*(1-3), 894–900. doi:10.1016/j.jhazmat.2009.06.081 PMID:19615818

Agbor, V. B., Cicek, N., Sparling, R., Berlin, A., & Levin, D. B. (2011). Biomass pretreatment: Fundamentals toward application. *Biotechnology Advances*, *29*(6), 675–685. doi:10.1016/j.biotechadv.2011.05.005 PMID:21624451

Ahalya, N., Kanamadi, R. D., & Ramachandra, T. V. (2005). Biosorption of chromium (VI) from aqueous solutions by the husk of Bengal gram (Cicer arientinum). *Electronic Journal of Biotechnology*, *8*(3).

Ahamed, M. I. N., & Kashif, P. M. (2014). Safety disposal of tannery effluent sludge: Challenges to researchers—a review. *International Journal of Pharma Sciences and Research*, *5*, 733–736.

Ahemad, M. (2012). Implications of bacterial resistance against heavy metals in bioremediation: A review. *Journal of Institute of Integrative Omics and Applied Biotechnology*, *3*(3).

Ahemad, M., & Khan, M. S. (2012). A Comparative analysis of the growth parameters of different legume crops grown in quizalafop-p-ethyl applied soils. *Insight Botany*, *2*(1), 1–6. doi:10.5567/BOTANY-IK.2012.1.6

Ahemad, M., & Malik, A. (2011). Bioaccumulation of heavy metals by zinc resistant bacteria isolated from agricultural soils irrigated with wastewater. *Journal of Bacteriology*, *2*(1), 12–21. doi:10.3923/bj.2012.12.21

Ahemad, M., & Malik, A. (2011). *Bioremediation Potentials of Heavy Metal Resistant Bacteria: A Practical Approach.* VDM Publishing.

Ahmad, J., & Ansari, T. A. (2013). Alkaline protease production using proteinaceous tannery solid waste. *Journal of Petroleum & Environmental Biotechnology*, *4*(1). doi:10.4172/2157-7463.1000136

Ahmad, Q., & Qazi, J. I. (2014). Thermophilic fermentations of lignocellulosic substrates and economics of biofuels: Prospects in Pakistan. *International Journal of Energy and Environmental Engineering*, *5*(2-3), 2–3. doi:10.100740095-014-0094-4

Ahmaruzzaman, M. (2010). A review on the utilization of fly ash. *Progress in Energy and Combustion Science*, *36*(3), 327–363. doi:10.1016/j.pecs.2009.11.003

Aiken, B. S., & Logan, B. E. (1989). Degradation of pentachlorophenol by the white rot fungus Phanerochaetechrysosporium grown in ammonium lignosulphonate media. *Biodegradation*, *7*(3), 175–182. doi:10.1007/BF00058177 PMID:8782389

Akansha, K., Prasad, A., Sukumaran, R. K., Nampoothiri, K. M., & Pandey, A. (2014). Dilute acid pretreatment and enzymatic hydrolysis of sorghum biomass for sugar recovery: A statistical approach. *Indian Journal of Experimental Biology*, *52*, 1082–1089. PMID:25434103

Akar, T., Tunali, S., & Çabuk, A. (2007). Study on the characterization of lead (II) biosorption by fungus Aspergillus parasiticus. *Applied Biochemistry and Biotechnology*, *136*(3), 389–405. doi:10.100712010-007-9032-8 PMID:17625240

Akpomie, O. O., & Ejechi, B. O. (2016). Bioremediation of Soil Contaminated with Tannery Effluent by Combined Treatment with Cow Dung and Microorganisms Isolated from Tannery Effluent. *Journal of Bioremediation & Biodegradation*, *7*, 354.

Aksu, Z., Sag, Y., & Kutsal, T. (1992). The biosorpnon of copperod by C. vulgaris and Z. ramigera. *Environmental Technology*, *13*(6), 579–586. doi:10.1080/09593339209385186

Alalewi, A., & Jiang, C. (2012). Bacterial influence on textile wastewater decolorization. *Journal of Environmental Protection*, *3*(08), 889–901. doi:10.4236/jep.2012.328104

Albertsson, A. C., Andersson, S. O., & Karlsson, S. (1987). The mechanism of biodegradation of polyethylene. *Polymer Degradation & Stability*, *18*(1), 73–87. doi:10.1016/0141-3910(87)90084-X

Albertsson, A.-C. (1980). The shape of the biodegradation curve for low and high density polyethenes in prolonged series of experiments. *European Polymer Journal*, *16*(7), 623–630. doi:10.1016/0014-3057(80)90100-7

Albertsson, A.-C., Barenstedt, C., Karlsson, S., & Lindberg, T. (1995). Degradation product pattern and morphology changes as means to differentiate abiotically and biotically aged degradable polyethylene. *Polymer*, *36*(16), 3075–3083. doi:10.1016/0032-3861(95)97868-G

Albertsson, A.-C., Erlandsson, B., Hakkarainen, M., & Karlsson, S. (1998). Molecular weight changes and polymeric matrix changes correlated with the formation of degradation products in biodegraded polyethylene. *Journal of Environmental Polymer Degradation*, *6*(4), 187–195. doi:10.1023/A:1021873631162

Albertsson, A.-C., & Karlsson, S. (1990). The influence of biotic and abiotic environments on the degradation of polyethylene. *Progress in Polymer Science*, *15*(2), 177–192. doi:10.1016/0079-6700(90)90027-X

Alcalde, M., Ferrer, M., Plou, F. J., & Ballesteros, A. (2006). Environmental biocatalysis: From remediation with enzymes to novel green processes. *Trends in Biotechnology*, *24*(6), 281–287. doi:10.1016/j.tibtech.2006.04.002 PMID:16647150

Alexander, K. T. W., Donohue, V. J., Cory, N. J., & Sykes, R. L. (1992). Environmental and safety issues-clean technology and environmental auditing. *Journal of the Society of Leather Technologists and Chemists*, *76*, 17–23.

Alexandratos, N., & Bruinsma, J. (2012). World Agriculture Towards 2030/2050 (The 2012 Revision) (ESA Working paper No. 12-03). FAO.

Alford, É. R., Pilon-Smits, E. A., Fakra, S. C., & Paschke, M. W. (2012). Selenium hyperaccumulation by *Astragalus* (Fabaceae) does not inhibit root nodule symbiosis. *American Journal of Botany*, *99*(12), 1930–1941. doi:10.3732/ajb.1200124 PMID:23204487

Alford, É. R., Pilon-Smits, E. A., & Paschke, M. W. (2010). Metallophytes—a view from the rhizosphere. *Plant and Soil*, *337*(1-2), 33–50. doi:10.100711104-010-0482-3

Ali, S. M., Pervaiz, A., Afzal, B., Hamid, N., & Yasmin, A. (2014). Open dumping of municipal solid waste and its hazardous impacts on soil and vegetation diversity at waste dumping sites of Islamabad city. *Journal of King Saud University - Science*, *26*(1), 59-65. doi:10.1016/j.jksus.2013.08.003

Ali, H., Khan, E., & Sajad, M. A. (2013). Phytoremediation of heavy metals—concepts and applications. *Chemosphere*, *91*(7), 869–881. doi:10.1016/j.chemosphere.2013.01.075 PMID:23466085

Ali, M. I., Perveen, Q., Ahmad, B., Javed, I., Razi-Ul-Hussnain, R., Andleeb, S., ... Hameed, A. (2009). Studies on biodegradation of cellulose blended polyvinyl chloride films. *International Journal of Agriculture and Biology*, *11*, 577–580.

Aljibouri, A. K., Turcotte, G., Wu, J., & Cheng, C. (2015). Ozone pretreatment of humid wheat straw for biofuel production. *Energy Science & Engineering*, *3*(6), 541–548. doi:10.1002/ese3.93

Alkorta, I., Hernández-Allica, J., Becerril, J. M., Amezaga, I., Albizu, I., & Garbisu, C. (2004). Recent findings on the phytoremediation of soils contaminated with environmentally toxic heavy metals and metalloids such as zinc, cadmium, lead, and arsenic. *Reviews in Environmental Science and Biotechnology*, *3*(1), 71–90. doi:10.1023/B:RESB.0000040059.70899.3d

Alleman, B. C., Logan, B. E., & Gilbertson, R. L. (1995). Degradation of pentachlorophenol by fixed films of white rot fungi in rotating tube bioreactors. *Water Research*, *29*(1), 61–67. doi:10.1016/0043-1354(94)E0106-G

Alloway, B. J. (1996). *Heavy Metal in Soils*. London: Halsted Press, John Wiley & Sons Inc.

Allpress, J. D., & Gowland, P. C. (1999). Biodegradation of chlorinated paraffins and long-chain chloroalkanes by Rhodococcus sp. S45-1. *International Biodeterioration & Biodegradation*, *43*(4), 173–179. doi:10.1016/S0964-8305(99)00050-5

Alluri, H. K., Ronda, S. R., Settalluri, V. S., Bondili, J. S., Suryanarayana, V., & Venkateshwar, P. (2007). Biosorption: An eco-friendly alternative for heavy metal removal. *African Journal of Biotechnology*, *6*(25).

Almodares, A., & Hadi, M. R. (2009). Production of bioethanol from sweet sorghum: A review. *African Journal of Agricultural Research*, *4*(9), 772–780.

Alok, S., Soni, R., Zaidi, M., Shouche, Y., & Goel, R. (2008). Comparative biodegradation of HDPE and LDPE using an indigenously developed microbial consortium. *Journal of Microbiology and Biotechnology*, *18*, 477–482. PMID:18388465

Alonzo, H. G. A., & Corrêa, C. L. (2014). Praguicidas. In S. Oga, M. M. A. Camargo, & J. A. Batistuzzo (Eds.), *Fundamentos de Toxicologia* (4th ed., pp. 323–341). Editora Atheneu, São Paulo.

Alvarenga, P., Mourinha, C., Farto, M., Santos, T., Palma, P., Sengo, J., ... Cunha-Queda, C. (2015). Sewage sludge, compost and other representative organic wastes as agricultural soil amendments: Benefits versus limiting factors. *Waste Management (New York, N.Y.)*, *40*, 44–52. doi:10.1016/j.wasman.2015.01.027 PMID:25708406

Amann, R. I., Ludwig, W., & Schleifer, K. H. (1995). Phylogenetic identification and in situ detection of individual microbial cells without cultivation. *Microbiological Reviews*, *59*(1), 143–169. PMID:7535888

Ameen, F., Moslem, M., Hadi, S., & Al-Sabri, A. E. (2016). Biodegradation of diesel fuel hydrocarbons by mangrove fungi from Red Sea Coast of Saudi Arabia. *Saudi Journal of Biological Sciences*, *23*(2), 211–218. doi:10.1016/j.sjbs.2015.04.005 PMID:26981002

Ameri, M., Ghobadian, B., & Baratian, I. (2008). Technical comparison of a CHP using various blends of gasohol in an IC engine. *Renewable Energy*, *33*(7), 1469–1474. doi:10.1016/j.renene.2007.09.015

Amrane, A., & Prigent, Y. (1998). Lactic acid production rates during the different growth phases of *Lactobacillus helveticus* cultivated on whey supplemented with yeast extract. *Biotechnology Letters*, *20*(4), 379–383. doi:10.1023/A:1005331430943

Añasco, N., Uno, S., Koyama, J., Matsuoka, T., & Kuwahara, N. (2010). Assessment of pesticide residues in freshwater areas affected by rice paddy effluents in Southern Japan. *Environmental Monitoring and Assessment*, *160*(1), 371–383. doi:10.100710661-008-0701-z PMID:19089593

Anderson, A. J., & Dawes, E. A. (1990). Occurrence, metabolism, metabolic role, and industrial uses of bacterial polyhydroxyalkanoates. *Microbiological Reviews*, *54*(4), 450–472. PMID:2087222

Anderson, T. A., Guthrie, E. A., & Walton, B. T. (1993). Bioremediation in the rhizosphere. *Environmental Science & Technology*, *27*(13), 2630–2636. doi:10.1021/es00049a001

Anderson, T. A., Kruger, E. L., & Coats, J. R. (1994). Enhanced degradation of a mixture of three herbicides in the rhizosphere of a herbicide-tolerant plant. *Chemosphere*, *28*(8), 1551–1557. doi:10.1016/0045-6535(94)90248-8

Angelova, B., Avramova, T., Stefanova, L., & Mutafov, S. (2008). Temperature effect on bacterial azo bond reduction kinetics: An Arrhenius plot analysis. *Biodegradation*, *19*(3), 387–393. doi:10.100710532-007-9144-4 PMID:17653820

Annepu, R. K. (2012, January 10). Report on Sustainable Solid Waste Management in India- Open Dumping of Solid Waste in India. Retrieved from http://www.seas.columbia.edu/earth/wtert/sofos/Sustainable%20 Solid%20Waste%20 Management %20in%20India_Final.pdf

Antunes, M. C., Agnelli, J. A. M., Babetto, A. S., Bonse, B. C., & Bettini, S. H. P. (2017). Abiotic thermos-oxidative degradation of high density polyethylene: Effect of manganese stearate concentration. *Polymer Degradation & Stability, 143*, 95–103. doi:10.1016/j.polymdegradstab.2017.06.012

APHA. (2005). *Standard methods for the examination of water and wastewater*. American Public Health Association.

Ariffin, H., Abdullah, N., Umi Kalsom, M. S., Shirai, Y., & Hassan, M. A. (2006). Production and characterisation of cellulase by bacillus pumilus. *IACSIT International Journal of Engineering and Technology, 3*(1), 47–53.

Arkatkar, A., Arutchelvi, J., Bhaduri, S., Uppara, P. V., & Doble, M. (2009). Degradation of unpretreated and thermally pretreated polypropylene by soil consortia. *International Biodeterioration & Biodegradation, 63*(1), 106–111. doi:10.1016/j.ibiod.2008.06.005

Armenta, S., & de la Guardia, M. (2016). Pollutants and Air Pollution. *Comprehensive Analytical Chemistry, 73*, 27–44. doi:10.1016/bs.coac.2016.03.002

Arshad, M., Saleem, M., & Hussain, S. (2007). Perspectives of bacterial ACC deaminase in phytoremediation. *Trends in Biotechnology, 25*(8), 356–362. doi:10.1016/j.tibtech.2007.05.005 PMID:17573137

Arshad, M., Zafar, M. N., Younis, S., & Nadeem, R. (2008). The use of Neem biomass for the biosorption of zinc from aqueous solutions. *Journal of Hazardous Materials, 157*(2), 534–540. doi:10.1016/j.jhazmat.2008.01.017 PMID:18289783

Artan, N. D., Orhon, D., & Tasli, R. (2002). Design of SBR systems for nutrient removal from wastewaters subject to seasonal fluctuations. *Water Science and Technology, 46*(8), 91–98. PMID:12420970

Artham, T., Sudhakar, M., Venkatesan, R., Nair, C. M., Murty, K. V. G. K., & Doble, M. (2009). Biofouling and stability of synthetic polymers in sea water. *International Biodeterioration & Biodegradation, 63*(7), 884–890. doi:10.1016/j.ibiod.2009.03.003

Arutchelvi, J., Sudhakar, M., Arkatkar, A., Doble, M., Bhaduri, S., & Uppara, P. V. (2008). Biodegradation of polyethylene and polypropylene. *Indian Journal of Biotechnology, 7*, 9–22.

Arvaniti, E., Bjerre, A. B., & Schmidt, J. E. (2012). Wet oxidation pretreatment of rape straw for ethanol production. *Biomass and Bioenergy, 39*, 94–105. doi:10.1016/j.biombioe.2011.12.040

Ashrafi, S. D., Rezaei, S., Forootanfar, H., Mahvi, A. H., & Faramarzi, M. A. (2013). The enzymatic decolorization and detoxification of synthetic dyes by the laccase from a soil-isolated ascomycete, *Paraconiothyrium variabile. International Biodeterioration & Biodegradation, 85*, 173–181. doi:10.1016/j.ibiod.2013.07.006

Asif, H. K., Ehsan, A., Kashaf, Z., Abeera, A. A., Azra, N., & Muneeb, Q. (2015). Comparative study of bioethanol production from sugarcane molasses by using *Zymomonas mobilis* and *Saccharomyces cerevisiae. African Journal of Biotechnology, 14*(31), 2455–2462. doi:10.5897/AJB2015.14569

Asnani, P. U. (2006). *Solid waste Management: India Infrastructure Report, 3iNetwork*. New Delhi: Oxford University Press.

Auerswald, K., Litz, N., Mayer, R., Stoy, A., Alef, K., & Kerndorff, H. (2000). Soil, 2. Damage and Soil Pollution. In Ullmann's Encyclopedia of Industrial Chemistry. Wiley-VCH Verlag GmbH & Co. KGaA.

Awasthi, S., Srivastava, N., Singh, T., Tiwary, D., Mishra, P. K. (2017). Biodegradation of thermally treated low density polyethylene by fungus *Rhizopus oryzae* NS5. *3 Biotech, 7*, 73.

Ayadi, M., Sarma, S. J., Pachapur, V. L., Brar, S. K., & Cheikh, R. B. (2016). History and Global Policy of Biofuels. In C.R. Soccol, S.K. Brar, C. Faulds et al. (Eds.), Green Fuels Technology. Switzerland: Springer International Publishing. doi:10.1007/978-3-319-30205-8_1

Ayed, L., Mahdhi, A., Cheref, A., & Bakhrouf, A. (2011). Decolorization and degradation of azo dye Methyl Red by an isolated *Sphingomonas paucimobilis*: Biotoxicity and metabolites characterization. *Desalination, 274*(1-3), 272–277. doi:10.1016/j.desal.2011.02.024

Aygan, A., Karcioglu, L., & Arikan, B. (2011). Alkaline thermostable and halophilic endoglucanase from Bacillus licheniformis C108. *African Journal of Biotechnology, 10*(5), 789–796.

Bafana, A., & Chakrabarti, T. (2008). Lateral gene transfer in phylogeny of azoreductase enzyme. *Computational Biology and Chemistry, 32*(3), 191–197. doi:10.1016/j.compbiolchem.2008.03.003 PMID:18440871

Bafana, A., Chakrabarti, T., Muthal, P., & Kanade, G. (2009). Detoxification of benzidine-based azo dye by *E. gallinarum*: Time-course study. *Ecotoxicology and Environmental Safety, 72*(3), 960–964. doi:10.1016/j.ecoenv.2007.11.013 PMID:18206233

Bafana, A., Devi, S. S., & Chakrabarti, T. (2011). Azo dyes: Past, present and the future. *Environmental Reviews, 19*(NA), 350–371. doi:10.1139/a11-018

Bahl, A., & Bahl, B. S. (2003). A text book of organic chemistry. S. Chand & Company Ltd.

Bailón, L., Nikolausz, M., Kästner, M., Veiga, M. C., & Kennes, C. (2009). Removal of dichloromethane from waste gases in one-and two-liquid-phase stirred tank bioreactors and biotrickling filters. *Water Research, 43*(1), 11–20. doi:10.1016/j.watres.2008.09.031 PMID:18945466

Bajpai, P. (2017). Anaerobic Reactors Used for Waste Water Treatment. In *Anaerobic Technology in Pulp and Paper Industry* (pp. 37–53). Springer Singapore. doi:10.1007/978-981-10-4130-3_5

Baker, K. H. (1994). Bioremediation of surface and surfaces soils. In K.H. Baker & D.S. Herson (Eds.), Bioremediation (pp. 9-60). New York, USA: McGraw-Hill.

Bak, J. S., Ko, J. K., Han, Y. H., Lee, B. C., Choi, I., & Kim, K. H. (2009). Improved enzymatic hydrolysis yield of rice straw using electron beam irradiation pretreatment. *Bioresource Technology, 100*(3), 1285–1290. doi:10.1016/j.biortech.2008.09.010 PMID:18930388

Balaji, V., Arulazhagan, P., & Ebenezer, P. (2014). Enzymatic bioremediation of polyaromatic hydrocarbons by fungal consortia enriched from petroleum contaminated soil and oil seeds. *Journal of Environmental Biology, 35*(3), 521–529. PMID:24813008

Balaji, V., & Ebenzer, P. (2008). Optimization of extracellular lipase production in Colletotrichum gloeosporioides by solid state fermentation. *Indian Journal of Science and Technology, 1*(7), 1–8.

Balakrishnan, B., Prasad, B., Rai, A. K., Velappan, S. P., Subbanna, M. N., & Narayan, B. (2011). In vitro antioxidant and antibacterial properties of hydrolysed proteins of delimed tannery fleshings: Comparison of acid hydrolysis and fermentation methods. *Biodegradation, 22*(2), 287–295. doi:10.100710532-010-9398-0 PMID:20680665

Balaria, A., & Schiewer, S. (2008). Assessment of biosorption mechanism for Pb binding by citrus pectin. *Separation and Purification Technology, 63*(3), 577–581. doi:10.1016/j.seppur.2008.06.023

Balasubramaniam, M. K., & Rajarathinam, R. (2013). Implementation of white rot fungal pretreated rice straw for sustainable bioethanol production by *Saccharomyces cerevisiae. International Journal of Engineering Research & Technology, 2*(11), 4047–4053.

Balasubramanian, M., & Dhulasi Birundha, V. (2012). An Economic Analysis of Solid Waste Management in Madurai District, Tamil Nadu. *Applied Journal of Hygiene, 1*(1).

Balasubramanian, V., Natarajan, K., Hemambika, B., Ramesh, N., Sumathi, C. S., Kottaimuthu, R., & Kannan, V. R. (2010). High-density polyethylene (HDPE)-degrading potential bacteria from marine ecosystem of Gulf of Mannar, India. *Letters in Applied Microbiology*, *51*, 205–211. PMID:20586938

Balat, M. (2009). Gasification of biomass to produce gaseous products. *Energy Sources*, *31*(6), 516–526. doi:10.1080/15567030802466847

Balat, M., Balat, H., & Cahide, O. Z. (2008). Progress in bioethanol processing. *Progress in Energy & Combustion*, *34*(5), 551–573. doi:10.1016/j.pecs.2007.11.001

Banat, I., Nigam, P., Singh, D., & Marchant, R. (1996). Microbial decolorization of textile dye containing effluents: A Review. *Bioresource Technology*, *58*(3), 217–227. doi:10.1016/S0960-8524(96)00113-7

Banerjee, S., Sen, R., Morone, A., Chakrabarti, T., Pandey, R. & Mudliar, S. (2012). Improved wet air oxidation pre-treatment for enhanced enzymatic hydrolysis of rice husk for bioethanol production. *Dynamic biochemistry, process biotechnology and molecular biology*, *6*, 43-45.

Bañuelos, G., Terry, N., LeDuc, D. L., Pilon-Smits, E. A., & Mackey, B. (2005). Field trial of transgenic Indian mustard plants shows enhanced phytoremediation of selenium-contaminated sediment. *Environmental Science & Technology*, *39*(6), 1771–1777. doi:10.1021/es049035f PMID:15819237

Banu, J. R., Esakkiraj, S., Nagendran, R., & Logakanthi, S. (2005). Biomanagement of petrochemical sludge using an exotic earthworm Eudrilus eugineae. *Journal of Environmental Biology*, *26*(1), 43–47. PMID:16114460

Barea, J. M., Pozo, M. J., Azcon, R., & Azcon-Aguilar, C. (2005). Microbial co-operation in the rhizosphere. *Journal of Experimental Botany*, *56*(417), 1761–1778. doi:10.1093/jxb/eri197 PMID:15911555

Barfield, B. J., Blevins, R. L., Flofle, A. W., Madison, C. E., Inamder, S., Carey, D. I., & Evangelou, V. P. (1992). Water quality impacts of natural riparian grasses: Empirical studies. American Society of Agricultural Engineers, St. Joseph, MI.

Barnes, D. K. A., Galgani, F., Thompson, R. C., & Barlaz, M. (2009). Accumulation and fragmentation of plastic debris in global environments. *Philosophical Transactions of the Royal Society of London. Series B, Biological Sciences*, *364*(1526), 1985–1998. doi:10.1098/rstb.2008.0205 PMID:19528051

Barr, D. B., & Needham, L. L. (2002). Analytical methods for biological monitoring of exposure to pesticides: A review. *Journal of Chromatography. A*, *778*, 5–29. PMID:12376114

Bayer, E. A., Lamed, R., & Himmel, M. E. (2007). The potential of cellulases and cellulosomes for cellulosic waste management. *Current Opinion in Biotechnology*, *18*(3), 237–245. doi:10.1016/j.copbio.2007.04.004 PMID:17462879

Bayley, M. L., Davison, L., & Headley, T. R. (2003). Nitrogen removal from domestic effluent using subsurface flow constructed wetlands: Influence of depth, hydraulic residence time and pre-nitrification. *Water Science and Technology*, *48*(5), 175–182. PMID:14621162

Beesley, L., Moreno-Jiménez, E., & Gomez-Eyles, J. L. (2010). Effects of biochar and green waste compost amendments on mobility, bioavailability and toxicity of inorganic and organic contaminants in a multi-element polluted soil. *Environmental Pollution*, *158*(6), 2282–2287. doi:10.1016/j.envpol.2010.02.003 PMID:20219274

Behki, R., Topp, E., Dick, W., & Germon, P. (1993). Metabolism of the herbicide atrazine by *Rhodococcus* strains. *Applied and Environmental Microbiology*, *59*(61), 955–1959. PMID:8328812

Bellin, C. A., & O'Connor, G. A. (1990). Plant uptake of pentachlorophenol from sludgeamended soils. *Journal of Environment Qualality*, *19*(3), 598–602. doi:10.2134/jeq1990.00472425001900030040x

Bending, G. D., Lincoln, S. D., & Edmondson, R. N. (2006). Spatial variation in the degradation rate of the pesticides isoproturon, azoxystrobin and diflufenican bioremediation and Phytoremediation of Pesticides in soil and its relationship with chemical and microbial properties. *Environmental Pollution, 139*(2), 279–287. doi:10.1016/j.envpol.2005.05.011 PMID:16043274

Bending, G. D., Turner, M. K., & Jones, J. E. (2002). Interactions between crop residue and soil organic matter quality and the functional diversity of soil microbial communities. *Soil Biology & Biochemistry, 34*(8), 1073–1082. doi:10.1016/S0038-0717(02)00040-8

Bensah, E. C., & Mensah, M. (2013). Chemical Pretreatment methods for the production of cellulosic ethanol: Technologies and innovations. *International Journal of Chemical Engineering, 2013*, 1–21. doi:10.1155/2013/719607

Berezinaa, N., Nysb, J., & Yadaa, B. (2013). Method for the Analysis of Grafted Cellulosic Materials. *Chemical Engineering (Albany, N.Y.)*, 32.

Bertazzi, P. A., & di Domenico, A. (1994). Chemical, Environmental, and Health Aspects of the Seveso, Italy, Accident. In A. Schecter (Ed.), *Dioxins and Health* (pp. 587–632). Boston, MA: Springer US. doi:10.1007/978-1-4899-1462-0_18

Berti, W. R., & Cunningham, S. D. (2000). *Phytostabilizat ion of metals* (pp. 71–88). New York: Wiley.

Besis, A., & Samara, C. (2012). Polybrominated diphenyl ethers (PBDEs) in the indoor and outdoor environments--a review on occurrence and human exposure. *Environmental Pollution, 169*, 217-229. doi:10.1016/j.envpol.2012.04.009

Beveridge, T. J. (1981). Ultrastructure, chemistry, and function of the bacterial wall. *International Review of Cytology, 72*, 229–317. doi:10.1016/S0074-7696(08)61198-5 PMID:6166584

Bharathiraja, B., Sudharsanaa, T., Bharghavi, A., Jayamuthunagai, J., & Praveenkumar, R. (2016). Biohydrogen and Biogas – An overview on feedstocks and enhancement process. *Fuel, 185*, 810–828. doi:10.1016/j.fuel.2016.08.030

Bhaskar, N., Sakhare, P. Z., Suresh, P. V., Gowda, L. R., & Mahendrakar, N. S. (2007). Biostabilization and preparation of protein hydrolysates from delimed leather fleshings. *Journal of Scientific and Industrial Research, 66*, 1054–1063.

Bhat, D., & Padmaja, P. (2014). Assessment of organic pesticides in Ground and surface water in Bhopal India. *IOSR journal of Environmental Science. Toxicology, and Food Technology, 8*(5), 51–52. doi:10.9790/2402-08535152

Bhatia, M., Girdhar, A., Tiwari, A., & Nayarisseri, A. (2014). Implications of a novel *Pseudomonas* species on low density polyethylene biodegradation: An *in vitro* to *in silico* approach. *SpringerPlus, 3*, 497.

Bhat, M. K. (2000). Cellulases and related enzymes in biotechnology. *Biotechnology Advances, 18*(5), 355–383. doi:10.1016/S0734-9750(00)00041-0 PMID:14538100

Bhatnagar, A., & Sillanpää, M. (2010). Utilization of agro-industrial and municipal waste materials as potential adsorbents for water treatment — A review. *Chemical Engineering Journal, 157*(2-3), 277–296. doi:10.1016/j.cej.2010.01.007

Bhatt, P., & Nailwal, T. K. (2018). Crop Improvement Through Microbial Technology:a Step Towards Sustainable Agriculture. Elsevier. doi:/10.1016/B978-0-444-63987-5.00011-6

Bhattacharya, S. S., Iftikar, W., Sahariah, B., & Chattopadhyay, G. N. (2012). Vermicomposting converts fly ash to enrich soil fertility and sustain crop growth in red and lateritic soils. *Resources, Conservation and Recycling, 65*, 100–106. doi:10.1016/j.resconrec.2012.05.008

Bhatt, P., Bhatt, R., & Bhatt, A. (2017). Soil Microbial Community affected by Climate change. *Scientific India Magazine, 5*(3), 38–39.

Bhatt, P., Kumar, M. S., Mudliar, S., & Chakrabarti, T. (2008). Enhanced biodegradation of hexachlorocyclohexane in upflow anaerobic sludge blanket reactor using methanol as an electron donor. *Bioresource Technology*, *99*(7), 2594–2602. doi:10.1016/j.biortech.2007.04.057 PMID:17574842

Białowiec, A. L., Davies, L., Albuquerque, A., & Randerson, P. F. (2012). The influence of plants on nitrogen removal from landfill leachate in discontinuous batch shallow constructed wetland with recirculating subsurface horizontal flow. *Ecological Engineering*, *40*, 44–52. doi:10.1016/j.ecoleng.2011.12.011

Bin, Y., Jiti, Z., Jing, W., Cuihong, D., Hongman, H., Zhiyong, S., & Yongming, B. (2004). Expression and characteristics of the gene encoding azoreductase from *Rhodobacter sphaeroides* AS1.1737. *FEMS Microbiology Letters*, *236*(1), 129–136. doi:10.1111/j.1574-6968.2004.tb09638.x PMID:15212802

Biswas, R., Uellendahl, H., & Ahring, B. K. (2015). Wet Explosion: A universal and efficient pretreatment process for lignocellulosic biorefineries. *BioEnergy Research*, *8*(3), 1101–1116. doi:10.100712155-015-9590-5

Bitter, G. A., & Egan, K. M. (1984). Expression of heterologous genes in *Saccharomyces cerevisiae* from vectors utilizing the glyceraldehyde-3-phosphate dehydrogenase gene promoter. *Gene*, *32*(3), 263–274. doi:10.1016/0378-1119(84)90002-7 PMID:6099318

Blümel, S., Knackmuss, H. J., & Stolz, A. (2002). Molecular cloning and characterization of the gene coding for the aerobic azoreductase from *Xenophilus azovorans* KF46F. *Applied and Environmental Microbiology*, *68*(8), 3948–3955. doi:10.1128/AEM.68.8.3948-3955.2002 PMID:12147495

Bocquené, G., & Franco, A. (2005). Pesticide contamination of the coastline of Martinique. *Marine Pollution Bulletin*, *51*(5-7), 612–619. doi:10.1016/j.marpolbul.2005.06.026 PMID:16045943

Bolaños, R. M. L., Damianovic, M. H. R. Z., Zaiat, M., & Foresti, E. (2005). Assessment of the ability of sludge to degrade PCP under anaerobic conditions. *Brazilian Journal of Chemical Engineering*, *22*(4), 611–617. doi:10.1590/S0104-66322005000400014

Bonfante, P., & Anca, I. A. (2009). Plants, mycorrhizal fungi, and bacteria: A network of interactions. *Annual Review of Microbiology*, *63*(1), 363–383. doi:10.1146/annurev.micro.091208.073504 PMID:19514845

Bonhomme, S., Cuer, A., Delort, A.-M., Lemaire, J., Sancelme, M., & Scott, G. (2003). Environmental biodegradation of polyethylene. *Polymer Degradation & Stability*, *81*(3), 441–452. doi:10.1016/S0141-3910(03)00129-0

Bonvin, D. (1998). Optimal operation of batch reactors—a personal view. *Journal of Process Control*, *8*(5), 355–368. doi:10.1016/S0959-1524(98)00010-9

Boonmee, A. (2009). Screening of rice straw degrading microorganisms and their cellulose activities. *KKU Sci. j.*, *37*, 83-88.

Boopathy, R.(2000). Factors limiting bioremediation technologies. *Bioresource technology.74.63*

Booth, G., Zollinger, H., McLaren, K., Sharpless, W. G., & Westwell, A. (2000). Dyes, General Survey. In *Ullmann's Encyclopedia of Industrial Chemistry*. Weinheim, Germany: Wiley-VCH. doi:10.1002/14356007.a09_073

Borbon, A., Fontaine, H., Veillerot, M., Locoge, N., Galloo, J. C., & Guillermo, R. (2001). An investigation into the traffic-related fraction of isoprene at an urban location. *Atmospheric Environment*, *35*(22), 3749–3760. doi:10.1016/S1352-2310(01)00170-4

Borghei, M., Karbassi, A., Khoramnejadian, S., Oromiehie, A., & Javid, A. H. (2010). Microbial biodegradable potato starch based low density polyethylene. *African Journal of Biotechnology*, *9*, 4075–4080.

Bossi, R., Skov, H., Vorkamp, K., Christensen, J., Rastogi, S. C., Egeløv, A., & Petersen, D. (2008). Atmospheric concentrations of organochlorine pesticides, polybrominated diphenyl ethers and polychloronaphthalenes in Nuuk, South-West Greenland. *Atmospheric Environment*, *42*(31), 7293–7303. doi:10.1016/j.atmosenv.2008.06.049

Bo, T., Zhu, X., Zhang, L., Tao, Y., He, X., Li, D., & Yan, Z. (2014). A new upgraded biogas production process: Coupling microbial electrolysis cell and anaerobic digestion in single-chamber, barrel-shape stainless steel reactor. *Electrochemistry Communications*, *45*, 67–70. doi:10.1016/j.elecom.2014.05.026

Bouallagui, H., Torrijos, M., Godon, J. J., Moletta, R., Cheikh, R. B., Touhami, Y., & Hamdi, M. (2004). Two-phases anaerobic digestion of fruit and vegetable wastes: Bioreactors performance. *Biochemical Engineering Journal*, *21*(2), 193–197. doi:10.1016/j.bej.2004.05.001

Bowen, E., Dolfing, J., Davenport, R., Read, F., & Curtis, T. (2014). Low-temperature limitation of bioreactor sludge in anaerobic treatment of domestic wastewater. *Water Science and Technology*, *69*(5), 1004. doi:10.2166/wst.2013.821 PMID:24622549

Bowmer, K. H. (1991). Atrazine persistence and toxicity in two irrigated soils in Australia. *Australian Journal of Soil Research*, *29*(2), 339–350. doi:10.1071/SR9910339

Boyd, D. R., Sharma, N. D., & Allen, C. C. R. (2001). Aromatic dioxygenases: Molecular biocatalysis and applications. *Current Opinion in Biotechnology*, *12*(6), 564–573. doi:10.1016/S0958-1669(01)00264-6 PMID:11849939

Brazil, G. M., Kenefick, L., Callanan, M., Haro, A., De Lorenzo, V., Dowling, D. N., & O'gara, F. (1995). Construction of a rhizosphere pseudomonad with potential to degrade polychlorinated biphenyls and detection of bph gene expression in the rhizosphere. *Applied and Environmental Microbiology*, *61*(5), 1946–1952. PMID:7646029

Breitling, R. (2010). What is Systems Biology? *Frontiers in Physiology*, *1*. doi:10.3389/fphys.2010.00009 PMID:21423352

Breslin, V. T., & Swanson, R. L. (1993). Deterioration of starch-plastic composites in the environment. *Journal of the Air & Waste Management Association*, *43*(3), 325–335. doi:10.1080/1073161X.1993.10467137

Breton-Deval, L., Galindez-Mayer, J., Moreno-Medina, C. U., Ríos-Leal, E., Rinderknecht-Seijas, N., & Poggi-Varaldo, H. M. (2012). Operation of hybrid fluidized bioreactors for on site bioremediation of water polluted with high concentration of PCE. *Environmental Engineering and Management Journal*, *11*(3), S25.

Bretón-Deval, L., Rios-Leal, E., Poggi-Varaldo, H. M., & Ponce-Noyola, T. (2016). Biodegradability of Nonionic Surfactant Used in the Remediation of Groundwaters Polluted with PCE. *Water Environment Research*, *88*(11), 2159–2168. doi:10.2175/106143016X14733681695564 PMID:28661332

Bretzel, F. C., & Calderisi, M. (2011). Contribution of a municipal solid waste incinerator to the trace metals in the surrounding soil. *Environmental Monitoring and Assessment*, *182*(1-4), 523–533. doi:10.100710661-011-1894-0 PMID:21327482

Brinza, L., Dring, M. J., & Gavrilescu, M. (2007). Marine micro-and macro-algal species as biosorbents for heavy metals. *Environmental Engineering and Management Journal*, *6*(3), 237–251.

Brisson, J., & Chazarenc, F. (2009). Maximizing pollutant removal in constructed wetlands: Should we pay more attention to macrophyte species selection? *The Science of the Total Environment*, *407*(13), 3923–3930. doi:10.1016/j.scitotenv.2008.05.047 PMID:18625516

Brix, H. (1997). Do macrophytes play a role in constructed treatment wetlands? *Water Science and Technology*, *35*(5), 11–17.

Bruggemann, F. J., & Westerhoff, H. V. (2006). The nature of systems biology. *Trends in Microbiology*, *15*(1), 45–50. doi:10.1016/j.tim.2006.11.003 PMID:17113776

Brydson, J. A. (1999). *Plastic materials*. Oxford, UK: Butterworth-Heinemann.

Bürger, S., & Stolz, A. (2010). Characterisation of the flavin-free oxygen-tolerant azoreductase from *Xenophilus azovorans* KF46F in comparison to flavin-containing azoreductases. *Applied Microbiology and Biotechnology*, *87*(6), 2067–2076. doi:10.100700253-010-2669-1 PMID:20508929

Burlein, G. A. D., & Rocha, M. C. G. (2014). LDPE/PHB blends filled with castor oil pressed cake. *Materials Research*, *17*(1), 203–212. doi:10.1590/S1516-14392013005000166

Buss, S. E., Butler, A. P., Sollars, C. J., Perry, R., & Johnston, P. M. (1995). Mechanisms of leakage through synthetic landfill liner materials. *Water and Environment Journal: the Journal / the Chartered Institution of Water and Environmental Management*, *9*(4), 353–359. doi:10.1111/j.1747-6593.1995.tb00952.x

Bustos, G., Moldes, A. B., Alonso, J. L., & Vázquez, M. (2004). Optimization of D-lactic acid production by *Lactobacillus coryniformis* using response surface methodology. *Food Microbiology*, *21*(2), 143–148. doi:10.1016/S0740-0020(03)00061-3

Button, M., Nivala, J., Weber, K. P., Aubron, T., & Müller, R. A. (2015). Microbial community metabolic function in subsurface flow constructed wetlands of different designs. *Ecological Engineering*, *80*, 162–171. doi:10.1016/j.ecoleng.2014.09.073

Byadgi, S. A., & Kalburgi, P. B. (2016). Production of bioethanol from waste newspaper. *Procedia Environmental Sciences*, *35*, 555–562. doi:10.1016/j.proenv.2016.07.040

Bychkov, A. L., Buchtoyarov, V. A., & Lomovsky, O. L. (2014). Mechanical pretreatment of corn straw in a centrifugal roller mill. *Cellulose Chemistry and Technology*, *48*(5-6), 545–551.

Cain, R. B. (1992). Microbial degradation of synthetic polymers. *Journal of Microbial control of pollution*.

Calheiros, C. S., Duque, A. F., Moura, A., Henriques, I. S., Correia, A., Rangel, A. O., & Castro, P. M. (2009). Substrate effect on bacterial communities from constructed wetlands planted with *Typha latifolia* treating industrial wastewater. *Ecological Engineering*, *35*(5), 744–753. doi:10.1016/j.ecoleng.2008.11.010

Campbell, A. W. (2014). Glyphosate: Its effects on humans. *Alternative Therapies in Health and Medicine*, *20*, 9–11. PMID:24755565

Campo, A. B. (2012). PID Control Design, MATLAB - A Fundamental Tool for Scientific Computing and Engineering Applications (Vol. 1). InTech. Retrieved 1 February 2017 from http://www.intechopen.com/books/matlab-a-fundamental-tool-for-scientific-computing-and-engineering-applications-volume-1/pid-control-design

Cardona, C. A., Quintero, J. A., & Paz, I. C. (2010). Production of bioethanol from sugarcane bagasse: Status and perspectives. *Bioresource Technology*, *101*(13), 4754–4766. doi:10.1016/j.biortech.2009.10.097 PMID:19945863

Cardona, C. A., & Sanchez, O. J. (2007). Fuel ethanol production: Process design trends and integration opportunities. *Bioresource Technology*, *98*(12), 2415–2457. doi:10.1016/j.biortech.2007.01.002 PMID:17336061

Carmen, Z., & Daniela, Z. (2012). Textile organic dyes - characteristics, polluting effects and separation/ elimination procedures from industrial effluents – A Critical Overview. In Organic Pollutants Ten Years After the Stockholm Convention - Environmental and Analytical Update. InTech.

Carmo, I. M. T. D. D. (2013). *Food waste valorization through the production of polyhydroxyalkanoates by mixed microbial cultures* [Doctoral dissertation]. Faculdade de Ciências e Tecnologia.

Cases, I., & de Lorenzo, V. (2005). Genetically modified organisms for the environment: Stories of success and failure and what we have learned from them. *International Microbiology, 8*, 213–222. PMID:16200500

Casey, E., Glennon, B., & Hamer, G. (1999). Review of membrane aerated biofilm reactors. *Resources, Conservation and Recycling, 27*(1), 203–215. doi:10.1016/S0921-3449(99)00007-5

Casida, J. E., & Durkin, K. A. (2013). Neuroactive insecticides: Targets, selectivity, resistance, and secondary effects. *Annual Review of Entomology, 58*(1), 99–117. doi:10.1146/annurev-ento-120811-153645 PMID:23317040

Castilho, L. R., Mitchell, D. A., & Freire, D. M. (2009). Production of polyhydroxyalkanoates (PHAs) from waste materials and by-products by submerged and solid-state fermentation. *Bioresource Technology, 100*(23), 5996–6009. doi:10.1016/j.biortech.2009.03.088 PMID:19581084

Cavaco-Paulo, A. (1998). Mechanism of cellulose action in textile processes. *Carbohydrate Polymers, 37*(3), 273–277. doi:10.1016/S0144-8617(98)00070-8

Cavalli, V. L. D. L. O., Cattani, D., Rieg, C. E. H., Pierozan, P., Zanatta, L., Parisotto, E. B., ... & Zamoner, A. (2013). Roundup disrupted male reproductive functions by triggering calcium-mediated cell death in rat testis and Sertoli cells. *Free Radicals, Biology and Medicine, 65*, 335–346.

Census of India. (2011). [Data file] Retrieved from www.censusindia.gov.in/2011census/PCA/PCA_Highlights/pca_highlights_file/India/Chapter-1.pdf

Cerejeira, M. J., Viana, P., Batista, S., Pereira, T., Silva, E., Valério, M. J., ... Silva-Fernandes, A. M. (2003). Pesticides in Portuguese surface and ground waters. *Water Research, 37*(5), 1055–1063. doi:10.1016/S0043-1354(01)00462-6 PMID:12553980

Cesare, A., Claudio, S., & Alberto, V. (2005). Influence of flooding on the degradation of linuron, isoproturon and metolachlor in soil. *Agronomy for Sustainable Development, 25*(3), 401–406. doi:10.1051/agro:2005030

Chagas, E. P., & Durrant, L. R. (2001). Decolorization of azo dyes by *Phanerochaete chrysosporium* and *Pleurotus sajorcaju. Enzyme and Microbial Technology, 29*(8-9), 473–477. doi:10.1016/S0141-0229(01)00405-7

Chakraborty, R., & Sarkar, S. K. (1998). Enzyme hydrolysis of solid tannery wastes: Solid state enzyme production. *Journal of the Society of Leather Technologists and Chemists, 82*(2), 56–58.

Chandel, A.K., E.S. C., Rudravaram, R., Narasu, M.L., Rao, L.V., & Ravindra, P. (2007). Economics and environmental impact of bioethanol production technologies: an appraisal. *Biotechnology & Molecular Biology Review, 2*(1), 014-032.

Chaney, R. L., Angle, J. S., Broadhurst, C. L., Peters, C. A., Tappero, R. V., & Sparks, D. L. (2007). Improved understanding of hyperaccumulation yields commercial phytoextraction and phytomining technologies. *Journal of Environmental Quality, 36*(5), 1429–1443. doi:10.2134/jeq2006.0514 PMID:17766822

Chaney, R. L., Malik, M., Li, Y. M., Brown, S. L., Angle, J. S., & Baker, A. J. M. (1997). Phytoremediation of soil metals. *Current Opinion in Biotechnology, 8*(3), 279–284. doi:10.1016/S0958-1669(97)80004-3 PMID:9206007

Chang, J.-S., Chou, C., Lin, Y. C., Lin, P. J., Ho, J. Y., & Hu, T. L. (2001). Kinetic characteristics of bacterial azo-dye decolorization by *Pseudomonas luteola. Water Research, 35*(12), 2841–2850. doi:10.1016/S0043-1354(00)00581-9 PMID:11471684

Chang, Y. C., Park, C. K., Jung, K., & Kikuchi, S. (2010). Dechlorination of High Concentrations of Tetrachloroethylene Using a Fixed-bed Reactor. *Korean Journal of Environmental Health Sciences, 36*(4), 323–336. doi:10.5668/JEHS.2010.36.4.323

Chan, S. Y., Tsang, Y. F., & Chua, H. (2008a). Domestic wastewater treatment using tidal-flow cinder bed with Cyperus alternifolius. *Aquatic Ecosystem Health & Management*, *11*(2), 206–211. doi:10.1080/14634980802100022

Chan, S. Y., Tsang, Y. F., & Chua, H. (2008b). Domestic wastewater treatment using batch-fed constructed wetland and predictive model development for NH_3-N removal. *Process Biochemistry*, *43*(3), 297–305. doi:10.1016/j.procbio.2007.12.009

Charnay, M. P., Tuis, S., Coquet, Y., & Barriuso, E. (2005). Spatial variability in 14C-herbicide degradation in surface and subsurface soils. *Pest Management Science*, *61*(9), 845–855. doi:10.1002/ps.1092 PMID:16003827

Chatterjee, S., Roy, B., Roy, D., & Banerjee, R. (2010). Enzyme-mediated biodegradation of heat treated commercial polyethylene by *Staphylococcal* species. *Polymer Degradation & Stability*, *95*(2), 195–200. doi:10.1016/j.polymdegradstab.2009.11.025

Chaudhry, G. R., & Chapalamadugu, S. (1991). Biodegradation of halogenated organic compounds. *Microbiological Reviews*, *55*(1), 59–79. PMID:2030673

Chaudhry, Q., Blom-Zandstra, M., Gupta, S. K., & Joner, E. (2005). Utilising the synergy between plants and rhizosphere microorganisms to enhance breakdown of organic pollutants in the environment. *Environmental Science and Pollution Research International*, *12*(1), 34–48. doi:10.1065/espr2004.08.213 PMID:15768739

Chaurasia, P. K., Bharati, S. L., & Singh, S. K. (2014e). Role of Laccase from *Coriolus versicolor* MTCC-138 in Selective Oxidation of Aromatic Methyl Group. *Russian Journal of Bioorganic Chemistry*, *40*(3), 288–292. doi:10.1134/S1068162014020034 PMID:25898738

Chaurasia, P. K., Bharati, S. L., Singh, S. K., & Yadava, S. (2015b). Synthetic Applications of Purified Laccase from *Pleurotus sajor caju* MTCC-141. *Russian Journal of General Chemistry*, *85*(1), 173–175. doi:10.1134/S1070363215010302

Chaurasia, P. K., Bharati, S. L., Yadava, S., & Yadav, R. S. S. (2015a). Purification, characterization and synthetic application of a thermally stable laccase from *Hexagonia tenuis* MTCC-1119. *Indian Journal of Biochemistry & Biophysics*, *52*(1), 60–67. PMID:26040112

Chaurasia, P. K., Yadav, A., Yadav, R. S. S., & Yadava, S. (2013a). Purification and Characterization of Laccase Secreted by *Phellinus linteus* MTCC-1175 and its Role in the selective Oxidation of aromatic Methyl group. *Applied Biochemistry and Microbiology*, *49*(6), 592–599. doi:10.1134/S000368381306006 PMID:25434183

Chaurasia, P. K., Yadava, S., Bharati, S. L., & Singh, S. K. (2014b). Syntheses of aromatic aldehydes by laccase of *Pleurotus ostreatus* MTCC-1801. *Synthetic Communications*, *44*(17), 2535–2544. doi:10.1080/00397911.2014.904879

Chaurasia, P. K., Yadava, S., Bharati, S. L., & Singh, S. K. (2014c). Selective Oxidation and N-coupling of Amines by Purified Laccase of *Xylaria polymorpha* MTCC-1100. *Russian Journal of Bioorganic Chemistry*, *40*(4), 455–460. doi:10.1134/S1068162014040025

Chaurasia, P. K., Yadava, S., Bharati, S. L., & Singh, S. K. (2014d). Syntheses of aromatic aldehydes by laccase without the help of mediators. *Green Chemistry Letters and Reviews*, *7*(1), 100–104. doi:10.1080/17518253.2014.895869

Chaurasia, P. K., Yadav, R. S. S., & Yadava, S. (2013b). Purification and Characterization of Laccase from *Coriolopsis floccosa* MTCC-1177 and its use in the selective oxidation of aromatic methyl group to aldehyde without mediators. *Journal of Chemical Sciences*, *125*(6), 1395–1403. doi:10.100712039-013-0525-4

Chaurasia, P. K., Yadav, R. S. S., & Yadava, S. (2014a). Purification and characterization of yellow laccase from *Trametes hirsuta* MTCC-1171 and its application in synthesis of aromatic aldehydes. *Process Biochemistry*, *49*(10), 1647–1655. doi:10.1016/j.procbio.2014.06.016

Cheng, K., Ning, Z., Zhang, X., Li, L., Liao, B., Mayne, J., ... Figeys, D. (2017). MetaLab: An automated pipeline for metaproteomic data analysis. *Microbiome*, *5*(1), 157. doi:10.118640168-017-0375-2 PMID:29197424

Cheng, S., Xing, D., Call, D., & Logan, B. (2009). Direct Biological Conversion of Electrical Current into Methane by Electromethanogenesis. *Environmental Science & Technology*, *43*(10), 3953–3958. doi:10.1021/es803531g PMID:19544913

Chen, H., & Cutright, T. (2001). EDTA and HEDTA effects on Cd, Cr, and Ni uptake by *Helianthus annuus*. *Chemosphere*, *45*(1), 21–28. doi:10.1016/S0045-6535(01)00031-5 PMID:11572587

Chen, H., Feng, J., Kweon, O., Xu, H., & Cerniglia, C. E. (2010). Identification and molecular characterization of a novel flavin-free NADPH preferred azoreductase encoded by *azoB* in *Pigmentiphaga kullae* K24. *BMC Biochemistry*, *11*(1), 13. doi:10.1186/1471-2091-11-13 PMID:20233432

Chen, H., Hopper, S. L., & Cerniglia, C. E. (2005). Biochemical and molecular characterization of an azoreductase from *Staphylococcus aureus*, a tetrameric NADPH-dependent flavoprotein. *Microbiology*, *151*(5), 1433–1441. doi:10.1099/mic.0.27805-0 PMID:15870453

Chen, H., Wang, R., & Cerniglia, C. E. (2004). Molecular cloning, overexpression, purification, and characterization of an aerobic FMN-dependent azoreductase from *Enterococcus faecalis*. *Protein Expression and Purification*, *34*(2), 302–310. doi:10.1016/j.pep.2003.12.016 PMID:15003265

Chen, H., Xu, H., Kweon, O., Chen, S., & Cerniglia, C. E. (2008). Functional role of Trp-105 of *Enterococcus faecalis* azoreductase (AzoA) as resolved by structural and mutational analysis. *Microbiology*, *154*(9), 2659–2667. doi:10.1099/mic.0.2008/019877-0 PMID:18757799

Chen, K. C., Wu, J. Y., Liou, D. J., & Hwang, S. C. (2003). Decolorization of the textile dyes by newly isolated bacterial strains. *Journal of Biotechnology*, *101*(1), 57–68. doi:10.1016/S0168-1656(02)00303-6 PMID:12523970

Chen, V. B., Arendall, W. B. III, Headd, J. J., Keedy, D. A., Immormino, R. M., Kapral, G. J., ... Richardson, D. C. (2010). MolProbity: All-atom structure validation for macromolecular crystallography. *Acta Crystallographica. Section D, Biological Crystallography*, *66*(1), 12–21. doi:10.1107/S0907444909042073 PMID:20057044

Chen, X., Xia, X., Zhao, Y., & Zhang, P. (2010). Heavy metal concentrations in roadside soils and correlation with urban traffic in Beijing, China. *Journal of Hazardous Materials*, *181*(1), 640–646. doi:10.1016/j.jhazmat.2010.05.060 PMID:20541319

Cherry J. R & Fidantsef A. L (2003). Directed evolution of industrial enzymes: an update. *Curr. Opin. Biotechnol.*, *14*, 438-443.

Chiellini, E., Corti, A., & Swift, G. (2003). Biodegradation of thermally-oxidized, fragmented low-density polyethylenes. *Polymer Degradation & Stability*, *81*(2), 341–351. doi:10.1016/S0141-3910(03)00105-8

Chiemchaisri, C., Wong, Y. K., Urase, T., & Yamamoto, A. K. (1992). Organic stabilization and nitrogen removal in membrane separation bioreactor for domestic wastewater treatment. *Water Science and Technology*, *25*(10), 231–240.

Chirila, T. V., Constable, I. J., Crawford, G. J., Vijayasekaran, S., Thompson, D. E., Chen, Y. C., & Griffin, B. J. (1993). Poly (2-hydroxyethyl methacrylate) sponges as implant materials: In vivo and in vitro evaluation of cellular invasion. *Biomaterials*, *14*(1), 26–38. doi:10.1016/0142-9612(93)90072-A PMID:7678755

Choudhary, R., Umagiliyage, A. L., Liang, Y., Siddaramu, T., Haddock, J., & Markevicius, G. (2012). Microwave pretreatment for enzymatic saccharification of sweet sorghum bagasse. *Biomass and Bioenergy*, *39*, 218–226. doi:10.1016/j.biombioe.2012.01.006

Chrysikou, L., Gemenetzis, P., Kouras, A., Manoli, E., Terzi, E., & Samara, C. (2008). Distribution of persistent organic pollutants, polycyclic aromatic hydrocarbons and trace elements in soil and vegetation following a large scale landfill fire in northern Greece. *Environment International*, *34*(2), 210–225. doi:10.1016/j.envint.2007.08.007 PMID:17900688

Chundawat, S. P., Venkatesh, B., & Dale, B. E. (2006). Effect of particle size based separation of milled corn stover on AFEX pretreatment and enzymatic digestibility. *Biotechnology and Bioengineering*, *96*(2), 219–231. doi:10.1002/bit.21132 PMID:16903002

Chung, J., Krajmalnik-Brown, R., & Rittmann, B. E. (2007). Bioreduction of trichloroethene using a hydrogen-based membrane biofilm reactor. *Environmental Science & Technology*, *42*(2), 477–483. doi:10.1021/es702422d PMID:18284150

Chunping, Y., Zhiqiang, S., Guoce, Y. U., & Jianlong, W. (2008). Effect and after effect of gama radiation pretreatment on enzymatic hydrolysis of wheat straw. *Bioresource Technology*, *99*(14), 6240–6245. doi:10.1016/j.biortech.2007.12.008 PMID:18248982

Cicek, A., & Koparal, A. S. (2004). Accumulation of sulfur and heavy metals in soil and tree leaves sampled from the surroundings of Tunçbilek Thermal Power Plant. *Chemosphere*, *57*(8), 1031–1036. doi:10.1016/j.chemosphere.2004.07.038 PMID:15488593

City Mayors Statistics. (2007). The Largest Cities in the World by Land Area, Population and Density. [Data file]. Retrieved from www.citymayors.com/statistics/largest-cities-density-125.html

Clarke, E. A., & Anliker, R. (1980). Organic dyes and pigments. In Anthropogenic Compounds. Springer.

Clauwaert, P., & Verstraete, W. (2008). Methanogenesis in membraneless microbial electrolysis cells. *Applied Microbiology and Biotechnology*, *82*(5), 829–836. doi:10.100700253-008-1796-4 PMID:19050859

Cohen, M. F., Williams, J., & Yamasaki, H. (2002). Biodegradation of diesel fuel by an *Azolla*-derived bacterial consortium. *Journal of Environmental Science and Health. Part A*, *37*(9), 1593–1606.

Cointreau, S. (1996). *Sanitary landfill design and sitting criteria*. World Bank .

Colberg, P. J. S., & Young, L. Y. (1995). Anaerobic degradation of no halogenated homocyclic aromatic compounds coupled with nitrate, iron, or sulfate reduction. In Microbial Transformation and Degradation of Toxic Organic Chemicals (pp. 307–330). New York: Wiley-Liss.

Collett, R. S., Oduyemi, K., & Lill, D. E. (1998). An investigation of environmental levels of cadmium and lead in airborne matter and surface soils within the locality of a municipal waste incinerator. *The Science of the Total Environment*, *209*(2-3), 157–167. doi:10.1016/S0048-9697(98)80107-1 PMID:9514037

Collins, G., Foy, C., McHugh, S., & O'Flaherty, V. (2005). Anaerobic treatment of 2, 4, 6-trichlorophenol in an expanded granular sludge bed-anaerobic filter (EGSB-AF) bioreactor at 15 C. *FEMS Microbiology Ecology*, *53*(1), 167–178. doi:10.1016/j.femsec.2004.10.008 PMID:16329938

Collins, P. J., Kotterman, M., Field, J. A., & Dobson, A. (1996). Oxidation of Anthracene and Benzo[a]pyrene by Laccases from *Trametes versicolor*. *Applied and Environmental Microbiology*, *62*(12), 4563–4567. PMID:16535468

Čolović, M. B., Krstić, D. Z., Lazarević-Pašti, T. D., Bondžić, A. M., & Vasić, V. M. (2013). Acetylcholinesterase Inhibitors: Pharmacology and Toxicology. *Current Neuropharmacology*, *11*(3), 315–335. doi:10.2174/1570159X11311030006 PMID:24179466

Comans, R. N., Haller, M., & De Preter, P. (1991). Sorption of cesium on illite: Non-equilibrium behaviour and reversibility. *Geochimica et Cosmochimica Acta*, *55*(2), 433–440. doi:10.1016/0016-7037(91)90002-M

Constantin, M., Răut, I., Vasilescu, G., Arsene, M. L., & Jecu, L. (2012). Colonization and degradation of polyethylene composites by fungal strains isolated. *Scientific Bulletin. Series F. Biotechnologies*, *16*, 109–112.

Cooper, P. (1999). A review of the design and performance of vertical-flow and hybrid reed bed treatment systems. *Water Science and Technology*, *40*(3). PMID:17802831

Cordovil, C. M., Cabral, F., & Coutinho, J. (2007). Potential mineralization of nitrogen from organic wastes to ryegrass and wheat crops. *Bioresource Technology*, *98*(17), 3265–3268. doi:10.1016/j.biortech.2006.07.014 PMID:16905311

Correia, B., Chen, Z., Mendes, S., Martins, L. O., & Bento, I. (2011). Crystallization and preliminary X-ray diffraction analysis of the azoreductase PpAzoR from *Pseudomonas putida* MET94. *Acta Crystallographica. Section F, Structural Biology and Crystallization Communications*, *67*(1), 121–123. doi:10.1107/S1744309110048220 PMID:21206041

Costa, L. G. (2008). Toxic Effects of Pesticides. In C. D. Klaassen (Ed.), *Cassarett and Doull's Toxicology The Basic Science of Poisons* (7th ed.). New York: McGraw-Hill.

Cox, H. H., & Deshusses, M. A. (1998). Biological waste air treatment in biotrickling filters. *Current Opinion in Biotechnology*, *9*(3), 256–262. doi:10.1016/S0958-1669(98)80056-6 PMID:9690989

Crist, R. H., Oberholser, K., Shank, N., & Nguyen, M. (1981). Nature of bonding between metallic ions and algal cell walls. *Environmental Science & Technology*, *15*(10), 1212–1217. doi:10.1021/es00092a010

Cui, D., Li, G., Zhao, D., Gu, X., Wang, C., & Zhao, M. (2012). Purification and characterization of an azoreductase from *Escherichia coli* CD-2 possessing quinone reductase activity. *Process Biochemistry*, *47*(3), 544–549. doi:10.1016/j.procbio.2011.12.013

Cui, L. H., Ouyang, Y., Chen, Y., Zhu, X. Z., & Zhu, W. L. (2009). Removal of total nitrogen by *Cyperus alternifolius* from wastewaters in simulated vertical-flow constructed wetlands. *Ecological Engineering*, *35*(8), 1271–1274. doi:10.1016/j.ecoleng.2009.05.017

Cui, L. H., Zhu, X. Z., Ouyang, Y., Chen, Y., & Yang, F. L. (2011). Total phosphorus removal from domestic wastewater with Cyperus Alternifolius in vertical-flow constructed wetlands at the microcosm level. *International Journal of Phytoremediation*, *13*(7), 692–701. doi:10.1080/15226514.2010.525552 PMID:21972496

Cui, L., Liu, Z., Hui, L. F., & Si, L. F. (2011). Effect of cellobiase and surfactant supplementation on enzymatic hydrolysis of pretreated wheat straw. *BioResources*, *6*(4), 3850–3858.

Cunliffe, D., Smart, C. A., Alexander, C., & Vulfson, E. N. (1999). Bacterial adhesion at synthetic surfaces. *Applied and Environmental Microbiology*, *65*, 4995–5002. PMID:10543814

Cunningham, S. D., Shann, J. R., Crowley, D. E., & Anderson, T. A. (1997). Phytoremediation of contaminated water and soil. In: E. Kruger et al. (Ed.), Phytoremediation of soil and water allocainants, (pp 2–17). ACS Symp. Ser. 664. Am. Chemical Societ, Washington, DC. doi:10.1021/bk-1997-0664.ch001

Cusick, R., Bryan, B., Parker, D., Merrill, M., Mehanna, M., Kiely, P., ... Logan, B. E. (2011). Performance of a pilot-scale continuous flow microbial electrolysis cell fed winery wastewater. *Applied Microbiology and Biotechnology*, *89*(6), 2053–2063. doi:10.100700253-011-3130-9 PMID:21305277

Cycon, M., & Piotrowska-Seget, Z. (2009). Changes in bacterial diversity and community structure following pesticides addition to soil estimated by cultivation technique. *Ecotoxicology (London, England)*, *18*(5), 632–642. doi:10.100710646-009-0321-6 PMID:19444604

Da Silva, A. S., Inoue, H., Endo, T., Yano, S., & Bon, E. P. (2010). Milling pretreatment of sugarcane bagasse and straw for enzymatic hydrolysis and ethanol fermentation. *Bioresource Technology*, *101*(19), 7402–7409. doi:10.1016/j.biortech.2010.05.008

Danh, L. T., Truong, P., Mammucari, R., Tran, T., & Foster, N. (2009). Vetiver grass, Vetiveria zizanioides: A choice plant for phytoremediation of heavy metals and organic wastes. *International Journal of Phytoremediation*, *11*(8), 664–691. doi:10.1080/15226510902787302 PMID:19810597

Danjaji, I. D., Nawang, R., Ishiaku, U. S., Ismail, H., & Ishak, Z. A. M. M. (2002). Degradation studies and moisture uptake of sago-starch-filled linear low-density polyethylene composites. *Polymer Testing*, *21*(1), 75–81. doi:10.1016/S0142-9418(01)00051-4

Daramwal, N., & Gaur, R. (2004). Single cell protein. *Environmental microbiology and biotechnology, 218.*

Dary, M., Chamber-Pérez, M. A., Palomares, A. J., & Pajuelo, E. (2010). "In situ" phytostabilisation of heavy metal polluted soils using Lupinus luteus inoculated with metal resistant plant-growth promoting rhizobacteria. *Journal of Hazardous Materials*, *177*(1), 323–330. doi:10.1016/j.jhazmat.2009.12.035 PMID:20056325

Das N, Chandran P (2011) Microbial Degradation of Petroleum Hydrocarbon Contaminants: An Overview. *Biotechnology Research International*. doi:10.4061/2011/941810

Das, N., and Chandran, P. (2011). Microbial degradation of petroleum hydrocarbon contaminants: An overview. *Biotechnology Research International.*

Dashtban, M., Schraft, H., & Qin, W. (2009). Fungal bioconversion of lignocellulosic residues; opportunities & perspectives. *International Journal of Biological Sciences*, *5*(6), 578–595. doi:10.7150/ijbs.5.578 PMID:19774110

Dashtban, M., Schraft, H., Syed, T. A., & Qin, W. (2010). Fungal biodegradation and enzymatic modification of lignin. *International Journal of Biochemistry and Molecular Biology*, *1*, 36–50. PMID:21968746

Das, S. K., & Guha, A. K. (2007). Biosorption of chromium by Termitomyces clypeatus. *Colloids and Surfaces. B, Biointerfaces*, *60*(1), 46–54. doi:10.1016/j.colsurfb.2007.05.021 PMID:17618091

Dave, H., Rao, P. V. C., & Desai, J. D. (1997). Biodegradation of starch-polyethylene films in soil and by microbial cultures. *World Journal of Microbiology & Biotechnology*, *13*(6), 655–658. doi:10.1023/A:1018566820659

Davies, B. E. (1983). A graphical estimation of the normal lead content of some British soils. *Geoderma*, *29*(1), 67–75. doi:10.1016/0016-7061(83)90031-9

Davis, T. A., Volesky, B., & Mucci, A. (2003). A review of the biochemistry of heavy metal biosorption by brown algae. *Water Research*, *37*(18), 4311–4330. doi:10.1016/S0043-1354(03)00293-8 PMID:14511701

Dawkar, V. V., Jadhav, U. U., Ghodake, G. S., & Govindwar, S. P. (2009). Effect of inducers on the decolorization and biodegradation of textile azo dye Navy blue 2GL by *Bacillus* sp. VUS. *Biodegradation*, *20*(6), 777–787. doi:10.100710532-009-9266-y PMID:19468842

Dayanandan, A., Rani, S., Shanmugavel, M., Gnanamani, A., & Rajakumar, G. S. (2013). Enhanced production of Aspergillus tamarii lipase for recovery of fat from tannery fleshings. *Brazilian Journal of Microbiology*, *44*(4), 1089–1095. doi:10.1590/S1517-83822013000400010 PMID:24688497

de Albuquerque Wanderley, M. C., Martín, C., de Moraes Rocha, G. J., & Gouveia, E. R. (2013). Increase in ethanol production from sugarcane bagasse based on combined pretreatments and fed-batch enzymatic hydrolysis. *Bioresource Technology*, *128*, 448–453. doi:10.1016/j.biortech.2012.10.131

de Best, J. H., Hunneman, P., Doddema, H. J., Janssen, D. B., & Harder, W. (1999). Transformation of carbon tetrachloride in an anaerobic packed-bed reactor without addition of another electron donor. *Biodegradation*, *10*(4), 287–295. doi:10.1023/A:1008309003059 PMID:10633544

De Haan, F. A. M., & Van Riemsdijk, W. H. (1986). Behaviour of Inorganic Contaminants in Soil. In J. W. Assink & W. J. Van Den Brink (Eds.), *Contaminated Soil: First International TNO Conference on Contaminated Soil, November 11–15, Utrecht, The Netherlands* (pp. 19-32). Dordrecht: Springer Netherlands. 10.1007/978-94-009-5181-5_3

De la Fuente, C., Clemente, R., Martínez-Alcalá, I., Tortosa, G., & Bernal, M. P. (2011). Impact of fresh and composted solid olive husk and their water-soluble fractions on soil heavy metal fractionation; microbial biomass and plant uptake. *Journal of Hazardous Materials*, *186*(2), 1283–1289. doi:10.1016/j.jhazmat.2010.12.004 PMID:21216095

Deans, J. R., & Dixon, B. G. (1992). Uptake of Pb2+ and Cu2+ by novel biopolymers. *Water Research*, *26*(4), 469–472. doi:10.1016/0043-1354(92)90047-8

Debarati, P., Gunjan, P., Janmejay, P., & Rakesh, V. J. K. (2005). Accessing Microbial Diversity for Bioremediation and Environmental Restoration. *Trends in Biotechnology*, *23*(3), 135–142. doi:10.1016/j.tibtech.2005.01.001 PMID:15734556

Dechamps, C., Roosens, N. H., Hotte, C., & Meerts, P. (2005). Growth and mineral element composition in two ecotypes of *Thlaspi caerulescens* on Cd contaminated soil. *Plant and Soil*, *273*(1), 327–335. doi:10.100711104-005-0099-0

Deller, S., Sollner, S., Trenker-El-Toukhy, R., Jelesarov, I., Gübitz, G. M., & Macheroux, P. (2006). Characterization of a thermostable NADPH:FMN oxidoreductase from the mesophilic bacterium *Bacillus subtilis*. *Biochemistry*, *45*(23), 7083–7091. doi:10.1021/bi052478r PMID:16752898

Demirbas, A. (2005). Bioethanol from Cellulosic Materials: A Renewable Motor Fuel from Biomass. *Energy Sources*, *27*(4), 327–337. doi:10.1080/00908310390266643

Demirbas, A. (2008). Biofuels sources, biofuel policy, biofuel economy and global biofuel projections. *Energy Conversion and Management*, *49*(8), 2106–2116. doi:10.1016/j.enconman.2008.02.020

Demirbas, A. (2008). The importance of bioethanol and biodiesel form biomass. *Energy Source. Part B*, *3*, 177–185.

Department of Economic Affairs, Ministry of Finance, Government of India. (2009, November). Position Paper on the Solid Waste Management Sector in India. *Public Private Parternships in India.* Retrieved from http://www.seas.columbia.edu/earth/wtert/sofos/Sustainable%20Solid%20Waste%20Management%20in%20India_Final.pdf

Devi, R., & Dahiya, R. (2006). Chemical oxygen demand (COD) reduction in domestic wastewater by fly ash and brick kiln ash. *Water, Air, and Soil Pollution*, *174*(1-4), 33–46. doi:10.100711270-005-9022-9

di Domenico, A., Silano, V., Viviano, G., & Zapponi, G. (1980). Accidental release of 2,3,7,8-tetrachlorodibenzo-p-dioxin (TCDD) at Seveso, Italy. IV. Vertical distribution of TCDD in soil. *Ecotoxicology and Environmental Safety*, *4*(3), 327–338. doi:10.1016/0147-6513(80)90034-2 PMID:7439102

Di Sioudi, B. D., Miller, C. E., Lai, K., Grimsley, J. K., & Wild, J. R. (1999). Rational design of organophosphorus hydrolase for altered substrate specificities. *Chemico-Biological Interactions*, *119*, 211–223. doi:10.1016/S0009-2797(99)00030-7 PMID:10421455

Dias, M. O. S., Ensinas, A. V., Nebra, S. A., Maciel Filho, R., Rossell, C. E. V., & Maciel, M. R. W. (2009). Production of bioethanol and other bio-based materials from sugarcane bagasse: Integration to conventional bioethanol production process. *Chemical Engineering Research & Design*, *87*(9), 1206–1216. doi:10.1016/j.cherd.2009.06.020

Dibble, J. T., & Bartha, R. (1979). *Effects of environmental parameters on the biodegradation of oil sludges*. Applied Environmental Biology.

Dickinson, N. M, Baker, A. J. M., Doronila, A., Laidlaw, S., and Reeves, R. D. (2009). Phytoremediation of inorganics: realism and synergies. *International Journal of Phytoremediation, 11*(2), 97-114.

Ding, A., Yang, Y., Sun, G., & Wu, D. (2016). Impact of applied voltage on methane generation and microbial activities in an anaerobic microbial electrolysis cell (MEC). *Chemical Engineering Journal, 283*, 260–265. doi:10.1016/j.cej.2015.07.054

Donlan, R. M. (2002). Biofilms: Microbial life on surfaces. *Emerging Infectious Diseases, 8*(9), 881–890. doi:10.3201/eid0809.020063 PMID:12194761

Dos Santos, A. B., Cervantes, F. J., & van Lier, J. B. (2007). Review paper on current technologies for decolourisation of textile wastewaters: Perspectives for anaerobic biotechnology. *Bioresource Technology, 98*(12), 2369–2385. doi:10.1016/j.biortech.2006.11.013 PMID:17204423

Douglas, M. R., & Tooker, J. F. (2015). Large-scale deployment of seed treatments has driven rapid increase in use of neonicotinoid insecticides and preemptive pest management in U.S. field crops. *Environmental Science & Technology, 49*(8), 5088–5097. doi:10.1021/es506141g PMID:25793443

Dragović, S., Ćujić, M., Slavković-Beškoski, L., Gajić, B., Bajat, B., Kilibarda, M., & Onjia, A. (2013). Trace element distribution in surface soils from a coal burning power production area: A case study from the largest power plant site in Serbia. *Catena, 104*, 288–296. doi:10.1016/j.catena.2012.12.004

Drennan, C. L., Pattridge, K. A., Weber, C. H., Metzger, A. L., Hoover, D. M., & Ludwig, M. L. (1999). Refined structures of oxidized flavodoxin from *Anacystis nidulans. Journal of Molecular Biology, 294*(3), 711–724. doi:10.1006/jmbi.1999.3151 PMID:10610791

Drzyzga, O., El Mamouni, R., Agathos, S. N., & Gottschal, J. C. (2002). Dehalogenation of chlorinated ethenes and immobilization of nickel in anaerobic sediment columns under sulfidogenic conditions. *Environmental Science & Technology, 36*(12), 2630–2635. doi:10.1021/es010184x PMID:12099458

Dubey, K. K., & Fulekar, M. H. (2011). Effect of pesticides on the seed germination of *Cenchrus setigerus* and *Pennisetum pedicellatum* as monocropping and co-cropping system: Implications for rhizospheric bioremediation. *Romanian Biotechnological Letters, 16*(1), 5908–5918.

Dubey, S. K., Dubey, J., Viswas, A. J., & Tiwari, P. (2011). Studies on cyanobacterial biodiversity in paper mill and pharmaceutical industrial effluents. *British Biotechnology Journal, 1*(3), 61–67. doi:10.9734/BBJ/2011/395

Dubey, S. K., Meena, R. K., Sao, S., Pate, J., Thakur, S., & Shukla, P. (2014). Isolation and Characterization of Cellulose Degrading Bacteria from Biogas Slurry and their RAPD profiling. *Current Research in Microbiology and Biotechnology., 2*(4), 416–442.

Dudhane, M., Borde, M., & Jite, P. K. (2012). Effect of aluminium toxicity on growth responses and antioxidant activities in Gmelina arborea Roxb. inoculated with AM fungi. *International Journal of Phytoremediation, 14*(7), 643–655. doi:10.1080/15226514.2011.619230 PMID:22908633

Dufey, A. (2006). *Biofuels production, trade and sustainable development: emerging issues.* London: International Institute for Environment and Development.

Duff, S. J. B., & Murray, W. D. (1996). Bioconversion of forest products industry waste cellulosics to fuel ethanol: A review. *Bioresource Technology, 55*(1). doi:10.1016/0960-8524(95)00122-0

Du, H., Kommalapati, R. R., & Huque, Z. (2016). Assessment of Bioethanol applications on transportation vehicles in houston. *JFundam Renewable Energy App, 6*, 207.

Duran, N., & Esposito, E. (2000). Potential applications of oxidative enzymes and phenoloxidase-like compounds in wastewater and soil treatment: A review. *Applied Catalysis B: Environmental*, *28*(2), 83–99. doi:10.1016/S0926-3373(00)00168-5

Dutton, J. A. (n.d.). 7.2 Sugarcane Ethanol Production. Retrieved from https://www.e-education.psu.edu/egee439/node/647

Du, Z., Li, H., & Gu, T. (2007). A state of the art review on microbial fuel cells: A promising technology for wastewater treatment and bioenergy. *Biotechnology Advances*, *25*(5), 464–482. doi:10.1016/j.biotechadv.2007.05.004 PMID:17582720

ECA. (2017). European Chemicals Agency. Retrieved from https://echa.europa.eu/regulations/reach/legislation

Economopoulos, A. P. (1993). *Assessment of Sources of Air Water and Land Pollution. A Guide to Rapid Source Inventory Techniques and Their Use in Formulating Environmental Control Strategies. Part One: Rapid Inventory Techniques in Environmental Pollution. Part Two: Approaches for Consideration in Formulating Environmental Control Strategies.* WHO.

Edens, L., Bom, I., Ledeboer, A. M., Maat, J., Toonen, M. Y., Visser, C., & Verrips, C. T. (1984). Synthesis and processing of the plant protein thaumatin in yeast. *Cell*, *37*(2), 629–633. doi:10.1016/0092-8674(84)90394-5 PMID:6327079

Eggeman, T., & Elander, R. T. (2005). Process and economic analysis of pretreatment technologies. *Bioresource Technology*, *96*(18), 2019–2025. doi:10.1016/j.biortech.2005.01.017 PMID:16112490

Elisangela, F., Andrea, Z., Fabio, D. G., de Menezes Cristiano, R., Regina, D. L., & Artur, C.-P. (2009). Biodegradation of textile azo dyes by a facultative *Staphylococcus arlettae* strain VN-11 using a sequential microaerophilic/aerobic process. *International Biodeterioration & Biodegradation*, *63*(3), 280–288. doi:10.1016/j.ibiod.2008.10.003

Eljarrat, E., Marsh, G., Labandeira, A., & Barcelo, D. (2008). Effect of sewage sludges contaminated with polybrominated diphenylethers on agricultural soils. *Chemosphere*, *71*(6), 1079–1086. doi:10.1016/j.chemosphere.2007.10.047 PMID:18068210

Ellenhorn, M. J., & Barceloux, D. G. (1988). Medical Toxicology. In Diagnosis and Treatment of Human Poisoning. New York: Elsevier.

El-Rehim, H. A. A., Hegazy, E.-S. A., Ali, A. M., & Rabie, A. M. (2004). Synergistic effect of combining UV-sunlight-soil burial treatment on the biodegradation rate of LDPE/starch blends. *Journal of Photochemistry and Photobiology A Chemistry*, *163*(3), 547–556. doi:10.1016/j.jphotochem.2004.02.003

El-Shafei, H. A., El-Nasser, N. H. A., Kansoh, A. L., & Ali, A. M. (1998). Biodegradation of disposable polyethylene by fungi and *Streptomyces* species. *Polymer Degradation & Stability*, *62*(2), 361–365. doi:10.1016/S0141-3910(98)00019-6

El-Sheekh, M. M., Gharieb, M. M., & Abou-El-Souod, G. W. (2009). Biodegradation of dyes by some green algae and cyanobacteria. *International Biodeterioration & Biodegradation*, *63*(6), 699–704. doi:10.1016/j.ibiod.2009.04.010

Endo, S., Kimura, S., Takatsuji, T., Nanasawa, K., Imanaka, T., & Shizuma, K. (2012). Measurement of soil contamination by radionuclides due to the Fukushima Dai-ichi Nuclear Power Plant accident and associated estimated cumulative external dose estimation. *Journal of Environmental Radioactivity*, *111*, 18–27. doi:10.1016/j.jenvrad.2011.11.006 PMID:22137554

Ensley, B. D. (2000). Rationale for use of phytoremediation. In I. Raskin & B. D. Ensley (Eds.), *Phy- toremediation of toxic metals. Using plants to clean up the environment* (pp. 3–11). New York: John Wiley & Sons.

Environment Agency. (2004). *Guidance on the management of landfill gas* (LFTGN03). Author.

Environmental Protection Agency. (2008, October 16). Control of Hazardous Air Pollutants From Mobile Sources: Early Credit Technology Requirement Revision. *Federal Register*, *73*(201), 61358–61363. Retrieved from https://www.gpo.gov/fdsys/pkg/FR-2008-10-16/pdf/E8-24591.pdf

EPA Victoria. (2009). *Solid industrial waste hazard categorization and management* (EPA publication IWRG631). Author.

EPA. (April 2015). What is a Pesticide? Retrieved from https://www.epa.gov/minimum-risk-pesticides/what-pesticide

Erable, B., Goubet, I., Lamare, S., Legoy, M. D., & Maugard, T. (2006). Bioremediation of halogenated compounds: Comparison of dehalogenating bacteria and improvement of catalyst stability. *Chemosphere*, *65*(7), 1146–1152. doi:10.1016/j.chemosphere.2006.04.007 PMID:16723151

Erkurt, E., Unyayar, A., & Kumbur, H. (2007). Decolorization of synthetic dyes by White Rot Fungi, involving laccase enzyme in the process. *Process Biochemistry*, *42*(10), 1429–1435. doi:10.1016/j.procbio.2007.07.011

Ertuğrul, S., Bakır, M., & Dönmez, G. (2008). Treatment of dye-rich wastewater by an immobilized thermophilic cyanobacterial strain: *Phormidium* sp. *Ecological Engineering*, *32*(3), 244–248. doi:10.1016/j.ecoleng.2007.11.011

Escapa, A., Mateos, R., Martínez, E., & Blanes, J. (2016). Microbial electrolysis cells: An emerging technology for wastewater treatment and energy recovery. From laboratory to pilot plant and beyond. *Renewable & Sustainable Energy Reviews*, *55*, 942–956. doi:10.1016/j.rser.2015.11.029

Eubeler, J. P., Bernhard, M., & Knepper, T. P. (2010). Environmental biodegradation of synthetic polymers II. Biodegradation of different polymer groups. *Trends in Analytical Chemistry*, *29*(1), 84–100. doi:10.1016/j.trac.2009.09.005

Eur-Lex. (2009). Title and reference. Retrieved from http://eur-lex.europa.eu/legal-content/EN/TXT/?uri=CELEX:32009L0030

European Commission. Directorate General JRC, Joint Research Centre, Institute for prospective technological Studies, Technologies for sustainable development, European IPPC Bureau, Integrated Pollution Prevention and Control. (2001). Reference document on best available techniques for the tanning of hides and skins (Final draft). Retrieved from http://eippcb.jrc.es.2001

European Environment Agency (EEA). (2010, July 9). European Union Emmision Inventory report 1990-2008 under the UNECE Convention on Long-Range Transboundary Air Pollution (LRTAP). Retrieved June 1, 2017 from https://www.eea.europa.eu/publications/european-union-emission-inventory-report/

European Environment Agency (EEA). (2010, November 8). Atmoshperic Greenhouse Gas Concentrations. Retrieved July 6, 2017 from https://www.eea.europa.eu/data-and-maps/indicators/atmospheric-greenhouse-gas-concentrations/atmospheric-greenhouse-gas-concentrations-assessment-3

Evangelista-Barreto, N. S., Albuquerque, C. D., Vieira, R. H. S. F., & Campos-Takaki, G. M. (2009). Cometabolic decolorization of the reactive azo dye orange II by *Geobacillus stearothermophilus* UCP 986. *Textile Research Journal*, *79*(14), 1266–1273. doi:10.1177/0040517508087858

Fan, C. Y., & Krishnamurthy, S. (1995). Enzymes for Enhancing Bioremediation of Petroleum-Contaminated Soils: A Brief Review. *Journal of the Air & Waste Management Association*, *45*(6), 453–460. doi:10.1080/10473289.1995.10467375 PMID:7788508

Fang, Y., Nie, Z., Die, Q., Tian, Y., Liu, F., He, J., & Huang, Q. (2017). Organochlorine pesticides in soil, air, and vegetation at and around a contaminated site in southwestern China: Concentration, transmission, and risk evaluation. *Chemosphere*, *178*, 340–349. doi:10.1016/j.chemosphere.2017.02.151 PMID:28334674

Fantroussi, S. E., & Agathos, S. N. (2005). Is bioaugmentation a feasible strategy for pollutant removal and site remediation? *Current Opinion in Microbiology*, *8*(3), 268–275. doi:10.1016/j.mib.2005.04.011 PMID:15939349

FAO. (2003). Report on State of the World's Forests, Rome. Retrieved from http://www.fao.org/docrep /005/ Y7581E/ y7581e04.htm

FAO. (2017). Retrieved from http://www.fao.org/faostat/en/#home

Farooq, U., Khan, M. A., Athar, M., & Kozinski, J. A. (2011). Effect of modification of environmentally friendly biosorbent wheat (Triticum aestivum) on the biosorptive removal of cadmium (II) ions from aqueous solution. *Chemical Engineering Journal*, *171*(2), 400–410. doi:10.1016/j.cej.2011.03.094

Farzenah, R., Joh, L. D., Kashima, H., Reddy, A. P., & VanderGheynst, J. S. (2011). Selection of condition for cellulase and xylanase extraction from switchgrass colonized by Acidothermus cellulolyticus. *Applied Biochemistry and Biotechnology*, *164*(6), 793–803. doi:10.100712010-011-9174-6 PMID:21318368

Fathima, N., Rao, R., & Nair, B. U. (2012). Tannery solid waste to treat toxic liquid wastes: A new holistic paradigm. *Environmental Engineering Science*, *29*(6), 363–372. doi:10.1089/ees.2010.0445

Fawzy, E. M., Abdel-Motaal, F. F., & El-zayat, S. A. (2017). Biosorption of heavy metals onto different eco-friendly substrates. *Journal of Toxicology and Environmental Health Sciences*, *9*(5), 35–44.

Feng, J., Kweon, O., Xu, H., Cerniglia, C. E., & Chen, H. (2012). Probing the NADH - and methyl red-binding site of a FMN-dependent azoreductase (AzoA) from *Enterococcus faecalis. Archives of Biochemistry and Biophysics*, *520*(2), 99–107. doi:10.1016/j.abb.2012.02.010 PMID:22387379

Feng, X., Pisula, W., & Mullen, K. (2009). Large polycyclic aromatic hydrocarbons: Synthesis and discotic organisation. *Pure and Applied Chemistry*, *81*(12), 2203–2224. doi:10.1351/PAC-CON-09-07-07

Fengxia, G., Min, B., Yijun, M., Zupei, L., Yawen, L., & Haifeng, S. (2016). Temporal and spatial characteristics of lightning-produced nitrogen oxides in China. *Journal of Atmospheric and Solar-Terrestrial Physics*, *149*, 100–107. doi:10.1016/j.jastp.2016.10.007

Feng, Y., Zhang, Y., Chen, S., & Quan, X. (2015). Enhanced production of methane from waste activated sludge by the combination of high-solid anaerobic digestion and microbial electrolysis cell with iron–graphite electrode. *Chemical Engineering Journal*, *259*, 787–794. doi:10.1016/j.cej.2014.08.048

Ferguson, S. J., Richardson, D. J., & van Spanning, R. J. M. (2007). *Biochemistry and Molecular Biology of Nitrification Biology of the Nitrogen Cycle* (pp. 209–222). Amsterdam: Elsevier. doi:10.1016/B978-044452857-5.50015-1

Fishbein, L. (1979). Potential halogenated industrial carcinogenic and mutagenic chemicals II. Halogenated saturated hydrocarbons. *The Science of the Total Environment*, *11*(2), 163–195. doi:10.1016/0048-9697(79)90026-3 PMID:373115

Fitch, M. W., Pearson, N., Richards, G., & Burken, J. G. (1998). Biological fixed-film systems. *Water Environment Research*, *70*(4), 495–518. doi:10.2175/106143098X134226

Fitzpatrick, J. J., & O'keeffe, U. (2001). Influence of whey protein hydrolysate addition to whey permeate batch fermentations for producing lactic acid. *Process Biochemistry*, *37*(2), 183–186. doi:10.1016/S0032-9592(01)00203-5

Flouty, R., & Estephane, G. (2012). Bioaccumulation and biosorption of copper and lead by a unicellular algae Chlamydomonas reinhardtii in single and binary metal systems: A comparative study. *Journal of Environmental Management*, *111*, 106–114. doi:10.1016/j.jenvman.2012.06.042 PMID:22835654

Foladori, P., Ruaben, J., & Ortigara, A. R. (2013). Recirculation or artificial aeration in vertical flow constructed wetlands: A comparative study for treating high load wastewater. *Bioresource Technology*, *149*, 398–405. doi:10.1016/j.biortech.2013.09.099 PMID:24128403

Folberth, G. A., Butler, T. M., Collins, W. J., & Rumbold, S. T. (2015). Megacities and climate change – A brief overview. *Environmental Pollution*, *203*(Suppl. C), 235–242. doi:10.1016/j.envpol.2014.09.004 PMID:25300966

Folch, A., Vilaplana, M., Amado, L., Vicent, R., & Caminal, G. (2013). Fungal permeable reactive barrier to remediate groundwater in an artificial aquifer. *Journal of Hazardous Materials*, *262*, 554–560. doi:10.1016/j.jhazmat.2013.09.004 PMID:24095995

Fontanella, S., Bonhomme, S., Koutny, M., Husarova, L., Brusson, J.-M., Courdavault, J.-P., ... Delort, A.-M. (2010). Comparison of the biodegradability of various polyethylene films containing pro-oxidant additives. *Polymer Degradation & Stability*, *95*(6), 1011–1021. doi:10.1016/j.polymdegradstab.2010.03.009

Foo, K., & Hameed, B. (2009). Value-added utilization of oil palm ash: A superior recycling of the industrial agricultural waste. *Journal of Hazardous Materials*, *172*(2), 523–531. doi:10.1016/j.jhazmat.2009.07.091 PMID:19695771

Forbes, P. B. C., & Garland, R. M. (2016). Outdoor Air Pollution. *Comprehensive Analytical Chemistry*, *73*, 73–96. doi:10.1016/bs.coac.2016.02.004

Forgacs, E., Cserhati, T., & Oros, G. (2004). Removal of synthetic dyes from wastewater: A Review. *Environment International*, *30*(7), 953–971. doi:10.1016/j.envint.2004.02.001 PMID:15196844

Forootanfar, H., Moezzi, A., Aghaie-Khozani, M., Mahmoudjanlou, Y., Ameri, A., Niknejad, F., & Faramarzi, M. A. (2012). Synthetic dye decolorization by three sources of fungal laccase. *Iranian Journal of Environmental Health Sciences & Engineering*, *9*(1), 27. doi:10.1186/1735-2746-9-27 PMID:23369690

Foyle, T., Jennings, L., & Mulcahy, P. (2007). Compositional analysis of lignocellulosic materials: Evaluation of methods used for sugar analysis of waste paper and straw. *Bioresource Technology*, *98*(16), 3026–3036. doi:10.1016/j.biortech.2006.10.013 PMID:17142038

Frascari, D., Fraraccio, S., Nocentini, M., & Pinelli, D. (2013). Aerobic/anaerobic/aerobic sequenced biodegradation of a mixture of chlorinated ethenes, ethanes and methanes in batch bioreactors. *Bioresource Technology*, *128*, 479–486. doi:10.1016/j.biortech.2012.10.026 PMID:23201903

Frascari, D., Zanaroli, G., & Danko, A. S. (2015). In situ aerobic cometabolism of chlorinated solvents: A review. *Journal of Hazardous Materials*, *283*, 382–399. doi:10.1016/j.jhazmat.2014.09.041 PMID:25306537

Fritsche, W., & Hofrichter, M. (2000). In J. Klein (Ed.), *Aerobic degradation by microorganisms in Environmental Processes- Soil Decontamination.* Weinheim, Germany: Wiley-VCH.

Fuchs, J. G. (2010). Interactions Between Beneficial and Harmful. Microorganisms: From the Composting Process to Compost Application. In Insam et al... (Eds.), *Microbes at Work. From waste to resources* (pp. 213–230). Heidelberg: Springer-Verlag. doi:10.1007/978-3-642-04043-6_11

Fu, D., Teng, Y., Luo, Y., Tu, C., Li, S., Li, Z., & Christie, P. (2012). Effects of alfalfa and organic fertilizer on benzo [a] pyrene dissipation in an aged contaminated soil. *Environmental Science and Pollution Research International*, *19*(5), 1605–1611. doi:10.100711356-011-0672-4 PMID:22134861

Fujisawa, M., Hiral, H., & Nishida, T. (2001). Degrdation of polyethylene and nylon-66 by the laccase-mediator system. *Journal of Polymers and the Environment*, *9*(3), 103–108. doi:10.1023/A:1020472426516

Fujiwara, T., Saito, T., Muroya, Y., Sawahata, H., Yamashita, Y., Nagasaki, S., ... Tanaka, S. (2012). Isotopic ratio and vertical distribution of radionuclides in soil affected by the accident of Fukushima Dai-ichi nuclear power plants. *Journal of Environmental Radioactivity*, *113*, 37–44. doi:10.1016/j.jenvrad.2012.04.007 PMID:22634028

Fukumori, F., & Hausinger, R. P. (1993). Alcaligenes eutrophus JMP134 "2,4-dichlorophenoxyacetate monooxygenase" is an alpha-ketoglutarate-dependent dioxygenase. *Journal of Bacteriology*, *175*(7), 2083–2086. doi:10.1128/jb.175.7.2083-2086.1993 PMID:8458850

Furukawa, K. (2003). 'Super bugs' for Bioremediation. *Trends in Biotechnology*, *21*(5), 187–190. doi:10.1016/S0167-7799(03)00054-4 PMID:12727376

Gadd, G. M. (2009). Biosorption: Critical review of scientific rationale, environmental importance and significance for pollution treatment. *Journal of Chemical Technology and Biotechnology (Oxford, Oxfordshire)*, *84*(1), 13–28. doi:10.1002/jctb.1999

Gadd, G. M., & White, C. (1993). Microbial treatment of metal pollution—a working biotechnology? *Trends in Biotechnology*, *11*(8), 353–359. doi:10.1016/0167-7799(93)90158-6 PMID:7764182

Gajaraj, S., Huang, Y., Zheng, P., & Hu, Z. (2017). Methane production improvement and associated methanogenic assemblages in bioelectrochemically assisted anaerobic digestion. *Biochemical Engineering Journal*, *117*, 105–112. doi:10.1016/j.bej.2016.11.003

Galliher, P. M., Hodge, G., Guertin, P., Chew, L., & Deloggio, T. (2011). Single-Use Bioreactor Platform for Microbial Fermentation. *Single-Use Technology in Biopharmaceutical Manufacture*, 241-250.

Gangola, S., Pankaj, Khati, P., & Sharma, A. (2015). Mycoremediation of Imidaclopid in the presence of different soil amendment using trichoderma longibrachiatum and Aspergillus oryzae isolated from pesticide contaminated agricultural field of Uttarakhand. *Journal of Bioremediation & Biodegradation*, *6*(5), 2–5.

Gangola, S., Pankaj, Negi, G., Srivastava, A., & Sharma, A. (2014). Enhanced Biodegradation of Endosulfan using a consortium of Aspergillus and Trichoderma spp. isolated from agriculture field of Kumaun region of Uttarakhand. *Pesticide Research Journal*, *27*(2), 223–230.

Gao, F., Yang, Z. H., Li, C., & Jin, W. H. (2013). Saline domestic sewage treatment in constructed wetlands: study of plant selection and treatment characteristics. *Desalination and Water Treatment.*

García-Aparicio, M. P., Ballesteros, I., González, A., Oliva, J., Ballesteros, M., & Negro, M. (2006). Effect of inhibitors released during steam-explosion pretreatment of barley straw on enzymatic hydrolysis. *Applied Biochemistry and Biotechnology*, *129*(1-3), 278–288. doi:10.1385/ABAB:129:1:278 PMID:16915647

García-Cubero, M. T., González-Benito, G., Indacoechea, I., Coca, M., & Bolado, S. (2009). Effect of ozonolysis pretreatment on enzymatic digestibility of wheat and rye straw. *Bioresource Technology*, *100*(4), 1608–1613. doi:10.1016/j.biortech.2008.09.012 PMID:18951781

García, I. L., Lopez, J. A., Dorado, M. P., Kopsahelis, N., Alexandri, M., Papanikolaou, S., & Koutinas, A. A. (2013). Evaluation of by-products from the biodiesel industry as fermentation feedstock for poly (3-hydroxybutyrate-co-3-hydroxyvalerate) production by *Cupriavidus necator. Bioresource Technology*, *130*, 16–22. doi:10.1016/j.biortech.2012.11.088 PMID:23280181

Garcia, M. A., Alonso, J., & Melgar, M. J. (2005). Agaricus macrosporus as a potential bioremediation agent for substrates contaminated with heavy metals. *Journal of Chemical Technology and Biotechnology (Oxford, Oxfordshire)*, *80*(3), 325–330. doi:10.1002/jctb.1203

Garcia-Ochoa, F., & Gomez, E. (2009). Bioreactor scale-up and oxygen transfer rate in microbial processes: An overview. *Biotechnology Advances*, *27*(2), 153–176. doi:10.1016/j.biotechadv.2008.10.006 PMID:19041387

Garde, A., Jonsson, G., Schmidt, A. S., & Ahring, B. K. (2002). Lactic acid production from wheat straw hemicellulose hydrolysate by *Lactobacillus pentosus* and *Lactobacillus brevis*. *Bioresource Technology*, *81*(3), 217–223. doi:10.1016/S0960-8524(01)00135-3 PMID:11800488

Garfield, S. (2002). *Mauve: how one man invented a color that changed the world.* W. W. Norton & Company.

Gavrilescu, M. (2005). Fate of Pesticides in the Environment and its Bioremediation. *Engineer in Life Science*, 5-6, 497-526.

GEOELEC. (n.d.). Geothermal Electricity: Potential for CO2 Mitigation. Retrieved on July 7, 2017 from http://www.geoelec.eu/wp-content/uploads/2013/11/Geoelec-CO2-mitigation.pdf

George, J., Masto, R. E., Ram, L. C., Das, T. B., Rout, T. K., & Mohan, M. (2015). Human exposure risks for metals in soil near a coal-fired power-generating plant. *Archives of Environmental Contamination and Toxicology*, *68*(3), 451–461. doi:10.100700244-014-0111-x PMID:25537921

Gevao, B., Semple, K. T., & Jones, K. C. (2000). Bound pesticide residues in soils: a review. *Environmental Pollution (Barking, Essex: 1987)*, *108*(1), 3-14.

Ghazali, K. A., Salleh, S. F., Riayatsyah, T. M., Aditiya, H. B., & Mahlia, T. M. (2016). The effect of dilute acid pre-treatment process in bioethanol production from durian (*Durio zibethinus*) seeds waste. In *Proceedings of IOP Conference Series: Earth and Environmental Science*. Malaysia: IOP Publishing. 10.1088/1755-1315/32/1/012058

Ghodake, G., Jadhav, U., Tamboli, D., Kagalkar, A., & Govindwar, S. (2011). Decolorization of textile dyes and degradation of mono-azo dye amaranth by *Acinetobacter calcoaceticus* NCIM 2890. *Indian Journal of Microbiology*, *51*(4), 501–508. doi:10.100712088-011-0131-4 PMID:23024414

Ghosh, A., Chakroborti, A., Das, F., Chattopadhyay, D., & Chakrabarti, K. (2004). Use of bacteria for utilization of solid waste from poultry and leather industry. *Journal of Indian Leather Technologists'. Association.*, *55*, 619–623.

Ghosh, P. G., Sawant, N. A., Patil, S. N., & Aglave, B. A. (2010). Microbial Biodegradation of Organophosphate Pesticides. *International Journal of Biotechnology and Biochemistry*, *6*(6), 871–876.

Gianfreda, L., & Rao, M. A. (2004). Potential of extra cellular enzymes in remediation of polluted soils: A review. *Enzyme and Microbial Technology*, *35*(4), 339–354. doi:10.1016/j.enzmictec.2004.05.006

Gibson, D. T., & Parales, R. E. (2000). Aromatic hydrocarbon dioxygenases in environmental biotechnology. *Current Opinion in Biotechnology*, *11*(3), 236–243. doi:10.1016/S0958-1669(00)00090-2 PMID:10851146

Gietz, R. D., & Schiest, R. H. (2007). High-efficiency yeast transformation using the LiAc/SS carrier DNA/PEG method. *Nature Protocols*, *2*(1), 31–34. doi:10.1038/nprot.2007.13 PMID:17401334

Gikas, G. D., & Tsihrintzis, V. A. (2012). A small-size vertical flow constructed wetland for on-site treatment of household wastewater. *Ecological Engineering*, *44*, 337–343. doi:10.1016/j.ecoleng.2012.04.016

Gilan (Orr), I., Hadar, Y., & Sivan, A. (2004). Colonization, biofilm formation and biodegradation of polyethylene by a strain of *Rhodococcus ruber*. *Applied Microbiology and Biotechnology*, *65*, 97–104. PMID:15221232

Gilliom, R. J. (2007). Pesticides in U.S. Streams and Groundwater. *Environmental Science & Technology*, *41*(10), 3408–3414. doi:10.1021/es072531u PMID:17547156

Glick, B. R., Cheng, Z., Czarny, J., & Duan, J. (2007). Promotion of plant growth by ACC deaminase-producing soil bacteria. *European Journal of Plant Pathology*, *119*(3), 329–339. doi:10.100710658-007-9162-4

Godia, F., & Sola, C. (1995). Fluidized-bed bioreactors. *Biotechnology Progress, 11*(5), 479–497. doi:10.1021/bp00035a001

Goel, A. K. (2015). Anthrax: A disease of biowarfare and public health importance. *World Journal of Clinical Cases: WJCC, 3*(1), 20–33. doi:10.12998/wjcc.v3.i1.20 PMID:25610847

Goh, C. S., Tan, K. T., Lee, K. T., & Bhatia, S. (2010). Bioethanol from lignocellulose: Status, perspectives and challenges in Malaysia. *Bioresource Technology, 101*(13), 4834–4841. doi:10.1016/j.biortech.2009.08.080 PMID:19762229

Goldberg, S. Z., Eisenberg, R., Miller, J. S., & Epstein, A. J. (1976). Tetrakis (methyl isocyanide) palladium (II) tetrakis (7, 7, 8, 8-tetracyano-p-quinodimethane), [Pd (CNMe) 4](TCNQ) 4.2 MeCN: Synthesis, structure, and physical properties. *Journal of the American Chemical Society, 98*(17), 5173–5182. doi:10.1021/ja00433a020

Gold, M. H., Youngs, H. L., & Gelpke, M. D. S. (2000). Manganese peroxidase. In A. Sigel & H. Sigel (Eds.), *Metal ions in biological systems* (pp. 559–586). New York: Marcel Dekker Inc.

Gomaa, E. Z. (2012). Production and characteristics of a heavy metals removing bioflocculant produced by Pseudomonas aeruginosa. *Polish Journal of Microbiology, 61*(4), 281–289. PMID:23484411

Gomashe, A. V., Gulhane, P. A., & Bezalwar, P. M. (2013). Isolation and Screening of Cellulose Degrading Microbes from Nagpur Region Soil. *International Journal of Life Sciences (Kathmandu), 1*(4), 291–293.

Gonçalves, A. M. D., Mendes, S., de Sanctis, D., Martins, L. O., & Bento, I. (2013). The crystal structure of *Pseudomonas putida* azoreductase — The active site revisited. *The FEBS Journal, 280*(24), 6643–6657. doi:10.1111/febs.12568 PMID:24127652

Gopinath, K. P., Murugesan, S., Abraham, J., & Muthukumar, K. (2009). *Bacillus* sp. mutant for improved biodegradation of Congo red: Random mutagenesis approach. *Bioresource Technology, 100*(24), 6295–6300. doi:10.1016/j.biortech.2009.07.043 PMID:19692233

Gordon, T., Stanek, L. W., & Brown, J. (2014). *Pollution. In Air in Encyclopedia of Toxicology A2 - Wexler, Philip Encyclopedia of Toxicology* (3rd ed., pp. 995–1002). Oxford: Academic Press.

Goszczynski, S., Paszczynski, A., Pasti-Grigsby, M. B., Crawford, R. L., & Crawford, D. L. (1994). New pathway for degradation of sulfonated azo dyes by microbial peroxidases of *Phanerochaete chrysosporium* and *Streptomyces chromofuscus*. *Journal of Bacteriology, 176*(5), 1339–1347. doi:10.1128/jb.176.5.1339-1347.1994 PMID:8113173

Grady, C. L. Jr, Daigger, G. T., Love, N. G., & Filipe, C. D. (2011). *Biological wastewater treatment*. CRC press.

Gray, K. A., Zhao, L., & Emptage, M. (2006). Bioethanol. *Current Opinion in Chemical Biology, 10*(2), 141–146. doi:10.1016/j.cbpa.2006.02.035 PMID:16522374

Gray, K. R., Sherman, K., & Biddlestone, A. J. (1971). A review of composting Part 2- The practical process. *Process Biochemistry, 6*(10), 22–28.

Grbic´-Galic, ´. D., Churchman-Eisel, N., & Mrakovic´, I. (1990). Microbial transformation of styrene by anaerobic consortia. *The Journal of Applied Bacteriology, 69*(2), 247–260. doi:10.1111/j.1365-2672.1990.tb01516.x PMID:2272946

Green, M., Friedler, E., & Safrai, I. (1998). Enhancing nitrification in vertical flow constructed wetland utilizing a passive air pump. *Water Research, 32*(12), 3513–3520. doi:10.1016/S0043-1354(98)00182-1

Greger, M., & Landberg, T. (1999). Use of willow in phytoextraction. *International Journal of Phytoremediation, 1*(2), 115–123. doi:10.1080/15226519908500010

Gribble, G. W. (1994). The natural production of chlorinated compounds. *Environmental Science & Technology, 28*(7), 310–319. doi:10.1021/es00056a712 PMID:22662801

Gribble, G. W. (2003). The diversity of naturally produced organohalogens. *Chemosphere*, *52*(2), 289–297. doi:10.1016/S0045-6535(03)00207-8 PMID:12738253

Gribble, G. W. (2004). Natural organohalogens: A new frontier for medicinal agents? *Journal of Chemical Education*, *81*(10), 1441. doi:10.1021/ed081p1441

Griffiths, A. D., & Tawfik, D. S. (2003). Directed evolution of an extremely fast phosphotriesterase by in vitro compartmentalization. *The EMBO Journal*, *22*(1), 24–35. doi:10.1093/emboj/cdg014 PMID:12505981

Griza, F. T., Ortiz, K. S., Geremias, G., & Thiesen, F. V. (2008). Avaliação da Contaminação por Organofosforados em Águas Superficiais no Município de Rondinha- Rio Grandedo Sul. *Quimica Nova*, *31*(7), 1631–1635. doi:10.1590/S0100-40422008000700006

Grostern, A., & Edwards, E. A. (2006). A 1, 1, 1-trichloroethane-degrading anaerobic mixed microbial culture enhances biotransformation of mixtures of chlorinated ethenes and ethanes. *Applied and Environmental Microbiology*, *72*(12), 7849–7856. doi:10.1128/AEM.01269-06 PMID:17056695

Gu, J. D., Ford, T. E., Mitton, D. B., & Mitchel, R. (2000). Microbial degradation of polymeric materials. In R. W. Revie (Ed.), *The Uhlig's corrosion handbook* (pp. 421–438). New York, NY: Wiley.

Guo, J., Xu, W., & Ma, M. (2012 a). The assembly of metals chelation by thiols and vacuolar compartmentalization conferred increased tolerance to and accumulation of cadmium and arsenic in transgenic Arabidopsis thaliana. *Journal of Hazardous Materials*, *199*, 309–313. doi:10.1016/j.jhazmat.2011.11.008 PMID:22119299

Guo, J., Zhou, J., Wang, D., Tian, C., Wang, P., Salah Uddin, M., & Yu, H. (2007). Biocalalyst effects of immobilized anthraquinone on the anaerobic reduction of azo dyes by the salt-tolerant bacteria. *Water Research*, *41*(2), 426–432. doi:10.1016/j.watres.2006.10.022 PMID:17129594

Guo, X. Y., Hao, L. J., Qiu, P. Z., Chen, R., Xu, J., Kong, X. J., ... Wang, N. (2016). Pollution characteristics of 23 veterinary antibiotics in livestock manure and manure-amended soils in Jiangsu province, China. *Journal of Environmental Science and Health. Part. B, Pesticides, Food Contaminants, and Agricultural Wastes*, *51*(6), 383–392. doi:10.1080/03601234.2016.1142743 PMID:26963628

Guo, Y.-M., Liu, Y.-G., Zeng, G.-M., Hu, X.-J., Xu, W.-H., Liu, Y.-Q., ... Huang, H.-J. (2014). An integrated treatment of domestic wastewater using sequencing batch biofilm reactor combined with vertical flow constructed wetland and its artificial neural network simulation study. *Ecological Engineering*, *64*, 18–26. doi:10.1016/j.ecoleng.2013.12.040

Gupta, P. K. (2004). Pesticide exposure—Indian scene. *Toxicology*, *198*(1–3), 83–90. doi:10.1016/j.tox.2004.01.021 PMID:15138033

Gupta, S., Mohan, K., Prasad, R., Gupta, S., & Kansal, A. (1998). Solid Waste Management in India, Options and Opportunity. *Resources, Conservation and Recycling*, *24*(2), 115–137. doi:10.1016/S0921-3449(98)00033-0

Gurav, A., Ghosh, J., & Kulkarni, G. (2011). Decolorization of anthroquinone based dye Vat Red 10 by *Pseudomonas desmolyticum* NCIM 2112 and *Galactomyces geotrichum* MTCC 1360. *International Journal for Biotechnology and Molecular Biology Research*, *2*, 93–97.

Gurgel, L. V., Pimenta, M. T., & Curvelo, A. A. (2014). Enhancing liquid hot water (LHW) pretreatment of sugarcane bagasse by high pressure carbon dioxide (HP-CO_2). *Industrial Crops and Products*, *57*, 141–149. doi:10.1016/j.indcrop.2014.03.034

Hadad, D., Geresh, S., & Sivan, A. (2005). Biodegradation of polyethylene by the thermophilic bacterium *Brevibacillus borstelensis*. *Journal of Applied Microbiology*, *98*(5), 1093–1100. doi:10.1111/j.1365-2672.2005.02553.x PMID:15836478

Hakkarainen, M., & Albertsson, A.-C. (2004). Environmental degradation of polyethylene. *Advances in Polymer Science, 169*, 177–199. doi:10.1007/b13523

Hamdi, H., Benzarti, S., Aoyama, I., & Jedidi, N. (2012). Rehabilitation of degraded soils containing aged PAHs based on phytoremediation with alfalfa (Medicago sativa L.). *International Biodeterioration & Biodegradation, 67*, 40–47.

Hamelinck, C. N., Hooijdonk, G. V., & Faaij, A. P. (2005). Ethanol from lignocellulosic biomass: Techno-economic performance in short-, middle- and long-term. *Biomass and Bioenergy, 28*(4), 384–410. doi:10.1016/j.biombioe.2004.09.002

Hamid, T. H. T. A., Hamid, A. A. A., & Huyop, F. (2013). A review on non-stereospecific haloalkanoic acid dehalogenases. *African Journal of Biotechnology, 10*(48), 9725–9736. doi:10.5897/AJB11.934

Harada, N., & Nonaka, M. (2012). Soil radiocesium distribution in rice fields disturbed by farming process after the Fukushima Dai-ichi Nuclear Power Plant accident. *The Science of the Total Environment, 438*, 242–247. doi:10.1016/j.scitotenv.2012.05.032 PMID:23000550

Hardman, D. J. (1991). Biotransformation of halogenated compounds. *Critical Reviews in Biotechnology, 11*(1), 1–40. doi:10.3109/07388559109069182 PMID:1863980

Harmsen, P., Huijgen, W., Bermudez, L., & Bakker, R. (2010). Literature review of physical and chemical pretreatment processes for lignocellulosic biomass (Tech. Rep. 1184). Biosynergy, Wageningen UR Food & Biobased Research.

Harshvardhan, K., & Jha, B. (2013). Biodegradation of low-density polyethylene by marine bacteria from pelagic waters, Arabian sea, India. *Marine Pollution Bulletin, 77*(1-2), 100–106. doi:10.1016/j.marpolbul.2013.10.025 PMID:24210946

Harvey, R. G. (1996). *Polycyclic aromatic hydrocarbons*. New York: Wiley-VCH.

Hasan, F., Shah, A. A., Hameed, A., & Ahmed, S. (2007). Synergistic effect of photo- and chemical treatment on the rate of biodegradation of low density polyethylene by Fusarium sp. AF4. *Journal of Applied Polymer Science, 105*(3), 1466–1470. doi:10.1002/app.26328

Hasan, S. E. (1996). *Geology and hazardous waste management*. Upper Saddle River, NJ: Prentice Hall.

Hattori, M., Yamaji-Tsukamoto, K. A., Kumagai, H., Feng, Y., & Takahashi, K. (1998). Antioxidative activity of soluble elastin peptides. *Journal of Agricultural and Food Chemistry, 46*(6), 2167–2170. doi:10.1021/jf970484r

Hayo, M. G., & Werf, V. D. (1996). Assessing the impact of pesticides on the environment. *Agriculture, Ecosystems & Environment, 60*(2-3), 81–96. doi:10.1016/S0167-8809(96)01096-1

Heidrich, E., Curtis, T., & Dolfing, J. (2011). Determination of the Internal Chemical Energy of Wastewater. *Environmental Science & Technology, 45*(2), 827–832. doi:10.1021/es103058w PMID:21142001

Heidrich, E., Edwards, S., Dolfing, J., Cotterill, S., & Curtis, T. (2014). Performance of a pilot scale microbial electrolysis cell fed on domestic wastewater at ambient temperatures for a 12month period. *Bioresource Technology, 173*, 87–95. doi:10.1016/j.biortech.2014.09.083 PMID:25285764

Heinfling, A., Martinez, M. J., Martínez, A. T., Bergbauer, M., & Szewzyk, U. (1998). Transformation of industrial dyes by manganese peroxidases from *Bjerkandera adusta* and *Pleurotus eryngii* in a manganese-independent reaction. *Applied and Environmental Microbiology, 64*, 2788–2793. PMID:9687431

Held, T., & Dörr, H. (2000). *In Situ* Remediation. *Biotechnology, 11b*, 350–370.

Hellmann, B., Zelles, L., Palojarvi, A., & Bai, Q. (1997). Emission of climate-relevant trace gases and succession of microbial communities during open-window composting. *Applied and Environmental Microbiology, 63*, 1011–1018. PMID:16535535

Hendriks, A., & Zeeman, G. (2009). Pretreatments to enhance the digestibility of lignocellulosic biomass. *Bioresource Technology*, *100*(1), 10–18. doi:10.1016/j.biortech.2008.05.027 PMID:18599291

Henniges, U., Hasani, M., Potthast, A., Westman, G., & Rosenau, T. (2013). Electron beam irradiation of cellulosic materials-opportunities and limitations. *Materials (Basel)*, *6*(5), 1584–1598. doi:10.3390/ma6051584 PMID:28809230

Hicks, B. (2013). Agricultural pesticides and human health. In: National Association of Geoscience Teachers. Retrieved from http://serc.carleton.edu/NAGTWorkshops/health/case_studies/pesticides.html

Higa, T. (1991). Effective Microorganisms: A biotechnology for mankind. In J.F. Parr, S.B. Hornick, & C.E. Whitman (Eds.), *Proceedings of the First International Conference on Kyusei Nature Farming* (pp.8-14). U.S. Department of Agriculture, Washington, DC.

Higa, T., & Parr, J. F. (1994). *Beneficial and Effective Microorganisms for a Sustainable Agriculture and Environment* (p. 16). Atami, Japan: International Nature Farming Research Center.

Hoagland, R. E., Zablotowicz, R. M., & Locke, M. A. (1993). Propanil metabolism by rhizosphere microflora. In J. A. Anderson & J. R. Coats (Eds.), *Bioremediation through Rhizosphere Technology* (pp. 160–183). Washington, D.C.: American Chemical Society.

Hocking, P. J. (1985). Effects of sodium and potassium chlorides on the growth and accumulation of mineral ions by Cyperus involucratus Rottb. *Aquatic Botany*, *21*(3), 201–217. doi:10.1016/0304-3770(85)90049-X

Hodaifa, G., Nieto, L. M., & Casanova, M. S. (2009). Elimination of pesticide residues from virgin olive oil by ultraviolet light: Preliminary results. *Journal of Hazardous Materials*, *168*(1), 555–559. doi:10.1016/j.jhazmat.2009.02.030 PMID:19272709

Hood, E. (2004). Nanotechnology: Looking As We Leap. *Environmental Health Perspectives*, *112*(13), A740–A749. doi:10.1289/ehp.112-a740 PMID:15345364

Howard, G. T. (2002). Biodegradation of polyurethane: A review. *International Biodeterioration & Biodegradation*, *49*(4), 245–252. doi:10.1016/S0964-8305(02)00051-3

Howard, R., Abotsi, E., Jansen, V. R., & Howard, S. (2003). Lignocellulose biotechnology: Issues of bioconversion and enzyme production. *African Journal of Biotechnology*, *2*(12), 602–619. doi:10.5897/AJB2003.000-1115

Hsueh, C.-C., Chen, B.-Y., & Yen, C.-Y. (2009). Understanding effects of chemical structure on azo dye decolorization characteristics by *Aeromonas hydrophila. Journal of Hazardous Materials*, *167*(1-3), 995–1001. doi:10.1016/j.jhazmat.2009.01.077 PMID:19237244

Huang, Q., Song, J., Zhong, Y., Peng, P., & Huang, W. (2014). Atmospheric depositional fluxes and sources apportionment of organochlorine pesticides in the Pearl River Delta region, South China. *Environmental Monitoring and Assessment*, *186*(1), 247–256. doi:10.100710661-013-3370-5 PMID:23943243

Hugenholtz, P., Goebel, B. M., & Pace, N. R. (1998). Impact of culture-independent studies on the emerging phylogenetic view of bacterial diversity. *Journal of Bacteriology*, *180*, 4765–4774. PMID:9733676

Hu, S., & Van Bruggen, A. H. C. (1997). Microbial dynamics associated with multiphasic decomposition of 14C-labeled cellulose in soil. *Microbial Ecology*, *33*(2), 134–143. doi:10.1007002489900015 PMID:9052647

Hussain, S., Siddique, T., Arshad, M., & Saleem, M. (2009). Bioremediation and Phytoremediation of Pesticides: Recent Advances. *Critical Reviews in Environmental Science and Technology*, *39*(10), 843–907. doi:10.1080/10643380801910090

Ibrahim, W. M., Hassan, A. F., & Azab, Y. A. (2016). Biosorption of toxic heavy metals from aqueous solution by Ulva lactuca activated carbon. *Egyptian Journal of Basic and Applied Sciences, 3*(3), 241–249. doi:10.1016/j.ejbas.2016.07.005

Idaka, E., Horitsu, H., & Ogawa, T. (1987). Some properties of azoreductase produced by *Pseudomonas cepacia. Bulletin of Environmental Contamination and Toxicology, 39*(6), 982–989. doi:10.1007/BF01689588 PMID:3440155

Ihedioha, J. N., Ukoha, P. O., & Ekere, N. R. (2017). Ecological and human health risk assessment of heavy metal contamination in soil of a municipal solid waste dump in Uyo, Nigeria. *Environmental Geochemistry and Health, 39*(3), 497–515. doi:10.100710653-016-9830-4 PMID:27170184

Iida, T., Sumita, T., Ohta, A., & Takagi, M. (2000). The cytochrome P450ALKmultigene family of an n-alkane-assimilating yeast, *Yarrowia lipolytica*: Cloning and characterization of genes coding for new CYP52 family members. *Yeast (Chichester, England), 16*(12), 1077–1087. doi:10.1002/1097-0061(20000915)16:12<1077::AID-YEA6013.0.CO;2-K PMID:10953079

Iiyoshi, Y., Tsutsumi, Y., & Nishida, T. (1998). Polyethylene degradation by lignin-degrading fungi and manganese peroxidaes. *Journal of Wood Science, 44*(3), 222–229. doi:10.1007/BF00521967

Ijah, U. J. J., & Antai, S. P. (2003). The potential use of chichen-drop micro-organisms for oil spill remediation. *The Environmentalist, 23*(1), 89–95. doi:10.1023/A:1022947727324

Ilhan, S., Nourbakhsh, M. N., Kiliçarslan, S., & Ozdag, H. (2004). Removal of chromium, lead and copper ions from industrial waste waters by Staphylococcus saprophyticus. *Turkish Electronic Journal of Biotechnology, 2*(2), 50–57.

Imandi, S. B., Bandaru, V. V. R., Somalanka, S. R., Bandaru, S. R., & Garapati, H. R. (2008). Application of statistical experimental designs for the optimization of medium constituents for the production of citric acid from pineapple waste. *Bioresource Technology, 99*(10), 4445–4450. doi:10.1016/j.biortech.2007.08.071 PMID:17936623

Imman, S., Arnthong, J., Laosiripojana, N., & Champreda, V. (2012). Liquid hot water pretreatment of rice straw for enzymatic hydrolysis. In *Proceedings of the4th International Conference on Sustainable Energy and Environment: A Paradigm Shift to Low Carbon Society, Bangkok.*

Immanuel, G., Bhagavath, C. M., Raj, I. P., Esakkiraj, P., & Palavesam, A. (2007). Production and Partial Purification of Cellulase by Aspergillus niger and A. fumigates Fermented in Coir waste and Sawdust. *The Internet Journal of Microbiology, 3*(1). doi:10.5580/49

Inamori, R., Gui, P., Dass, P., Matsumura, M., Xu, K. Q., Kondo, T., ... Inamori, Y. (2007). Investigating CH_4 and N_2O emissions from eco-engineering wastewater treatment processes using constructed wetland microcosms. *Process Biochemistry, 42*(3), 363–373. doi:10.1016/j.procbio.2006.09.007

Ingram, T., Rogalinski, T., Bockemühl, V., Antranikian, G., & Brunner, G. (2009). Semi-continuous liquid hot water pretreatment of rye straw. *The Journal of Supercritical Fluids, 48*(3), 238–246. doi:10.1016/j.supflu.2008.10.023

Irfan, M., Safdar, A., Syed, Q., & Nadeem, M. (2012). Isolation and screening of cellulolytic bacteria from soil and optimization of cellulose production and activity. *Turkish Journal of Biochemistry–Turk. Journal of Biochemistry, 37*(3), 287–293.

Ito, K., Nakanishi, M., Lee, W. C., Sasaki, H., Zenno, S., Saigo, K., ... Tanokura, M. (2006). Three-dimensional structure of AzoR from *Escherichia coli*: An oxidereductase conserved in microorganisms. *The Journal of Biological Chemistry, 281*(29), 20567–20576. doi:10.1074/jbc.M513345200 PMID:16684776

Ito, K., Nakanishi, M., Lee, W. C., Zhi, Y., Sasaki, H., Zenno, S., ... Tanokura, M. (2008). Expansion of substrate specificity and catalytic mechanism of azoreductase by X-ray crystallography and site-directed mutagenesis. *The Journal of Biological Chemistry, 283*(20), 13889–13896. doi:10.1074/jbc.M710070200 PMID:18337254

Jackson, A. V. (2007). *Sources of Air Pollution Handbook of Atmospheric Science* (pp. 124–155). Blackwell Science Ltd.

Jacobsen, C. S. (1997). Plant protection and rhizosphere colonization of barley by seed inoculated herbicide degrading Burkholderia (*Pseudomonas*) cepacia DBO1 (pRO101) in 2, 4-D contaminated soil. *Plant and Soil, 189*(1), 139–144. doi:10.1023/A:1004296615446

Jadhav, G., & Ghangrekar, M. (2009). Performance of microbial fuel cell subjected to variation in pH, temperature, external load and substrate concentration. *Bioresource Technology, 100*(2), 717–723. doi:10.1016/j.biortech.2008.07.041 PMID:18768312

Jadhav, J. P., Parshetti, G. K., Kalme, S. D., & Govindwar, S. P. (2007). Decolourization of azo dye methyl red by *Saccharomyces cerevisiae* MTCC 463. *Chemosphere, 68*(2), 394–400. doi:10.1016/j.chemosphere.2006.12.087 PMID:17292452

Jadhav, J. P., Phugare, S. S., Dhanve, R. S., & Jadhav, S. B. (2010). Rapid biodegradation and decolorization of Direct Orange 39 (Orange TGLL) by an isolated bacterium *Pseudomonas aeruginosa* strain BCH. *Biodegradation, 21*(3), 453–463. doi:10.100710532-009-9315-6 PMID:19937265

Jadhav, S. B., Surwase, S. N., Kalyani, D., Gurav, R. G., & Jadhav, J. P. (2012). Biodecolorization of azo dye remazol orange by *Pseudomonas aeruginosa* BCH and toxicity (Oxidative stress) reduction in allium cepa root cells. *Applied Biochemistry and Biotechnology, 168*(5), 1319–1334. doi:10.100712010-012-9860-z PMID:22948606

Jadhav, S. U., Jadhav, M. U., Kagalkar, A. N., & Govindwar, S. P. (2008). Decolorization of Brilliant Blue G dye mediated by degradation of the microbial consortium of *Galactomyces geotrichum* and *Bacillus* sp. *Journal of the Chinese Institute of Chemical Engineers, 39*(6), 563–570. doi:10.1016/j.jcice.2008.06.003

Jadhav, S. U., Kalme, S. D., & Govindwar, S. P. (2008). Biodegradation of methyl red by *Galactomyces geotrichum* MTCC 1360. *International Biodeterioration & Biodegradation, 62*(2), 135–142. doi:10.1016/j.ibiod.2007.12.010

Jadia, C. D., & Fulekar, M. H. (2008). Vermicomposting of vegetable waste: A biophysicochemical process based on hydro-operating bioreactor. *African Journal of Biotechnology, 7*(20), 3723–3730.

Jafari, N., Soudi, M. R., & Kasra-Kermanshahi, R. (2014). Biodecolorization of textile azo dyes by isolated yeast from activated sludge: *Issatchenkia orientalis* JKS6. *Annals of Microbiology, 64*(2), 475–482. doi:10.100713213-013-0677-y

Jaisamut, K., Paulova, L., Patakova, P., Rychtera, M., & Melzoch, K. (2013). Optimization of alkali pretreatment of wheat straw to be used as substrate for biofuels production. *Plant, Soil and Environment, 59*(12), 537–542. doi:10.17221/7129-PSE

Jamal, F., Qidwai, T., Pandey, P., Singh, R., & Singh, S. (2011). Azo and anthraquinone dye decolorization in relation to its molecular structure using soluble *Trichosanthes dioica* peroxidase supplemented with redox mediator. *Catalysis Communications, 12*(13), 1218–1233. doi:10.1016/j.catcom.2011.04.012

Janssen, D. B., Pries, F., & Van der Ploeg, J. R. (1994). Genetics and biochemistry of dehalogenating enzymes. *Annual Review of Microbiology, 48*(1), 163–191. doi:10.1146/annurev.mi.48.100194.001115 PMID:7826004

Javed, A. (2012). *The Hydrogenation of Pyrolosis Gasoline (PyGas) over Nickel and Palladium catalysts* (PhD Thesis). University of Glasgow. Retrieved from http://theses.gla.ac.uk/3542/]

Jayasekara, R., Harding, I., Bowater, I., & Lonergan, G. (2005). Biodegradability of a selected range of polymers and polymer blends and standard methods for assessment of biodegradation. *Journal of Polymers and the Environment, 13*(3), 231–251. doi:10.100710924-005-4758-2

Jendrossek, D., Knoke, I., Habibian, R. B., Steinbüchel, A., & Schlegel, H. G. (1993). Degradation of poly (3-hydroxybutyrate), PHB, by bacteria and purification of a novel PHB depolymerase from *Comamonas sp. Journal of Polymers and the Environment*, *1*(1), 53–63.

Jesus, H. E., Peixoto, R. S., & Rosado, A. S. (2015). Bioremediation in Antarctic Soils. *Journal of Petroleum & Environmental Biotechnology*, *6*(06), 248. doi:10.4172/2157-7463.1000248

Jeyaratnam, J. (1990). Acute pesticide poisoning: A major global health problem. *World Health Statistics Quarterly*, *43*(3), 139–144. PMID:2238694

Jiang, W., & Fang, B. S. (2016). Construction and evaluation of a novel bifunctional phenylalanine–formate dehydrogenase fusion protein for bienzyme system with cofactor regeneration. *Journal of Industrial Microbiology & Biotechnology*, *43*(5), 577–584. doi:10.100710295-016-1738-6 PMID:26819086

Jiang, Y. F., Wang, X. T., Jia, Y., Wang, F., Wu, M.-H., Sheng, G.-Y., & Fu, J.-M. (2009). Occurrence, distribution and possible sources of organochlorine pesticides in agricultural soil of Shanghai, China. *Journal of Hazardous Materials*, *170*(2-3), 989–997. doi:10.1016/j.jhazmat.2009.05.082 PMID:19535203

Jini, R., Bijinu, B., Baskaran, V., & Bhaskar, N. (2016). Utilization of Solid Wastes from Tanneries as Possible Protein Source for Feed Applications: Acute and Sub-acute Toxicological Studies to Assess Safety of Products Prepared from Delimed Tannery Fleshings. *Waste and Biomass Valorization*, *7*(3), 439–446. doi:10.100712649-015-9463-3

Jin, R., Yang, H., Zhang, A., Wang, J., & Liu, G. (2009). Bioaugmentation on decolorization of C.I. Direct Blue 71 by using genetically engineered strain *Escherichia coli* JM109 (pGEX-AZR). *Journal of Hazardous Materials*, *163*(2-3), 1123–1128. doi:10.1016/j.jhazmat.2008.07.067 PMID:18755538

Jin, X., Liu, G., Xu, Z., & Tao, W. (2007). Decolourisation of a dye industry effluent by *Aspergillus fumigatus* XC6. *Applied Microbiology and Biotechnology*, *74*(1), 239–243. doi:10.100700253-006-0658-1 PMID:17086413

Johannes, C., Majcherczyk, A., & Huttermann, A. (1998). Oxidation of acenaphthene and acenaphthylene by laccase of Trametes versicolor in a laccase-mediator system. *Journal of Biotechnology*, *61*(2), 151–156. doi:10.1016/S0168-1656(98)00030-3

Johansson, H. E., Johansson, M. K., Wong, A. C., Armstrong, E. S., Peterson, E. J., Grant, R. E., ... Cook, R. M. (2011). BTI1, an azoreductase with pH-dependent substrate specificity. *Applied and Environmental Microbiology*, *77*(12), 4223–4225. doi:10.1128/AEM.02289-10 PMID:21531830

John, R. P., Sukumaran, R. K., Nampoothiri, K. M., & Pandey, A. (2007). Statistical optimization of simultaneous saccharification and L (+)-lactic acid fermentation from cassava bagasse using mixed culture of lactobacilli by response surface methodology. *Biochemical Engineering Journal*, *36*(3), 262–267. doi:10.1016/j.bej.2007.02.028

Joshi, B., Bhatt, M. R., Sharma, D., Joshi, J., Malla, R., & Sreerama, L. (2011). Lignocellulosic ethanol production: Current practices and recent developments. *Biotechnology & Molecular Biology Review*, *6*(8), 172–182.

Joshi, V., & Srivastava, A., Pankaj, Sharma, A., & Srivastava, P, C. (2015). Efficacy of cereal straw and its conjoint use with microbial consortium in reducing the leaching of chlorpyrifos: A soil column study. *Research Journal of Chemical Sciences.*, *5*(6), 9–14.

Joutey, N. T., Bahafid, W., Sayel, H., & Ghachtouli, N. E. (2013). Biodegradation: Involved microorganisms and genetically engineered microorganisms. In R. Chamy & F. Rosenkranz (Eds.), *Biodegradation – Life of science* (pp. 289–320). InTech.

Kadi, M. W. (2009). "Soil Pollution Hazardous to Environment": A case study on the chemical composition and correlation to automobile traffic of the roadside soil of Jeddah city, Saudi Arabia. *Journal of Hazardous Materials*, *168*(2-3), 1280–1283. doi:10.1016/j.jhazmat.2009.03.015 PMID:19345483

Kadlec, R. H., & Wallace, S. (2008). *Treatment wetlands*. CRC press. doi:10.1201/9781420012514

Kala. (2014). Rhizoremediation: A Promising Rhizosphere Technology. *IOSR-JESTFT, 8*(8), 23-27.

Kalme, S., Jadhav, S., Jadhav, M., & Govindwar, S. (2009). Textile dye degrading laccase from *Pseudomonas desmolyticum* NCIM 2112. *Enzyme and Microbial Technology*, *44*(2), 65–71. doi:10.1016/j.enzmictec.2008.10.005

Kalyani, D. C., Telke, A. A., Dhanve, R. S., & Jadhav, J. P. (2009). Ecofriendly biodegradation and detoxification of Reactive Red 2 textile dye by newly isolated *Pseudomonas* sp. SUK1. *Journal of Hazardous Materials*, *163*(2-3), 735–742. doi:10.1016/j.jhazmat.2008.07.020 PMID:18718713

Kalyani, D., Dhiman, S. S., Kim, H., Jeya, M., Kim, I. W., & Lee, J. K. (2012). Characterization of a novel laccase from the isolated *Coltricia perennis* and its application to detoxification of biomass. *Process Biochemistry*, *47*(4), 671–678. doi:10.1016/j.procbio.2012.01.013

Kamini, N. R., Hemachander, C., Mala, J. G. S., & Puvanakrishnan, R. (1999). Microbial enzyme technology as an alternative to conventional chemicals in leather industry. *Current Science*, *77*(1), 80–86.

Kanagaraj, J., Velappan, K. C., Babu, N. K., & Sadulla, S. (2006). Solid wastes generation in the leather industry and its utilization for cleaner environment-A review. *Journal of Scientific and Industrial Research*, *65*, 541–548.

Kandasamy, M., Hamawand, I., Bowtell, L., Seneweera, S., Chakrabarty, S., Yusaf, T., ... Eberhard, F. (2017). Investigation of ethanol production potential from lignocellulosic material without enzymatic hydrolysis using the ultrasound technique. *Energies*, *10*(1), 62. doi:10.3390/en10010062

Kang, S. Y., Lee, J. U., & Kim, K. W. (2007). Biosorption of Cr (III) and Cr (VI) onto the cell surface of Pseudomonas aeruginosa. *Biochemical Engineering Journal*, *36*(1), 54–58. doi:10.1016/j.bej.2006.06.005

Kantawanichkul, S., & Neamkam, P. (2003). Optimum recirculation ratio for nitrogen removal in a combined system: vertical flow vegetated bed over horizontal flow sand bed.

Kantawanichkul, S., Kladprasert, S., & Brix, H. (2009). Treatment of high-strength wastewater in tropical vertical flow constructed wetlands planted with Typha angustifolia and Cyperus involucratus. *Ecological Engineering*, *35*(2), 238–247. doi:10.1016/j.ecoleng.2008.06.002

Kantawanichkul, S., Sattayapanich, S., & Van Dien, F. (2013). Treatment of domestic wastewater by vertical flow constructed wetland planted with umbrella sedge and Vetiver grass. *Water Science and Technology*, *68*(6), 1345–1351. doi:10.2166/wst.2013.379 PMID:24056433

Kant, R. (2011). *Report on Sustainable Solid Waste management in India*. Ramky Enviro Engineers Ltd. Retrieved from ramkyenviroengineers.com/.../enviro/enviro%20engeneering%20brochure_latest.pdf

Kant, R. (2012). Textile dyeing industry an environmental hazard. *Nature and Science*, *4*(01), 22–26. doi:10.4236/ns.2012.41004

Karataş, S., Hasar, H., Taşkan, E., Özkaya, B., & Şahinkaya, E. (2014). Bio-reduction of tetrachloroethen using a H2-based membrane biofilm reactor and community fingerprinting. *Water Research*, *58*, 21–28. doi:10.1016/j.watres.2014.03.053 PMID:24731873

Karigar, C.S., & Rao, S.S. (2011). Role of Microbial Enzymes in the Bioremediation of Pollutants: A Review. *Enzyme Research*. doi:10.4061/2011/805187

Karlsson, S., Ljungquist, O., & Albertsson, A.-C. (1988). Biodegradation of polyethylene and the influence of surfactants. *Polymer Degradation & Stability*, *21*(3), 237–250. doi:10.1016/0141-3910(88)90030-4

Karmakar, M., & Ray, R. (2011). Current trends in research and application of microbial cellulases. *Research Journal of Microbiology*, *6*(1), 41–53. doi:10.3923/jm.2011.41.53

Karpouzas, D. G., Morgan, J. A., & Walker, A. (2000). Isolation and Characterization of 23 carbofuran degrading bacteria from soils from distant geographical areas. *Letters in Applied Microbiology*, *31*(5), 353–358. doi:10.1046/j.1472-765x.2000.00823.x PMID:11069636

Karthikeyan, K., Nanthakumar, K., Shanthi, K., & Lakshmanaperumalsamy, P. (2010). Response surface methodology for optimization of culture conditions for dye decolorization by a fungus, *Aspergillus niger* HM11 isolated from dye affected soil. *Iranian Journal of Microbiology*, *2*, 213–222. PMID:22347575

Kasassi, A., Rakimbei, P., Karagiannidis, A., Zabaniotou, A., Tsiouvaras, K., Nastis, A., & Tzafeiropoulou, K. (2008). Soil contamination by heavy metals: Measurements from a closed unlined landfill. *Bioresource Technology*, *99*(18), 8578–8584. doi:10.1016/j.biortech.2008.04.010 PMID:18508262

Kathiresan, K. (2003). Polythene and plastic degrading microbes from mangrove soil. *Revista de Biología Tropical*, *51*, 629–633. PMID:15162769

Katsifas, E. A., Giannoutsou, E., Lambraki, M., Barla, M., & Karagouni, A. D. (2004). Chromium recycling of tannery waste through microbial fermentation. *Journal of Industrial Microbiology & Biotechnology*, *31*(2), 57–62. doi:10.100710295-004-0115-z PMID:14767674

Kaur, I., & Gautam, N. (2010). Starch grafted polyethylene evincing biodegradation behaviour. *Malaysian Polymer Journal*, *5*, 26–38.

Kaur, M., & Arora, S. (2012). Isolation and Screening of Cellulose Degrading Bacteria in Kitchen Waste and Detecting Their Degrading Potential. *IOSR Journal of Mechanical and Civil Engineering*, *1*(2), 33–35.

Kawai, F., Watanabe, M., Shibata, M., Yokoyama, S., Sudate, Y., & Hayashi, S. (2004). Comparative study on biodegradability of polyethylene wax by bacteria and fungi. *Polymer Degradation & Stability*, *86*(1), 105–114. doi:10.1016/j.polymdegradstab.2004.03.015

Kay, P., Blackwell, P. A., & Boxall, A. B. (2004). Fate of veterinary antibiotics in a macroporous tile drained clay soil. *Environmental Toxicology and Chemistry*, *23*(5), 1136–1144. doi:10.1897/03-374 PMID:15180364

Kekre, A., Bhushan, A., Kumar, P., & Kalia, V. C. (2015). Genome Wide Analysis for Searching Novel Markers to Rapidly Identify Clostridium Strains. *Indian Journal of Microbiology*, *55*(3), 250–257. doi:10.100712088-015-0535-7 PMID:26063934

Kelley, S. L., Aitchison, E. W., Deshpande, M., Schnoor, J. L., & Alvarez, P. J. (2001). Biodegradation of 1, 4-dioxane in planted and unplanted soil: Effect of bioaugmentation with *Amycolata* sp. CB1190. *Water Research*, *35*(16), 3791–3800. doi:10.1016/S0043-1354(01)00129-4 PMID:12230161

Kennes, C., & Veiga, M. C. (2013). Introduction to Air Pollution. *Air Pollution Prevention and Control; the Official Journal of the World Heart Federation.*

Kerle, E. A., Jenkins, J. J., & Vogue, P. A. (2007). Understanding pesticide persistence and mobility for groundwater and surface water protection. Oregon State University.

Keshwani, D. R., Cheng, J. J., Burns, J. C., Li, L., & Chiang, V. (2007). Microwave pretreatment of switchgrass to enhance enzymatic hydrolysis. *Paper presented at the 2007 ASAE Annual Meeting.*

Khabbaz, F., & Albertsson, A. C. (2001). Rapid test methods for analyzing degradable polyolefins with a pro-oxidant system. *Journal of Applied Polymer Science, 79*(12), 2309–2316. doi:10.1002/1097-4628(20010321)79:12<2309::AID-APP10383.0.CO;2-E

Khan, M.A. (2014). Microbiological Solution to Environmental Problems- A Review on Bioremediation. *International Journal of Pure and Applied Bioscience, 2*(6), 295-303.

Khani, M. H. (2011). Uranium biosorption by Padina sp. algae biomass: Kinetics and thermodynamics. *Environmental Science and Pollution Research International, 18*(9), 1593–1605. doi:10.100711356-011-0518-0 PMID:21607729

Khan, R., Bhawana, P., & Fulekar, M. H. (2013). Microbial decolorization and degradation of synthetic dyes: A review. *Reviews in Environmental Science and Biotechnology, 12*(1), 75–97. doi:10.100711157-012-9287-6

Khati, P., Parul, Gangola, S., Bhatt, P., & Sharma, A. (2017). *Nanochitosan induced growth of Zea Mays with soil health maintenance.* Biotech. DOI 10.1007/s13205-017-0668-y

Khati, P., Sharma, A., Gangola, S., Kumar, R., Bhatt, P., & Kumar, G. (2017). Impact of agriusable nanocompounds on soil microbial activity: an indicator of soil health. *Soil, Air, Water, 45*(5). DOI: 10.1002/clen.201600458

Khehra, M. S., Saini, H. S., Sharma, D. K., Chadha, B. S., & Chimni, S. S. (2006). Biodegradation of azo dye C.I. Acid Red 88 by an anoxic–aerobic sequential bioreactor. *Dyes and Pigments, 70*(1), 1–7. doi:10.1016/j.dyepig.2004.12.021

Kikuchi, R. (2001). Environmental management of sulfur trioxide emission: Impact of SO3 on human health. *Environmental Management, 27*(6), 837–844. doi:10.1007002670010192 PMID:11393318

Kılıç, N. K., Nielsen, J. L., Yüce, M., & Dönmez, G. (2007). Characterization of a simple bacterial consortium for effective treatment of wastewaters with reactive dyes and Cr(VI). *Chemosphere, 67*(4), 826–831. doi:10.1016/j.chemosphere.2006.08.041 PMID:17217991

Kim, M., Li, L. Y., Gorgy, T., & Grace, J. R. (2017). Review of contamination of sewage sludge and amended soils by polybrominated diphenyl ethers based on meta-analysis. *Environmental Pollution, 220*(Pt B), 753-765. doi:10.1016/j.envpol.2016.10.053

Kim, S. Y., Jo, Y. A., Choi, J., & Choi, M. J. (2001). Characterization of s-Triazine Antibodies and Comparison of Enzyme Immunoassay for the Determination of s-Triazine. *Microchemical Journal, 68*, 163–172. doi:10.1016/S0026-265X(00)00143-0

Kim, S., & Dale, B. E. (2004). Global potential bioethanol production from wasted crops and crop residues. *Biomass and Bioenergy, 26*(4), 361–375. doi:10.1016/j.biombioe.2003.08.002

Kim, S., Krajmalnik-Brown, R., Kim, J.-O., & Chung, J. (2014). Remediation of petroleum hydrocarbon-contaminated sites by DNA diagnosis-based bioslurping technology. *The Science of the Total Environment, 497-498*, 250–259. doi:10.1016/j.scitotenv.2014.08.002 PMID:25129160

Kim, Y., Yeo, S., Kum, J., Song, H. G., & Choi, H. T. (2005). Cloning of a manganese peroxidase cDNA gene repressed by manganese in *Trametes versicolor. Journal of Microbiology (Seoul, Korea), 43*, 569–571. PMID:16410775

King, A. M., & Aaron, C. K. (2015). Organophosphate and carbamate poisoning. Organophosphate and carbamate poisoning. *Emergency Medicine Clinics, 33*(1), 133–151.

King, P., Anuradha, K., Lahari, S. B., Kumar, Y. P., & Prasad, V. S. R. K. (2008). Biosorption of zinc from aqueous solution using Azadirachta indica bark: Equilibrium and kinetic studies. *Journal of Hazardous Materials*, *152*(1), 324–329. doi:10.1016/j.jhazmat.2007.06.101 PMID:17681426

Kingsley, M. T., Fredrickson, J. K., Metting, F. B., & Seidler, R. J. (1994). Environmental restoration using plant-microbe bioaugmentation. In Bioremediation of chlorinated and polyaromatic hydrocarbon compounds (pp. 287-292). Boca Raton, FL: Lewis Publishers.

Kinoshita, H., Sohma, Y., Ohtake, F., Ishida, M., Kawai, Y., Kitazawa, H., ... Kimura, K. (2013). Biosorption of heavy metals by lactic acid bacteria and identification of mercury binding protein. *Research in Microbiology*, *164*(7), 701–709. doi:10.1016/j.resmic.2013.04.004 PMID:23603782

Kiran, I., Akar, T., & Tunali, S. (2005). Biosorption of Pb (II) and Cu (II) from aqueous solutions by pretreated biomass of Neurospora crassa. *Process Biochemistry*, *40*(11), 3550–3558. doi:10.1016/j.procbio.2005.03.051

Kiraye, M., John, W., & Gabriel, K. (2016). Bioremediation Rate of Total Petroleum Hydrocarbons from Contaminated Water by *Pseudomonas aeruginosa* Case Study: Lake Albert, Uganda. *Journal of Bioremediation & Biodegradation*, *7*, 335.

Kirk, D. W., Chan, C. C., & Marsh, H. (2002). Chromium behavior during thermal treatment of MSW fly ash. *Journal of Hazardous Materials*, *90*(1), 39–49. doi:10.1016/S0304-3894(01)00328-4 PMID:11777591

Kirk, T. K., & Farrell, R. L. (1987). Enzymatic "combustion": The microbial degradation of lignin. *Annual Review of Microbiology*, *41*(1), 465–505. doi:10.1146/annurev.mi.41.100187.002341 PMID:3318677

Kishore, S. (2000). Response of Effective Micro-organisms (EM) and other Nutrients to Rice and Wheat under Field conditions at Khumaltar, Nepal. *World Journal of Emergency Medicine*, *1*(1), 40–44.

Klamer, M., & Baath, E. (1998). Microbial community dynamics during composting of straw material studied using phospholipid fatty acid analysis. *FEMS Microbiology Ecology*, *27*, 9–20. doi:10.1111/j.1574-6941.1998.tb00521.x

Ko, P. C., & Yu, Y. (1968). Production of SCP from hydrocarbons: Taiwan. *Single Cell Protein*, 225.

Kodam, K. M., Soojhawon, I., Lokhande, P. D., & Gawai, K. R. (2005). Microbial decolorization of reactive azo dyes under aerobic conditions. *World Journal of Microbiology & Biotechnology*, *21*(3), 367–370. doi:10.100711274-004-5957-z

Kolb, H. C., Van Nieuwenhze, M. S., & Sharpless, K. B. (1994). Catalytic asymmetric dihydroxylation. *Chemical Reviews*, *94*(8), 2483–2547. doi:10.1021/cr00032a009

Koltermann, A., Kraus, M., Rarbach, M., Reisinger, C., Zavrel, M., & Soltl, Y. (2014). Cellulosic ethanol from agricultural residues–an advanced biofuel and biobased chemical platform. *JSM Biotechnology Bioengineering*, *2*(1), 1024.

Konduri, M. K. R., Anupam, K. S., Vivek, J. S., & Kumar, D. B. R. (2010). Synergistic effect of chemical and photo treatment on the rate of biodegradation of high density polyethylene by indigenous fungal isolates. *International Journal of Biotechnology and Biochemistry*, *6*, 157–174.

Koning, M., Hupe, K., & Stegmann, R. (2000). Thermal Processes, Scrubbing/Extraction, Bioremediation and Disposal. *Biotechnology*, *11b*, 306–317.

Koning, M., Hupe, K., & Stegmann, R. (2008). *Thermal processes, scrubbing/extraction, bioremediation and disposal. In Biotechnology Set* (2nd ed., pp. 304–317).

Konstantinou, I. K., Hela, D. G., & Albanis, T. A. (2006). The status of pesticide pollution in surface waters (rivers and lakes) of Greece. Part I. Review on occurrence and levels. *Environmental Pollution*, *141*(3), 555–570. doi:10.1016/j.envpol.2005.07.024 PMID:16226830

Korade, D. L., & Fulekar, M. H. (2009). Rhizosphere remediation of chlorpyrifos in mycorrhizospheric soil using ryegrass. *Journal of Hazardous Materials*, *172*(2), 1344–1350. doi:10.1016/j.jhazmat.2009.08.002 PMID:19720454

Koutny, M., Amato, P., Muchova, M., Ruzicka, J., & Delort, A.-M. (2009). Soil bacterial strains able to grow on the surface of oxidized polyethylene film containing prooxidant additives. *International Biodeterioration & Biodegradation*, *63*(3), 354–357. doi:10.1016/j.ibiod.2008.11.003

Koutny, M., Lemaire, J., & Delort, A.-M. (2006a). Biodegradation of polyethylene films with prooxidant additives. *Chemosphere*, *64*(8), 1243–1252. doi:10.1016/j.chemosphere.2005.12.060 PMID:16487569

Koutny, M., Sancelme, M., Dabin, C., Pichon, N., Delort, A.-M., & Lemaire, J. (2006b). Acquired biodegradability of polyethylenes containing pro-oxidant additives. *Polymer Degradation & Stability*, *91*(7), 1495–1503. doi:10.1016/j.polymdegradstab.2005.10.007

Kramer, U. (2010). Metal hyperaccumulation in plants. *Annual Review of Plant Biology*, *61*(1), 517–534. doi:10.1146/annurev-arplant-042809-112156 PMID:20192749

Kratochvil, D., & Volesky, B. (1998). Advances in the biosorption of heavy metals. *Trends in Biotechnology*, *16*(7), 291–300. doi:10.1016/S0167-7799(98)01218-9

Krause, A., Hancock, W. G., Minard, R. D., Freyer, A. J., Honeycutt, R. C., LeBaron, H. M., ... Bollang, J. M. (1985). Microbial transformation of the herbicide metolachlor by a soil *actinomycete*. *Journal of Agricultural and Food Chemistry*, *33*(4), 584–589. doi:10.1021/jf00064a008

Krishnan, C. S., Sakinah, A. M. M., Singh, L., & Wahid, Z. A. (2016). Production of biodiesel using tannery fleshing as a feedstock via solid state fermentation. *International Journal of Chemical, Molecular, Nuclear. Materials and Metallurgical Engineering*, *10*(5), 591–596.

Krishnan, C., Sousa, L. D., Jin, M., Chang, L., Dale, B. E., & Balan, V. (2010). Alkali-based AFEX pretreatment for the conversion of sugarcane bagasse and cane leaf residues to ethanol. *Biotechnology and Bioengineering*, *107*(3), 441–450. doi:10.1002/bit.22824 PMID:20521302

Kröpfelová, L. (2008). *Wastewater treatment in constructed wetlands with horizontal sub-surface flow*. Springer.

Krueger, M. C., Harms, H., & Schlosser, D. (2015). Prospects for microbiological solutions to environmental pollution with plastics. *Applied Microbiology and Biotechnology*, *99*(21), 8857–8874. doi:10.100700253-015-6879-4 PMID:26318446

Krupp, M., Weinmann, A., Galle, P. R., & Teufel, A. (2006). Actin binding LIM protein 3 (abLIM3). *International Journal of Molecular Medicine*, *17*(1), 129–133. PMID:16328021

Kuhad, R. C., Gupta, R., & Singh, A. (2011). Microbial cellulases and their industrial applications. *Enzyme Research*. doi:.10.4061/2011/280696

Kuila, A., Mukhopadhyay, M., Tuli, D. K., & Banerjee, R. (2011). Production of ethanol from lignocellulosics: An enzymatic venture. *EXCLI Journal*, *10*, 85–96. PMID:27857667

Kuiper, I., Bloemberg, G. V., & Lugtenberg, B. J. (2001). Selection of a plant-bacterium pair as a novel tool for rhizostimulation of polycyclic aromatic hydrocarbon-degrading bacteria. *Molecular Plant-Microbe Interactions*, *14*(10), 1197–1205. doi:10.1094/MPMI.2001.14.10.1197 PMID:11605959

Kuiper, I., Lagendijk, E. L., Bloemberg, G. V., & Lugtenberg, B. J. (2004). Rhizoremediation: A beneficial plant-microbe interaction. *Molecular Plant-Microbe Interactions*, *17*(1), 6–15. doi:10.1094/MPMI.2004.17.1.6 PMID:14714863

Kulkarni, A. G., & Vedamurthy, A. B. (2015). Isolation and Characterization of Cellulolytic Bacteria from Soil. *International Journal of Environmental Biology*, *5*(3), 57–65.

Kulozik, U., & Wilde, J. (1999). Rapid lactic acid production at high cell concentrations in whey ultrafiltrate by *Lactobacillus helveticus*. *Enzyme and Microbial Technology*, *24*(5), 297–302. doi:10.1016/S0141-0229(98)00122-7

Kumar, S., & Chakrabarti, T. (2010). Effective Municipal Solid Waste Management in India, Waste Management. InTech. Retrieved from http://www.intechopen.com/books/waste-management/effective-municipal-solid-waste-management-in-india

Kumar, A. G., Kamatchi, P., Umashankari, J., Vidhya, S., Murthy, P. S., & Sekaran, G. (2008a). Acidogenic fermentation of proteinaceous solid waste and characterization of different bioconversion stages and extracellular products. *Biodegradation*, *19*(4), 535–543. doi:10.100710532-007-9159-x PMID:17952609

Kumar, A. G., Nagesh, N., Prabhakar, T. G., & Sekaran, G. (2008b). Purification of extracellular acid protease and analysis of fermentation metabolites by Synergistes sp. utilizing proteinaceous solid waste from tanneries. *Bioresource Technology*, *99*(7), 2364–2372. doi:10.1016/j.biortech.2007.05.001 PMID:17601727

Kumar, A. G., Swarnalatha, S., Sairam, B., & Sekaran, G. (2008c). Production of alkaline protease by Pseudomonas aeruginosa using proteinaceous solid waste generated from leather manufacturing industries. *Bioresource Technology*, *99*(6), 1939–1944. doi:10.1016/j.biortech.2007.03.025 PMID:17481889

Kumar, A. G., Venkatesan, R., Kirubagaran, R., Prabhakar, T. G., & Sekaran, G. (2008d). Effects of nonionic surfactant on hydrolysis and fermentation of protein rich tannery solid waste. *Biodegradation*, *19*(5), 739–748. doi:10.100710532-008-9178-2 PMID:18288576

Kumar, A. K., & Sharma, S. (2017). Recent updates on different methods of pretreatment of lignocellulosic feedstocks: A review. *Bioresources and Bioprocessing*, *4*(1), 7. doi:10.118640643-017-0137-9 PMID:28163994

Kumar, D., Jain, V. K., Shanker, G., & Srivastava, A. (2003). Utilisation of fruits waste for citric acid production by solid state fermentation. *Process Biochemistry*, *38*(12), 1725–1729. doi:10.1016/S0032-9592(02)00253-4

Kumar, G., Arya, K., Verma, A., Pankaj, P., Khati, P., Gangola, S., ... Singh, H. (2017). Bioremediation of Petrol Engine Oil Polluted Soil Using Microbial Consortium and Wheat Crop. *Journal of Pure & Applied Microbiology*, *11*(3), 1583–1588. doi:10.22207/JPAM.11.3.45

Kumar, P. B. A. N., Dushenkov, V., Motto, H., & Raskin, I. (1995). Phytoextraction: The use of plants to remove heavy metals from soils. *Environmental Science & Technology*, *29*(5), 1232–1238. doi:10.1021/es00005a014 PMID:22192016

Kumar, P. D., Barrett, D. M., Delwiche, M. J., & Stroeve, P. (2009). Methods for pretreatment of lignocellulosic biomass for efficient hydrolysis and biofuel production. *Industrial & Engineering Chemistry Research*, *48*(8), 3713–3729. doi:10.1021/ie801542g

Kumar, P. S., Sounderraj, L., Sivakumar, K., & Rajakumar, S. (2012). Enzymatic hydrolyses of chrome shaving waste from tannery through Aspergillus tamerii. *Journal of Biological and Environmental Sciences*, *6*(18), 285–292.

Kumar, R., Joshi, S. R., & Acharya, C. (2008b). Metal tolerant Bacillus and Pseudomonas from uranium rich soils of Meghalaya. In *Proceedings of ISBT 2008* (pp. 345–350).

Kumar, R., Singh, S., & Singh, O. (2008). Bioconversion of lignocellulosic biomass: Biochemical and molecular perspectives. *Journal of Industrial Microbiology & Biotechnology*, *35*(5), 377–391. doi:10.100710295-008-0327-8 PMID:18338189

Kumar, S., Hatha, A. A. M., & Christi, K. S. (2007). Diversity and effectiveness of tropical mangrove soil microflora on the degradation of polythene carry bags. *Revista de Biología Tropical*, *55*(3-4), 777–786. doi:10.15517/rbt.v55i3-4.5954 PMID:19086383

Kumar, S., Sharma, K. P., Sharma, S., Grover, R., Kumar, P., Soni, P., & Sharma, S. (2006). Optimization of microbial degradation of an azo dye (methyl red) in fixed film bioreactors. *Indian Journal of Biotechnology*, *5*, 68–75.

Kurade, M. B., Waghmode, T. R., & Govindwar, S. P. (2011). Preferential biodegradation of structurally dissimilar dyes from a mixture by *Brevibacillus laterosporus*. *Journal of Hazardous Materials*, *192*(3), 1746–1755. doi:10.1016/j.jhazmat.2011.07.004 PMID:21803494

Kurade, M., Waghmode, T., Kagalkar, A., & Govindwar, S. (2012). Decolorization of textile industry effluent containing disperse dye Scarlet RR by a newly developed bacterial-yeast consortium BL-GG. *Chemical Engineering Journal*, *184*, 33–41. doi:10.1016/j.cej.2011.12.058

Kurbanoglu, E. B., & Kurbanoglu, N. I. (2003). Utilization for lactic acid production with a new acid hydrolysis of ram horn waste. *FEMS Microbiology Letters*, *225*(1), 29–34. doi:10.1016/S0378-1097(03)00472-5 PMID:12900017

Kutty, S. R. M., Isa, M. H., & Leong, L. C. (2011). Removal of Ammonia-Nitrogen (NH_3-N) and Nitrate by Modified Conventional Activated-Sludge System to Meet New DOE Regulations.

Kuyicak, N., & Volesky, B. (1990). Biosorption by fungal biomass. In B. Volesky (Ed.), *Biosorption of heavy metals* (pp. 173–198). CRC Press.

Kwon, E. E., Jean, Y. J., & Yi, H. (2012). New candidate for biofuel feedstock beyond terrestrial biomass for thermo-chemical process (pyrolysis/gasification) enhanced by carbon dioxide (CO_2). *Bioresource Technology*, *123*, 673–677. doi:10.1016/j.biortech.2012.07.035 PMID:22939597

Kwon, H. K., Jeon, J. Y., & Oh, S. J. (2017). Potential for heavy metal (copper and zinc) removal from contaminated marine sediments using microalgae and light emitting diodes. *Ocean Science Journal*, *52*(1), 57–66. doi:10.100712601-017-0001-z

Kyaw, B. M., Champakalakshmi, R., Sakharkar, M. K., Lim, C. S., & Sakharkar, K. R. (2012). Biodegradation of low density polythene (LDPE) by *Pseudomonas* species. *Indian Journal of Microbiology*, *52*(3), 411–419. doi:10.100712088-012-0250-6 PMID:23997333

Kyrikou, I., & Briassoulis, D. (2007). Biodegradation of agricultural plastic films: A critical review. *Journal of Polymers and the Environment*, *15*(2), 125–150. doi:10.100710924-007-0053-8

Lai, K., Stolowich, N. J., & Wild, J. R. (1995). Characterization of PS bond hydrolysis in organophosphorothioate pesticides by organophosphorus hydrolase. *Archives of Biochemistry and Biophysics*, *318*(1), 59–64. doi:10.1006/abbi.1995.1204 PMID:7726573

Langergraber, G. (2005). The role of plant uptake on the removal of organic matter and nutrients in subsurface flow constructed wetlands: A simulation study. *Water Science and Technology*, *51*(9), 213–223. PMID:16042261

Langer, S., Nakanishi, S., Mathes, T., Knaus, T., Binter, A., Macheroux, P., ... Mack, M. (2013). The flavoenzyme azobenzene reductase AzoR from Escherichia coli binds roseoflavin mononucleotide (RoFMN) with high affinity and is less active in its RoFMN form. *Biochemistry*, *52*(25), 4288–4295. doi:10.1021/bi400348d PMID:23713585

Laohaprapanon, S., Marques, M., & Hogland, W. (2010). Removal of Organic Pollutants from Wastewater Using Wood Fly Ash as a Low-Cost Sorbent. Clean–Soil, Air. *Water (Basel)*, *38*(11), 1055–1061.

Laskowski, D. A. (2002). Physical and chemical properties of pyrethroids. *Reviews of Environmental Contamination and Toxicology*, *174*, 49–170. doi:10.1007/978-1-4757-4260-2_3 PMID:12132343

Laughlin, R. B. Jr, & Neff, J. M. (1979). Interactive effects of salinity, temperature and polycyclic aromatic hydrocarbons on the survival and development rate of larvae of the mud crab *Rhithropanopeus harrisii. Marine Biology*, *53*(3), 281–291. doi:10.1007/BF00952437

Lau, M. W., & Dale, B. E. (2009). Cellulosic ethanol production from AFEX-treated corn stover using *Saccharomyces cerevisiae* 424A(LNH-ST). *Proceedings of the National Academy of Sciences of the United States of America*, *106*(5), 1368–1373. doi:10.1073/pnas.0812364106 PMID:19164763

Leal, R. M., Figueira, R. F., Tornisielo, V. L., & Regitano, J. B. (2012). Occurrence and sorption of fluoroquinolones in poultry litters and soils from Sao Paulo State, Brazil. *The Science of the Total Environment*, *432*, 344–349. doi:10.1016/j.scitotenv.2012.06.002 PMID:22750180

Lee, B.-H., & Scholz, M. (2007). What is the role of *Phragmites australis* in experimental constructed wetland filters treating urban runoff? *Ecological Engineering*, *29*(1), 87–95. doi:10.1016/j.ecoleng.2006.08.001

Lee, B., Pometto, A. L. III, Fratzke, A., & Bailey, T. B. Jr. (1991). Biodegradation of degradable plastic polyethylene by *Phanerochaete* and *Streptomyces* species. *Applied and Environmental Microbiology*, *57*, 678–685. PMID:16348434

Lee, J. (1997). Biological conversion of lignocellulosic biomass to ethanol. *Journal of Biotechnology*, *56*(1), 1–24. doi:10.1016/S0168-1656(97)00073-4 PMID:9246788

Leelakriangsak, M. (2013). Molecular approaches for bacterial azoreductases. *Songklanakarin Journal of Science and Technology*, *35*, 647–657.

Leelakriangsak, M., & Borisut, S. (2012). Characterization of the decolorizing activity of azo dyes by *Bacillus subtilis* azoreductase AzoR1. *Songklanakarin Journal of Science and Technology*, *34*, 509–516.

Lee, S. Y. (1996). Bacterial polyhydroxyalkanoates. *Biotechnology and Bioengineering*, *49*(1), 1–14. doi:10.1002/(SICI)1097-0290(19960105)49:1<1::AID-BIT13.0.CO;2-P PMID:18623547

Lee, S. Y., Choi, J. I., & Wong, H. H. (1999). Recent advances in polyhydroxyalkanoate production by bacterial fermentation: Mini-review. *International Journal of Biological Macromolecules*, *25*(1), 31–36. doi:10.1016/S0141-8130(99)00012-4 PMID:10416647

Leitão, A. L. (2009). Potential of Penicillium species in the bioremediation field. *International Journal of Environmental Research and Public Health*, *6*(4), 1393–1417. doi:10.3390/ijerph6041393 PMID:19440525

Lei, W., Wang, D. D., Dou, T. Y., Hou, J., Feng, L., Yin, H., ... Yang, L. (2017). Assessment of the inhibitory effects of pyrethroids against human carboxylesterases. *Toxicology and Applied Pharmacology*, *321*, 48–56. doi:10.1016/j.taap.2017.02.018 PMID:28242322

Lemoigne, M. (1926). Produits de deshydration et de polymerisation de l'acide boxybutyrique. *Biological science, 8*, 770-782.

Letcher, T. M. (2014). *Introduction with a Focus on Atmospheric Carbon Dioxide and Climate Change Future Energy* (2nd ed., pp. 3–16). Boston: Elsevier.

Leto, C., Tuttolomondo, T., Bella, S. L., Leone, R., & Licata, M. (2013). Growth of Arundo donax L. and *Cyperus alternifolius* L. in a horizontal subsurface flow constructed wetland using pre-treated urban wastewater—a case study in Sicily (Italy). *Desalination and Water Treatment*, *51*(40-42), 7447–7459. doi:10.1080/19443994.2013.792134

Leto, C., Tuttolomondo, T., La Bella, S., Leone, R., & Licata, M. (2013). Effects of plant species in a horizontal subsurface flow constructed wetland–phytoremediation of treated urban wastewater with *Cyperus alternifolius* L. and *Typha latifolia* L. in the West of Sicily (Italy). *Ecological Engineering*, *61*, 282–291. doi:10.1016/j.ecoleng.2013.09.014

Lettinga, G. (1995). Anaerobic digestion and wastewater treatment systems. *Antonie van Leeuwenhoek*, *67*(1), 3–28. doi:10.1007/BF00872193 PMID:7741528

Li, X., Lu, J., Zhao, J., & Qu, Y. (2014). Characteristics of corn stover pretreated with liquid hot water and fed-batch semi-simultaneous saccharification and fermentation for bioethanol production. *PLoS ONE, 9*(4).

Liao, X., Luo, S., Wu, Y., & Wang, Z. (2003). Studies on the abilities of *Vetiveria zizanioides* and *Cyperus alternifolius* for pig farm wastewater treatment. In *International conference on vetiver and exhibition.*

Liao, X., Luo, S., Wu, Y., & Wang, Z. (2005). Comparison of nutrient removal ability between *Cyperus alternifolius* and Vetiveria zizanioides in constructed wetlands. *Ying yong sheng tai xue bao [The journal of applied ecology]*, *16*(1), 156-160.

Li, C., Wang, L., Chen, Z., Li, Y., Wang, R., Luo, X., & Lu, J. (2015). Ozonolysis pretreatment of maize stover: The interactive effect of sample particle size and moisture on ozonolysis process. *Bioresource Technology*, *183*, 240–247. doi:10.1016/j.biortech.2015.01.042 PMID:25746300

Lifeng, C., & Jian, Y. (2008). Biochemical Characterization of the Tetrachlorobenzoquinone Reductase Involved in the Biodegradation of Pentachlorophenol. *International Journal of Molecular Sciences*, *9*(3), 198–212. doi:10.3390/ijms9030198 PMID:19325743

Liguori, R., Amore, A., & Faraco, V. (2013). Waste valorization by biotechnological conversion into added value products. *Applied Microbiology and Biotechnology*, *97*(14), 6129–6147. doi:10.100700253-013-5014-7 PMID:23749120

Li, L. L., McCorkle, S. R., Monchy, S., Taghavi, S., & van der Lelie, D. (2009). Bioprospecting metagenomes: Glycosyl hydrolases for converting biomass. *Biotechnology for Biofuels*, *128*, 231–236. PMID:19450243

Li, L., Yang, Y., Tam, N. F., Yang, L., Mei, X. Q., & Yang, F. J. (2013). Growth characteristics of six wetland plants and their influences on domestic wastewater treatment efficiency. *Ecological Engineering*, *60*, 382–392. doi:10.1016/j.ecoleng.2013.09.044

Li, M. T., Hao, L. L., Sheng, L. X., & Xu, J. B. (2008). Identification and degradation characterization of hexachlorobutadiene degrading strain Serratia marcescens HL1. *Bioresource Technology*, *99*(15), 6878–6884. doi:10.1016/j.biortech.2008.01.048 PMID:18337093

Limayem, A., & Ricke, S. C. (2012). Lignocellulosic biomass for bioethanol production: Current perspectives, potential issues and future prospects. *Progress in Energy and Combustion Science*, *38*(4), 449–467. doi:10.1016/j.pecs.2012.03.002

Lin, C., Yang, L., Xu, G., & Wu, J. (2011). Biodegradation and metabolic pathway of β-chlorinated aliphatic acid in Bacillus sp. CGMCC no. 4196. *Applied Microbiology and Biotechnology*, *90*(2), 689–696. doi:10.100700253-010-3081-6 PMID:21210101

Linde, M., Galbe, M., & Zacchi, G. (2008). Bioethanol production from non-starch carbohydrate residues in process streams from a dry-mill ethanol plant. *Bioresource Technology*, *99*(14), 6505–6511. doi:10.1016/j.biortech.2007.11.032 PMID:18281213

Lin, H. (2011). Three principles of soil change and pedogenesis in time and space. *Soil Science Society of America Journal*, *75*(6), 2049–2070. doi:10.2136ssaj2011.0130

Lin, J., Zhang, X., Li, Z., & Lei, L. (2010). Biodegradation of Reactive blue 13 in a two-stage anaerobic/aerobic fluidized beds system with a *Pseudomonas* sp. isolate. *Bioresource Technology, 101*(1), 34–40. doi:10.1016/j.biortech.2009.07.037 PMID:19713103

Lin, L., Kan, X., Yan, H., & Wang, D. (2012). Characterization of extracellular cellulose degrading enzymes from Bacillus thuringiensis strains. *Electronic Journal of Biotechnology, 15*(3). doi:10.2225/vol15-issue3-fulltext-1

Lin, M., Chen, Y., Villeneuve, P. J., Burnett, R. T., Lemyre, L., Hertzman, C., ... Krewski, D. (2004). Gaseous air pollutants and asthma hospitalization of children with low household income in Vancouver, British Columbia, Canada. *American Journal of Epidemiology, 159*(3), 294–303. doi:10.1093/aje/kwh043 PMID:14742290

Lin, Y., & Tanaka, S. (2005). Ethanol fermentation from biomass resources: Current state and prospects. *Applied Microbiology and Biotechnology, 69*(6), 627–642. doi:10.100700253-005-0229-x PMID:16331454

Li, R., Wu, H., Ding, J., Fu, W., Gan, L., & Li, Y. (2017). Mercury pollution in vegetables, grains and soils from areas surrounding coal-fired power plants. *Scientific Reports, 7*, 46545. doi:10.1038rep46545 PMID:28484233

Liszka, M. J., Clark, M. E., Schneider, E., & Clark, D. S. (2012). Nature Versus Nurture: Developing Enzymes That Function Under Extreme Conditions. *Annual Review of Chemical and Biomolecular Engineering, 3*(1), 77–102. doi:10.1146/annurev-chembioeng-061010-114239 PMID:22468597

Li, T., Fang, Q., Chen, H., Qi, F., Ou, X., Zhao, X., & Liu, D. (2017). Solvent-based delignification and decrystallization of wheat straw for efficient enzymatic hydrolysis of cellulose and ethanol production with low cellulase loadings. *RSC Advances, 7*(17), 10609–10617. doi:10.1039/C6RA28509K

Liu, D., Zou, J., Wang, M., & Jiang, W. (2008). Hexavalent chromium uptake and its effects on mineral uptake, antioxidant defence system and photosynthesis in Amaranthus viridis L. *Bioresource Technology, 99*(7), 2628–2636. doi:10.1016/j.biortech.2007.04.045 PMID:17570658

Liu, G., Niu, J., Guo, W., Zhao, L., Zhang, C., Wang, M., ... Guo, G. (2017). Assessment of terrain factors on the pattern and extent of soil contamination surrounding a chemical industry in Chongqing, Southwest China. *Catena, 156*, 237–243. doi:10.1016/j.catena.2017.04.005

Liu, G., Zhou, J., Fu, Q. S., & Wang, J. (2009). The *Escherichia coli* azoreductase AzoR is involved in resistance to thiol-specific stress caused by electrophilic quinones. *Journal of Bacteriology, 191*(20), 6394–6400. doi:10.1128/JB.00552-09 PMID:19666717

Liu, G., Zhou, J., Jin, R., Zhou, M., Wang, J., Lu, H., & Qu, Y. (2008A). Enhancing survival of *Escherichia coli* by expression of azoreductase AZR possessing quinone reductase activity. *Applied Microbiology and Biotechnology, 80*(3), 409–416. doi:10.100700253-008-1555-6 PMID:18548247

Liu, G., Zhou, J., Lv, H., Xiang, X., Wang, J., Zhou, M., & Qv, Y. (2007). Azoreductase from *Rhodobacter sphaeroides* AS1.1737 is a flavodoxin that also functions as nitroreductase and flavin mononucleotide reductase. *Applied Microbiology and Biotechnology, 76*(6), 1271–1279. doi:10.100700253-007-1087-5 PMID:17846764

Liu, G., Zhou, J., Wang, J., Yan, B., Li, J., Lu, H., ... Jin, R. (2008B). Site-directed mutagenesis of substrate binding sites of azoreductase from *Rhodobacter sphaeroides*. *Biotechnology Letters, 30*(5), 869–875. doi:10.100710529-007-9627-8 PMID:18165868

Liu, Y., & Hua, X. (2014). Degradation of acenaphthylene and anthracene by chemically modified laccase from *Trametes versicolor*. *RSC Advances, 4*(59), 31120–31122. doi:10.1039/C4RA02807D

Li, W. C., Ye, Z. H., & Wong, M. H. (2007). Effects of bacteria on enhanced metal uptake of the Cd/Zn-hyperaccumulating plant, *Sedum alfredii*. *Journal of Experimental Botany*, *58*(15-16), 4173–4182. doi:10.1093/jxb/erm274 PMID:18039737

Li, X., Li, A., Long, M., & Tian, X. (2015). Equilibrium and kinetic studies of copper biosorption by dead Ceriporia lacerata biomass isolated from the litter of an invasive plant in China. *Journal of Environmental Health Science & Engineering*, *13*(1), 37. doi:10.118640201-015-0191-1 PMID:25954509

Li, X., Liang, D., Bai, Y., Fan, Y., & Hou, H. (2014). Enhanced H2 production from corn stalk by integrating dark fermentation and single chamber microbial electrolysis cells with double anode arrangement. *International Journal of Hydrogen Energy*, *39*(17), 8977–8982. doi:10.1016/j.ijhydene.2014.03.065

Li, X., Wang, W., Wang, J., Cao, X., Wang, X., Liu, J., ... Jiang, X. (2008). Contamination of soils with organochlorine pesticides in urban parks in Beijing, China. *Chemosphere*, *70*(9), 1660–1668. doi:10.1016/j.chemosphere.2007.07.078 PMID:17869322

Li, Y., Xia, L., Huang, R., Xia, C., & Song, S. (2017). Algal biomass from the stable growth phase as a potential biosorbent for Pb (II) removal from water. *RSC Advances*, *7*(55), 34600–34608. doi:10.1039/C7RA06749F

Lobelle, D., & Cunliffe, M. (2011). Early microbial biofilm formation on marine plastic debris. *Marine Pollution Bulletin*, *62*(1), 197–200. doi:10.1016/j.marpolbul.2010.10.013 PMID:21093883

Loman, N. J., Constantinidou, C., Chan, J. Z., Halachev, M., Sergeant, M., Penn, C. W., ... Pallen, M. J. (2012). High-throughput bacterial genome sequencing: An embarrassment of choice, a world of opportunity. *Nature Reviews. Microbiology*, *10*(9), 599–606. doi:10.1038/nrmicro2850 PMID:22864262

Lone, M. I., He, Z. L., Stoffella, P. J., & Yang, X. E. (2008). Phytoremediation of heavy metal polluted soils and water: Progresses and perspectives. *Journal of Zhejiang University. Science. B.*, *9*(3), 210–220. doi:10.1631/jzus.B0710633 PMID:18357623

Looman, A. C., & Kuivenhoven, J. A. (1993). Influence of the three nucleotides upstream of the initiation codon on expression of the *Escherichia coli* lacZ gene in *Saccharomyces cerevisiae*. *Nucleic Acids Research*, *21*(18), 4268–4271. doi:10.1093/nar/21.18.4268 PMID:8414981

Lorah, M. M., Walker, C., & Graves, D. (2015). Performance of an anaerobic, static bed, fixed film bioreactor for chlorinated solvent treatment. *Biodegradation*, *26*(5), 341–357. doi:10.100710532-015-9738-1 PMID:26154697

Loredo-Trevino, A., Gutierrez-Sanchez, G., Rodriguea-Herrera, R., & Aguilar, C. N. (2012). Microbial enzymes involved in polyurethane biodegradation: A review. *Journal of Polymers and the Environment*, *20*(1), 258–265. doi:10.100710924-011-0390-5

Lourival, C., Sonia, C. N., Debora, R. C., & Maria, L. (2011). Risk Simulation of Soil Contamination by Polycyclic Aromatic Hydrocarbons from Sewage Sludge used as Fertilizers. *Journal of the Brazilian Chemical Society*, *22*(6), 1156–1163. doi:10.1590/S0103-50532011000600022

Lucas, N., Bienaime, C., Belloy, C., Quenedec, M., Silvestre, F., & Nava-Saucedo, J.-E. (2008). Polymer biodegradation: Mechanism and estimation techniques-a review. *Chemosphere*, *73*(4), 429–442. doi:10.1016/j.chemosphere.2008.06.064 PMID:18723204

Luckachan, G. E., & Pillai, C. K. S. (2011). Biodegradable polymers – A review on recent trends and emerging perspectives. *Journal of Polymers and the Environment*, *19*(3), 637–676. doi:10.100710924-011-0317-1

Luederitz, V., Eckert, E., Lange-Weber, M., Lange, A., & Gersberg, R. M. (2001). Nutrient removal efficiency and resource economics of vertical flow and horizontal flow constructed wetlands. *Ecological Engineering*, *18*(2), 157–171. doi:10.1016/S0925-8574(01)00075-1

Lu, L., & Ren, Z. (2016). Microbial electrolysis cells for waste biorefinery: A state of the art review. *Bioresource Technology*, *215*, 254–264. doi:10.1016/j.biortech.2016.03.034 PMID:27020129

Lu, W.-J., Wang, H.-T., Yang, S.-J., Wang, Z.-C., & Nie, Y.-F. (2005). Isolation and characterization of mesophilic cellulose degrading bacteria from flower stalks-vegetable waste co-composting system. *The Journal of General and Applied Microbiology*, *51*(6), 353–360. doi:10.2323/jgam.51.353 PMID:16474195

Lyand, L. R., Weimer, P. J., Vanzyl, W. H., & Pretorious, I. S. (2002). Microbial cellulose utilization. Fundamentals and Biotechnology. *Microbiology and Molecular Biology Reviews*, *66*(3), 506–577. doi:10.1128/MMBR.66.3.506-577.2002 PMID:12209002

Lynch, J. M., Slater, J. H., Bennett, J., & Harper, S. H. T. (1981). Cellulase activities of some aerobic microorganisms isolated from soil. *Journal of General Microbiology*, *127*, 231–236.

Lynd, L., Zyl, W., Mcbride, J., & Laser, M. (2005). Consolidated bioprocessing of cellulosic biomass: An update. *Current Opinion in Biotechnology*, *16*(5), 577–583. doi:10.1016/j.copbio.2005.08.009 PMID:16154338

Lyyra, S., Meagher, R. B., Kim, T., Heaton, A., Montello, P., Balish, R. S., & Merkle, S. A. (2007). Coupling two mercury resistance genes in Eastern cottonwood enhances the processing of organomercury. *Plant Biotechnology Journal*, *5*(2), 254–262. doi:10.1111/j.1467-7652.2006.00236.x PMID:17309680

Maas, R., & Chaudhari, S. (2005). Adsorption and biological decolourization of azo dye Reactive Red 2 in semicontinuous anaerobic reactors. *Process Biochemistry*, *40*(2), 699–705. doi:10.1016/j.procbio.2004.01.038

Macek, T., Mackova, M., & Káš, J. (2000). Exploitation of plants for the removal of organics in environmental remediation. *Biotechnology Advances*, *18*(1), 23–34. doi:10.1016/S0734-9750(99)00034-8 PMID:14538117

Machado, K. M. G., Luciana, C. A., Compart, L. C. A., Morais, R. O., Luiz, H., Rosa, L. H., & Santos, M. H. (2006). Biodegradation of reactive textile dyes by basidiomycetous fungi from Brazilian ecosystems. *Brazilian Journal of Microbiology*, *37*(4), 481–487. doi:10.1590/S1517-83822006000400015

Maciel, M. J. M., Silva, A. C., & Ribeiro, H. C. T. (2010). Industrial and biotechnological applications of ligninolytic enzymes of the basidiomycota: A review. *Electronic Journal of Biotechnology*, *13*. doi:10.2225/vol13-issue6-fulltext-2

Maeng, J. H. O., Sakai, Y., Tani, Y., & Kato, N. (1996). Isolation and characterization of a novel oxygenase that catalyzes the first step of n-alkane oxidation in *Acinetobacter* sp. strain M-1. *Journal of Bacteriology*, *178*(13), 3695–3700. doi:10.1128/jb.178.13.3695-3700.1996 PMID:8682768

Ma, H., Liu, W., Chen, X., Wu, Y., & Yu, Z. (2009). Enhanced enzymatic saccharification of rice straw by microwave pretreatment. *Bioresource Technology*, *100*(3), 1279–1284. doi:10.1016/j.biortech.2008.08.045 PMID:18930389

Mai, A., Bryan, J., Goodarzi, N., & Kantas, A. (2006). *Insights Into Non-Thermal Recovery of Heavy Oil. World Heavy Oil Conference (WHOC)*, Calgary, Alberta.

Makan, A., Assobhei, O., & Mountadar, M. (2013). Effect of initial moisture content on the in-vessel composting under air pressure of organic fraction of municipal solid waste in Morocco. *Iranian Journal of Environmental Health Sciences & Engineering*, *10*(1), 3. doi:10.1186/1735-2746-10-3 PMID:23369502

Ma, L. Q., Komar, K. M., Zhang, W., Cai, Y., & Kennelley, E. D. (2001). A fern that hyperaccumulates arsenic. *Nature*, *409*(6820), 579. doi:10.1038/35054664 PMID:11214308

Malherbe, S., & Cloete, T. E. (2003). Lignocellulose biodegradation: Fundamentals and applications. *Reviews in Environmental Science and Biotechnology*, *1*(2), 105–114. doi:10.1023/A:1020858910646

Malik, A. (2004). Metal bioremediation through growing cells. *Environment International*, *30*(2), 261–278. doi:10.1016/j.envint.2003.08.001 PMID:14749114

Mandenius, C. F. (2016). *Bioreactors: Design, Operation and Novel Applications*. John Wiley & Sons. doi:10.1002/9783527683369

Mane, U. V., Gurav, P. N., Deshmukh, A. M., & Govindwar, S. P. (2008). Degradation of Textile Dye Reactive Navy-Blue Rx (Reactive Blue 59) by an Isolated Actinomycete *Streptomyces krainskii* SUK-5. *Malaysian Journal of Microbiology*, *4*, 1–5.

Mani, D., Sharma, B., & Kumar, C. (2007). Phytoaccumulation, interaction, toxicity and remediation of cadmium from *Helianthus annuus* L. (sunflower). *Bulletin of Environmental Contamination and Toxicology*, *79*(1), 71–79. doi:10.100700128-007-9153-3 PMID:17549427

Mani, D., Sharma, B., Kumar, C., & Balak, S. (2012c). Cadmium and lead bioaccumulation during growth stages alters sugar and vitamin C content in dietary vegetables. *Proceedings of the National Academy of Sciences. India. Section B, Biological Sciences*, *82*(4), 477–488. doi:10.100740011-012-0057-6

Mani, D., Sharma, B., Kumar, C., Pathak, N., & Balak, S. (2012b). Phytoremediation potential of Helianthus annuus L. in sewageirrigated Indo-Gangetic alluvial soils. *International Journal of Phytoremediation*, *14*(3), 235–246. doi:10.1080/15226514.2010.498844 PMID:22567708

Manzur, A., Cuamatzi, F., & Favela, E. (1997). Effect of the growth of *Phanerochaete chrysosporium* in a blend of low density polyethylene and sugar cane bagasse. *Journal of Applied Polymer Science*, *66*(1), 105–111. doi:10.1002/(SICI)1097-4628(19971003)66:1<105::AID-APP123.0.CO;2-0

Manzur, A., Limón-González, M., & Favela-Torres, E. (2004). Biodegradation of physicochemically treated LDPE by a consortium of filamentous fungi. *Journal of Applied Polymer Science*, *92*(1), 265–271. doi:10.1002/app.13644

Marchiol, L., Fellet, G., Perosa, D., & Zerbi, G. (2007). Removal of trace metals by *Sorghum bicolor and Helianthus annuus* in a site polluted by industrial wastes: A field experience. *Plant Physiology and Biochemistry*, *45*(5), 379–387. doi:10.1016/j.plaphy.2007.03.018 PMID:17507235

Markandey, D. K., & Rajvaidya, N. (2004). *Environmental Biotechnology*. New Delhi: APH Publishing Corporation.

Marszalek, J., & Kaminski, W. (2008). Environmental impact of bioethanol production. In *Proceedings of ECOpole*. Wolczanska.

Martínková, L., Uhnáková, B., Pátek, M., Nesvera, J., & Kren, V. (2009). Biodegradation potential of the genus *Rhodococcus*. *Environment International*, *35*(1), 162–177. doi:10.1016/j.envint.2008.07.018 PMID:18789530

Martin, P., & Dombrowski, C. (2008). Air and Water Pollution. In R. P. Martin & S. C. Dombrowski (Eds.), *Prenatal Exposures: Psychological and Educational Consequences for Children* (pp. 185–199). Boston, MA: Springer US.

Martin-Reina, J., Duarte, J. A., Cerrillos, L., Bautista, J. D., & Moreno, I. (2017). Insecticide Reproductive Toxicity Profile: Organophosphate, Carbamate and Pyrethroids. *J. Toxins*, *4*(1), 7.

Ma, T., Luo, Y., Christie, P., Teng, Y., & Liu, W. (2012). Removal of phthalic esters from contaminated soil using different cropping systems: A field study. *European Journal of Soil Biology*, *50*, 76–82. doi:10.1016/j.ejsobi.2011.12.001

Mathiyazhagan, N., & Natarajan, D. (2011). Bioremediation on effluents from Magnesite and Bauxite mines using *Thiobacillus* Spp and *Pseudomonas* Spp. *Journal of Bioremediation & Biodegradation*, *2*(01), 115. doi:10.4172/2155-6199.1000115

Mathur, G., Mathur, A., & Prasad, R. (2011). Colonization and degradation of thermally oxidized high-density polyethylene by Aspergillus niger (ITCC No. 6052) isolated from plastic waste dumpsite. *Bioremediation Journal*, *15*(2), 69–76. doi:10.1080/10889868.2011.570281

Matsumoto, K., Mukai, Y., Ogata, D., Shozui, F., Nduko, J. M., Taguchi, S., & Ooi, T. (2010). Characterization of thermostable FMN-dependent NADH azoreductase from the moderate thermophile *Geobacillus stearothermophilus*. *Applied Microbiology and Biotechnology*, *86*(5), 1431–1438. doi:10.100700253-009-2351-7 PMID:19997911

McCutcheon, S. C., & Schnoor, J. L. (2003). Overview of phytotransformation and control of wastes. *Phytoremediation: Transformation and control of contaminants, 358.*

McDonald, I. R., Miguez, C. B., Rogge, G., Bourque, D., Wendlandt, K. D., Groleau, D., & Murrell, J. C. (2006). Diversity of solublemethane monooxygenase-containing methanotrophs isolated from polluted environments. *FEMS Microbiology Letters*, *255*(2), 225–232. doi:10.1111/j.1574-6968.2005.00090.x PMID:16448499

McGovern, P. E., & Michel, R. H. (1985). Royal Purple Dye: Tracing chemical origins of the industry. *Analytical Chemistry*, *57*, 1514A–1522A.

McMullan, G., Meehan, C., Conneely, A., Kirby, N., Robinson, T., Nigam, P., ... Smyth, W. F. (2001). Microbial decolourisation and degradation of textile dyes. *Applied Microbiology and Biotechnology*, *56*(1-2), 81–87. doi:10.1007002530000587 PMID:11499950

Meagher, R. (2000). Phytoremediation of toxic elements and organic pollutants. *Current Opinion in Plant Biology*, *3*(2), 153–162. doi:10.1016/S1369-5266(99)00054-0 PMID:10712958

Medvedeva, N., Polyak, Y., Kuzikova, I., Orlova, O., & Zharikov, G. (2008). The effect of mustard gas on the biological activity of soil. *Environmental Research*, *106*(3), 289–295. doi:10.1016/j.envres.2007.04.003 PMID:17537425

Melnyk, A., Dettlaff, A., Kuklińska, K., Namieśnik, J., & Wolska, L. (2015). Concentration and sources of polycyclic aromatic hydrocarbons (PAHs) and polychlorinated biphenyls (PCBs) in surface soil near a municipal solid waste (MSW) landfill. *The Science of the Total Environment*, *530*, 18–27. doi:10.1016/j.scitotenv.2015.05.092 PMID:26026405

Mendes, S., Pereira, L., Batista, C., & Martins, L. O. (2011). Molecular determinants of azo reduction activity in the strain *Pseudomonas putida* MET94. *Applied Microbiology and Biotechnology*, *92*(2), 393–405. doi:10.100700253-011-3366-4 PMID:21655981

Menezes, E. G., Do Carmo, J. R., Alves, J. G., Menezes, A. G., Guimaraes, I. C., Queiroz, F., & Pimenta, C. J. (2014). Optimization of alkaline pretreatment of coffee pulp for production of bioethanol. *Biotechnology Progress*, *30*(2), 451–462. doi:10.1002/btpr.1856 PMID:24376222

Menone, M. L., De Moreno, J. A., Moreno, V. J., Lanfranchi, A. L., Metcalfe, T. L., & Metcalfe, C. D. (2001). Organochlorine Pesticides and PCBs in a Southern Atlantic Coastal Lagoon Watershed, Argentina. *Archives of Environmental Contamination and Toxicology*, *40*(3), 355–362. doi:10.1007002440010183 PMID:11443366

Menzel, R. G. (1965). Soil-plant relationships of radioactive elements. *Health Physics*, *11*(12), 1325–1332. doi:10.1097/00004032-196512000-00009 PMID:5884542

Mervat, S. M. (2009). Degradation of methomyl by the novelbacterial strain *Stenotrophomonas maltophilia* M1. *Electronic Journal of Biotechnology*, *12*(4).

Messias, J. M., da Costa, B. Z., de Lima, V. M. G., Dekker, R. F. G., Rezende, M. I., Krieger, N., & Barbosa, A. M. (2009). Screening Botryosphaeria species for lipase: Production of lipase by Botryosphaeria ribis EC-01 grown on soyabean oil and other carbon sources. *Enzyme and Microbial Technology*, *45*(6-7), 426–431. doi:10.1016/j.enzmictec.2009.08.013

Mester, T., & Field, J. A. (1998). Characterization of a novel manganese peroxidase-lignin peroxidase hybrid isozyme produced by *Bjerkandera* species strain BOS55 in the absence of manganese. *The Journal of Biological Chemistry*, *273*(25), 15412–15417. doi:10.1074/jbc.273.25.15412 PMID:9624124

Miller, J. P., & Logan, B. E. (2000). Sustained perchlorate degradation in an autotrophic, gas-phase, packed-bed bioreactor. *Environmental Science & Technology*, *34*(14), 3018–3022. doi:10.1021/es991155d

Ministry of Environment. (1999). *Generation and discharge of municipal waste*. Author.

Ministry of New and Renewable Energy, Government of India. (2009). *Annual report.*

Minnikova, T. V., Denisova, T. V., Mandzhieva, S. S., Kolesnikov, S. I., Minkina, T. M., Chaplygin, V. A., ... Bauer, T. V. (2017). Assessing the effect of heavy metals from the Novocherkassk power station emissions on the biological activity of soils in the adjacent areas. *Journal of Geochemical Exploration*, *174*, 70–78. doi:10.1016/j.gexplo.2016.06.007

Minton, N. A., & Murray, V. S. (1988). A review of organophosphate poisoning. *Medical Toxicologyand Adverse Drug Experiment*, *3*, 350–375. PMID:3057326

Mireles, F., Davila, J. I., Pinedo, J. L., Reyes, E., Speakman, R. J., & Glascock, M. D. (2012). Assessing urban soil pollution in the cities of Zacatecas and Guadalupe, Mexico by instrumental neutron activation analysis. *Microchemical Journal*, *103*, 158–164. doi:10.1016/j.microc.2012.02.009

Mishra, A., Kumar, S., & Pandey, A. K. (2011). Laccase production and simultaneous decolorization of synthetic dyes in unique inexpensive medium by new isolates of White Rot Fungus. *International Biodeterioration & Biodegradation*, *65*(3), 487–493. doi:10.1016/j.ibiod.2011.01.011

Mitton, F. M., Gonzalez, M., Peña, A., & Miglioranza, K. S. (2012). Effects of amendments on soil availability and phytoremediation potential of aged p, p′-DDT, p, p′-DDE and p, p′-DDD residues by willow plants (*Salix* sp.). *Journal of Hazardous Materials*, *203*, 62–68. doi:10.1016/j.jhazmat.2011.11.080 PMID:22188788

Moghimi, H., Tabar, R. H., & Hamedi, J. (2017). Assessing the biodegradation of polycyclic aromatic hydrocarbons and laccase production by new fungus *Trematophoma* sp. UTMC 5003. *World Journal of Microbiology & Biotechnology*, *33*(7), 136. doi:10.100711274-017-2304-8 PMID:28585171

Mohammed, A. K., Ali, S. A., & Jiyad, J. H. (2015). Production of bioethanol from waste potatoes. Al-Khawarizmi Engineering Journal, 10(3), 62-67.

Mohanty, A. K., Misra, M., & Drzal, L. T. (2002). Sustainable bio-composites from renewable resources: Opportunities and challenges in the green materials world. *Journal of Polymers and the Environment*, *10*(1), 19–26. doi:10.1023/A:1021013921916

Mohanty, B., & Abdullahi, I. I. (2016). Bioethanol production from lignocellulosic waste-a review. *Biosciences Biotechnology Research Asia*, *13*(2), 1153–1161. doi:10.13005/bbra/2146

Moiser, N. (2005b). Features of promising technologies for pretreatment of lignocellulosic biomass. *Bioresource Technology*, *96*(6), 673–686. doi:10.1016/j.biortech.2004.06.025 PMID:15588770

Moore, B. C., Lafer, J. E., & Funk, W. H. (1994). Influence of aquatic macrophytes on phosphorus and sediment porewater chemistry in a freshwater wetland. *Aquatic Botany*, *49*(2), 137–148. doi:10.1016/0304-3770(94)90034-5

Moore, C. J. (2008). Synthetic polymers in the marine environment: A rapidly increasing, long-term threat. *Environmental Research, 108*(2), 131–139. doi:10.1016/j.envres.2008.07.025 PMID:18949831

Moosvi, S., Kher, X., & Madamwar, D. (2007). Isolation, characterization and decolorization of textile dyes by a mixed bacterial consortium JW-2. *Dyes and Pigments, 74*(3), 723–729. doi:10.1016/j.dyepig.2006.05.005

Moreira, L. R., & Filho, E. X. (2016). Insights into the mechanism of enzymatic hydrolysis of xylan. *Applied Microbiology and Biotechnology, 100*(12), 5205–5214. doi:10.100700253-016-7555-z PMID:27112349

Moreno-Medina, C. U., Poggi-Varaldo, H. M., Breton-Deval, L., & Rinderknecht-Seijas, N. (2016). Effect of sudden addition of PCE and bioreactor coupling to ZVI filters on performance of fluidized bed bioreactors operated in simultaneous electron acceptor modes. *Environmental Science and Pollution Research International*, 1–16. PMID:27498752

Moreno, R., San-Martín, M., Escapa, A., & Morán, A. (2016). Domestic wastewater treatment in parallel with methane production in a microbial electrolysis cell. *Renewable Energy, 93*, 442–448. doi:10.1016/j.renene.2016.02.083

Morgenroth, E., & Wilderer, P. (1998). Sequencing Batch Reactor Technology: Concepts, Design and Experiences (Abridged). *Water and Environment Journal: the Journal / the Chartered Institution of Water and Environmental Management, 12*(5), 314–320. doi:10.1111/j.1747-6593.1998.tb00192.x

Morrison, J. M., Wright, C. M., & John, G. H. (2012). Identification, isolation and characterization of a novel azoreductase from *Clostridium perfringens. Anaerobe, 18*(2), 229–234. doi:10.1016/j.anaerobe.2011.12.006 PMID:22182443

Mortan, S. H., Martín-González, L., Vicent, T., Caminal, G., Nijenhuis, I., Adrian, L., & Marco-Urrea, E. (2017). Detoxification of 1, 1, 2-trichloroethane to ethene in a bioreactor co-culture of Dehalogenimonas and Dehalococcoides mccartyi strains. *Journal of Hazardous Materials, 331*, 218–225. doi:10.1016/j.jhazmat.2017.02.043 PMID:28273571

MoUD. (2000). Annual Report on Municipal Solid Waste Management on a Regional Basis, Government of India [Data file]. Retrieved from www.asean.org/uploads/archive/files/UNEP.pdf

Mrema, E. J., Rubino, F. M., Mandic-Rajcevic, S., Sturchio, E., Turci, R., Osculati, A., ... Colosio, C. (2013). Exposure to priority organochlorine contaminants in the Italian general population. Part 1. Eight priority organochlorinated pesticides in blood serum. *Human and Experimental Toxicology, 32*(12), 1323–1339. doi:10.1177/0960327113485255 PMID:23857032

Mtui, G.Y.S. (2009). Recent advances in pretreatment of lignocellulosic wastes and production of value added products. *African Journal of Biotechnology, 8*(8).

Mudaliyar, P., Sharma, L., & Kulkarni, C. (2012). Food waste management-lactic acid production by *Lactobacillus species. International Journal of Advanced Biological Research, 2*(1), 34–38.

Mueller, R. J. (2006). Biological degradation of synthetic polyesters—enzymes as potential catalysts for polyester recycling. *Process Biochemistry, 41*(10), 2124–2128. doi:10.1016/j.procbio.2006.05.018

Muhammad, H., Iqbal, N., Ahmed, I., Zia, M. A., & Irfan, M. (2012). Purification and characterization of the kinetic parameters of cellulase produced from wheat straw by Trichoderma viride under SSF and its detergent compatibility. *Advances in Bioscience and Biotechnology, 2*, 149–156.

Muir, D. C., & Howard, P. H. (2006). Are there other persistent organic pollutants? A challenge for environmental chemists. *Environmental Science & Technology, 40*(23), 7157–7166. doi:10.1021/es061677a PMID:17180962

Mukhopadhyay, S., & Maiti, S. K. (2010). Phytoremediation of metal mine waste. *Applied Ecology and Environmental Research, 8*(3), 207–222.

Mulbry, W. W., & Karns, J. S. (1989). Parathion hydrolase specified by the Flavobacterium opd gene: Relationship between the gene and protein. *Journal of Bacteriology*, *171*(12), 6740–6746. doi:10.1128/jb.171.12.6740-6746.1989 PMID:2556372

Mulchandani, A., Kaneva, I., & And Chen, W. (1999). Detoxification of organophosphate pesticides by immobilized Escherichia coli expressing organophosphorus hydrolase on cell surface. *Biotechnology and Bioengineering*, *63*, 216–223. doi:10.1002/(SICI)1097-0290(19990420)63:2<216::AID-BIT103.0.CO;2-0 PMID:10099598

Muller, E. E., Bringel, F., & Vuilleumier, S. (2011). Dichloromethane-degrading bacteria in the genomic age. *Research in Microbiology*, *162*(9), 869–876. doi:10.1016/j.resmic.2011.01.008 PMID:21288483

Mumtaz, T., Khan, M. R., & Hassan, M. A. (2010). Study of environmental biodegradation of LDPE films in soil using optical and scanning electron microscopy. *Micron (Oxford, England)*, *41*(5), 430–438. doi:10.1016/j.micron.2010.02.008 PMID:20207547

Murrell, J. N. (1973). *The theory of the electronic Spectra of Organic Molecules*. John Wiley & Sons.

Mussatto, S. I., & Teixeira, J. A. (2010). Lignocellulose as raw material in fermentation processes. In A. Mendez-Vilas (Ed.), *Current Research, Technology and Education, Topics in Applied Microbiology and Microbial Biotechnology* (pp. 897–907). Badajoz: Formatex Research Center.

Mustapha, M. U., & Halimoon, N. (2015). Screening and isolation of heavy metal tolerant bacteria in industrial effluent. *Procedia Environmental Sciences*, *30*, 33–37. doi:10.1016/j.proenv.2015.10.006

Nachiyar, C. V., & Rajakumar, G. S. (2005). Purification and characterization of an oxygen insensitive azoreductase from *Pseudomonas aeruginosa*. *Enzyme and Microbial Technology*, *36*(4), 503–509. doi:10.1016/j.enzmictec.2004.11.015

Nadeau, L. J., Menn, F. M., Breen, A., & Sayler, G. S. (1994). Aerobic degradation of 1, 1, 1-trichloro-2, 2-bis (4-chlorophenyl) ethane (DDT) by Alcaligenes eutrophus A5. *Applied and Environmental Microbiology*, *60*(1), 51–55. PMID:8117093

Nagrale, S. D., Hajare, H., & Modak, P. R. (2012). Utilization of Rice Husk Ash. *Carbon*, *2*, 6=42.

Naja, G., & Volesky, B. (2011). The mechanism of metal cation and anion biosorption. In *Microbial biosorption of metals* (pp. 19–58). Springer Netherlands. doi:10.1007/978-94-007-0443-5_3

Nam, I. H., Kim, Y. M., Schmidt, S., & Chang, Y. S. (2006). Biotransformation of 1, 2, 3-tri-and 1, 2, 3, 4, 7, 8-hexachlorodibenzo-p-dioxin by *Sphingomonas wittichii* strain RW1. *Applied and Environmental Microbiology*, *72*(1), 112–116. doi:10.1128/AEM.72.1.112-116.2006 PMID:16391032

Nanda, S., Sahu, S. S., & Abraham, J. (2010). Studies on the biodegradation of natural and synthetic polyethylene by Pseudomonas spp. *Journal of Applied Science & Environmental Management*, *14*(2), 57–60. doi:10.4314/jasem.v14i2.57839

Narihiro, T., Takebayashi, S., & Hiraishi, A. (2004). Activity and phylogenetic composition of proteolytic bacteria in mesophilic fedbatch garbage composters. *Microbes and Environments*, *19*(4), 292–300. doi:10.1264/jsme2.19.292

Nasseri, A. T., Rasoul-Amini, S., Morowvat, M. H., & Ghasemi, Y. (2011). Single cell protein: Production and process. *American. Journal of Food Technology*, *6*(2), 103–116. doi:10.3923/ajft.2011.103.116

Nath, A., Dixit, M., Bandiya, A., Chavda, S., & Desai, A. J. (2008). Enhanced PHB production and scale up studies using cheese whey in fed batch culture of Methylobacterium sp. ZP24. *Bioresource Technology*, *99*(13), 5749–5755. doi:10.1016/j.biortech.2007.10.017 PMID:18032031

National Oceanic & Atmospheric Administration (NOAA). (n.d.). Trends in Atmospheric Carbon Dioxide). Retrieved June 30, 2017 from https://www.esrl.noaa.gov/gmd/ccgg/trends/weekly.html

Neethu, C. S., Mujeeb Rahiman, K. M., Saramma, A. V., & Mohamed Hatha, A. A. (2015). Heavy-metal resistance in Gram-negative bacteria isolated from Kongsfjord, Arctic. *Canadian Journal of Microbiology, 61*(6), 429–435. doi:10.1139/cjm-2014-0803 PMID:25942102

Negi, G., Pankaj, Srivastava, A., & Sharma, A. (2014). In situ Biodegradation of Endosulfan, Imidacloprid, and Carbendazim Using Indigenous Bacterial Cultures of Agriculture Fields of Uttarakhand, India. *Int. J. of Biological biomolecular agriculture food & biotech. Engineering, 8*(9), 953-961.

Negi, G., Pankaj, Srivastava, A., & Sharma, A. (2016). Optimization of Endosulfan biodegradation using indigenous bacterial isolate *Bacillus aryabhatti* through Response surface methodology. *Journal of Industrial Pollution Control, 32*(2), 638-644.

Neu, T. R., & Lawrence, J. R. (2009). Extracellular polymeric substances in microbial biofilms. In Microbial Glycobiology (pp. 733-758). Academic Press.

Nichols, T. D., Wolf, D. C., Rogers, H. B., Beyrouty, C. A., & Reynolds, C. M. (1997). Rhizosphere microbial populations in contaminated soils. *Water, Air, and Soil Pollution, 95*(1), 165–178. doi:10.1007/BF02406163

Nigam, J. N. (2000). Cultivation of Candida langeronii in sugar cane bagasse hemicellulosic hydrolyzate for the production of single cell protein. *World Journal of Microbiology & Biotechnology, 16*(4), 367–372. doi:10.1023/A:1008922806215

Nivala, J., Hoos, M. B., Cross, C., Wallace, S., & Parkin, G. (2007). Treatment of landfill leachate using an aerated, horizontal subsurface-flow constructed wetland. *The Science of the Total Environment, 380*(1), 19–27. doi:10.1016/j.scitotenv.2006.12.030 PMID:17241653

Nouri, J., Khorasani, N., Lorestani, B., Karami, M., Hassani, A. H., & Yousefi, N. (2009). Accumulation of heavy metals in soil and uptake by plant species with phytoremediation potential. *Environmental Earth Sciences, 59*(2), 315–323. doi:10.100712665-009-0028-2

Nowak, B., Pająk, J., Drozd-Bratkowicz, M., & Rymarz, G. (2011). Microorganisms participating in the biodegradation of modified polyethylene films in different soils under laboratory conditions. *International Biodeterioration & Biodegradation, 65*(6), 757–767. doi:10.1016/j.ibiod.2011.04.007

Nunan, F. S. L. (2000). Urban Organic Waste Markets: Responding to Change in Hubli-Dharwad, India. *Habitat International, 24*(3), 347–360. doi:10.1016/S0197-3975(00)00002-3

Nwoko, C. O. (2010). Trends in phytoremediation of toxic elemental and organic pollutants. *African Journal of Biotechnology, 9*(37), 6010–6016.

Nwosu-Obieogu, K., Chiemenem, L. I., & Adekunle, K. F. (2016). Utilization of agricultural waste for bioethanol production- a review. *International Journal of Current Research & Review, 8*(19).

Obasi, H. C., & Onuegbu, G. C. (2013). Biodegradability and mechanical properties of low density polyethylene/waste maize cob flour blends. *International Journal of Applied Sciences and Engineering Research, 2*, 2277–9442.

Obreque-Contreras, J., Pérez-Flores, D., Gutiérrez, P., & Chávez-Crooker, P. (2015). Acid mine drainage in Chile: An opportunity to apply bioremediation technology. *Hydrology: Current Research, 6*(3), 1. doi:10.4172/2157-7587.1000215

Oga, S. (2003). Fundamentos de toxicologia (2nd ed.). São Paulo: Atheneu Editora.

Oh, H., Wee, Y. J., Yun, J. S., Han, S. H., Jung, S., & Ryu, H. W. (2005). Lactic acid production from agricultural resources as cheap raw materials. *Bioresource Technology, 96*(13), 1492–1498. doi:10.1016/j.biortech.2004.11.020 PMID:15939277

Ohtaki, A., Sato, N., & Nakasaki, K. (1998). Biodegradation of poly-Ɛ-caprolactone under controlled composting conditions. *Polymer Degradation & Stability*, *61*(3), 499–505. doi:10.1016/S0141-3910(97)00238-3

Oikeh, E. I., Oriakhi, K., & Omoregie, E. S. (2013). Proximate analysis and phytochemical screening of *Citrus sinensis* fruit wastes. *The Bioscientist*, *1*(2), 164–170.

Ojumu, T. V., Yu, J., & Solomon, B. O. (2004). Production of polyhydroxyalkanoates, a bacterial biodegradable polymers. *African Journal of Biotechnology*, *3*(1), 18–24. doi:10.5897/AJB2004.000-2004

Okedeyi, O. O., Dube, S., Awofolu, O. R., & Nindi, M. M. (2014). Assessing the enrichment of heavy metals in surface soil and plant (Digitaria eriantha) around coal-fired power plants in South Africa. *Environmental Science and Pollution Research International*, *21*(6), 4686–4696. doi:10.100711356-013-2432-0 PMID:24352550

Okurut, T. O., Rijs, G. B. J., & Van Bruggen, J. J. A. (1999). Design and performance of experimental constructed wetlands in Uganda, planted with *Cyperus papyrus* and *Phragmites mauritianus. Water Science and Technology*, *40*(3), 265–271.

Olofsson, K., Bertilsson, M., & Liden, G. (2008). A short review on SSF – an interesting process option for ethanol production from lignocellulosic feedstocks. *Biotechnology for Biofuels*, *1*(1), 7. doi:10.1186/1754-6834-1-7 PMID:18471273

Olsson, L., & Hahn-Hagerdal, B. (1996). Fermentation of lignocellulosic hydrolysates for ethanol production. *Enzyme and Microbial Technology*, *18*(5), 312–331. doi:10.1016/0141-0229(95)00157-3

Ontiveros-Valencia, A., Tang, Y., Krajmalnik-Brown, R., & Rittmann, B. E. (2013). Perchlorate reduction from a highly contaminated groundwater in the presence of sulfate-reducing bacteria in a hydrogen-fed biofilm. *Biotechnology and Bioengineering*, *110*(12), 3139–3147. doi:10.1002/bit.24987 PMID:23797735

Onuki, S., Koziel, J. A., Leeuwen, V. J., Jenks, W. S., Grewell, D., & Cai, L. (2008). Ethanol production, purification, and analysis techniques: a review. In *Proceedings of the* ASABE Annual International Meeting, *Rhode Island.*

Ooi, T., Shibata, T., Sato, R., Ohno, H., Kinoshita, S., Thuoc, T. L., & Taguchi, S. (2007). An azoreductase, aerobic NADH-dependent flavoprotein discovered from *Bacillus* sp.: Functional expression and enzymatic characterization. *Applied Microbiology and Biotechnology*, *75*(2), 377–386. doi:10.100700253-006-0836-1 PMID:17546472

Oreb, M., Dietz, H., Farwick, A., & Boles, E. (2012). Novel strategies to improve co-fermentation of pentoses with D-glucose by recombinant yeast strains in lignocellulosic hydrolysates. *Bioengineered*, *3*(6), 347–351. doi:10.4161/bioe.21444 PMID:22892590

Organization for the Prohibition of Chemical Weapons. (2017, December 10). Genesis and Historical Development. Retrieved from https://www.opcw.org/chemical-weapons-convention/genesis-and-historical-development/

Orhan, Y., & Büyükgüngör, H. (2000). Enhancement of biodegradability of disposable polyethylene in controlled biological soil. *International Biodeterioration & Biodegradation*, *45*(1-2), 49–55. doi:10.1016/S0964-8305(00)00048-2

Orłowska, E., Przybyłowicz, W., Orlowski, D., Turnau, K., & Mesjasz-Przybyłowicz, J. (2011). The effect of mycorrhiza on the growth and elemental composition of Ni-hyperaccumulating plant, *Berkheya coddii Roessler. Environmental Pollution*, *159*(12), 3730–3738. doi:10.1016/j.envpol.2011.07.008 PMID:21835516

Osman, K. T. (2014). Soil Pollution. In K. T. Osman (Ed.), *Soil Degradation, Conservation and Remediation* (pp. 149–226). Dordrecht: Springer Netherlands. doi:10.1007/978-94-007-7590-9_6

Otake, Y., Kobayashi, T., Asabe, H., Murakami, N., & Ono, K. (1995). Biodegradation of low-density polyethylene, polystyrene, polyvinyl chloride, and urea formaldehyde resin buried under soil for over 32 years. *Journal of Applied Polymer Science*, *56*(13), 1789–1796. doi:10.1002/app.1995.070561309

Ouellet-Plamondon, C., Chazarenc, F., Comeau, Y., & Brisson, J. (2006). Artificial aeration to increase pollutant removal efficiency of constructed wetlands in cold climate. *Ecological Engineering*, *27*(3), 258–264. doi:10.1016/j.ecoleng.2006.03.006

Oyeyiola, A. O., Fatunsin, O. T., Akanbi, L. M., Fadahunsi, D. E., & Moshood, O. M. (2017). Human Health Risk of Organochlorine Pesticides in Foods Grown in Nigeria. *Journal of Health & Pollution*, *7*(15), 63–70. doi:10.5696/2156-9614-7.15.63

Ozdemir, C., Sen, N., Dursun, S., & Kalipci, E. (2010). Removal of dichloromethane in up-flow anaerobic sludge bed reactors and methane production. *Asian Journal of Chemistry*, *22*(8), 6423.

Ozkul, C. (2016). Heavy metal contamination in soils around the Tuncbilek Thermal Power Plant (Kutahya, Turkey). *Environmental Monitoring and Assessment*, *188*(5), 284. doi:10.100710661-016-5295-2 PMID:27071662

Page, V., & Schwitzguébel, J. P. (2009). The role of cytochromes P450 and peroxidases in the detoxification of sulphonated anthraquinones by rhubarb and common sorrel plants cultivated under hydroponic conditions. *Environmental Science and Pollution Research International*, *16*(7), 805–816. doi:10.100711356-009-0197-2 PMID:19529968

Pai, S. G., Riley, M. B., & Camper, N. D. (2001). Microbial degradation of mefenoxam in rhizosphere of *Zinnia angustifolia. Chemosphere*, *44*(4), 577–582. doi:10.1016/S0045-6535(00)00368-4 PMID:11482644

Palaniswamy, R., & Veluchamy, C. (2017). *Removal of chromium from effluent by biosorption using spirulina platensis*. Academic Press.

Pampanin, D. M., & Sydnes, M. O. (2013). Polycyclic Aromatic Hydrocarbons a Constituent of Petroleum: Presence and Influence in the Aquatic Environment. Academic Press. Doi:10.5772/48176

Pandey, A., Singh, P., & Iyengar, L. (2007). Bacterial decolorization and degradation of azo dyes. *International Biodeterioration & Biodegradation*, *59*(2), 73–84. doi:10.1016/j.ibiod.2006.08.006

Pandey, A., Soccol, C. R., Nigam, P., & Soccol, V. T. (2000). Biotechnological potential of agro-industrial residues. I: Sugarcane bagasse. *Bioresource Technology*, *74*(1), 69–80. doi:10.1016/S0960-8524(99)00142-X

Pankaj & Sharma, A. (2018). *Microbial Biotechnology in Environmental Monitoring and Cleanup*. Hershey, PA: IGI Global. Doi:10.4018/978-1-5225-3126-5

Pankaj, Negi, G., Gangola, S., Khati, P., Kumar, G., Srivastava, A., & Sharma, A. (2016). Differential expression and characterization of cypermethrindegrading potential proteins in Bacillus thuringiensis strain, SG4. *Biotech, 6*, 225. DOI: 10.1007/s13205-016-0541-4

Pankaj, Sharma, A., Gangola, S., Khati, P., Kumar, G., & Srivastava, A. (2016). Novel pathway of cypermethrin biodegradation in a Bacillus sp. strain SG2 isolated from cypermethrin-contaminated agriculture field. *Biotech, 6*(1), 45. doi: 10.1007/s13205-016-0372-3

Pankaj, Tiwari, S., Gangola, S., Khati, P., Kumar, G., & Sharma, A. (2017). Removal of Xenobiotics from environment using microbial metabolism. *Scientific India Magzine, 5*(1), 33-34.

Pankaj, Bisht, T.S., Pathak, V.M., Barh, A., & Chandra, D. (2015). Optimization Of Amylase Production From The Fungal Isolates of Himalayan Region Uttarakhand. *Ecology. Environmental Conservation*, *21*(3). 1517-1521.

Pankaj, Nailwal, T.K., Singh, L., & Amit, P. (2014). Isolation and characterization of Rhizobial isolates from the Rhizospheric soil of an Endangered plant Meizotropis pellita. *Asian Journal of Microbiology, Biotechnology & Environmental Science, 16*(2), 301–306.

Pankaj, Negi, G., Gangola, S., Khati, P., Srivastava, A., & Sharma, A. (2015). Optimization of Sulfosulfuron Biodegradation through Response Surface Methodology using Indigenous Bacterial Strain Isolated from Contaminated Agriculture Field. *International Journal of Current Microbiology and Applied Sciences*, *4*(8), 105–112.

Paoli, D., Giannandrea, F., Gallo, M., Turci, R., Cattaruzza, M. S., & Lombardo, F. (2015). Exposure to polychlorinated biphenyls and hexachlorobenzene, semen quality and testicular cancer risk. *The Journal of Endocrinology*. PMID:25770454

Parajó, J. C., Santos, V., Domínguez, H., & Vázquez, M. (1995). NH4 OH-Based pretreatment for improving the nutritional quality of single-cell protein (SCP). *Applied Biochemistry and Biotechnology*, *55*(2), 133–149. doi:10.1007/BF02783554

Paterson, S., Mackay, D., Tam, D., & Shiu, W. Y. (1990). Uptake of organic chemicals by plants: A review of processes, correlations and models. *Chemosphere*, *21*(3), 297–331. doi:10.1016/0045-6535(90)90002-B

Pathak, V. M., Navneet, Bisht, T. S., & Pandey, A. (2018). Biodegradation of Natural and Synthetic Polymer. Microbial Biotechnology in Environmental Monitoring and Cleanup, 106.

Pathak, V, M. & Navneet. (2017a). Review on the current status of polymer degradation: A microbial approach. *Bioresour Bioprocess. Springer Berlin Heidelberg*, *4*, 15.

Pathak, V. M., & Navneet. (2017b). Dataset on the impact of UV, nitric acid and surfactant treatments on low-density polyethylene biodegradation. *Data in Brief*, *14*, 393.

Pathak, V. M., & Navneet. (2017c). Biodegradation of Polyurethane. *Life Science Journal*, *14*(1), 60–62.

Pathak, V. M., & Navneet. (2017d). Implications of SiO 2 nanoparticles for in vitro biodegradation of low-density polyethylene with potential isolates of Bacillus, Pseudomonas, and their synergistic effect on Vigna mungo growth. Energy. *Ecology & Environment*, *2*(6), 418–427.

Pathak, V. M., & Navneet. (2017). Navneet. Review on the current status of polymer degradation: A microbial approach. *Bioresources and Bioprocessing*, *4*(1), 15. doi:10.118640643-017-0145-9

Patnaik, P. (2003). *A Comprehensive Guide to the Hazardous Properties of Chemical Substances*. New Jersey, USA: John Wiley & Sons.

Patterson, A. T., Kaffenberger, B. H., Keller, R. A., & Elston, D. M. (2016). Skin diseases associated with Agent Orange and other organochlorine exposures. *Journal of the American Academy of Dermatology*, *74*(1), 143–170. doi:10.1016/j.jaad.2015.05.006 PMID:26210237

Pearce, C. I., Christie, R., Boothman, C., von Canstein, H., Guthrie, J. T., & Lloyd, J. R. (2006). Reactive azo dye reduction by Shewanella strain J18 143. *Biotechnology and Bioengineering*, *95*(4), 692–703. doi:10.1002/bit.21021 PMID:16804943

Pegram, J. E., & Andrady, A. L. (1989). Outdoor weathering of selected polymeric materials under marine exposure conditions. *Polymer Degradation & Stability*, *26*(4), 333–345. doi:10.1016/0141-3910(89)90112-2

Perocco, P., Bolognesi, S., & Alberghini, W. (1983). Toxic activity of seventeen industrial solvents and halogenated compounds on human lymphocytes cultured in vitro. *Toxicology Letters*, *16*(1), 69–75. doi:10.1016/0378-4274(83)90012-7 PMID:6836616

Persson, J., Somes, N. L. G., & Wong, T. H. F. (1999). Hydraulics efficiency of constructed wetlands and ponds. *Water Science and Technology*, *40*(3), 291–300.

Petre, M., Zarnea, G., Adrian, P., & Gheorghiu, E. (1999). Biodegradation and bioconversion of cellulose wastes using bacterial and fungal cells immobilized in radio polymerized hydrogels. *Resources, Conservation and Recycling*, *27*(4), 309–332. doi:10.1016/S0921-3449(99)00028-2

Petticrew, E. L., & Kalff, J. (1992). Water flow and clay retention in submerged macrophyte beds. *Canadian Journal of Fisheries and Aquatic Sciences*, *49*(12), 2483–2489. doi:10.1139/f92-274

Pew, J. C. (1957). Properties of powdered wood and isolation of lignin by cellulolytic enzymes. *TAPPI*, *40*, 553–558.

Philp, J. C., & Atlas, R. M. (2005). Bioremediation of contaminated soils and aquifers. In Atlas RM, Philp JC, editors. Bioremediation: applied microbial solutions for real-world environmental cleanup (pp. 139–236). Washington: American Society for Microbiology (ASM) Press. doi:10.1128/9781555817596.ch5

Phua, S. K., Castillo, E., Anderson, J. M., & Hiltner, A. (1987). Biodegradation of polyurethane in vitro. *Journal of Biomedical Materials Research*, *21*(2), 231–246. doi:10.1002/jbm.820210207 PMID:3818683

Pieper, D. H., & Reineke, W. (2000). Engineering bacteria for bioremediation. *Current Opinion in Biotechnology*, *11*(3), 262–270. doi:10.1016/S0958-1669(00)00094-X PMID:10851148

Pierzynski, G. M., Schnoor, J. L., Banks, M. K., Tracy, J. C., Licht, L. A., & Erickson, L. E. (1994). Vegetative remediation at Superfund Sites. Mining and It's Environ. Impact. *Environmental Science & Technology*, *1*, 49–69.

Pilon-Smits, E. (2005). Phytoremediation. *Annual Review of Plant Biology*, *56*(1), 15–39. doi:10.1146/annurev.arplant.56.032604.144214 PMID:15862088

Pimentel, D., & Patzek, T. W. (2005). Ethanol Production Using Corn, Switch grass and wood; biodiesel production using soybean and sunflower. *Natural Resources Research*, *14*(1), 65–76. doi:10.100711053-005-4679-8

Pink, D.H. (2006). *Investing in Tomorrow's liquid gold.* Yahoo.

PlasticsEuorope. (2016). *Plastics-The facts 2016*. Brussels, Belgium: PlasticsEurope. Retrieved from http://www.plasticseurope.org/Document/plastics-the-facts-2016-15787.aspx?Page=DOCUMENT&FolID=2

Platzer, C. (1999). Design recommendations for subsurface flow constructed wetlands for nitrification and denitrification. *Water Science and Technology*, *40*(3), 257–263.

Pohland, F., & Kim, J. (1999). In situ anaerobic treatment of landfills for optimum Stabilization and biogas production. *Water Science and Technology*, *40*(8), 203–210.

Pometto, A. L. III, Lee, B., & Johnson, K. E. (1992). Production of an extracellular polyethylene-degrading enzyme(s) by *Streptomyces* species. *Applied and Environmental Microbiology*, *58*, 731–733. PMID:1610196

Popat, S. C., & Deshusses, M. A. (2009). Reductive dehalogenation of trichloroethene vapors in an anaerobic biotrickling filter. *Environmental Science & Technology*, *43*(20), 7856–7861. doi:10.1021/es901305x PMID:19921905

Popat, S. C., Zhao, K., & Deshusses, M. A. (2012). Bioaugmentation of an anaerobic biotrickling filter for enhanced conversion of trichloroethene to ethene. *Chemical Engineering Journal*, *183*, 98–103. doi:10.1016/j.cej.2011.12.026

Potts, J. E. (1978). Biodegradation. In H. H. G. Jelink (Ed.), *Aspects of degradation and stabilization of polymers* (pp. 617–658). New York, NY: Elsevier.

Pramila, R., & Ramesh, K. V. (2011a). Biodegradation of low density polyethylene (LDPE) by fungi isolated from municipal landfill area. *Journal of Microbiology and Biotechnology Research*, *1*, 131–136.

Pramila, R., & Ramesh, K. V. (2011b). Biodegradation of low density polyethylene (LDPE) by fungi isolated from marine water – a SEM analysis. *African Journal of Microbiological Research*, *5*, 5013–5018. doi:10.5897/AJMR11.670

Prasad, M. N. V. (2007). Phytoremediation in India. *Phytoremediation: Methods and Reviews*, 435-454.

Prasad, S., Singh, A., Jain, N., & Joshi, H. C. (2007a). Ethanol production from sweet sorghum syrup for utilization as automotive fuel in India. *Energy & Fuels*, *21*(4), 2415–2420. doi:10.1021/ef060328z

Prasad, S., Singh, A., & Joshi, H. C. (2007b). Ethanol as an alternative fuel from agricultural, industrial and urban residues. *Conservation & Recycling*, *50*(1). doi:10.1016/j.resconrec.2006.05.007

Preethi, K., Anand, M., & Thazeem, B. (2015). Isolation and Identification of keratinolytic bacteria from tannery effluent: A Study on their biodegradative and dehairing activity. *International Journal of Multidisciplinary Research and Development*, *2*(10), 227–234.

Preethi, K., & Maha Lakshmi, G., Mridul Umesh, Priyanka, K., & Thazeem, B. (2017). Fruit peels: A potential substrate for acetic acid production using *Acetobacter aceti*. *International Journal of Applied Research*, *2*(4), 286–291.

Prescott, M., Harley, J. P., & Klein, D. A. (2002). *Industrial Microbiology and Biotechnology in General Microbiology*. Academic Press.

Priya, B., Uma, L., Ahamed, A. K., Subramanian, G., & Prabaharan, D. (2011). Ability to use the diazo dye, C.I. Acid Black 1 as a nitrogen source by the marine cyanobacterium *Oscillatoria curviceps* BDU92191. *Bioresource Technology*, *102*(14), 7218–7223. doi:10.1016/j.biortech.2011.02.117 PMID:21571528

Putnam, R. A., Nelson, J. O., & Clark, J. M. (2003). The persistence and degradation of chlorothalonil and chlorpyrifos in a cranberry bog. *Journal of Agricultural and Food Chemistry*, *51*(1), 170–176. doi:10.1021/jf020744r PMID:12502403

Puvaneswari, N., Muthukrishnan, J., & Gunasekaran, P. (2006). Toxicity assesment and microbial degradation of azo dyes. *Indian Journal of Experimental Biology*, *44*, 618–626. PMID:16924831

Qi, J., Anke, M. K., Szymańska, K., & Tischler, D. (2017A). Immobilization of *Rhodococcus opacus* 1CP azoreductase to obtain azo dye degrading biocatalysts operative at acidic pH. *International Biodeterioration & Biodegradation*, *118*, 89–94. doi:10.1016/j.ibiod.2017.01.027

Qi, J., Paul, C. E., Hollmann, F., & Tischler, D. (2017B). Changing the electron donor improves azoreductase dye degrading activity at neutral pH. *Enzyme and Microbial Technology*, *100*, 17–19. doi:10.1016/j.enzmictec.2017.02.003 PMID:28284307

Qi, J., Schlömann, M., & Tischler, D. (2016). Biochemical characterization of an azoreductase from *Rhodococcus opacus* 1CP possessing methyl red degradation ability. *Journal of Molecular Catalysis. B, Enzymatic*, *130*, 9–17. doi:10.1016/j.molcatb.2016.04.012

Qing, Q., Yang, B., & Wyman, C. E. (2010). Impact of surfactants on pretreatment of corn stover. *Bioresource Technology*, *101*(15), 5941–5951. doi:10.1016/j.biortech.2010.03.003 PMID:20304637

Qiu, X., Shah, S. I., Kendall, E. W., Sorensen, D. L., Sims, R. C., & Engelke, M. C. (1994). Grass-enhanced bioremediation for clay soils contaminated with polynuclear aromatic hydrocarbons. In T.A. Anderson & J.R. Coats (Eds.), Bioremediation Through RhizosphereTechnology (pp. 142-157). American Chemical Society, Washington, DC. 10.1021/bk-1994-0563.ch013

Raaman, N., Rajitha, N., Jayshree, A., & Jegadeesh, R. (2012). Biodegradation of plastic by *Aspergillus* spp. isolated from polythene polluted sites around Chennai. *Journal of Academic and Industrial Research*, *1*, 313–316.

Rader, G., & Logan, B. (2010). Multi-electrode continuous flow microbial electrolysis cell for biogas production from acetate. *International Journal of Hydrogen Energy*, *35*(17), 8848–8854. doi:10.1016/j.ijhydene.2010.06.033

Raghavan, D., & Torma, A. E. (1992). DSC and FTIR characterization of biodegradation of polyethylene. *Polymer Engineering and Science*, *32*(6), 438–442. doi:10.1002/pen.760320609

Rahman, M. S., & Sathasivam, K. V. (2015). Heavy metal adsorption onto kappaphycus sp. from aqueous solutions: The use of error functions for validation of isotherm and kinetics models. *BioMed Research International*. PMID:26295032

Rai, A. K., Bhaskar, N., Halami, P. M., Indirani, K., Suresh, P. V., & Mahendrakar, N. S. (2009). Characterization and application of a native lactic acid bacterium isolated from tannery fleshings for fermentative bioconversion of tannery fleshings. *Applied Microbiology and Biotechnology*, *83*(4), 757–766. doi:10.100700253-009-1970-3 PMID:19333593

Rai, A. K., General, T., Bhaskar, N., Suresh, P. V., Sakhare, P. Z., Halami, P. M., ... Mahendrakar, N. S. (2010). Utilization of tannery fleshings: Optimization of conditions for fermenting delimed tannery fleshings using Enterococcus faecium HAB01 by response surface methodology. *Bioresource Technology*, *101*(6), 1885–1891. doi:10.1016/j.biortech.2009.10.015 PMID:19892548

Rai, J., Bhattacharya, M., Singh, J., Bansal, T. K., Vats, P., & Banerjee, U. C. (2005). Removal of dyes from the effluent of textile and dyestuff manufacturing industry: A review of emerging techniques with reference to biological treatment. *Critical Reviews in Environmental Science and Technology*, *35*(3), 219–238. doi:10.1080/10643380590917932

Rajandas, H., Parimannan, S., Sathasivam, K., Ravichandran, M., & Yin, L. S. (2012). A novel FTIR-ATR spectroscopy based technique for the estimation of low-density polyethylene biodegradation. *Polymer Testing*, *31*(8), 1094–1099. doi:10.1016/j.polymertesting.2012.07.015

Raja, R., Nayak, A. K., Shukla, A. K., Rao, K. S., Gautam, P., Lal, B., ... Patra, D. K. (2015). Impairment of soil health due to fly ash-fugitive dust deposition from coal-fired thermal power plants. *Environmental Monitoring and Assessment*, *187*(11), 679. doi:10.100710661-015-4902-y PMID:26450689

Ralebits, T. K., Senior, E., & Van Versevell, H. W. (2002). Microbial aspects of atrazine degradation in natural environments. *Biodegradation*, *13*(1), 11–19. doi:10.1023/A:1016329628618 PMID:12222950

Ramakrishnan, B. (2016). *Bioremediation with Simultaneous Recovery and Reuse of Resources. Journal of Bioremediation and Biodegradation*, *7*, 172.

Ramos, J. L., Díaz, E., Dowling, D., de Lorenzo, V., Molin, S., O'Gara, F., ... Timmis, K. N. (1994). The behavior of bacteria designed for biodegradation. *Nature Biotechnology*, *12*(12), 1349–1356. doi:10.1038/nbt1294-1349 PMID:7765565

Ran, Z., Gefu, Z., Kumar, J., Chaoxiang, L., Xu, H., & Lin, L. (2014). Hydrogen and methane production in a bio-electrochemical system assisted anaerobic baffled reactor. *International Journal of Hydrogen Energy*, *39*(25), 13498–13504. doi:10.1016/j.ijhydene.2014.02.086

Rashad, M. M., Moharib, S. A., & Jwanny, E. W. (1990). Yeast conversion of mango waste or methanol to single cell protein and other metabolites. *Biological Wastes*, *32*(4), 277–284. doi:10.1016/0269-7483(90)90059-2

Raskin, I., Smith, R. D., & Salt, D. E. (1997). Phytoremediation of metals: Using plants to remove pollutants from the environment. *Current Opinion in Biotechnology*, *8*(2), 221–226. doi:10.1016/S0958-1669(97)80106-1 PMID:9079727

Ratnavathi, C. V., Suresh, K., Kumar, B. S. V., Pallavi, M., Komala, V. V., & Seetharama, N. (2010). Study on genotypic variation for ethanol production from sweet sorghum juice. *Biomass and Bioenergy*, *34*(7), 947–952. doi:10.1016/j.biombioe.2010.02.002

Ravindran, B., Gayathri, V., Priya, D. M., Kanimozhi, S., & Sekaran, G. (2012). Hydrolysis of proteinaceous tannery solid waste for the production of extracellular acidic protease by Selenomonas ruminantium. *African Journal of Biotechnology*, *11*(56), 11978–11990. doi:10.5897/AJB11.2484

Ravindran, B., Kumar, A. G., Bhavani, P. A., & Sekaran, G. (2011). Solid-state fermentation for the production of alkaline protease by Bacillus cereus 1173900 using proteinaceous tannery solid waste. *Current Science (Bangalore), 100*(5), 726–730.

Ravindran, B., Nagalakshmi, C., Devi, S. G., Subramanian, K. S., Mandal, A. B., & Sekaran, G. (2013). Efficiency of hydrolyzed tannery solid waste for the growth and yield potentials on tomato. *Journal of Plant Nutrition, 36*(5), 795–810. doi:10.1080/01904167.2012.756893

Ravindran, B., & Sekaran, G. (2010). Bacterial composting of animal fleshing generated from tannery industries. *Waste Management (New York, N.Y.), 30*(12), 2622–2630. doi:10.1016/j.wasman.2010.07.013 PMID:20727727

Ray, M. J., Leak, D. J., Spanu, P. D., & Murphy, R. J. (2010). Brown rot fungal early stage decay mechanism as a biological pretreatment for softwood biomass in biofuel production. *Biomass and Bioenergy, 34*(8), 1257–1262. doi:10.1016/j.biombioe.2010.03.015

Rayu, S., Nielsen, U. N., Nazaries, L., & Singh, B. K. (2017). Isolation and Molecular Characterization of Novel Chlorpyrifos and 3,5,6- trichloro-2-pyridinol-degrading Bacteria from Sugarcane Farm Soils. *Frontiers in Microbiology, 8*, 518. doi:10.3389/fmicb.2017.00518 PMID:28421040

Reddy, C. S. K., Ghai, R., & Kalia, V. (2003). Polyhydroxyalkanoates: An overview. *Bioresource Technology, 87*(2), 137–146. doi:10.1016/S0960-8524(02)00212-2 PMID:12765352

Rees, H. C., Oswald, S. E., Banwart, S. A., Pickup, R. W., & Lerner, D. N. (2007). Biodegradation processes in a laboratory-scale groundwater contaminant plume assessed by fluorescence imaging and microbial analysis. *Applied and Environmental Microbiology, 73*(12), 3865–3876. doi:10.1128/AEM.02933-06 PMID:17468279

Rehman, M. S. U., Kim, I., Chisti, Y., & Han, J. I. (2013). Use of ultrasound in the production of bioethanol from lignocellulosic biomass. *Energy Education Science and Technology, Part A. Energy Science and Research, 30*(2), 1391–1410.

Reigart, J. R., & Roberts, J. R. (2014). *Recognition and Management of Pesticide Poisonings* (5^a ed.). United States Environmental Protection Agency. Retrieved from http://www.epa.gov/oppfead1/safety/healthcare/handbook/handbook.pdf

Resnick, S. M., & Gibson, D. T. (1996). Regio- and stereospecific oxidation of fluorene, dibenzofuran, and dibenzothiophene by naphthalene dioxygenase from *Pseudomonas* sp. strain NCIB 9816-4. *Applied and Environmental Microbiology, 62*(11), 4073–4080. PMID:8899998

Resnick, S. M., Lee, K., & Gibson, D. T. (1996). Diverse reactions catalyzed by naphthalene dioxygenase from *Pseudomonas* sp strain NCIB 9816. *Journal of Industrial Microbiology & Biotechnology, 17*(5), 438–457. doi:10.1007/BF01574775

Revankar, M., & Lele, S. (2007). Synthetic dye decolorization by White Rot Fungus, *Ganoderma* sp. WR-1. *Bioresource Technology, 98*(4), 775–780. doi:10.1016/j.biortech.2006.03.020 PMID:16730976

Richter, A., Burrows, J. P., Nusz, H., Granier, C., & Niemeier, U. (2005). Increase in tropospheric nitrogen dioxide over China observed from space. *Nature, 437*(7055), 129-132. Retrieved from http://www.nature.com/nature/journal/v437/n7055/suppinfo/nature04092_S1.html

Rigden, D. J., Fernández-Suárez, X. M., & Galperin, M. Y. (2016). The 2016 database issue of *Nucleic Acids Research* and an updated molecular biology database collection. *Nucleic Acids Research, 44*(D1), D1–D6. doi:10.1093/nar/gkv1356 PMID:26740669

Rincon, J., Gonzalez, F., Ballester, A., Blazquez, M. L., & Munoz, J. A. (2005). Biosorption of heavy metals by chemically-activated alga Fucus vesiculosus. *Journal of Chemical Technology and Biotechnology (Oxford, Oxfordshire), 80*(12), 1403–1407. doi:10.1002/jctb.1342

Riser-Roberts, E. (1998). *Remediation of petroleum contaminated soils. Biological, physical and chemical process.* USA: Lewis Publishers Inc, CRC Press LLC. doi:10.1201/9781420050578

Rivard, C., Moens, L., Brigham, K. R., & Kelley, S. (1995). Starch esters as biodegradable plastics: Effects of ester group chain length and degree of substitution on anaerobic biodegradation. *Enzyme and Microbial Technology, 17*(9), 848–852. doi:10.1016/0141-0229(94)00120-G

Robinson, T., McMullan, G., Marchant, R., & Nigam, P. (2001). Remediation of dyes in textile effluent: A critical review on current treatment technologies with a proposed alternative. *Bioresource Technology, 77*(3), 247–255. doi:10.1016/S0960-8524(00)00080-8 PMID:11272011

Rodríguez Martín, J. A., & Nanos, N. (2016). Soil as an archive of coal-fired power plant mercury deposition. *Journal of Hazardous Materials, 308*, 131–138. doi:10.1016/j.jhazmat.2016.01.026 PMID:26808251

Rodriguez-Iruretagoiena, A., Fdez-Ortiz de Vallejuelo, S., Gredilla, A., Ramos, C. G., Oliveira, M. L. S., Arana, G., ... Silva, L. F. O. (2015). Fate of hazardous elements in agricultural soils surrounding a coal power plant complex from Santa Catarina (Brazil). *The Science of the Total Environment, 508*, 374–382. doi:10.1016/j.scitotenv.2014.12.015 PMID:25497677

Roehr, M. (2000). *Biotechnology of Ethanol, Classic and future application. Publication.* Wiley. doi:10.1002/3527602348

Rojo, F. (2010). Enzymes for aerobic degradation of alkanes. In Handbook of Hydrocarbon and Lipid Microbiology (pp. 781-798). Berlin: Springer. doi:10.1007/978-3-540-77587-4_59

Romera, E., Gonzalez, F., Ballester, A., Blazquez, M. L., & Munoz, J. A. (2006). Biosorption with algae: A statistical review. *Critical Reviews in Biotechnology, 26*(4), 223–235. doi:10.1080/07388550600972153 PMID:17095433

Ronchel, M. C., & Ramos, J. L. (2001). Dual system to reinforce biological containment of recombinant bacteria designed for rhizoremediation. *Applied and Environmental Microbiology, 67*(6), 2649–2656. doi:10.1128/AEM.67.6.2649-2656.2001 PMID:11375176

Roy, P. K., Titus, S., Surekha, P., Tulsi, E., Deshmukh, C., & Rajagopal, C. (2008). Degradation of abiotically aged LDPE films containing pro-oxidant by bacterial consortium. *Polymer Degradation & Stability, 93*(10), 1917–1922. doi:10.1016/j.polymdegradstab.2008.07.016

Rstrepo-Florez, J.-M., Bassi, A., & Thompson, M. R. (2014). Microbial degradation and deterioration of polyethylene – a review. *International Biodeterioration & Biodegradation, 88*, 83–90. doi:10.1016/j.ibiod.2013.12.014

Ruiz-Arias, A., Juárez-Ramírez, C., de los Cobos-Vasconcelos, D., Ruiz-Ordaz, N., Salmerón-Alcocer, A., Ahuatzi-Chacón, D., & Galíndez-Mayer, J. (2010). Aerobic biodegradation of a sulfonated phenylazonaphthol dye by a bacterial community immobilized in a multistage packed-bed BAC reactor. *Applied Biochemistry and Biotechnology, 162*(6), 1689–1707. doi:10.100712010-010-8950-z PMID:20376575

Ruiz-Dueñas, F. J., & Martínez, A. T. (2009). Microbial degradation of lignin: How a bulky recalcitrant polymer is efficiently recycled in nature and how we can take advantage of this. *Microbial Biotechnology, 2*(2), 164–177. doi:10.1111/j.1751-7915.2008.00078.x PMID:21261911

Ruiz, O. N., Alvarez, D., Gonzalez-Ruiz, G., & Torres, C. (2011). Characterization of mercury bioremediation by transgenic bacteria expressing metallothionein and polyphosphate kinase. *BMC Biotechnology, 11*(1), 82. doi:10.1186/1472-6750-11-82 PMID:21838857

Ryan, A., Kaplan, E., Nebel, J. C., Polycarpoum, E., Crescente, V., Lowe, E., ... Sim, E. (2014). Identification of NAD(P)H quinone oxidoreductase activity in azoreductases from *P. aeruginosa*: Azoreductases and NAD(P)H quinone oxidoreductases belong to the same FMN-dependent superfamily of enzymes. *PLoS One*, *9*(6), e98551. doi:10.1371/journal.pone.0098551 PMID:24915188

Ryan, A., Laurieri, N., Westwood, I., Wang, C. J., Lowe, E., & Sim, E. (2010). A novel mechanism for azoreduction. *Journal of Molecular Biology*, *400*(1), 24–37. doi:10.1016/j.jmb.2010.04.023 PMID:20417637

Ryckeboer, J., Mergaert, J., Coosemans, J., Deprins, K., & Swings, J. (2003a). Microbiological aspects of bio waste during composting in a monitored compost bin. *Journal of Applied Microbiology*, *94*(1), 127–137. doi:10.1046/j.1365-2672.2003.01800.x PMID:12492933

Ryckeboer, J., Mergaert, J., Vaes, K., Klammer, S., De Clerq, D., Coosemans, J., ... Swings, J. (2003b). Survey of bacteria and fungi occurring during composting and self-heating processes. *Annals of Microbiology*, *53*(4), 349–410.

Ryckebosch, E., Drouillon, M., & Vervaeren, H. (2011). Techniques for transformation of biogas to biomethane. *Biomass and Bioenergy*, *35*(5), 1633–1645. doi:10.1016/j.biombioe.2011.02.033

Rynk, R. (Ed.). (1992). *On-Farm Composting Handbook*. Ithaca, NY: Northeast Regional Agricultural Engineering Service Cooperative Extension.

Ryoo, D., Shim, H., Arenghi, F., Barbieri, P., & Wood, T. (2001). Tetrachloroethylene, trichloroethylene, and chlorinated phenols induce toluene-o-xylene monooxygenase activity in Pseudomonas stutzeri OX1. *Applied Microbiology and Biotechnology*, *56*(3-4), 545–549. doi:10.1007002530100675 PMID:11549035

Sack, U., Hofrichter, M., & Fritsche, W. (1997). Degradation of phenanthrene and pyrene by *Nematoloma frowardii. Journal of Basic Microbiology*, *37*(4), 287–293. doi:10.1002/jobm.3620370408 PMID:9379367

Saeed, T., & Sun, G. (2011). Enhanced denitrification and organics removal in hybrid wetland columns: Comparative experiments. *Bioresource Technology*, *102*(2), 967–974. doi:10.1016/j.biortech.2010.09.056 PMID:20934326

Saeed, T., & Sun, G. (2012). A review on nitrogen and organics removal mechanisms in subsurface flow constructed wetlands: Dependency on environmental parameters, operating conditions and supporting media. *Journal of Environmental Management*, *112*, 429–448. doi:10.1016/j.jenvman.2012.08.011 PMID:23032989

Sag, Y., & Kutsal, T. (2001). Recent trends in the biosorption of heavy metals: A review. *Biotechnology and Bioprocess Engineering; BBE*, *6*(6), 376–385. doi:10.1007/BF02932318

Saha, B. C., & Cotta, M. A. (2007). Enzymatic saccharification and fermentation of alkaline peroxide pretreated rice hulls to ethanol. *Enzyme and Microbial Technology*, *41*(4), 528–532. doi:10.1016/j.enzmictec.2007.04.006

Saha, S., Roy, R., Sen, S. K., & Ray, A. K. (2006). Characterization of cellulase-producing bacteria from the digestive tract of tilapia, Oreochromis mossambica (Peters) and grass carp, Ctenopharyngodon idella (Valenciennes). *Aquaculture Research*, *37*(4), 380–388. doi:10.1111/j.1365-2109.2006.01442.x

Sakaguchi, T., Nakajima, A., & Horikoshi, T. (1981). Studies on the accumulation of heavy metal elements in biological systems. *European Journal of Applied Microbiology and Biotechnology*, *12*(2), 84-89.

Salleh, A. B., Ghazali, F. M., Rahman, R. N. Z. A., & Basri, M. (2003). Bioremediation of Petroleum hydrocarbon pollution. *Indian Journal of Biotechnology*, *2*, 411–425.

Sambrook, J., Fritsch, E. F., & Maniatis, T. (1989). *Molecular cloning: a laboratory manual*. New York: Cold Spring Harbor Press.

Samin, G., & Janssen, D. B. (2012). Transformation and biodegradation of 1, 2, 3-trichloropropane (TCP). *Environmental Science and Pollution Research International*, *19*(8), 3067–3078. doi:10.100711356-012-0859-3 PMID:22875418

Sampogna, R. V., & Cunard, R. (2007). Roundup intoxication and a rationale for treatment. *Clinical Nephrology*, *68*(3), 190–196. doi:10.5414/CNP68190 PMID:17915625

Samsel, A., & Seneff, S. (2013). Glyphosate, pathways to modern diseases II: Celiac sprue and gluten intolerance. *Interdisciplinary Toxicology*, *6*(4), 159–184. doi:10.2478/intox-2013-0026 PMID:24678255

Sanchez, C. (2009). Lignocellulosic residues: Biodegradation and bioconversion by fungi. *Biotechnology Advances*, *27*(2), 185–194. doi:10.1016/j.biotechadv.2008.11.001 PMID:19100826

Sanchez, G., Pilcher, L., Roslander, C., Modig, T., Galbe, M., & Liden, G. (2004). Dilute-acid hydrolysis for fermentation of the bolivian straw material pajabrava. *Bioresource Technology*, *93*(3), 249–256. doi:10.1016/j.biortech.2003.11.003 PMID:15062819

Sánchez, M. A., & González, B. (2007). Genetic characterization of 2, 4, 6-trichlorophenol degradation in Cupriavidus necator JMP134. *Applied and Environmental Microbiology*, *73*(9), 2769–2776. doi:10.1128/AEM.02584-06 PMID:17322325

Sandhya, S. (2010). Biodegradation of azo dyes under anaerobic condition: role of azoreductase. Handbook of Environmental Chemistry – Biodegradation of Azo Dyes Eds Erkurt, 39-57. doi:10.1007/698_2009_43

Sandmann, E. R. I. C., & Loos, M. A. (1984). Enumeration of 2, 4-D-degrading microorganisms in soils and crop plant rhizospheres using indicator media; high populations associated with sugarcane (*Saccharum officinarum). Chemosphere*, *13*(9), 1073–1084. doi:10.1016/0045-6535(84)90066-3

Sangkharak, K., Vangsirikul, P., & Janthachat, S. (2013). Isolation of novel cellulase from agricultural soil and application for ethanol production. *International Journal of Advanced Biotechnology and Research*, *2*(2), 230-239.

Sani, R. K., & Banerjee, U. C. (1999). Decolorization of triphenylmethane dyes and textile and dye-stuff effluent by *Kurthia* sp. *Enzyme and Microbial Technology*, *24*(7), 433–437. doi:10.1016/S0141-0229(98)00159-8

Santhoskumar, A. U., Palanivelu, K., Sharma, S. K., & Nayak, S. K. (2010). A new synthesis of nickel 12-hydroxy oleate formulation to improve polyolefin's degradation. *Journal of Bioremediation and Biodegradation, 1.*

Santo, M., Weitsman, R., & Sivan, A. (2013). The role of the copper-binding enzyme – laccase – in the biodegradation of polyethylene by the actinomycete *Rhodococcus ruber. International Biodeterioration & Biodegradation*, *84*, 204–210. doi:10.1016/j.ibiod.2012.03.001

Saratale, R. G., Saratale, G. D., Chang, J. S., & Govindwar, S. P. (2009). Decolorization and biodegradation of textile dye Navy blue HER by *Trichosporon beigelii* NCIM-3326. *Journal of Hazardous Materials*, *166*(2-3), 1421–1428. doi:10.1016/j.jhazmat.2008.12.068 PMID:19157708

Saratale, R. G., Saratale, G. D., Chang, J. S., & Govindwar, S. P. (2011). Bacterial decolorization and degradation of azo dyes: A Review. *Journal of the Taiwan Institute of Chemical Engineers*, *42*(1), 138–157. doi:10.1016/j.jtice.2010.06.006

Saravanakumar, K., & Kathiresan, K. (2013). Bioconversion of lignocellulosic waste to bioethanol by *Trichoderma* and yeast fermentation. *Biotech*, *4*(5), 493–499. PMID:28324381

Saritha, M., Arora, A., & Lata. (2011). Biological pretreatment of lignocellulosic substrates for enhanced delignification and enzymatic digestibility. *Indian Journal of Microbiology*, *52*(2), 122–130. doi:10.100712088-011-0199-x PMID:23729871

Sarkar, N., Ghosh, S. K., Bannerjee, S., & Aikat, K. (2011). Bioethanol production from agricultural wastes: An overview. *Renewable Energy*, *37*(1), 19–27. doi:10.1016/j.renene.2011.06.045

Sato, T., Kimura, Y., & Takamizawa, K. (2002). Modeling transport and biodegradation of PCE in sandy soil. *WIT Transactions on Ecology and the Environment, 55*.

Sato, K. (1989). In F. Kunc (Ed.), *Interrelationships between Microorganisms and Plants in Soil. Vancura, V* (pp. 335–342). New York: Elsevier. doi:10.1016/S0166-2481(08)70234-4

Satpathy, R., Konkimalla, V. B., & Ratha, J. (2015). Application of bioinformatics tools and databases in microbial dehalogenation research: A review. *Applied Biochemistry and Microbiology*, *51*(1), 11–20. doi:10.1134/S0003683815010147 PMID:25842899

Satyapal, G. K., Rani, S., & Kumar, M. (2016). Potential Role of Arsenic Resistant Bacteria in Bioremediation: Current Status and Future Prospects. *Journal of Microbial & Biochemical Technology*, *8*(3), 256–258. doi:10.4172/1948-5948.1000294

Savchenko, V. K. (1995). *The ecology of the Chernobyl catastrophe Scientific outlines of an international programme of collaborative research*. UNESCO France.

Sawangjit, S. (2016). Isolation and Characterization of Atrazine-Degrading Xanthomonas sp. ARB2 and Its Use in Bioremediation of Contaminated Soils. *International Journal of Environmental Sciences and Development*, *7*(5), 351–354. doi:10.7763/IJESD.2016.V7.798

Saxena, K. K., & Seth, N. (2002). Toxic effects of cypermethrin on certain hematological aspects of fresh water fish *Channa punctatus*. *Bulletin of Environmental Contamination and Toxicology*, *69*(3), 364–369. doi:10.100700128-002-0071-0 PMID:12177757

Schepers, A. W., Thibault, J., & Lacroix, C. (2002). *Lactobacillus helveticus* growth and lactic acid production during pH-controlled batch cultures in whey permeate/yeast extract medium. Part I. multiple factor kinetic analysis. *Enzyme and Microbial Technology*, *30*(2), 176–186. doi:10.1016/S0141-0229(01)00465-3

Scheuer, U., Zimmer, T., Becher, D., Schauer, F., & Schunck, W.-H. (1998). Oxygenation cascade in conversion of n-alkanes to α,ω-dioic acids catalyzed by cytochrome P450 52A3. *The Journal of Biological Chemistry*, *273*(49), 32528–32534. doi:10.1074/jbc.273.49.32528 PMID:9829987

Schmid, A., Sonnleitner, B., & Witholt, B. (1998). Medium chain length (mcl) alkane mass transfer rates in two-liquid phase *Pseudomonas oleovorans* cultivations. *Biotechnology and Bioengineering*, *60*(1), 10–23. doi:10.1002/(SICI)1097-0290(19981005)60:1<10::AID-BIT23.0.CO;2-3 PMID:10099401

Schnoor, J. L., Aitchison, E. W., Kelley, S. L., & Alvarez, P. J. J. (1998). Phytoremediation and bioaugmentation of 1, 4-dioxane. In *Abstr. Adv. Biol. Syst. Rem. Conf.* (pp. 87-91).

Schofield, D. A., & DiNovo, A. A. (2010). Generation of a mutagenized organophosphorus hydrolase for the biodegradation of the organophosphate pesticides malathion and demeton-S. *Journal of Applied Microbiology*, *109*(2), 548–557. PMID:20132373

Schultz, A., Jonas, U., Hammer, E., & Schauer, F. (2001). Dehalogenation of chlorinated hydroxybiphenyls by fungal laccase. *Applied and Environmental Microbiology*, *67*(9), 4377–4381. doi:10.1128/AEM.67.9.4377-4381.2001 PMID:11526052

Scientific Research. (2000). Phytoremediation: An Environmentally Sound Technology for Pollution Prevention, Control and Remediation: An Introductory Guide to Decision-Makers. United Nation Environment Program.

Scott, D. S., & Legge, R. L. (1999). Bioplastics. CIGR handbook of agricultural engineering, 5, 305-310.

Secchi, E. R., & Zarzur, S. (1999). Plastic debris ingested by a Blainville's beaked whale, *Mesoplodon densirostris*, washed ashore in Brazil. *Aquatic Mammals*, *25*, 21–24.

Seibert, K., Fuehr, F., & Cheng, H. H. (1981). In *Theory and Practical Use of Soil-Applied Herbicides Symposium* (pp. 137-146). European Weed Resource Society, Paris.

Sekaran, G., Swarnalatha, S., & Srinivasulu, T. (2007). Solid waste management in leather sector. *Journal on Design and Manufacturing Technologies*, *1*(1), 47–52. doi:10.18000/ijodam.70008

Seki, H., Suzuki, A., & Mitsueda, S. I. (1998). Biosorption of heavy metal ions onRhodobacter sphaeroidesandAlcaligenes eutrophusH16. *Journal of Colloid and Interface Science*, *197*(2), 185–190. doi:10.1006/jcis.1997.5284 PMID:9466859

Selvam, A., & Wong, J. W. C. (2008). Phytochelatin systhesis and cadmium uptake of *Brassica napus*. *Environmental Technology*, *29*(7), 765–773. doi:10.1080/09593330801987079 PMID:18697518

Seneviratne, G., Tennalkoon, N., & Nandasena, K. (2006). Polyethylene biodegradation by a developed Penicillium-Bacillus biofilm. *Current Science*, *90*, 20–21.

Sengupta, S., Chatterjee, T., Ghosh, P. B., & Saha, T. (2010). Heavy metal accumulation in agricultural soils around a coal fired thermal power plant (Farakka) in India. *Journal of Environmental Science & Engineering*, *52*(4), 299–306. PMID:22312798

Sen, K., Pakshirajan, K., & Santra, S. B. (2012). Modelling the biomass growth and enzyme secretion by the White Rot Fungus *Phanerochaete chrysosporium* in presence of a toxic pollutant. *Journal of Environmental Protection*, *3*(01), 114–119. doi:10.4236/jep.2012.31014

Sen, S. K., & Raut, S. (2015). Microbial degradation of low density polyethylene (LDPE): A review. *Journal of Environmental Chemical Engineering*, *3*(1), 462–473. doi:10.1016/j.jece.2015.01.003

Seymour, R. B. (1989). Polymer science before & after 1899: Notable developments during the lifetime of Maurtis Dekker. *Journal of Macromolecular Science. Chemistry*, *26*(8), 1023–1032. doi:10.1080/00222338908052032

Shah, A. A., Hasan, F., Hameed, A., & Ahmed, S. (2008). Biological degradation of plastics: A comprehensive review. *Biotechnology Advances*, *26*(3), 246–265. doi:10.1016/j.biotechadv.2007.12.005 PMID:18337047

Shah, K. L. (2000). *Basics of solid and hazardous waste management technology*. Upper Saddle River, NJ: Prentice Hall.

Shah, V., Garg, N., & Madamwar, D. (1999). Exopolysaccharide production by a marine cyanobacterium *Cyanothece* sp. application in dye removal by its gelatin phenomenon. *Applied Biochemistry and Biotechnology*, *82*(2), 81–90. doi:10.1385/ABAB:82:2:81

Shanahan, P. (2004). *Bioremediation. Waste Containment and RemediationTechnology*. Springer.

Shanmugam, P., & Horan, N. J. (2009). Optimising the biogas production from leather fleshing waste by co-digestion with MSW. *Bioresource Technology*, *100*(18), 4117–4120. doi:10.1016/j.biortech.2009.03.052 PMID:19395254

Sharholy, M., Ahmad, K., Vaishya, R. C., & Gupta, R. D. (2007). Municipal Solid Waste Characteristics and Management in Allahabad, India. *Journal of Waste Management*, *27*(4), 490–496. doi:10.1016/j.wasman.2006.03.001 PMID:16766176

Sharma, A., & Bhatt, P. (2016). Bioremediation: A microbial technology for improvising wildlife. Wildlife Management: Concept Analysis and Conservation. edited by Amita Saxena. Daya Publisher. Chapter-3, pp29-39.

Sharma, A., Pankaj, Khati, P., Gangola, S., & Kumar, G. (2016). Microbial degradation of pesticides for environmental cleanup. Bioremediation of Industrial Pollutants. edited by Ram Naresh Bharagava and Gaurav Saxena. Write and Print Publication. Chapter-6.pp178-205.

Sharma, A., & Sharma, A. (2004). Degradation assessment of low density polythene (LDP) and polythene (PP) by an indigenous isolate of *Pseudomonas stutzeri*. *Journal of Scientific and Industrial Research*, *63*, 293–296.

Sharma, M., Chaurasia, P. K., Yadav, A., Yadav, R. S. S., Yadava, S., & Yadav, K. D. S. (2016). Purification and characterization of a thermally stable yellow laccase from *Daedalea flavida* MTCC-145 with higher catalytic performance towards selective synthesis of substituted benzaldehydes. *Russian Journal of Bioorganic Chemistry, 42*(1), 59–68. doi:10.1134/S1068162016010143

Shegunova, P., Klánová, J., & Holoubek, I. (2007). Residues of organochlorinated pesticides in soils from the Czech Republic. *Environmental Pollution, 146*(1), 257–261. doi:10.1016/j.envpol.2006.03.057 PMID:17045372

Sheikh, M. M. I., Kim, C., Lee, J., Kim, S., Kim, G., Shim, S., & Kim, J. (2013). Production of bioethanol from waste money bills – A new cellulosic material for biofuels. *Food and Bioproducts Processing, 91*(1), 60–65. doi:10.1016/j.fbp.2012.09.001

Shelef, O., Gross, A., & Rachmilevitch, S. (2013). Role of Plants in a Constructed Wetland: Current and New Perspectives. *Water (Basel), 5*(2), 405–419. doi:10.3390/w5020405

Shen, C., Tang, X., Cheema, S. A., Zhang, C., Khan, M. I., Liang, F., ... Chen, Y. (2009). Enhanced phytoremediation potential of polychlorinated biphenyl contaminated soil from e-waste recycling area in the presence of randomly methylated-β-cyclodextrins. *Journal of Hazardous Materials, 172*(2), 1671–1676. doi:10.1016/j.jhazmat.2009.08.064 PMID:19748180

Shimao, M. (2001). Biodegradation of plastics. *Current Opinion in Biotechnology, 12*(3), 242–247. doi:10.1016/S0958-1669(00)00206-8 PMID:11404101

Shimp, J. F., Tracy, J. C., Davis, L. C., Lee, E., Huang, W., Erickson, L. E., & Schnoor, J. L. (1993). Beneficial effects of plants in the remediation of soil and groundwater contaminated with organic materials. *Critical Reviews in Environmental Science and Technology, 23*(1), 41–77. doi:10.1080/10643389309388441

Shong, J., Jimenez Diaz, M., & Collins, C. (2012). Towards synthetic microbial consortia for bioprocessing. *Current Opinion in Biotechnology, 23*(5), 798–802. doi:10.1016/j.copbio.2012.02.001 PMID:22387100

Shukla, V. B., Zhou, S., Yomano, L. P., Shanmugam, K. T., Preston, J. F., & Ingram, L. O. (2004). Production of d (–)-lactate from sucrose and molasses. *Biotechnology Letters, 26*(9), 689–693. doi:10.1023/B:BILE.0000024088.36803.4e PMID:15195965

Sidiras, D. K., & Salapa, L. S. (2015). Organosolv pretreatment as a major step of lignocellulosic biomass refining. In *Proceedings of the International Engineering Conferences.*

Simcik, M., Ruzicka, M., Karaszova, M., Sedlakova, Z., Vejrazka, J., Vesely, M., ... Izak, P. (2016). Polyamide thin-film composite membranes for potential raw biogas purification: Experiments and modeling. *Separation and Purification Technology, 167*, 163–173. doi:10.1016/j.seppur.2016.05.008

Singh, A., & Lal, R. (2009). Sphingobium ummariense sp. nov., a hexachlorocyclohexane (HCH)-degrading bacterium, isolated from HCH-contaminated soil. *International Journal of Systematic and Evolutionary Microbiology, 59*(1), 162–166. doi:10.1099/ijs.0.65712-0 PMID:19126742

Singhal, S. (2000). Solid Waste Management in India, Status and Future direction, TERI. *Information Monitor on Environ. Sci, 6*(1), 1–4.

Singhania, R. R., Sukumaran, R. K., Patel, A. K., Larroche, C., & Pandey, A. (2010). Advancement and comparative profiles in the production technologies using solid-state and submerged fermentation for microbial cellulases. *Enzyme and Microbial Technology, 46*(7), 541–549. doi:10.1016/j.enzmictec.2010.03.010

Singh, C. P. (1987). Preparation of high grade compost by enrichment on organic matter decomposition. *Bio. Agric. Host*, *5*, 41–49.

Singh, D. K. (2008). Biodegradation and bioremediation of pesticide in soil:concept, method and recent developments. *Indian Journal of Medical Microbiology*, *48*(1), 35–40. doi:10.100712088-008-0004-7 PMID:23100698

Singh, D. P., & Trivedi, R. K. (2013). Acid and Alkaline pretreatment of lignocellulosic biomass to produce ethanol as biofuel. *International Journal of Chemtech Research*, *5*(2), 727–734.

Singh, J., Agrawal, M., & Narayan, D. (1995). Changes in soil characteristics around coal-fired power plants. *Environment International*, *21*(1), 93–102. doi:10.1016/0160-4120(94)00035-6

Singh, P. (2007). Sequential anaerobic and aerobic treatment of pulp and paper mill effluent in pilot scale bioreactor. *Journal of Environmental Biology*, *28*(1), 77–82. PMID:17717990

Singh, R. L., Singh, P. K., & Singh, R. P. (2015). Enzymatic decolorization and degradation of azo dyes - A review. *International Biodeterioration & Biodegradation*, *104*, 21–31. doi:10.1016/j.ibiod.2015.04.027

Singh, R., Singh, R., & Singh, P. (2014). Bacterial decolorization of textile azo dye acid orange by *Staphylococcus hominis* RMLRT03. *Toxicology International*, *21*(2), 160–166. doi:10.4103/0971-6580.139797 PMID:25253925

Singh, S. P., & Prerna, P. (2009). Review of recent advances in anaerobic packed-bed biogas reactors. *Renewable & Sustainable Energy Reviews*, *13*(6), 1569–1575. doi:10.1016/j.rser.2008.08.006

Singh, V., Chauhan, P. K., Kanta, R., Dhewa, T., & Kumar, V. (2010). Isolation and characterization of pseudomonas resistant to heavy metals contaminants. *International Journal of Pharmaceutical Sciences Review and Research*, *3*, 164–167.

Singh, V., Dubey, M., & Bhadauria, S. (2012). Microbial degradation of polyethylene (low density) by *Aspergillus fumigatus* and *Penicillium* sp. *Asian Journal of Experimental Biological Sciences*, *3*, 498–501.

Singh, Z. J., Kaur, R., & Hundal, S. S. (2016). Toxic effects of organochlorine pesticides: A review. *American Journal of Bioscience*, *4*(3), 11–18. doi:10.11648/j.ajbio.s.2016040301.13

Sinha, C. (1997). Open burning of municipal solid waste: State level Analysis. *TERI Monitoring on Environmental Science*, *2*(2), 71–75.

Sinumvayo, J. P., Ishimwe, N., Komera, I., & Niyomukiza, S. (2015). Ethanol biofuel production from lignocellulosic biomass by Engineered *Saccharomyces cerevisiae*. *Journal of Academia & Industrial Research*, *3*(10), 463–471.

Sivan, A., Szanto, M., & Pavlov, V. (2006). Biofilm development of the polyethylene-degrading bacterium *Rhodococcus ruber*. *Applied Microbiology and Biotechnology*, *72*(2), 346–352. doi:10.100700253-005-0259-4 PMID:16534612

Sklarz, M. Y., Gross, A., Yakirevich, A., & Soares, M. I. M. (2009). A recirculating vertical flow constructed wetland for the treatment of domestic wastewater. *Desalination*, *246*(1–3), 617–624. doi:10.1016/j.desal.2008.09.002

Soccol, C. R., Vandenbarghe, L. P. S., Medeiros, A. B. P., Krap, S. G., Buckeridge, M., Ramos, L. P., ... Torres, F. A. G. (2010). Bioethanol from lignocelluloses: Status and perspectives in Brazil. *Bioresource Technology*, *101*(13), 4820–4825. doi:10.1016/j.biortech.2009.11.067 PMID:20022746

Sogorb, M. A., & Vilanova, E. (2002). Enzymes Involved in the Detoxification of, Organophosphorus, Carbamate and Pyrethroid Insecticides Through Hydrolysis. *Toxicology Letters*, *128*(1-3), 215–228. doi:10.1016/S0378-4274(01)00543-4 PMID:11869832

Solis, M., Solis, A., Perez, H. I., Manjarrez, N., & Flores, M. (2012). Microbial decolouration of azo dyes: A Review. *Process Biochemistry*, *47*(12), 1723–1748. doi:10.1016/j.procbio.2012.08.014

Solomons, G. L. (1985). Production of biomass by filamentous fungi. *Comprehensive biotechnology: the principles, applications, and regulations of biotechnology in industry, agriculture, and medicine.*

Song, B., Palleroni, N. J., & Häggblom, M. M. (2000). Isolation and characterization of diverse halobenzoate-degrading denitrifying bacteria from soils and sediments. *Applied and Environmental Microbiology*, *66*(8), 3446–3453. doi:10.1128/AEM.66.8.3446-3453.2000 PMID:10919805

Sorensen, S., Holtez, M. S., Simosen, A., & Aamand, J. (2007). Degradation and Mineralization of Nanomolar Concentrations of the Herbicide Dichlobenil and Its Persistent Metabolite 2,6-Dichlorobenzamide by *Aminobacter* spp. Isolated from Dichlobenil-Treated Soils. *Applied and Environmental Microbiology*, *73*(2), 399–406. doi:10.1128/AEM.01498-06 PMID:17114317

Soriano, J. M., Jiménez, B., Font, G., & Moltó, J. C. (2001). Analysis of Carbamate Pesticides and Their Metabolites in Water by Solid Phase Extraction and Liquid Chromatography: A Review. *Critical Reviews in Analytical Chemistry*, *31*(1), 19–5. doi:10.1080/20014091076677

Sowmya, H. V., Ramalingappa, M., & Krishnappa, M. (2012). Degradation of polyethylene by *Chaetomium* sp. and *Aspergillus flavus*. *International Journal of Recent Science Research*, *3*, 513–517.

Spear, L. B., Ainley, D. G., & Ribic, C. A. (1995). Incidence of plastic in seabirds from the tropical pacific, 1984-91: Relation with distribution of species, sex, age, season, year and body weight. *Marine Environmental Research*, *40*(2), 123–146. doi:10.1016/0141-1136(94)00140-K

Spear, R. (1991). Recognised and possible exposure to pesticides. In W. J. Hayes & E. R. Laws (Eds.), *Handbook of pesticide toxicology* (pp. 245–274). San Diego, CA: Academic.

Spencer, R. C. (2003). Bacillus anthracis. *Journal of Clinical Pathology*, *56*(3), 182–187. doi:10.1136/jcp.56.3.182 PMID:12610093

Sreenath, H. K., Moldes, A. B., Koegel, R. G., & Straub, R. J. (2001). Lactic acid production from agriculture residues. *Biotechnology Letters*, *23*(3), 179–184. doi:10.1023/A:1005651117831

Srinivasan, A., & Viraraghavan, T. (2010). Decolorization of dye wastewaters by biosorbents: A review. *Journal of Environmental Management*, *91*(10), 1915–1929. doi:10.1016/j.jenvman.2010.05.003 PMID:20627542

Sriprang, R., Hayashi, M., Yamashita, M., Ono, H., Saeki, K., & Murooka, Y. (2002). A novel bioremediation system for heavy metals using the symbiosis between leguminous plant and genetically engineered rhizobia. *Journal of Biotechnology*, *99*(3), 279–293. doi:10.1016/S0168-1656(02)00219-5 PMID:12385715

Stamatis, N., Konstantinou, I., & Hela, D. (2010). Pesticide Inputs from the Sewage Treatment Plant of Agrinio to River Acheloos, Western Greece: Occurrence and Removal. *Water Science and Technology*, *62*(5), 1098–1105. doi:10.2166/wst.2010.932 PMID:20818051

Stefanakis, A., Akratos, C. S., & Tsihrintzis, V. A. (2014). *Vertical flow constructed wetlands : eco-engineering systems for wastewater and sludge treatment*. Elsevier Science.

Steinbüchel, A., & Lütke-Eversloh, T. (2003). Metabolic engineering and pathway construction for biotechnological production of relevant polyhydroxyalkanoates in microorganisms. *Biochemical Engineering Journal*, *16*(2), 81–96. doi:10.1016/S1369-703X(03)00036-6

Stokes, P. M., & Lindsay, J. E. (1979). Copper tolerance and accumulation in Penicillium ochro-chloron isolated from copper-plating solution. *Mycologia*, *71*(4), 796–806. doi:10.2307/3759190 PMID:530290

Stolz, A. (2001). Basic and applied aspects in the microbial degradation of azo dyes. *Applied Microbiology and Biotechnology, 56*(1-2), 69–80. doi:10.1007002530100686 PMID:11499949

Stomp, A., Han, K.-H., Wilbert, S., Gordon, M. P., & Cunningham, S. D. (1994). Genetic strategies for enhancing phytoremediation. *Annals of the New York Academy of Sciences, 721*(1), 481–491. doi:10.1111/j.1749-6632.1994.tb47418.x PMID:8010696

Stompor, M., Potaniec, B., Szumny, A., Zielinski, P., & Anna Katarzyna Żołnierczyk, A. K. (2013). Microbial synthesis of dihydrochalcones using *Rhodococcus* and *Gordonia* species. *Journal of Molecular Catalysis. B, Enzymatic, 97*, 283–288. doi:10.1016/j.molcatb.2013.09.009

Stottmeister, U., Wießner, A., Kuschk, P., Kappelmeyer, U., Kästner, M., Bederski, O., ... Moormann, H. (2003). Effects of plants and microorganisms in constructed wetlands for wastewater treatment. *Biotechnology Advances, 22*(1), 93–117. doi:10.1016/j.biotechadv.2003.08.010 PMID:14623046

Strantz, A. A., & Zottola, E. A. (1992). Stability of cornstarch-containing polyethylene films to starch-degrading enzymes. *Journal of Food Protection, 55*(9), 736–738. doi:10.4315/0362-028X-55.9.736

Strevett, K., Vieth, R., & Grasso, D. (1995). Chemo-autotrophic biogas purification for methane enrichment: Mechanism and kinetics. *The Chemical Engineering Journal and the Biochemical Engineering Journal, 58*(1), 71–79. doi:10.1016/0923-0467(95)06095-2

Sudhakar, M., Doble, M., Murthy, P. S., & Venkatesan, R. (2008). Marine microbe-mediated biodegradation of low- and high-density polyethylenes. *International Biodeterioration & Biodegradation, 61*(3), 203–213. doi:10.1016/j.ibiod.2007.07.011

Sugiura, W., Yoda, T., Matsuba, T., Tanaka, Y., & Suzuki, Y. (2006). Expression and characterization of the genes encoding azoreductases from *Bacillus subtilis* and *Geobacillus stearothermophilus. Bioscience, Biotechnology, and Biochemistry, 70*(7), 1655–1665. doi:10.1271/bbb.60014 PMID:16861800

Sukumaran, R. K., Singhania, R. R., & Pandey, A. (2005). Microbial Cellulases Production, Applications and Challenges. *Journal of Scientific and Industrial Research, 64*(11), 832–844.

Sukumaran, R. K., Singhania, R. R., & Pandey, A. (2005). Microbial cellulases- Production, Applications and challenges. *Journal of Scientific and Industrial Research, 64*, 832–844.

Sulaymon, A. H., Mohammed, A. A., & Al-Musawi, T. J. (2013). Competitive biosorption of lead, cadmium, copper, and arsenic ions using algae. *Environmental Science and Pollution Research International, 20*(5), 3011–3023. doi:10.100711356-012-1208-2 PMID:23054774

Suman, S., Sinha, A., & Tarafdar, A. (2016). Polycyclic aromatic hydrocarbons (PAHs) concentration levels, pattern, source identification and soil toxicity assessment in urban traffic soil of Dhanbad, India. *The Science of the Total Environment, 545-546*, 353–360. doi:10.1016/j.scitotenv.2015.12.061 PMID:26747999

Sumathi, C., Jayashree, S., & Sekaran, G. (2012a). Screening and evaluation of antimicrobial activity of tannery fleshing utilizing fish gut bacteria. *International Journal of Pharmacology, 8*(5), 373–381. doi:10.3923/ijp.2012.373.381

Sumathi, C., Mohanapriya, D., Mandal, A. B., & Sekaran, G. (2012b). Production of different proteases from fish gut microflora utilizing tannery fleshing. *Engineering in Life Sciences, 12*(2), 223–237. doi:10.1002/elsc.201100097

Sumathi, S., & Sekaran, G. (2010). Nutritional Evaluation of Animal Fleshing as a Fish Meal Replacer in Labeo rohita. *Journal of Aquaculture Feed Science and Nutrition, 2*(2-4), 6–10.

Sundar, V. J., Gnanamani, A., Muralidharan, C., Chandrababu, N. K., & Mandal, A. B. (2011). Recovery and utilization of proteinous wastes of leather making: A review. *Reviews in Environmental Science and Biotechnology*, *10*(2), 151–163. doi:10.100711157-010-9223-6

Sun, Y., & Cheng, J. (2002). Hydrolysis of lignocellulosic materials for ethanol production: A review. *Bioresource Technology*, *83*(1). doi:10.1016/S0960-8524(01)00212-7 PMID:12058826

Sun, Y., Zhou, Q., Xu, Y., Wang, L., & Liang, X. (2011). Phytoremediation for co-contaminated soils of benzo [a] pyrene (B [a] P) and heavy metals using ornamental plant *Tagetes patula*. *Journal of Hazardous Materials*, *186*(2), 2075–2082. doi:10.1016/j.jhazmat.2010.12.116 PMID:21269763

Surekha, R. M., Lakshmi, P. K. L., Suvarnalatha, D., Jaya, M., Aruna, S., Jyothi, K., ... Venkateswarlu, K. (2008). Isolation and characterization of a chlorpyrifos degrading bacterium from agricultural soil and its growth response. *African Journal of Microbiological Research*, *2*, 26–31.

Suzuki, Y., Yoda, T., Ruhul, A., & Sugiura, W. (2001). Molecular cloning and characterization of the gene coding for azoreductase from *Bacillus* sp. OY1-2 isolated from soil. *The Journal of Biological Chemistry*, *276*(12), 9059–9065. doi:10.1074/jbc.M008083200 PMID:11134015

Syed, S., & Chinthala, P. (2015). Heavy metal detoxification by different Bacillus species isolated from solar salterns. *Scientifica*. PMID:26525498

Taherzadeh, M. J., & Karimi, K. (2008). Pretreatment of lignocellulosic wastes to improve ethanol and biogas production: A review. *International Journal of Molecular Sciences*, *9*(9), 1621–1651. doi:10.3390/ijms9091621 PMID:19325822

Talebnia, F., Karakashev, D., & Angelidaki, I. (2010). Production of bioethanol from wheat straw: An overview on pretreatment, hydrolysis and fermentation. *Bioresource Technology*, *101*(13), 4744–4753. doi:10.1016/j.biortech.2009.11.080 PMID:20031394

Tamura, T., & Jacobs, D. G. (1960). Structural implications in cesium sorption. *Health Physics*, *2*(4), 391–398. doi:10.1097/00004032-195910000-00009 PMID:13836791

Tan, L., Sun, Z. Y., Okamoto, S., Takaki, M., Tang, Y. Q., Morimura, S., & Kida, K. (2015). Production of ethanol from raw juice and thick juice of sugar beet by continuous ethanol fermentation with flocculating yeast strain KF-7. *Biomass and Bioenergy*, *81*, 265–272. doi:10.1016/j.biombioe.2015.07.019

Tarazona, J. V. (2014). *Pollution, Soil A2 - Wexler, Philip Encyclopedia of Toxicology* (3rd ed., pp. 1019–1023). Oxford: Academic Press.

Taşeli, B. K., Gökçay, C. F., & Taşeli, H. (2004). Up flow column reactor design for dechlorination of chlorinated pulping wastes by Penicillium camemberti. *Journal of Environmental Management*, *72*(3), 175–179. doi:10.1016/j.jenvman.2004.04.008 PMID:15251223

Tee, H.-C., Lim, P.-E., Seng, C.-E., Mohd Nawi, M. A., & Adnan, R. (2015). Enhancement of azo dye Acid Orange 7 removal in newly developed horizontal subsurface-flow constructed wetland. *Journal of Environmental Management*, *147*, 349–355. doi:10.1016/j.jenvman.2014.09.025 PMID:25284799

Tee, H.-C., Lim, P.-E., Seng, C.-E., & Nawi, M.-A. M. (2012). Newly developed baffled subsurface-flow constructed wetland for the enhancement of nitrogen removal. *Bioresource Technology*, *104*, 235–242. doi:10.1016/j.biortech.2011.11.032 PMID:22130081

Telke, A. A., Ghodake, G. S., Kalyani, D. C., Dhanve, R. S., & Govindwar, S. P. (2011). Biochemical characteristics of a textile dye degrading extracellular laccase from a *Bacillus* sp. ADR. *Bioresource Technology, 102*(2), 1752–1756. doi:10.1016/j.biortech.2010.08.086 PMID:20855194

Thapa, B., Ajay Kumar, K. C., & Ghimire, A. (2012). A review on bioremediation of petroleum hydrocarbon contaminants in soil. Kathmandu University Journal of Science. *Engineering and Technology., 8*(1), 164–170.

Thassitou, P.K., & Arvanitonyannis, I S. (2001). Bioremediation: A novel approach to food waste management. *Trends in Food Science & Technology, 12*(185).

Thazeem, B., Umesh, M., & Vikas, O. V. (2016). Isolation and identification of proteolytic *Bacillus pumilus* from tannery lime effluent and its dehairing activity. *International Journal of Academic Research and Development, 1*(2), 5-7.

Thazeem, B., Beryl, G. P., & Umesh, M. (2017a). A comparative study on alkaline protease production from *Bacillus* spp. and their biodegradative, dehairing and destaining activity. *International Journal of Academic Research and Development, 2*(2), 74–79.

Thazeem, B., Preethi, K., Umesh, M., & Radhakrishnan, S. (2017b). Nutritive Characterization of Delimed Bovine Tannery Fleshings for Their Possible Use as a Proteinaceous Aqua Feed Ingredient. *Waste and Biomass Valorization.*

Thompson, R. C., Swan, S. H., Moore, C. J., & vom Saal, F. S. (2009). Our plastic age. *Philosophical Transactions of the Royal Society of London. Series B, Biological Sciences, 364*(1526), 1973–1976. doi:10.1098/rstb.2009.0054 PMID:19528049

Thongkred, P., Lotrakul, P., Prasongsuk, S., Imai, T., & Punnapayak, H. (2011). Oxidation of polycyclic aromatic hydrocarbons by a tropical isolate of Pycnoporus coccineus and its laccase. *ScienceAsia, 37*(3), 225–233. doi:10.2306cienceasia1513-1874.2011.37.225

Tian, H., Cheng, K., Wang, Y., Zhao, D., Lu, L., Jia, W., & Hao, J. (2012). Temporal and spatial variation characteristics of atmospheric emissions of Cd, Cr, and Pb from coal in China. *Atmospheric Environment, 50*, 157–163. doi:10.1016/j.atmosenv.2011.12.045

Tiwari, S., Kumari, B., & Singh, S. N. (2008). Evaluation of metal mobility/immobility in fly ash induced by bacterial strains isolated from the rhizospheric zone of Typha latifolia growing on fly ash dumps. *Bioresource Technology, 99*(5), 1305–1310. doi:10.1016/j.biortech.2007.02.010 PMID:17382536

Toan, V., Thao, V., Walder, J., Schmutz, H.-R., & Ha, C. (2007). Contamination by Selected Organochlorine Pesticides (OCPs) in Surface Soils in Hanoi, Vietnam. *Bulletin of Environmental Contamination and Toxicology, 78*(3), 195–200. doi:10.100700128-007-9149-z PMID:17476441

Tobin, J. M., Cooper, D. G., & Neufeld, R. J. (1984). Uptake of metal ions by Rhizopus arrhizus biomass. *Applied and Environmental Microbiology, 47*(4), 821–824. PMID:16346521

Togna, M. T., Kafkewitz, D., & Armenante, P. M. (1995). Rapid dehalogenation of 2, 4, 6-trichlorophenol at alkaline pH by an anaerobic enrichment culture. *Letters in Applied Microbiology, 20*(2), 113–116. doi:10.1111/j.1472-765X.1995.tb01299.x

Tomizawa, M., & Casida, J. E. (2005). Neonicotinoid insecticide toxicology: Mechanisms of selective action. *Annual Review of Pharmacology and Toxicology*, 24. PMID:15822177

Tony, B. D., Goyal, D., & Khanna, S. (2009). Decolorization of textile azo dyes by aerobic bacterial consortium. *International Biodeterioration & Biodegradation, 63*(4), 462–469. doi:10.1016/j.ibiod.2009.01.003

Topp, E. (2001). A comparison of three atrazine-degrading bacteria for soil bioremediation. *Biology and Fertility of Soils*, *33*(6), 529–534. doi:10.1007003740100371

Topp, E., Hanson, R. S., Ringelberg, D. B., White, D. C., & Wheatcroft, R. (1993). Isolation and characterization of an N-methylcarbamate insecticide-degrading *methylotrophic* bacterium. *Applied and Environmental Microbiology*, *59*, 3339–3349. PMID:7504430

Torres-Jimenez, E., Jerman, M. S., Gregorc, A., Lisec, I., Dorado, M. P., & Kegl, B. (2011). Physical and chemical properties of ethanol-diesel fuel blends. *Fuel*, *90*(2), 795–802. doi:10.1016/j.fuel.2010.09.045

Travaini, R., Martin-Juarez, J., Lorenzo-Hernando, A., & Bolado-Rodriguez, S. (2016). Ozonolysis: An advantageous pretreatment for lignocellulosic biomass revisited. *Bioresource Technology*, *199*, 2–12. doi:10.1016/j.biortech.2015.08.143 PMID:26409859

Tresse, O., Mounien, F., Lévesque, M. J., & Guiot, S. (2005). Comparison of the microbial population dynamics and phylogenetic characterization of a CANOXIS reactor and a UASB reactor degrading trichloroethene. *Journal of Applied Microbiology*, *98*(2), 440–449. doi:10.1111/j.1365-2672.2004.02475.x PMID:15659198

Tribedi, P., & Sil, A. K. (2013). Low-density polyethylene degradation by *Pseudomonas* sp. AKS2 biofilm. *Environmental Science and Pollution Research International*, *20*(6), 4146–4153. doi:10.100711356-012-1378-y PMID:23242625

Trinelli, M. A., Areco, M. M., & dos Santos Afonso, M. (2013). Co-biosorption of copper and glyphosate by Ulva lactuca. *Colloids and Surfaces. B, Biointerfaces*, *105*, 251–258. doi:10.1016/j.colsurfb.2012.12.047 PMID:23376752

Tripathi, V., Fraceto, L. F., & Abhilash, P. C. (2015). Sustainable clean-up technologies for soils contaminated with multiple pollutants: Plant-microbe-pollutant and climate nexus. *Ecological Engineering*, *82*, 330–335. doi:10.1016/j.ecoleng.2015.05.027

Tsai, W.-T. (2010). Current Status and Regulatory Aspects of Pesticides Considered to be Persistent Organic Pollutants (POPs) in Taiwan. *International Journal of Environmental Research and Public Health*, *7*(10), 3615–3627. doi:10.3390/ijerph7103615 PMID:21139852

Tsezos, M. (2001). Biosorption of metals. The experience accumulated and the outlook for technology development. *Hydrometallurgy*, *59*(2), 241–243. doi:10.1016/S0304-386X(99)00056-0

Tsezos, M., & Keller, D. M. (1983). Adsorption of radium-226 by biological origin absorbents. *Biotechnology and Bioengineering*, *25*(1), 201–215. doi:10.1002/bit.260250116 PMID:18548548

Tsezos, M., & Volesky, B. (1981). Biosorption of uranium and thorium. *Biotechnology and Bioengineering*, *23*(3), 583–604. doi:10.1002/bit.260230309

Tuomela, M., Vikman, M., Hatakka, A., & Itavaara, M. (2000). Biodegradation of lignin in a compost environment: A review. *Bioresource Technology*, *72*(2), 169–183. doi:10.1016/S0960-8524(99)00104-2

Turner, D., Xu, H., Cracknell, R. F., Natarajan, V., & Chen, X. (2011). Combustion performance of bio-ethanol of various blend ratios in a gasoline direct injection engine. *Fuel*, *90*(5), 1999–2006. doi:10.1016/j.fuel.2010.12.025

Turrio-Baldassarri, L., Abate, V., Alivernini, S., Battistelli, C. L., Carasi, S., Casella, M., ... La Rocca, C. (2007). A study on PCB, PCDD/PCDF industrial contamination in a mixed urban-agricultural area significantly affecting the food chain and the human exposure. Part I: Soil and feed. *Chemosphere*, *67*(9), 1822–1830. doi:10.1016/j.chemosphere.2006.05.124 PMID:17234238

Tutt, M., Kikas, T., Kahr, H., Pointner, M., Kuttner, P., & Olt, J. (2014). Using steam explosion pretreatment method for bioethanol production from floodplain meadow hay. *Agronomy Research (Tartu)*, *12*(2), 417–424.

U.S. Dept. of Energy. (2016). Alternative fuels data center: Maps and data. Retrieved 1 February 2017 from http://www.afdc.energy.gov/data/

U.S. Environment Protection Agency. (1994). *Composting Yard Trimmings and Municipal Solid Waste*. Solid Waste and Emergency Response (Report No. EPA530-R- 94-003).

U.S. Environmental Protection Agency. (2004). *Monitoring approaches for landfill bioreactors*. EPA/600/R-04/301.

Uchiyama, I., Mihara, M., Nishide, H., & Chiba, H. (2015). MBGD update: Microbial genome database for flexible ortholog analysis utilizing a diverse set of genomic data. *Nucleic Acids Research*, *43*(D1), D270–D276. doi:10.1093/nar/gku1152 PMID:25398900

Umesh, M., & Preethi, K. (2014). Fermentative Utilization of fruit peel waste for Lactic Acid production by *Lactobacillus plantarum. Indian Journal of Applied Research*, *4*(9), 449–451.

UNEP Stockholm Convention. (2008). All POPs listed in the Stockholm Convention. Retrieved on June 30, 2017 from http://chm.pops.int/TheConvention/ThePOPs/AllPOPs/tabid/2509/Default.aspx

UNEP. (2004). Report on State of Waste Management in South East Asia, Types of Wastes - Sources and Composition [Data file]. Retrieved from www.asean.org/uploads/archive/files/UNEP.pdf

United Nations Environment Program, Ozone Secretariat. (2011). The London Amendment (1990): The amendment to the Montreal Protocol agreed by the Second Meeting of the Parties (London, 27-29 June 1990) Retrieved June 30, 2017 from https://ozone.unep.org/new_site/en/Treaties/treaties_decisions-hb.php?dec_id_anx_auto=780

United States Department of Agriculture. (2009). Colombia: a new ethanol producer on the rise. Economic Research Service: USDA.

United States Environmental Protection Agency (EPA). (2017a). Air Emissions Inventories. Retrieved June 1, 2017 from https://www.epa.gov/air-emissions-inventories/air-emissions-sources

United States Environmental Protection Agency (EPA). (2017b). Greenhouse Gas Emissions: Carbon Dioxide Emissions. Retrieved June 30, 2017 from https://www.epa.gov/ghgemissions/overview-greenhouse-gases#carbon-dioxide

United States Environmental Protection Agency (EPA). (2017c). Greenhouse Gas Emissions: Methane Emissions. Retrieved June 30, 2017 from https://www.epa.gov/ghgemissions/overview-greenhouse-gases#methane

Uqab, B., Mudasir, S., & Nazir, R. (2016). Review on Bioremediation of Pesticides. *Journal of Bioremediation & Biodegradation*, *7*, 343.

Uqab, B., Mudasir, S., & Qayoom, A. (2016). Bioremediation: A Management Tool. *Journal of Bioremediation & Biodegradation*, *7*, 331.

Urrutia, M. M. (1997). General bacterial sorption processes. *Biosorbents for metal ions*, 39-66.

US Environmental Protection Agency. (1994). Review of liner and cap regulations for landfills. EPA/600/A-94/246., Risk Reduction Engineering Laboratory, Office of Research and Development, U.S. EPA.

Vaccari, D. A., Strom, P. F., & Alleman, J. E. (2005). Environmental Biology for Engineers and Scientists.

Van Beilen, J. B., & Funhoff, E. G. (2007). Alkane hydroxylases involved in microbial alkane degradation. *Applied Microbiology and Biotechnology*, *74*(1), 13–21. doi:10.100700253-006-0748-0 PMID:17216462

Van Beilen, J. B., Funhoff, E. G., Van Loon, A., Just, A., Kaysser, L., Bouza, M., ... Witholt, B. (2006). Cytochrome P450 alkane hydroxylases of the CYP153 family are common in alkane-degrading eubacteria lacking integral membrane alkane hydroxylases. *Applied and Environmental Microbiology, 72*(1), 59–65. doi:10.1128/AEM.72.1.59-65.2006 PMID:16391025

Van Beilen, J. B., Kingma, J., & Witholt, B. (1994). Substrate specificity of the alkane hydroxylase system of *Pseudomonas oleovorans* GPo1. *Enzyme and Microbial Technology, 16*(10), 904–911. doi:10.1016/0141-0229(94)90066-3

Van den Wijngaard, A. J., Van Der Kleij, R. G., Doornweerd, R. E., & Janssen, D. B. (1993). Influence of organic nutrients and cocultures on the competitive behavior of 1, 2-dichloroethane-degrading bacteria. *Applied and Environmental Microbiology, 59*(10), 3400–3405. PMID:8250561

Van der Heul, R. M. (2011). *Environmental degradation of petroleum hydrocarbons* (Unpublished master's dissertation). Utrecht University, Netherlands.

Van Deuren, J., Lloyd, T., Chhetry, S., Raycharn, L., & And Peck, J. (2002). Remediation Technologies Screening Matrix and Reference Guide. In *Federal Remediation Technologies Roundtable.*

Van Deuren, J., Lloyd, T., Chhetry, S., Raycharn, L., & Peck, J. (2002). Remediation Technologies Screening Matrix and Reference Guide. In *Federal Remediation Technologies Roundtable.*

Van Hamme, J. D., Singh, A., & Ward, O. P. (2003). Recent Advances in Petroleum Microbiology. *Microbiology and Molecular Biology Reviews, 67*(4), 503–549. doi:10.1128/MMBR.67.4.503-549.2003 PMID:14665675

Van Kempen-Fryling, R. J., & Camper, A. K. (2017). Escherichia coli O157:H7 attachment and persistence within root biofilm of common treatment wetlands plants.

Van Pee, K. H., & Unversucht, S. (2003). Biological dehalogenation and halogenation reactions. *Chemosphere, 52*(2), 299–312. doi:10.1016/S0045-6535(03)00204-2 PMID:12738254

Vandevivere, P. C., Bianchi, R., & Verstraete, W. (1998). Treatment and reuse of wastewater from the textile wet-processing industry: Review of Emerging Technologies. *Journal of Chemical Technology and Biotechnology (Oxford, Oxfordshire), 72*(4), 289–302. doi:10.1002/(SICI)1097-4660(199808)72:4<289::AID-JCTB9053.0.CO;2-#

Vanholme, R., Demedts, B., Morreel, K., Ralph, J., & Boerjan, W. (2010). Lignin biosynthesis and structure. *Plant Physiology, 153*(3), 895–905. doi:10.1104/pp.110.155119 PMID:20472751

Varga, E., Schmidt, A. S., Reczey, K., & Thomsen, A. B. (2003). Pretreatment of corn stover using wet oxidation to enhance enzymatic digestibility. *Applied Biochemistry and Biotechnology, 104*(1), 37–50. doi:10.1385/ABAB:104:1:37 PMID:12495204

Vasile, C. (1993). Degradation and decomposition. In C. Vasile & R. B. Seymour (Eds.), *Handbook of polyolefins synthesis and properties* (pp. 479–509). New York, NY: Marcel Dekker Inc.

Vassilev, S. V., Vassileva, C. G., & Vassilev, V. S. (2015). Advantages and disadvantages of composition and properties of biomass in comparison with coal: An overview. *Fuel, 158*, 330–350. doi:10.1016/j.fuel.2015.05.050

Vasudevan, N., & Ravindran, A. D. (2007). Biotechnological process for the treatment of fleshing from tannery industries for methane generation. *Current Science, 93*(11), 1492–1494.

Vasudevan, N., & Ravindran, A. D. (2009). Bioconversion of tannery solid waste into energy. *International Journal of Integrated Energy Systems, 1*(2), 97–101.

Vasudevan, P., Padmavathy, V., Tewari, N., & Dhingra, S. C. (2001). Biosorption of heavy metal ions. *Journal of Scientific and Industrial Research (New Delhi, India)*, *60*, 112–120.

Veiga, M. M., Silva, D. M., & Veiga, L. B. E., and De CastroFaria, M. V. (. (2006). Pesticide pollution in water systems in a small rural community in Southeast Brazil. *Cadernos de Saude Publica*, *22*(11), 2391–2399. doi:10.1590/S0102-311X2006001100013 PMID:17091176

Velazquez, F., Peak-Chew, S. Y., Fernández, I. S., Neumann, C. S., & Kay, R. R. (2011). Identification of a eukaryotic reductive dechlorinase and characterization of its mechanism of action on its natural substrate. *Chemistry & Biology*, *18*(10), 1252–1260. doi:10.1016/j.chembiol.2011.08.003 PMID:22035794

Venkatesagowda, B., Ponugupaty, E., Barbisa, A. M., & Dekker, R. F. (2012). Diversity of plant oil seed-associated fungi isolated from seven oil bearing seeds and their potential for the production of lipolytic enzymes. *World Journal of Microbiology & Biotechnology*, *28*(1), 71–80. doi:10.100711274-011-0793-4 PMID:22806781

Verkleij, J. A., Golan-Goldhirsh, A., Antosiewisz, D. M., Schwitzguébel, J. P., & Schröder, P. (2009). Dualities in plant tolerance to pollutants and their uptake and translocation to the upper plant parts. *Environmental and Experimental Botany*, *67*(1), 10–22. doi:10.1016/j.envexpbot.2009.05.009

Verma, A., Kumar, S., & Jain, P. K. (2011). Key pretreatment technologies on cellulosic ethanol production. *Journal of Scientific Research*, *55*, 57–63.

Verma, P., & Madamwar, D. (2002). Production of ligninolytic enzymes for dye decolorization by cocultivation of white-rot fungi *Pleurotus ostreatus* and *Phanerochaete chrysosporium* under solid-state fermentation. *Applied Biochemistry and Biotechnology*, *102*(1-6), 109–118. doi:10.1385/ABAB:102-103:1-6:109 PMID:12396115

Verma, P., & Madamwar, D. (2003). Decolourization of synthetic dyes by a newly isolated strain of *Serratia marcescens. World Journal of Microbiology & Biotechnology*, *19*(6), 615–618. doi:10.1023/A:1025115801331

Veselý, T., Tlustoš, P., & Száková, J. (2012). Organic acid enhanced soil risk element (Cd, Pb and Zn) leaching and secondary bioconcentration in water lettuce (Pistia stratiotes L.) in the rhizofiltration process. *International Journal of Phytoremediation*, *14*(4), 335–349. doi:10.1080/15226514.2011.620650 PMID:22567715

Viard, B., Pihan, F., Promeyrat, S., & Pihan, J.-C. (2004). Integrated assessment of heavy metal (Pb, Zn, Cd) highway pollution: Bioaccumulation in soil, Graminaceae and land snails. *Chemosphere*, *55*(10), 1349–1359. doi:10.1016/j.chemosphere.2004.01.003 PMID:15081778

Vidali, M. (2001). Bioremediation. an overview. *Pure and Applied Chemistry*, *73*(7), 1163–1172. doi:10.1351/pac200173071163

Vidal, P. F., & Molinier, J. (1988). Ozonolysis of lignin improvement of in vitro digestibility of poplar sawdust. *Biomass*, *16*(1), 1–17. doi:10.1016/0144-4565(88)90012-1

Vijayaraghavan, K., & Yun, Y. S. (2008). Bacterial biosorbents and biosorption. *Biotechnology Advances*, *26*(3), 266–291. doi:10.1016/j.biotechadv.2008.02.002 PMID:18353595

Vikas, O.V., & Mridul, U. (2014). Bioconversion of papaya peel waste in to vinegar by *Acetobacter aceti. International Journal of Scientific Research*, *3*(11), 409–411.

Villano, M., Aulenta, F., Beccari, M., & Majone, M. (2010). Microbial generation of H_2 or CH_4 coupled to wastewater treatment in bioelectrochemical systems. *Chemical Engineering Transactions*, *20*, 163–168. doi:10.3303/CET1020028

Villano, M., Monaco, G., Aulenta, F., & Majone, M. (2011). Electrochemically assisted methane production in a biofilm reactor. *Journal of Power Sources*, *196*(22), 9467–9472. doi:10.1016/j.jpowsour.2011.07.016

Villano, M., Scardala, S., Aulenta, F., & Majone, M. (2013). Carbon and nitrogen removal and enhanced methane production in a microbial electrolysis cell. *Bioresource Technology*, *130*, 366–371. doi:10.1016/j.biortech.2012.11.080 PMID:23313682

Vobis, G., & Miyadoh, S. (1997). Atlas of actinomycetes. Academic Press.

Vohra, M., Manwar, J., Manmode, R., Padgilwar, S., & Patil, S. (2014). Bioethanol production: Feedstock and current technologies. *Journal of Environmental Chemical Engineering*, *2*(1), 573–584. doi:10.1016/j.jece.2013.10.013

Volesky, B. (2007). Biosorption and me. *Water Research*, *41*(18), 4017–4029. doi:10.1016/j.watres.2007.05.062 PMID:17632204

Volke-Sepúlveda, T., Saucedo-Castañeda, G., Gutiérrez-Rojas, M., Manzur, A., & Favela-Torres, E. (2002). Thermally treated low density polyethylene biodegradation by *Penicillium pinophilum* and *Aspergillus niger*. *Journal of Applied Polymer Science*, *83*(2), 305–314. doi:10.1002/app.2245

Volova, T. G. (2004). *Polyhydroxyalkanoates--plastic materials of the 21st century: production, properties, applications*. Nova publishers.

Vymazal, J. (2011) Plants used in constructed wetlands with horizontal subsurface flow: a review. *Hydrobiologia*, *674*(1), 133–156.

Vymazal, J. (2007). Removal of nutrients in various types of constructed wetlands. *The Science of the Total Environment*, *380*(1), 48–65. doi:10.1016/j.scitotenv.2006.09.014 PMID:17078997

Vymazal, J. (2013). The use of hybrid constructed wetlands for wastewater treatment with special attention to nitrogen removal: A review of a recent development. *Water Research*, *47*(14), 4795–4811. doi:10.1016/j.watres.2013.05.029 PMID:23866134

Wahman, D. G., Katz, L. E., & Speitel, G. E. (2005). Cometabolism of trihalomethanes by Nitrosomonas europaea. *Applied and Environmental Microbiology*, *71*(12), 7980–7986. doi:10.1128/AEM.71.12.7980-7986.2005 PMID:16332776

Walker, G. M. (2011). Fuel alcohol: Current production and future challenges. *Journal of the Institute of Brewing*, *117*(1), 3–22. doi:10.1002/j.2050-0416.2011.tb00438.x

Wanderley, K. J., Torres, F. A., Moraes, L. M., & Ulhoa, C. J. (2004). Biochemical characterization of Î±-amylase from the yeast Cryptococcus flavus. *Microbiology Letters*, *231*(2), 165–169. doi:10.1016/S0378-1097(03)00955-8

Wang, C. J., Hagemeier, C., Rahman, N., Lowe, E., Noble, M., Coughtrie, M., ... Westwood, I. (2007). Molecular cloning, characterisation and ligand-bound structure of an azoreductase from *Pseudomonas aeruginosa*. *Journal of Molecular Biology*, *373*(5), 1213–1228. doi:10.1016/j.jmb.2007.08.048 PMID:17904577

Wang, C.-J., Laurieri, N., Abuhammad, A., Lowe, E., Westwood, I., Ryan, A., & Sim, E. (2010). Role of tyrosine 131 in the active site of paAzoR1, an azoreductase with specificity for the inflammatory bowel disease prodrug balsalazide. *Acta Crystallographica. Section F, Structural Biology and Crystallization Communications*, *66*(1), 2–7. doi:10.1107/S1744309109044741 PMID:20057057

Wang, F., Grundmann, S., Schmid, M., Dörfler, U., Roherer, S., Munch, J. C., & Schroll, R. (2007). Isolation and characterization of 1, 2, 4-trichlorobenzene mineralizing Bordetella sp. and its bioremediation potential in soil. *Chemosphere*, *67*(5), 896–902. doi:10.1016/j.chemosphere.2006.11.019 PMID:17204305

Wang, H., Su, J., Zheng, X., Tian, Y., Xiong, X., & Zheng, T. L. (2009). Bacterial decolorization and degradation of the reactive dye Reactive Red 180 by *Citrobacter* sp. SK3. *International Biodeterioration & Biodegradation*, *63*(4), 395–399. doi:10.1016/j.ibiod.2008.11.006

Wang, H., Zheng, X. W., Su, J. Q., Tian, Y., Xiong, X. J., & Zheng, T. L. (2009). Biological decolorization of the reactive dyes Reactive Black 5 by a novel isolated bacterial strain *Enterobacter* sp. EC3. *Journal of Hazardous Materials, 171*(1-3), 654–659. doi:10.1016/j.jhazmat.2009.06.050 PMID:19577842

Wang, J., & Chen, C. (2009). Biosorbents for heavy metals removal and their future. *Biotechnology Advances, 27*(2), 195–226. doi:10.1016/j.biotechadv.2008.11.002 PMID:19103274

Wang, J., Qiao, M., Wei, K., Ding, J., Liu, Z., Zhang, K. Q., & Huang, X. (2011). Decolorizing activity of malachite green and its mechanism involved in dye biodegradation by *Achromobacter xylosoxidans* MG1. *Journal of Molecular Microbiology and Biotechnology, 20*(4), 220–227. doi:10.1159/000330669 PMID:21865764

Wang, J., Zhang, X., Ling, W., Liu, R., Liu, J., Kang, F., & Gao, Y. (2017). Contamination and health risk assessment of PAHs in soils and crops in industrial areas of the Yangtze River Delta region, China. *Chemosphere, 168*, 976–987. doi:10.1016/j.chemosphere.2016.10.113 PMID:27816287

Wang, L. K., Tay, J. H., Tay, S. T. L., & Hung, Y. T. (Eds.). (2010). *Environmental bioengineering* (Vol. 11). Springer Science & Business Media. doi:10.1007/978-1-60327-031-1

Wang, M., Saricks, C., & Santini, D. (1999). *Effect of fuel ethanol use on fuel-cycle energy and greenhouse gas emission.* Argonne, IL: Argonne National Laboratory.

Wang, Q., & Zhu, S. (2010). Genetically modified lignocellulosic biomass for improvement of ethanol production. *BioResources, 5*(1), 3–4.

Wang, T. C., Weissman, J. C., Ramesh, G., Varadarajan, R., & Benemann, J. R. (1998). Heavy metal binding and removal by Phormidium. *Bulletin of Environmental Contamination and Toxicology, 60*(5), 739–744. doi:10.1007001289900688 PMID:9595189

Ward, M. M. (2002). Composting: A beginner's guide. The Robert A. Macoskey Center for Sustainable Systems Education and Research, Slippery Rock University, PA.

Wariishi, H., Valli, K., & Gold, M. H. (1989). Oxidative cleavage of a phenolic diarylpropane lignin model dimer by manganese peroxidase from *Phanerochaete chrysosporium. Biochemistry, 28*(14), 6017–6023. doi:10.1021/bi00440a044

Wasserbauer, R., Beranová, M., Vancurová, D., & Doležel, B. (1990). Biodegradation of polyethylene foils by bacterial and liver homogenates. *Biomaterials, 11*(1), 36–40. doi:10.1016/0142-9612(90)90049-V PMID:2302448

Watanabe, T., Ohtake, Y., Asabe, H., Murakami, N., & Furukawa, M. (2009). Biodegradability and degrading microbes of low-density polyethylene. *Journal of Applied Polymer Science, 111*(1), 551–559. doi:10.1002/app.29102

Weber, K. P., & Legge, R. L. (2013). Comparison of the catabolic activity and catabolic profiles of rhizospheric, gravel-associated and interstitial microbial communities in treatment wetlands. *Water Science and Technology, 67*(4), 886–893. doi:10.2166/wst.2012.637 PMID:23306269

Weber, W. J. (1972). *Physicochemical processes for water quality control.* Wiley Interscience.

Wee, Y. J., Kim, J. N., Yun, J. S., & Ryu, H. W. (2004). Utilization of sugar molasses for economical L (+)-lactic acid production by batch fermentation of Enterococcus faecalis. *Enzyme and Microbial Technology, 35*(6), 568–573. doi:10.1016/j.enzmictec.2004.08.008

Wei, R., & Zimmermann, W. (2017). *Microbial enzymes for the recycling of recalcitrant petroleum-based plastics: how far are we?* Academic Press. DOI: 10.1111/1751-7915.12710

Weiland, P. (2010). Biogas production: Current state and perspectives. *Applied Microbiology and Biotechnology*, *85*(4), 849–860. doi:10.100700253-009-2246-7 PMID:19777226

Wei, R., Ge, F., Zhang, L., Hou, X., Cao, Y., Gong, L., ... Bao, E. (2016). Occurrence of 13 veterinary drugs in animal manure-amended soils in Eastern China. *Chemosphere*, *144*, 2377–2383. doi:10.1016/j.chemosphere.2015.10.126 PMID:26610297

Weis, J. S., & Weis, P. (2004). Metal uptake, transport and release by wetland plants: Implications for phytoremediation and restoration. *Environment International*, *30*(5), 685–700. doi:10.1016/j.envint.2003.11.002 PMID:15051245

Wei, Y., Han, I.-K., Hu, M., Shao, M., Zhang, J., & Tang, X. (2010). Personal exposure to particulate PAHs and anthraquinone and oxidative DNA damages in humans. *Chemosphere*, *81*(10), 1280–1285. doi:10.1016/j.chemosphere.2010.08.055 PMID:20869742

Wenzel, W. W. (2009). Rhizosphere processes and management in plant-assisted bioremediation (phytoremediation) of soils. *Plant and Soil*, *321*(1-2), 385–408. doi:10.100711104-008-9686-1

West, L. (2006). *World water day: A billion people worldwide lack safe drinking water*. About.com.

Whelan, M. J., Coulon, F., Hince, G., Rayner, J., McWatters, R., Spedding, T., & Snape, I. (2015). Fate and transport of petroleum hydrocarbons in engineered biopiles in polar regions. *Chemosphere*, *131*, 232–240. doi:10.1016/j.chemosphere.2014.10.088 PMID:25563162

White, K. D. (1995). Enhancement of nitrogen removal in subsurface flow constructed wetlands employing a 2-stage configuration, an unsaturated zone, and recirculation. *Water Science and Technology*, *32*(3), 59–67.

WHO. 9/2017). Agrochemicals, health and environment: directory of resources. Retrieved from http://www.who.int/heli/risks/toxics/chemicalsdirectory/en/index1.html

Wididana, G. N., & Higa, T. (1999). Integrated Recycle System of Organic Urban Waste with EM-Technology. *Presented in 6th International Conference on Kyusei Nature Farming*, South Africa.

Wikipedia. (2016). *Pyrolosis Gasoline*. Retrieved from https://en.wikipedia.org/wiki/Pyrolysis_gasoline#cite_note-1

Wikipedia. (2017a). *Crude*. Retrieved from https://en.wikipedia.org/wiki/Crude

Wikipedia. (2017b). *Brent Crude*. Retrieved from https://en.wikipedia.org/wiki/Brent_Crude

Wikipedia. (2017c). *Light Crude Oil*. Retrieved from https://en.wikipedia.org/wiki/Light_crude_oil

Wikipedia. (2018a). *Water Pollution*. Retrieved from https://en.wikipedia.org/wiki/Water_pollution

Wikipedia. (2018b). *Petroleum*. Retrieved from https://en.wikipedia.org/wiki/Petroleum

Wikipedia. (2018c). *Petrochemical*. Retrieved from https://en.wikipedia.org/wiki/Petrochemical

Wikipedia. (2018d). *Pollution*. Retrieved from https://en.wikipedia.org/wiki/Pollution#Pollutants

Wikipedia. (2018e). *Gasoline*. Retrieved from https://en.wikipedia.org/wiki/Gasoline

Wikipedia. (2018f). *Oil Refinery*. Retrieved from https://en.wikipedia.org/wiki/Oil_refinery#Major_products

Wikipedia. (2018g). *Heavy Crude Oil*. Retrieved from https://en.wikipedia.org/wiki/Heavy_crude_oil

Wikipedia. (2018h). *Sweet Crude Oil*. Retrieved from https://en.wikipedia.org/wiki/Sweet_crude_oil

Wilda, P. J. D., Huijgena, W. J. J., & Heeres, H. J. (2012). Pyrolysis of wheat straw-derived organosolv lignin. *Journal of Analytical and Applied Pyrolysis*, *93*, 95–103. doi:10.1016/j.jaap.2011.10.002

Wilke, A., Buchholz, R., & Bunke, G. (2006). Selective biosorption of heavy metals by algae. *Environmental Biotechnology*, *2*, 47–56.

Williams, W. A., Lobos, J. H., & Cheetham, W. E. (1997). Notes; A Phylogenetic Analysis of Aerobic Polychlorinated Biphenyl-Degrading Bacteria. *International Journal of Systematic and Evolutionary Microbiology*, *47*(1), 207–210. PMID:8995824

Wilson, D. B. (2011). Microbial diversity of cellulose hydrolysis. *Current Opinion in Microbiology*, *14*(3), 1–5. doi:10.1016/j.mib.2011.04.004 PMID:21531609

Wilson, J. H., Leung, W. Y., Bosco, G., Dieu, D., & Haber, J. E. (1994). The frequency of gene targeting in yeast depends on the number of target copies. *Proceedings of the National Academy of Sciences of the United States of America*, *91*(1), 177–181. doi:10.1073/pnas.91.1.177 PMID:8278360

Wojas, S., Hennig, J., Plaza, S., Geisler, M., Siemianowski, O., Skłodowska, A., & Antosiewicz, D. M. (2009). Ectopic expression of Arabidopsis ABC transporter MRP7 modifies cadmium root-to-shoot transport and accumulation. *Environmental Pollution*, *157*(10), 2781–2789. doi:10.1016/j.envpol.2009.04.024 PMID:19467746

Wong, K. S., Huang, Q., Au, C. H., Wang, J., & Kwan, H. S. (2012). Biodegradation of dyes and polyaromatic hydrocarbons by two allelic forms of *Lentinula edodes* laccase expressed from *Pichia pastoris*. *Bioresource Technology*, *104*, 157–164. doi:10.1016/j.biortech.2011.10.097 PMID:22130082

Wong, R. C., & Solomon, A. R. (1985). Acquired dermal smooth-muscle hamartoma. *Cutis*, *35*(4), 369–370. PMID:3996041

Wong, Y. C., & Sanggari, V. (2014). Bioethanol production from sugarcane bagasse using fermentation process. *Oriental Journal of Chemistry*, *30*(2), 507–513. doi:10.13005/ojc/300214

World Bank. (n.d.). *Guidance Gact Sheet: Landfill Design*. Retrieved from http://siteresources.worldbank.org/INTCARBONFINANCE/Resources/MSW_SCS_FactSheet_Landfill_Design_Final.pdf

World Health Organization. (1990). *Public health impact of pesticides used in agriculture*.

Woudneh, M. B., Ou, Z., Sekela, M., Tuominen, T., & Gledhill, M. (2009). Pesticide Multiresidues in Waters of the Lower Fraser Valley, British Columbia, Canada. Part I. Surface Water. *Journal of Environmental Quality*, *38*(3), 940–947. doi:10.2134/jeq2007.0524 PMID:19329682

Wrona, M., Salafranca, J., & Nerin, C. (2017). Fast assessment of oxo-biodegradable polyethylene film oxidation by surface-enhanced Raman scattering with *in situ* formation of a silver nanoparticle substrate. *Journal of Materials Chemistry. C, Materials for Optical and Electronic Devices*, *5*(2), 463–469. doi:10.1039/C6TC04401H

Wuebbles, D. J. (2013). *Atmospheric Gases A2. In A Encyclopedia of Biodiversity* (2nd ed., pp. 281–290). Waltham: Academic Press. doi:10.1016/B978-0-12-384719-5.00275-6

Wu, H., Tang, S., Zhang, X., Guo, J., Song, Z., Tian, S., & Smith, D. L. (2009). Using elevated CO 2 to increase the biomass of a Sorghum vulgare× Sorghum vulgare var. sudanense hybrid and Trifolium pratense L. and to trigger hyperaccumulation of cesium. *Journal of Hazardous Materials*, *170*(2), 861–870. doi:10.1016/j.jhazmat.2009.05.069 PMID:19523760

Wu, L., Li, Z., Akahane, I., Liu, L., Han, C., Makino, T., ... Christie, P. (2012). Effects of organic amendments on Cd, Zn and Cu bioavailability in soil with repeated phytoremediation by Sedum plumbizincicola. *International Journal of Phytoremediation*, *14*(10), 1024–1038. doi:10.1080/15226514.2011.649436 PMID:22908662

Wyman, C. E. (1994). Ethanol from lignocellulosic biomass: Technology, economics, and opportunities. *Bioresource Technology*, *50*(1), 3–15. doi:10.1016/0960-8524(94)90214-3

Wyman, C. E. (2007). What is (and is not) vital to advancing cellulosic ethanol. *Trends in Biotechnology*, *25*(4), 153–157. doi:10.1016/j.tibtech.2007.02.009 PMID:17320227

Xafenias, N., & Mapelli, V. (2014). Performance and bacterial enrichment of bioelectrochemical systems during methane and acetate production. *International Journal of Hydrogen Energy*, *39*(36), 21864–21875. doi:10.1016/j.ijhydene.2014.05.038

Xenia, M. E., & Refugio, R. V. (2016). Microorganisms Metabolism during Bioremediation of Oil Contaminated Soils. *Journal of Bioremediation & Biodegradation*, *7*(340), 2.

Xinshan, S., Qin, L., & Denghua, Y. (2010). Nutrient removal by hybrid subsurface flow constructed wetlands for high concentration ammonia nitrogen wastewater. *Procedia Environmental Sciences*, *2*, 1461–1468. doi:10.1016/j.proenv.2010.10.159

Xu, M. X., Yan, J. H., Lu, S. Y., Li, X. D., Chen, T., Ni, M. J., ... Cen, K. F. (2009). Agricultural soil monitoring of PCDD/Fs in the vicinity of a municipal solid waste incinerator in Eastern China: Temporal variations and possible sources. *Journal of Hazardous Materials*, *166*(2-3), 628–634. doi:10.1016/j.jhazmat.2008.11.072 PMID:19135306

Xun, L., Topp, E., & Orser, C. S. (1992). Purification and characterization of a tetrachloro-*p*-hydroquinone reductive dehalogenase from a *Flavobacterium* sp. *Journal of Bacteriology*, *174*(24), 8003–8007. doi:10.1128/jb.174.24.8003-8007.1992 PMID:1459949

Yabannavar, A. V., & Bartha, R. (1994). Methods for assessment of biodegradability of plastic films in soil. *Applied and Environmental Microbiology*, *60*, 3608–3614. PMID:16349408

Yablokov, A. V., Nesterenko, V. B., & Nesterenko, A. V. (2009). 8. Atmospheric, water, and soil contamination after Chernobyl. *Annals of the New York Academy of Sciences*, *1181*, 223–236. doi:10.1111/j.1749-6632.2009.04831.x PMID:20002050

Yadav, J. S., & Reddy, C. A. (1993). Degradation of benzene, toluene, ethylbenzene, and xylenes (BTEX) by the lignin-degrading basidiomycete *Phanerochaete chrysosporium. Applied and Environmental Microbiology*, *59*, 756–762. PMID:8481002

Yamada-Onodera, K., Mukumoto, H., Katsuyaya, Y., Saiganji, A., & Tani, Y. (2001). Degradation of polyethylene by a fungus, *Penicillium simplicissimum* YK. *Polymer Degradation & Stability*, *72*(2), 323–327. doi:10.1016/S0141-3910(01)00027-1

Yamasaki, S., Nomura, N., Nakajima, T., & Uchiyama, H. (2012). Cultivation-independent identification of candidate dehalorespiring bacteria in tetrachloroethylene degradation. *Environmental Science & Technology*, *46*(14), 7709–7716. doi:10.1021/es301288y PMID:22708499

Yang, C. F., Lee, C. M., & Wang, C. C. (2005). Degradation of chlorophenols using pentachlorophenol-degrading bacteria Sphingomonas chlorophenolica in a batch reactor. *Current Microbiology*, *51*(3), 156–160. doi:10.100700284-005-4481-1 PMID:16086107

Yang, J., Yang, Y., Wu, W.-M., Zhao, J., & Jiang, L. (2014). Evidence of polyethylene biodegradation by bacterial strains from the guts of plastic-eating waxworms. *Environmental Science & Technology*, *48*(23), 13776–13784. doi:10.1021/es504038a PMID:25384056

Yang, P. Y., Su, R., & Kim, S. J. (2003). EMMC process for combined removal of organics, nitrogen and an odor producing substance. *Journal of Environmental Management*, *69*(4), 381–389. doi:10.1016/j.jenvman.2003.09.015 PMID:14680899

Yang, Q. X., Tao, L. X., Yang, M., & Zhang, H. (2008). Effects of glucose on the decolorization of Reactive Black 5 by yeast isolates. *Journal of Environmental Sciences (China)*, *20*(1), 105–108. doi:10.1016/S1001-0742(08)60016-9 PMID:18572531

Yao, F., Shen, G. X., Li, X. L., Li, H. Z., Hu, H., & Ni, W. Z. (2011). A comparative study on the potential of oxygen release by roots of selected wetland plants. *Physics and Chemistry of the Earth Parts A/B/C*, *36*(9), 475–478. doi:10.1016/j.pce.2010.11.001

Yeh, T. Y., Wu, M. H., Cheng, C. Y., & Hsu, Y. H. (2015). *A Study and Analysis on the Physical Shading Effect of Water Quality Control in Constructed Wetlands in Research and Practices in Water Quality*. InTech.

Yelda, S. (2005). Modified landfill design for sustainable waste management. *International Journal of Global Energy Issues*, *23*(1), 93–106. doi:10.1504/IJGEI.2005.006412

Yemendzhiev, H., Alexieva, Z., & Krastanov, A. (2009). Decolorization of synthetic dye reactive blue 4 by mycelial culture of white-rot fungi *Trametes versicolor* 1. *Biotechnology, Biotechnological Equipment*, *23*(3), 1337–1339. doi:10.1080/13102818.2009.10817665

Yin, L. J., Huang, P. S., & Lin, H. H. (2010). Isolation of Cellulase-Producing Bacteria and Characterization of the Cellulase from the Isolated Bacterium *Cellulomonas* Sp. YJ5. *Journal of Agricultural and Food Chemistry*, *58*(17), 9833–9837. doi:10.1021/jf1019104 PMID:20687562

Yin, L. J., Lin, H. H., & Xiao, Z. R. (2010). Purification and characterization of a cellulase from Bacillus subtilis YJ1. *Journal of Marine Science and Technology*, *18*(3), 466–471.

Yin, Q., Zhu, X., Zhan, G., Bo, T., Yang, Y., Tao, Y., ... Yan, Z. (2016). Enhanced methane production in an anaerobic digestion and microbial electrolysis cell coupled system with co-cultivation of *Geobacter* and *Methanosarcina*. *Journal of Environmental Sciences (China)*, *42*, 210–214. doi:10.1016/j.jes.2015.07.006 PMID:27090713

Yoon, M. G., Jeon, H. J., & Kim, M. N. (2012). Biodegradation of polyethylene by a soil bacterium and AlkB cloned recombinant cell. *Journal of Bioremediation & Biodegradation*, *3*, 145–151.

You, S.-J., & Teng, J.-Y. (2009). Anaerobic decolorization bacteria for the treatment of azo dye in a sequential anaerobic and aerobic membrane bioreactor. *Journal of the Taiwan Institute of Chemical Engineers*, *40*(5), 500–504. doi:10.1016/j.jtice.2009.01.007

Yu, J., Ogata, D., Gai, Z., Taguchi, S., Tanaka, I., Ooi, T., & Yao, M. (2014). Structures of AzrA and of AzrC complexed with substrate or inhibitor: Insight into substrate specificity and catalytic mechanism. *Acta Crystallographica. Section D, Biological Crystallography*, *70*(2), 553–564. doi:10.1107/S1399004713030988 PMID:24531489

Yu, W. W., Chang, E., Falkner, J. C., Zhang, J., Al-Somali, A. M., Sayes, C. M., & Colvin, V. L. (2007). Forming biocompatible and nonaggregated nanocrystals in water using amphiphilic polymers. *Journal of the American Chemical Society*, *129*(10), 2871–2879. doi:10.1021/ja067184n PMID:17309256

Yu, Z., & Wen, X. (2005). Screening and identification of yeasts for decolorizing synthetic dyes in industrial wastewater. *International Biodeterioration & Biodegradation*, *56*(2), 109–114. doi:10.1016/j.ibiod.2005.05.006

Zablotowicz, R. M., Hoagland, R. E., & Locke, M. A. (1994). Glutathione S-transferase activity in rhizosphere bacteria and the potential for herbicide detoxification. In T. A. Anderson & J. R. Coats (Eds.), *Bioremediation Through Rhizosphere Technology* (pp. 184–198). Washington, DC: American Chemical Society. doi:10.1021/bk-1994-0563.ch015

Zan, L., Fa, W., & Wang, S. (2006). Novel photodegradable low-density polyethylene-TiO_2 nanocomposite film. *Environmental Science & Technology*, *40*(5), 1681–1685. doi:10.1021/es051173x PMID:16568787

Zeeman, G., Kujawa, K., De Mes, T., Hernandez, L., De Graaff, M., Abu-Ghunmi, L., ... Van Lier, J. (2008). Anaerobic treatment as a core technology for energy, nutrients and water recovery from source-separated domestic waste (water). *Water Science and Technology*, *57*(8), 1207. doi:10.2166/wst.2008.101 PMID:18469391

Zeng, J., Lin, X., Zhang, J., Li, X., & Wong, M. H. (2011). Oxidation of polycyclic aromatic hydrocarbons by the bacterial laccase CueO from E. coli. *Applied Microbiology and Biotechnology*, *89*(6), 1841–1849. doi:10.100700253-010-3009-1 PMID:21120471

Zenteno, A., Lieberwirth, I., Catalina, F., Corrales, T., Guerrero, S., Vasco, D. A., & Zapata, P. A. (2017). Study of the effect of the incorporation of TiO2 nanotubes on the mechanical and photodegradation properties of polyethylenes. *Composites. Part B, Engineering*, *112*, 66–73. doi:10.1016/j.compositesb.2016.12.007

Zerdansi, I., Faid, M., & Malki, A. (2004). Digestion of solid tannery wastes by strains of Bacillus species isolated from compost in Morocco. *International Journal of Agriculture and Biology*, *6*(5), 758–761.

Zhai, J., Zou, J., He, Q., Ning, K., & Xiao, H. (2012). Variation of dissolved oxygen and redox potential and their correlation with microbial population along a novel horizontal subsurface flow wetland. *Environmental Technology*, *33*(17), 1999–2006. doi:10.1080/09593330.2012.655320 PMID:23240193

Zhang, H., Zhou, Y., Huang, Y., Wu, L., Liu, X., & Luo, Y. (2016). Residues and risks of veterinary antibiotics in protected vegetable soils following application of different manures. *Chemosphere*, *152*, 229–237. doi:10.1016/j.chemosphere.2016.02.111 PMID:26971176

Zhang, H., Ziv-El, M., Rittmann, B. E., & Krajmalnik-Brown, R. (2010). Effect of dechlorination and sulfate reduction on the microbial community structure in denitrifying membrane-biofilm reactors. *Environmental Science & Technology*, *44*(13), 5159–5164. doi:10.1021/es100695n PMID:20524654

Zhang, P., Chen, Y., & Zhou, Q. (2009). Waste activated sludge hydrolysis and short-chain fatty acids accumulation under mesophilic and thermophilic conditions: Effect of pH. *Water Research*, *43*(15), 3735–3742. doi:10.1016/j.watres.2009.05.036 PMID:19555988

Zhang, Y. H., Himmel, M. E., & Mielenz, J. R. (2006). Outlook for cellulase improvement: Screening and selection strategies. *Biotechnology Advances*, *24*(5), 452–481. doi:10.1016/j.biotechadv.2006.03.003 PMID:16690241

Zhang, Y., & Angelidaki, I. (2014). Microbial electrolysis cells turning to be versatile technology: Recent advances and future challenges. *Water Research*, *56*, 11–25. doi:10.1016/j.watres.2014.02.031 PMID:24631941

Zhang, Y., Mu, Y., Liu, J., & Mellouki, A. (2012). Levels, sources and health risks of carbonyls and BTEX in the ambient air of Beijing, China. *Journal of Environmental Sciences (China)*, *24*(1), 124–130. doi:10.1016/S1001-0742(11)60735-3 PMID:22783623

Zhao, S., Ding, C., & He, J. (2015). Detoxification of 1, 1, 2-trichloroethane to ethene by Desulfitobacterium and identification of its functional reductase gene. *PLoS One*, *10*(4), e0119507. doi:10.1371/journal.pone.0119507 PMID:25835017

Zhao, Y. Q., Sun, G., & Allen, S. J. (2004). Purification capacity of a highly loaded laboratory scale tidal flow reed bed system with effluent recirculation. *The Science of the Total Environment*, *330*(1), 1–8. doi:10.1016/j.scitotenv.2004.03.002 PMID:15325153

Zhao, Z., Zhang, Y., Quan, X., & Zhao, H. (2016). Evaluation on direct interspecies electron transfer in anaerobic sludge digestion of microbial electrolysis cell. *Bioresource Technology*, *200*, 235–244. doi:10.1016/j.biortech.2015.10.021 PMID:26492177

Zhao, Z., Zhang, Y., Wang, L., & Quan, X. (2015). Potential for direct interspecies electron transfer in an electric-anaerobic system to increase methane production from sludge digestion. *Scientific Reports*, *5*(1), 11094. doi:10.1038rep11094 PMID:26057581

Zheng, Y. I., Pan, Z., & Zhang, R. (2009). Overview of biomass pretreatment for cellulosic ethanol production. *International Journal of Agricultural and Biological Engineering*, *2*(3), 51–68.

Zhuang, P., Yang, Q. W., Wang, H. B., & Shu, W. S. (2007). Phytoextraction of heavy metals by eight plant species in the field. *Water, Air, and Soil Pollution*, *184*(1-4), 235–242. doi:10.100711270-007-9412-2

Zhuang, X., Chen, J., Shim, H., & Bai, Z. (2007). New advances in plant growth-promoting rhizobacteria for bioremediation. *Environment International*, *33*(3), 406–413. doi:10.1016/j.envint.2006.12.005 PMID:17275086

Ziegler, F., Nitenberg, G., Coudray-Lucas, C., Lasser, P., Giboudeau, J., & Cynober, L. (1998). Pharmacokinetic assessment of an oligopeptide-based enteral formula in abdominal surgery patients. *The American Journal of Clinical Nutrition*, *67*(1), 124–128. doi:10.1093/ajcn/67.1.124 PMID:9440386

Zimmermann, T., Gasser, F., Kulla, H. G., & Leisinger, T. (1984). Comparison of two bacterial azoreductases acquired during adaptation to growth on azo dyes. *Archives of Microbiology*, *138*(1), 37–43. doi:10.1007/BF00425404 PMID:6742955

Zimmermann, T., Kulla, H. G., & Leisinger, T. (1982). Properties of purified Orange II azoreductase, the enzyme initiating azo dye degradation by *Pseudomonas* KF46. *European Journal of Biochemistry*, *129*(1), 197–203. doi:10.1111/j.1432-1033.1982.tb07040.x PMID:7160382

Ziv-El, M., Popat, S. C., Cai, K., Halden, R. U., Krajmalnik-Brown, R., & Rittmann, B. E. (2012). Managing methanogens and homoacetogens to promote reductive dechlorination of trichloroethene with direct delivery of H2 in a membrane biofilm reactor. *Biotechnology and Bioengineering*, *109*(9), 2200–2210. doi:10.1002/bit.24487 PMID:22392141

Zumarraga, M., Plou, F.J., Garcia-Arellano, H., Ballesteros, A., & Alcalde, M. (2007). *Bioremediation of polycyclic aromatic hydrocarbons by fungal laccases engineered by directed evolution*. Academic Press.

About the Contributors

Vinay Mohan Pathak is currently working as research associate in VPKAS-ICAR Almora, India. He earned B.Sc. Biotechnology and M.Sc. Microbiology Degree in 2007 and 2010 respectively from Kumaun University, Nainital. He was awared UGC BSR Junior Research Fellowship in 2014 and Senior Research Fellowship in 2016. He was also qulified ICAR NET in 2016. He has submitted his Ph.D. thesis in 2017 under the guidance of Prof. Navneet. He has published research papers and book chapters in peer reviewed journals and reputed publications respectively.

Navneet is an Professor in the Department of Botany and Microbiology Gurukul Kangri University Haridwar, India. He earned M.Sc. and M.Phil. Degree in 1985 and 1987 respectively and Ph.D. in 1991 from Kurukshetra University, Kurukshetr. He joined Gurukul Kangri University Haridwar as Lecturer in 1992. He has been the Head of the Department of Department of Botany and Microbiology from July 2013 to June 2016. He has guided 12 Ph.D. students and is supervising 7 research students on varied aspects of microbiology. He had been the Dean Student Welfare and Coordinator, National Service Scheme of Gurukul Kangri University. He has published several papers in National and International journals.

* * *

Akash is presently working as research scholar in Department of Botany and Microbiology, Gurukul Kangri University Haridwar, Uttarakhand, India.

Sophie Anderson is a graduate student researcher at Teesside University.

Neelesh Babu is presently working as research scholar in Department of Botany and Microbiology, Gurukul Kangri University Haridwar, Uttarakhand, India.

Thazeem Basheer is pursuing Ph.D. in Microbial Biotechnology at Department of Microbial Biotechnology School of Biotechnology and Genetic Engineering, Coimbatore, Tamil Nadu, India.

Shashi Lata Bharati is working as Assistant Professor in Department of Chemistry, NERIST, Nirjuli, Arunachal Pradesh, India.

Pankaj Bhatt did his master degree in Microbiology from DBT Bhimtal, Kumaun University Nainital and Doctorate degree in Microbiology from G.B Pant University of Agriculture and Technology Pantnagar U.S Nagar in 2015. He has worked as Post Doctoral Research Associate at Indian Veterinary Research Institute Izatnagar, Bareilly, India. Currently he is working as Assistant Professor at Department of Microbiology, Dolphin (P.G) Institute of Biomedical and natural Sciences Dehradun, India. He has published 20 Research articles in reputed journals. Presently he is doing research on the Bioinformatics of Bioremediation.

Pankaj Kumar Chaurasia was working as Research Associate-CSIR in Department of Biotechnology, MNNIT, Allahabad, UP, India and presently working as Assistant Professor in Department of Chemistry, LS College Muzaffarpur, Bihar.

Arvind Chel is a PhD from IIT Delhi with more than 30 international papers.

Kingshuk Dutta is currently working as an Indo-U.S. Postdoctoral Fellow at the Cornell University, NY. Prior to this appointment, he worked as a National Post-Doctoral Fellow at Indian Institute of Technology, Kharagpur. Earlier, Dr. Dutta has obtained his B. Sc. in Chemistry, and B. Tech., M. Tech. and Ph. D. (Tech.) in Polymer Science and Technology, all from University of Calcutta, Kolkata. Until date, he has published 35 articles, 1 edited book and 7 book chapters in internationally reputed platforms.

Soumita Dutta Laha has completed her B.Sc. in Biotechnology from West Bengal University of Technology in 2012 and M.Sc. in Biotechnology from Bangalore University in 2014. She is currently working as a project student under Prof. Kundu.

Hossein Farraji is a Ph.D candidate in Civil(Environmental) Engineering at Universiti Sains Malaysia (USM) from 2012.Author completed his Master of (Horticulture)Engineering at Azad University (AU), central unit of science and researches in 1999.Author has completed his Bachelor of (Horticulture) Engineering in 1997.He has 11 years governmental job experiences in Institute of Standard and Industrial Researches of Iran (ISIRI).His research area is application green technology (phytoremediation and constructed wetland)for wastewater treatment. Designing vertical, horizontal and combined constructed wetlands for higher decontamination efficiency especially for high strength agroindustrial wastewater following traditional treatment methods is major of his studies.

Pooja Sharma Joshi is working as Asssistant Professor in the Department of Chemistry at Northern India Engineering College (affiliated to GGSIPU), Delhi. She cleared CSIR-UGC (NET) JRF exam in Chemical Sciences and completed her Ph.D in 2007. She has more than 10 years teaching and research experience She has several research papers to her credit in various reputed journals. She has also published books nationally and internationally. She is an active life member of various professional societies (APA, ISAS etc). She has presented papers and participated in many seminars and conferences.

Geetanjali Kaushik completed her PhD in Environmental Sciences, from the prestigious Indian Institute of Technology Delhi with 45 papers in International Journals of repute. She has participated in International Conferences held in Germany (won first prize) and USA (received American Society of Nutrition travel grant). Subsequent to her PhD she has also worked with UC Berkley on an air quality monitoring project for over 40 locations across Delhi. She has also edited 4 International books. Her taught assignments include stints as Assistant Professor with the University of Petroleum and Energy Studies, Dehradun and Amity University, Noida. In 2012 Dr. Kaushik received full scholarship from School of Management, Swansea University UK to undertake MBA. She topped the MBA class and has worked in UK and Turkey in various positions. She returned to India in 2015 and worked as Associate Professor in Environmental Management before joining Clean Air Asia. In Clean Air Asia as an Air Quality Specialist she has worked on creating Air Quality Profiles for 30 Indian cities with the application of the ADB's Clean Air Scorecard Tool.

Abderrezak Khelfi got Doctorate in Biochemistry-Immunology applied to envenomations, Institute of Biology, University Houari boumediene "USTHB", diploma of specialized medical study in toxicology, Department of Pharmacy, Faculty of Medicine of Algiers and diploma in Biology Graduate Option Biochemistry, Department of Biology, University Saad Dahleb, Blida, Algeria. His Bachelor in life science, lyceum Ibn El Khettab Blida, Algeria.

Amritha G. Kulkarni is working in P.G. Department of Biotechnology and Microbiology, Karnatak University, Dharwad, India.

Patit Paban Kundu is currently a Professor in the Department of Chemical Engineering, Indian Institute of Technology – Roorkee, India, since 2016. After completing his B. Sc. In Chemistry (1986), and B. Tech. (1989) and M. Tech. (1992) in Plastics and Rubber Technology from University of Calcutta, he joined the group of Prof. D. K. Tripathy at the Indian Institute of Technology – Kharagpur's Rubber Technology Center for carrying out his doctoral research. Following his Ph.D. in 1997, he joined as a faculty in the Department of Chemical Technology, Sant Longowal Institute of Engineering and Technology, India. Then he joined as a Professor in the Department of Polymer Science and Technology, University of Calcutta, India, in 2009. He did his post-doctoral studies in South Korea at the Inha University with Prof. S. Choe (2001-2002) and at the Yonsei University with Prof. Y. G. Shul (2006), and in USA at the Iowa State University with Prof. R. C. Larock (2003). He is a Polymer chemist and technologist, and has been actively working in the field of fuel cells as one of his areas of expertise. His interest in fuel cells include microbial, direct methanol and hydrogen fuel cells, and spans from finding novel materials for fabricating membranes, electrodes, catalysts and catalyst supports to designing of membrane electrode assemblies and flow channels.

Ashutosh Mani is working as Assistant Professor in Department of Biotechnology MNNIT, Allahabad, UP, India.

Mercy Manyuchi is a professional Chemical Engineer by training and holds a Doctor of Technology, a Master of Science and a Bachelor of Engineering Honours in Chemical Engineering. Dr. Manyuchi's research interest lies in Waste to Energy Technologies and the Value Addition of Waste to Bio Products. To date she has published more than 30 research articles in peer reviewed journals and conferences. Dr. Manyuchi's work in green initiatives has won several awards including the prestigious German based Green Talents, the Royal Academy of Engineering UK Africa Innovates and Africa Award for AgriTech Innovators. Dr. Manyuchi is driven by applying research to make the lives of people better.

Charles Mbohwa is a Professor of Sustainable Engineering and Energy Systems at the University of Johannesburg.

Edison Muzenda is a Professor in Chemical and Petrochemical Engineering at the Botswana University of Science and Technology. He is also a visiting professor at the University of Johannesburg.

Meenakshi Nandal is currently working as Assistant Professor in the Department of Environmental Sciences, pursued her Doctorate degree from Maharshi Dayanand University. She is working for last 14 years in the field of Environment science. Dr. Meenakshi Nandal also published 9 international publications and a book chapter. Currently 4 Ph. D. scholars working with her and working on salinity stress on crops, wastewater treatment, Solid waste management using aerobic, Anaerobic and microbial treatments.

Anna Christina R. Ngo received M.Sc. major in Microbiology from University of Santo Tomas, Manila.

Jingxian Qi is working in Institute of Biosciences, TU Bergakademie Freiberg, Germany.

Pattanathu Rahman is a Course Leader for MSc Food Sci & Biotechnology, Senior Lecturer in Process Engineering and Biotechnology at Teesside University and Director of TeeGene Biotech Ltd. with 20 years' research experience on novel biotechnological approach in bioproduct development. He is a recipient of TU Enterprise Project of the Year – 2016 award and SAB Award of Excellence in Microbial Biotechnology. He is a Fellow of Society for Applied Biotechnology, Member of the Institute of Environmental Science and IChemE.

Mansi Rastogi is currently working as research scholar in Department of Environmental Sciences, Maharshi Dayanand University Rohtak, Haryana, India.

Raghunath Satpathy received his M.Sc in Botany (with specialisation in Biotechnology) from Berhampur University science, India; (Post M.Sc) Advanced P.G Diploma in Bioinformatics from University of Hyderabad and M. Tech. degree in Biotechnology from VIT University, Vellore India. He has submitted the thesis for the degree for the doctor of philosophy in Biotechnology of Sambalpur University, Odisha India. Currently he is working as an Assistant professor in M.I.T.S Engineering College Odisha, India. He is having more than 9 years of experiences in both teaching and research. He has authored many national and international journal papers and six book chapters to his credit. In addition to this Mr. Satpathy is associated with many professional bodies like ISRD, APCBEES, SCIEI, SABT, OBA. His current research interests include computational analysis of protein structure and environmental applications, data mining, biological database and analytical tool development. In 2017 he has received Adarsh Vidya Saraswati Rashtriya Puraskar and senior educator and researcher award.

Michael Schlömann is working as Professor and Head of Institute, Research Group Environmental Microbiology, TU Bergakademie Freiberg, Germany.

Shruti Shukla is currently working as research scholar in Department of Microbiology, Kanya Gurukul Campus, Haridwar, India.

Poonam Singhal has a PhD from IIT Delhi with 20 International papers and is currently teaching in the University of Hyderabad.

Dirk Tischler is working in TU Bergakademie Freiberg. He received his Doctorate from Environmental Microbiology Group, TU Bergakademie Freiberg, Germany.

Mridul Umesh is pursuing Ph.D. in Microbial Biotechnology at Department of Microbial Biotechnology School of Biotechnology and Genetic Engineering, Coimbatore, Tamil Nadu, India.

Ankala Vedamurthy is Professor and Chairman, P.G. Department of Biotechnology and Microbiology, Karnatak University, Dharwad, India.

Ellie Vipond is working as Graduate student at Teesside University, UK.

Verinder Wahla is working as Assistant Professor in the Microbiology Department, Kanya Gurukul Campus, Haridwar, India.

Index

E

F

G

H

I

L

M

N

O

P

R

S

T

V

W

X

www.ingramcontent.com/pod-product-compliance
Lightning Source LLC
LaVergne TN
LVHW062103300926
849161LV00010B/2368